Random Vibrations

Analysis of Structural and Mechanical Systems

Random Vibrations

Analysis of Structural and Mechanical Systems

Loren D. Lutes
Texas A&M University
College Station, Texas

Shahram Sarkani
The George Washington University
Washington, D.C.

ELSEVIER
BUTTERWORTH HEINEMANN

AMSTERDAM • BOSTON • HEIDELBERG • LONDON
NEW YORK • OXFORD • PARIS • SAN DIEGO
SAN FRANCISCO • SINGAPORE • SYDNEY • TOKYO

Elsevier Butterworth–Heinemann
200 Wheeler Road, Burlington, MA 01803, USA
Linacre House, Jordan Hill, Oxford OX2 8DP, UK

Library of Congress Cataloging-in-Publication Data
Application submitted.

British Library Cataloguing-in-Publication Data
A catalogue record for this book is available from the British Library.

ISBN: 0-7506-7765-1

For information on all Butterworth–Heinemann publications visit our website at www.bh.com

To our children
Douglas, Daniel, Alan, David, Richard,
Laura, Steven, Rebekah, and Caroline

Contents

Chapter 4 Analysis of Stochastic Processes

Chapter 5 Time Domain Linear Vibration Analysis

Chapter 6 Frequency Domain Analysis

Chapter 7 Frequency, Bandwidth, and Ampitude

Chapter 8 Matrix Analysis of Linear Systems

Chapter 9 Direct Stochastic Analysis of Linear Systems

Chapter 10 Introduction to Nonlinear Stochastic Vibration

Chapter 11 Failure Analysis

Chapter 12 Effect of Parameter Uncertainty

PREFACE

This book has much in common with our earlier book (Lutes and Sarkani, 1997). In fact, a few of the chapters are almost unchanged. At the same time, we introduce several concepts that were not included in the earlier book and reorganize and update the presentation on several other topics.

The book is designed for use as a text for graduate courses in random vibrations or stochastic structural dynamics, such as might be offered in departments of civil engineering, mechanical engineering, aerospace engineering, ocean engineering, and applied mechanics. It is also intended for use as a reference for graduate students and practicing engineers with a similar level of preparation. The focus is on the determination of response levels for dynamical systems excited by forces that can be modeled as stochastic processes.

Because many readers will be new to the subject, our primary goal is clarity, particularly regarding the fundamental principles and relationships. At the same time, we seek to make the presentation sufficiently thorough and rigorous that the reader will be able to move on to more advanced work. We believe that the book can meet the needs of both those who wish to apply existing stochastic procedures to practical problems and those who wish to prepare for research that will extend the boundaries of knowledge.

In the hopes of meeting the needs of a broad audience, we have made this book relatively self-contained. We begin with a fairly extensive review of probability, random variables, and stochastic processes before proceeding to the analysis of dynamics problems. We do presume that the reader has a background in deterministic structural dynamics or mechanical vibration, but we also give a brief review of these methods before extending them for use in stochastic problems. Some knowledge of complex functions is necessary for the understanding of important frequency domain concepts. However, we also present time domain integration techniques that provide viable alternatives to the calculus of residues. Because of this, the book should also be useful to engineers who do not have a strong background in complex analysis.

The choice of prerequisites, as well as the demands of brevity, sometimes makes it necessary to omit mathematical proofs of results. We do always try to give mathematically rigorous definitions and results even when mathematical details are omitted. This approach is particularly important for the reader who wishes to pursue further study. An important part of the book is the inclusion of a number of worked examples that illustrate the modeling of physical problems as well as the proper application of theoretical solutions. Similar problems are also presented as exercises to be solved by the reader.

We attempt to introduce engineering applications of the material at the earliest possible stage, because we have found that many engineering students become impatient with lengthy study of mathematical procedures for which they do not know the application. Thus, we introduce linear vibration problems immediately after the introductory chapter on the modeling of stochastic problems. Time-domain interpretations are emphasized throughout the book, even in the presentation of important frequency-domain concepts. This includes, for example, the time history implications of bandwidth, with situations varying from narrowband to white noise.

One new topic added in this book is the use of evolutionary spectral density and the necessary time-domain and frequency-domain background on modulated processes. The final chapter is also new, introducing the effect of uncertainty about parameter values. Like the rest of the book, this chapter focuses on random vibration problems. The discussion of fatigue has major revisions and is grouped with first passage in an expanded chapter on the analysis of failure.

We intentionally include more material than can be covered in the typical one-semester or one-quarter format, anticipating that different instructors will choose to include different topics within an introductory course. To promote this flexibility, the crucial material is concentrated in the early portions of the book. In particular, the fundamentals of stochastic modeling and analysis of vibration problems are presented by the end of Chapter 6. From this point the reader can proceed to most topics in any of the other chapters. The book, and a modest number of readings on current research, could also form the basis for a two-semester course. It should be noted that Chapters 9–12 include topics that are the subjects of ongoing research, with the intent that these introductions will equip the reader to use the current literature, and possibly contribute to its future.

Loren D. Lutes
Shahram Sarkani

Chapter 1
Introduction

1.1 Why Study Random Vibration?

Most structural and mechanical engineers who study probability do so specifically so that they may better estimate the likelihood that some engineering system will provide satisfactory service. This is often stated in the complementary way as estimating the probability of unsatisfactory service or failure. Thus, the study of probability generally implies that the engineer accepts the idea that it is either impossible or infeasible to devise a system that is absolutely sure to perform satisfactorily. We believe that this is an honest acceptance of the facts in our uncertain world, but it is somewhat of a departure from the philosophy of much of past engineering education and of the explicit form of many engineering design codes. Of course, engineers have always known that there was a possibility of failure, but they have not always made an effort to quantify the likelihood of that event and to use it in assessing the adequacy of a design. We believe that more rational design decisions will result from such explicit study of the likelihood of failure, and that is our motivation for the study of probabilistic methods.

The characterization of uncertainty in this book will always be done by methods based on probability theory. This is a purely pragmatic choice, based on the fact that these methods have been shown to be useful for a great variety of problems. Methods based on fundamentally different concepts, such as fuzzy sets, have also been demonstrated for some problems, but they will not be investigated here.

The engineering systems studied in this book are dynamical systems. Specifically, they are systems for which the dynamic motion can be modeled by a differential or integral equation or a set of such equations. Such systems usually consist of elements having mass, stiffness, and damping and exhibiting vibratory dynamic behavior. The methods presented are general and can be applied to a great variety of problems of structural and mechanical vibration. Examples will vary from simple mechanical oscillators to buildings or other large structures, with excitations that can be either forces or base motion. The primary emphasis will be on problems with linear models, but we will also include some study of nonlinear problems. For nonlinear problems, we will particularly emphasize methods that are direct extensions of linear methods.

Throughout most of this book, the uncertainty studied will be limited to that in the excitation of the system. Only in Chapter 12 will we introduce the topic of uncertainty about the parameters of the system. Experience has shown that there are indeed many problems in which the uncertainty about the input to the system is a key factor determining the probability of system failure. This is particularly true when the inputs are such environmental loads as earthquakes, wind, or ocean waves, but it also applies to numerous other situations such as the pressure variations in the exhaust from a jet engine. Nonetheless, there is almost always additional uncertainty about the system parameters, and this also can affect the probability of system failure.

The response of a dynamical system of the type studied here is a time history defined on a continuous set of time values. The field of probability that is applicable to such problems is called stochastic (or random) processes. Thus, the applications presented here involve the use of stochastic processes to model problems involving the behavior of dynamical systems. An individual whose background includes both a course in stochastic processes and a course in either structural dynamics or mechanical vibrations might be considered to be in an ideal situation to study stochastic vibrations, but this is an unreasonably high set of prerequisites for the beginning of such study. In particular, we will not assume prior knowledge of stochastic processes and will develop the methods for analysis of such processes within this book. The probability background needed for the study of stochastic processes is a fairly thorough understanding of the fundamental methods for investigating probability and, especially, random variables. This is because a stochastic process is generally viewed as a family of random variables. For the benefit of readers lacking the necessary random variable background, Chapters 2 and 3 give a relatively comprehensive introduction to the topic, focusing on the aspects that are most important for the understanding of stochastic processes. This material may be bypassed by readers with a strong background in probability and random variables, although some review of the notation used may be helpful, because it is also the notation of the remainder of this book.

We expect the reader to be familiar with deterministic approaches to vibration problems by using superposition methods such as the Duhamel convolution integral and, to a lesser extent, the Fourier transform. We will present brief reviews of the principal ideas involved in these methods of vibration analysis, but the reader without a solid background in this area will probably need to do some outside reading on these topics.

1.2 Probabilistic Modeling and Terminology

Within the realm of probabilistic methods, there are several terms related to uncertainty that warrant some comment. The term *random* will be used here for any variable about which we have some uncertainty. This does not mean that no knowledge is available but rather that we have less than perfect knowledge. As indicated in the previous section, we will particularly use results from the area of random variables. The word *stochastic* in common usage is essentially synonymous with *random*, but we will use it in a somewhat more specialized way. In particular, we will use the term *stochastic* to imply that there is a time history involved. Thus, we will say that the dynamic response at one instant of time t is a random variable $X(t)$ but that the uncertain history of response over a range of time values is a stochastic process $\{X(t)\}$. The practice of denoting a stochastic process by putting the notation for the associated random variables in braces will be used to indicate that the stochastic process is a family of random variables—one for each t value. The term *probability*, of course, will be used in the sense of fundamental probability theory. The probability of any event is a number in the range of zero to unity that models the likelihood of the event occurring. We can compute the probabilities of events that are defined in terms of random variables having certain values, or in terms of stochastic processes behaving in certain ways.

One can view the concepts of event, random variable, and stochastic process as forming a hierarchy, in order of increasing complexity. One can always give all the probabilistic information about an event by giving one number—the probability of occurrence for the event. To have all the information about a random variable generally requires knowledge of the probability of many events. In fact, we will be most concerned with so-called continuous random variables, and one must know the probabilities of infinitely many events to completely describe the probabilities of a continuous random variable. As mentioned before, a stochastic process is a family of random variables, so its probabilistic description will always require significantly more information than does the description of any one of those random variables. We will be most concerned with the case in which the stochastic process consists of infinitely many random variables, so the additional information required will be much more than for a random variable. One can also extend this hierarchy further, with the next step being stochastic fields, which are families of stochastic processes. Within this book we will use events, random variables, and especially stochastic processes, but we will avoid stochastic fields and further generalizations.

****

Example 1.1: Let t denote time in seconds and the random variable $X(t)$, for any fixed t value, be the magnitude of the wind speed at a specified location at that time. Furthermore, let the family of $X(t)$ random variables for all nonnegative t values be a stochastic process, $\{X(t)\}$, and let A be the event $\{X(10) \le 5\text{m/s}\}$. Review the amount of information needed to give complete probabilistic descriptions of the event A, the random variable $X(t)$, and the stochastic process $\{X(t)\}$.

All the probabilistic information about the event A is given by one number—its probability of occurrence. Thus, we might say that $p = P(A)$ is that probability of occurrence, and the only other probabilistic statement that can be made about A is the almost trivial affirmation that $1 - p = P(A^c)$, in which A^c denotes the event of A not occurring, and is read as "A complement" or "not A."

We expect there to be many possible values for $X(10)$. Thus, it takes much more information to give its probabilistic description than it did to describe A. In fact, one of the simpler comprehensive ways of describing the random variable $X(10)$ is to give the probability of infinitely many events like A. That is, if we know $P[X(10) \le u]$ for all possible u values, then we have a complete probabilistic description of the random variable $X(10)$. Thus, in going from an event to a random variable we have moved from needing one number to needing many (often infinitely many) numbers to describe the probabilities.

The stochastic process $\{X(t): t \ge 0\}$ is a family of random variables, of which $X(10)$ is one particular member. Clearly, it takes infinitely more information to give the complete probability description for this stochastic process than it does to describe any one member of the family. In particular, we would need to know the probability of events such as $[X(t_1) \le u_1, X(t_2) \le u_2, \cdots, X(t_j) \le u_j]$ for all possible choices of j, $t_1,\cdots,t_j$, and $u_1,\cdots,u_j$.

If one chooses to extend this hierarchy further, then a next step could be a stochastic field giving the wind speed at many different locations, with the speed at any particular location being a stochastic process like $\{X(t)\}$.

****

It should also be noted that there exist special cases that somewhat blur the boundaries between the various levels of complexity in the common classification system based on the concepts of event, random variable, stochastic process, stochastic field, and so forth. In particular, there are random variables that can be described in terms of the probabilities of only a few events, or even only one event. Similarly, one can define stochastic processes that are families of only a few random variables. Within this book, we will generally use the concept of a vector random variable to describe any finite family of random variables and reserve the term *stochastic process* for an infinite (usually uncountable) family of random variables. Finally, we will treat a finite family of stochastic processes as a vector stochastic process, even though it could be considered a stochastic field.

****

Example 1.2: Let the random variable X denote the maintenance cost for an antenna subjected to the wind, and presume that $X=0$ if the antenna is undamaged and \$5,000 (replacement cost) if it is damaged. How much information is needed to describe all probabilities of X?

Because X has only two possible values in this simplified situation, one can describe all its probabilities with only one number—$p=P(X=5{,}000)=P(D)$, in which D denotes the event of antenna damage occurring. The only other information that can be given about the random variable X is $P(X=0)=P(D^c)=1-p$.

****

Example 1.3: Let the random variable X denote the maintenance cost for an antenna structure subjected to the wind, and presume that there are two possible types of damage. Event A denotes damage to the structure that supports the antenna dish, and it costs \$2,000 to repair, while event B denotes damage to the dish itself, and costs \$3,000 to repair. Let the random variable X denote the total maintenance cost. How much information is needed to describe all probabilities of X?

In this problem, X may take on any of four values: zero if neither the structure or the dish is damaged, 2,000 if only the structure is damaged, 3,000 if only the dish is damaged, and 5,000 if both structure and dish are damaged. Thus, one can give all the probability information about X with no more than the four numbers giving the probability that X takes on each of its possible values. These are easily described by using the events A and B and the operations of complement and intersection. For example, we might write $p_1=P(X=5{,}000)=P(AB)$, $p_2=P(X=3{,}000)=P(A^cB)$, $p_3=P(X=2{,}000)=P(AB^c)$, and $p_4=P(X=0)=P(A^cB^c)$. Even this is somewhat redundant because we also know

that $p_1 + p_2 + p_3 + p_4 = 1$, so knowledge of only three of the probabilities, such as p_1, p_2, and p_3, would be sufficient to describe the problem.
****

Example 1.4: Consider the permanent displacement of a rigid 10 meter square foundation slab during an earthquake that causes some sliding of the underlying soil. Let X, Y, and Z denote the east-west, north-south, and vertical translations of the center of the slab, and let θ_x, θ_y, and θ_z be the rotations (in radians) about the three axes. What type of probability information is required to describe this foundation motion?

Because $\{X,Y,Z,\theta_x,\theta_y,\theta_z\}$ is a family of random variables, one could consider this to be a simple stochastic process. The family has only a finite number of members, though, so we can equally well consider it to be a vector random variable. We will denote vectors by putting an arrow over them and treat them as column matrices. Thus we can write $\vec{V} = (X,Y,Z,\theta_x,\theta_y,\theta_z)^T$, in which the T superscript denotes the matrix transpose operation, and this column vector $\vec{V}$ gives the permanent displacement of the foundation. Knowledge of all the probability information about $\vec{V}$ would allow us to write the probability of any event that was defined in terms of the components of $\vec{V}$. That is, we want to be able to give $P(A)$ for any event A that depends on $\vec{V}$ in the sense that we can tell whether A has or has not occurred if we know the value of the vector $\vec{V}$. Clearly we must have information such as $P(X \le 100\,\text{mm})$ and $P(\theta_z > 0.05\,\text{rad})$, but we must also know probabilities of intersections like $P(X \le 100\,\text{mm},\ \theta_z < 0.05\,\text{rad},\ \theta_y < 0.1\,\text{rad})$, and so forth.
****

Example 1.5: Consider the permanent deformation of a system consisting of a rigid building 20 meters high resting on the foundation of Example 1.4. Let a new random variable W denote the translation to the west of a point at the top of the north face of the building, as shown in the sketch. Show the relationship between W and the vector $\vec{V}$ of Example 1.4.

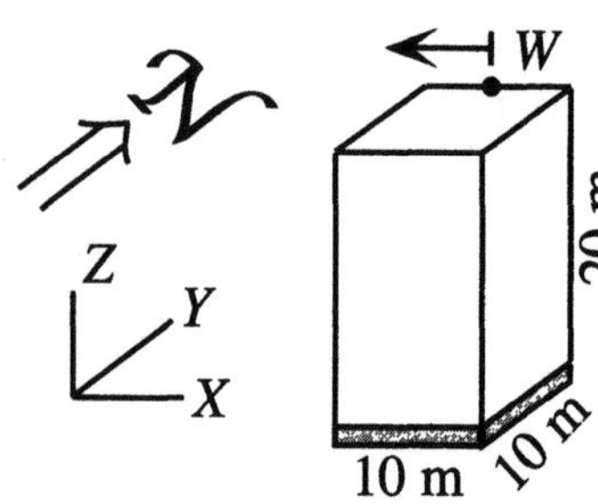

In order to describe the random variable W, we need to be able to calculate probabilities of the sort $P(W \le 200\,\text{mm})$. We can see, though, that W is related to the components of our vector $\vec{V}$ by $W = -X + 5\theta_z - 20\theta_y$, so $P(W \le 200\,\text{mm}) = P(-X + 5\theta_z - 20\theta_y \le 200\,\text{mm})$. It can be shown that one has sufficient information to compute all such terms as this if one knows $P(X \le u,$

$Y \le v, Z \le w, \theta_x \le \alpha, \theta_y \le \beta, \theta_z \le \gamma)$ for all values of the six parameters $(u,v,w,\alpha,\beta,\gamma)$.

****

Example 1.6: For the same situation described in Example 1.4, consider the foundation motion at any time during the earthquake. That is, rather than simply considering permanent displacement, let

$$\vec{V}(t) = [X(t), Y(t), Z(t), \theta_x(t), \theta_y(t), \theta_z(t)]^T$$

Identify the appropriate probabilistic model for this problem.

Now the description of the motion at any one particular time t has the same complexity as the vector $\vec{V}$ in Example 1.4. A complete description of the motion at all times, though, is much more complicated and requires information about events related to $\vec{V}(t)$ for any t value and/or to several t values. This is a problem in which we need the probabilistic description of the vector stochastic process $\{\vec{V}(t)\}$.

****

1.3 Approach to the Study of Failure Probability

Unsatisfactory performance or "failure" of a structural or mechanical dynamical system can usually be classified as being due to either "first passage" or "fatigue," and both of these failure modes will be studied within this book. Study of first-passage failure is appropriate when we consider the system to have performed unsatisfactorily if some measure of response has ever reached some critical value. Thus, for example, first-passage failure might be considered to have occurred if the stress or strain at some critical location had ever exceeded the yield level or if the displacement had exceeded some preselected value regarded as a boundary of a region of instability due to buckling.

In order to calculate the probability of first-passage failure during a given time interval $0 \le t \le T$, we will need terms such as $P[\max X(t) > x_{critical}]$, in which the maximum is over the set of values $0 \le t \le T$. Obviously, such probabilistic analysis of the maximum of a time history requires considerable knowledge about how the response $X(t_j)$ at one time relates to the response at some other time. In fact, we will need to consider the relationship of $X(t_j)$ to X at every other time within the interval of interest. Fatigue failure differs from first passage inasmuch as it involves an accumulation of damage over a stress or strain time history, rather than the maximum of that time history. It is similar to first passage, though, in that one cannot calculate the accumulated fatigue damage without knowing the relationship of the response X at any one time to X

at every other time. As mentioned previously, analyzing how $X(t_j)$ relates to X at every other time necessitates our study of a stochastic process $\{X(t)\}$ defined on a continuous set of t values. We begin this study in Chapter 4, after a review of the fundamentals of probability and random variables in Chapters 2 and 3.

Exercises

****

1.1 Assume that a given machine part may fail either due to gradual wear or due to brittle fracture. Let $X(t)$ denote the amount of gradual wear at time t. Parts (a)–(d) list various levels of detail that might be studied for this problem. For each of these, indicate whether the required model would be an event, a scalar random variable, a vector random variable, a scalar stochastic process, a vector stochastic process, or a stochastic field. Give the answer that is the simplest adequate model

(a) The likelihood that the part will fail during 10,000 hours of service
(b) The likelihood that there will be brittle fracture during 10,000 hours of service
(c) The possible time histories of $X(t)$ on the set $0 \le t \le 10{,}000$ hours
(d) The set of possible values of $X(t)$ at $t = 10{,}000$ hours

****

1.2 Let $X_j(t)$ for j = 1 to 20 denote the time history of the shear distortion in story j of a 20-story building that is subjected to an earthquake. Let Y_j denote the maximum value of $X_j(t)$ throughout the duration of the earthquake. Parts (a)–(f) list various levels of detail that might be studied for this problem. For each of these, indicate whether the required model would be an event, a scalar random variable, a vector random variable, a scalar stochastic process, a vector stochastic process, or a stochastic field. Give the answer that is the simplest adequate model.

(a) The likelihood that $X_5(t)$ will exceed 100 mm at any time during the earthquake
(b) The likelihood that the translation at the top of the building will exceed 200 mm at any time during the earthquake
(c) The likelihoods of all the possible values for Y_5
(d) The likelihoods of all the possible values for the combination of all Y_j terms
(e) The likelihoods of all the possible time histories of $X_5(t)$
(f) The likelihoods of all the possible combinations of $X_j(t)$ time histories

****

1.3 Let $X(t,u)$ denote the bending moment in a cantilever beam at time t and a distance u from the fixed end. Parts (a)–(g) list various levels of detail that might be studied for this problem. For each of these, indicate whether the required model would be an event, a scalar random variable, a vector random variable, a scalar stochastic process, a vector stochastic process, or a stochastic field. Give the answer that is the simplest adequate model.

(a) The likelihood that the bending moment is exactly $3\text{kN}\cdot\text{m}$ when $t=5$ seconds and $u=2$ m

(b) The likelihood that the bending moment exceeds $2\text{kN}\cdot\text{m}$ when $t=3$ seconds and $u=2$ m

(c) The likelihoods of all the possible values of the bending moment when $t=5$ seconds and $u=2$ m

(d) The possible time histories on the set $0\le t\le 5$ seconds of the bending moment at $u=3$ m

(e) The possible values of the set of bending moments at $u=0\text{m}$, $u=1\text{m}$, $u=2\text{m}$, and $u=3\text{m}$, all observed at $t=3$ seconds

(f) The possible time histories on the set $0\le t\le 5$ seconds of the bending moments at $u=0\text{m}$, $u=1\text{m}$, $u=2\text{m}$, and $u=3\text{m}$

(g) The possible time histories on the set $0\le t\le 5$ seconds of the bending moments at every u value for $0\le u\le 3\text{m}$

****

1.4 Let $X(t)$ and $Y(t)$ denote the tensile stress and the temperature, respectively, at time t at a given location in a critical element. Parts (a)–(g) list various levels of detail that might be studied for this problem. For each of these, indicate whether the required model would be an event, a scalar random variable, a vector random variable, a scalar stochastic process, a vector stochastic process, or a stochastic field. Give the answer that is the simplest adequate model.

(a) The likelihood that the tensile stress exceeds 70 MPa and the temperature exceeds 300°C at time $t=120$ seconds

(b) The likelihoods of all the possible values of the tensile stress at time $t=120$ seconds

(c) The likelihoods of all the possible combinations of tensile stress and temperature at time $t=120$ seconds

(d) The possible time histories of the temperature on the set of values $0\le t\le 120$ seconds

(e) The possible combinations of time histories of the tensile stress and temperature on the set of values $0\le t\le 120$ seconds

(f) The likelihood that $W(t)\equiv X(t)+0.2Y(t)>200$ at time $t=120$ seconds

(g) The possible time histories of $W(t)$ on the set of values $0\le t\le 120$ seconds

****

1.5 Let $X(t)$ denote the load applied to a crane cable at time t and $Y(t)$ denote the strength of the cable at that time. The strength is a decreasing function of time, because of wear on the cable. Parts (a)–(e) list various levels of detail that might be studied for this problem. For each of these, indicate whether the required model would be an event, a scalar random variable, a vector random variable, a scalar stochastic process, a vector stochastic process, or a stochastic field. Give the answer that is the simplest adequate model.

(a) The likelihoods of all the possible combinations of load and strength at time $t = 15{,}000$ hours

(b) The likelihoods of all possible values of the load at time $t = 15{,}000$ hours

(c) The likelihood that the load will exceed the strength prior to $t = 15{,}000$ hours

(d) The possible time histories of the load on the set of values $0 \le t \le 15{,}000$ hours

(e) The possible combinations of time histories of the load and strength on the set of values $0 \le t \le 15{,}000$ hours

****

Chapter 2 Fundamentals of Probability and Random Variables

2.1 Use of Probability and Random Variable Theory

As mentioned in Chapter 1, most of the problems in this book are modeled using the concept of stochastic processes. The concepts and definitions of stochastic process theory, though, are extensions of the concepts of random variables, which are, in turn, based on the fundamentals of probability theory. This chapter and Chapter 3 give an introduction or review of the concepts that are essential for the study of stochastic processes, which begins in Chapter 4. If the reader already has a firm grounding in probability and random variable theory, then the primary usefulness of this chapter will be to establish items of notation and nomenclature that will be used throughout the book and to review mathematical concepts.

This chapter begins with a very brief review of the fundamentals of probability theory, followed by a more extensive coverage of the concepts and mathematics of random variables. There are two reasons for the difference in depth of the coverage of the two areas. The primary reason is that a thorough understanding of random variables is essential to the understanding of the presentation given here for stochastic processes. Even though the probability of events is a more fundamental concept, which forms the basis for random variable theory, essentially all of the events of interest in this book can be formulated in terms of the values of random variables or stochastic processes. The other reason that the fundamental theory of the probability of events is given only cursory treatment here is the assumption that the typical reader will come equipped with an understanding of set theory and the fundamental concepts of probability.

The reader should always remember that there is a clear distinction between probability theory and any particular physical problem being modeled. That is, probability is a field of mathematics, not of physics. Probability theory alone can never tell us the likelihood of a particular outcome of a physical problem. We must always choose some particular probability model, including the values of certain probabilities, in order to begin analyzing a problem. Probability theory then allows us to compute the likelihood of other events of interest, consistent with those initial assumptions. The answers, of course, are generally no better than the initial assumptions.

Many of the key terms in probability nomenclature are also used in common conversation—for example, *probability*, *expected value*, and *independence*. This presents both advantages and disadvantages for the student. The advantage is that the mathematical term is generally closely related to an intuitive concept that is already known. The disadvantage is the danger of concentrating on the intuitive concept to the extent of ignoring the mathematical definition. Each of us may have a slightly different understanding of a particular intuitive concept, but confusion will result unless we all agree to use a unique mathematical definition of each term.

2.2 Fundamental Concepts of Probability Theory

The basic concepts of probability theory are quite simple. In particular, one must be able to define a space of all possible outcomes for some operation or experiment, an acceptable family of *events* (i.e., sets) on the space of possible outcomes, and a probability measure that assigns a probability to each of the events. The restrictions on the events are minor and rarely present any difficulty in applications of probability. In particular, the family of events must be Boolean, which means that it is closed under the Boolean operations of union, intersection, and complement.

To be more specific, let ω_j denote the *j*th possible outcome and Ω be the space of all possible ω_js. Any event A is then a particular subset of Ω. Often the event is defined as the collection of all possible outcomes meeting some physically meaningful condition, such as $A = \{\omega:$ wind speed ≤ 40 km/h$\}$. Similarly, another event might be defined as $B = \{\omega:$ wind direction is from the south$\}$. The Boolean restriction then means that for any such events A and B, one knows that $A \cup B$, $A \cap B$, A^c, and B^c are also events, in which $A \cup B = \{\omega : \omega \in A$, or $\omega \in B$, or both$\}$, $A \cap B = \{\omega : \omega \in A$ and $\omega \in B\}$, and $A^c = \{\omega : \omega \notin A\}$. One can define a number of new events by using the Boolean operations. For example, with the definitions of A and B suggested previously, the event $A^c \cap B = \{\omega:$ wind speed > 40 km/h and wind direction is from the south$\}$. We will commonly use a simplified notation for the intersection operation in which the intersection sign is omitted in a mathematical expression and the word *and* is replaced by a comma in a textual expression—for example, $A^c B = \{\omega:$ wind speed > 40 km/h, wind direction is from the south$\}$. No such simplified notation is used for the union operation, except that *and/or* can be used in the textual definition—for example, $A^c \cup B = \{\omega:$ wind speed > 40 km/h and/or wind direction is from the south$\}$.

Given a space of possible outcomes Ω and a family of events on Ω, a probability measure $P(\cdot)$ is a function of the events meeting three conditions:

1. $P(A) \ge 0$ for any event A
2. $P(\Omega) = 1$
3. $P\left(\bigcup_{i \in I} A_i\right) = \sum_{i \in I} P(A_i)$ if $A_i A_j = \phi$ for all i and $j \in I$, in which ϕ is the empty set

These three conditions are called the axioms of probability. It is important to note that Axiom 3 must apply for I containing either a finite or an infinite number of terms. The condition of $A_i A_j = \phi$, which appears in Axiom 3, is also commonly stated as either "A_i and A_j are disjoint" or "A_i and A_j are mutually exclusive." Also, it may be noted that $\phi = \Omega^c$, so it definitely is included in any family of events on Ω. Axiom 3 (which applies not only to probability but also to any other mathematical measure of events) simply says that the measure of a union of events is the sum of the measures of the individual events if there is no "overlap" between the events. This often seems intuitively obvious to engineers, because they are familiar with quantities such as area, volume, and mass, each of which is an example of a mathematical measure.

Probability nomenclature includes calling Ω the *sure event* and ϕ the *impossible event*, for obvious reasons. In some cases it is important to distinguish between these terms and an *almost sure event*, which is one with unit probability, and an *almost impossible event*, which is one with zero probability. For example, if A is an almost sure event, then $P(A) = P(\Omega) = 1$. This implies that $P(A^c) = 0$, so the almost impossible event A^c is a collection of ω outcomes with zero probability. At first glance, the distinction between sure and almost sure or impossible and almost impossible may appear to be trivial, but this is often not the case. For example, assume that all outcomes in Ω are *equally likely*. That is, if Ω contains a finite number N of possible ω outcomes, then there is a probability of $1/N$ assigned to each ω and ϕ is the only event with probability zero. If N is allowed to tend to infinity, however, the probability of any particular outcome tends to zero, so any event containing only one ω outcome (or any finite number of ω outcomes) has probability zero, although it is generally not an impossible event.

One of the most useful concepts in applied probability is that of *conditional probability*. The mathematical definition of conditional probability is quite simple, as illustrated by

$$P(B|A) \equiv \frac{P(AB)}{P(A)} \tag{2.1}$$

The left-hand side of this expression is read as "the conditional probability of B given A" and the definition, as shown, is simply the probability of the intersection of the two events divided by the probability of the "conditioning" or "given" event. The only limitation on this definition is that $P(A) > 0$. If $P(A) = 0$, then the ratio is undefined (zero over zero), so we say that $P(B|A)$ is undefined if A is an almost impossible event. The intuitive concept of conditional probability relates to the informational connection between two events. In particular $P(B|A)$ is the probability that the outcome is in event B given that it is in event A. In applied probability we often want to estimate the likelihood of some nonobservable event based on one or more observable outcomes. For example, we want to estimate the remaining load capacity of a structural member given that we have observed certain cracks on its surface. In this case, of course, the remaining load capacity can be said to be observable, but generally not without destroying the member.

Based on Eq. 2.1, one can write a *product rule* as

$$P(AB) = P(A)P(B|A) \quad \text{or} \quad P(AB) = P(B)P(A|B) \tag{2.2}$$

Both of these expressions are true for any pair of events, provided that the conditional probabilities are defined. That is, the first form is true provided that $P(A) \neq 0$ and the second is true provided that $P(B) \neq 0$. Such product rules are very useful in applications, and they also form the basis for the definition of another important concept. Events A and B are said to be *independent* if

$$P(AB) = P(A)P(B) \tag{2.3}$$

Comparing this product rule with that in Eq. 2.2, we see that A and B are independent if and only if $P(B|A) = P(B)$ if $P(A) \neq 0$ and also if and only if $P(A|B) = P(A)$ if $P(B) \neq 0$. It is these latter relationships that give independence its important intuitive value. Assume for the moment that $P(A) \neq 0$ and $P(B) \neq 0$. We then say that A is independent of B if knowledge that B has occurred gives us absolutely no new information about the likelihood that A has occurred. That is, $P(A|B)$ (i.e., the probability of A given that B has occurred) is exactly the same as the unconditional $P(A)$ that applied before we had any information about whether the outcome was in A or A^c. Furthermore, if B is independent of A, then A is also independent of B, so knowledge about the occurrence or nonoccurrence of B gives no new information about the likelihood of the occurrence of A. It should also be noted that disjoint events are almost never independent. That is, if A and B have no outcomes in common, then $P(AB)=0$, because the intersection is empty. This contradicts Eq. 2.3 except in the almost trivial special cases with $P(A)=0$, or $P(B)=0$, or both.

2.3 Random Variables and Probability Distributions

A real random variable is a mathematical tool that we can use to describe an entity that must take on some real value but for which we are uncertain regarding what that value will be.[1] For example, in a structural dynamics problem, the uncertain quantity of interest might be the force that will occur on a structure at some specified future instant of time, or it might be some measure of response such as displacement, velocity, or acceleration at the specified time. Because we are uncertain about the value of the uncertain entity, the best description we can hope to find is one that gives the probability of its taking on particular values or values in any particular subset of the set of possible values. Thus, the description of a random variable is simply a description of its probabilities. It should perhaps be noted that there is nothing to be gained by debating whether a certain physical quantity truly is a random variable. The more pertinent question is whether our uncertainty about the value of the quantity can be usefully modeled by a random variable. As in all other areas of applied mathematics, it is safe to assume that our mathematical model is never identical to a physical quantity, but that does not necessarily preclude our using the model to obtain meaningful results.

As presented in Section 2.2, probabilities are always defined for sets of possible outcomes, called events. For a problem described by a single real random variable X, any event of interest is always equivalent to the event of X belonging to some union (finite or infinite) of disjoint intervals of the real line. In some problems we are most interested in events of the type $X=u$, in which u is a particular real number. In order to include this situation within the idea of events corresponding to intervals, we can consider the event to be $X\in I_u$ with $I_u=[u,u]$ being a closed interval that includes only the single point u. In other problems, we are more interested in intervals of finite length. Probably the most general way to describe the probabilities associated with a given random variable is with the use of the cumulative distribution function, which will be written as $F_X(\cdot)$. The argument of this function is always a real number, and the domain of definition of the function is the entire real line. That is, for any real random variable the argument of the cumulative distribution function can be any real number. The definition of the $F_X(\cdot)$ function is in terms of a probability of X being smaller than or equal to a

[1]Later we will also use complex random variables and vector random variables, but these are unnecessary at this stage of the development.

given number. Arbitrarily choosing u to denote the argument, we can write the definition as

$$F_X(u) \equiv P(X \le u) \tag{2.4}$$

in which the notation on the right-hand side is somewhat of an abbreviation. In particular, the actual event for which we are calculating the probability is more accurately represented as $\{\omega: X(\omega) \le u\}$, with $X(\omega)$ representing a mapping from the space Ω of possible outcomes to the real line. For most of our problems, though, one can use the alternative interpretation of considering Ω to be the real line, so a particular outcome ω can be replaced by an outcome u on the real line. This abbreviated notation will be used throughout this book.

Based on Eq. 2.4, $F_X(u)$ is exactly the probability of X being within the semi-infinite interval to the left of u on the real line: $F_X(u) = P\big(X \in (-\infty,u]\big) \equiv P(-\infty < X \le u)$. Again, it should be kept in mind that u can be any real number on the real line. Also, it should be kept in mind that u is not the random variable. It is a given real number denoting one of the possible outcomes for X. The random variable is X and our uncertainty about its value is represented by the values of the function $F_X(u)$ for given values of u.

****

Example 2.1: Consider X to be a random variable that is the numerical value of the result from a single roll of a standard gaming die. Thus, X has six possible values, $\{1,2,3,4,5,6\}$, and is equally likely to have any one of these values: $P(X=1) = P(X=2) = P(X=3) = P(X=4) = P(X=5) = P(X=6) = 1/6$. Find the cumulative distribution function for X.

The cumulative distribution function for this random variable is found simply by summing outcomes that fall within the interval $(-\infty,u]$:

$F_X(u) = 0$ for $-\infty < u < 1$

$F_X(u) = 1/6$ for $1 \le u < 2$

$F_X(u) = 2/6$ for $2 \le u < 3$

$F_X(u) = 3/6$ for $3 \le u < 4$

$F_X(u) = 4/6$ for $4 \le u < 5$

$F_X(u) = 5/6$ for $5 \le u < 6$

$F_X(u) = 6/6 = 1$ for $6 \le u < \infty$

Note that this $F_X(u)$ function is defined over the entire real line $-\infty < u < \infty$, even though the random variable X has a very limited set of possible values. For example, $F_X(\pi) = 3/6 = 0.5$.

****

Example 2.2: Let X denote a real number chosen "at random" from the interval $[0,10]$. Find the cumulative distribution function of X.

In this case there is a continuous set of possible values for the random variable, so there are infinitely many values that the random variable X might assume and it is equally likely that X will take on any one of these values. Obviously, this requires that the probability of X being equal to any particular one of the possible values must be zero, because the total probability assigned to the set of all possible values is always unity for any random variable (Axiom 2 of probability theory, also called *total probability*). Thus, it is not possible to define the probabilities of this random variable by giving the probability of events of the type $\{X = u\}$, but there is no difficulty in using the cumulative distribution function. It is given by

$$F_X(u) = 0 \qquad \text{for } -\infty < u < 0$$

$$F_X(u) = 0.1\,u \qquad \text{for } 0 \le u < 10$$

$$F_X(u) = 1 \qquad \text{for } 10 \le u < \infty$$

For example, $P(0 \le X \le 4) = F_X(4) - F_X(0) = 0.4$.

****

In both of the examples, note that the function $F_X(u)$ starts at zero for $u \to -\infty$ and increases monotonically to unity for $u \to +\infty$. These limits of zero and unity and the property of being monotonically increasing (or, more precisely, "monotonically nondecreasing") are characteristic of the cumulative distribution function of any random variable. They follow directly from the axioms of probability theory. In fact, one can define a random variable with a distribution function equal to $F(u)$ for any real function $F(u)$ that approaches zero for $u \to -\infty$, approaches unity for $u \to +\infty$, and is monotonically nondecreasing and continuous from the right for all finite u values.

Example 2.1 is an illustration of the category of *discrete random variables*, whereas Example 2.2 gives a *continuous random variable*. Precisely, a random variable is discrete if it may assume only values within a discrete set. (It has zero probability of being outside the discrete set.) The $F_X(u)$ function for a discrete random variable is always of the "stairstep" form, with the magnitude of the discontinuity at any particular point u being the probability that $X = u$. In Example 2.1 the number of possible values for X was finite, but this is not necessary for a discrete random variable. For example, $P(X=j)=2^{-j}$ for $j\in$ {set of positive integers} gives a well-defined random variable X. In this case the steps in $F_X(u)$ become smaller and smaller as u becomes larger and larger, so $F_X(u)$ does approach unity for $u\to\infty$. A random variable is said to be continuous if its cumulative distribution function is continuous. Although the designations "discrete" and "continuous" are useful, they are not comprehensive; that is, there are also random variables that are neither discrete nor continuous. These are sometimes called *mixed random variables*.

****

Example 2.3: Let the random variable X represent the DC voltage output from some force transducer, and let the distribution be uniform on the set $[-4,16]$ such that $F_X(u)=0.05(u+4)$ for $-4\le u\le 16$. Of course, $F_X(u)$ is zero to the left and unity to the right of the $[-4,16]$ interval. Now let another (mixed) random variable Y represent the output from a voltmeter that reads only from zero to 10 volts and that has X as the input. Whenever $0\le X\le 10$ we will have $Y=X$, but we will get $Y=0$ whenever $X<0$, and $Y=10$ whenever $X>10$. Find the cumulative distribution function for Y.

From $P(Y\le u)$ we get

$$F_Y(u)=0 \qquad \text{for } u<0$$
$$F_Y(u)=0.05(u+4) \qquad \text{for } 0\le u<10$$
$$F_Y(u)=1 \qquad \text{for } u\ge 10$$

This $F_Y(u)$ function for this mixed random variable has discontinuities of 0.2 at $u=0$ and 0.3 at $u=10$, representing the finite probabilities that Y takes on these two particular values.

****

In describing cumulative distribution functions, it is often convenient to use the simple discontinuous function called the unit step function.[2] We will use the notation $U(\cdot)$ to denote this function and define it as

$$U(x)=0 \qquad \text{for } x<0$$
$$U(x)=1 \qquad \text{for } x\ge 0 \tag{2.5}$$

[2]This widely used function is sometimes called the Heaviside function or the Heaviside step function. More detail on the unit step function and its relationship to the Dirac delta function is given in Appendix A.

Using the unit step function, it is possible to define a "stairstep" function as a summation. In particular, the cumulative distribution function for a discrete random variable such as that given in Example 2.1 can be written as

$$F_X(u) = \sum_j p_j U(u - x_j) \tag{2.6}$$

in which p_j is the magnitude of the discontinuity at $u = x_j$. Note that for any u value the summation in Eq. 2.6 is over all the possible values of j, whether that number is finite or infinite. However, for a given u value, some of the terms may contribute nothing. Specifically, there is no contribution to the summation for any j value for which $x_j > u$, because that corresponds to a $U(\cdot)$ function with a negative argument.

****

Example 2.4: Use the unit step function to write a single expression for the cumulative distribution function of the discrete random variable of Example 2.1, having $P(X=1)=P(X=2)=P(X=3)=P(X=4)=P(X=5)=P(X=6)=1/6$.

The summation in Eq. 2.6 is simply

$$F_X(u) = \frac{1}{6}\left[U(u-1)+U(u-2)+U(u-3)+U(u-4)+U(u-5)+U(u-6)\right]$$

****

Example 2.5: Use the unit step function to write a single expression for the cumulative distribution function for the mixed random variable Y of Example 2.3, having

$$\begin{aligned} F_Y(u) &= 0 && \text{for } u < 0 \\ F_Y(u) &= 0.05(u+4) && \text{for } 0 \le u < 10 \\ F_Y(u) &= 1 && \text{for } u \ge 10 \end{aligned}$$

The discontinuous $F_Y(u)$ is exactly given by

$$F_Y(u) = 0.05(u+4)\left[U(u-0) - U(u-10)\right] + U(u-10)$$

or

$$F_Y(u) = 0.05(u+4)\,U(u) + (0.8 - 0.05u)\,U(u-10)$$

****

These examples illustrate how the unit step function can be used to write the cumulative distribution function for a discrete or mixed random variable as a single equation that is valid everywhere on the real line, rather than using a piecewise description of the function, as was done in Examples 2.1, 2.2, and 2.3. This is strictly a matter of convenience, but it will often be useful. Note that the fact that $U(\cdot)$ was defined to be continuous from the right ensures that a cumulative distribution function written according to Eq. 2.6 will have the proper value at any point of discontinuity.

2.4 Probability Density Functions

For many of our calculations, it will be more convenient to describe the probability distribution of a random variable *X* by using what is called the probability density function $p_X(u)$ rather than the cumulative distribution function $F_X(u)$. The $p_X(u)$ function gives the probability per unit length along the line of possible values for a continuous random variable. Using $p_X(u)$ is analogous to describing the mass distribution of a nonuniform rod by giving the mass per unit length at each location *u*, whereas using $F_X(u)$ is like describing the rod by telling how much mass is located to the left of *u* for any particular *u* value. If $F_X(u)$ is continuous and differentiable everywhere, then one can define the probability density function as

$$p_X(u) = \frac{d}{du} F_X(u) \tag{2.7}$$

The inverse of this relationship is

$$F_X(u) = \int_{-\infty}^{u} p_X(v)\, dv \tag{2.8}$$

because this gives both Eq. 2.7 and the limiting value of $F_X(-\infty) = 0$. Equation 2.8 illustrates the fundamental nature of probability density functions—the integral of $p_X(u)$ over an interval gives the probability that *X* lies within that interval. In Eq. 2.8, of course, the interval is $(-\infty, u]$. A special case of Eq. 2.8 is that

$$\int_{-\infty}^{\infty} p_X(v)\, dv = 1 \tag{2.9}$$

This is equivalent to $F_X(\infty) = 1$ and is another form of the axiom of total probability. The probability of *X* being within any finite interval $[a,b]$ is generally given by

$$P(a < X \le b) = \int_a^b p_X(v)\,dv = F_X(b) - F_X(a) \tag{2.10}$$

Note that the expression in Eq. 2.10 actually excludes $P(X = a)$. This is identical to the probability $P(X \in [a,b]) \equiv P(a \le X \le b)$ for a continuous random variable, though, because $P(X = a)$ is the magnitude of the discontinuity in $F_X(u)$ at $u = a$, and this must be zero for a continuous function. Equation 2.10 can also be used to write an infinitesimal expression that may help illustrate the meaning of the probability density function. In particular, if $p_X(u)$ is continuous on the infinitesimal interval $[u, u+du]$, then it can be considered to be constant across the interval so that we have

$$P(u \le X \le u + du] = \int_u^{u+du} p_X(w)\,dw = p_X(u)\,du \tag{2.11}$$

Thus, at any point u for which $p_X(u)$ is continuous, the probability density gives the probability of X being in the neighborhood of u, in the sense that $p_X(u)\,du$ is the probability of being in the infinitesimal increment of length du.

The fact that any cumulative distribution function is nondecreasing tells us that any probability density function determined from Eq. 2.7 will be nonnegative. In fact, nonnegativity and Eq. 2.9 are the only restrictions on a probability density function. For any function $p(\cdot)$ that satisfies these two conditions, one can define a random variable having $p(\cdot)$ as its probability density function. Note, in particular, that there is no requirement that $p(\cdot)$ be continuous. A discontinuity in $p(\cdot)$ corresponds to an instantaneous change in the slope of $F(\cdot)$. As long as the number of points of discontinuity is countable, one can show that probabilities of X are uniquely defined by $p_X(u)$, even if that function is not uniquely defined at the points of discontinuity.

****

Example 2.6: Find the probability density function for the random variable X of Example 2.2, for which

$F_X(u) = 0$ for $-\infty < u < 0$

$F_X(u) = 0.1\,u$ for $0 \le u < 10$

$F_X(u) = 1$ for $10 \le u < \infty$

Differentiating the cumulative distribution function, the probability density function is found to be

$p_X(u)=0.1$ for $0 \le u < 10$

$p_X(u)=0$ otherwise

or, equivalently,

$$p_X(u)=0.1U(u)U(10-u) \quad \text{or} \quad p_X(u)=0.1[U(u)-U(u-10)]$$

Note that the final two forms differ only in their values at the single point $u=10$. Such a finite difference in a probability density function at a single point (or a finite number of points) can be considered trivial, because it has no effect on the cumulative distribution function, or any other integral of the density function.

****

Example 2.7: Let the random variable X have the Gaussian (or "normal") distribution with probability density function

$$p_X(u)=\frac{1}{(2\pi)^{1/2}\sigma}\exp\left[-\frac{1}{2}\left(\frac{u-\mu}{\sigma}\right)^2\right]$$

in which μ and $\sigma>0$ are constants. Find the cumulative distribution function $F_X(u)$.

Using Eq. 2.8 gives

$$F_X(u)=\int_{-\infty}^{u} p_X(v)\,dv=\frac{1}{(2\pi)^{1/2}}\int_{-\infty}^{(u-\mu)/\sigma} e^{-w^2/2}\,dw \equiv \Phi\left(\frac{u-\mu}{\sigma}\right)$$

in which the change of variables $w=(v-\mu)/\sigma$ has been used, and the function $\Phi(\cdot)$ is defined by the integral:

$$\Phi(r)\equiv\frac{1}{(2\pi)^{1/2}}\int_{-\infty}^{r} e^{-w^2/2}\,dw$$

Note that $\Phi(u)=F_X(u)$ for the special case with $\mu=0$ and $\sigma=1$, so it has the form of a cumulative distribution function. In particular, $\Phi(-\infty)=0$, $\Phi(\infty)=1$, and it is monotonically increasing. Also, $\Phi(0)=0.5$, because the probability density function for this special case is $p_X(u)=(2\pi\sigma^2)^{-1/2}e^{-u^2/2}$, which is symmetric about $u=0$. The $\Phi(\cdot)$ function is tabulated in many probability books, and it can also be related to the error function by using the change of variables $s=w/2^{1/2}$:

$$\Phi(r)\equiv\frac{1}{2}+\frac{1}{(2\pi)^{1/2}}\int_{0}^{r/2^{1/2}} e^{-s^2}\,ds=\frac{1+\text{erf}(r/2^{1/2})}{2}$$

The error function is also tabulated in many mathematical handbooks (for example, Abramowitz and Stegun, 1965). In addition, many types of computer software are readily available for evaluating the $\Phi(\cdot)$ and $\text{erf}(\cdot)$ functions.

****

Example 2.8: Let the random variable X have a probability density function of

$$p_X(u) = A\exp(-au^2 - bu - c)$$

in which A, a, b, and c are constants, with the limitations that $a > 0$, $A > 0$, and A has a value such that the integral of $p_X(u)$ over the entire real line is unity. Show that X is Gaussian, as in Example 2.7.

First let us rewrite the exponent in $p_X(u)$ as

$$(-au^2 - bu - c) = -a(u + a^{-1}b/2)^2 + a^{-1}b^2/4 - c$$

Now if we choose $\mu = a^{-1}b/2$ and $\sigma = (2a)^{-1/2}$ then we have

$$-a(u + a^{-1}b/2)^2 = (u-\mu)^2/(2\sigma^2)$$

which is the exponent in $p_X(u)$ in Example 2.7. This choice of μ and σ now gives the $p_X(u)$ as

$$p_X(u) = A\exp\left(-\frac{1}{2}\left[\frac{u-\mu}{\sigma}\right]^2\right)\exp\left(\frac{b^2}{4a} - c\right)$$

Presuming that $p_X(u)$ in Example 2.7 is a legitimate probability density function, then its integral over the entire real line is unity, so the $p_X(u)$ in the current example is legitimate only if

$$A\exp\left(\frac{b^2}{4a} - c\right) = \frac{1}{(2\pi)^{1/2}\sigma}$$

Thus, we have demonstrated that any random variable is Gaussian if its $p_X(u)$ has the form of a constant multiplying an exponential of a quadratic form of u. The coefficient on the u^2 term in the exponent ($-a$ in this example) must be negative so that $p_X(u)$ has a finite integral over the real line.

****

Any random variable *X*, such as that in Example 2.6, for which $p_X(u)$ has a constant value over all the possible values of *X*, is said to have a uniform distribution. Note that for any random variable, including one with a uniform distribution, the $p_X(u)$ function is defined on the entire real line (except possibly at points of discontinuity), even though it will be exactly zero at any *u* value that is not a possible value of *X*. Note also that Example 2.6 illustrates the use of the unit step function to simplify the form of the probability density function for a random variable that would otherwise require the use of a piecewise description.

The definition of the probability density function according to Eqs. 2.7 and 2.8 is applicable only to continuous random variables, unless one extends the boundaries of ordinary calculus. In particular, no bounded function $p_X(u)$ can satisfy Eq. 2.8 if $F_X(u)$ contains discontinuities, as occurs for a discrete or mixed random variable. This limitation of the probability density function can be removed by using the Dirac delta function $\delta(\cdot)$, which is defined by the following properties:[3]

$$\begin{aligned} \delta(x) &= 0 \qquad \text{for } x \neq 0 \\ \delta(x) &= \infty \qquad \text{for } x = 0 \end{aligned} \tag{2.12}$$

and

$$\int_{-\infty}^{\infty} \delta(x - x_0)\, f(x)\, dx \equiv \int_{x_0-\varepsilon}^{x_0+\varepsilon} \delta(x - x_0)\, f(x)\, dx = f(x_0) \tag{2.13}$$

for any function that is finite and continuous at the point $x = x_0$. The $\delta(\cdot)$ function can also be thought of as the formal derivative of the unit step function (see Appendix A):

$$\delta(x) = \frac{d}{dx} U(x) \tag{2.14}$$

so it allows one to formulate the derivative of a cumulative distribution function that contains discontinuities.

Using the Dirac delta function, one can formally describe any random variable by using a probability density function. For a discrete random variable described by Eq. 2.6, in particular, one obtains

$$p_X(u) = \sum_j p_j\, \delta(u - x_j) \tag{2.15}$$

The p_j multiplier of a term $\delta(u - x_j)$ in such a description of $p_X(u)$ for a discrete or mixed random variable always gives the finite probability that $X = x_j$.

****

Example 2.9: Use the Dirac delta function to write an expression for the probability density function of the discrete random variable X of Examples 2.1 and 2.4, having

$$P(X=1) = P(X=2) = P(X=3) = P(X=4) = P(X=5) = P(X=6) = 1/6$$

This probability density function can be written as

---

[3]Strictly speaking, $\delta(\cdot)$ is not a function, but it can be thought of as a limit of a sequence of functions, as considered in Appendix A.

$$p_X(u) = \frac{1}{6}\left[\delta(u-1) + \delta(u-2) + \delta(u-3) + \delta(u-4) + \delta(u-5) + \delta(u-6)\right]$$

This expression is exactly the formal derivative of the cumulative distribution function in Example 2.4.

****

Example 2.10: Find the probability density function for the mixed random variable Y of Examples 2.3 and 2.5, having

$$F_Y(u) = 0.05(u+4)\,U(u) + (0.8 - 0.05u)\,U(u-10)$$

Differentiating this cumulative distribution function gives

$$p_Y(u) = 0.05\,U(u) + 0.05(u+4)\,\delta(u) - 0.05\,U(u-10) + (0.8 - 0.05u)\,\delta(u-10)$$

which can be rewritten as

$$p_Y(u) = 0.05\left[U(u) - U(u-10)\right] + 0.2\delta(u) + 0.3\delta(u-10)$$

or

$$p_Y(u) = 0.05U(u)\,U(10-u) + 0.2\delta(u) + 0.3\delta(u-10)$$

but these final forms require using some properties of the Dirac delta function. In particular, $0.05(u+4)\delta(u) = 0.05\,\delta(u)$, because the terms are equal at $u = 0$ and both are zero everywhere else. Similarly, $(0.8 - 0.05u)\,\delta(u-10) = 0.3\delta(u-10)$, because they match at the sole nonzero point of $u = 10$.

****

Example 2.10 illustrates an important feature of Dirac delta functions. Any term $g(x)\delta(x - x_0)$ can be considered to be identical to $g(x_0)\delta(x - x_0)$, because both are zero for $x \neq x_0$. A related property is that

$$g(x)\delta(x - x_0) \equiv 0 \qquad \text{if } g(x_0) = 0 \tag{2.16}$$

That is, $g(x_0)\delta(x - x_0)$ can be considered to be identically zero for all x values if $g(x_0) = 0$. Clearly $g(x_0)\delta(x - x_0)$ is zero for $x \neq 0$, but its value seems to be indeterminate at $x = 0$, because it is then the product of zero and infinity. However, the integral of the expression is zero, by the definition of the Dirac delta function in Eq. 2.12. Because Dirac delta functions are useful only as they contribute to integrals, this demonstrates that $g(x_0)\delta(x - x_0)$ with $g(x_0) = 0$ can never contribute to any finite expression. Thus, it can be considered as identically zero and dropped from calculations. A particular example of this is the term $x\delta(x)$, which can always be considered to be zero.

2.5 Joint and Marginal Distributions

In many problems, we must use more than one random variable to describe the uncertainty that exists about various outcomes of a given activity. Example 2.3 represents one very simple situation involving two random variables. In that particular case, Y is a function of X, so if one knows the value of X, then one knows exactly the value of Y. If one thinks of the (X,Y) plane, for this example, then all the possible outcomes lie on a simple (piecewise linear) curve. In other problems, there may be a less direct connection between the random variables of interest. For example, the set of possible values and/or the probability of any particular value for one random variable Y may depend on the value of another random variable X.

As with a single random variable, the probabilities of two or more random variables can always be described by using a cumulative distribution function. For two random variables X and Y, this can be written as

$$F_{XY}(u,v) \equiv P(X \le u, Y \le v) \tag{2.17}$$

in which the comma within the parentheses on the right-hand side of the expression represents the intersection operation. That is, the probability denoted is for the joint event that $X \le u$ and $Y \le v$. The function $F_{XY}(u,v)$ is defined on the two-dimensional space of all possible (X,Y) values, and it is called the joint cumulative distribution function. When we generalize to more than two or three random variables, it will often be more convenient to use a vector notation. In particular, we will use an arrow over a symbol to indicate that the quantity involved is a vector (which may also be viewed as a matrix with only one column). Thus, we will write

$$\vec{X} = \begin{pmatrix} X_1 \\ X_2 \\ \vdots \\ X_n \end{pmatrix}, \qquad \vec{u} = \begin{pmatrix} u_1 \\ u_2 \\ \vdots \\ u_n \end{pmatrix}$$

and use the notation

$$F_{\vec{X}}(\vec{u}) \equiv F_{X_1 X_2 \cdots X_n}(u_1, u_2, \cdots, u_n) \equiv P\left(\bigcap_{j=1}^{n} (X_j \le u_j \right) \tag{2.18}$$

for the general joint cumulative distribution function of n random components.

As with a single random variable, one can convert from a cumulative distribution function to a probability density function by differentiation. The situation is slightly more complicated now, because we must use multiple partial derivatives:

$$p_{XY}(u,v) = \frac{\partial^2}{\partial u\, \partial v} F_{XY}(u,v), \qquad p_{\vec{X}}(\vec{u}) = \frac{\partial^n}{\partial u_1\, \partial u_2 \cdots \partial u_n} F_{\vec{X}}(\vec{u}) \qquad (2.19)$$

To obtain probabilities from an n-dimensional joint probability density function, one must use an n-fold integration such as

$$F_{\vec{X}}(\vec{u}) = \int_{-\infty}^{u_n} \cdots \int_{-\infty}^{u_1} p_{\vec{X}}(\vec{w})\, dw_1 \cdots dw_n \qquad (2.20)$$

for the probability defined as the joint cumulative distribution function, or

$$P[(X,Y) \in A] = \int\int_A p_{XY}(u,v)\, du\, dv \qquad (2.21)$$

for the probability of (X,Y) falling within any arbitrary region A on the plane of possible values for the pair.

****

Example 2.11: Let the joint distribution of (X,Y) be such that the possible outcomes (u,v) are the points within the rectangle $(-1 \le u \le 2, -1 \le v \le 1)$. Furthermore, let every one of these possible outcomes be "equally likely." Note that we are not dealing with discrete random variables in this example, because the set of possible values is not discrete. As with a single random variable on a continuous set of possible values, the term *equally likely* denotes a constant value of the probability density function. Thus, we say that

$p_{XY}(u,v) = C$ for $-1 \le u \le 2, -1 \le v \le 1$

$p_{XY}(u,v) = 0$ otherwise

in which C is some constant. Find the value of the constant C and the joint cumulative distribution function $F_{XY}(u,v)$ for all values of (u,v).

We find the value of C by using the *total probability* property that (X,Y) must fall somewhere within the space of possible values and that the probability of being within a set is the double integral over that set, as in Eq. 2.21. Thus, we obtain $C = 1/6$, because the rectangle of possible values has an area of 6 and the joint probability density function has the constant value C everywhere in the rectangle:

$$1 = P(-1 \le X \le 2, -1 \le Y \le 1) = \int_{-1}^{1} \int_{-1}^{2} p_{XY}(u,v)\, du\, dv = 6C$$

We may now calculate the joint cumulative distribution of these two random variables by integrating as in Eq. 2.20, with the result that

$$F_{XY}(u,v) = C(u+1)(v+1) = (u+1)(v+1)/6 \qquad \text{for } -1 \le u \le 2, -1 \le v \le 1$$
$$F_{XY}(u,v) = 3C(v+1) = (v+1)/2 \qquad \text{for } u > 2, -1 \le v \le 1$$
$$F_{XY}(u,v) = 2C(u+1) = (u+1)/3 \qquad \text{for } -1 \le u \le 2, v > 1$$
$$F_{XY}(u,v) = 6C = 1 \qquad \text{for } u > 2, v > 1$$
$$F_{XY}(u,v) = 0 \qquad \text{otherwise}$$

One can verify that this cumulative distribution function is continuous. Thus, the random variables X and Y are said to have a continuous joint distribution. Clearly, for this particular problem the description given by the density function is much simpler than that given by the cumulative distribution function.

****

Example 2.12: Let the joint probability density function of X and Y be equal to the constant C on the set of possible values, as in Example 2.11, but let the space of possible values be a triangle in the (u,v) plane such that

$$p_{XY}(u,v) = C \qquad \text{for } 1 \le v \le u \le 5$$
$$p_{XY}(u,v) = 0 \qquad \text{otherwise}$$

A sketch of this joint probability density function is shown. Find the value of the constant C and the joint cumulative distribution function $F_{XY}(u,v)$ for all values of (u,v).

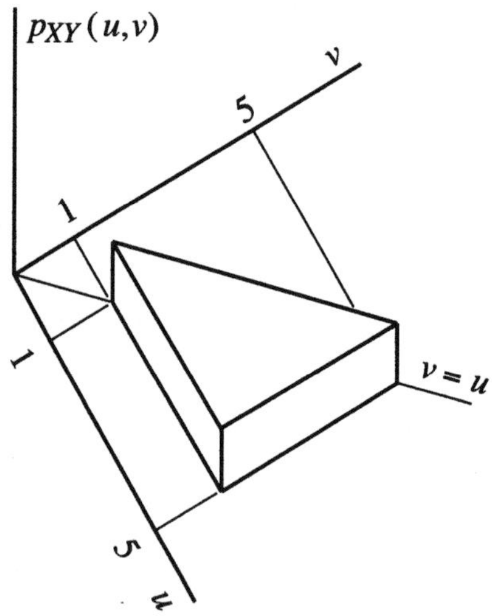

Again we use total probability to find that $C = 1/8$, because the triangle of possible values has an area of 8 and the density function is constant on that triangle. Integration gives the joint cumulative distribution function as the continuous function

$F_{XY}(u,v) = (2u - v + 1)(v - 1)/16$ for $1 \le v \le u \le 5$

$F_{XY}(u,v) = (u - 1)^2/16$ for $v > u, 1 \le u \le 5$

$F_{XY}(u,v) = (9 - v)(v - 1)/16$ for $1 \le v \le 5, u > 5$

$F_{XY}(u,v) = 1$ for $u > 5, v > 5$

$F_{XY}(u,v) = 0$ otherwise

****

Example 2.13: Consider two random variables X and Y with the joint cumulative distribution given by

$$F_{XY}(u,v) = \left(\frac{(1-e^{-4u})(1-e^{-4v})}{2} + \frac{(1-e^{-4u})(1-e^{-3v})}{4} + \frac{(1-e^{-3u})(1-e^{-4v})}{4} \right) U(u)U(v)$$

in which the unit step function has been used to convey the information that the nonzero function applies only in the first quadrant of the (u,v) plane. Find the joint probability density function for X and Y.

Taking the mixed partial derivative according to Eq. 2.19 gives

$$p_{XY}(u,v) = \left(8e^{-4u-4v} + 3e^{-4u-3v} + 3e^{-3u-4v} \right) U(u)U(v)$$

Note that no $\delta(u)$ or $\delta(v)$ Dirac delta functions have been included in the derivative, even though such terms do appear as derivatives of the $U(u)$ and $U(v)$ unit step functions. However, it is found that each Dirac delta function is multiplied by an expression that is zero at the single point where the Dirac delta function is nonzero, so they can be ignored, as in Eq. 2.16.

****

Some Properties of Joint Probability Distributions

It is left to the reader to verify that for any problem involving two random variables *X* and *Y*, the joint cumulative distribution function and joint probability distribution function must satisfy the following properties:

$$F_{XY}(-\infty,-\infty) \equiv \lim_{\substack{u \to -\infty \\ v \to -\infty}} F_{XY}(u,v) = 0 \tag{2.22}$$

$$F_{XY}(u,-\infty) \equiv \lim_{v \to -\infty} F_{XY}(u,v) = 0 \tag{2.23}$$

$$F_{XY}(-\infty,v) \equiv \lim_{u \to -\infty} F_{XY}(u,v) = 0 \tag{2.24}$$

$$F_{XY}(\infty,\infty) \equiv \lim_{\substack{u \to \infty \\ v \to \infty}} F_{XY}(u,v) = 1 \tag{2.25}$$

$$F_{XY}(u,\infty) \equiv \lim_{v \to \infty} F_{XY}(u,v) = F_X(u) \tag{2.26}$$

$$F_{XY}(\infty,v) \equiv \lim_{u \to \infty} F_{XY}(u,v) = F_Y(v) \tag{2.27}$$

$$p_{XY}(u,v) \ge 0 \text{ for all } (u,v) \text{ values} \tag{2.28}$$

$$\int_{-\infty}^{\infty} p_{XY}(u,v)\, dv = p_X(u) \tag{2.29}$$

$$\int_{-\infty}^{\infty} p_{XY}(u,v)\, du = p_Y(v) \tag{2.30}$$

$$\int_{-\infty}^{\infty}\int_{-\infty}^{\infty} p_{XY}(u,v)\, du\, dv = 1 \tag{2.31}$$

These properties can also be extended to problems with any number of random variables. A few of these extensions are:

$$F_{\vec{X}}(-\infty,u_2,\cdots,u_n) = 0 \tag{2.32}$$

$$F_{\vec{X}}(\infty,\cdots,\infty) = 1 \tag{2.33}$$

$$F_{\vec{X}}(u_1,\infty,\cdots,\infty) = F_{X_1}(u_1) \tag{2.34}$$

$$F_{\vec{X}}(u_1,u_2,\infty,\cdots,\infty) = F_{X_1X_2}(u_1,u_2) \tag{2.35}$$

$$\int_{-\infty}^{\infty}\cdots\int_{-\infty}^{\infty} p_{\vec{X}}(\vec{u})\, du_1 \cdots du_n = 1 \tag{2.36}$$

$$\int_{-\infty}^{\infty} \cdots \int_{-\infty}^{\infty} p_{\vec{X}}(\vec{u})\, du_2 \cdots du_n = p_{X_1}(u_1) \tag{2.37}$$

$$\int_{-\infty}^{\infty} \cdots \int_{-\infty}^{\infty} p_{\vec{X}}(\vec{u})\, du_3 \cdots du_n = p_{X_1 X_2}(u_1, u_2) \tag{2.38}$$

In problems involving more than one random variable, the one-dimensional distributions of single random variables are commonly referred to as the *marginal distributions*. Using this terminology, Eqs. 2.26, 2.27, 2.29, 2.30, 2.34, and 2.37 give formulas for deriving marginal distributions from joint distributions. Similarly, Eqs. 2.35 and 2.38 can be considered to give two-dimensional marginal distributions for problems involving more than two random variables.

****

Example 2.14: Find the marginal distributions for X and Y for the joint probability distribution of Example 2.11 with

$p_{XY}(u,v) = 1/6$ for $-1 \le u \le 2, -1 \le v \le 1$

$p_{XY}(u,v) = 0$ otherwise

Note that Eq. 2.29 gives $p_X(u) \equiv 0$ for $u < -1$ or $u > 2$, because the integrand of that formula is zero everywhere on the specified interval. For the nontrivial situation of $-1 \le u \le 2$, the integral gives

$$p_X(u) = \int_{-1}^{1} \frac{dv}{6} = \frac{1}{3}$$

Thus, this marginal probability density function can be written as

$$p_X(u) = \frac{1}{3} U(u+1)\, U(2-u)$$

We can now obtain the marginal cumulative distribution function of X either by integrating $p_X(u)$ or by applying Eq. 2.26 to the joint cumulative distribution function given in Example 2.11. Either way, we obtain

$$F_X(u) = \left(\frac{u+1}{3}\right)[U(u+1) - U(u-2)] + U(u-2)$$

Note that replacing $[U(u-1) - U(u-2)]$ in this expression with the almost equivalent form of $U(u-1)\,U(2-u)$ is unacceptable, because it would give $F_X(2) = 2$, instead of the correct value of $F_X(2) = 1$. Proceeding in the same way to find the marginal distribution of Y gives

$$p_Y(v) = \frac{1}{2} U(v+1)\, U(1-v)$$

and

$$F_Y(v) = \left(\frac{v+1}{2}\right)[U(v+1) - U(v-1)] + U(v-1)$$

****

Example 2.15: Find the marginal distributions for X and Y for the probability distribution of Example 2.12 with

$p_{XY}(u,v) = 1/8$ for $1 \le v \le u \le 5$

$p_{XY}(u,v) = 0$ otherwise

This $p_{XY}(u,v)$ function can be rewritten as

$$p_{XY}(u,v) = U(u-1)U(5-u)U(v-1)U(u-v)/8$$

which slightly simplifies the manipulations. First integrating $p_{XY}(u,v)$ with respect to v, in order to apply Eq. 2.29, gives

$$p_X(u) = \left(\frac{u-1}{8}\right)U(u-1)U(5-u)$$

which gives a cumulative distribution of

$$F_X(u) = \left(\frac{(u-1)^2}{16}\right)[U(u-1) - U(u-5)] + U(u-5)$$

This can be shown to be exactly the same as is obtained by taking $v = \infty$ in $F_{XY}(u,v)$ from Example 2.12, confirming Eq. 2.26. Similarly

$$p_Y(v) = \left(\frac{5-v}{8}\right)U(v-1)U(5-v)$$

and

$$F_Y(v) = \left(\frac{5(v-1)}{8} - \frac{(v^2-1)}{16}\right)[U(v-1) - U(v-5)] + U(v-5)$$

****

Example 2.16: Find the marginal probability distribution for X for the random variables of Example 2.13 with

$$F_{XY}(u,v) = \left(\frac{(1-e^{-4u})(1-e^{-4v})}{2} + \frac{(1-e^{-4u})(1-e^{-3v})}{4} + \frac{(1-e^{-3u})(1-e^{-4v})}{4}\right)U(u)U(v)$$

Because $F_{XY}(u,v)$ is given in the original statement of the problem, it is convenient to set $v=\infty$ in that equation and obtain $F_X(u)$ from Eq. 2.26 as

$$F_X(u)=\left(\frac{(1-e^{-4u})}{2}+\frac{(1-e^{-4u})}{4}+\frac{(1-e^{-3u})}{4}\right)U(u)=\left(1-\frac{3e^{-4u}}{4}-\frac{e^{-3u}}{4}\right)U(u)$$

Differentiating this expression gives

$$p_X(u)=\left(3e^{-4u}+\frac{3e^{-3u}}{4}\right)U(u)$$

It is left to the reader to verify that this agrees with the result of integrating $p_{XY}(u,v)$ from Example 2.13 with respect to v. We will not derive the marginal distribution for Y, in this example, because we can see from symmetry that it will have the same form as the marginal distribution for X.

****

Example 2.17: Let the components of the random vector $\vec{X}$ be jointly Gaussian, which means that the probability density function can be written as

$$p_{\vec{X}}(\vec{u})=\frac{1}{(2\pi)^{n/2}|\mathbf{K}|^{1/2}}\exp\left(-\frac{1}{2}(\vec{u}-\vec{\mu})^T\mathbf{K}^{-1}(\vec{u}-\vec{\mu})\right)$$

in which $\vec{\mu}$ is a vector of constants; $\mathbf{K}$ is a square, symmetric, positive-definite matrix of constants; and $\mathbf{K}^{-1}$ and $|\mathbf{K}|$ denote the inverse and determinant, respectively, of $\mathbf{K}$. Show that any subset of the components of $\vec{X}$ also is jointly Gaussian.

First note that the exponent in $p_{\vec{X}}(\vec{u})$ is a quadratic form in all the u_j terms, making this joint distribution consistent with the scalar Gaussian distribution investigated in Examples 2.7 and 2.8—the probability density function is a constant multiplying an exponential of a quadratic form. It can be shown that any joint probability density function meeting this condition can be written in the standard form given here.

Next let us find the probability density function of components X_1 to X_{n-1}. To do this we need to integrate $p_{\vec{X}}(\vec{u})$ over all possible values of X_n. This integration will be easier if we first rearrange the exponent in $p_{\vec{X}}(\vec{u})$ as follows

$$-\frac{1}{2}(\vec{u}-\vec{\mu})^T\mathbf{K}^{-1}(\vec{u}-\vec{\mu})\equiv-\frac{1}{2}\sum_{j=1}^{n}\sum_{k=1}^{n}K_{jk}^{-1}(u_j-\mu_j)(u_k-\mu_k)=$$

$$-\frac{K_{nn}^{-1}}{2}(u_n-\mu_n)^2-(u_n-\mu_n)\sum_{k=1}^{n-1}K_{jn}^{-1}(u_j-\mu_j)-\frac{1}{2}\sum_{j=1}^{n-1}\sum_{k=1}^{n-1}K_{jk}^{-1}(u_j-\mu_j)(u_k-\mu_k)$$

in which the symmetry of $\mathbf{K}^{-1}$ (which follows from the symmetry of $\mathbf{K}$) has been used to simplify the terms that are linear in $(u_n - \mu_n)$. Note that K_{nn}^{-1} (as well as each of the other K_{jj}^{-1} terms) must be greater than zero so that the probability density function has a finite integral. Now one can use

$$\sigma = \frac{1}{K_{nn}^{-1/2}}, \qquad \mu = \mu_n - \frac{1}{K_{nn}^{-1}} \sum_{k=1}^{n-1} K_{jn}^{-1}(u_j - \mu_j)$$

and rewrite the exponent as

$$-\frac{1}{2}(\vec{u}-\vec{\mu})^T \mathbf{K}^{-1}(\vec{u}-\vec{\mu}) = -\frac{1}{2}\left(\frac{u_n - \mu}{\sigma}\right)^2 + \frac{1}{2\sigma^2}\left(\sum_{k=1}^{n-1} K_{jn}^{-1}(u_j - \mu_j)\right)^2 - \frac{1}{2}\sum_{j=1}^{n-1}\sum_{k=1}^{n-1} K_{jk}^{-1}(u_j - \mu_j)(u_k - \mu_k)$$

This separation of the terms depending on u_n from the other terms allows the desired probability density function to be written as

$$p_{X_1 \cdots X_{n-1}}(u_1,\cdots,u_{n-1}) = C\int_{-\infty}^{\infty} \exp\left(-\frac{1}{2}\left[\frac{u_n - \mu}{\sigma}\right]^2\right) du_n = C(2\pi)^{1/2}\sigma$$

in which the value of the integral has been determined by observation from comparison with the probability density function of Example 2.7. Substituting for the term C then gives

$$p_{X_1 \cdots X_{n-1}}(u_1,\cdots,u_{n-1}) = \frac{(2\pi)^{1/2}\sigma}{(2\pi)^{n/2}\sqrt{|\mathbf{K}|}} \exp\left[\frac{1}{2\sigma^2}\left(\sum_{k=1}^{n-1} K_{jn}^{-1}(u_j - \mu_j)\right)^2 - \frac{1}{2}\sum_{j=1}^{n-1}\sum_{k=1}^{n-1} K_{jk}^{-1}(u_j - \mu_j)(u_k - \mu_k)\right]$$

Note that the exponent in this probability density function is a quadratic form in all the u_j terms for $j = 1,\cdots,n-1$. This is sufficient, without further simplification, to demonstrate that $p_{X_1 \cdots X_{n-1}}(u_1,\cdots,u_{n-1})$ is a jointly Gaussian distribution. This procedure can be extended to show that if X_1 to X_n are jointly Gaussian, then any subset of those random variables is also jointly Gaussian, including the limiting case that each individual random variable in the set is Gaussian. This is a very important property of Gaussian distributions.

****

2.6 Distribution of a Function of a Random Variable

For any reasonably smooth function $g(\cdot)$, we can define a new random variable $Y = g(X)$ in terms of a given random variable X. We now want to consider how the probability distribution of Y relates to that of X. First, let us consider the simplest case of g being a monotonically increasing function. Then g has a unique inverse $g^{-1}(\cdot)$, and the event $\{Y \le v\}$ is identical to the event $\{X \le g^{-1}(v)\}$. Thus, $F_Y(v) = F_X[g^{-1}(v)]$, and taking the derivative with respect to v of both sides of this equation gives the probability density function as

$$p_Y(v) = p_X[g^{-1}(v)] \frac{d}{dv} g^{-1}(v)$$

or

$$p_Y(v) = \frac{p_X[g^{-1}(v)]}{\left(\frac{dg(u)}{du}\right)_{u=g^{-1}(v)}} \qquad \text{for } g \text{ monotonically increasing}$$

Similarly, if g is decreasing, then $\{Y \le v\} = \{X \ge g^{-1}(v)\}$, so $F_Y(v) = 1 - F_X[g^{-1}(v)]$ and

$$p_Y(v) = \frac{p_X[g^{-1}(v)]}{-\left(\frac{dg(u)}{du}\right)_{u=g^{-1}(v)}} \qquad \text{for } g \text{ monotonically decreasing}$$

That the derivative of g is positive if g is monotonically increasing and negative if g is monotonically decreasing allows these two probability density function results to be combined as

$$p_Y(v) = \frac{p_X[g^{-1}(v)]}{\left|\frac{dg(u)}{du}\right|_{u=g^{-1}(v)}} \qquad \text{if } g \text{ is monotonic} \tag{2.39}$$

Note that the point $X = g^{-1}(v)$ maps into $Y = v$. Thus, it is surely logical that the probability of Y being in the neighborhood of v depends only on the probability of X being in the neighborhood of $g^{-1}(v)$. The derivative of g appearing in the probability density function reflects the fact that an increment of length du

does not map into an increment dv of equal length. Thus, the expression in Eq. 2.39 must include the dv/du ratio given in the denominator. An important special case of the monotonic function is the linear relationship $Y = c + bX$, for which $p_Y(v) = p_X[(v-c)/b]/|b|$.

For a general g function, there may be many inverse points $u = g^{-1}(v)$. In this situation, Eq. 2.39 becomes

$$p_Y(v) = \sum_j \frac{p_X[g_j^{-1}(v)]}{\left| \frac{d\,g(u)}{du} \right|_{u=g_j^{-1}(v)}} \tag{2.40}$$

with the summation being over all points $X = g_j^{-1}(v)$ that map into $Y = v$.

The results given here can also be generalized to situations involving vectors $\vec{X}$ and $\vec{Y} = \vec{g}(\vec{X})$. The situation corresponding to the monotonic scalar g function is when $\vec{g}$ has a unique inverse. This can happen only if the dimensions of $\vec{X}$ and $\vec{Y}$ are the same, and in such cases one can obtain a result that resembles Eq. 2.39:

$$p_{\vec{Y}}(\vec{v}) = \frac{p_{\vec{X}}[\vec{g}^{-1}(\vec{v})]}{\left\| \frac{d\,\vec{g}(\vec{u})}{d\vec{u}} \right\|_{\vec{u}=\vec{g}^{-1}(\vec{v})}} \quad \text{if the inverse is unique} \tag{2.41}$$

in which the derivative term in the denominator denotes a matrix

$$\frac{d\,\vec{g}(\vec{u})}{d\vec{u}} = \begin{pmatrix} \frac{\partial g_1(\vec{u})}{\partial u_1} & \frac{\partial g_1(\vec{u})}{\partial u_2} & \cdots & \frac{\partial g_1(\vec{u})}{\partial u_n} \\ \frac{\partial g_2(\vec{u})}{\partial u_1} & \frac{\partial g_2(\vec{u})}{\partial u_2} & \cdots & \frac{\partial g_2(\vec{u})}{\partial u_n} \\ \vdots & \vdots & \ddots & \vdots \\ \frac{\partial g_n(\vec{u})}{\partial u_1} & \frac{\partial g_n(\vec{u})}{\partial u_2} & \cdots & \frac{\partial g_n(\vec{u})}{\partial u_n} \end{pmatrix} \tag{2.42}$$

and the double bars denote the absolute value of the determinant of the matrix. It may also be noted that the determinant of the matrix is called the Jacobian of the transformation, so the denominator in Eq. 2.41 is the absolute value of the Jacobian.

As with the scalar situation, the linear transformation is an important special case. This can be written for the vectors as $\vec{Y}=\vec{C}+\mathbf{B}\vec{X}$, in which **B** is a square matrix. It is then easily shown that the matrix in the denominator of Eq. 2.41 is exactly **B**, so

$$p_{\vec{Y}}(\vec{v})=\frac{p_{\vec{X}}[\vec{g}^{-1}(\vec{v})]}{\|\mathbf{B}\|}=\frac{p_{\vec{X}}[\mathbf{B}^{-1}(\vec{v}-\vec{C})]}{\|\mathbf{B}\|} \quad \text{for } \vec{Y}=\vec{C}+\mathbf{B}\vec{X}$$

provided that **B** is not singular.

There are, of course, also many situations in which we have a function of multiple random variables, but in which the inverse is not unique. For example, we might have a scalar random variable Y defined as $Y=g(X_1,\cdots,X_n)$. In this case we can use the same principle as was used in finding Eq. 2.39. That is, we can write

$$F_Y(v)=\underset{g(\vec{u})\le v}{\int\cdots\int} p_{\vec{X}}(\vec{u})\,du_1\cdots du_n$$

then find $p_Y(v)$ by taking the derivative with respect to v.

****

Example 2.18: Let X have the probability density function

$$p_X(u)=(1-u)\,U(u+1)\,U(1-u)/2$$

and let $Y=X^2$. Find the $p_Y(v)$ probability density function.

Clearly we can write $Y=g(X)$ with $g(u)=u^2$. The inverse function then is double valued, with $u=g^{-1}(v)=\pm v^{1/2}$, and the possible values of Y are in the range of 0 to 1. For $0<v\le 1$, using Eq. 2.40 gives

$$p_Y(v)=\frac{p_X(v^{1/2})}{2v^{1/2}}+\frac{p_X(-v^{1/2})}{2v^{1/2}}=\frac{1-v^{1/2}}{4v^{1/2}}+\frac{1+v^{1/2}}{4v^{1/2}}=\frac{1}{2v^{1/2}}$$

so that $p_Y(v)=1/(2v^{1/2})\,U(v)\,U(1-v)$. Note that $p_Y(v)$ tends to infinity for $v\to 0$, but it is integrable at this singularity.

****

Example 2.19: Let the components of $\vec{X}$ be jointly Gaussian, as in Example 2.17 with

$$p_{\vec{X}}(\vec{u})=\frac{1}{(2\pi)^{n/2}\,|\mathbf{K}|^{1/2}}\exp\left(-\frac{1}{2}(\vec{u}-\vec{\mu})^T\,\mathbf{K}^{-1}(\vec{u}-\vec{\mu})\right)$$

and let $\vec{Y} = \mathbf{B}\vec{X}$, in which $\mathbf{B}$ is a nonsingular square matrix. Show that the components of $\vec{Y}$ are jointly Gaussian.

Because the transformation is linear, we can use Eq. 2.42 to obtain

$$p_{\vec{Y}}(\vec{v}) = \frac{p_{\vec{X}}(\mathbf{B}^{-1}\vec{v})}{\|\mathbf{B}\|} = \frac{1}{(2\pi)^{n/2}|\mathbf{K}|^{1/2}\|\mathbf{B}\|}\exp\left(-\frac{1}{2}(\mathbf{B}^{-1}\vec{v}-\vec{\mu})^T\mathbf{K}^{-1}(\mathbf{B}^{-1}\vec{v}-\vec{\mu})\right)$$

Without further calculation, we can note that $p_{\vec{Y}}(\vec{v})$ has the form of a constant multiplying an exponential of a quadratic form in the v_j components, and this is sufficient to assure that $\vec{Y}$ is Gaussian. This is an example of a very important property of Gaussian distributions—a linear combination of jointly Gaussian random variables is always Gaussian.

****

2.7 Conditional Probability Distributions

Recall that the conditional probability of one event given another event is defined as the probability of the intersection divided by the probability of the conditioning event, as in Eq. 2.1. This has an intuitive interpretation as the likelihood that one event will occur simultaneously with a known occurrence of the other event.

For random variable problems, we will usually be interested in conditioning the probability distribution of one random variable based on certain events involving the outcome for that same random variable or a different random variable. If this conditioning event has a nonzero probability, then Eq. 2.1 provides an adequate tool for defining the conditional distribution. For example, for a random variable X having the entire real line as possible values, we can define a conditional cumulative distribution function given $X \le 10$ as

$$F_X(u \mid X \le 10) \equiv P(X \le u \mid X \le 10) = \frac{P[X \le \min(u,10)]}{P(X \le 10)} = \frac{F_X[\min(u,10)]}{F_X(10)}$$

Thus,

$$F_X(u \mid X \le 10) = \frac{F_X(u)}{F_X(10)} \qquad \text{for } u \le 10$$

$$F_X(u \mid X \le 10) = 1 \qquad \text{for } u > 10$$

The $P(X \le u)$ and $P(X \le 10)$ terms can be evaluated directly from the unconditional cumulative distribution function $F_X(u)$, if that is known, or from integration of the unconditional probability density function $p_X(u)$. One can then define a conditional probability density function for X as

$$p_X(u|X \le 10) \equiv \frac{d}{du} F_X(u|X \le 10)$$

giving

$$p_X(u|X \le 10) = \frac{p_X(u)}{F_X(10)} \qquad \text{for } u \le 10$$

$$p_X(u|X \le 10) = 0 \qquad \text{for } u > 10$$

Similarly, in a problem involving the two random variables X and Y, one might need to compute a conditional distribution for X given some range of outcomes for Y. For example, one particular conditional cumulative distribution function is

$$F_X(u|Y \le 10) \equiv P(X \le u|Y \le 10) = \frac{P(X \le u, Y \le 10)}{P(Y \le 10)} = \frac{F_{XY}(u,10)}{F_Y(10)}$$

and the corresponding conditional probability density function is

$$p_X(u|Y \le 10) \equiv \frac{d}{du} F_X(u|Y \le 10) = \frac{1}{F_Y(10)} \frac{d}{du} F_{XY}(u,10)$$

These procedures can easily be extended to situations with more than two random variables. In summary, whenever the conditioning event has a nonzero probability, the recommended procedure for finding the conditional probability distribution for some random variable X is first to write the conditional cumulative distribution according to Eq. 2.1, as was done in the two examples. After that, one can obtain the conditional probability density function by taking a derivative:

$$F_X(u|B) \equiv \frac{P(\{X \le u\} \cap B)}{P(B)} \qquad \text{for } P(B) > 0 \tag{2.43}$$

and

$$p_X(u|B) \equiv \frac{d}{du} F_X(u|B) \qquad \text{for } P(B) > 0 \tag{2.44}$$

****

Example 2.20: Reconsider the random variables X and Y of Examples 2.11 and 2.14, with joint probability distribution

$$p_{XY}(u,v) = 1/6 \qquad \text{for } -1 \le u \le 2,\ -1 \le v \le 1$$

$$p_{XY}(u,v) = 0 \qquad \text{otherwise}$$

Find the conditional cumulative distribution function and conditional density function for the random variable X given that $Y \ge 0.5$.

To calculate the conditional probability of any event A given the event $Y \ge 0.5$, we need the probability of $Y \ge 0.5$ and the probability of the intersection of A with $Y \ge 0.5$. Thus, to compute the conditional cumulative distribution function for X, we need

$$P(Y > 0.5) = \int_{0.5}^{\infty} \int_{-\infty}^{\infty} p_{XY}(u,v)\, du\, dv = \int_{0.5}^{1} \int_{-1}^{2} \left(\frac{1}{6}\right) du\, dv = 0.25$$

and

$$P(X \le u, Y > 0.5) = \int_{0.5}^{\infty} \int_{-\infty}^{u} p_{XY}(u,v)\, du\, dv$$

which gives

$$P(X \le u, Y > 0.5) = 0 \qquad \text{for } u < -1$$

$$P(X \le u, Y > 0.5) = \frac{u+1}{12} \qquad \text{for } -1 \le u \le 2$$

$$P(X \le u, Y > 0.5) = \frac{1}{4} \qquad \text{for } u > 2$$

Taking the ratio of probabilities, as in Eq. 2.43, gives the conditional cumulative distribution function as

$$F_X(u \mid Y > 0.5) = 0 \qquad \text{for } u < -1$$

$$F_X(u \mid Y > 0.5) = \frac{u+1}{3} \qquad \text{for } -1 \le u \le 2$$

$$F_X(u \mid Y > 0.5) = 1 \qquad \text{for } u > 2$$

****

Example 2.21: For the random variables of Example 2.12 for which the probability density function may be written as

$$p_{XY}(u,v) = U(u-1)\,U(5-u)\,U(v-1)\,U(u-v)/8$$

find the conditional cumulative distribution function and conditional density function for the random variable Y given that $X \le 3$.

Proceeding as in the previous example:

$$P(X \le 3) = \int_{-\infty}^{\infty} \int_{v}^{3} \left(\frac{1}{8}\right) du\, dv = \frac{1}{4}$$

and

$$P(X \le 3, Y \le v) = 0 \qquad \text{for } v < 1$$

$$P(X \le 3, Y \le v) = \int_{1}^{v} \int_{w}^{3} \left(\frac{1}{8}\right) du\, dw = \frac{(v-1)(5-v)}{16} \qquad \text{for } 1 \le v \le 3$$

$$P(X \le 3, Y \le v) = 1/4 \qquad \text{for } v > 3$$

so

$$F_Y(v \mid X \le 3) = 0 \qquad \text{for } v < 1$$
$$F_Y(v \mid X \le 3) = (v-1)(5-v)/4 \qquad \text{for } 1 \le v \le 3$$
$$F_Y(v \mid X \le 3) = 1 \qquad \text{for } v > 3$$

Differentiating this expression gives the conditional probability density function as

$$p_Y(v \mid X \le 3) = [(3-v)/2] U(v-1) U(3-v)$$

****

A slightly different sort of conditioning arises in many random variable problems. In particular, one may wish to condition by an event with zero probability, and this requires a new definition, because Eq. 2.1 no longer applies. In particular, one often is interested in the conditional distribution of one random variable given a precise value of another random variable. If the conditioning random variable has a continuous distribution, then this gives a conditioning event with zero probability. The definition of the conditional distribution for the situation in which X and Y have a continuous joint distribution is in terms of the conditional probability density function:

$$p_X(u \mid Y = v) \equiv \frac{p_{XY}(u,v)}{p_Y(v)} \tag{2.45}$$

which is defined only for v values giving $p_Y(v) \ne 0$. Note that Eq. 2.45 is very similar in form to Eq. 2.1, but the terms in Eq. 2.45 are all probability density functions, whereas those in 2.1 are probabilities. Of course, Eq. 2.45 is consistent with a limit of Eq. 2.1, and one can convert Eq. 2.45 into infinitesimal increments of probability by multiplying the density terms by infinitesimal increments of length:

$$p_X(u \mid Y = v)\, du = \frac{p_{XY}(u,v)\, du\, dv}{p_Y(v)\, dv}$$

which is essentially a statement that

$$P(u < X < u + du \mid Y = v) = \frac{P(u < X < u + du, v < Y < v + dv)}{P(v < Y < v + dv)}$$

which is of the form of Eq. 2.1 for the infinitesimal probabilities. Because the conditional distribution in this situation is defined by a conditional probability density function, one must integrate if the conditional cumulative distribution function is needed:

$$F_X(u \mid Y = v) = \int_{-\infty}^{u} p_X(w \mid Y = v)\, dw \tag{2.46}$$

****

Example 2.22: Find the conditional probability distribution of X given the event $Y = v$ and the conditional distribution of Y given $X = u$ for the joint probability distribution of Examples 2.11 and 2.14 with

$p_{XY}(u,v) = 1/6$ for $-1 \le u \le 2, -1 \le v \le 1$

$p_{XY}(u,v) = 0$ otherwise

We note that $p_X(u \mid Y = v)$ is defined only for $-1 \le v \le 1$, because $p_Y(v)$ is zero otherwise. For $-1 \le v \le 1$ we use Eq. 2.45 and the marginal probability density function derived in Example 2.14 to obtain

$$p_X(u \mid Y = v) = \frac{p_{XY}(u,v)}{p_Y(v)} = \left(\frac{1}{3}\right) U(u+1)U(2-u)$$

Integrating this expression gives the conditional cumulative distribution function as

$$F_X(u \mid Y = v) = \left(\frac{u+1}{3}\right)[U(u+1) - U(u-2)] + U(u-2)$$

Similarly,

$$p_Y(v \mid X = u) = \frac{p_{XY}(u,v)}{p_Y(v)} = \left(\frac{1}{2}\right) U(u+1)U(2-u)$$

and

$$F_Y(v \mid X = u) = \left(\frac{v+1}{2}\right)[U(v+1) - U(v-1)] + U(v-1)$$

for $-1 \le u \le 2$, and the conditional distribution is undefined for other values of u.

****

Example 2.23: For the random variables of Examples 2.12 and 2.15 with

$$p_{XY}(u,v) = U(u-1)U(5-u)U(v-1)U(u-v)/8$$

find the conditional distribution of X given the event $Y = v$ and the conditional distribution of Y given $X = u$.

In Example 2.15, we found that

$$p_X(u) = (u-1)U(u-1)U(5-u)/8, \qquad p_Y(v) = (5-v)U(v-1)U(5-v)/8$$

Thus, we can now take the ratio of joint and marginal probability density functions, according to Eq. 2.45, to obtain

$$p_Y(v \mid X = u) = \frac{p_{XY}(u,v)}{p_X(u)} = \left(\frac{1}{u-1}\right) U(v-1)U(u-v) \qquad \text{for } 1 \le u \le 5$$

and

$$p_X(u \mid Y = v) = \frac{p_{XY}(u,v)}{p_Y(v)} = \left(\frac{1}{5-v}\right) U(u-v)U(5-u) \qquad \text{for } 1 \le v \le 5$$

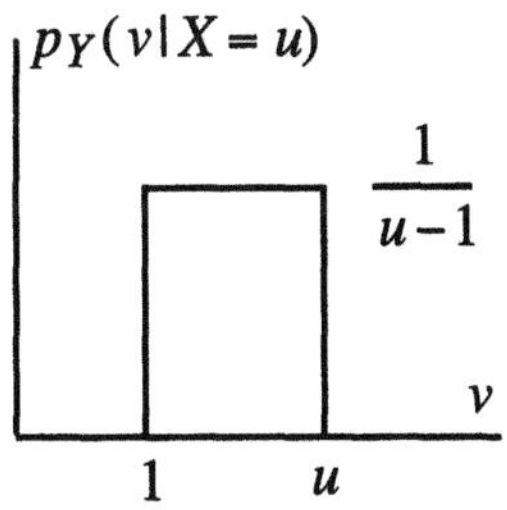

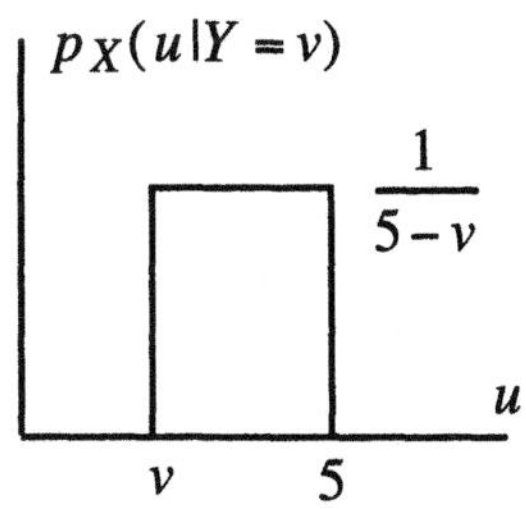

These conditional density functions are shown in the sketches. They show that when $X = u$ is known, the set of possible values of Y is limited to the interval $[1,u]$, and $[v,5]$ gives the set of possible values for X when $Y = v$ is known. On these sets of possible values, both of the conditional distributions are uniform (each of the conditional density functions is a constant). Integrating these conditional probability density functions gives the conditional cumulative distribution functions as

$$F_Y(v \mid X = u) = \left(\frac{v-1}{u-1}\right) U(v-1)U(u-v) \qquad \text{for } 1 \le u \le 5$$

and

$$F_X(u \mid Y = v) = \left(\frac{u-v}{5-v}\right) U(u-v)U(5-u) \qquad \text{for } 1 \le v \le 5$$

These are also sketched.

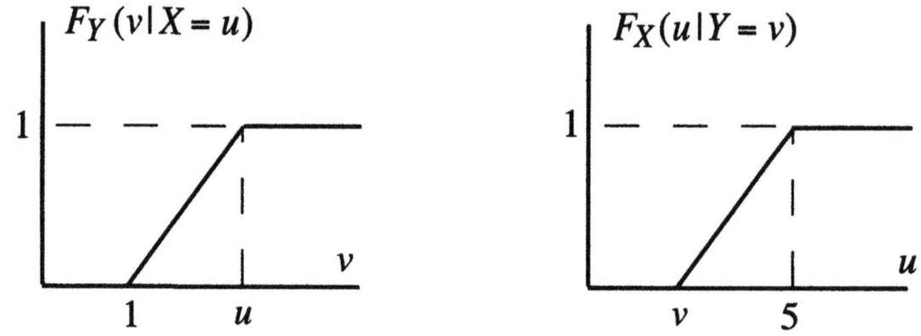

****

Example 2.24: Find the conditional distribution of X given $Y=v$ for the probability distribution of Examples 2.13 and 2.16 with

$$p_{XY}(u,v)=\left(8e^{-4u-4v}+3e^{-4u-3v}+3e^{-3u-4v}\right)U(u)U(v)$$

We can use symmetry and the marginal probability density function derived in Example 2.16 to write

$$p_Y(v)=3\left(e^{-4v}+e^{-3v}/4\right)U(v)$$

Then Eq. 2.45 gives

$$p_X(u|Y=v)=\frac{\left(8e^{-4u-4v}+3e^{-4u-3v}+3e^{-3u-4v}\right)}{3\left(e^{-4v}+e^{-3v}/4\right)}U(u) \quad \text{for } v\ge 0$$

As a more specific example,

$$p_X(u|Y=4)=\left(3.91e^{-4u}+0.0683e^{-3u}\right)U(u)$$

****

Example 2.25: Let the components of $\vec{X}$ be jointly Gaussian, as in Example 2.17 with

$$p_{\vec{X}}(\vec{u})=\frac{1}{(2\pi)^{n/2}|\mathbf{K}|^{1/2}}\exp\left(-\frac{1}{2}(\vec{u}-\vec{\mu})^T\mathbf{K}^{-1}(\vec{u}-\vec{\mu})\right)$$

Show that the joint conditional distribution of X_1 to X_{n-1} given $X_n=u_n$ has a jointly Gaussian form.

In Example 2.17 we showed that the components are individually Gaussian, so we can write

$$p_{X_n}(u_n) = \frac{1}{(2\pi)^{1/2}\sigma_n} \exp\left[-\frac{1}{2}\left(\frac{u_n - \mu_n}{\sigma_n}\right)^2\right]$$

Using this along with $p_{\vec{X}}(\vec{u})$ gives

$$p_{X_1 \cdots X_{n-1}}(u_1, \cdots, u_{n-1} \mid X_n = u_n) = \frac{\exp\left[-\frac{1}{2}\sum_{j=1}^{n}\sum_{k=1}^{n} K_{jk}^{-1}(u_j - \mu_j)(u_k - \mu_k)\right]}{\frac{(2\pi)^{n/2}|\mathbf{K}|^{1/2}}{(2\pi)^{1/2}\sigma_n}\exp\left[-\frac{1}{2}\left(\frac{u_n - \mu_n}{\sigma_n}\right)^2\right]}$$

We now note that this conditional probability density function has the form of a constant multiplying an exponential of a quadratic form in the u_1 to u_{n-1} components. This is sufficient to ensure that the conditional distribution is Gaussian. This is an example of another general property of Gaussian distributions—if the components of $\vec{X}$ are jointly Gaussian, then the conditional distribution of any subset of these components is also Gaussian.

****

It is important to remember that any conditional cumulative distribution function or conditional probability density function has the same mathematical characteristics as any other cumulative distribution function or probability density function. Specifically, any conditional cumulative distribution function, such as that given by Eq. 2.43 or 2.46, satisfies the condition of being monotonically increasing from zero to unity as one considers all possible arguments of the function (i.e., as u is increased from negative infinity to positive infinity). Similarly, any conditional probability density function, such as that given by Eq. 2.44 or 2.45, satisfies the conditions of nonnegativity and having a unit integral (Eq. 2.9), which are necessary for any probability density function.

In many situations, it is convenient to give the initial definition of a problem in terms of conditional distributions rather than in terms of joint distributions. We can always rewrite Eq. 2.45 as

$$p_{XY}(u,v) = p_X(u \mid Y = v)\, p_Y(v) \tag{2.47}$$

or

$$p_{XY}(u,v) = p_Y(v \mid X = u)\, p_X(u) \tag{2.48}$$

so if we know the marginal distribution for one random variable and the conditional distribution for a second random variable given the value of the first, then we also know the joint distribution.

****

Example 2.26: Let a random variable X be uniform on the set $[0,10]$ so that its probability density function is given by

$$p_X(u) = U(u)U(10-u)/10$$

Let another random variable Y be uniform on the set $[0,X]$; that is, if we are given the information that $X = u$, then Y is uniform on $[0,u]$:

$$p_Y(v \mid X = u) = \left(\frac{1}{u}\right) U(v)U(u-v)$$

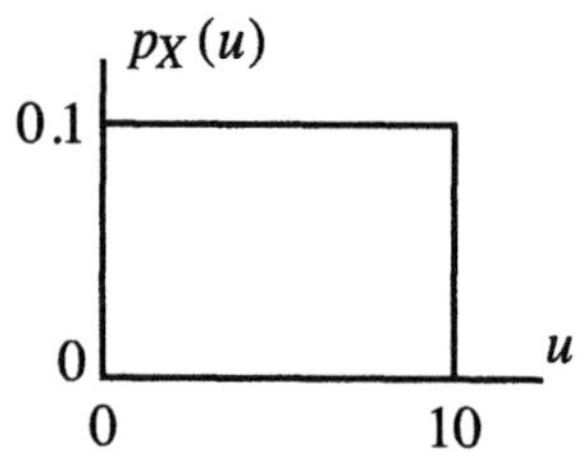

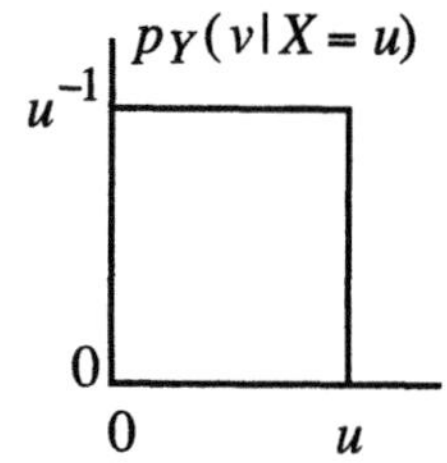

Find the joint probability density function of X and Y, and identify the domain on which this density function is nonzero.

Substituting into Eq. 2.48 gives

$$p_{XY}(u,v) = \left(\frac{1}{10\,u}\right) U(u)U(10-u)U(v)U(u-v)$$

and this function is nonzero on the triangular region described by $0 \le v \le u \le 10$. This region is shown on a sketch of the (u,v) plane.

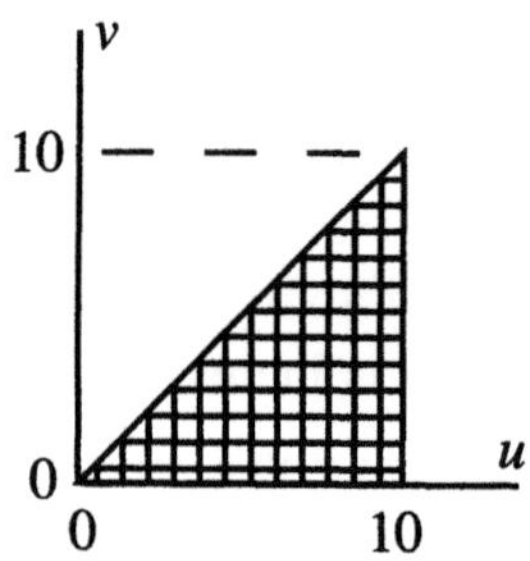

****

Example 2.27: Let the random variable X again be uniform with probability density function

$$p_X(u) = U(u)U(10-u)/10$$

Let Y be a biased estimate of X such that it is always greater than X. In particular, let the conditional distribution be of the exponential form

$$p_Y(v \mid X = u) = b e^{-b(v-u)} U(v-u)$$

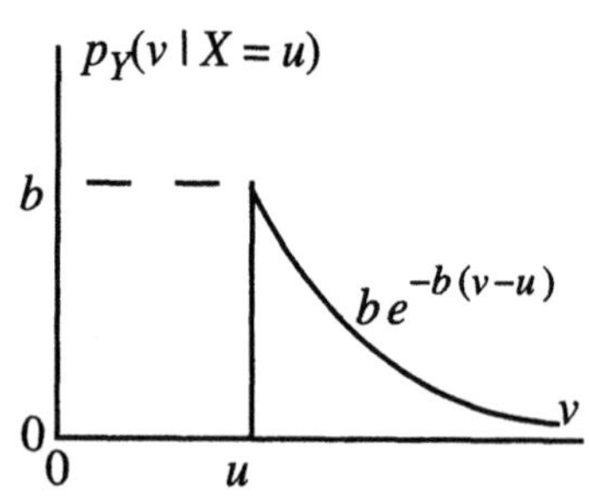

in which $b>0$ is a known constant. Find the joint probability density function of X and Y, and identify the domain on which this density function is nonzero.

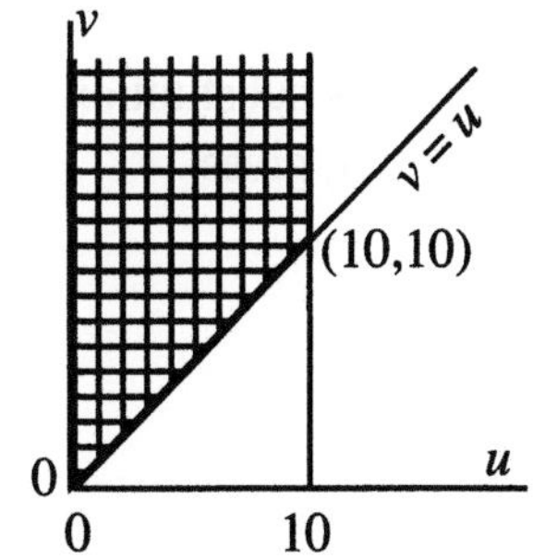

Multiplying the marginal and conditional probability densities gives

$p_{XY}(u,v) = (b/10)e^{-b(u-v)}U(u)U(10-u)U(v-u)$
and this is nonzero on the semi-infinite strip shaded in the sketch.

****

It is also possible to use the idea of conditional and joint probability distributions to describe a problem in which one random variable is a function of another random variable. However, such functional relationships always give degenerate conditional and joint distributions. For example, let *Y* be a function of *X*, say $Y = g(X)$. Now if we are given the event $X = u$, then the only possible value of *Y* is $g(u)$. By using the Dirac delta function, though, we can write a conditional probability density function as $p_Y(v \mid X = u) = \delta[v - g(U)]$. This function is an acceptable probability density function, because it does give unity when integrated over the real line and it is nonzero only when *v* is equal to the possible value for *Y*. By using Eq. 2.48, we can now obtain a joint probability density function as $p_{XY}(u,v) = p_X(u)\delta[v - g(U)]$. Clearly this joint probability density function is nonzero only on a one-dimensional subset of the two-dimensional (u,v) plane, and it is infinite on this one-dimensional subset. This degeneracy is typical of what one obtains when a joint distribution is used to describe the relationship between one random variable and a function of that random variable. It is also possible to use this relationship to find $p_Y(v)$ for $Y = g(X)$, but this requires an additional property of the Dirac delta function. To develop this property, note that

$$U[v - g(u)] = U[g^{-1}(v) - u] \quad \text{if } g'[g^{-1}(u)] > 0$$

$$U[v - g(u)] = U[u - g^{-1}(v)] \quad \text{if } g'[g^{-1}(u)] < 0$$

provided that the inverse is unique [i.e., that $g(\cdot)$ is a monotonic function]. Taking the derivative with respect to *v* then gives

$$\delta[v - g(u)] = \left| \frac{d\, g^{-1}(v)}{d\, v} \right| \delta[g^{-1}(v) - u] = \frac{\delta[u - g^{-1}(v)]}{|\, g'[g^{-1}(v)]\,|} \tag{2.49}$$

Making this substitution in $p_{XY}(u,v) = p_X(u)\delta\big[v - g(U)\big]$, then integrating with respect to u, gives

$$p_Y(v) = \frac{p_X[g^{-1}(v)]}{|g'[g^{-1}(v)]|}$$

which is identical with Eq. 2.39.

2.8 Independence of Random Variables

The concept of independent random variables is essentially the same as that of independent events, as outlined in Section 2.1. The intuitive idea is that a random variable Y is independent of another random variable X if knowledge of the value of X gives absolutely no information about the possible values of Y, or about the likelihood that Y will take on any of those possible values. Because conditional probability has been defined precisely for the purpose of conveying this type of informational connection between random variables, it would be natural to use conditional probabilities in defining the concept of independence. However, there are minor mathematical difficulties in using this approach for the definition. Thus, we will use an alternative statement as the definition of independence, then show that the definition is consistent with the intuitive concept. Specifically, we will say that two random variables X and Y are defined to be independent if and only if

$$p_{XY}(u,v) = p_X(u)\, p_Y(v) \qquad \text{for all } u \text{ and } v \qquad (2.50)$$

or, equivalently,

$$F_{XY}(u,v) = F_X(u)\, F_Y(v) \qquad \text{for all } u \text{ and } v \qquad (2.51)$$

Thus, the definition is in terms of joint and marginal probability distributions rather than conditional distributions.

Using Eq. 2.50 along with Eq. 2.45 to compute the conditional distributions for independent X and Y gives

$$p_Y(v \mid X = u) = p_Y(v) \qquad \text{provided that } p_X(u) \neq 0 \qquad (2.52)$$

and

$$p_X(u \mid Y = v) = p_X(u) \qquad \text{provided that } p_Y(v) \neq 0 \qquad (2.53)$$

One can also show that

$$F_Y(v \mid X = u) = F_Y(v) \qquad \text{provided that } p_X(u) \neq 0 \tag{2.54}$$

and

$$F_X(u \mid Y = v) = F_X(u) \qquad \text{provided that } p_Y(v) \neq 0 \tag{2.55}$$

and various other statements such as

$$p_Y(v \mid X \le u) = p_Y(v) \qquad \text{provided that } P(X \le u) \neq 0 \tag{2.56}$$

or

$$F_X(u \mid Y > v) = F_X(u) \qquad \text{provided that } P(Y > v) \neq 0 \tag{2.57}$$

Equations 2.52–2.57 are examples of the intuitive idea of independence discussed in the first paragraph of this section. Specifically, independence of X and Y implies that knowledge of the value of X gives absolutely no information about the probability distribution of Y, and knowledge of Y gives absolutely no information about the distribution of X. Expressions such as Eqs. 2.52–2.55 have not been used as the definition of independence simply because, for many problems, each of these equations may be defined only for a subset of all possible (u,v) values, whereas Eqs. 2.50 and 2.51 hold for all (u,v) values.

****

Example 2.28: Reconsider the probability distribution of Examples 2.11, 2.14, 2.20, and 2.22 with

$$p_{XY}(u,v) = U(u+1)\,U(2-u)\,U(v+1)\,U(1-v)/6$$

Are X and Y independent?

We have already found the marginal distributions for this problem in Example 2.14, so they can easily be used to check for independence. In particular, we can use the previously evaluated functions $p_X(u) = U(u+1)\,U(2-u)/3$ and $p_Y(v) = U(v+1)\,U(1-v)/2$ to verify that for every choice of u and v we get $p_{XY}(u,v) = p_X(u)\,p_Y(v)$. Thus, Eq. 2.50 is satisfied, and X and Y are independent.

We could, equally well, have demonstrated independence by using the joint and marginal cumulative distribution functions to show that Eq. 2.51 was satisfied. We can also use the conditional probabilities that have been derived for this problem in Examples 2.20 and 2.22 to verify the intuitive idea of independence—that knowledge of one of the random variables gives no information about the distribution of the other random variable. For example, we found $F_X(u \mid Y > 0.5)$ in Example

2.20, and it is easily verified that this is identical to the marginal cumulative distribution function $F_X(u)$ for all values of u. Also, we found $p_X(u|Y=v)$, and this is identical to $p_X(u)$, provided that $-1 \le v \le 1$ so that the conditional distribution is defined. These are two of many equations that we could write to confirm that knowledge of Y gives no information about either the possible values of X, or the probability distribution on those possible values.

****

Example 2.29: Determine whether X and Y are independent for the probability distribution of Examples 2.12, 2.15, 2.21, and 2.23 with

$$p_{XY}(u,v) = U(u-1)U(5-u)U(v-1)U(u-v)/8$$

Using the marginal distributions derived for this problem in Example 2.15, we have

$$p_X(u)p_Y(v) = \left(\frac{u-1}{8}\right)\left(\frac{5-v}{8}\right)U(u-1)U(5-u)U(v-1)U(5-v)$$

and this is clearly not the same as $p_{XY}(u,v)$. Thus, X and Y are not independent. In comparing $p_X(u)p_Y(v)$ with $p_{XY}(u,v)$, it is worthwhile to note two types of differences. Probably the first discrepancy that the reader will note is that on the domain where both are nonzero the functions are $(u-1)(5-v)/64$ and $1/8$, and these clearly are not the same functions of u and v. One can also note, though, that $p_X(u)p_Y(v)$ is nonzero on the square defined by $1 \le u \le 5$, $1 \le v \le 5$, whereas $p_{XY}(u,v)$ is nonzero on only the lower right half of this square. This difference of the domain of nonzero values of the functions is sufficient to prove that the functions are not the same. This latter comparison is sometimes an easy way to show that $p_X(u)p_Y(v) \ne p_{XY}(u,v)$ in a problem in which both the joint and marginal probability density functions are complicated.

Recall also that in Examples 2.21 and 2.23 we derived conditional probability density functions for this problem:

$$p_Y(v|X \le 3) = \left(\frac{3-v}{2}\right)U(v-1)U(3-v)$$

and

$$p_Y(v|X=u) = \left(\frac{1}{u-1}\right)U(v-1)U(u-v)$$

If X and Y were independent, then both of these conditional probability density functions for Y would be the same as its marginal probability density function, which we found in Example 2.15 to be

$$p_Y(v) = \left(\frac{5-v}{8}\right)U(v-1)U(5-v)$$

Clearly neither of these conditional functions is the same as the marginal. Showing that any one conditional function for Y given information about X is not the same as the corresponding marginal function for X is sufficient to prove that X and Y are not independent. Knowing something about the value of X (such as $X = u$ or $X \le 3$) does give information about the distribution of Y in this example. In particular, the information given about X restricts the range of possible values of Y, as well as the probability distribution on those possible values. The following example will illustrate a case in which the probabilities for one random variable are modified by knowledge of another random variable, even though the domain is unchanged.

****

Example 2.30: Determine whether X and Y are independent for the probability distribution of Examples 2.13, 2.16, and 2.24 with

$$p_{XY}(u,v) = \left(8e^{-4u-4v} + 3e^{-4u-3v} + 3e^{-3u-4v}\right)U(u)U(v)$$

Using the marginal distribution from Example 2.16, we have

$$p_X(u)\, p_Y(v) = 9\left(e^{-4u} + e^{-3u}/4\right)\left(e^{-4v} + e^{-3v}/4\right)U(u)U(v)$$

This time we do find that the domain of nonzero values is the same for $p_X(u)\, p_Y(v)$ as for $p_{XY}(u,v)$ (i.e., the first quadrant), but the functions of u and v are not the same. Thus, X and Y are not independent.

Looking at conditional probabilities, we found in Examples 2.16 and 2.24 that the marginal distribution for X could be described by

$$p_X(u) = 3\left(e^{-4u} + e^{-3u}/4\right)U(u)$$

and the conditional distribution had

$$p_X(u \mid Y = v) = \frac{\left(8e^{-4u-4v} + 3e^{-4u-3v} + 3e^{-3u-4v}\right)}{3\left(e^{-4v} + e^{-3v}/4\right)}U(u)$$

Thus, the set of possible values of X is the semi-infinite interval $X > 0$, and knowledge of the value of Y does not change this set. Knowledge of the value of Y, though, does change the probability distribution on this set of possible values, confirming that X and Y are not independent.

****

Example 2.31: Let X_1 to X_n be independent and Gaussian with

$$p_{X_j}(u_j) = \frac{1}{(2\pi)^{1/2}\sigma_j}\exp\left[-\frac{1}{2}\left(\frac{u_j - \mu_j}{\sigma_j}\right)^2\right]$$

Show that $p_{\vec{X}}(\vec{u})$ is a special case of the jointly Gaussian distribution as given in Example 2.17.

Because the components are independent we can write

$$p_{\vec{X}}(\vec{u}) = \prod_{j=1}^{n} p_{X_j}(u_j) = \frac{1}{(2\pi)^{n/2}\sigma_1 \cdots \sigma_n} \exp\left(-\frac{1}{2} \sum_{j=1}^{n} \left(\frac{u_j - \mu_j}{\sigma_j} \right)^2 \right)$$

This, though, is identical to

$$p_{\vec{X}}(\vec{u}) = \frac{1}{(2\pi)^{n/2} |\mathbf{K}|^{1/2}} \exp\left(-\frac{1}{2} (\vec{u} - \vec{\mu})^T \mathbf{K}^{-1} (\vec{u} - \vec{\mu}) \right)$$

if we take $\mathbf{K}$ to be the diagonal matrix with $K_{jj} = \sigma_j^2$ terms on the diagonal, which gives $\mathbf{K}^{-1}$ as the diagonal matrix with σ_j^{-2} terms on the diagonal and $|\mathbf{K}| = \sigma_1^2 \cdots \sigma_n^2$.

****

Example 2.32: Let X_1 and X_2 be independent and Gaussian with the identical mean-zero probability density function

$$p_{X_j}(u_j) = \frac{1}{(2\pi)^{1/2}\sigma} e^{-u_j^2/(2\sigma^2)}$$

Find the probability distribution of $Y = (X_1^2 + X_2^2)^{1/2}$.

As in Section 2.6, we write

$$F_Y(v) = \iint_{u_1^2 + u_2^2 \le v^2} p_{X_1 X_2}(u_1, u_2)\, du_1\, du_2$$

$$= \frac{1}{2\pi\sigma^2} \iint_{u_1^2 + u_2^2 \le v^2} e^{-(u_1^2 + u_2^2)/(2\sigma^2)}\, du_1\, du_2$$

This integral is most easily evaluated by making a change of variables to polar coordinates: $u_1 = w\cos(\theta)$, $u_2 = w\sin(\theta)$. Then we have

$$F_Y(v) = \frac{1}{2\pi\sigma^2} \int_0^{2\pi} \int_0^{v} e^{-w^2/(2\sigma^2)}\, w\, dw\, d\theta = 1 - e^{-v^2/(2\sigma^2)}$$

Taking the derivative with respect to v then gives

$$p_Y(v) = \frac{v\, e^{-v^2/(2\sigma^2)}}{\sigma^2}$$

This is called the *Rayleigh distribution*. It plays an important role in many applications, and we will use it later in this book.

****

Exercises

****

Distribution of One Random Variable

****

2.1 Consider a random variable X with cumulative distribution function

$$F_X(u) = 0 \qquad \text{for } u < 0$$

$$F_X(u) = u^3 \qquad \text{for } 0 \le u < 1$$

$$F_X(u) = 1 \qquad \text{for } u \ge 1$$

Find the probability density function for X.

****

2.2 Consider a random variable X with cumulative distribution function

$$F_X(u) = (1 - u^2/2)U(u-1)$$

Find the probability density function for X.

****

2.3 Consider a random variable X with probability density function

$$p_X(u) = [2/(u+1)^3]U(u)$$

Find the cumulative distribution function for X.

****

2.4 Consider a random variable X with probability density function

$$p_X(u) = \left(\frac{3}{4}\right) u(2-u)U(u)U(2-u)$$

Find the cumulative distribution function for X.

****

2.5 Let the Gaussian random variable X, denoting the wind velocity (in m/s) at time t at a given location, have the probability density function

$$p_X(u) = \frac{1}{5(2\pi)^{1/2}} \exp\left[-\frac{1}{2}\left(\frac{u-10}{5}\right)^2\right]$$

Find the probability that X will exceed 25. [Hint: See Example 2.7.]

****

Joint and Marginal Distributions

****

2.6 Let X and Y be two random variables with the joint cumulative distribution function

$$F_{XY}(u,v) = \left((1-e^{-2u})(1-e^{-3v}) + \frac{6}{5}(u+v)e^{-2u-3v} - ue^{-2u} - ve^{-3v}\right)U(u)U(v)$$

(a) Find the joint probability density function $p_{XY}(u,v)$.
(b) Find the marginal cumulative distribution functions $F_X(u)$ and $F_Y(v)$.
(c) Find the marginal probability density functions $p_X(u)$ and $p_Y(v)$, and verify that they satisfy Eqs. 2.29 and 2.30, as well as Eq. 2.7.

****

2.7 Let X and Y be two random variables with the joint cumulative distribution function

$$
\begin{aligned}
&F_{XY}(u,v) = u^3(v-1)^2 && \text{for } 0 \le u \le 1,\ 1 \le v \le 2 \\
&F_{XY}(u,v) = 1 && \text{for } u > 1,\ v > 2 \\
&F_{XY}(u,v) = u^3 && \text{for } 0 \le u \le 1,\ v > 2 \\
&F_{XY}(u,v) = (v-1)^2 && \text{for } u > 1,\ 1 \le v \le 2 \\
&F_{XY}(u,v) = 0 && \text{otherwise}
\end{aligned}
$$

(a) Find the joint probability density function $p_{XY}(u,v)$.
(b) Find the marginal cumulative distribution functions $F_X(u)$ and $F_Y(v)$.
(c) Find the marginal probability density functions $p_X(u)$ and $p_Y(v)$, and verify that they satisfy Eqs. 2.29 and 2.30, as well as Eq. 2.7.

****

2.8 Let the joint probability density function of two random variables X and Y be given by

$p_{XY}(u,v) = C$ for (u,v) inside the shaded area on the sketch

$p_{XY}(u,v) = 0$ otherwise

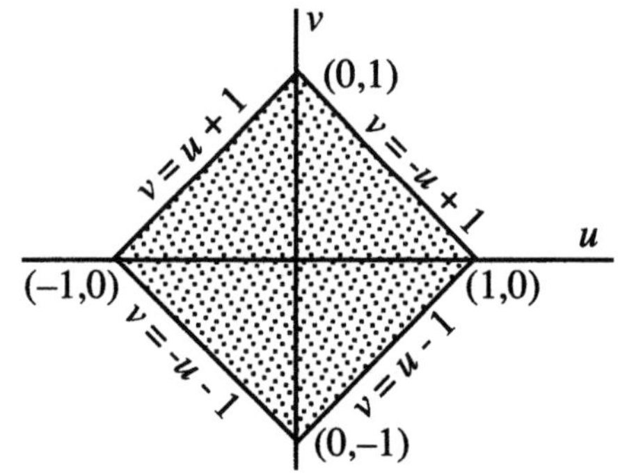

(a) Find the value of the constant C.
(b) Find both marginal probability density functions: $p_X(u)$ and $p_Y(v)$.
(c) Find $F_{XY}(0.5,0.75)$; that is, find the joint cumulative distribution function $F_{XY}(u,v)$ only for arguments of $u = 0.5$, $v = 0.75$.

****

2.9 Let X and Y be two random variables with joint probability density function

$$p_{XY}(u,v) = C\, v\, e^{-2uv-5v}\, U(u)\, U(v)$$

(a) Find the value of the constant C.
(b) Find the marginal probability density function $p_X(u)$.

(c) Find the joint cumulative distribution function $F_{XY}(u,v)$. [Hint: It is easier to do the integration in the X direction first.]

(d) Find the marginal cumulative distribution function $F_X(u)$, and verify that it satisfies Eq. 2.27, as well as Eqs. 2.7 and 2.8.

****

2.10 Let X and Y be two random variables with joint probability density function

$$p_{XY}(u,v) = C\,e^{-u-4v}\,U(u)U(v-u)$$

(a) Find the value of the constant C.

(b) Find the marginal probability density function $p_X(u)$.

(c) Find the joint cumulative distribution function $F_{XY}(u,v)$.

(d) Find the marginal cumulative distribution function $F_X(u)$, and verify that it satisfies Eq. 2.27, as well as Eqs. 2.7 and 2.8.

****

2.11 Consider two random variables X and Y with the joint probability density function

$$p_{XY}(u,v) = \frac{1}{12\pi}\exp\left[-\frac{1}{2}\left(\frac{u-1}{2}\right)^2 - \frac{1}{2}\left(\frac{v-2}{3}\right)^2\right]$$

Find the marginal probability density functions: $p_X(u)$ and $p_Y(v)$.

[Hint: See Example 2.7.]

****

2.12 Consider two random variables with a joint probability density of

$$p_{XY}(u,v) = \frac{1}{2\pi\sigma_1\sigma_2}\exp\left(-\frac{u^2}{2\sigma_1^2} - \frac{v^2}{2\sigma_2^2}\right)\left(1 + uv\exp\left(-\frac{u^2v^2}{2}\right)\right)$$

(a) Find the marginal probability density functions $p_X(u)$ and $p_Y(v)$.

(b) Does X have the Gaussian distribution?

(c) Does Y have the Gaussian distribution?

(d) Are X and Y jointly Gaussian?

[Hint: Symmetry and antisymmetry can be used to simplify the integrals. Also see Examples 2.7 and 2.17.]

****

2.13 Consider two random variables with a joint probability density of

$$p_{XY}(u,v) = \frac{1}{(2\pi)^{1/2}}\exp\left(-\frac{u^2}{2\sigma_1^2} - |v|\right)U(uv)$$

(a) Find the marginal probability density functions $p_X(u)$ and $p_Y(v)$.
(b) Does X have the Gaussian distribution?
[Hint: Symmetry and antisymmetry can be used to simplify the integrals. Also see Example 2.7.]
****

Conditional Distributions

****

2.14 Let X and Y be the two random variables described in Exercise 2.6 with

$$F_{XY}(u,v)=\left((1-e^{-2u})(1-e^{-3v})+\frac{6}{5}(u+v)e^{-2u-3v}-ue^{-2u}-ve^{-3v}\right)U(u)U(v)$$

(a) Find the conditional cumulative distribution function $F_X(u\,|\,A)$ given the event $A=\{0\le Y\le 2\}$.
(b) Find the conditional probability density function $p_X(u\,|\,A)$.
****

2.15 Let X and Y be the two random variables described in Exercise 2.7 with

$$\begin{aligned}
&F_{XY}(u,v)=u^3(v-1)^2 && \text{for } 0\le u\le 1,\ 1\le v\le 2\\
&F_{XY}(u,v)=1 && \text{for } u>1,\ v>2\\
&F_{XY}(u,v)=u^3 && \text{for } 0\le u\le 1,\ v>2\\
&F_{XY}(u,v)=(v-1)^2 && \text{for } u>1,\ 1\le v\le 2\\
&F_{XY}(u,v)=0 && \text{otherwise}
\end{aligned}$$

(a) Find the conditional cumulative distribution function $F_Y(v\,|\,A)$ given the event $A=\{0\le X\le 0.75\}$.
(b) Find the conditional probability density function $p_Y(v\,|\,A)$.
****

2.16 Let X and Y be the two random variables described in Exercise 2.8, for which we can write

$$p_{XY}(u,v)=C\,U(v+u+1)U(u+1-v)U(u+1)U(-u)]+ \\ C\,U(v-u+1)U(1-u-v)U(u)U(1-u)$$

(a) Find the conditional cumulative distribution function $F_Y(v\,|\,A)$ given the event $A=\{Y\le 0.5\}$.
(b) Find the conditional probability density function $p_Y(v\,|\,A)$.
****

2.17 Let X and Y be the two random variables described in Exercise 2.9 with

$$p_{XY}(u,v)=C\,v\,e^{-2uv-5v}\,U(u)U(v)$$

(a) Find the conditional cumulative distribution function $F_X(u\,|\,A)$ given the event $A=\{X\ge 3\}$.

(b) Find the conditional probability density function $p_X(u|A)$.

****

2.18 Let X and Y be the two random variables described in Exercise 2.8, for which we can write

$$p_{XY}(u,v) = C\,U(v+u+1)U(u+1-v)U(u+1)U(-u)] + \\ C\,U(v-u+1)U(1-u-v)U(u)U(1-u)$$

(a) Find the conditional probability density function $p_Y(v|X=u)$.
(b) Find the conditional cumulative distribution function $F_Y(v|X=u)$.

****

2.19 Let X and Y be the two random variables described in Exercise 2.9 with

$$p_{XY}(u,v) = C\,v\,e^{-2uv-5v}\,U(u)U(v)$$

(a) Find the conditional probability density function $p_Y(v|X=u)$.
(b) Find the conditional cumulative distribution function $F_Y(v|X=u)$.

****

2.20 Let X and Y be the two random variables described in Exercise 2.10 with

$$p_{XY}(u,v) = C\,e^{-u-4v}\,U(u)U(v-u)$$

(a) Find the conditional probability density function $p_Y(v|X=u)$.
(b) Find the conditional cumulative distribution function $F_Y(v|X=u)$.

****

2.21 Let X and Y be the two random variables described in Exercise 2.6 with

$$F_{XY}(u,v) = \left((1-e^{-2u})(1-e^{-3v}) + \frac{6}{5}(u+v)e^{-2u-3v} - ue^{-2u} - ve^{-3v}\right)U(u)U(v)$$

(a) Find the conditional probability density functions $p_X(u|Y=v)$ and $p_Y(v|X=u)$.
(b) Find the conditional cumulative distribution functions $F_X(u|Y=v)$ and $F_Y(v|X=u)$.

****

2.22 Let X and Y be the two random variables described in Exercise 2.7, for which we can write

$$F_{XY}(u,v) = U(u-1)U(v-2) + u^3(v-1)^2[U(u)-U(u-1)][U(v-1)-U(v-2)] + \\ u^3[U(u)-U(u-1)]U(v-2) + (v-1)^2U(u-1)[U(v-1)-U(v-2)]$$

(a) Find the conditional probability density functions $p_X(u|Y=v)$ and $p_Y(v|X=u)$.
(b) Find the conditional cumulative distribution functions $F_X(u|Y=v)$ and $F_Y(v|X=u)$.

****

2.23 Consider the two random variables X and Y of Exercise 2.11 with the joint probability density function

$$p_{XY}(u,v)=\frac{1}{12\pi}\exp\left(-\frac{1}{2}\left(\frac{u-1}{2}\right)^2-\frac{1}{2}\left(\frac{v-2}{3}\right)^2\right)$$

Find the $p_X(u\,|\,Y=v)$ and $p_Y(v\,|\,X=u)$ conditional probability density functions.
****

2.24 Consider the random variables of Exercise 2.12 with

$$p_{XY}(u,v)=\frac{1}{2\pi\sigma_1\sigma_2}\exp\left(-\frac{u^2}{2\sigma_1^2}-\frac{v^2}{2\sigma_2^2}\right)\left(1+u\,v\exp\left[-\frac{u^2v^2}{2}\right]\right)$$

Find the conditional probability density functions $p_X(u\,|\,Y=v)$ and $p_Y(v\,|\,X=u)$.
****

2.25 Consider the random variables of Exercise 2.13 with

$$p_{XY}(u,v)=\frac{1}{(2\pi)^{1/2}}\exp\left(-\frac{u^2}{2\sigma_1^2}-|\,v\,|\right)U(u\,v)$$

Find the conditional probability density functions $p_X(u\,|\,Y=v)$ and $p_Y(v\,|\,X=u)$.
****

Independence
****

2.26 Are the random variables X and Y of Exercises 2.6 and 2.21 independent?
****

2.27 Are the random variables X and Y of Exercises 2.7 and 2.22 independent?
****

2.28 Are the random variables X and Y of Exercises 2.8 and 2.18 independent?
****

2.29 Are the random variables X and Y of Exercises 2.9 and 2.19 independent?
****

2.30 Are the random variables X and Y of Exercises 2.10 and 2.20 independent?
****

2.31. Are the random variables X and Y of Exercises 2.11 and 2.23 independent?
****

2.32 Are the random variables X and Y of Exercises 2.12 and 2.24 independent?
****

2.33 Are the random variables X and Y of Exercises 2.13 and 2.25 independent?
****

Chapter 3
Expected Values of Random Variables

3.1 Concept of Expected Values

As noted in Chapter 2, any nontrivial random variable has more than one possible value. In fact, there are infinitely many possible values for many random variables. A complete description of a real random variable requires the probability of its being within any finite or infinitesimal interval on the real line. In some cases this is more information than we are able to obtain conveniently. In other situations, this complete information can be obtained from knowledge of only a few parameters, because the general form of the probability distribution is known or assumed. In either of these instances, it is very valuable to give at least a partial characterization of the random variable by using quantities that are averages over all the possible values. The particular types of averages that we will be considering are called expected values, and they form the basis for this chapter.

3.2 Definition of Expected Values

For any random variable X, the expected value is defined as

$$E(X) \equiv \int_{-\infty}^{\infty} u\, p_X(u)\, du \tag{3.1}$$

This expression can be viewed as a weighted average over all the possible values of X. The weighting function is the probability density function. Thus, u represents a particular value of X, $p_X(u)\, du$ is the probability of X being in the vicinity of u, and the integration gives us a "sum" over all such terms. In a weighted integral we usually expect to find an integral such as that in Eq. 3.1 normalized by an integral of the weighting function. When the weighting function is a probability density, though, that normalization integral is exactly unity, so it can be omitted. Thus, the expected value is the probability weighted average of all the possible values of X. It is also called the expectation or, most commonly, the mean value of X.

One must be careful to remember the meaning of $E(X)$. The phrase *expected value* could easily lead to the misinterpretation that $E(X)$ is something like the most likely value of X, but this is generally not true; in fact, we will find in some situations that $E(X)$ is not even a possible value of X.

****

Example 3.1: Find the mean value of the random variable X having the exponential distribution with $p_X(u) = 2e^{-2u}\, U(u)$.

Applying Eq. 3.1 gives

$$E(X) = \int_{-\infty}^{\infty} u\, p_X(u)\, du = 2\int_0^{\infty} u\, e^{-2u}\, du = 0.5$$

One would generally say that the most likely value of this random variable is zero, because $p_X(u)$ is larger for $u=0$ than for any other u value, indicating that there is higher probability of X being in the neighborhood of zero than in the neighborhood of any other possible value. The fact that $E(X)=0.5$, though, reinforces the idea that $E(X)$ is an average of all the possible outcomes and need not coincide with a large value of the probability density function.

****

Example 3.2: Find the expected value of a random variable X that is uniformly distributed on the two intervals $[-2,-1]$ and $[1,2]$ so that the probability density function is as shown on the sketch.

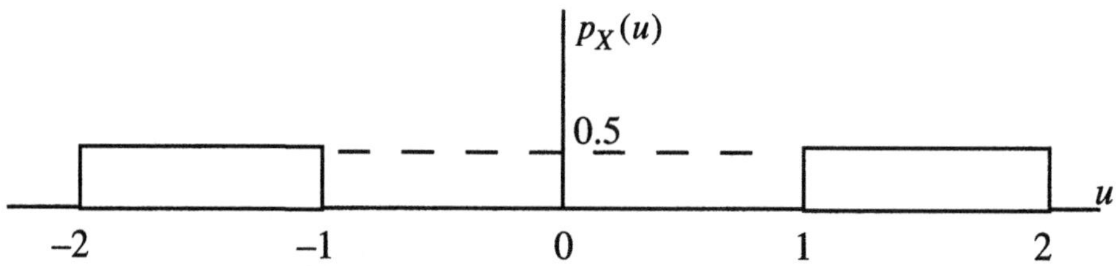

Integration of Eq. 3.1 gives

$$E(X) = \int_{-\infty}^{\infty} u\, p_X(u)\, du = 0.5\int_{-2}^{-1} u\, du + 0.5\int_1^2 u\, du = 0$$

In this case, $E(X)$ is not within the set of possible values for X. It is, however, the average of all the equally likely possible values. Similarly, for the discrete distribution of Example 2.1 with $P(X=1)$ = $P(X=2)$ = $P(X=3)$ = $P(X=4)$ = $P(X=5)$ = $P(X=6)=1/6$, we find that $E(X)=3.5$, which is surely not a possible value for the outcome of a die roll but is the average of the six equally likely outcomes.

****

Similar to Eq. 3.1, we now define the expected value of any function of the random variable X as

$$E[g(X)] \equiv \int_{-\infty}^{\infty} g(u)\, p_X(u)\, du \tag{3.2}$$

As before, we have $E[g(X)]$ as a weighted average of all the possible values of $g(X)$, because $p_X(u)\,du$ gives the probability of $g(X)$ being in the neighborhood of $g(u)$ due to X being in the neighborhood of u. Integrating over all the possible values of X gives $g(u)$ varying over all the possible values of $g(X)$, and $E[g(X)]$ is the probability weighted average of $g(X)$.

****

Example 3.3: Find $E(X^2)$ and $E[\sin(3X)]$ for the random variable X of Examples 2.2 and 2.6, for which the probability density can be written as

$$p_X(u) = 0.1\, U(u)\, U(10-u)$$

Using the probability density function in the appropriate integrals, we have

$$E(X^2) = \int_{-\infty}^{\infty} u^2 p_X(u)\, du = 0.1 \int_0^{10} u^2\, du = 33.3$$

and

$$E[\sin(3X)] = \int_{-\infty}^{\infty} \sin(3u)\, p_X(u)\, du = 0.1 \int_0^{10} \sin(3u)\, du = \frac{1-\cos(30)}{30}$$

****

Because any function of a random variable is a random variable, we could write $Y = g(X)$ for the random quantity considered in Eq. 3.2 and derive the probability distribution of Y from that of X. In particular, Eq. 2.40 gives $p_Y(v)$ in terms of $p_X(u)$. After having found $p_Y(v)$, we could use Eq. 3.1 to write

$$E[g(X)] \equiv E(Y) = \int_{-\infty}^{\infty} v\, p_Y(v)\, dv \tag{3.3}$$

It can be shown that Eq. 3.2 will always give the same result as Eq. 3.3. The difference between the two is that Eq. 3.3 prescribes calculation of the weighted average by summing over all possible values of $Y = g(X)$ in the order of increasing Y, whereas Eq. 3.2 involves summing the same terms in the order of increasing X. Both Eqs. 3.2 and 3.3 are exactly correct for calculating $E[g(X)]$, but in many situations we will find it easier to apply Eq. 3.2.

****

Example 3.4: For the random variables of Examples 2.3 and 2.10 one can write

$$F_X(u) = 0.05(u+4)U(u+4)U(16-u) + U(u-16)$$

and $Y = X[U(X) - U(X-10)] + 10U(X-10)$. Find $E(Y)$ both by using Eq. 3.2 and by using Eq. 3.3.

In Eq. 3.2 we need $p_X(u)$, which is $p_X(u) = 0.05U(u+4)U(16-u)$, along with the function $g(u)$ given by $g(u) = u[U(u) - U(u-10)] + 10U(u-10)$. The integral is then

$$E(Y) = \int_{-\infty}^{\infty} g(u)\, p_X(u)\, du = \int_{-4}^{0} (0)(0.05)\, du + \int_{0}^{10} u(0.05)\, du + \int_{10}^{16} (10)(0.05)\, du$$

which gives $E(Y) = 5.5$.

In order to use Eq. 3.3, we use the probability density obtained for Y in Example 2.10 to write

$$E(Y) = \int_{-\infty}^{\infty} v[0.05U(v)U(10-v) + 0.2\delta(v) + 0.3\delta(v-10)]\, dv$$

giving

$$E(Y) = 0.05\int_{0}^{10} v\, dv + 0.2\int_{-\infty}^{\infty} v\,\delta(v)\, dv + 0.3\int_{-\infty}^{\infty} v\,\delta(v-10)\, dv = 5.5$$

Of course, the two results are identical.

****

An important generalization of Eq. 3.2 is needed for the situation in which the new random quantity is a function of more than one random variable, such as $g(X,Y)$. The idea of a probability weighted average over all possible values of the function is maintained and the joint probability density of X and Y is used to write

$$E[g(X,Y)] \equiv \int_{-\infty}^{\infty}\int_{-\infty}^{\infty} g(u,v)\, p_{XY}(u,v)\, du\, dv \tag{3.4}$$

For a function of many random variables, we can use the corresponding vector notation

$$E[g(\vec{X})] \equiv \int_{-\infty}^{\infty}\cdots\int_{-\infty}^{\infty} g(\vec{u})\, p_{\vec{X}}(\vec{u})\, du_1 \cdots du_n \tag{3.5}$$

in which n is the number of components of the random vector $\vec{X}$ and of the deterministic vector $\vec{u}$ representing possible values of $\vec{X}$.

****

Example 3.5: Find $E(X^2)$ and $E(XY)$ for the distribution of Examples 2.12 and 2.15, for which the probability density can be written as

$$p_{XY}(u,v) = U(u-1)\,U(5-u)\,U(v-1)\,U(u-v)/8$$

We have two ways to calculate $E(X^2)$. We can use Eq. 3.4 with $g(u,v) = u^2$ and the given joint density function $p_{XY}(u,v)$, or we can use Eq. 3.1 with $g(u) = u^2$ and the marginal probability density function $p_X(u)$ that we derived in Example 2.15. First, using Eq. 3.4, we have

$$E(X^2) = \int_{-\infty}^{\infty}\int_{-\infty}^{\infty} u^2\, p_{XY}(u,v)\,du\,dv = \frac{1}{8}\int_1^5\int_v^5 u^2\,du\,dv = 14.33$$

The corresponding result from Eq. 3.1 is

$$E(X^2) = \int_{-\infty}^{\infty} u^2\, p_X(u)\,du = \int_1^5 u^2\left(\frac{u-1}{8}\right)du\,dv = 14.33$$

The result from Eq. 3.1 is simpler when the marginal probability density is known. Equation 3.4 involves a double integration, and one can consider that one of these integrations (the one with respect to v) has already been performed in going from $p_{XY}(u,v)$ to $p_X(u)$. Thus, there is only a single integration (with respect to u) remaining in applying Eq. 3.1.
For the function $g(X,Y) = XY$, we naturally use Eq. 3.4 and obtain

$$E(XY) = \int_{-\infty}^{\infty}\int_{-\infty}^{\infty} u\,v\, p_{XY}(u,v)\,du\,dv = \frac{1}{8}\int_1^5\int_v^5 u\,v\,du\,dv = 9$$

****

One very important property of expectation is linearity. This follows directly from the fact that the expected value is defined as an integral, and integration is a linear operation. The consequences of this are that $E(X+Y) = E(X) + E(Y)$ and $E(bX) = b\,E(X)$ for any random variables X and Y. It should be noted that we cannot reverse the order of the expectation operation and nonlinear functions such as squaring or multiplication of random variables. For example $E(X^2) \neq [E(X)]^2$, and $E(XY)$ generally does not equal $E(X)E(Y)$. We are allowed to reverse only the order of expectation and other linear operations.

****

Example 3.6: Find $E[X(X-Y)]$ for the distribution of Examples 2.12 and 3.5 with

$$p_{XY}(u,v) = U(u-1)\,U(5-u)\,U(v-1)\,U(u-v)/8$$

We could do this directly from Eq. 3.4 by using $g(u,v) = u(u-v)$, but we can find the result more simply by noting that $E[X(X-Y)] = E(X^2) - E(XY)$. Using the results obtained in Example 3.5 gives $E[X(X-Y)] = 14.33 - 9 = 5.33$.

****

Note that there is no assurance that $E[g(X)]$ will exist in general. In particular, if $|X|$ is not bounded and $p_X(u)$ does not decay sufficiently rapidly, then the integral of $g(u)\,p_X(u)$ in Eq. 3.2 may not exist. We do know that $p_X(u)$ is integrable, because it cannot be a probability density function unless its integral is unity, but this does not ensure that $g(u)\,p_X(u)$ will be integrable unless $g(u)$ is bounded. For example, $p_X(u) = U(u-1)/u^2$ is an acceptable probability density function, but it gives an infinite value for $E(X)$ because the integrand does not decay fast enough to give a finite value for the integral of $u\,p_X(u)$. Also, $E(X^2)$ and many other expected values are infinite for this random variable.

3.3 Moments of Random Variables

The quantities called moments of a random variable are simply special cases of the expectation of a function of the random variable. In particular, the jth moment of X is defined to be $E(X^j)$, corresponding to a special case of $E[g(X)]$. We will find it convenient to use a special name (mean) and notation (μ_X) for the first moment, which is simply the expected value of the random variable:

$$\text{mean value: } \mu_X \equiv E(X) = \int_{-\infty}^{\infty} u\, p_X(u)\, du \tag{3.6}$$

The reason for this special treatment of the first moment is that it plays an especially important role in many of our calculations. Similarly, the second moment $E(X^2)$ is commonly called the mean-squared value of the random variable, but we will not use any special notation for it.

One can also define cross moments of any pair of random variables (X,Y). The term $E(X^j Y^k)$ is the cross moment of order (j,k). The most commonly studied of these cross moments is $E(XY)$. This cross moment of order (1,1) is called the cross-product of the random variables.

In some situations it is more convenient to use a special form of the moments, in which one first subtracts the mean of each random variable. These moments of the form $E[(X-\mu_X)^j]$ and $E[(X-\mu_X)^j (Y-\mu_Y)^k]$ are called central moments. Two of these central moments are given special names and notations. They are

$$\text{variance: } \sigma_X^2 \equiv E[(X-\mu_X)^2] = E[X^2 - 2\mu_X E(X) + \mu_X^2] = E(X^2) - \mu_X^2 \quad (3.7)$$

and

$$\text{covariance: } K_{XY} \equiv E[(X-\mu_X)(Y-\mu_Y)] = E(XY) - \mu_X \mu_Y \quad (3.8)$$

The square root of the second moments of a random variable are also given special names. In particular, $[E(X^2)]^{1/2}$ is called the root-mean-square value, or rms, of X, and the standard deviation of X is σ_X.

Note that Eqs. 3.7 and 3.8, in addition to giving the definitions of variance and covariance as central moments, also show how these quantities can be evaluated in terms of ordinary noncentral moments. It is easily shown that this sort of expansion can be found for any central moment. Specifically, the central moment of order j can be written as a linear combination of noncentral moments of order 1 to j, and $E[(X-\mu_X)^j (Y-\mu_Y)^k]$ can be written as a linear combination of noncentral cross moments involving orders 1 to j of X and orders 1 to k of Y. In some situations it is useful to realize that the converse is also true—that noncentral moments can be written as linear combinations of central moments of lower and equal order. For the moments of X, the explicit relationships can be written as

$$E[(X-\mu_X)^j] = \sum_{i=0}^{j} \frac{(-1)^i \, j!}{i!(j-i)!} \mu_X^i \, E(X^{j-i})$$

or

$$E[(X-\mu_X)^j] = \sum_{i=0}^{j-2} \frac{(-1)^i \, j!}{i!(j-i)!} \mu_X^i \, E(X^{j-i}) - (-1)^j (j-1) \mu_X^j \quad (3.9)$$

and

$$E(X^j) = E\left([(X-\mu_X)+\mu_X]^j\right) = \sum_{i=0}^{j-2} \frac{j!}{i!(j-i)!} \mu_X^i \, E[(X-\mu_X)^{j-i}] + \mu_X^j \quad (3.10)$$

One other form of the covariance deserves special attention. This form is called the correlation coefficient and can be viewed either as a normalized form of covariance or as a cross-product of a modified form of the original random variables. It is defined as

$$\text{correlation coefficient:} \quad \rho_{XY} \equiv \frac{K_{XY}}{\sigma_X \sigma_Y} = E\left(\left[\frac{X-\mu_X}{\sigma_X}\right]\left[\frac{Y-\mu_Y}{\sigma_Y}\right]\right) \tag{3.11}$$

This form of covariance information is commonly used, and its significance will be investigated somewhat following Example 3.8. The random variables $(X-\mu_X)/\sigma_X$ and $(Y-\mu_Y)/\sigma_Y$ appearing in Eq. 3.11 are sometimes called standardized forms of X and Y. The standardized form of any random variable, obtained by subtracting the mean value then dividing by the standard deviation, is a new random variable that is a linear function of the original random variable. The linear function has been chosen such that the standardized random variable always has zero mean and unit variance. The next two moments of the standardized random variable are given special names

$$\text{skewness:} \quad E\left(\left[\frac{X-\mu_X}{\sigma_X}\right]^3\right) = \frac{E[(X-\mu_X)^3]}{\sigma_X^3} \tag{3.12}$$

and

$$\text{kurtosis:} \quad E\left(\left[\frac{X-\mu_X}{\sigma_X}\right]^4\right) = \frac{E[(X-\mu_X)^4]}{\sigma_X^4} \tag{3.13}$$

These quantities are used much less frequently than the first and second moments.

****

Example 3.7: Let the random variable X have the exponential distribution with probability density function $p_X(u) = be^{-bu}\, U(u)$, in which b is a positive constant. Find a general formula for $E(X^j)$.

We can simplify the integral

$$E(X^j) = \int_{-\infty}^{\infty} u^j p_X(u)\, du = b\int_0^{\infty} u^j e^{-bu}\, du$$

by using the change of variables $v = bu$, giving

$$E(X^j) = b^{-j}\int_0^{\infty} v^j e^{-v}\, dv$$

This gives us an integral that does not depend on the parameter b. Furthermore, it is a common integral. It is called the *gamma function* and is commonly written as

$$\Gamma(a)=\int_0^\infty v^{a-1}e^{-v}\,dv$$

The gamma function is a continuous real function of a on the range $a>0$, and at integer values of a it is equal to the factorial for a shifted argument: $\Gamma(a)=(a-1)!$. Thus, we have $E(X^j)=b^{-j}\,\Gamma(j-1)$ and $E(X^j)=b^{-j}\,j!$ for j = an integer. Note that there is no difficulty with the existence of $E(X^j)$ for any j value: the exponential decay of the probability density function assures the existence of all moments of X.

****

Example 3.8: Let the random variable X have the Gaussian distribution with probability density function

$$p_X(u)=\frac{1}{(2\pi)^{1/2}\sigma}\exp\left(-\frac{1}{2}\left[\frac{u-\mu}{\sigma}\right]^2\right)$$

in which μ and σ are constants. Find general formulas for μ_X and the central moments $E[(X-\mu_X)^j]$.

First we note that the probability density function is symmetric about the line $u=\mu$, so we immediately know that $\mu_X=\mu$ and

$$E[(X-\mu_X)^j]=\int_{-\infty}^\infty (u-\mu)^j p_X(u)\,du$$

If j is an odd integer, then $(u-\mu)^j$ is antisymmetric about $u=\mu$, so the integral must be zero, giving $E[(X-\mu_X)^j]=0$ for j odd. For j even, $(u-\mu)^j$ is symmetric about $u=\mu$, so we can use only the integral on one side of this point of symmetry, then double it:

$$E[(X-\mu_X)^j]=2\int_0^\infty (u-\mu)^j p_X(u)\,du \qquad \text{for } j \text{ even}$$

Using the change of variables $v=(u-\mu)^2/(2\sigma^2)$ gives

$$E[(X-\mu_X)^j]=\frac{(2\sigma^2)^{j/2}}{\pi^{1/2}}\int_0^\infty v^{(j-1)/2}e^{-v}\,dv=\frac{(2\sigma^2)^{j/2}}{\pi^{1/2}}\Gamma\left(\frac{j+1}{2}\right) \qquad \text{for } j \text{ even}$$

in terms of the gamma function introduced in Example 3.7. Applying this relationship will require knowledge of the gamma function for arguments that are half-integers. All such terms can be obtained from the fact that $\Gamma(1/2)=\pi^{1/2}$ and then using a general recursive property of the gamma function: $\Gamma(a+1)=a\,\Gamma(a)$. Thus,

$$\Gamma\left(\frac{j+1}{2}\right)=\left(\frac{j-1}{2}\right)\,\Gamma\left(\frac{j-1}{2}\right)=\left(\frac{(j-1)(j-3)\cdots(1)}{2^{j/2}}\right)\,\Gamma\left(\frac{1}{2}\right)$$

or

$$\Gamma\left(\frac{j+1}{2}\right)=\left(\frac{(1)(3)\cdots(j-3)(j-1)}{2^{j/2}}\right)\pi^{1/2}$$

and

$$E[(X-\mu_X)^j]=(1)(3)\cdots(j-3)(j-1)\sigma^j \quad \text{for } j \text{ even}$$

Thus, $j=2$ shows us that the variance $\sigma_X^2 \equiv E[(X-\mu_X)^2]$ is equal to the parameter σ^2. Similarly, $j=4$ gives $E[(X-\mu_X)^4]=3\sigma^4$, which shows that $kurtosis=3$ for a Gaussian random variable. From knowledge of the mean and variance we now know that we can write the probability density function of any Gaussian random variable as

$$p_X(u)=\frac{1}{(2\pi)^{1/2}\sigma_X}\exp\left(-\frac{1}{2}\left[\frac{u-\mu_X}{\sigma_X}\right]^2\right)$$

If we let the random variable Z be the standardized form of X:

$$Z=\frac{X-\mu_X}{\sigma_X}$$

then $\mu_Z=0$ and $\sigma_Z=1$. From Example 2.19 we know that this linear combination is also Gaussian, so its probability density function can be written as

$$p_Z(w)=\frac{1}{(2\pi)^{1/2}}e^{-w^2/2}$$

This is called the unit Gaussian random variable, and any Gaussian random variable can be written in terms of such a unit Gaussian random variable as

$$X=\mu_X+\sigma_X Z$$

****

To illustrate the significance of covariance and correlation coefficient, we will now investigate some results called Schwarz inequalities. Consider any two random variables W and Z for which the joint probability distribution is known. Now let b and c be two real numbers, and we will investigate the following mean-squared value of a linear combination of the random variables

$$E[(bW+cZ)^2]=b^2E(W^2)+2bc\,E(W\,Z)+c^2E(Z^2)$$

If $E(W^2)\ne 0$, we can rewrite the equation as

$$E[(bW+cZ)^2]=\left(b[E(W^2)]^{1/2}+c\frac{E(WZ)}{[E(W^2)]^{1/2}}\right)^2+c^2\left(E(Z^2)-\frac{[E(WZ)]^2}{E(W^2)}\right)$$

Note that the first term in this expression can be made to be zero by an appropriate choice of b and c but that $E[(bW+cZ)^2]\geq 0$ for all choices of b and c. Thus, we also know that the last term in the equation must be nonnegative:

$$E(Z^2)-\frac{[E(WZ)]^2}{E(W^2)}\geq 0 \qquad \text{or} \qquad [E(WZ)]^2\leq E(W^2)E(Z^2)$$

Furthermore, if we choose $b/c=-E(WZ)/E(W^2)$, then $E[(bW+cZ)^2]=0$ if and only if $[E(WZ)]^2=E(W^2)E(Z^2)$. For the situation with $E(W^2)\neq 0$, this is a proof of one form of the Schwarz inequality:

Schwarz Inequality I: For any two random variables W and Z

$$[E(WZ)]^2\leq E(W^2)E(Z^2) \tag{3.14}$$

and equality holds if and only if there exist constants b and c, not both zero, such that

$$E[(bW+cZ)^2]=0 \tag{3.15}$$

Note that Eq. 3.15 is simply a condition that all the probability of the joint distribution of W and Z lies on the straight line described by $(bW+cZ)=0$ so that there is a simple linear functional relationship between W and Z. If $E(W^2)=0$ but $E(Z^2)\neq 0$, then we prove the inequality by simply modifying this procedure to choose $c/b=-E(WZ)/E(Z^2)$. If both $E(W^2)=0$ and $E(Z^2)=0$, the result is trivial.

A more general form of the Schwarz inequality is found by letting W and Z be standardized random variables. In particular, let

$$W=\frac{X-\mu_X}{\sigma_X}, \qquad Z=\frac{Y-\mu_Y}{\sigma_Y}$$

In terms of X and Y, the Schwarz inequality can be written as

Schwarz Inequality II: For any two random variables X and Y

$$\rho_{XY}^2 \le 1 \tag{3.16}$$

and equality holds if and only if there exist constants $\tilde{a}$, $\tilde{b}$, and $\tilde{c}$, not all zero, such that

$$E[(\tilde{b}\,X+\tilde{c}\,Y-\tilde{a})^2]=0 \tag{3.17}$$

indicating a linear relationship between X and Y.

Note that the values of the coefficients in Eq. 3.17 can be related to those in Eq. 3.15 as $\tilde{b}=b/\sigma_X$, $\tilde{c}=c/\sigma_Y$, and $\tilde{a}=\tilde{b}\,\mu_X/\sigma_X+\tilde{c}\,\mu_Y/\sigma_Y$.

The correlation coefficient ρ_{XY} (or perhaps ρ_{XY}^2) can be considered to give the extent to which there is a linear relationship between the random variables X and Y. The limits of the correlation coefficient are $\rho_{XY}=\pm 1$ (so that $\rho_{XY}^2=1$), and at these limits there is a perfect linear functional relationship between X and Y of the form $\tilde{b}\,X+\tilde{c}\,Y=\tilde{a}$. That is, in these limiting cases one can find a linear function of X that is exactly equal to Y and/or a linear function of Y that is exactly equal to X. The other extreme is when $\rho_{XY}=0$, which requires that $K_{XY}=0$. At this extreme, X and Y are called uncorrelated random variables. One can say that there is no linear relationship between uncorrelated random variables. For other values of ρ_{XY} (i.e., $0<\rho_{XY}^2<1$) one may say that there is a partial linear relationship between X and Y, in the sense that some linear function of X is an imperfect approximation of Y and some linear function of Y is an imperfect approximation of X. Linear regression presents a slightly different way of looking at this matter. In linear regression of Y on X, one compares Y with a linear function $(a+b\,X)$, with a and b chosen to minimize $E[(a+b\,X-Y)^2]$. When this is done, it is found that the minimum value of $E[(a+b\,X-Y)^2]$ is $\sigma_Y^2(1-\rho_{XY}^2)$, indicating that the mean-squared error in this best linear fit is directly proportional to $(1-\rho_{XY}^2)$. Again we see that ρ_{XY}^2 gives the extent to which there is a linear relationship between X and Y.

The preceding paragraph notes that $\rho_{XY}=1$ and $\rho_{XY}=-1$ both give a perfect linear relationship between X and Y, but it says nothing about the significance of the sign of ρ_{XY}. Reviewing the proof of the Schwarz inequality, it is easily seen that the slope of the linear approximate relationship between Y and X or Z and W has the same sign as $E(WZ)$, which is the same sign as ρ_{XY}. Thus, $\rho_{XY} > 0$ indicates a positive slope for the linear approximation, whereas $\rho_{XY} < 0$ indicates a negative slope. In particular, the slope of the linear regression of Y on X is $\rho_{XY}\sigma_Y / \sigma_X$.

It should be kept in mind that having random variables X and Y uncorrelated does not generally mean that they are independent. Rather, $\rho_{XY} = 0$ (or $K_{XY} = 0$) implies only that there is not a linear relationship between X and Y. However, if X and Y are independent, they must also be uncorrelated. This is easily shown as follows

$$E(XY) = \int_{-\infty}^{\infty} \int_{-\infty}^{\infty} u\, v\, p_{XY}(u,v)\, du\, dv$$

so that

$$E(XY) = \int_{-\infty}^{\infty} u\, p_X(u)\, du \int_{-\infty}^{\infty} v\, p_Y(v)\, dv = E(X)\, E(Y)$$

if X and Y are independent. Thus, the covariance of X and Y is

$$K_{XY} = E(XY) - E(X)\, E(Y) = 0$$

if X and Y are independent. Independence of X and Y is generally a much stronger condition than is their being uncorrelated. Independence implies no relationship of any kind (linear or nonlinear) between X and Y.

****

Example 3.9: Let the probability density of X be uniform on $[-2,-1]$ and $[1,2]$, as shown in Example 3.2, and let Y be a simple square of X: $Y = X^2$. Investigate the correlation and dependence between X and Y.

Clearly X and Y are not independent. In fact, the joint probability density of X and Y is degenerate with $p_{XY}(u,v) \neq 0$ only on the curve $v = u^2$. Nonetheless, $E(X) = 0$ and $E(XY) = E(X^3) = 0$, so $E(XY) = E(X)\, E(Y)$, proving that X and Y are uncorrelated even though functionally dependent. They are uncorrelated because there is no linear relationship between them that is even approximately correct. Linear regression of Y on X for this problem can be shown to give a line of zero slope, again affirming that Y cannot be approximated (to any degree of accuracy at all) by a linear function of X.

****

The engineering reader may find it useful to relate the concept of moments of random variables to a more familiar topic of moments of areas or of rigid bodies. In particular, the expression for finding the mean of a random variable is exactly the same as that for finding the centroid of the area under the curve describing the probability density function. In general, to find the centroid, one takes this first moment integral and then divides by the total area. For a probability density function, though, the area is unity, so the first moment is exactly the centroidal location. This will often prove to be an easy way to detect errors in the evaluation of a mean value. If it is obvious that the computed value could not be the centroid of the area under the probability density curve, then a mistake has been made. Also, note that if the probability density is symmetric about some particular value, then that point of symmetry must be the mean value, just as it is the centroid of the area.

Similarly, for the second moments, $E(X^2)$ is the same as the moment of inertia of the area under the probability density curve about the line at $u=0$, and the variance σ_X^2 is the same as the centroidal moment of inertia. Extending these mechanics analogies to joint distributions shows that the point in two-dimensional space $u=\mu_X$ and $v=\mu_Y$ is at the centroid of the volume under the joint probability density function $p_{XY}(u,v)$, and the cross-product $E(X,Y)$ and covariance K_{XY} are the same as products of inertia of the volume under the probability density function.

We previously noted that it is convenient to use vector notation when we are dealing with many random variables. Consistent with this, we can use a matrix to organize all the information related to various cross-products and covariances of n random variables. In particular, the expression

$$E(\vec{X}\vec{X}^T)=E\left(\begin{bmatrix} X_1 \\ X_2 \\ \vdots \\ X_n \end{bmatrix}\begin{bmatrix} X_1 & X_2 & \cdots & X_n \end{bmatrix}\right)$$

gives a square symmetric matrix in which the (j,k) component is the cross-product of X_j and X_k. Similarly, if we let $\vec{\mu}_X$ denote $E(\vec{X})$, then

$$\mathbf{K}_{XX}=E\Big((\vec{X}-\vec{\mu}_X)(\vec{X}-\vec{\mu}_X)^T\Big) \tag{3.18}$$

defines a square symmetric covariance matrix in which the (j,k) component is the covariance of X_j and X_k. Note that the diagonal elements of $\mathbf{K}_{XX}$ are simply the variance values of the components. One can also show that the matrices $E(\vec{X}\vec{X}^T)$ and $\mathbf{K}_{XX}$ both have the property of being *nonnegative definite*. This property says that none of the eigenvalues of either matrix is negative. More directly relevant to our purposes is the fact that $\mathbf{K}_{XX}$, for example, is nonnegative definite if the scalar $\vec{v}^T\mathbf{K}_{XX}\vec{v}$ is nonnegative for every vector $\vec{v}$. It is easy to prove that $\mathbf{K}_{XX}$ has this property because $\vec{v}^T\mathbf{K}_{XX}\vec{v}$ is exactly the variance of a new scalar random variable $Y=\vec{v}^T\vec{X}$ and variance is always nonnegative. Similarly, the nonnegative definite property of $E(\vec{X}\vec{X}^T)$ follows from $\vec{v}^T E(\vec{X}\vec{X}^T)\vec{v}=E(Y^2)\ge 0$. It may be noted that for the special case of $n=2$, the random variable Y is exactly equivalent to the linear combination used in proving the Schwarz inequalities of Eqs. 3.14–3.17.

The reason for the double subscript on the covariance matrix in Eq. 3.18 is that we will sometimes also want to consider a matrix of covariances of different vectors defined by

$$\mathbf{K}_{XY}=E\left((\vec{X}-\vec{\mu}_X)(\vec{Y}-\vec{\mu}_Y)^T\right) \tag{3.19}$$

Note that $\mathbf{K}_{XY}$ generally need not be symmetric, or even square.

****

Example 3.10: Let $\vec{X}$ be a general Gaussian random vector, for which

$$p_{\vec{X}}(\vec{u})=\frac{1}{(2\pi)^{n/2}\,|\mathbf{K}|^{1/2}}\exp\left(-\frac{1}{2}(\vec{u}-\vec{\mu})^T\mathbf{K}^{-1}(\vec{u}-\vec{\mu})\right)$$

in which $\vec{\mu}$ is a vector of constants; $\mathbf{K}$ is a square, symmetric, positive definite matrix of constants; and $\mathbf{K}^{-1}$ and $|\mathbf{K}|$ denote the inverse and determinant, respectively, of $\mathbf{K}$. Show that $\vec{\mu}$ and $\mathbf{K}$ give the mean and covariance of $\vec{X}$; that is, $\vec{\mu}_X=\vec{\mu}$ and $\mathbf{K}_{XX}=\mathbf{K}$.

A convenient way to do this is to let $\mathbf{D}$ be a matrix with columns that are the unit length, orthogonal eigenvectors of $\mathbf{K}$, meaning that $\mathbf{K}\mathbf{D}=\mathbf{D}\mathbf{\Lambda}$, in which $\mathbf{\Lambda}$ is a diagonal matrix containing the eigenvalues of $\mathbf{K}$. Then $\mathbf{D}^T\mathbf{D}=\mathbf{I}$ and $\mathbf{D}^T\mathbf{K}\mathbf{D}=\mathbf{\Lambda}$. Manipulating these expressions also shows that $\mathbf{D}^{-1}=\mathbf{D}^T$, $\mathbf{D}^T\mathbf{K}^{-1}\mathbf{D}=\mathbf{\Lambda}^{-1}$, and $|\mathbf{K}|=|\mathbf{\Lambda}|/|\mathbf{D}|^2=|\mathbf{\Lambda}|=\Lambda_{11}\cdots\Lambda_{nn}$. Now consider a new random vector $\vec{Y}$ defined as $\vec{Y}=\mathbf{D}^T(\vec{X}-\vec{\mu})$. Using Eq. 2.41 then gives

$$p_{\vec{Y}}(\vec{v}) = \frac{1}{(2\pi)^{n/2}(\Lambda_{11}\cdots\Lambda_{nn})^{1/2}} \exp\left(-\frac{1}{2}\vec{v}^T \mathbf{D}^T \mathbf{K}^{-1} \mathbf{D} \vec{v}\right)$$

Using the fact that $\mathbf{D}^T \mathbf{K}^{-1} \mathbf{D} = \mathbf{\Lambda}^{-1}$ is diagonal allows this to be expanded as

$$p_{\vec{Y}}(\vec{v}) = \frac{\exp[(-\Lambda_{11}^{-1} v_1^2 -, \cdots, -\Lambda_{nn}^{-1} v_n^2)/2]}{(2\pi)^{n/2}(\Lambda_{11}\cdots\Lambda_{nn})^{1/2}} = \prod_{j=1}^{n} \frac{e^{-\Lambda_{jj}^{-1} v_j^2/2}}{(2\pi \Lambda_{jj})^{1/2}}$$

The fact that $p_{\vec{Y}}(\vec{v})$ is a product of individual functions of v_j terms indicates that the Y_j terms are independent with

$$p_{Y_j}(v_j) = \frac{e^{-\Lambda_{jj}^{-1} v_j^2/2}}{(2\pi \Lambda_{jj})^{1/2}}$$

Noting that the (j,j) component of $\mathbf{\Lambda}^{-1}$ is $\Lambda_{jj}^{-1} = 1/\Lambda_{jj}$, and comparing $p_{Y_j}(v_j)$ with the standard form for a scalar Gaussian random variable, as in Example 2.7, shows that

$$E(Y_j) = 0, \quad \sigma_{Y_j}^2 = \Lambda_{jj}$$

Thus, we can say that the mean vector and covariance matrix for $\vec{Y}$ are $\vec{\mu}_Y = \vec{0}$ and $\mathbf{K}_{YY} = E(\vec{Y}\vec{Y}^T)$. The independence of the Y_j components, along with their known variance values, gives $\mathbf{K}_{YY} = \mathbf{\Lambda}$. Next we note that the definition of $\vec{Y}$ and the fact that $\mathbf{D}^{-1} = \mathbf{D}^T$ imply that $\vec{X} = \vec{\mu} + \mathbf{D}\vec{Y}$, so

$$\vec{\mu}_X = \vec{\mu}, \quad \mathbf{K}_{XX} \equiv E\left((\vec{X}-\vec{\mu})(\vec{X}-\vec{\mu})^T\right) = \mathbf{D}\mathbf{K}_{YY}\mathbf{D}^T = \mathbf{D}\mathbf{\Lambda}\mathbf{D}^T = \mathbf{K}$$

Thus, the desired result is obtained without evaluating any expectation integrals involving the jointly Gaussian probability density function.

****

Example 3.11: Let $\vec{X}$ be a Gaussian random vector with

$$p_{\vec{X}}(\vec{u}) = \frac{1}{(2\pi)^{n/2} |\mathbf{K}_{XX}|^{1/2}} \exp\left(-\frac{1}{2}(\vec{u}-\vec{\mu}_X)^T \mathbf{K}_{XX}^{-1} (\vec{u}-\vec{\mu}_X)\right)$$

and $g(\vec{X})$ be a nonlinear function of the *n* components of $\vec{X}$. Show that

$$E[(\vec{X}-\vec{\mu}_X) g(\vec{X})] = \mathbf{K}_{XX} E\left(\left[\frac{\partial g(\vec{X})}{\partial X_1}, \cdots, \frac{\partial g(\vec{X})}{\partial X_n}\right]^T\right)$$

for any $g(\cdot)$ function such that $g(\vec{u})\, p_{\vec{X}}(\vec{u}) \rightarrow 0$ as the length of the $\vec{u}$ vector tends to infinity.

We begin by considering the expected value of one component of the derivative vector:

$$E\left(\frac{\partial g(\vec{X})}{\partial X_j}\right)=\int_{-\infty}^{\infty}\cdots\int_{-\infty}^{\infty}\frac{\partial g(\vec{u})}{\partial u_j}\,p_{\vec{X}}(\vec{u})\,du_1\cdots du_n$$

$$=-\int_{-\infty}^{\infty}\cdots\int_{-\infty}^{\infty}g(\vec{u})\frac{\partial}{\partial u_j}\,p_{\vec{X}}(\vec{u})\,du_1\cdots du_n$$

in which the second form has been obtained by integration by parts, taking advantage of the fact that $g(\vec{u})\,p_{\vec{X}}(\vec{u})\to 0$ at the limits of the integral. Expanding the exponent in the probability density function as

$$(\vec{u}-\vec{\mu}_X)^T\,\mathbf{K}_{XX}^{-1}(\vec{u}-\vec{\mu}_X)=\sum_{l=1}^{n}(u_j-\mu_{X_j})[\mathbf{K}_{XX}^{-1}]_{jl}(u_l-\mu_{X_l})$$

now gives

$$\frac{\partial}{\partial u_j}p_{\vec{X}}(\vec{u})=-\sum_{k=1}^{n}(u_k-\mu_{X_k})[\mathbf{K}_{XX}^{-1}]_{jk}\,p_{\vec{X}}(\vec{u})$$

and substitution yields

$$E\left(\frac{\partial g(\vec{X})}{\partial X_j}\right)=\sum_{k=1}^{n}E(X_k-\mu_{X_k})[\mathbf{K}_{XX}^{-1}]_{jk}$$

This, though, is exactly the jth row of the vector expression

$$E\left(\left[\frac{\partial g(\vec{X})}{\partial X_1},\cdots,\frac{\partial g(\vec{X})}{\partial X_n}\right]^T\right)=\mathbf{K}_{XX}^{-1}\,E[(\vec{X}-\vec{\mu}_X)\,g(\vec{X})]$$

Multiplying this expression by $\mathbf{K}_{XX}$ provides the desired result.

The expression derived here will be used in Chapter 10 in the process of "equivalent" linearization of nonlinear dynamics problems. It was apparently first used in this context by Atalik and Utku (1976), but it was presented earlier by Kazakov (1965). Falsone and Rundo Sotera (2003) have pointed out that, in addition to its use in nonlinear dynamics, the expression can also be used to obtain recursive relationships for the moments of Gaussian random variables. First, let us consider central moments by letting

$$g(\vec{X})=(X_1-\mu_{X_1})^{\alpha_1}\cdots(X_n-\mu_{X_n})^{\alpha_n}$$

The formula then gives

$$E[(X_j-\mu_{X_j})\,g(\vec{X})]=\sum_{l=1}^{n}\alpha_l\,\mathbf{K}_{X_jX_l}E\left(\frac{g(\vec{X})}{(X_l-\mu_{X_l})}\right)$$

which allows computation of higher-order moment expressions from lower-order moment expressions. In particular, the expression has $(X_j - \mu_{X_j})$ to the power $\alpha_j + 1$ on the left-hand side and each term on the right-hand side is either to its original power or to a lower power. Repeated use of this relationship allows computation of any order central moments from lower order moments, such as mean and covariance. For the special case of $n = 1$ we have

$$E[(X - \mu_X)^{\alpha+1}] = \alpha \sigma_X^2 \, E[(X - \mu_X)^{\alpha-1}]$$

For example,

$$E[(X - \mu_X)^4] = 3\sigma_X^2 \, E[(X - \mu_X)^2] = 3\sigma_X^4$$

and

$$E[(X - \mu_X)^6] = 5\sigma_X^2 \, E[(X - \mu_X)^4] = 15\sigma_X^6$$

These relationships agree exactly with those in Example 3.8. Furthermore, given the fact that the central moment of order unity is zero, the recursive relationship also shows that all odd central moments are zero, as found in Example 3.8. This recursive relationship can also be extended to noncentral moments by choosing the nonlinear function as $g(\vec{X}) = X_1^{\alpha_1} \cdots X_n^{\alpha_n}$. This gives

$$E[(X_j - \mu_{X_j})\, g(\vec{X})] = \sum_{l=1}^{n} \alpha_l \, \mathbf{K}_{X_j X_l} E\big(g(\vec{X}) / X_l\big)$$

or

$$E[X_j \, g(\vec{X})] = \mu_{X_j} E[g(\vec{X})] + \sum_{l=1}^{n} \alpha_l \, \mathbf{K}_{X_j X_l} E\big(g(\vec{X}) / X_l\big)$$

For the special case of $n = 1$ we have

$$E(X^{\alpha+1}) = \mu_X E(X^{\alpha}) + \alpha \sigma_X^2 \, E(X^{\alpha-1})$$

For example,

$$E(X^2) = \mu_X E(X^1) + \sigma_X^2 E(X^0) = \mu_X^2 + \sigma_X^2$$

$$E(X^3) = \mu_X E(X^2) + 2\sigma_X^2 E(X^1) = \mu_X^3 + 3\mu_X \sigma_X^2$$

$$E(X^4) = \mu_X E(X^3) + 3\sigma_X^2 E(X^2) = \mu_X^4 + 6\mu_X^2 \sigma_X^2 + 3\sigma_X^4$$

and so forth. Clearly these recursive relationships can be very useful when higher-order moments are needed for Gaussian random variables.

****

Example 3.12: Find $\mathbf{K}_{XX}$ and a simplified form for $p_{\vec{X}}(\vec{u})$ for a Gaussian random vector $\vec{X}$ with only two components.

The covariance matrix is written directly as

$$\mathbf{K}_{XX} \equiv E\left((\vec{X}-\vec{\mu})(\vec{X}-\vec{\mu})^T\right) = \begin{pmatrix} \sigma_{X_1}^2 & K_{X_1X_2} \\ K_{X_1X_2} & \sigma_{X_2}^2 \end{pmatrix}$$

or

$$\mathbf{K}_{XX} = \begin{pmatrix} \sigma_{X_1}^2 & \rho_{X_1X_2}\sigma_{X_1}\sigma_{X_2} \\ \rho_{X_1X_2}\sigma_{X_1}\sigma_{X_2} & \sigma_{X_2}^2 \end{pmatrix}$$

From this we find that $|\mathbf{K}_{XX}| = \sigma_{X_1}^2 \sigma_{X_2}^2 (1-\rho_{X_1X_2}^2)$ and

$$\mathbf{K}_{XX}^{-1} = \frac{1}{(1-\rho_{X_1X_2}^2)} \begin{pmatrix} 1/\sigma_{X_1}^2 & -\rho_{X_1X_2}/(\sigma_{X_1}\sigma_{X_2}) \\ -\rho_{X_1X_2}/(\sigma_{X_1}\sigma_{X_2}) & 1/\sigma_{X_2}^2 \end{pmatrix}$$

Substituting these expressions into the general form for $p_{\vec{X}}(\vec{u})$ for a Gaussian random vector, then expanding the matrix multiplication, gives

$$p_{\vec{X}}(\vec{u}) = \frac{1}{(2\pi)\sigma_{X_1}\sigma_{X_2}(1-\rho_{X_1X_2}^2)^{1/2}} \exp\left(-\frac{1}{2(1-\rho_{X_1X_2}^2)}\left[\left(\frac{u_1-\mu_{X_1}}{\sigma_{X_1}}\right)^2 - 2\rho_{X_1X_2}\left(\frac{u_1-\mu_{X_1}}{\sigma_{X_1}}\right)\left(\frac{u_2-\mu_{X_2}}{\sigma_{X_2}}\right) + \left(\frac{u_2-\mu_{X_2}}{\sigma_{X_2}}\right)^2\right]\right)$$

****

Example 3.13: Let $\vec{X}$ be a Gaussian random vector, with uncorrelated components: $\rho_{X_jX_k} = 0$ for all j and k. Show that the components are also independent.

For the special case of $n = 2$, this is evident in Example 3.12. That is, for $\rho_{X_1X_2} = 0$, the $p_{\vec{X}}(\vec{u})$ probability density becomes $p_{\vec{X}}(\vec{u}) = p_{X_1}(u_1)\, p_{X_2}(u_2)$, with $p_{X_1}(u_1)$ and $p_{X_2}(u_2)$ each having the standard scalar Gaussian form, as in Example 2.7. The more general result is easily obtained by noting that $\mathbf{K}_{XX}$ and $\mathbf{K}_{XX}^{-1}$ are diagonal for the uncorrelated components and that this results in $p_{\vec{X}}(\vec{u}) = p_{X_1}(u_1)\cdots p_{X_n}(u_n)$. As noted earlier, any random variables are uncorrelated if they are independent, but it is a very special property of the Gaussian distribution that jointly Gaussian random variables are uncorrelated if and only if they are independent.

****

3.4 Conditional Expectation

For any given event A we will define the conditional expected value of any random variable X given A, written $E(X|A)$, to be exactly the expectation defined in Eq. 3.1, but using the conditional probability density function $p_X(u|A)$ in place of $p_X(u)$. Thus,

$$E(X|A)=\int_{-\infty}^{\infty} u\, p_X(u|A)\, du \tag{3.20}$$

Similarly

$$E[g(X)|A]=\int_{-\infty}^{\infty} g(u)\, p_X(u|A)\, du \tag{3.21}$$

and

$$E[g(\vec{X})|A]=\int_{-\infty}^{\infty}\cdots\int_{-\infty}^{\infty} g(\vec{u})\, p_{\vec{X}}(\vec{u}|A)\, du_1\cdots du_n \tag{3.22}$$

The key idea is that a conditional probability density function truly is a probability density function. It can be used in all the same ways as any other probability density function, including as a weighting function for calculating expected values. The conditional probability density function given an event A provides the revised probabilities of all possible outcomes based on the known or hypothesized occurrence of A. Similarly the expected value calculated using this conditional probability density function provides the revised probability weighted average of all possible outcomes based on the occurrence of A.

Recall the special case of independent random variables. We know that if X and Y are independent, then the conditional distribution of Y given any information about X is the same as the marginal distribution of Y—that is, it is the same as if no information were given about X. Thus, if X and Y are independent, then the conditional expectation of Y given any information about X will be the same as if the information had not been given. It will be simply $E(Y)$.

****

Example 3.14: For the distribution of Examples 2.12, 2.21, and 2.23 with

$$p_{XY}(u,v)=U(u-1)U(5-u)U(v-1)U(u-v)/8$$

find $E(Y|X\le 3)$, $E(Y^2|X\le 3)$, and $E(Y|X=u)$ for all u for which it is defined.

In Example 2.21 we found that $p_Y(v|X\le 3)=[(3-v)/2]U(v-1)U(3-v)$, so we can now integrate to obtain

$$E(Y|X\le 3)=\int_1^3 v\left(\frac{3-v}{2}\right)dv=\frac{5}{3}$$

and

$$E(Y^2|X\le 3)=\int_1^3 v^2\left(\frac{3-v}{2}\right)dv=3$$

We also know from Example 2.23 that the conditional distribution of Y given $X=u$ is defined for $1\le u\le 5$ and is given by

$$p_Y(v|X=u)=\left(\frac{1}{u-1}\right)U(v-1)U(u-v)$$

so we now obtain

$$E(Y|X=u)=\int_1^u\left(\frac{v}{u-1}\right)dv=\frac{u+1}{2}\quad \text{for } 1\le u\le 5$$

Actually we could have used symmetry to write this last expected value without performing integration. The $p_Y(v|X=u)$ conditional probability density function is uniform on the set $[1,u]$, so the conditional mean must be at the center of this range.

****

Example 3.15: Find $E(X|Y=v)$ and $E(X^2|Y=v)$ for all v values for which they are defined for the random variables of Examples 2.13 and 2.24 with

$$p_{XY}(u,v)=\left(8e^{-4u-4v}+3e^{-4u-3v}+3e^{-3u-4v}\right)U(u)U(v)$$

In Example 2.24 we found that

$$p_X(u|Y=v)=\frac{8e^{-4u-4v}+3e^{-4u-3v}+3e^{-3u-4v}}{3\left(e^{-4v}+e^{-3v}/4\right)}U(u)\quad \text{for } v\ge 0$$

so we now obtain

$$E(X|Y=v)=\int_0^\infty u\frac{8e^{-4u-4v}+3e^{-4u-3v}+3e^{-3u-4v}}{3\left(e^{-4v}+e^{-3v}/4\right)}du=\frac{9+40e^{-v}}{36(1+4e^{-v})}\quad \text{for } v\ge 0$$

and

$$E(X^2|Y=v)=\int_0^\infty u^2\frac{8e^{-4u-4v}+3e^{-4u-3v}+3e^{-3u-4v}}{3\left(e^{-4v}+e^{-3v}/4\right)}du=\frac{27+136e^{-v}}{216(1+4e^{-v})}$$

for $v\ge 0$

One can perform a simple check on these results by looking at the limiting case for v tending to infinity. In this situation we will have $e^{-4v} << e^{-3v}$, so we can neglect the e^{-4v} terms in the conditional probability density function, reducing it to $4e^{-4u}$. This is an exponential distribution for which the mean and mean-squared values are $1/4$ and $1/8$. It is easy to verify that our conditional expected values do tend to these limits as v tends to infinity so that the e^{-v} terms become negligible.

****

Example 3.16: Let X and Y be jointly Gaussian. Find the conditional mean, conditional variance, given $X = u$, and $E(Y^2 | X = u)$.

From the definition of the conditional probability density function and the forms given in Examples 2.17 and 3.12 we obtain

$$p_Y(v|X=u) = \frac{1}{(2\pi)^{1/2}\sigma_Y(1-\rho^2)^{1/2}} \exp\left(-\frac{1}{2(1-\rho^2)}\left[\left(\frac{u-\mu_X}{\sigma_X}\right)^2 - 2\rho\left(\frac{u-\mu_X}{\sigma_X}\right)\left(\frac{v-\mu_Y}{\sigma_Y}\right)+\left(\frac{v-\mu_Y}{\sigma_Y}\right)^2\right]+\frac{1}{2}\left(\frac{u-\mu_X}{\sigma_X}\right)^2\right)$$

in which $\rho \equiv \rho_{XY}$. Defining new quantities as

$$\tilde{\sigma}_Y = \sigma_Y(1-\rho^2)^{1/2}, \qquad \tilde{\mu}_Y = \mu_Y + \rho\frac{\sigma_Y}{\sigma_X}(u-\mu_X)$$

allows this to be rewritten as

$$p_Y(v|X=u) = \frac{1}{(2\pi)^{1/2}\tilde{\sigma}_Y}\exp\left(-\frac{1}{2}\left(\frac{v-\tilde{\mu}_Y}{\tilde{\sigma}_Y}\right)^2\right)$$

Note that this expression shows that a jointly Gaussian distribution implies that the components are conditionally Gaussian. Comparing this with the standard scalar Gaussian form also shows that the conditional mean and variance are

$$E(Y|X=u) = \tilde{\mu}_Y = \mu_Y + \rho\frac{\sigma_Y}{\sigma_X}(u-\mu_X), \quad \text{Var}(Y|X=u) = \tilde{\sigma}_Y^2 = \sigma_Y^2(1-\rho^2)$$

We can now say that

$$E(Y^2|X=u) = [E(Y^2|X=u)]^2 + \text{Var}(Y|X=u)$$

or

$$E(Y^2|X=u) = \left(\mu_Y + \rho\frac{\sigma_Y}{\sigma_X}(u-\mu_X)\right)^2 + \sigma_Y^2(1-\rho^2)$$

****

Example 3.17: For the distribution of Examples 2.11, 2.14, and 2.22 with

$$p_{XY}(u,v) = U(u+1)U(2-u)U(v+1)U(1-v)/6$$

find the conditional expected values $E(Y|X=u)$ and $E(X|Y=v)$ for all u and v for which they are defined; also find $E(X|Y>0.5)$.

Because the random variables are independent, as shown in Example 2.28, we know that $E(Y|X=u)=E(Y)$ and $E(X|Y=v)=E(X)$ for u and v values for which they are defined. Similarly, $E(X|Y>0.5)=E(X)$. Using the marginal probability density functions derived in Example 2.14 then gives $E(Y|X=u)=0$ for $-1 \le u \le 2$, $E(X|Y=v)=0.5$ for $-1 \le v \le 1$, and $E(X|Y>0.5)=0.5$.

****

3.5 Generalized Conditional Expectation

To this point, our discussion of conditional expectation has involved the expectation given some specific event. These conditional expectations are always deterministic, being derived from integrals of conditional probability density functions. Now we want to define a conditional expectation that is, itself, a random variable. The uncertainty about the conditional expectation will be introduced by introducing the idea of uncertainty about the given event. In order to be more specific about this idea, let us consider the expected value of a random variable *X* given a value of a random variable *Y*: $E(Y|X=u)$. In general, this conditional expectation will be different for each value of the deterministic variable u. Thus, the conditional expectation can be considered to define a function of u. For the moment, let us write this as

$$f(u) \equiv E(Y|X=u)$$

Now that we have the definition of this function, we might also consider the same function with a random variable for an argument. In particular, we will be interested in taking the random variable *X* to be the argument of $f(\cdot)$. Now we can investigate the expected value of $f(X)$, just as we might for any other function of a random variable. We write

$$E[f(X)] = \int_{-\infty}^{\infty} f(u)\, p_X(u)\, du$$

but we also know from its definition that $f(u)$ is given by

$$f(u) = \int_{-\infty}^{\infty} v\, p_Y(v \mid X = u)\, dv$$

Substituting $f(u)$ into the integral for $E[f(X)]$ gives

$$E[f(X)] = \int_{-\infty}^{\infty}\int_{-\infty}^{\infty} v\, p_Y(v \mid X = u)\, p_X(u)\, dv\, du$$

but the product of probability density functions in this integrand is exactly the joint probability density function for X and Y: $p_{XY}(u,v) = p_X(u)\, p_Y(v \mid X = u)$. Thus, the integral for $E[f(X)]$ becomes exactly the same as one for evaluating $E[g(X,Y)]$ with $g(X,Y) = Y$, and

$$E[f(X)] = E(Y)$$

The usual notation for the conditional expectation random variable defined earlier as $f(X)$ is simply $E(Y \mid X)$. That is, $E(Y \mid X)$ is the function of u defined by $E(Y \mid X = u)$ when the deterministic quantity u is replaced by the random variable X. The result that we have just proved can then be written as

$$E[E(Y \mid X)] = E(Y) \tag{3.23}$$

which can be stated in words as follows: The expected value of the conditional expected value is the unconditional expected value. This statement may sound rather confusing, but the meaning of Eq. 3.23 should be quite understandable. The left-hand side of Eq. 3.23 can be considered to depend on the joint distribution of X and Y. Evaluating $E(Y \mid X)$ corresponds to finding the probability weighted average over all possible values of Y for a given value of X, then the second expectation gives the probability weighted average over all possible values of X. Equation 3.23 can also be generalized to a form

$$E\Big(g(X)\, E\big[h(Y) \mid X\big]\Big) = E[g(X)\, h(Y)] \tag{3.24}$$

for any functions g and h for which the expectations exist. The definition of $E[h(Y) \mid X]$ in Eq. 3.24 is precisely $f(X)$ with the function $f(\cdot)$ defined by $f(u) = E[h(Y) \mid X = u]$. One can verify Eq. 3.24 by substituting

$$E[h(Y)|X=u]=\int_{-\infty}^{\infty} h(v)\, p_Y(v|X=u)\, dv$$

into

$$E\big(g(X)\,E[h(Y)|X]\big)=\int_{-\infty}^{\infty} g(u)\, E[h(Y)|X=u]\, p_X(u)\, du$$

In some problems it will be found to be quite convenient to use Eqs. 3.23 and 3.24 to evaluate the expectations given on the right-hand sides of the equations. In particular, using Eq. 3.23 may be simpler than finding and using $p_Y(v)$ to find $E(Y)$, and using 3.24 may be simpler than finding and using $p_{XY}(u,v)$ to find $E[g(X)\,h(Y)]$. Most important, there are situations in which Eq. 3.23 can be used when one is given insufficient information to describe $p_Y(v)$, and others in which 3.24 can be used when one is given insufficient information to describe $p_{XY}(u,v)$.

****

Example 3.18: Find $E(Y)$ for the probability distribution of Example 2.26 with

$$p_X(u)=0.1U(u)U(10-u)$$

and

$$p_Y(v|X=u)=U(v)U(u-v)/u$$

From the fact that the conditional distribution of Y given $X=u$ is uniform on the set $[0,u]$, we can immediately note that $E(Y|X=u)=u/2$. Thus, $E(Y|X)=X/2$ is the conditional expectation random variable obtained by substituting X for u in $E(Y|X=u)$. Now using Eq. 3.23 gives $E(Y)=E[E(Y|X)]=E(X)/2$. Using the fact that the distribution of X is uniform on the set $[0,10]$ gives $E(X)=5$, so we have $E(Y)=2.5$ without ever having evaluated $p_Y(v)$ or explicitly worked with $p_{XY}(u,v)$.

One can, of course, do this problem by first evaluating $p_Y(v)$, but that is less convenient. Using the joint probability density function we can write

$$p_Y(v)=\int_{-\infty}^{\infty} p_{XY}(u,v)\, du=\int_{v}^{10}\left(\frac{0.1}{u}\right) du\ U(v)U(10-v)$$

giving

$$p_Y(v)=0.1\log(10/v)\ U(v)U(10-v)$$

The reader can verify that this probability density function does give $E(Y)=2.5$, but it seems clear that the approach using Eq. 3.23 is the simpler procedure for evaluating the expectation.

****

Example 3.19: Let the random variable X have an exponential distribution with

$$p_X(u) = 2e^{-2u}\, U(u)$$

and let the conditional mean and mean-squared values of Y be given as

$$E(Y|X=u) = 3u \quad \text{and} \quad E(Y^2|X=u) = 10u^2 + 2u$$

Find $E(Y)$, $E(Y^2)$, and $E(X^2Y^2)$.

This is a situation in which there clearly is not enough information given to allow us to write out $p_Y(v)$ or $p_{XY}(u,v)$. It is not unusual to have practical problems in which one is given only partial information of this sort, particularly when the information about the Y random variable has been obtained strictly from statistical analysis of measured data. We can use Eqs. 3.23 and 3.24, though. We can say that $E(Y|X) = 3X$ so that Eq. 3.23 gives $E(Y) = 3E(X)$. Now using the distribution of X we find that $E(X) = 0.5$, so $E(Y) = 1.5$. Similarly, $E(Y^2|X) = 10X^2 + 2X$, so $E(Y^2) = 10E(X^2) + 2E(X)$. From $p_X(u)$ we find that $E(X^2) = 0.5$, and finally $E(Y^2) = 6$. This result for $E(Y^2)$ can be considered either as an application of Eq. 3.24 with $g(X) = 1$ and $h(Y) = Y^2$ or as an application of Eq. 3.23 with the second random variable being taken as Y^2 rather than the Y that was written previously.

For the final expectation we need to use Eq. 3.24 with $g(X) = X^2$ and $h(Y) = Y^2$. Substituting gives

$$E(X^2Y^2) = E\Big[E(X^2)\, E(Y^2|X)\Big] = 10E(X^4) + 2E(X^3)$$

Integrals using $p_X(u)$ now give $E(X^4) = 1.5$ and $E(X^3) = 0.75$, so $E(X^2Y^2) = 16.5$.

****

3.6 Characteristic Function of a Random Variable

The characteristic function provides an alternative form for giving a complete description of a probability distribution. Recall that we have previously noted that either the cumulative distribution function or the probability density function gives such a complete description. We will now show that the function called the *characteristic function* is also adequate because it gives all the information that is included within the probability density function. The characteristic function for a random variable X will be denoted by $M_X(\theta)$ and is defined as

$$M_X(\theta) \equiv E\left(e^{i\theta X}\right) = \int_{-\infty}^{\infty} e^{i\theta u}\, p_X(u)\, du \tag{3.25}$$

Thus, $M_X(\theta)$ is the expected value of a complex function of X, and it can be evaluated from an integral with a complex integrand. One can, of course, convert this to the evaluation of two real integrals by using $e^{i\theta u} = \cos(\theta u) + i \sin(\theta u)$. Those familiar with Fourier analysis will note that Eq. 3.25 gives the characteristic function $M_X(\theta)$ as a form of the Fourier transform of the probability density function $p_X(u)$.[1] Noting that this is a Fourier transform allows us to use known results to write the corresponding inverse Fourier transform formula, giving

$$p_X(u) = \frac{1}{2\pi} \int_{-\infty}^{\infty} e^{-i\theta u} M_X(\theta)\, d\theta \tag{3.26}$$

Actually, this inverse formula is an equality only at points where $p_X(u)$ is continuous, but this is sufficient for our purposes. Changing the value of $p_X(u)$ only at points of discontinuity would not change the value of an integral containing $p_X(u)$ in the integrand. Thus, knowledge of $p_X(u)$ everywhere except at a countable number of points of discontinuity is sufficient to give the cumulative distribution function, which is a complete description of the probability distribution. This proves our earlier assertion that knowing $M_X(\theta)$ gives all the information necessary for a complete description of the probability distribution.

There are several reasons why one might choose to use the characteristic function of a random variable. Many times, the motivation for using $M_X(\theta)$ will be that it simplifies some analytical development or proof. There is one property of characteristic functions, though, that sometimes proves to be very useful for simple calculations. This is the so-called moment generating property, and it is obtained by differentiating Eq. 3.25 with respect to θ. In particular, if we take the jth derivative we obtain

$$\frac{d^j}{d\theta^j} M_X(\theta) = i^j\, E\left(X^j e^{i\theta X}\right)$$

and letting $\theta = 0$ in this expression gives

$$E(X^j) = i^{-j} \left(\frac{d^j}{d\theta^j} M_X(\theta) \right)_{\theta=0} \tag{3.27}$$

[1]Appendix B gives a brief introduction to Fourier analysis.

Thus, if the characteristic function for a random variable is known, then the moments of the random variable can be found by differentiating that characteristic function, then evaluating the derivatives at $\theta = 0$. If many moments are needed, it may be easier to find them this way instead of by performing an integral for each moment according to Eq. 3.2. Of course, if the probability distribution of X is given as $p_X(u)$, then the integration of Eq. 3.25 must be performed before one can begin to evaluate moments according to Eq. 3.27. This is only one integration, though, rather than one for each moment being calculated.

Some readers may be acquainted with the real function $E(e^{-rX})$, which is a Laplace transform of $p_X(u)$ and is called the moment generating function. The moment equations from this function are simpler than those from Eq. 3.27 inasmuch as all terms are real, because $E(e^{-rX})$ is real for all real values of r. The disadvantage of $E(e^{-rX})$ is that it does not exist for some probability density functions or for all values of r for other probability density functions. The condition for the existence of a Fourier transform, though, is simply that the original function be absolute value integrable. This means that the characteristic function $M_X(\theta)$ exists for all values of θ if

$$\int_{-\infty}^{\infty} |p_X(u)| du < \infty$$

but this condition is met for any random variable. In fact, the integral shown is exactly unity. Thus, one never needs to be concerned about the existence of the characteristic function.

****

Example 3.20: Find the characteristic function for the random variable with the general exponential distribution with

$$p_X(u) = b e^{-bu} U(u)$$

and verify that the mean and mean-squared values obtained by taking derivatives of $M_X(\theta)$ agree with those obtained in Example 3.7 by integration of the probability density function.

In Example 3.7 we used a change of variables of $v = bu$ to obtain

$$E(X) = b\int_0^{\infty} u e^{-bu} du = \frac{1}{b}\int_0^{\infty} v e^{-v} dv = \frac{1}{b}$$

and

$$E(X^2) = b \int_0^\infty u^2 e^{-bu} \, du = \frac{1}{b^2} \int_0^\infty v^2 e^{-v} \, dv = \frac{2}{b^2}$$

Similarly, using $v = (b + i\theta) u$ gives the characteristic function as

$$M_X(\theta) = b \int_0^\infty e^{iu\theta} e^{-bu} \, du = \frac{b}{b - i\theta} \int_{C_1} e^{-v} \, dv$$

The contour C_1 in the final integral is the straight line shown in the sketch. The fact that the contour is made up of complex v values causes the evaluation of the integral to be nontrivial. It can be found from the study of complex functions, though, that for this integrand the integral over C_1 is identical to the integral from zero to infinity along the real axis.[2] Thus, one obtains

$$M_X(\theta) = \frac{b}{b - i\theta}$$

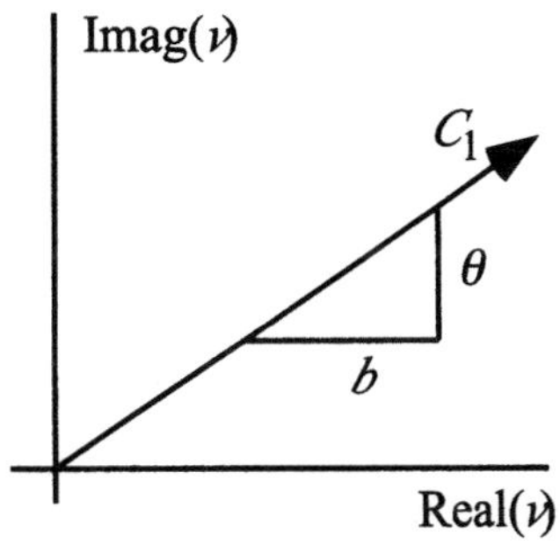

Note that $M_X(0) = 1$, as is always the case. The first two derivatives are

$$M'_X(\theta) = \frac{ib}{(b - i\theta)^2}, \qquad M''_X(\theta) = \frac{2i^2 b}{(b - i\theta)^3}$$

giving $M'_X(0) = i/b$ and $M''_X(0) = 2i^2/b^2$. Using Eq. 3.27 then gives $E(X) = 1/b$ and $E(X^2) = 2/b^2$, confirming our results from integration.

****

Example 3.21: Find the characteristic function and calculate the mean and mean-squared values by taking derivatives of $M_X(\theta)$ for the probability distribution of Examples 2.1, 2.4, and 2.9 with

$$p_X(u) = \left[\delta(u-1) + \delta(u-2) + \delta(u-3) + \delta(u-4) + \delta(u-5) + \delta(u-6)\right]/6$$

Using this probability density function, we obtain

$$M_X(\theta) = \int_{-\infty}^{\infty} \frac{e^{i\theta u}}{6} \left[\delta(u-1) + \delta(u-2) + \delta(u-3) + \delta(u-4) + \delta(u-5) + \delta(u-6)\right] du$$

which is easily integrated by the property of the Dirac delta function to give

---

[2]The reader is cautioned that considerable care should be used in making changes of variables or changing the contours of integration in problems involving complex functions. The complex analysis theorems governing these matters are not covered in this book.

$$M_X(\theta)=\frac{1}{6}\left(e^{i\theta}+e^{2i\theta}+e^{3i\theta}+e^{4i\theta}+e^{5i\theta}+e^{6i\theta}\right)$$

From this one obtains

$$M'_X(0)=\frac{1}{6}\left(i+2i+3i+4i+5i+6i\right)=\frac{7i}{2}$$

and

$$M''_X(0)=\frac{1}{6}\left((i)^2+(2i)^2+(3i)^2+(4i)^2+(5i)^2+(6i)^2\right)=\frac{91i^2}{6}$$

Thus, Eq. 3.27 gives $E(X)=7/2$ and $E(X^2)=91/6$.

****

Example 3.22: Find the characteristic function for a Gaussian random variable with

$$p_X(u)=\frac{1}{(2\pi)^{1/2}\sigma_X}\exp\left(-\frac{1}{2}\left[\frac{u-\mu_X}{\sigma_X}\right]^2\right)$$

One of the simplest ways to do this is first to note that $Y=(X-\mu_X)/\sigma_X$ is a Gaussian random variable with mean zero and unit variance. The characteristic function for Y is then

$$M_Y(\theta)\equiv E(e^{i\theta Y})=\frac{1}{(2\pi)^{1/2}}\int_{-\infty}^{\infty}\exp(i\theta v-v^2/2)\,dv$$

using the change of variables $w=v-i\theta$ allows this to be rewritten as

$$M_Y(\theta)\equiv E(e^{i\theta Y})=e^{-\theta^2/2}\int_{C_1}\frac{e^{-w^2/2}}{(2\pi)^{1/2}}\,dw$$

in which the contour C_1 is from $-\infty-i\theta$ to $+\infty-i\theta$ along a line parallel to the real axis in the space of complex w values. From the theory of complex variables one can show that the integral is the same as if it were along the real axis, though, which requires that the integral is unity so that $M_Y(\theta))=e^{-\theta^2/2}$.
Note the remarkable similarity between $M_Y(\theta)$ and $p_Y(v)$. Each is an exponential of the square of the argument. Only for this "standardized" Gaussian probability density function does the Fourier transform of the function have the same form as the original function. Now we can use $X=\mu_X+\sigma_X Y$ to obtain

$$M_X(\theta)\equiv E(e^{i\theta X})=e^{i\theta\mu_X}E(e^{i\theta\sigma_X Y})=e^{i\theta\mu_X}M_Y(\sigma_X\theta)=\exp\left(i\theta\mu_X-\frac{\sigma_X^2\theta^2}{2}\right)$$

Note that the characteristic function $M_X(\theta)$ for any Gaussian random variable X is an exponential of a quadratic form in the variable θ. In fact, this can be used as a definition of a Gaussian random variable, in lieu of saying that the probability density function $p_X(u)$ is a constant multiplying an exponential of a quadratic form in u.

****

One can easily generalize the characteristic function formulas to joint distributions of random variables. Using vector notation, one can write

$$M_{\vec{X}}(\vec{\theta}) = E\left(e^{i\vec{\theta}^T \vec{X}}\right) = \int_{-\infty}^{\infty} \cdots \int_{-\infty}^{\infty} e^{i\vec{\theta}^T \vec{u}}\, p_{\vec{X}}(\vec{u})\, du_1 \cdots du_n \tag{3.28}$$

in which the superscript T denotes the transpose of the vector so that $\vec{\theta}^T \vec{X} = (\theta_1 X_1 + \cdots + \theta_n X_n)$. The inverse formula is

$$p_{\vec{X}}(\vec{u}) = \frac{1}{(2\pi)^n} \int_{-\infty}^{\infty} \cdots \int_{-\infty}^{\infty} e^{-i\vec{\theta}^T \vec{u}}\, M_{\vec{X}}(\vec{\theta})\, d\theta_1 \cdots d\theta_n \tag{3.29}$$

and moments are found from formulas such as

$$E\left(X_1^j X_2^k\right) = i^{-(j+k)} \left(\frac{\partial^{j+k}}{\partial\theta_1^j\, \partial\theta_2^k} M_{\vec{X}}(\vec{\theta}) \right)_{\vec{\theta}=\vec{0}}$$

or

$$E\left(X_1^{j_1} \cdots X_n^{j_n}\right) = i^{-(j_1+\cdots+j_n)} \left(\frac{\partial^{j_1+\cdots+j_n}}{\partial\theta_1^{j_1} \cdots \partial\theta_n^{j_n}} M_{\vec{X}}(\vec{\theta}) \right)_{\vec{\theta}=\vec{0}} \tag{3.30}$$

A particularly convenient property of characteristic functions relates to the finding of marginal distributions from joint distributions. It is easily shown that a marginal characteristic function is found from a joint characteristic function simply by setting one or more of the θ arguments equal to zero. For example, if $M_{XYZ}(\theta_1,\theta_2,\theta_3) \equiv E(e^{i\theta_1 X} + e^{i\theta_2 Y} + e^{i\theta_3 Z})$ is known, then $M_{XY}(\theta_1,\theta_2) \equiv E(e^{i\theta_1 X} + e^{i\theta_2 Y}) = M_{XYZ}(\theta_1,\theta_2,0)$, $M_X(\theta_1) \equiv E(e^{i\theta_1 X}) = M_{XYZ}(\theta_1,0,0)$, and so forth.

****

Example 3.23: Find the characteristic function for a Gaussian random vector with

$$p_{\vec{X}}(\vec{u}) = \frac{1}{(2\pi)^{n/2}(\|\mathbf{K}_{XX}\|)^{1/2}} \exp\left(-\frac{1}{2}(\vec{u}-\vec{\mu}_X)^T \mathbf{K}_{XX}^{-1}(\vec{u}-\vec{\mu}_X)\right)$$

Recall the $\mathbf{D}$ defined in Example 3.10, for which $\mathbf{D}^T\mathbf{D} = \mathbf{I}$ and $\mathbf{D}^T\mathbf{K}\mathbf{D} = \mathbf{\Lambda}$, in which $\mathbf{\Lambda}$ is a diagonal matrix. Now let $\vec{Y} = \mathbf{D}^T(\vec{X} - \vec{\mu}_X)$. From Example 3.10 we know that the components of $\vec{Y}$ are independent and Gaussian, with $\sigma_{Y_j}^2 = \Lambda_{jj}$. The independence property allows us to write

$$M_{\vec{Y}}(\vec{\theta}) \equiv E(e^{i\vec{\theta}^T\vec{Y}}) = E\left(\exp\left[i\sum_{j=1}^{n}\theta_j Y_j\right]\right) = \prod_{j=1}^{n} E[e^{i\theta_j Y_j}] = \prod_{j=1}^{n} M_{Y_j}(\theta_j)$$

and using the results of Example 3.22 gives

$$M_{\vec{Y}}(\vec{\theta}) = \prod_{j=1}^{n}\exp[-\sigma_{Y_j}^2\theta_j^2] = \prod_{j=1}^{n}\exp[-\Lambda_{jj}\theta_j^2] = \exp(-\vec{\theta}^T\mathbf{\Lambda}\vec{\theta})$$

As shown in Example 3.10, $\mathbf{\Lambda} = \mathbf{D}^T\mathbf{K}_{XX}\mathbf{D}$, so this can be rewritten as

$$M_{\vec{Y}}(\vec{\theta}) = \exp(-\vec{\theta}^T\mathbf{D}^T\mathbf{K}_{XX}\mathbf{D}\vec{\theta})$$

Using $\vec{X} = \vec{\mu}_X + \mathbf{D}\vec{Y}$ now gives

$$M_{\vec{X}}(\vec{\theta}) \equiv E(e^{i\vec{\theta}^T\vec{X}}) = e^{i\vec{\theta}^T\vec{\mu}_X} E(e^{i\vec{\theta}^T\mathbf{D}Y}) = e^{i\vec{\theta}^T\vec{\mu}_X} M_{\vec{Y}}(\mathbf{D}^T\vec{\theta})$$

or

$$M_{\vec{X}}(\vec{\theta}) = \exp(i\vec{\theta}^T\vec{\mu}_X - \vec{\theta}^T\mathbf{D}\mathbf{\Lambda}\mathbf{D}^{\mathbf{T}}\vec{\theta})$$

From the results in Example 3.10, $\mathbf{D}\mathbf{\Lambda}\mathbf{D}^T = \mathbf{K}_{XX}$, so we now have

$$M_{\vec{X}}(\vec{\theta}) = \exp(i\vec{\theta}^T\vec{\mu}_X - \vec{\theta}^T\mathbf{K}_{XX}\vec{\theta})$$

Consistent with the results in Example 3.22, this shows that the characteristic function for any Gaussian random vector is an exponential of a quadratic form of the arguments. It may also be noted that this form of the Gaussian characteristic function does not involve the inverse of $\mathbf{K}_{XX}$, so it can be written even in the degenerate case with a singular covariance matrix.

****

3.7 Power Series for Characteristic Function

Recall that Eq. 3.27 gives a simple relationship between the moments of a random variable and the derivatives of its characteristic function when evaluated at the origin of θ space. Thus, if all the moments of X exist, one can use them in a Taylor series expansion for $M_X(\theta)$ about the origin (i.e., a Maclaurin expansion):

$$M_X(\theta) = \sum_{j=0}^{\infty} \frac{\theta^j}{j!} \left(\frac{d^j M_X(\theta)}{d\theta^j} \right)_{\theta=0} = \sum_{j=0}^{\infty} \frac{(i\theta)^j}{j!} E(X^j) \tag{3.31}$$

This equation represents a very important property of many probability distributions. Knowledge of all the moments of a random variable is usually sufficient information to provide a complete description of its probability distribution. In particular, from knowledge of all the $E(X^j)$ terms, one can write $M_X(\theta)$ according to Eq. 3.31, then Eq. 3.26 allows determination of the probability density function. We are not suggesting that this is a practical method for finding $p_X(u)$ in an applied problem. Rather, it is an illustration of how much information is needed for a complete description of most probability distributions. It should also be noted, though, that this procedure does not always work. The log-normal distribution is probably the most commonly used distribution for which knowledge of all the moments does not uniquely define the probability distribution (Kendall and Stuart, 1977).

Sometimes it is convenient to work with the natural logarithm of $M_X(\theta)$ instead of with $M_X(\theta)$ itself. This function $\log[M_X(\theta)]$ is called the *log-characteristic function.* One can also expand the log-characteristic function as a Taylor series like the one in Eq. 3.31, but the coefficients are different, of course. Specifically, we will write

$$\log[M_X(\theta)] = \sum_{j=0}^{\infty} \frac{(i\theta)^j}{j!} \kappa_j(X) \tag{3.32}$$

in which

$$\kappa_j(X) \equiv i^{-j} \left(\frac{d^j}{d\theta^j} \log[M_X(\theta)] \right)_{\theta=0} \tag{3.33}$$

The term $\kappa_j(X)$ is called the jth *cumulant* of X. Note that $\kappa_0(X) = 0$, so the first term of the summation in Eq. 3.32 makes no contribution. We introduce the cumulant terms because it is sometimes convenient to use them, rather than moments, to describe a probability distribution.

For joint distributions, one can use joint cumulants of the form

$$\kappa_J(\underbrace{X_1,\cdots X_1}_{j_1},\cdots \underbrace{X_n,\cdots X_n}_{j_n}) \equiv i^{-J}\left(\frac{\partial^J}{\partial\theta_1^{j_1}\cdots\partial\theta_n^{j_n}}\log[M_{\vec{X}}(\vec{\theta})]\right)_{\vec{\theta}=\vec{0}} \tag{3.34}$$

in which the order of the cumulant is $J = j_1 + \cdots + j_n$. This gives the series expansion

$$\log[M_{\vec{X}}(\vec{\theta})] = \sum_{j_1=0}^{\infty}\cdots\sum_{j_n=0}^{\infty}\frac{(i\theta_1)^{j_1}\cdots(i\theta_n)^{j_n}}{j_1!\cdots j_n!}\kappa_J(\underbrace{X_1,\cdots,X_1}_{j_1},\cdots,\underbrace{X_n,\cdots,X_n}_{j_n}) \tag{3.35}$$

just as the joint characteristic function can be expanded in terms of the joint moments as

$$M_{\vec{X}}(\vec{\theta}) = \sum_{j_1=0}^{\infty}\cdots\sum_{j_n=0}^{\infty}\frac{(i\theta_1)^{j_1}\cdots(i\theta_n)^{j_n}}{j_1!\cdots j_n!}E(X_1^{j_1}\cdots X_n^{j_n}) \tag{3.36}$$

Any particular cumulant of order J can be written in terms of the moments up the Jth order. The procedure for doing this involves writing derivatives of the log-characteristic function in terms of the characteristic function and its derivatives, then using Eqs. 3.38 and 3.30. Alternatively, one can use Eqs. 3.33 and 3.27 for the special case of a single random variable. Even though the procedure is straightforward, the general relationship is quite complicated. The first few cumulants are given by

$$\kappa_1(X) = E(X) = \mu_X \tag{3.37}$$

$$\kappa_2(X_1,X_2) = E(X_1X_2) - \mu_{X_1}\mu_{X_2} = K_{X_1X_2} \tag{3.38}$$

$$\begin{aligned}\kappa_3(X_1,X_2,X_3) &= E(X_1X_2X_3) - \mu_{X_1}E(X_1X_3) - \mu_{X_2}E(X_1X_3) - \\ &\qquad \mu_{X_3}E(X_1X_2) + 2\mu_{X_1}\mu_{X_2}\mu_{X_3} \\ &= E[(X_1 - \mu_{X_1})(X_2 - \mu_{X_2})(X_3 - \mu_{X_3})]\end{aligned} \tag{3.39}$$

$$\begin{aligned}\kappa_4(X_1,X_2,X_3,X_4) = E(X_1X_2X_3X_4) - E(X_1X_2)E(X_3X_4) - E(X_1X_3)E(X_2X_4) - \\ E(X_1X_4)E(X_2X_3) - \mu_{X_1}E(X_2X_3X_4) - \mu_{X_2}E(X_1X_3X_4) - \\ -\mu_{X_3}E(X_1X_2X_4) - \mu_{X_4}E(X_1X_2X_3) + 2\mu_{X_1}\mu_{X_2}E(X_3X_4) + \\ 2\mu_{X_1}\mu_{X_3}E(X_2X_4) + 2\mu_{X_1}\mu_{X_4}E(X_2X_3) + 2\mu_{X_2}\mu_{X_3}E(X_1X_4) + \\ 2\mu_{X_2}\mu_{X_4}E(X_1X_3) + 2\mu_{X_3}\mu_{X_4}E(X_1X_2) - 6\mu_{X_1}\mu_{X_2}\mu_{X_3}\mu_{X_4}\end{aligned}$$

or

$$\begin{aligned}\kappa_4(X_1,X_2,X_3,X_4) = E[(X_1-\mu_{X_1})(X_2-\mu_{X_2})(X_3-\mu_{X_3})(X_4-\mu_{X_4})] - \\ K_{X_1X_2}K_{X_3X_4} - K_{X_1X_3}K_{X_2X_4} - K_{X_1X_4}K_{X_2X_3}\end{aligned} \tag{3.40}$$

For the special case of a single random variable, the final three equations simplify to

$$\kappa_2(X) = E[(X-\mu_X)^2] = \sigma_X^2 \tag{3.41}$$

$$\kappa_3(X) = E[(X-\mu_X)^3] = \sigma_X^3 \times (skewness) \tag{3.42}$$

and

$$\kappa_4(X) = E[(X-\mu_X)^4] - 3\sigma_X^4 = \sigma_X^4 \times (kurtosis - 3) \tag{3.43}$$

One can also use the log-characteristic function to prove a property of joint cumulants that is sometimes of considerable importance in applications. In particular, we will prove that

$$\kappa_{n+1}\left(X_1,\cdots,X_n,\sum a_jY_j\right) = \sum a_j\kappa_{n+1}\left(X_1,\cdots,X_n,Y_j\right) \tag{3.44}$$

which can be considered a property of linearity and includes the distributive property. To show this, we use Eq. 3.34 to write

$$\kappa_{n+1}\left(X_1,\cdots,X_n,\sum a_jY_j\right) = \frac{1}{i^{n+1}}\left(\frac{\partial^{n+1}}{\partial\theta_1\cdots\partial\theta_n\partial\phi}\log\left[E(e^{i\vec{\theta}^T\vec{X}+i\phi\sum a_jY_j})\right]\right)_{\substack{\vec{\theta}=\vec{0}\\ \phi=0}}$$

Performing the differentiation with respect to ϕ gives

$$\kappa_{n+1}\Big(X_1,\cdots,X_n,\sum a_j Y_j\Big)=\frac{1}{i^n}\left(\frac{\partial^n}{\partial\theta_1\cdots\partial\theta_n}\frac{E\Big(\sum a_j Y_j e^{i\vec{\theta}^T\vec{X}+i\phi\sum a_j Y_j}\Big)}{E\Big(e^{i\vec{\theta}^T\vec{X}+i\phi\sum a_j Y_j}\Big)}\right)_{\substack{\vec{\theta}=\vec{0}\\ \phi=0}}$$

and setting $\phi=0$, then rearranging terms allows this to be written as

$$\kappa_{n+1}\Big(X_1,\cdots,X_n,\sum a_j Y_j\Big)=\frac{1}{i^n}\left(\sum a_j\frac{E[Y_j e^{i\vec{\theta}^T\vec{X}}]}{E[e^{i\vec{\theta}^T\vec{X}}]}\right)_{\vec{\theta}=\vec{0}} \tag{3.45}$$

This expression, though, is precisely what one would obtain by performing exactly the same operations on each of the cumulant terms appearing on the right-hand side of Eq. 3.44. Thus, Eq. 3.44 is proved to be true. One can also let some of the X_j arguments in Eq. 3.44 be identical, in which case the expression applies to any sort of joint cumulant term rather than being restricted to being first order in each of the X_j arguments.

****

Example 3.24: Find the log-characteristic function and all cumulants for a Gaussian random vector $\vec{X}$ with known mean vector $\vec{\mu}_X$ and covariance matrix $\mathbf{K}_{XX}$.

From Example 3.23 we know that $M_{\vec{X}}(\vec{\theta})=\exp(i\vec{\theta}^T\vec{\mu}_X-\vec{\theta}^T\mathbf{K}_{XX}\vec{\theta})$. Thus,

$$\log[M_{\vec{X}}(\vec{\theta})]=i\vec{\theta}^T\vec{\mu}_X-\vec{\theta}^T\mathbf{K}_{XX}\vec{\theta}$$

From Eq. 3.38 we then have

$$\kappa_1(X_j)=\mu_{X_j} \qquad \text{for } j=1,\cdots,n$$

$$\kappa_2(X_j,X_k)=K_{X_jX_k} \qquad \text{for } j=1,\cdots,n;\ k=1,\cdots,n$$

$$\kappa_J(X_{j_1},\cdots,X_{j_J})=0 \qquad \text{for } J\ge 2$$

Of course, the first- and second-order cumulants agree with Eqs. 3.38 and 3.39. All the higher-order cumulants are zero, because the log-characteristic function is a quadratic in the θ terms and a cumulant of order J involves a Jth order derivative. In fact, a Gaussian random vector or scalar variable may be defined to be one that has all zero cumulants beyond the second order.

****

3.8 Importance of Moment Analysis

In many random variable problems (and in much of the analysis of stochastic processes), one performs detailed analysis of only the first and second moments of the various quantities, with occasional consideration of skewness and/or kurtosis. One reason for this is surely the fact that analysis of mean, variance, or mean-squared value is generally much easier than analysis of probability distributions. Furthermore, in many problems, one has some idea of the shape of the probability density functions, so knowledge of moment information may allow evaluation of the parameters in that shape, thereby giving an estimate of the complete probability distribution. If the shape has only two parameters to be chosen, in particular, then knowledge of mean and variance will generally suffice for this procedure. In addition to these pragmatic reasons, though, the results in Eqs. 3.31, 3.32, 3.35, and 3.36 give a theoretical justification for focusing attention on the lower-order moments. Specifically, mean, variance, skewness, kurtosis, and so forth, in that order, are the first items in an infinite sequence of information that would give a complete description of the problem. In most situations it is impossible to achieve the complete description, but it is certainly logical for us to focus our attention on the first items in the sequence.

Exercises

****

Mean and Variance of One Random Variable

****

3.1 Consider a random variable X with probability density function

$$p_X(u) = C\,u^4\,U(u+1)U(2-u)$$

(a) Find the value of the constant C.
(b) Find the mean value μ_X.
(c) Find the mean-squared value $E(X^2)$.
(d) Find the variance σ_X^2.

****

3.2 Consider the random variable X of Exercise 2.4 with

$$p_X(u) = 0.75\,u(2-u)U(u)U(2-u)$$

(a) Find the mean value μ_X.
(b) Find the mean-squared value $E(X^2)$.
(c) Find the variance σ_X^2.

****

3.3 Consider a random variable X with probability density function

$$p_X(u) = a/(b+u^6)$$

in which a and b are positive constants. For what positive integer j values does $E(X^j) < \infty$ exist?

****

Moments of Jointly Distributed Random Variables

****

3.4 Let X and Y be the two random variables described in Exercise 2.8, for which we can write

$$p_{XY}(u,v) = C\,U(v+u+1)\,U(1+u-v)\,U(u+1)\,U(-u) + C\,U(v-u+1)\,U(1-u-v)\,U(u)\,U(1-u)$$

(a) Find $E(X)$ and $E(Y)$.
(b) Find $E(X^2)$ and $E(Y^2)$.
(c) Find the variances and covariance σ_X^2, σ_Y^2, and K_{XY}.
(d) Are X and Y correlated?

****

3.5 Let X and Y be the two random variables described in Exercise 2.9 with

$$p_{XY}(u,v) = C\,v\,e^{-2uv-5v}\,U(u)\,U(v)$$

(a) Find $E(X)$ and $E(Y)$.
(b) Find $E(X^2)$ and $E(Y^2)$.
(c) Find the variances and covariance σ_X^2, σ_Y^2, and K_{XY}.
(d) Are X and Y correlated?

****

3.6 Let X and Y be the two random variables described in Exercise 2.10 with

$$p_{XY}(u,v) = C\,e^{-u-4v}\,U(u)\,U(v-u)$$

(a) Find $E(X)$ and $E(Y)$.
(b) Find $E(X^2)$ and $E(Y^2)$.
(c) Find the variances and covariance σ_X^2, σ_Y^2, and K_{XY}.
(d) Are X and Y correlated?

****

3.7 Let two random variables (X,Y) be uniformly distributed on the set $X^2+Y^2 \le 1$:

$$p_{XY}(u,v) = U(1-u^2-v^2)/\pi$$

(a) Are X and Y uncorrelated?
(b) Are X and Y independent?

****

3.8 Let the probability density function of a random variable X be given by

$$p_X(u) = C\,U(u)\,U(5-u)$$

and let the conditional probability density function of another random variable Y be given by

$$p_Y(v \mid X=u) = B(u)\,U(v)\,U(2-v) \qquad \text{for } 0 \le u \le 5$$

(a) Find the value of the constant C and the function $B(u)$.
(b) Find the joint probability density function $p_{XY}(u,v)$ and indicate on a sketch of the (u,v) plane the region on which $p_{XY}(u,v) \ne 0$.
(c) Find $E(X^2)$.
(d) Find $E(XY)$.
(e) Are X and Y independent?

****

3.9 Consider a random variable X with probability density function

$$p_X(u) = u\,e^{-u}\,U(u)$$

Let another random variable Y have the conditional probability density function

$$p_Y(v \mid X=u) = u^{-1}\,U(v)\,U(u-v) \qquad \text{for } u \ge 0$$

(a) Find the joint probability density function $p_{XY}(u,v)$ and indicate on a sketch of the (u,v) plane the region on which $p_{XY}(u,v) \ne 0$.
(b) Find $E(X^2)$.
(c) Find $E(Y)$.
(d) Find $E(X^2 Y)$.

****

3.10 Let the random variables X and Y denote the displacement and the velocity of response at a certain time t of a dynamic system. For a certain nonlinear oscillator the joint probability density function is $p_{XY}(u,v) = C\exp(-\alpha\,|u^3| - \gamma\,v^2)$ in which C, α, and γ are positive constants.

(a) Find the probability density function for X, including evaluation of C in terms of α and/or γ.
(b) Find $E(X^2)$ in terms of α and/or γ.

(Methods developed in Chapter 10 will show that this problem corresponds to the response to Gaussian white noise of an oscillator with linear damping and a quadratic hardening spring.)

****

3.11 Consider a discrete element model of a beam in which the random sequence $\{S_1, S_2, S_3, \cdots\}$ represents normalized average slopes in successive segments of the beam. Each segment is of unit length. The normalized deflections of the beam are then given by

$$X_1 = S_1, \qquad X_2 = S_1 + S_2, \qquad X_3 = S_1 + S_2 + S_3, \qquad \text{and so forth}$$

For $j \le 3$, $k \le 3$ it has been found that $E(S_j) = 0$, $E(S_j^2) = 1$, and $E(S_j S_k) = 0.8$ for $j \ne k$.

(a) Find $E(X_1^2)$, $E(X_2^2)$, and their correlation coefficient $\rho_{X_1X_2}$.
(b) Let $W_1 = S_2 - S_1$ and $W_2 = S_3 - S_2$ (which are proportional to bending moments in the beam). Find $E(W_1^2)$, $E(W_2^2)$, and their correlation coefficient $\rho_{W_1W_2}$.
(c) Let $V_2 = W_2 - W_1 = S_3 - 2S_2 + S_1$ (proportional to shear in the beam). Find $E(V_2^2)$.

****

3.12 Let X_j denote the east-west displacement at the top of the jth story of a four-story building at some particular instant of time during an earthquake. Each displacement is measured relative to the moving base of the building. Presume that these story motions are mean-zero and that their covariance matrix is

$$\mathbf{K} = E(\vec{X}\vec{X}^T) = \begin{pmatrix} 100 & 180 & 223 & 231 \\ 180 & 400 & 540 & 594 \\ 223 & 540 & 900 & 1080 \\ 231 & 594 & 1080 & 1600 \end{pmatrix} \text{mm}^2$$

Find the standard deviation σ_{Y_j} $(j = 1 \text{ to } 4)$ of each story shear deformation, $Y_j = X_j - X_{j-1}$, in which $X_0 \equiv 0$.

****

3.13 Consider two random variables with the joint probability density function

$$p_{X_1X_2}(u_1,u_2) = A\exp\left(-\frac{2}{3}\left[\frac{u_1-1}{0.5}\right]^2 + \frac{2}{3}\left[\frac{u_1-1}{0.5}\right]\left[\frac{u_2-2}{0.25}\right] - \frac{2}{3}\left[\frac{u_2-2}{0.25}\right]^2 \right)$$

(a) Find the mean and variance values for both X_1 and X_2.
(b) Find the $\rho_{X_1X_2}$ correlation coefficient.
(c) Find the value of the constant A.
(d) Find $P(X_2 > 2.5)$.
[Hint: See Examples 3.12 and 2.7.]

****

3.14 Consider two random variables with the joint probability density function

$$p_{X_1X_2}(u_1,u_2) = A\exp\left(-\frac{8}{15}\left[\frac{u_1-2}{2}\right]^2 - \frac{4}{15}\left[\frac{u_1-2}{2}\right]\left[\frac{u_2-3}{4}\right] - \frac{8}{15}\left[\frac{u_2-3}{4}\right]^2 \right)$$

(a) Find the mean and variance values for both X_1 and X_2.
(b) Find the $\rho_{X_1X_2}$ correlation coefficient.
(c) Find the value of the constant A.
(d) Find $P(X_2 > 13)$.
[Hint: See Examples 3.12 and 2.7.]

****

3.15 Consider the two random variables of Exercise 3.13 with the joint probability density function

$$p_{X_1X_2}(u_1,u_2) = A\exp\left(-\frac{2}{3}\left[\frac{u_1-1}{0.5}\right]^2 + \frac{2}{3}\left[\frac{u_1-1}{0.5}\right]\left[\frac{u_2-2}{0.25}\right] - \frac{2}{3}\left[\frac{u_2-2}{0.25}\right]^2\right)$$

Let $Y_1 = X_1 + X_2$ and $Y_2 = X_1 + 3X_2$.

(a) Find the mean and variance values for both Y_1 and Y_2.

(b) Find the $\rho_{Y_1Y_2}$ correlation coefficient.

(c) Find the joint probability density function $p_{Y_1Y_2}(v_1,v_2)$.

[Hint: See Example 3.12.]

****

3.16 Consider the two random variables of Exercise 3.14 with the joint probability density function

$$p_{X_1X_2}(u_1,u_2) = A\exp\left(-\frac{8}{15}\left[\frac{u_1-2}{2}\right]^2 - \frac{4}{15}\left[\frac{u_1-2}{2}\right]\left[\frac{u_2-3}{4}\right] - \frac{8}{15}\left[\frac{u_2-3}{4}\right]^2\right)$$

Let $Y_1 = 2X_1 - X_2$ and $Y_2 = X_1 + 3X_2$.

(a) Find the mean and variance values for both Y_1 and Y_2.

(b) Find the $\rho_{Y_1Y_2}$ correlation coefficient.

(c) Find the joint probability density function $p_{Y_1Y_2}(v_1,v_2)$.

[Hint: See Example 3.12.]

****

Conditional Expectations

****

3.17 Consider two random variables X and Y with the joint probability density function

$$p_{XY}(u,v) = e^{-u}\,U(v)\,U(u-v)$$

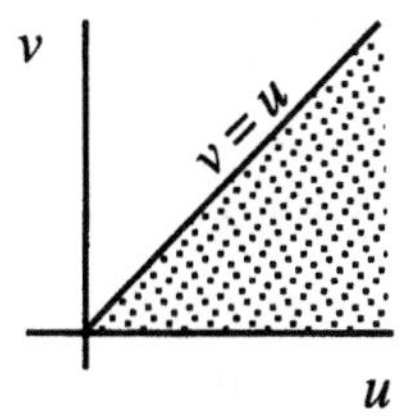

(a) Find both marginal probability density functions: $p_X(u)$ and $p_Y(v)$.

(b) Find the conditional probability density functions: $p_X(u|Y=v)$ and $p_Y(v|X=u)$.

(c) Find $E(Y|X=u)$.

****

3.18 Consider a random variable X with probability density function

$$p_X(u) = U(u)\,U(1-u)$$

Let another random variable Y have the conditional probability density function

$$p_Y(v \mid X=u) = U(v-u)U(1+u-v) \quad \text{for } 0 \le u \le 1$$

(a) Find the joint probability density function $p_{XY}(u,v)$ and indicate on a sketch of the (u,v) plane the region on which $p_{XY}(u,v) \ne 0$.
(b) Find $p_Y(v)$ and sketch this probability density function versus v.
(c) Find the conditional expected value $E(Y \mid X=u)$ for $0 \le u \le 1$.

****

3.19 Consider a random variable *X* with probability density function

$$p_X(u) = 2e^{-2u}\, U(u)$$

Let another random variable *Y* have the conditional probability density function

$$p_Y(v \mid X=u) = e^{u}\, U(v)U(e^{-u} - v) \quad \text{for } u \ge 0$$

(a) Find the joint probability density function $p_{XY}(u,v)$ and indicate on a sketch of the (u,v) plane the region on which $p_{XY}(u,v) \ne 0$.
(b) Find $p_Y(v)$ and sketch this probability density function versus v.
(c) Find the conditional expected value $E(Y \mid X=u)$ for $u \ge 0$.
(d) Find $E(Y)$.

****

3.20 Consider a random variable *X* with probability density function

$$p_X(u) = 0.75\, u(2-u)U(u)U(2-u)$$

Let another random variable *Y* have the conditional probability density function

$$p_Y(v \mid X=u) = u^{-1}\, U(v)U(u-v) \quad \text{for } 0 \le u \le 2$$

(a) Find the joint probability density function $p_{XY}(u,v)$ and indicate on a sketch of the (u,v) plane the region on which $p_{XY}(u,v) \ne 0$.
(b) Find $p_Y(v)$ and sketch this probability density function versus v.
(c) Find the conditional expected value $E(Y \mid X=u)$ for $0 \le u \le 2$.
(d) Find $E(X^2 Y)$.

****

3.21 Consider a random variable *X* with probability density function

$$p_X(u) = 3u^2\, U(u)U(1-u)]$$

Let another random variable *Y* have the conditional probability density function

$$p_Y(v \mid X=u) = u^{-1}\, U(v)U(u-v) \quad \text{for } 0 \le u \le 1$$

(a) Find the joint probability density function $p_{XY}(u,v)$ and indicate on a sketch of the (u,v) plane the region on which $p_{XY}(u,v) \ne 0$.
(b) Find $p_Y(v)$ and sketch this probability density function versus v.
(c) Find the conditional expected value $E(Y \mid X=u)$ for $0 \le u \le 1$.
(d) Find $E(Y)$.

****

3.22 Consider a random variable X with probability density function

$$p_X(u) = 4\,u^3\,U(u)U(1-u)$$

Let another random variable Y have the conditional probability density function

$$p_Y(v\,|\,X=u) = u^{-2}\,U(v)U(u^2-v) \quad \text{for } 0 \le u \le 1$$

(a) Find the joint probability density function $p_{XY}(u,v)$ and indicate on a sketch of the (u,v) plane the region on which $p_{XY}(u,v) \ne 0$.

(b) Find $p_Y(v)$ and sketch this probability density function versus v.

(c) Find the conditional expected value $E(Y\,|\,X=u)$ for $0 \le u \le 1$.

(d) Find $E(Y)$.

****

3.23 Consider a random variable X with probability density function

$$p_X(u) = \frac{\pi}{4}\cos\left(\frac{\pi u}{2}\right) U(1-|u|)$$

Let another random variable Y have the conditional probability density function

$$p_Y(v\,|\,X=u) = \left[\cos\left(\frac{\pi u}{2}\right)\right]^{-1} U(v)\,U\left[\cos\left(\frac{\pi u}{2}\right)-v\right] \quad \text{for } |u| \le 1$$

(a) Find the joint probability density function $p_{XY}(u,v)$ and indicate on a sketch of the (u,v) plane the region on which $p_{XY}(u,v) \ne 0$.

(b) Find the conditional expected value $E(Y\,|\,X=u)$ for $|u| \le 1$.

(c) Find $E(Y)$.

(d) Find $p_Y(v)$ and sketch this probability density function versus v.

****

3.24 Consider the two random variables of Exercise 3.13 with the joint probability density function

$$p_{X_1X_2}(u_1,u_2) = A\exp\left(-\frac{2}{3}\left[\frac{u_1-1}{0.5}\right]^2 + \frac{2}{3}\left[\frac{u_1-1}{0.5}\right]\left[\frac{u_2-2}{0.25}\right] - \frac{2}{3}\left[\frac{u_2-2}{0.25}\right]^2\right)$$

Find the conditional mean and variance: $E(X_2\,|\,X_1=3)$ and $\mathrm{Var}(X_2\,|\,X_1=3)$. [Hint: See Example 3.16.]

****

3.25 Consider the two random variables of Exercise 3.14 with the joint probability density function

$$p_{X_1X_2}(u_1,u_2) = A\exp\left(-\frac{8}{15}\left[\frac{u_1-2}{2}\right]^2 - \frac{4}{15}\left[\frac{u_1-2}{2}\right]\left[\frac{u_2-3}{4}\right] - \frac{8}{15}\left[\frac{u_2-3}{4}\right]^2\right)$$

Find the conditional mean and variance: $E(X_2\,|\,X_1=5)$ and $\mathrm{Var}(X_2\,|\,X_1=5)$. [Hint: See Example 3.16.]

****

General Conditional Expectation

****

3.26 Let X be uniform on $[-1,1]$: $p_X(u)=0.5$ for $|u|\le 1$. Let $E(Y|X=u)=u$, $E(Y^2|X=u)=2u^2$, and $Z=XY$.
Find the mean and variance of Z.
[Hint: Begin with $E(Z|X=u)$ and $E(Z^2|X=u)$, then find the unconditional expectations.]

****

3.27 Let X be uniform on $[-1,1]$: $p_X(u)=0.5$ for $|u|\le 1$. Let $E(Y|X=u)=2u^2$, $E(Y^2|X=u)=4u^4+2u^2$, and $Z=XY$.
Find the mean and variance of Z.
[Hint: Begin with $E(Z|X=u)$ and $E(Z^2|X=u)$, then find the unconditional expectations.]

****

Characteristic Function

****

3.28 Let X have a distribution that is uniform on $[c-h/2, c+h/2]$:

$$p_X(u)=h^{-1}U(u-c+h/2)U(c+h/2-u)$$

(a) Determine the mean and variance of X from integration of $p_X(u)$.
(b) Find the characteristic function of X.
(c) Check the mean and variance of X by using derivatives of the characteristic function.

****

3.29 Let the random variable X have a discrete distribution with

$$p_X(u)=a\,\delta(u-1)+a\,\delta(u-2)/4+a\,\delta(u-3)/9$$

That is, $P(X=k)=a/k^2$ for $k=$ 1, 2, and 3.
(a) Find the value of a.
(b) Find the mean and variance of X from integration of $p_X(u)$.
(c) Find the characteristic function of X.
(d) Check the mean and variance of X by using derivatives of the characteristic function.

****

Chapter 4
Analysis of Stochastic Processes

4.1 Concept of a Stochastic Process

As noted in Chapter 1, a stochastic process can be viewed as a family of random variables. It is common practice to use braces to denote a set or collection of items, so we write $\{X(t)\}$ for a stochastic process that gives a random variable $X(t)$ for any particular value of *t*. The parameter *t* may be called the index parameter for the process, and the set of possible *t* values is then the index set. The basic idea is that for every possible *t* value there is a random variable $X(t)$. In some situations we will be precise and include the index set within the notation for the process, such as $\{X(t): t \ge 0\}$ or $\{X(t): 0 \le t \le 20\}$. It should be kept in mind that such a specification of the index set within the notation is nothing more than a statement of the range of *t* for which $X(t)$ is defined. For example, writing a process as $\{X(t): t \ge 0\}$ means that $X(t)$ is not defined for $t < 0$. We will often simplify our notation by omitting the specification of the possible values of *t*, unless that is apt to cause confusion. In this book we will always treat the index set of possible *t* values as being continuous rather than discrete, which is precisely described as a *continuously parametered* or *continuously indexed* stochastic process. If we consider a set of $X(t)$ random variables with *t* chosen from a discrete set such as $\{t_1, t_2, \cdots, t_j, \cdots\}$, then we will use the nomenclature and notation of a vector random variable rather than of a stochastic process.

Another useful way to conceive of a stochastic process is in terms of its possible time histories—its variation with *t* for a particular observation of the process. For example, any earthquake ground acceleration record might be thought of as one of the many time histories that could have occurred for an earthquake with the same intensity at that site. In this example, as in many others, the set of possible time histories must be viewed as an infinite collection, but the concept is still useful. This idea of the set of all possible time histories may be viewed as a direct generalization of the concept of a random variable. The generalization can be illustrated by the idea of statistical sampling. A statistical sample from a random variable *Y* is a set of independent observations of the value of *Y*. Each observation

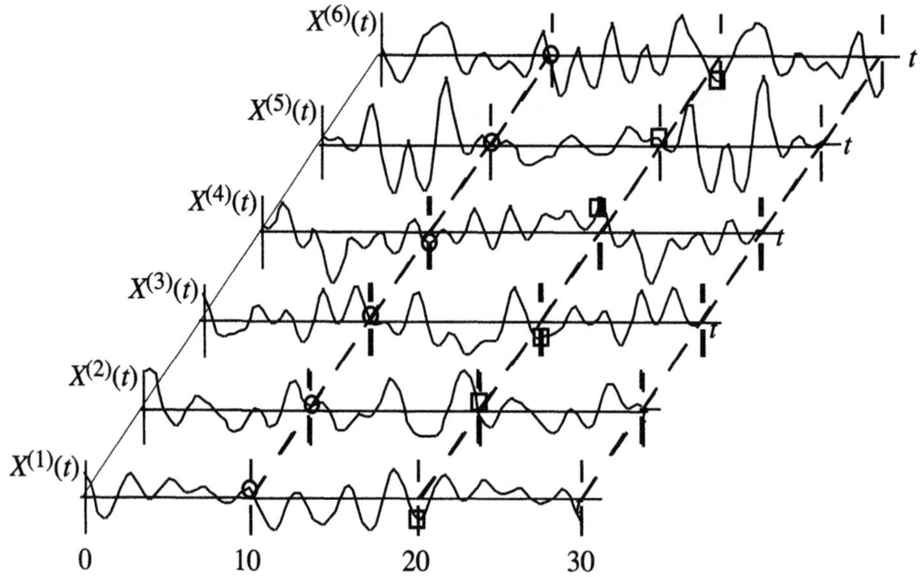

Figure 4.1 Ensemble of time histories of $\{X(t)\}$.

is simply a number from the set of possible values of Y. After a sufficient number of such observations, one will get an idea of the likelihood of various outcomes and can estimate the probability distribution of Y and/or expected values of functions of Y. For the stochastic process $\{X(t)\}$, each observation will give an observed time history rather than simply a number. Again, a sufficient number of observations will allow us to estimate probabilities and expected values related to the process. A collection of time histories for a stochastic process is typically called an *ensemble*.

Figure 4.1 illustrates the idea of a statistical sample, or ensemble, from a stochastic process, using the notation $X^{(j)}(t)$ for the jth sample time history observed for the process. Of course, the ensemble shown in Fig. 4.1 is for illustration only and is too small (i.e., it has too few time histories) to allow one to estimate probabilities or expected values with any confidence. It shows six time histories observed in separate, independent observations of the particular $\{X(t)\}$ process. A "section" across this ensemble at any particular time gives a statistical sample for the random variable corresponding to that t value. Thus, we have observed a sample of six values for $X(10)$, a sample of six values for $X(20)$, and so forth. The plot illustrates the fact that there is a sort of "orthogonality" between the idea that a stochastic process is characterized by an ensemble of time histories and the idea that it is a family of random variables. A time history is a single observation including many values of t. Many observations at a single value of t give a statistical sample of the random variable $X(t)$.

In most practical problems, it is not feasible to describe a stochastic process in terms of the probability of occurrence of particular time histories (although Examples 4.1 to 4.3 will illustrate a few situations in which this is feasible). However, we can always characterize the process by using information about the joint probability distribution of the random variables that make it up. Nonetheless, it is often useful to think of the process in terms of an ensemble (usually with infinitely many members) of all its possible time histories and to consider the characteristics of these time histories. The term *ensemble average* is often used for statistical averages across an ensemble of observed time histories. Thus, one might calculate the average value of $X(10)$ from the sample shown in Fig. 4.1 as $[X^{(1)}(10)+\cdots+X^{(6)}(10)]/6$, and this can be classified either as a statistical average of the observed values for the random variable $X(10)$ or as an ensemble average of the process $\{X(t)\}$ at time $t=10$. It should also be noted that the term *ensemble average* is also sometimes used to refer to mathematical expectation. This is generally legitimate if one considers an average over an infinite ensemble. Basically, we expect a statistical average to converge to the underlying expected value if we can make our sample size infinitely large, and an ensemble average is simply a statistical average for a stochastic process. One must always remember, though, that actual numerical ensemble averages are always statistical averages across a finite sample; they are not the same as expected values.

4.2 Probability Distribution

In order to have a complete probabilistic description of a stochastic process $\{X(t)\}$, one must know the probability distribution for every set of random variables belonging to that process. This can be simply stated as knowing the probability distribution for every set $\{X(t_1),X(t_2),\cdots,X(t_n)\}$, for all possible n values, and all possible choices of $\{t_1,t_2,\cdots,t_n\}$ for each n value. Because the probability distribution of a random variable or a set of random variables can always be given by a probability density function (or a cumulative distribution function), the necessary information can be written as

$$
\begin{array}{ll}
p_{X(t_1)}(u_1) & \text{for all choices of } t_1 \text{ and } u_1 \\
p_{X(t_1)X(t_2)}(u_1,u_2) & \text{for all choices of } t_1,\ t_2,\ u_1, \text{ and } u_2 \\
\vdots & \\
p_{X(t_1)\cdots X(t_n)}(u_1,\cdots,u_n) & \text{for all choices of } t_1,\cdots,t_n,u_1,\cdots,u_n \\
\vdots &
\end{array}
$$

Note, though, that this list is redundant. If one knows the joint distribution of the random variables $\{X(t_1),\cdots,X(t_k)\}$ corresponding to $n=k$, then the corresponding information for any n value that is less than k is simply a marginal distribution that can be easily derived from this joint distribution, as in Section 2.5. Thus, the function $p_{X(t_1)\cdots X(t_k)}(u_1,\cdots,u_k)$ contains all the information for $p_{X(t_1)}(u_1)$, $p_{X(t_1),X(t_2)}(u_1,u_2)$, $\cdots$, $p_{X(t_1)\cdots X(t_{k-1})}(u_1,\cdots,u_{k-1})$. All of these statements could equally well be written in terms of cumulative distribution functions like $F_{X(t_1)\cdots X(t_n)}(u_1,\cdots,u_n)$, but there is no point in repeating them because the forms would be the same.

The probability density $p_{X(t)}(u)$ can be viewed as a function of the two variables t and u, and it is sometimes written as $p_X(u,t)$ or $p_X(u;t)$, with similar notation for the higher-order density functions. There is nothing wrong with this notation, but we choose to keep any t variable explicitly stated as an index of X to remind us that the sole function of t is to specify a particular random variable $X(t)$. Stated another way, we do not want a notation such as $p_X(u,t)$ to lead the reader to consider this probability density function for the stochastic process $\{X(t)\}$ to be somehow different from $p_X(u)$ for a random variable X. The probability densities for a stochastic process must be identical in concept to those for random variables, because they are simply the descriptions of the random variables that make up the process. In some cases we will find it convenient to use alternative notations that are simpler, such as $X_k \equiv X(t_k)$, $\vec{X} \equiv [X(t_1),\cdots,X(t_n)]^T$, and $p_{\vec{X}}(\vec{u}) \equiv p_{X(t_1)\cdots X(t_n)}(u_1,\cdots,u_n)$, but we will always keep index parameters [such as t in $p_{X(t)}(u)$] specifying particular random variables separate from the dummy variables denoting possible values of the random variables [e.g., u in $p_{X(t)}(u)$].

We noted earlier that our discussion of stochastic processes will be limited to the situation in which the index set of possible t values in $\{X(t)\}$ is continuous. This is very different from saying that the set of possible values for $X(t)$ is continuous. In fact, we will usually be considering situations in which both the set of t values and the set of u values in $p_{X(t)}(u)$ is continuous, but our definitions certainly allow the possibility of having $\{X(t)\}$ defined on a continuous index set of t values, but with each random variable $X(t)$ belonging to a discrete set. Example 4.1 is a very simple case of this situation; because in this example $X(t)$ has only three possible values for any particular t value. Recall that a random variable X may not be a continuous random variable even if it has a continuous set of possible values. If a random variable X has a continuous set of possible values but a discontinuous cumulative distribution on those values, then X has a mixed rather than continuous distribution (like the random variable Y in Example 2.3). Similarly, saying that the index set of possible t values is continuous does not imply that the stochastic process $\{X(t)\}$ varies continuously on that set, either in terms of time history continuity or some sort of probabilistic continuity. This issue of continuity of $\{X(t)\}$ over the continuous index set will be considered in Section 4.6.

In many problems we will need to consider more than one stochastic process. This means that we will need information related to the joint distribution of random variables from two or more stochastic processes. For example, to have a complete probabilistic description of the two stochastic processes $\{X(t)\}$ and $\{Y(s)\}$, we will need to know the joint probability distribution for every set of random variables $\{X(t_1),\cdots,X(t_n),Y(s_1),\cdots,Y(s_m)\}$ for all possible n and m values, and all possible choices of $\{t_1,\cdots,t_n,s_1,\cdots,s_m\}$ for each (n,m) combination. Of course, this joint distribution could be described by a probability density function like $p_{X(t_1)\cdots X(t_n)Y(s_1)\cdots Y(s_m)}(u_1,\cdots,u_n,v_1,\cdots,v_m)$. The concept can be readily extended to more than two stochastic processes.

In summary, the fundamental definition of one or more stochastic processes is in terms of the underlying probability distribution, as given by probability density functions or cumulative distribution functions. Of course, we can never explicitly write out all of these functions because there are infinitely many of them for processes with continuous index sets. Furthermore, we will often find that we need to write out few or none of them in calculating or estimating the probabilities of interest for failure calculations. In many cases we can gain the information that we need for stochastic processes from considering their moments (as in Chapter 3), and the following section considers this characterization.

4.3 Moment and Covariance Functions

One can characterize any random variable X by moments in the form of mean value, mean-squared value, variance, and possibly higher moments or cumulants giving information like skewness and kurtosis, as in Section 3.7. Similarly, if we have more than one random variable, then we can use cross-products, covariance, and other moments or expected values involving two or more random variables. The material presented here for a stochastic process $\{X(t)\}$ is a direct application of the concepts in Section 3.3 to the set of random variables that compose the process.

For the mean value, or expected value, of a process we will use the notation

$$\mu_X(t) \equiv E[X(t)] = \int_{-\infty}^{\infty} u\, p_{X(t)}(u)\, du \tag{4.1}$$

That is, $\mu_X(t)$ is a function defined on the index set of possible values of t in $\{X(t)\}$, and at any particular t value, this mean value function is identical to the mean of the random variable $X(t)$. Similarly, we define a function called the autocorrelation function of $\{X(t)\}$ as a cross-product of two random variables from the same process

$$\phi_{XX}(t,s) \equiv E[X(t)\,X(s)] = \int_{-\infty}^{\infty}\int_{-\infty}^{\infty} u\,v\,p_{X(t)X(s)}(u,v)\,du\,dv \tag{4.2}$$

and this function is defined on a two-dimensional space with t and s each varying over all values in the index set for $\{X(t)\}$. The double subscript notation on $\phi_{XX}(t,s)$ is to distinguish autocorrelation from the corresponding concept of cross-correlation that applies when the two random variables in the cross-product are drawn from two different stochastic processes as

$$\phi_{XY}(t,s) \equiv E[X(t)\,Y(s)] = \int_{-\infty}^{\infty}\int_{-\infty}^{\infty} u\,v\,p_{X(t)Y(s)}(u,v)\,du\,dv \tag{4.3}$$

The cross-correlation function is defined on the two-dimensional space with t being any value from the index set for $\{X(t)\}$ and s being any value from the index set for $\{Y(t)\}$. These two index sets may be identical or quite different.

The reader should note that the use of the term *correlation* in the expressions autocorrelation and cross-correlation is not consistent with the use of the term for random variables. For two random variables the correlation coefficient is a normalized form of the covariance, whereas autocorrelation and cross-correlation simply correspond to cross-products of two random variables. This inconsistency is unfortunate but well established in the literature on random variables and stochastic processes.

We will also use the covariances of random variables drawn from one or two stochastic processes and refer to them as autocovariance

$$K_{XX}(t,s) \equiv E\big([X(t)-\mu_X(t)][X(s)-\mu_X(s)]\big) \tag{4.4}$$

and cross-covariance

$$K_{XY}(t,s) \equiv E\big([X(t)-\mu_X(t)][Y(s)-\mu_Y(s)]\big) \tag{4.5}$$

Of course, the two-dimensional domain of definition of $K_{XX}(t,s)$ is identical to that for $\phi_{XX}(t,s)$, and for $K_{XY}(t,s)$ it is the same as for $\phi_{XY}(t,s)$. We can rewrite these second central moment expressions for the stochastic processes in terms of mean and cross-product values and obtain

$$K_{XX}(t,s) = \phi_{XX}(t,s) - \mu_X(t)\,\mu_X(s) \tag{4.6}$$

and

$$K_{XY}(t,s) = \phi_{XY}(t,s) - \mu_X(t)\,\mu_Y(s) \tag{4.7}$$

Note that we can also define a mean-squared function for the $\{X(t)\}$ stochastic process as a special case of the autocorrelation function, and an ordinary variance function as a special case of the autocovariance function:

$$E[X^2(t)] = \phi_{XX}(t,t) \tag{4.8}$$

and

$$\sigma_X^2(t) = K_{XX}(t,t) \tag{4.9}$$

One can also extend the idea of the correlation coefficient for random variables and write

$$\rho_{XX}(t,s) = \frac{K_{XX}(t,s)}{\sigma_X(t)\,\sigma_X(s)} = \frac{K_{XX}(t,s)}{[K_{XX}(t,t)\,K_{XX}(s,s)]^{1/2}} \tag{4.10}$$

and

$$\rho_{XY}(t,s) = \frac{K_{XY}(t,s)}{\sigma_X(t)\,\sigma_Y(s)} = \frac{K_{XY}(t,s)}{[K_{XX}(t,t)\,K_{YY}(s,s)]^{1/2}} \tag{4.11}$$

Sometimes it is convenient to modify the formulation of a problem in such a way that one can analyze a mean-zero process, even though the physical process requires a model with a mean value function that is not zero. This is simply done by introducing a new stochastic process $\{Z(t)\}$ defined by $Z(t) = X(t) - \mu_X(t)$. This new process is mean-zero [i.e., $\mu_Z(t) \equiv 0$], and the autocovariance function for the original $\{X(t)\}$ process is given by $K_{XX}(t,s) = \phi_{ZZ}(t,s)$.

****

Example 4.1: Consider a process $\{X(t): t \ge 0\}$ that represents the position of a body that is minimally stable against overturning. In particular, $\{X(t)\}$ has only three possible time histories:

$X^{(1)}(t) = 0$

$X^{(2)}(t) = \alpha \sinh(\beta t)$

$X^{(3)}(t) = -\alpha \sinh(\beta t)$

in which α and β are constants, and the three time histories represent "not falling," "falling to the right," and "falling to the left." Let the probabilities be

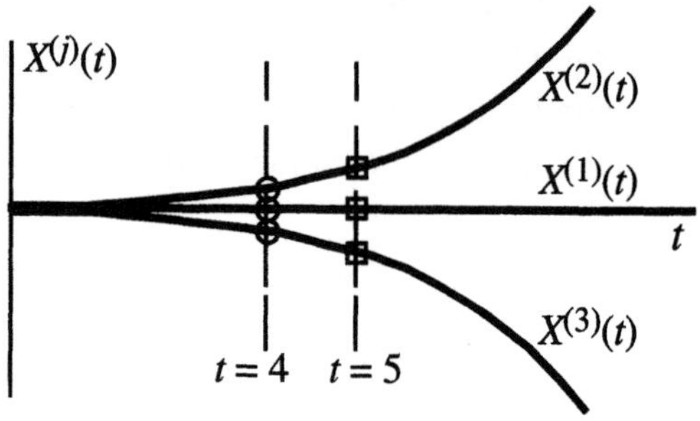

$$P[X(t)=X^{(1)}(t)]=0.50$$

$$P[X(t)=X^{(2)}(t)]=0.25$$

$$P[X(t)=X^{(3)}(t)]=0.25$$

Find:

(a) The first-order probability distribution of any random variable $X(t)$ from $\{X(t)\}$

(b) The joint probability distribution of any two random variables $X(t)$ and $X(s)$

(c) The mean value function for $\{X(t)\}$

(d) The covariance function for $\{X(t)\}$ and the variance function $\sigma_X^2(t)\equiv\sigma_{X(t)}^2$

(e) The correlation coefficient relating the random variables $X(t)$ and $X(s)$

(a) At any time there are only three possible values for $X(t)$, and the probability distribution for such a simple discrete random variable can be given as the probability of each possible outcome. Thus, we have

$$P[X(t)=0]=0.50,\quad P[X(t)=\alpha\sinh(\beta t)]=0.25,\quad P[X(t)=-\alpha\sinh(\beta t)]=0.25$$

By using the Dirac delta function, as in Section 2.4 (see also Appendix A), this can be described by a probability density function of

$$p_{X(t)}(u)=0.50\,\delta(u)+0.25\,\delta[u-\alpha\sinh(\beta t)]+0.25\,\delta[u+\alpha\sinh(\beta t)]$$

(b) Clearly there are three possible values for $X(t)$ and three possible values for $X(s)$, but in this case these result in only three possible values for the pair. In particular, for any t and s not equal to zero there is one point with $X(t)=X^{(1)}(t),X(s)=X^{(1)}(s)$, one with $X(t)=X^{(2)}(t),X(s)=X^{(2)}(s)$, and one with $X(t)=X^{(3)}(t),X(s)=X^{(3)}(s)$. The joint probability distribution of the two random variables is completely given by

$$P[X(t)=0,X(s)=0]=0.50$$

$$P[X(t)=\alpha\sinh(\beta t),X(s)=\alpha\sinh(\beta s)]=0.25$$

$$P[X(t)=-\alpha\sinh(\beta t),X(s)=-\alpha\sinh(\beta s)]=0.25$$

or as a joint probability density function as

$$p_{X(t),X(s)}(u,v)=0.50\,\delta(u)\,\delta(v)+0.25\,\delta[u-\alpha\sinh(\beta t)]\,\delta[v-\alpha\sinh(\beta s)]+$$
$$0.25\,\delta[u+\alpha\sinh(\beta t)]\,\delta[v+\alpha\sinh(\beta s)]$$

(c) The mean value of the random variable is the probability weighted average over the possible values, so

$$\mu_X(t)=(0.50)(0)+(0.25)[\alpha\sinh(\beta t)]+(0.25)[-\alpha\sinh(\beta t)]=0$$

(d) Because the mean is zero, the covariance function is the same as the autocorrelation function. Thus,

$$\phi_{XX}(t,s) \equiv E[X(t)X(s)] = (0.50)(0)(0) + (0.25)[\alpha \sinh(\beta t)][\alpha \sinh(\beta s)] + (0.25)[-\alpha \sinh(\beta t)][-\alpha \sinh(\beta s)]$$

gives

$$K_{XX}(t,s) = \phi_{XX}(t,s) = 0.50\alpha^2 \sinh(\beta t)\sinh(\beta s)$$

Choosing $s = t$ gives the variance and mean-squared values as

$$\sigma_X^2(t) = E[X^2(t)] = 0.50\alpha^2 \sinh^2(\beta t)$$

(e) The correlation coefficient of the two random variables is of the form

$$\rho_{XX}(t,s) = \frac{K_{XX}(t,s)}{\sigma_{X(t)}\sigma_{X(s)}} = \frac{0.50\alpha^2 \sinh(\beta t)\sinh(\beta s)}{0.50[\alpha^2 \sinh^2(\beta t)\alpha^2 \sinh(\beta s)]^{1/2}} = 1$$

Thus, we see that $X(t)$ and $X(s)$ are perfectly correlated. There is a perfect linear relationship between them that can be written as

$$X(s) = \frac{\sinh(\beta s)}{\sinh(\beta t)} X(t)$$

It should be noted that such a situation with only a few possible time histories is not typical of most stochastic problems of interest.

****

Example 4.2: Consider a stochastic process $\{X(t)\}$ that may have infinitely many possible time histories, but with these time histories depending on only one random variable—a random initial condition:

$$X(t) = A\cos(\omega t)$$

in which ω is a deterministic circular frequency and A is a random variable for which the probability distribution is known. Find expressions for the mean value function, the autocorrelation function, and the covariance function for $\{X(t)\}$.

Because only A is random in $X(t)$, we know that

$$\mu_X(t) \equiv E[X(t)] = E(A)\cos(\omega t) \equiv \mu_A \cos(\omega t)$$

$$\phi_{XX}(t,s) \equiv E[X(t)\,X(s)] = E(A^2)\cos(\omega t)\cos(\omega s)$$

and

$$K_{XX}(t,s) = \phi_{XX}(t,s) - \mu_X(t)\,\mu_X(s) = [E(A^2) - \mu_A^2]\cos(\omega t)\cos(\omega s)$$

$$= \sigma_A^2 \cos(\omega t)\cos(\omega s)$$

The number of possible time histories in this example depends entirely on the number of possible values for the random variable A. If , for example, A has only five possible values, then $X(t)$ has only five possible time histories. At the other extreme, if A is a continuous random variable, then there are infinitely many (in fact uncountably many) possible time histories for $X(t)$. Even in this case, though, we know the shape of each time history because they are all cosine functions with a given frequency. Furthermore, any two time histories are either exactly in phase or 180° out of phase, but there are many possible amplitudes.

****

Example 4.3: Consider a stochastic process $\{X(t)\}$ that has time histories depending on only two random variables:

$$X(t) = A\cos(\omega t + \theta)$$

in which ω is a deterministic circular frequency and A and θ are random variables for which the joint probability distribution is known. In particular, let A and θ be independent and θ be uniformly distributed on the range $[0,2\pi]$. Find expressions for the mean value function, the autocorrelation function, and the covariance function for $\{X(t)\}$.

Because A and θ are independent, it can be shown that A and $\cos(\omega t+\theta)$ are also independent. The gist of this idea can be stated as follows: Independence of A and θ tells us that knowledge of the value of A will give no information about the possible values or probability of θ, and this then guarantees that knowledge of the value of A will give no information about the possible values or probability of $\cos(\omega t+\theta)$; this is the condition of independence. Using this independence, we can write

$$\mu_X(t) \equiv \mu_A\, E[\cos(\omega t+\theta)]$$

but we can use the given uniform distribution for θ to obtain

$$E[\cos(\omega t+\theta)] = \int_{-\infty}^{\infty} p_\theta(\eta)\, \cos(\omega t+\eta)\, d\eta = \int_0^{2\pi} \frac{1}{2\pi} \cos(\omega t+\eta)\, d\eta = 0$$

Thus, $\mu_X(t) = 0$ regardless of the distribution of A, provided only that μ_A is finite. We use exactly the same ideas to find the other quantities asked for:

$$K_{XX}(t,s) = \phi_{XX}(t,s) = E(A^2)\, E[\cos(\omega t+\theta)\cos(\omega s+\theta)]$$

Using the identity $\cos(\alpha)\cos(\beta) = [\cos(\alpha+\beta)+\cos(\alpha-\beta)]/2$ allows us to obtain

$$E[\cos(\omega t+\theta)\cos(\omega s+\theta)] = \frac{1}{2}\left(\int_0^{2\pi} \frac{1}{2\pi} \cos[\omega(t+s)+2\eta)]\, d\eta + \cos[\omega(t-s)]\right)$$

$$= \frac{1}{2}\cos[\omega(t-s)]$$

so that

$$K_{XX}(t,s) = \phi_{XX}(t,s) = \frac{E(A^2)}{2}\cos[\omega(t-s)]$$

Note that in this problem, as in Example 4.2, the shape of each time history is a cosine function with a given frequency. The distinguishing factor is that now the phase difference between any two time histories can be any angle, and all possible values are equally likely. Thus, for any distribution of A, there are infinitely many possible time histories. In fact, even if A has only one value [i.e., $p_A(u) = \delta(u - A_0)$] there are still infinitely many possible time histories but they differ only in phase angle.

****

Example 4.4: Let $\{X(t)\}$ be the stochastic process of Example 4.3, and let $\{Y(t)\}$ be a process that also has trigonometric time histories. The time histories of $\{Y(t)\}$, though, are 45° out of phase with those of $\{X(t)\}$ and they are offset by a fixed amount of 5:

$$Y(t) = 5 + A\cos(\omega t + \theta + \pi/4)$$

Note that the same two random variables A and θ define $\{X(t)\}$ and $\{Y(t)\}$. Find the cross-correlation function and the cross-covariance function for the stochastic processes.

Because the mean value function of $\{X(t)\}$ is exactly zero, Eq. 4.7 tells us that the cross-covariance function is identical to the cross-correlation function. We find this function as

$$K_{XY}(t,s) = \phi_{XY}(t,s) = E\left(\left[A\cos\left(\omega t + \theta\right)\right]\left[5 + A\cos\left(\omega s + \theta + \frac{\pi}{4}\right)\right]\right)$$

or

$$K_{XY}(t,s) = 5\,E(A)\,E\left[\cos\left(\omega t + \theta\right)\right] + E(A^2)\,E\left(\cos\left(\omega t + \theta\right)\cos\left(\omega t + \theta + \frac{\pi}{4}\right)\right)$$

or

$$K_{XY}(t,s) = E(A^2)\,E\left(\cos\left(\omega t + \theta\right)\cos\left(\omega s + \theta + \frac{\pi}{4}\right)\right)$$

Using the identity $\cos(\alpha)\cos(\beta) = [\cos(\alpha + \beta) + \cos(\alpha - \beta)]/2$ now gives

$$K_{XY}(t,s) = \frac{E(A^2)}{2} E\left(\cos\left[\omega(t+s) + 2\theta + \frac{\pi}{4}\right] + \cos\left[\omega(t-s) - \frac{\pi}{4}\right]\right)$$

which reduces to

$$K_{XY}(t,s) = \phi_{XY}(t,s) = \frac{E(A^2)}{2}\cos\left(\omega(t-s) - \frac{\pi}{4}\right)$$

$$= \frac{E(A^2)}{2^{3/2}}\left(\cos\left[\omega(t-s)\right] + \sin\left[\omega(t-s)\right]\right)$$

****

4.4 Stationarity of Stochastic Processes

The property of stationarity (or homogeneity) of a stochastic process $\{X(t)\}$ always refers to some aspect of the description of the process being unchanged by any arbitrary shift along the t axis. There are many types of stationarity depending on what characteristic of the process has this property of being invariant under a time shift.

The simplest type of stationarity involves only invariance of the mean value function for the process. We say that $\{X(t)\}$ is mean-value stationary if

$$\mu_X(t+r) = \mu_X(t) \quad \text{for any value of the shift parameter } r \tag{4.12}$$

Clearly this can be true only if $\mu_X(t)$ is the same for all t values, so we can say that $\{X(t)\}$ is mean-value stationary if

$$\mu_X(t) = \mu_X \tag{4.13}$$

in which the absence of a t argument on the right-hand side conveys the information that the mean value is independent of time. Although the notation on the right-hand side of Eq. 4.13 is the same as for the mean value of a random variable, Eq. 4.13 does refer to the mean value function of a stochastic process. Of course, having the mean value be independent of t does not imply that the $X(t)$ random variables are all the same at different values of t or that the probability distributions of these random variables are all the same—only that they all have the same mean value.

As a next step in specifying more rigorous stationarity, let us say that $\{X(t)\}$ is second-moment stationary if the second moment function (i.e., the autocorrelation function) is invariant under a time shift:

$$E[X(t+r)\,X(s+r)] = E[X(t)\,X(s)] \quad \text{for any value of the shift parameter } r$$

or

$$\phi_{XX}(t+r,s+r) = \phi_{XX}(t,s) \quad \text{for any value of the shift parameter } r \tag{4.14}$$

Because Eq. 4.14 is specified to be true for any value of r, it must be true, in particular, for $r = -s$; using this particular value gives $\phi_{XX}(t,s) = \phi_{XX}(t-s,0)$. This shows that the autocorrelation function for a second-moment stationary process is a function of only one time argument—the $(t-s)$ difference between the two time arguments in $\phi_{XX}(t,s)$. We will use an alternative notation of R in place of ϕ for such a situation in which autocorrelation can be written as a function of a single time argument. Thus, we define $R_{XX}(t-s) \equiv \phi_{XX}(t-s,0)$ so that

$$\phi_{XX}(t,s) = R_{XX}(t-s) \quad \text{for any values of } t \text{ and } s \tag{4.15}$$

or, using $\tau = t - s$

$$R_{XX}(\tau) = \phi_{XX}(t+\tau,t) \equiv E[X(t+\tau)\,X(t)] \quad \text{for any values of } t \text{ and } \tau \tag{4.16}$$

We similarly say that two stochastic processes $\{X(t)\}$ and $\{Y(t)\}$ are jointly second-moment stationary if

$$\phi_{XY}(t+r,s+r) = \phi_{XY}(t,s) \quad \text{for any value of the shift parameter } r \tag{4.17}$$

and we define $R_{XY}(t-s) \equiv \phi_{XY}(t-s,0)$ so that

$$\phi_{XY}(t,s) = R_{XY}(t-s) \quad \text{for any values of } t \text{ and } s \tag{4.18}$$

or

$$R_{XY}(\tau) = \phi_{XY}(t+\tau,t) \equiv E[X(t+\tau)\,Y(t)] \quad \text{for any values of } t \text{ and } \tau \tag{4.19}$$

A slight variation on second-moment stationarity is obtained by using the autocovariance and cross-covariance functions in place of autocorrelation and cross-correlation functions. Thus, we say that a stochastic process is covariant stationary if

$$K_{XX}(t,s) = G_{XX}(t-s) \quad \text{for any values of } t \text{ and } s \tag{4.20}$$

or, equivalently,

$$G_{XX}(\tau) = K_{XX}(t+\tau,t) \quad \text{for any values of } t \text{ and } \tau \tag{4.21}$$

in which Eq. 4.21 gives the definition of the new stationary autocovariance function $G_{XX}(\tau)$. Similarly, $\{X(t)\}$ and $\{Y(t)\}$ are jointly covariant stationary if

$$K_{XY}(t,s) = G_{XY}(t-s) \quad \text{for any values of } t \text{ and } s \tag{4.22}$$

and

$$G_{XY}(\tau) = K_{XY}(t+\tau,t) \quad \text{for any values of } t \text{ and } \tau \tag{4.23}$$

Note that one can also say that $G_{XX}(\tau) = R_{ZZ}(\tau)$ if $\{Z(t)\}$ is defined as the mean-zero shifted version of $\{X(t)\}$: $Z(t) = X(t) - \mu_X(t)$.

Clearly one can extend this concept to definitions of third-moment stationarity, fourth-moment stationarity, skewness stationarity, kurtosis stationarity, and so forth. We will look explicitly only at one generic term in the sequence of moment stationary definitions. We will say that $\{X(t)\}$ is *j*th-moment stationary if

$$E[X(t_1+r)\,X(t_2+r)\cdots X(t_j+r)] = E[X(t_1)\,X(t_2)\cdots X(t_j)] \tag{4.24}$$

for all values of *r* and all values of $\{t_1,t_2,\cdots,t_j\}$. Using the notation $\tau_k = t_k - t_j$ for $k=1$ to $(j-1)$ allows this to be rewritten as

$$E[X(t_1)\,X(t_2)\cdots X(t_j)] = E[X(t_j+\tau_1)\,X(t_j+\tau_2)\cdots X(t_j+\tau_{j-1})\,X(t_j)] \tag{4.25}$$

with the expression on the right-hand side being independent of t_j. Thus, the *j*th moment function of a *j*th-moment stationary process is a function of only the $(j-1)$ time arguments $\{\tau_1,\tau_2,\cdots,\tau_{j-1}\}$ giving increments between the original time values of $\{t_1,t_2,\cdots,t_j\}$. Stationarity always reduces the number of necessary time arguments by one. Of course, we had already explicitly demonstrated this fact for first and second moments, showing that $\mu_X(t)$ depends on zero time arguments (is independent of time) and $\phi_{XX}(t,s)$ depends on the one time argument $\tau = t-s$.

Although a given stochastic process may simultaneously have various types of moment stationarity, this is not necessary. One of the most commonly considered combinations of moment stationarity involves the first and second moments. If a process $\{X(t)\}$ does have both mean-value and second-moment stationarity, then it is easily shown from Eq. 4.6 that it is also covariant stationary. Often such a process is described in the alternative way of saying that it is mean and covariant stationary, and Eq. 4.6 then shows that it is also second-moment stationary.

There are also forms of stationarity that are not defined in terms of moment functions. Rather, they are defined in terms of probability distributions being invariant under a time shift. The general relationship is that $\{X(t)\}$ is *j*th-order stationary if

$$p_{X(t_1+r)\cdots X(t_j+r)}(u_1,\cdots,u_j) = p_{X(t_1)\cdots X(t_j)}(u_1,\cdots,u_j) \tag{4.26}$$

for all values of $\{t_1,\cdots,t_j,u_1,\cdots,u_j\}$ and the shift parameter *r*. This includes, as special cases, $\{X(t)\}$ being first-order stationary if

$$p_{X(t+r)}(u) = p_{X(t)}(u) \tag{4.27}$$

for all values of t and u and the shift parameter r, and second-order stationary if

$$p_{X(t_1+r)X(t_2+r)}(u_1,u_2) = p_{X(t_1)X(t_2)}(u_1,u_2) \tag{4.28}$$

for all values of $\{t_1,t_2,u_1,u_2\}$ and the shift parameter r.

There are strong interrelationships among the various sorts of stationarity. One of the simplest of these involves the stationarities defined in terms of the probability distribution, as in Eqs. 4.26–4.28. Specifically, from consideration of marginal distributions it is easy to show that if $\{X(t)\}$ is jth-order stationary, then it is also first-order stationary, second-order stationary, and up to $(j-1)$-order stationary. Note that the same hierarchy does not apply to moment stationarity. It is quite possible to define a process that is second-moment stationary but not mean-value stationary, although there may be little practical usefulness for such a process. A slight variation on this is to have a process that is covariant stationary, but not mean-value or second-moment stationary; this process certainly does approximate various physical phenomena for which the mean value changes with time, but the variance is constant.

It is also important to note that jth-order stationarity always implies jth-moment stationarity, because a jth moment function can be calculated by using a jth-order probability distribution. Thus, for example, second-order stationarity implies second-moment stationarity. However, second-order stationarity also implies first-order stationarity, and this then implies first-moment (mean-value) stationarity. In general, we can say that jth-order stationarity implies stationarity of all moments up to and including the jth.

When comparing moment and order stationarity definitions, it is not always possible to say which is stronger. For example, consider the question of whether second-moment or first-order stationarity is stronger. Because first-order stationarity states that $p_{X(t)}(u)$ has time shift invariance, it implies that any moment that can be calculated from that probability density function also has time shift invariance. Thus, first-order stationarity implies that the jth moment for the random variable $X(t)$ at one instant of time has time shift invariance for any value of j. This seems to be a form of jth-moment stationarity that is certainly not implied by second-moment stationarity. On the other hand, first-order stationarity says nothing about the relationship of $X(t)$ to $X(s)$ for $t \neq s$, whereas second-moment stationarity does. Thus, there is no answer as to which condition is stronger. In general, jth-order stationarity implies time shift invariance of moments of any order, so long as they only depend on the values of $X(t)$ at no more than j different times. Similarly, *k*th-moment stationarity implies time shift invariance only of *k*th moment functions, which may involve values of $X(t)$ at up to *k* different time values. For $k > j$, one cannot say whether *j*th-order or *k*th-moment stationarity is stronger, because each implies certain behavior that the other does not.

The most restrictive type of stationarity is called strict stationarity. We say that $\{X(t)\}$ is strictly stationary if it is *j*th-order stationary for any value of *j*. This implies that any order probability density function has time shift invariance and any order moment function has time shift invariance. In the stochastic process literature, one also frequently encounters the terms *weakly stationary* and/or *wide-sense* stationary. It appears that both of these terms are usually used to mean a process that is both mean-value and covariant stationary, but caution is advised because there is some variation in usage, with meanings ranging up to second-order stationary. When we refer to a stochastic process as being stationary and give no qualification as to type of stationarity, we will generally mean strictly stationary. In some situations, though, the reader may find that some weaker form of stationarity is adequate for the calculations being performed. For example, if the analytical development or problem solution involves only second-moment calculations, then strict stationarity will be no more useful than second-moment stationarity. Thus, one can also say that the word *stationary* without qualifier simply means that all moments and/or probability distributions being used in the given problem are invariant under a time shift.

****

Example 4.5: Identify applicable types of stationarity for the stochastic process with $X^{(1)}(t)=0$, $X^{(2)}(t)=\alpha\sinh(\beta t)$, $X^{(3)}(t)=-\alpha\sinh(\beta t)$, and $P[X(t)=X^{(1)}(t)]=0.50$, $P[X(t)=X^{(2)}(t)]=P[X(t)=X^{(3)}(t)]=0.25$.

In Example 4.1 we found the first-order probability density function to be

$$p_{X(t)}(u)=0.50\,\delta(u)+0.25\,\delta[u-\alpha\sinh(\beta t)]+0.25\,\delta[u+\alpha\sinh(\beta t)]$$

Clearly this function is not invariant under a time shift because $\sinh(\beta t+\beta r)\neq\sinh(\beta t)$. Thus, $\{X(t)\}$ is not first-order stationary, and this precludes the possibility of its having any jth-order stationarity. However, we did find that the mean value function for this process is a constant [$\mu_X(t)=0$], so it is time shift invariant. Thus, the process is mean-value stationary. The autocorrelation function $\phi_{XX}(t,s)=(0.50)\alpha^2\sinh(\beta t)\sinh(\beta s)$ determined in Example 4.1 clearly demonstrates that the process is not second-moment stationary, because its dependence on t and s is not of the form $(t-s)$.

****

Example 4.6: Identify applicable types of stationarity for the stochastic process with $X(t) = A\cos(\omega t)$ in which ω is a constant and A is a random variable.

The mean value function $\mu_X(t) = \mu_A \cos(\omega t)$ and autocorrelation function $\phi_{XX}(t,s) = E(A^2)\cos(\omega t)\cos(\omega s)$ of this process were found in Example 4.2. These functions clearly show that this is neither mean-value or second-moment stationary.

We can investigate first-order stationarity by deriving the probability density function for the random variable $X(t)$. For any fixed *t*, this is a special case of $X = c\,A$, in which *c* is a deterministic constant. This form gives $F_X(u) = F_A(u/c)$ for $c > 0$ and $F_X(u) = 1 - F_A(u/c)$ for $c < 0$. Taking a derivative with respect to u then gives $p_X(u) = |c|^{-1}\, p_A(u/c)$. Thus, for $c = \cos(\omega t)$ we have

$$p_X(u) = \frac{1}{|\cos(\omega t)|}\, p_A\left(\frac{u}{\cos(\omega t)}\right)$$

and this is generally not invariant under a time shift. Thus, one cannot conclude that the process is first-order stationary, or jth-order stationary for any j.

****

Example 4.7: Investigate mean-value, second-moment, and covariance stationarity for the stochastic process with $X(t) = A\cos(\omega t + \theta)$ in which ω is a constant and A and θ are independent random variables with θ being uniformly distributed on the range $[0,2\pi]$.

In Example 4.3 we found the mean value, covariance, and autocorrelation functions for this stochastic process to be $\mu_X(t) = 0$ and $K_{XX}(t,s) = \phi_{XX}(t,s) = E(A^2)\cos[\omega(t-s)]/2$. These functions are all invariant under a time shift, because a time shift will change t and s by an equal amount in the K and ϕ functions. Thus, $\{X(t)\}$ does have stationarity of the mean value, the second moment, and the covariance functions. We can rewrite the autocorrelation and autocovariance functions as

$$G_{XX}(\tau) = R_{XX}(\tau) = \frac{E(A^2)}{2}\cos(\omega\tau)$$

****

Example 4.8: Consider the two stochastic processes $\{X(t)\}$ and $\{Y(t)\}$ defined in Example 4.4 with

$$K_{XY}(t,s) = \phi_{XY}(t,s) = \frac{E(A^2)}{2^{3/2}}\left(\cos[\omega(t-s)] + \sin[\omega(t-s)]\right)$$

Are $\{X(t)\}$ and $\{Y(t)\}$jointly covariant stationary?

The covariance function does have the time shift property, so $\{X(t)\}$ and $\{Y(t)\}$are jointly covariant stationary. We can rewrite the cross-correlation and cross-covariance functions as

$$G_{XY}(\tau) = R_{XY}(\tau) = \frac{E(A^2)}{2^{3/2}}\left[\cos(\omega\tau) + \sin(\omega\tau)\right]$$

****

4.5 Properties of Autocorrelation and Autocovariance

A number of properties of the autocorrelation and autocovariance functions apply for any stochastic process in any problem. We will list some of those here, giving both general versions and the forms that apply for stationary processes. Probably one of the more obvious, but nonetheless significant, properties is symmetry:

$$\phi_{XX}(s,t) = \phi_{XX}(t,s) \quad \text{and} \quad K_{XX}(s,t) = K_{XX}(t,s) \tag{4.29}$$

Rewriting these equations for a process with the appropriate stationarity (second-moment and covariant stationarity, respectively) gives

$$R_{XX}(-\tau) = R_{XX}(\tau) \quad \text{and} \quad G_{XX}(-\tau) = G_{XX}(\tau) \tag{4.30}$$

Next we recall the nonnegative definite property described in Section 3.3 for cross-product and covariance matrices for random variables. In particular, $\vec{v}^T E(\vec{X}\vec{X}^T)\vec{v} \ge 0$ and $\vec{v}^T \mathbf{K}_{XX}\vec{v} \ge 0$ for any real vector $\vec{v}$, in which $\vec{X}$ is a vector of real scalar components. In order to apply this result to a stochastic process $\{X(t)\}$, let

$$\vec{X} = \begin{pmatrix} X(t_1) \\ X(t_2) \\ \vdots \\ X(t_n) \end{pmatrix}$$

so that the $E(\vec{X}\vec{X}^T)$ and $\mathbf{K}_{XX}$ matrices have (j,k) components of $\phi_{XX}(t_j,t_k)$ and $K_{XX}(t_j,t_k)$, respectively. The ϕ_{XX} and K_{XX} functions, then, must be such that these matrices are nonnegative definite for all values of n and all possible choices of $\{t_1,t_2,\cdots,t_n\}$. It can be shown that any real square matrix $\mathbf{A}$ of dimension n is nonnegative definite if

$$\begin{vmatrix} A_{11} & A_{12} & \cdots & A_{1m} \\ A_{21} & A_{22} & \cdots & A_{2m} \\ \vdots & \vdots & & \vdots \\ A_{m1} & A_{m2} & \cdots & A_{mm} \end{vmatrix} \ge 0$$

for all integers m in the range $1 \le m \le n$, in which the bars denote the determinant of the submatrix shown. Because $E(\vec{X}\,\vec{X}^T)$ and $\mathbf{K}_{XX}$ must be nonnegative definite for all values of n, we can say that

$$\begin{vmatrix} \phi_{XX}(t_1,t_1) & \cdots & \phi_{XX}(t_1,t_n) \\ \vdots & & \vdots \\ \phi_{XX}(t_1,t_n) & \cdots & \phi_{XX}(t_n,t_n) \end{vmatrix} \ge 0, \qquad \begin{vmatrix} K_{XX}(t_1,t_1) & \cdots & K_{XX}(t_1,t_n) \\ \vdots & & \vdots \\ K_{XX}(t_1,t_n) & \cdots & K_{XX}(t_n,t_n) \end{vmatrix} \ge 0 \quad (4.31)$$

for $n \ge 1$. For $n = 1$, Eq. 4.31 reduces to the rather obvious requirements that $\phi_{XX}(t_1,t_1) \equiv E[X^2(t_1)] \ge 0$ and $K_{XX}(t_1,t_1) \equiv \sigma_X^2(t_1) \ge 0$. Expanding the determinants for $n = 2$ gives

$$|\phi_{XX}(t_1,t_2)| \le [\phi_{XX}(t_1,t_1)\,\phi_{XX}(t_2,t_2)]^{1/2} \quad (4.32)$$

and

$$|K_{XX}(t_1,t_2)| \le [K_{XX}(t_1,t_1)\,K_{XX}(t_2,t_2)]^{1/2} \quad (4.33)$$

Note that Eqs. 4.32 and 4.33 are converted into the Schwarz inequalities in Eqs. 3.14 and 3.15, respectively, by writing $\phi_{XX}(t,s) = E[X(t)\,X(s)]$ and $K_{XX}(t,s) = \rho_{XX}(t,s)\,\sigma_X(t)\,\sigma_X(s)$ with $\rho_{XX}(t,t) = 1$. Equations 4.32 and 4.33 provide bounds on the value of $\phi_{XX}(t,s)$ and $K_{XX}(t,s)$ anywhere off the 45° line of the (t,s) plane in terms of the values on that diagonal line. One particular implication of this is the fact that $|\phi_{XX}(t,s)| < \infty$ everywhere if $E[X^2(t)]$ is bounded for all values of t, and $|K_{XX}(t,s)| < \infty$ everywhere if $\sigma_X(t) < \infty$ for all t. The standard nomenclature is that $\{X(t)\}$ is a second-order process if its autocorrelation function is always finite.[1] If $\{X(t)\}$ has the appropriate stationarity, these Schwartz inequality relationships become

$$|R_{XX}(\tau)| \le R_{XX}(0) = E[X^2], \qquad |G_{XX}(\tau)| \le G_{XX}(0) = \sigma_X^2 \quad (4.34)$$

---

[1]The reader is cautioned that the word *order* in this term refers only to the order of the moment being considered and not to the order of the probability distribution, as in our definition of *j*th order stationarity or ergodicity.

The additional conditions that Eq. 4.31 must also apply for $m \ge 3$ are equally important, but less obvious.

We will next investigate certain continuity properties of the autocorrelation and autocovariance functions. It is not necessary that these functions be continuous everywhere, but there are some limitations on the types of discontinuities that they can have. The basic result is that if either of the functions is continuous in the neighborhood of a point (t,t) and in the neighborhood of a point (s,s), then it must also be continuous in the neighborhood of the point (t,s). The first step in clarifying this result is a statement of precisely what we mean by continuity for these functions of two arguments. We say that $\phi(t,s)$ is continuous at the point (t,s) if and only if

$$\lim_{\substack{\varepsilon_1 \to 0 \\ \varepsilon_2 \to 0}} \phi(t+\varepsilon_1, s+\varepsilon_2) = \phi(t,s) \tag{4.35}$$

meaning that we obtain the same limit as ε_1 and ε_2 both tend to zero along any route in the two-dimensional space. A special case of this relationship is that $\phi(t,s)$ is continuous at the point (t,t) if

$$\lim_{\substack{\varepsilon_1 \to 0 \\ \varepsilon_2 \to 0}} \phi(t+\varepsilon_1, t+\varepsilon_2) = \phi(t,t)$$

The reader should carefully consider the distinction between this last statement and the much weaker condition of $\phi(t+\varepsilon, t+\varepsilon)$ tending to $\phi(t,t)$ as ε tends to zero, which corresponds only to approaching the point (t,t) along a 45° diagonal in the (t,s) space. This seemingly minor issue has confused many students in the past.

To derive conditions for the continuity of $\phi(t,s)$ at the point (t,s), we write the autocorrelation function as an expected value and look at the difference between the value at (t,s) and at $(t+\varepsilon_1, s+\varepsilon_2)$:

$$\phi_{XX}(t+\varepsilon_1, s+\varepsilon_2) - \phi_{XX}(t,s) = E[X(t+\varepsilon_1)\, X(s+\varepsilon_2) - X(t)\, X(s)]$$

which can be rewritten as

$$\begin{aligned}\phi_{XX}(t+\varepsilon_1, s+\varepsilon_2) - \phi_{XX}(t,s) = E\big([X(t+\varepsilon_1) - X(t)][X(s+\varepsilon_2) - X(s)] + \\ [X(t+\varepsilon_1) - X(t)]\, X(s) + X(t)[X(s+\varepsilon_2) - X(s)]\big)\end{aligned} \tag{4.36}$$

Thus, we are assured of continuity of ϕ at the point (t,s) if the expected value of each of the three terms in the parentheses is zero. We can easily find sufficient conditions for this to be true by using the Schwarz inequality. For example,

$$E\Big([X(t+\varepsilon_1)-X(t)]\,X(s)\Big)\le\Big[E\Big([X(t+\varepsilon_1)-X(t)]^2\Big)E\Big([X(s)]^2\Big)\Big]^{1/2}$$

and converting the right-hand side back into autocorrelation functions gives

$$E\Big([X(t+\varepsilon_1)-X(t)]\,X(s)\Big)\le$$
$$[\phi_{XX}(t+\varepsilon_1,t+\varepsilon_1)-2\phi_{XX}(t+\varepsilon_1,t)+\phi_{XX}(t,t)]^{1/2}\,[\phi_{XX}(s,s)]^{1/2}$$

Now we can see that the first of these two square root terms tends to zero as $\varepsilon_1\to 0$ if ϕ_{XX} is continuous at the point (t,t), regardless of whether it is continuous anywhere else in the (t,s) plane. Thus, continuity of ϕ_{XX} at the point (t,t) will ensure that the second term in Eq. 4.36 goes to zero as $\varepsilon_1\to 0$. One can show the same result for the first term in Eq. 4.36 by exactly the same set of manipulations. Similarly, the third term in Eq. 4.36 goes to zero as $\varepsilon_2\to 0$ if ϕ_{XX} is continuous at the point (s,s). The sum of these three results shows that Eq. 4.35 holds for the function ϕ_{XX} if it is continuous at both the point (t,t) and the point (s,s). That is, if ϕ_{XX} is continuous at both the point (t,t) and the point (s,s), then it is also continuous at the point (t,s). Exactly the same result applies for the autocovariance function: If K_{XX} is continuous at both the point (t,t) and the point (s,s), then it is also continuous at the point (t,s).

We again emphasize that there is no requirement that ϕ_{XX} or K_{XX} be continuous. The requirement is that if either function is discontinuous at a point (t,s), then it must also be discontinuous at point (t,t) and/or point (s,s). One implication of this result is that if ϕ_{XX} or K_{XX} is continuous everywhere along the (t,t) diagonal of the (t,s) plane, it must be continuous everywhere in the plane.

The general continuity relationship just derived can be applied to the special case of a stationary process by simply replacing ϕ_{XX} and K_{XX} by R_{XX} and G_{XX} and noting that the points (t,t) and (s,s) both correspond to $\tau=0$. The result is that continuity of $R_{XX}(\tau)$ or $G_{XX}(\tau)$ at $\tau=0$ is sufficient to prove that the function is continuous everywhere. Alternatively, we can say that if $R_{XX}(\tau)$ or $G_{XX}(\tau)$ is discontinuous for any value of the argument τ, then it must also be discontinuous at $\tau=0$.

****

Example 4.9: Consider a very erratic stochastic process $\{X(t)\}$ that has discontinuities in its time histories at discrete values of t:

$$X(t) = A_j \qquad \text{for} \qquad j\Delta \le t < (j+1)\Delta$$

or

$$X(t) = \sum_j A_j \big(U[t - j\Delta] - U[t-(j+1)\Delta] \big)$$

in which the random variables are independent and identically distributed with mean zero and variance $E(A^2)$, and $U(\cdot)$ is the unit step function (see Eq. 2.5). The time histories of this process have discontinuities at the times $t = j\Delta$ as the process jumps from the random variable A_{j-1} to A_j. Find the $\phi_{XX}(t,s)$ autocorrelation function and identify the (t,s) points at which the function is discontinuous.

Using the definition $\phi_{XX}(t,s) = E[X(t)X(s)]$, we find that

$$\phi_{XX}(t,s) = E(A^2) \qquad \text{if t and s are in the same time interval (of length } \Delta)$$
$$\phi_{XX}(t,s) = 0 \qquad \text{otherwise}$$

This process is uncorrelated with itself at times t and s unless t and s are in the same time interval, varying from $j\Delta$ to $(j+1)\Delta$. The accompanying sketch indicates the region in which $\phi_{XX}(t,s) \neq 0$.

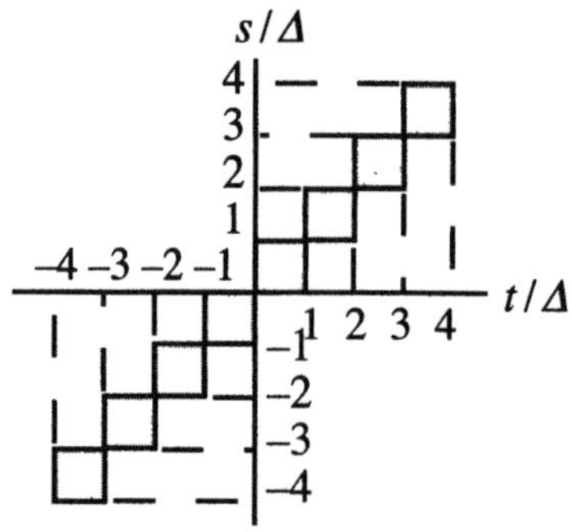

Clearly the function is discontinuous along the boundaries of each of the size Δ squares shown. Any one of these boundaries corresponds to either t being an integer times Δ or s being an integer times Δ (with the corners of the squares having both t and s being integers times Δ). The location of these discontinuities is consistent with the fact that along the 45° diagonal of the plane we have discontinuities of $\phi_{XX}(t,s)$ at (t,t) whenever t is an integer times Δ.

This example illustrates a situation in which $\phi_{XX}(t,s)$ is discontinuous at some points (t,s), even though $\phi_{XX}(t,t) = E(A^2)$ is a continuous one-dimensional function everywhere along the line of possible values. The apparent discrepancy between these two statements is simply due to the difference in definition of one-dimensional and two-dimensional continuity. Looking at one-dimensional continuity of $\phi_{XX}(t,t)$ at $t = 2\Delta$, for example, corresponds to approaching $(2\Delta, 2\Delta)$ along the 45° diagonal of the plane, whereas the discontinuity of $\phi_{XX}(t,s)$ at the point

$(2\Delta,2\Delta)$ is due to the fact that a different limit is achieved if the point is approached along some other paths, such as along a line with a negative 45° slope.

****

Example 4.10: Consider an even more erratic stochastic process for which

$$E[X^2(t)] = E(A^2)$$

but

$$E[X(t)X(s)] = 0 \qquad \text{if } t \neq s$$

This process is the limit of the process in Example 4.9 as Δ goes to zero. It is second-moment stationary and is uncorrelated with itself at any two distinct times, no matter how close together they are. Find the autocorrelation function and identify points at which it is discontinuous.

Using the definition that $R_{XX}(\tau) = E[X(t+\tau)X(s)]$, we have the stationary autocorrelation function as

$$R_{XX}(\tau) = E(A^2) \qquad \text{if } \tau = 0$$

and

$$R_{XX}(\tau) = 0 \qquad \text{if } \tau \neq 0$$

Clearly this function is discontinuous at the origin—at $\tau = 0$. It is continuous everywhere else. Note that this is the only possible type of $R_{XX}(\tau)$ function with discontinuity at only one point. Discontinuity at any point other than $\tau = 0$ would imply that there was also discontinuity at $\tau = 0$. Discontinuity at $\tau = 0$, though, implies only the possibility (not the fact) of discontinuity at other values of τ. Stating the result for this problem in terms of the general (rather than stationary) form of autocorrelation function, we have the rather odd situation of a $\phi_{XX}(t,s)$ function that is discontinuous everywhere along the major diagonal $t = s$ and continuous everywhere else; this is consistent with Example 4.9 as $\Delta \to 0$.

****

Example 4.11: Give a reason why each of the following $\phi(t,s)$ and $R(\tau)$ functions could not be the autocorrelation function for any stochastic process. That is, there could be no $\{X(t)\}$ process with $\phi_{XX}(t,s) = \phi(t,s)$ or $R_{XX}(\tau) = R(\tau)$:

(a) $\phi(t,s) = (t^2 - s)^{-2}$

(b) $\phi(t,s) = 2 - \cos(t-s)$

(c) $\phi(t,s) = U[4 - (t^2 + s^2)]$

(d) $R(\tau) = \tau^2 e^{-\tau}$

(e) $R(\tau) = e^{-\tau(\tau-1)}$

(f) $R(\tau) = U(1 - |\tau|)$

The general conditions are symmetry, nonnegative definiteness, and the conditions that $\phi(t,s)$ must be continuous in the neighborhood of the point (t,s) if it is continuous in the neighborhood of point (t,t) and in the neighborhood of point (s,s) and $R_{XX}(\tau)$ must be continuous for all values of τ if it is continuous at $\tau=0$.

Note that there might be various possible violations of necessary conditions for a particular candidate function, and it takes only one violation to prove that the function cannot be an autocorrelation function. Nonetheless, we will investigate at least two conditions for each of the six functions. In particular, we will investigate symmetry and the Schwartz inequality for each suggested autocorrelation function. We will investigate the continuity condition only for (c) and (f), because the other four are clearly continuous everywhere—they contain no discontinuous terms. We will not investigate the more complicated higher-order terms needed for proving nonnegative definiteness.

(a) Checking symmetry: We have $\phi(s,t)\neq\phi(t,s)$; therefore, this function violates the symmetry required of an autocorrelation function, so $\phi(t,s)$ cannot be an autocorrelation function.

Checking the Schwartz inequality: We see that $\phi(t,s)$ is unbounded whenever $s=t^2$, but $\phi(t,t)$ is bounded everywhere except at the single point $t=0$. Thus, this function also violates the necessary condition of having $\phi(t,s)$ bounded by $[\phi(t,t)\,\phi(s,s)]^{1/2}$. For example, $\phi(2,4)=\infty>[\phi(2,2)\,\phi(4,4)]^{1/2}=1/24$.

(b) Checking symmetry: Yes, $\phi(s,t)=\phi(t,s)$, so it does have the necessary symmetry.

Checking the Schwartz inequality: We have $\phi(t,t)=\phi(s,s)=1$, but $\phi(t,s)>1$ for some (t,s) values. Thus, this function violates the necessary condition of having $\phi(t,s)$ bounded by $[\phi(t,t)\,\phi(s,s)]^{1/2}$.

(c) Checking symmetry: Yes, $\phi(s,t)=\phi(t,s)$, so it does have the necessary symmetry.

Checking the Schwartz inequality: We have $\phi(t,t)=U(4-2t^2)=U(2-t^2)$ and $\phi(s,s)=U(2-s^2)$ so that $[\phi(t,t)\,\phi(s,s)]^{1/2}=[U(2-t^2)\,U(2-s^2)]^{1/2}$, which is unity on the square shown in the sketch, and zero outside the square. However, $\phi(t,s)=U[4-(t^2+s^2)]$ is unity on the circle in the sketch, and zero outside the circle. We have a situation with $\phi(t,s)>[\phi(t,t)\,\phi(s,s)>1]^{1/2}$ whenever (t,s) is in one of the portions of the circle that is outside the square. For example, $\phi(1.8,0)=U(0.76)=1$, $\phi(1.8,1.8)=U(-2.48)=0$,

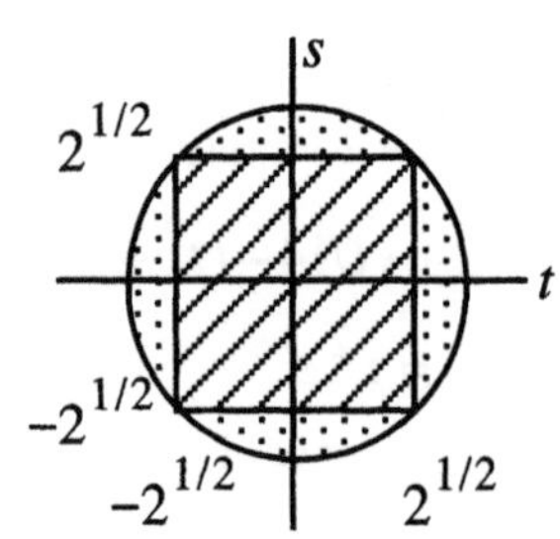

and $\phi(0,0) = U(4) = 1$. Thus, we have $\phi(1.8,0) = 1 > [\phi(1.8,1.8)\phi(0,0)]^{1/2} = 0$, and the Schwartz inequality is violated.
Checking the continuity condition: Clearly $\phi(t,s)$ is discontinuous on the circle $t^2 + s^2 = 4$. This gives discontinuity along the diagonal of the (t,s) plane only at the points $(2^{1/2},2^{1/2})$ and $(-2^{1/2},-2^{1/2})$. Clearly there are many points for which $\phi(t,s)$ is discontinuous even though the function is continuous at (t,t) and (s,s). For example, $(t,s) = (2,0)$ gives such a point. Thus, the discontinuity of this function is not of the type allowed by the Schwarz inequality. Note that it might have been more reasonable to start by checking to see whether the continuity condition was satisfied, because this $\phi(t,s)$ does have discontinuities.
(d) Checking symmetry: Now we have a stationary form of $R(\tau)$, so the necessary symmetry condition is that $R(-\tau) = R(\tau)$. Clearly this is satisfied.
Checking the Schwartz inequality: We need $|R(\tau)| < R(0)$. Clearly this is violated because $R(0) = 0$, but $R(\tau)$ is not zero everywhere.
(e) Checking symmetry: $R(-\tau) \neq R(\tau)$, so symmetry is violated.
Checking the Schwartz inequality: $R(0) = 1$ but $R(\tau) > 1$ whenever $\tau(\tau - 1) < 0$, which occurs for $0 < \tau < 1$. Thus, $R(\tau)$ does not meet the necessary condition of $|R(\tau)| < R(0)$.
(f) Checking symmetry: $R(-\tau) = R(\tau)$, so symmetry is satisfied.
Checking the Schwartz inequality: We do have $|R(\tau)| \le R(0) = 1$, so there is no problem with the Schwartz inequality.
Checking the continuity condition: $R(\tau)$ is continuous at $\tau = 0$, so the necessary condition is that it be continuous everywhere. Clearly this is violated.

****

4.6 Limits of Stochastic Processes

The next two major issues with which we wish to deal are (1) ergodicity and (2) differentiability of a process $\{X(t)\}$. Both of these involve taking a limit of a stochastic process as the index set approaches some value, and this is closely related to the simpler idea of continuity of a process. In particular, continuity involves the behavior of $\{X(t)\}$ as t approaches some particular value t_1, ergodicity involves evaluating an expected value of a function of the process from a limit of a time average over a time history of length T and finding the derivative $\dot{X}(t)$ involves taking a limit involving $X(t+h) - X(t)$ as $h \to 0$. Because continuity is probably the easiest to visualize, we will use it in this section to illustrate the idea of the limit of a stochastic process, even though the other two properties have more overall importance to the application of stochastic process methods to vibration problems. It should also be noted, though, that some sort of continuity of a process must be a condition for differentiability.

The concept of the limit of a deterministic function must be generalized somewhat to handle situations in which one wishes to find a limit of a stochastic process. The difficulty, of course, is that a stochastic process is a different random variable at each value of the index argument, so the limit of a stochastic process is like the limit of a sequence of random variables. There are a number of definitions of convergence of a stochastic process, in all of which one expects the limit to be some random variable. Rather than give a general discussion of this topic, we will focus on the particular sort of convergence to a limit that we wish to use in the three situations cited in the preceding paragraph.

We say that $\{X(t)\}$ *converges in probability* to the random variable Y as t approaches t_1 if

$$\lim_{t\to t_1} P\Big[|X(t)-Y|\geq\varepsilon\Big]=0 \quad \text{for any } \varepsilon>0 \tag{4.37}$$

Note that the limit in Eq. 4.37 is on the probability of an event, so this is an unambiguous deterministic limit. The well-known Chebyshev inequality provides a very useful tool for any problem that requires bounding of the probability of a random variable exceeding some particular value. This can be written in a quite general form as

$$P(|Z|\geq b)\leq\frac{E(|Z|^c)}{b^c} \tag{4.38}$$

for any random variable Z and any nonnegative numbers b and c. Thus, condition 4.37 will be satisfied if

$$\lim_{t\to t_1} E\Big(|X(t)-Y|^c\Big)=0 \quad \text{for some } c>0 \tag{4.39}$$

Note that Eq. 4.39 is very general, with convergence in probability ensured if Eq. 4.39 holds for any positive real exponent c. Unfortunately it is difficult to use the relationship in Eq. 4.39 unless the exponent is an even integer such that the power can be expanded in terms of moments of the random variables. Any even integer will serve this purpose, but choosing $c=2$ allows us to use the lowest possible moments. Thus, we will generally choose a much more restrictive condition than Eq. 4.39 and say that we are assured of convergence in probability to Y if we have *mean-square convergence*:

$$\lim_{t\to t_1} E\left([X(t)-Y]^2\right)=0 \tag{4.40}$$

Turning now to the issue of continuity of $\{X(t)\}$ at $t=t_1$, we simply need to replace the variable Y in the preceding equations and say that $\{X(t)\}$ is continuous in probability at $t=t_1$ if

$$\lim_{t\to t_1} E\left(|X(t)-X(t_1)|^c\right)=0 \qquad \text{for some } c>0 \tag{4.41}$$

and, in particular, if it is mean-square continuous, so

$$\lim_{t\to t_1} E\left([X(t)-X(t_1)]^2\right)=0$$

Expanding the quadratic expression inside the expected value allows us to rewrite this condition for mean-square continuity as

$$\lim_{t\to t_1} [\phi_{XX}(t,t)-2\phi_{XX}(t,t_1)+\phi_{XX}(t_1,t_1)]=0 \tag{4.42}$$

Clearly Eq. 4.42 is satisfied if $\phi_{XX}(t,s)$ is continuous at the point (t_1,t_1). Thus, we see that mean-square continuity of a process depends only on the continuity of the autocorrelation function of the process. The process $\{X(t)\}$ is mean-square continuous at time $t=t_1$ if and only if $\phi_{XX}(t,s)$ is continuous at (t_1,t_1).

It is easy to show that any process for which all time histories are continuous will have mean-square continuity. However, the mean-square continuity condition is considerably weaker than a requirement that all possible time histories of the process be continuous. This will be illustrated in Example 4.13, but first we will consider a simpler situation.

****

Example 4.12: Identify the t values corresponding to mean-square continuity of the process $\{X(t)\}$ with $X(t)=A_j$ for $j\Delta\le t<(j+1)\Delta$ with the random variables

A_j being independent and identically distributed with mean zero and variance $E(A^2)$.

We have already shown in Example 4.9 that $\phi_{XX}(t,s)$ for this process is $E(A^2)$ on the squares with dimension Δ along the 45° line and is zero outside those squares. Thus, this autocorrelation is discontinuous along the boundaries of those squares and continuous elsewhere. This means that $\phi_{XX}(t,s)$ is continuous at (t_1,t_1) if $t_1 \neq j\Delta$ for any j value. The process is mean-square continuous except at the times $t_1 = j\Delta$. Note that $t_1 = j\Delta$ describes the instants of time when the time histories of $\{X(t)\}$ are almost sure to have discontinuities. Thus, in this example, $\{X(t)\}$ is mean-square continuous when the time histories are continuous and it is mean-square discontinuous when the time histories are discontinuous.

****

Example 4.13: Let $\{X(t): t \geq 0\}$ be what is called a Poisson process defined by $X(0) = 0$ and

$$P[X(t) - X(s) = k] = \frac{e^{-b(t-s)}\, b^k\, (t-s)^k}{k!} \quad \text{for } k = 0,1,\cdots,\infty \text{ and } 0 \leq s \leq t$$

with $[X(t) - X(s)]$ being independent of $[X(s) - X(r)]$ for $0 \leq r \leq s \leq t$. Note that $X(t)$ is always integer valued. This distribution arises in many areas of applied probability, with $\{X(t)\}$ representing the number of occurrences (the count) of some event during the time interval $[0,t]$. It corresponds to having the number of occurrences in two nonoverlapping (i.e., disjoint) time intervals be independent of each other and with their being identically distributed if the time intervals are of the same length. The parameter b represents the mean rate of occurrence. Consider the continuity of $\{X(t)\}$.

Note that any time history of this process is a "stairstep" function. It is zero until the time of first occurrence, stays at unity until the time of second occurrence, and so forth, as shown in the accompanying sketch of the kth sample time history. We can write

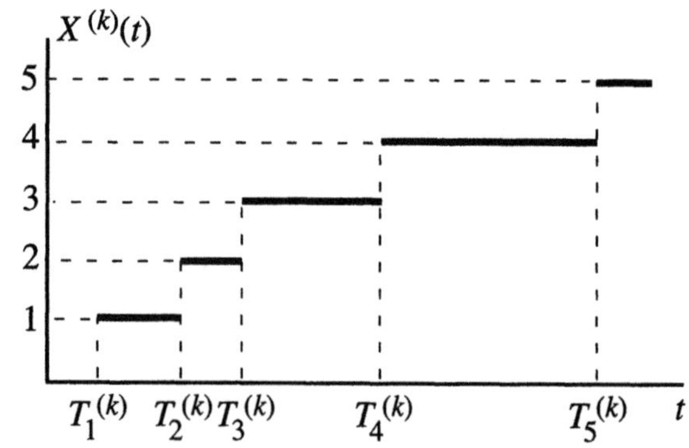

$$X(t) = \sum_{j=1}^{\infty} U(t - T_j)$$

with the random variable T_j denoting the time of jth occurrence. Clearly these time histories are not continuous at the times of occurrence. Furthermore, any time t is a time of occurrence for some of the possible time histories, so there is no time at which we can say that all time histories are continuous.

Let us now find the autocorrelation function for $\{X(t)\}$. For $s \le t$ we can define $\Delta X = X(t) - X(s)$ and write

$$\phi_{XX}(t,s) = E\big([X(s) + \Delta X]\, X(s)\big) = E\big(X^2(s)\big) + \mu_{\Delta X}\, \mu_X(s)$$

in which the independence of ΔX and $X(s)$ has been used. To proceed further we need the mean value function of $\{X(t)\}$:

$$\mu_{X(t)} = \sum_{k=0}^{\infty} k\, P[X(t) = k] = \sum_{k=1}^{\infty} \frac{e^{-bt} (bt)^k}{(k-1)!} = bt$$

Now we use the fact that the distribution of ΔX is the same as that of $X(t-s)$ to obtain

$$\phi_{XX}(t,s) = \sum_{k=0}^{\infty} k^2\, P[X(s) = k] + b^2 (t-s)\, s = \sum_{k=0}^{\infty} k^2 \frac{e^{-bs} (bs)^k}{k!} + b^2 (ts - s^2) =$$

$$\sum_{k=2}^{\infty} \frac{e^{-bs} (bs)^k}{(k-2)!} + \sum_{k=1}^{\infty} \frac{e^{-bs} (bs)^k}{(k-1)!} + b^2 (ts - s^2) = bs(bt + 1)$$

for $s \le t$. The general relationship can be written as $\phi_{XX}(t,s) = b^2 ts + b \min(t,s)$. It is easily verified that this function is continuous everywhere on the plane of (t,s) values, so $\{X(t)\}$ is mean-square continuous everywhere. This is despite the fact that there is nowhere at which all its time histories are continuous, nor does it have any continuous time histories, except the trivial one corresponding to no arrivals.

****

Example 4.14: Identify the t values giving mean-square continuity of the process $\{X(t)\}$ of Example 4.10 with the autocorrelation of $\phi_{XX}(t,s) = E(A^2)$ for $t = s$ and $\phi_{XX}(t,s) = 0$ off this diagonal line.

Because $\phi_{XX}(t,s)$ is discontinuous at every point on the 45° line, this $\{X(t)\}$ process is not mean-square continuous anywhere.

****

4.7 Ergodicity of a Stochastic Process

The concept of ergodicity has to do with using a time average obtained from one time history of a stochastic process as a substitute for a mathematical expectation. The stochastic process is said to be ergodic (in some sense) if the two are the same. Recall the idea of an ensemble of possible time histories of the process, as illustrated in Fig. 4.1. As noted before, an expected value can be thought of as a statistical average across an infinite ensemble of time histories, and this is an average on a section orthogonal to a particular time history. Thus, there is no obvious reason why the two should be the same, even though they are both averages across infinitely many values of $\{X(t)\}$. To illustrate the idea we begin with consideration of the simplest expected value related to a stochastic process $\{X(t)\}$—the mean value. A truncated time average over sample j corresponding to the mean of the process can be written as

$$\frac{1}{T}\int_{-T/2}^{T/2} X^{(j)}(t)\,dt \tag{4.43}$$

We say that $\{X(t)\}$ is ergodic in mean value if it is mean-value stationary and the time average of expression 4.43 tends to $\mu_X \equiv E[X(t)]$ as T tends to infinity, regardless of the value of j. We impose the stationarity condition, because taking a time average like that in expression 4.43 could not possibly approximate the mean value unless that value was independent of time. Similarly, we say that $\{X(t)\}$ is ergodic in second moment if it is second-moment stationary and

$$R_{XX}(\tau) \equiv E[X(t+\tau)\,X(t)] = \lim_{T\to\infty}\frac{1}{T}\int_{-T/2}^{T/2} X^{(j)}(t+\tau)\,X^{(j)}(t)\,dt \tag{4.44}$$

There are as many types of ergodicity as there are stationarity. For any stationary characteristic of the stochastic process, one can define a corresponding time average, and ergodicity of the proper type will ensure that the two are the same. The only additional type of ergodicity that we will list is the one related to the first-order probability distribution of the process. We will say that $\{X(t)\}$ is first-order ergodic if it is first-order stationary and

$$F_{X(t)}(u) \equiv E\big(U[u - X(t)]\big) = \lim_{T\to\infty}\frac{1}{T}\int_{-T/2}^{T/2} U[u - X^{(j)}(t)]\,dt \tag{4.45}$$

Note that the unit step function in the integrand is always zero or unity, such that the integral gives exactly the amount of time for which $X^{(j)}(t) \le u$ within the interval $[-T/2, T/2]$. Thus, this type of ergodicity is a condition that the probability that $X(t) \le u$ at any time t is the same as the fraction of the time that any time history $X^{(j)}(t)$ is less than or equal to u. We mention Eq. 4.45, both because it illustrates an idea that is somewhat different from the moment ergodicity of expression 4.43 and Eq. 4.44 and because it gives another property that one often wishes to determine in practical problems.

The reason that people usually want to consider stochastic processes to be ergodic really has to do with the issue of statistics rather than probability theory. That is, they wish to determine information about the moments and/or probability distribution of a process $\{X(t)\}$ from observed values of the process. If a sufficiently large ensemble of observed time histories is available, then one can use ensemble averages and there is no need to invoke a condition of ergodicity. In most situations, though, it is impossible or impractical to obtain a very large ensemble of observed time histories from a given physical process. Thus, for example, one may wish to determine moments and/or the probability distribution of the wind speed at Easterwood Airport on April 27 from one long time history obtained on that date in 2003, rather than wait many years to obtain an ensemble of time histories, all recorded on April 27. Furthermore, even if it is possible to obtain the many-year sample, one may suspect that the process may not be stationary over such a long time, so it may be necessary to consider each annual sample as being from a different process or a process with different parameter values. Thus, the use of the ergodicity property is almost essential in many situations involving data observed from physical processes.

At least two cautions should be noted regarding ergodicity. The first is that there may be some difficulty in proving that a physical process is ergodic. Once we have chosen a mathematical model, we can usually show that it has a particular form of ergodicity, but this avoids the more fundamental issue of whether this ergodic model is the appropriate model for the physical problem. Furthermore, even if we know that Eq. 4.45, for example, is satisfied, we will not have an infinite time history available to use in evaluating the cumulative distribution function from the time average. The best that we can ever hope for in a physical problem is a long finite sample time history. If we could be sure that a process is ergodic in the proper way, then we would probably feel more confident in using a time average such as in expression 4.43, or Eq. 4.44 or 4.45 to estimate the corresponding expected value, but it will still only be an estimate.

To illustrate the idea of ergodicity theorems, consider the proof of convergence of expression 4.43 to μ_X. What we wish to prove is convergence in probability:

$$\lim_{T\to\infty} P\big(|Q_T| \ge \varepsilon\big) = 0 \qquad \text{for any } \varepsilon > 0 \tag{4.46}$$

in which

$$Q_T = \mu_X - \frac{1}{T}\int_{-T/2}^{T/2} X(t)\,dt \tag{4.47}$$

As noted in Section 4.6, this is ensured if $E[|Q_T|^c] \to 0$ as $T\to\infty$ for some $c>0$, and in particular if we have mean-square convergence

$$\lim_{T\to\infty} E[|Q_T|^2] = 0 \tag{4.48}$$

One can find conditions to ensure that Eq. 4.48 is true by rewriting Q_T^2 as

$$Q_T^2 = \left(\frac{1}{T}\int_{-T/2}^{T/2}[X(t)-\mu_X(t)]\,dt\right)^2$$
$$= \frac{1}{T^2}\int_{-T/2}^{T/2}\int_{-T/2}^{T/2}[X(t_1)-\mu_X(t_1)][X(t_2)-\mu_X(t_2)]\,dt_1\,dt_2$$

Taking the expected value of this quantity, rewriting the integrand as $G_{XX}(t_1-t_2)$, and eliminating t_1 by the change of variables $\tau = t_1 - t_2$ gives

$$E(Q_T^2) = \frac{1}{T^2}\int_{-T/2}^{T/2}\int_{-T/2}^{T/2} G_{XX}(t_1-t_2)\,dt_1\,dt_2$$
$$= \frac{1}{T^2}\int_{-T/2}^{T/2}\int_{-T/2-t_2}^{T/2-t_2} G_{XX}(\tau)\,d\tau\,dt_2$$

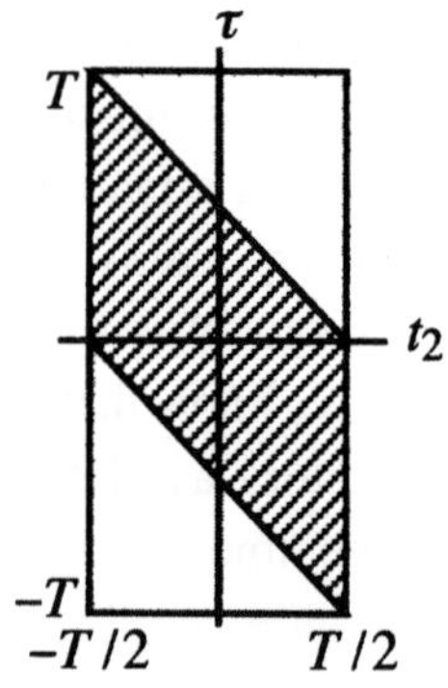

Figure 4.2 Domain of integration.

The parallelogram in Fig. 4.2 shows the domain of integration in the (t_2,τ) plane. We now replace this integral by the one over the rectangle shown, minus the integrals over the two triangular corners. Using the symmetry of $G_{XX}(\tau)$, it is easily shown that the two triangles give the same integral, so the result can be written as

$$E(Q_T^2) = \frac{1}{T^2}\left(\int_{-T/2}^{T/2} \int_{-T}^{T} G_{XX}(\tau)\, d\tau\, dt_2 - 2\int_{-T/2}^{T/2} \int_{T/2-t_2}^{T} G_{XX}(\tau)\, d\tau\, dt_2 \right) \quad (4.49)$$

We now easily perform the integration with respect to t_2 in the first term. We can also reverse the order of integration in the second term and then perform its integration with respect to t_2 to obtain a sufficient condition for ergodicity in mean value as

$$\lim_{T\to\infty} = \frac{1}{T} \int_{-T}^{T} \left(1 - \frac{|\tau|}{T}\right) G_{XX}(\tau)\, d\tau = 0 \quad (4.50)$$

Note that the development of Eq. 4.50 from 4.48 is exact. Thus, Eq. 4.50 is necessary as well as sufficient for ergodicity in mean value to hold in the sense of convergence in mean square. This does not imply, however, that Eq. 4.50 is necessary for ergodicity in mean value to hold in the sense of convergence in probability.

As an alternative to performing the integration with respect to t_2 in the second term in Eq. 4.49, we can introduce the new variable $t = T/2 - t_2$ to obtain a result that can be written as

$$E(Q_T^2) = \frac{1}{T} \int_{-T}^{T} G_{XX}(\tau)\, d\tau - \frac{1}{T} \int_{0}^{T} \frac{2}{T} \int_{T/2-t_2}^{T} G_{XX}(\tau)\, d\tau\, dt$$

om this form one can show that $E(Q_T^2) \to 0$ as $T \to \infty$ if

(.51)

this also gives a sufficient condition for ergodicity in mean value. The condition of Eq. 4.51 is certainly satisfied if the integral of $G_{XX}(\tau)$ over the entire real line is finite, but this is a somewhat more restrictive condition than Eq. 4.51. The essential feature of the finiteness of the integrals, of course, relates to the behavior of

$|G_{XX}(\tau)|$ for large values of $|\tau|$. To illustrate the difference between the integrability of $G_{XX}(\tau)$ and Eq. 4.51, note that the integral of $G_{XX}(\tau)$ is finite if $|G_{XX}(\tau)|$ tends to zero like $|\tau|^{-(1+\varepsilon)}$ for some small $\varepsilon > 0$, but Eq. 4.51 is satisfied provided only that $|G_{XX}(\tau)|$ tends to zero in any fashion as $|\tau|$ tends to infinity.

It may also be useful to note a situation that does not meet the conditions of ergodicity in mean value. If $G_{XX}(\tau)$ is a constant (other than zero) for all τ values, then the process violates the condition in Eq. 4.50, which is necessary for convergence in mean square. Consider what it means, though, for a process to be covariant stationary with a constant autocovariance function $G_{XX}(\tau)$. This would give $X(t+\tau)$ and $X(t)$ as being perfectly correlated, for any value of τ, and it can be shown (see Section 3.3) that this implies that there is a linear relationship of the form $X(t+\tau) = a + b\,X(t)$ for some constants a and b. Combining this condition with covariant stationarity requires that each time history of $\{X(t)\}$ be simply a constant value. This is an almost trivial case of a stochastic process that is always equal to a random variable. Each possible time history $X^{(j)}(t)$ is a constant $X^{(j)}$ independent of time, but each random sample may give a different value for that constant. It should be no surprise that this process is not ergodic in mean value. The time average in expression 4.43 is simply the constant $X^{(j)}$, and there is no convergence to μ_X as T tends to infinity. A necessary condition for ergodicity in mean value is that the random variables $X(t+\tau)$ and $X(t)$ not be perfectly correlated as $\tau \to \infty$.

The condition for ergodicity in mean value can be extended to the ergodicity in second moment of Eq. 4.44 by defining a new random process $\{Y(t)\}$ as

$$Y(t) = X(t+\tau)\,X(t)$$

for a particular τ value. Then Eq. 4.44 holds for that particular τ value if $\{Y(t)\}$ is ergodic in mean value. The autocovariance for this $\{Y(t)\}$ process can be written in terms of the second and fourth moment functions of the $\{X(t)\}$ process as

$$G_{YY}(s) = E[X(t+\tau+s)\,X(t+s)\,X(t+\tau)\,X(t)] - R_{XX}^2(\tau) \tag{4.52}$$

Thus, we can say that the $\{X(t)\}$ process is ergodic in second moment if it is fourth-moment stationary and the $G_{YY}(s)$ function satisfies a condition like Eq. 4.50 or 4.51, with the integration being over the variable s. Similarly, we can define a $\{Z(t)\}$ process as

$$Z(t) = U[u - X(t)]$$

with autocovariance function

$$G_{ZZ}(\tau) = E\big(U[u - X(t+\tau)]U[u - X(t)]\big) - F_X^2(u)$$
$$= F_{X(t+\tau)X(t)}(u,u) - F_X^2(u) \qquad (4.53)$$

for a given u value. The process $\{X(t)\}$ is first-order ergodic (Eq. 4.45) if it is second-order stationary and the $G_{ZZ}(\tau)$ function of Eq. 4.53 satisfies a condition like Eq. 4.50.

We will not proceed further with proofs of ergodicity. Equations 4.48–4.51 illustrate that it is not necessarily difficult to perform the manipulations to obtain conditions for convergence in mean square and that those conditions are not generally very restrictive. Note, though, that applying Eqs. 4.50–4.52 to show ergodicity in second moment requires conditions on the fourth-moment function of $\{X(t)\}$, and the corresponding relationship to show jth-moment ergodicity in general would involve conditions on the $(2j)$th moment function of $\{X(t)\}$. This need for consideration of higher moments can be seen to follow directly from the use of mean-square convergence. If one could derive simple conditions to show $E(|Q_T|^c)$ tending to zero for an appropriate Q_T error term and some c that is only slightly greater than zero, then one could also prove ergodicity without consideration of such high-moment functions. Unfortunately, such simple conditions are not available.

It is our observation that in common practice people assume that processes are ergodic unless there is some obvious physical reason why that would be inappropriate. For example, if $\{X(t)\}$ denotes the position of a body that has a finite probability of instability, then no one time history will be representative of the entire process because it will represent only one observation of either stable or unstable behavior of the system. Example 4.1 is of this type. We previously saw that the process used in that example is mean-value stationary, but it is clearly not mean-value ergodic. In terms of checking the ergodicity conditions presented here, one obvious difficulty is that the process is not covariant stationary.

Even for clearly nonstationary processes, it is sometimes necessary to use time averages as estimates of expected values. For example, one might modify expression 4.43 and use

$$\frac{1}{T}\int_{t-T/2}^{t+T/2} X^{(j)}(s)\,ds$$

as an estimate of $\mu_X(t)$ for a process that was known not to be mean-value stationary. Intuitively, this seems reasonable if $\mu_X(t)$ varies relatively slowly, but we will not go into any analysis of the error of such an approximation.

4.8 Stochastic Derivative

Most deterministic dynamics problems are governed by differential equations. Our goal is to analyze such systems when the deterministic excitation and response time histories are replaced by stochastic processes. This means that we want to analyze stochastic differential equations, but the first step in doing this must be the identification of an acceptable concept of the derivative of a stochastic process $\{X(t)\}$ with respect to its index parameter t:

$$\dot{X}(t) = \frac{d}{dt}X(t)$$

At each t value $X(t)$ is a different random variable, so the idea of a Riemann definition of the derivative gives $\dot{X}(t)$ as being like a limit as h goes to zero of

$$Y(t,h) = \frac{X(t+h)-X(t)}{h} \qquad (4.54)$$

with the numerator of the expression being a different random variable for each h value. We will take a very simple approach and say that if any property of $Y(t,h)$ exists and has a limit as $h \to 0$, then that property also describes $\dot{X}(t)$. That is, if $E[Y(t,h)]$ exists and has a limit as $h \to 0$, then that limit is $\mu_{\dot{X}}(t) \equiv E[\dot{X}(t)]$; if $E[Y(t,h)\,Y(s,h)]$ exists and has a limit as $h \to 0$, then that limit is $\phi_{\dot{X}\dot{X}}(t,s) \equiv E[\dot{X}(t)\,\dot{X}(s)]$; if the probability distribution of $Y(t,h)$ exists and has a limit as $h \to 0$, then that limit is the probability distribution of $\dot{X}(t)$; and so forth.

Let us now investigate the moments of $\{\dot{X}(t)\}$ by evaluating those for $Y(t,h)$. For the mean value function we have

$$\mu_{\dot{X}}(t) = \lim_{h\to 0} E\big[Y(t,h)\big] = \lim_{h\to 0} E\left(\frac{X(t+h)-X(t)}{h}\right) = \lim_{h\to 0}\frac{\mu_X(t+h)-\mu_X(t)}{h}$$

or

$$\mu_{\dot{X}}(t) = \frac{d}{dt}\mu_X(t) \tag{4.55}$$

This important result can be stated in words as "the mean of the derivative is the derivative of the mean." This certainly sounds reasonable because expectation is a linear operation and we want stochastic differentiation also to be linear, so as to be consistent with the deterministic definition of the derivative. Because we are defining the stochastic derivative by a limit of the linear operation in Eq. 4.54, though, we are ensured of having this linearity.

Before proceeding to find the autocorrelation function for the derivative, let us consider the intermediate step of finding the cross-correlation function between $\{X(t)\}$ and $\{\dot{X}(t)\}$. We say that

$$\phi_{X\dot{X}}(t,s) = \lim_{h\to 0} E\big[X(t)Y(s,h)\big] = \lim_{h\to 0} E\left(X(t)\frac{X(s+h)-X(s)}{h}\right)$$

which gives

$$\phi_{X\dot{X}}(t,s) = \lim_{h\to 0}\left[\frac{\phi_{XX}(t,s+h)-\phi_{XX}(t,s)}{h}\right] \equiv \frac{\partial}{\partial s}\phi_{XX}(t,s) \tag{4.56}$$

showing that this cross-correlation function is obtained by taking a partial derivative of the autocorrelation function. Similarly, the autocorrelation function for the derivative process

$$\phi_{\dot{X}\dot{X}}(t,s) = \lim_{\substack{h_1\to 0\\ h_2\to 0}} E\big[Y(t,h_1)Y(t,h_2)\big] = \lim_{\substack{h_1\to 0\\ h_2\to 0}} E\left(\frac{X(t+h_1)-X(t)}{h_1}\,\frac{X(s+h_2)-X(s)}{h_2}\right)$$

gives

$$\phi_{\dot{X}\dot{X}}(t,s) = \lim_{\substack{h_1\to 0\\ h_2\to 0}} \frac{\phi_{XX}(t+h_1,s+h_2)-\phi_{XX}(t,s+h_2)-\phi_{XX}(t+h_1,s)+\phi_{XX}(t,s)}{h_1\,h_2}$$

or

$$\phi_{\dot{X}\dot{X}}(t,s) = \frac{\partial^2}{\partial t\,\partial s}\phi_{XX}(t,s) \tag{4.57}$$

Conditions for higher-order moment functions could be derived in the same way. Also, a first-order probability distribution for $\{\dot{X}(t)\}$ could be derived from the second-order joint probability distribution for $X(t)$ and $X(t+h)$, which governs the probability distribution of $Y(t,h)$.

Note that our approach implies that $\mu_{\dot{X}}(t)$ exists if and only if $\mu_X(t)$ exists and has a derivative at the point *t*. Furthermore, $\mu_X(t)$ cannot be differentiable at *t* unless it is continuous at that point. Similarly, existence of the cross-correlation function $\phi_{X\dot{X}}(t,s)$ requires that $\phi_{XX}(t,s)$ be continuous at (t,s) and that the partial derivative exist at that point. The corresponding condition that $\phi_{XX}(t,s)$ be twice differentiable in order that the autocorrelation function $\phi_{\dot{X}\dot{X}}(t,s)$ exist requires that the first partial derivative of $\phi_{XX}(t,s)$ be continuous at the point of interest. This final point is sometimes slightly confusing and will be illustrated in Examples 4.16 and 4.17.

If the process $\{X(t)\}$ is second-moment stationary, then one can rewrite the conditions of Eqs. 4.56 and 4.57 in stationary notation. First note that using $\phi_{XX}(t,s) = R_{XX}(t-s)$ gives

$$\phi_{X\dot{X}}(t,s) = \frac{\partial}{\partial s} R_{XX}(t-s) = \left(-\frac{d}{d\tau} R_{XX}(\tau) \right)_{\tau=t-s}$$

and

$$\phi_{\dot{X}\dot{X}}(t,s) = \frac{\partial^2}{\partial t\, \partial s} R_{XX}(t-s) = \left(-\frac{d^2}{d\tau^2} R_{XX}(\tau) \right)_{\tau=t-s}$$

These two equations show that $\phi_{X\dot{X}}(t,s)$ and $\phi_{\dot{X}\dot{X}}(t,s)$, if they exist, are functions only of the time difference $\tau = t-s$. Thus, the derivative process is also second-moment stationary and $\{X(t)\}$ and $\{\dot{X}(t)\}$ are jointly second-moment stationary so that one can also write the results as

$$R_{X\dot{X}}(\tau) = -\frac{d}{d\tau} R_{XX}(\tau) \tag{4.58}$$

and

$$R_{\dot{X}\dot{X}}(\tau) = -\frac{d^2}{d\tau^2} R_{XX}(\tau) \tag{4.59}$$

Using the results given for the mean value function and correlation functions involving the derivative of $\{X(t)\}$, one can easily show that corresponding results hold for the covariance functions describing $\{\dot{X}(t)\}$:

$$K_{X\dot{X}}(t,s) = \frac{\partial}{\partial s} K_{XX}(t,s), \qquad G_{X\dot{X}}(\tau) = -\frac{\partial}{\partial \tau} G_{XX}(\tau) \tag{4.60}$$

$$K_{\dot{X}\dot{X}}(t,s) = \frac{\partial^2}{\partial t \partial s} K_{XX}(t,s), \qquad G_{\dot{X}\dot{X}}(\tau) = -\frac{d^2}{d\tau^2} G_{XX}(\tau) \tag{4.61}$$

Let us now consider the derivative of a function of a stochastic process. Specifically, we will define a new stochastic process $\{Z(t)\}$ by $Z(t) = g[X(t)]$ and investigate the derivative process $\{\dot{Z}(t)\}$. By the procedure of Eq. 4.54, the behavior of $\dot{Z}(t)$ must be the same as that of

$$\frac{Z(t+h) - Z(t)}{h} = \frac{g[X(t+h)] - g[X(t)]}{h}$$

in the limit as h goes to zero. In particular,

$$E[\dot{Z}(t)] = \lim_{h \to 0} E\left(\frac{Z(t+h) - Z(t)}{h} \right) = \lim_{h \to 0} \frac{E(g[X(t+h)]) - E(g[X(t)])}{h}$$
$$= \frac{d}{dt} E(g[X(t)])$$

provided that $E(g[X(t)])$ is a differentiable function of t. This can be rewritten as

$$E\left(\frac{d}{dt} g[X(t)] \right) = \frac{d}{dt} E(g[X(t)]) \tag{4.62}$$

to emphasize the very important result that the order of differentiation and expectation can generally be reversed. The results presented in Eqs. 4.55–4.62 can be considered as special cases of this general relationship.

The result in Eq. 4.62 will be very useful to us in our study of stochastic vibrations, particularly in the method of state space analysis. The derivative of $g[X(t)]$ can be written out in the same way as for a deterministic function giving

$$\frac{d}{dt}E\big(g[X(t)]\big)=E\big(\dot{X}(t)\,g'[X(t)]\big) \tag{4.63}$$

in which $g'(\cdot)$ denotes the derivative of $g(\cdot)$ with respect to its total argument. For example,

$$\frac{d}{dt}E[X^2(t)]=E[2\,X(t)\,\dot{X}(t)]=2\phi_{X\dot{X}}(t,t) \tag{4.64}$$

and

$$\frac{d}{dt}E[X^j(t)]=j\,E[X^{j-1}(t)\,\dot{X}(t)] \tag{4.65}$$

These relationships become particularly simple when the processes involved are stationary. Thus, if $\{X(t)\}$ is second-moment stationary, we can say that $E[X^2(t)]$ is a constant and Eq. 4.64 then tells us that $E[X(t)\,\dot{X}(t)]=0$. This particular result could also be seen by noting that $E[X(t)\,\dot{X}(t)]=R_{X\dot{X}}(0)$, using Eq. 4.58 to show that this is the negative of the slope of the $R_{XX}(\tau)$ function at $\tau=0$, and arguing that this slope at the origin must be zero (if it exists) because $R_{XX}(\tau)$ is a symmetric function. Equation 4.65 shows the somewhat less obvious fact that $E[X^{j-1}(t)\,\dot{X}(t)]=0$ for any process $\{X(t)\}$ that is jth-moment stationary. If $\{X(t)\}$ is strictly stationary, then $E(g[X(t)])$ must be a constant for any function $g(\cdot)$ and Eq. 4.63 then requires that $E(\dot{X}(t)\,g'[X(t)])=0$ for any function $g'(\cdot)$.

The procedure can also be extended to expected values involving more than one stochastic process, giving expressions such as

$$\frac{d}{dt}E[X^j(t)\,Z^k(t)]=j\,E[X^{j-1}(t)\,\dot{X}(t)\,Z^k(t)]+k\,E[X^j(t)\,Z^{k-1}(t)\,\dot{Z}(t)] \tag{4.66}$$

which includes such special cases as

$$\frac{d}{dt}E[X(t)\,\dot{X}(t)]=E[\dot{X}^2(t)]+E[X(t)\,\ddot{X}(t)]$$

and

$$\frac{d}{dt}E[X(t)\,\ddot{X}(t)] = E[\dot{X}(t)\,\ddot{X}(t)] + E[X(t)\,\dddot{X}(t)]$$

If the processes are second-moment stationary, then the derivatives of expected values must be zero and these two equations give us $E[X(t)\,\ddot{X}(t)] = -E[\dot{X}^2(t)]$ and $E[X(t)\,\dddot{X}(t)] = -E[\dot{X}(t)\,\ddot{X}(t)]$ but this latter equation gives $E[X(t)\,\dddot{X}(t)] = 0$ because $E[\dot{X}(t)\,\ddot{X}(t)]$ is one-half of the derivative of $E[\dot{X}^2(t)]$. Clearly, one can derive many such relationships between different expected values, and those given are only illustrative examples.

****

Example 4.15: Consider the differentiability of a stochastic process $\{X(t)\}$ with mean value and autocovariance functions of

$$\mu_X(t) = e^{3t}, \qquad G_{XX}(\tau) = K_{XX}(t+\tau,t) = 2e^{-5\tau^2}$$

Both of these functions are differentiable everywhere, so there is no problem with the existence of the mean and covariance of the derivative process $\{\dot{X}(t)\}$. The mean value function is given by

$$\mu_{\dot{X}}(t) = \frac{d}{dt}\mu_X(t) = 3e^{3t}$$

Because $\{X(t)\}$ is covariant stationary, we can use

$$K_{X\dot{X}}(t+\tau,t) = G_{X\dot{X}}(\tau) = -\frac{d}{d\tau}G_{XX}(\tau) = 20\,\tau\,e^{-5\tau^2}$$

and

$$K_{\dot{X}\dot{X}}(t+\tau,t) = G_{\dot{X}\dot{X}}(\tau) = -\frac{d^2}{d\tau^2}G_{XX}(\tau) = 20(1-10\tau^2)e^{-5\tau^2}$$

Alternatively, we could have found the general autocorrelation function for $\{X(t)\}$ as

$$\phi_{XX}(t,s) = K_{XX}(t,s) + \mu_X(t)\,\mu_X(s) = 2e^{-5(t-s)^2} + e^{3(t+s)}$$

and used

$$\phi_{X\dot{X}}(t,s) = \frac{\partial}{\partial s}\phi_{XX}(t,s) = 20(t-s)e^{-5(t-s)^2} + 3e^{3(t+s)}$$

and

$$\phi_{\dot{X}\dot{X}}(t,s) = \frac{\partial^2}{\partial t\,\partial s}\phi_{XX}(t,s) = 20[1-10(t-s)^2]e^{-5(t-s)^2} + 9e^{3(t+s)}$$

It is easy to verify that this gives

$$\phi_{X\dot{X}}(t,s) = K_{X\dot{X}}(t,s) + \mu_X(t)\,\mu_{\dot{X}}(s)$$

and

$$\phi_{X\dot{X}}(t,s) = K_{X\dot{X}}(t,s) + \mu_{\dot{X}}(t)\,\mu_{\dot{X}}(s)$$

This is a confirmation of something that we know must be true, because these relationships hold for any stochastic processes, as given in Eqs. 4.6 and 4.7.

****

Example 4.16: Consider the differentiability of a mean-zero stationary stochastic process $\{X(t)\}$ with autocorrelation function

$$R_{XX}(\tau) = e^{-a|\tau|}$$

in which a is a positive constant.

Because $\mu_X(t) = 0$ everywhere, its derivative is also zero and we can say that $\mu_{\dot{X}}(t) = 0$. Next, we can take the derivative of Eq. 4.58 and obtain

$$R_{X\dot{X}}(\tau) = e^{-a|\tau|}\frac{d}{d\tau}(a\,|\,\tau\,|) = a\,e^{-a|\tau|}\,\mathrm{sgn}(\tau) = a\,e^{-a|\tau|}\big[2U(\tau)-1\big]$$

This shows that the cross-correlation of $\{X(t)\}$ and $\{\dot{X}(t)\}$ exists (i.e., is finite) for all τ values. However, this first derivative is discontinuous at $\tau = 0$, so the autocorrelation for $\{\dot{X}(t)\}$

$$R_{\dot{X}\dot{X}}(\tau) = e^{-a|\tau|}\left(2\,a\,\tau\,\delta(\tau) - a^2\right)$$

is infinite for $\tau = 0$. Note that the derivative process has an infinite mean-squared value. This example illustrates that caution must be used in evaluating the second partial derivative of Eq. 4.57. If we had simply investigated the situations with $\tau < 0$ and $\tau > 0$, we would have found $R_{\dot{X}\dot{X}}(\tau) = -a^2\,e^{-a|\tau|}$ in both those regions. Thus, we find that $R_{\dot{X}\dot{X}}(\tau)$ approaches the same limit as we approach from either side toward the condition that should give mean-squared value. The fact that this limit is negative (namely, $-a^2$), though, is a sure indication that this could not be the mean-squared value. The reader is urged to use considerable caution in investigating the second moment properties of the derivatives of a stochastic process, unless the autocorrelation function for the original process is an analytic function, ensuring that all its derivatives exist.

****

Example 4.17: Consider the differentiability of the Poisson process $\{X(t): t \ge 0\}$ of Example 4.13, for which $\mu_X(t) = b\,t$ and

$$\phi_{XX}(t,s) = b^2\,t\,s + b\min(t,s) = b^2\,t\,s + b\,s\,U(t-s) + b\,t\,U(s-t)$$

Based on the mean-value function of $\{X(t)\}$, we can say that the mean of the derivative process exists and is equal to $\mu_{\dot{X}}(t) = b$. Similarly, the cross-correlation of $\{X(t)\}$ and $\{\dot{X}(t)\}$ should be given by

$$\phi_{X\dot{X}}(t,s) = \frac{\partial}{\partial s}\phi_{XX}(t,s) = b^2\, t + bU(t-s) + b(t-s)\delta(s-t)$$

in which $\delta(\cdot)$ denotes the Dirac delta function (see Section 2.4). The last term of this expression is always zero, though, so the cross-correlation always exists (i.e., is always finite) and is given by

$$\phi_{X\dot{X}}(t,s) = b^2\, t + bU(t-s)$$

Note that this first derivative function is discontinuous along the line $t = s$, so the second partial derivative of Eq. 4.57 does not exist. We can write it formally as

$$\phi_{\dot{X}\dot{X}}(t,s) = \frac{\partial^2}{\partial t\, \partial s}\phi_{XX}(t,s) = b^2 + b\delta(t-s)$$

which again emphasizes that the autocorrelation of $\{\dot{X}(t)\}$ is finite only for $t \ne s$. The mean-squared value of the derivative process, which is the autocorrelation function for $t = s$, is infinite.

This example illustrates a situation in which the time histories of $\{X(t)\}$ are not differentiable, but one can still consider a derivative process $\{\dot{X}(t)\}$, with the limitation that it has an infinite mean-squared value. As in Example 4.16, the second derivative exists both for $t < s$ and for $t > s$, and the limit is the same (i.e., b^2) as one approaches $t = s$ from either side. Nonetheless, the second derivative does not exist for $t = s$. A discontinuity in a function can always be expected to contribute such a Dirac delta function term to the derivative of the function.

****

Example 4.18: Consider the differentiability of the stochastic process $\{X(t)\}$ of Example 4.9, which has $X(t)$ equal to some A_j in each finite time interval, and with the A_j identically distributed, mean-zero, and independent.

Taking the expected value gives $\mu_X(t) = \mu_A$, so we can say that $\mu_{\dot{X}}(t) = 0$, and in Example 4.9 we found the autocorrelation function for $\{X(t)\}$, which can be written as

$$\phi_{XX}(t,s) = E(A^2)\sum_j \big(U[t-j\Delta] - U[t-(j+1)\Delta]\big)\big(U[s-j\Delta] - U[s-(j+1)\Delta]\big)$$

This function is discontinuous on the boundaries of the dimension Δ squares along the diagonal. Thus, the $\phi_{X\dot{X}}(t,s)$ cross-correlation function is infinite for some (t,s) values:

$$\phi_{X\dot{X}}(t,s) = E(A^2)\sum_j \big(U[t-j\Delta] - U[t-(j+1)\Delta]\big)\big(\delta[s-j\Delta] - \delta[s-(j+1)\Delta]\big)$$

In particular, it is infinite for $s = k\,\Delta$ for any integer k, if $(k-1)\,\Delta < t < (k+1)\,\Delta$ (i.e., along the boundaries of the dimension Δ squares along the diagonal). Similarly,

$\phi_{X\dot{X}}(t,s)$ is infinite along all the $t = k\,\Delta$ boundaries of the same squares along the diagonal. Everywhere, except on the boundaries of these squares, one obtains $\phi_{X\dot{X}}(t,s)=0$ and $\phi_{\dot{X}\dot{X}}(t,s)=0$, reflecting the fact that this process has a zero derivative except at certain discrete times at which the derivative is infinite. This agrees with the behavior of the time histories of the process.

****

Example 4.19: Consider the differentiability of the stochastic process $\{X(t):t\ge 0\}$ of Example 4.1 with $X^{(1)}(t)=0$, $X^{(2)}(t)=\alpha\sinh(\beta t)$, $X^{(3)}(t)=-\alpha\sinh(\beta t)$, and $P[X(t)=X^{(1)}(t)]=0.50$, $P[X(t)=X^{(2)}(t)]=P[X(t)=X^{(3)}(t)]=0.25$.

In Example 4.1 we showed that $\phi_{XX}(t,s)=(0.50)\alpha^2\sinh(\beta t)\sinh(\beta s)$ and $\mu_X(t)=0$. Thus, we can now take derivatives of these functions to obtain

$$\mu_{\dot{X}}(t)=0, \qquad \phi_{X\dot{X}}(t,s)=(0.50)\alpha^2\beta\sinh(\beta t)\cosh(\beta s)$$

and

$$\phi_{\dot{X}\dot{X}}(t,s)=(0.50)\alpha^2\beta^2\cosh(\beta t)\cosh(\beta s)$$

Inasmuch as the functions are analytic everywhere on the domain for which $\{X(t)\}$ is defined, there is no problem of existence of the moments of the derivative. This $\{X(t)\}$ was defined in terms of its time histories, so one can also differentiate those time histories and obtain corresponding time history relationships for the derivative process $\{\dot{X}(t)\}$. This should give the same results as shown for the moments of $\{\dot{X}(t)\}$. If the results are not consistent, it must mean that our definition of the stochastic derivative is defective. Checking this, we find that

$$\dot{X}^{(1)}(t)=0, \qquad \dot{X}^{(2)}(t)=\alpha\,\beta\cosh(\beta t), \qquad \dot{X}^{(3)}(t)=-\alpha\,\beta\cosh(\beta t)$$

which gives

$$\mu_{\dot{X}}(t)\equiv E[\dot{X}(t)]=(0.50)(0)+(0.25)[\alpha\,\beta\cosh(\beta t)]+(0.25)[-\alpha\,\beta\cosh(\beta t)]=0$$

$$\phi_{X\dot{X}}(t,s)=(0.25)\big([\alpha\sinh(\beta t)][\alpha\,\beta\cosh(\beta s)]+[-\alpha\sinh(\beta t)][-\alpha\,\beta\cosh(\beta s)]\big)$$

$$\phi_{\dot{X}\dot{X}}(t,s)=0.25\big([\alpha\,\beta\cosh(\beta t)][\alpha\,\beta\cosh(\beta s)]+[-\alpha\,\beta\cosh(\beta t)][-\alpha\,\beta\cosh(\beta s)]\big)$$

These results agree exactly with those obtained from differentiating the moment functions of $\{X(t)\}$, confirming that our definition of the stochastic derivative is consistent.

****

Example 4.20: For the stochastic process $\{X(t)\}$ of Example 4.2 with $X(t)=A\cos(\omega t)$ in which A is a random variable, find $\mu_{\dot{X}}(t)$, $\phi_{X\dot{X}}(t,s)$, and $\phi_{\dot{X}\dot{X}}(t,s)$. Confirm that differentiating the moment functions for $\{X(t)\}$ and analyzing the time histories of the derivative process $\{\dot{X}(t)\}$ give the same results.

From Example 4.2, we know that $\phi_{XX}(t,s) = E(A^2)\cos(\omega t)\cos(\omega s)$ and $\mu_X(t) = \mu_A \cos(\omega t)$. Taking derivatives of these functions gives

$$\mu_{\dot{X}}(t) = -\mu_A\, \omega \sin(\omega t), \qquad \phi_{X\dot{X}}(t,s) = -E(A^2)\,\omega \cos(\omega t)\sin(\omega s)$$

and

$$\phi_{\dot{X}\dot{X}}(t,s) = E(A^2)\,\omega^2 \sin(\omega t)\sin(\omega s)$$

The relationship for the time histories of $\{\dot{X}(t)\}$ is

$$\dot{X}(t) = -A\omega \sin(\omega t)$$

and it is obvious that this relationship gives the same moment functions for $\{\dot{X}(t)\}$ as were already obtained by differentiating the moment functions for $\{X(t)\}$.

****

In many cases, analysts choose to use a definition of stochastic derivative that is more precise but also more restrictive than the one we use here. In particular, if

$$\lim_{h \to 0} E\left(\left[\dot{X}(t) - Y(t,h)\right]^2\right) = 0 \tag{4.67}$$

then $\{\dot{X}(t)\}$ is said to be the mean-square derivative of $\{X(t)\}$. Expanding this expression, one can show that the mean-square derivative exists if Eqs. 4.56 and 4.57 are both satisfied. In general, one must say that the mean-square derivative does not exist if $E[\dot{X}^2(t)]$ is infinite. This situation is illustrated in Examples 4.16 and 4.17, in which the mean-square derivative does not exist for any t value, and in Example 4.18, in which the same is true for certain t values. These examples demonstrate that it may be overly restrictive to limit attention only to the mean-square derivative.

Note that we have considered only the first derivative $\{\dot{X}(t)\}$ of a stochastic process $\{X(t)\}$, whereas our dynamics equations will usually require at least two derivatives. In particular, we will usually need to include displacement, velocity, and acceleration terms in our differential equations, and these can be modeled as three stochastic processes $\{X(t)\}$, $\{\dot{X}(t)\}$, and $\{\ddot{X}(t)\}$. This presents no difficulty, though, because we can simply say that $\{\ddot{X}(t)\}$ is the derivative of $\{\dot{X}(t)\}$, reusing the concept of the first derivative. In this way, one can define the general jth-order derivative of $\{X(t)\}$ by j applications of the derivative procedure.

Finally, it should also be noted that the various expressions given here for autocorrelation or cross-correlation functions could equally well be written in terms of autocovariance or cross-covariance functions. This follows directly from the fact that an autocovariance or cross-covariance function is identical to an autocorrelation or cross-correlation function for the special situation in which the mean-value functions are zero. One cross-covariance result of particular significance is the fact that the random variables $X(t)$ and $\dot{X}(t)$ are uncorrelated at any time t for a covariant stationary $\{X(t)\}$ process, which is the generalization of $E[X(t)\,\dot{X}(t)]$ being zero for a second-moment stationary process.

4.9 Stochastic Integral

To complete our idea of stochastic calculus, we need to define stochastic integrals. In fact, we will consider three slightly different types of stochastic integrals that we will use in applications. The conditions for existence of a stochastic integral are usually easier to visualize than for a stochastic derivative, which is basically the same as for deterministic functions. Whereas existence of a derivative depends on smoothness conditions on the original function, an integral generally exists if the integrand is bounded and tends to zero sufficiently rapidly at any infinite limits of integration.

The simplest integral of a stochastic process $\{X(t)\}$ is the simple definite integral of the form

$$Z = \int_a^b X(t)\,dt \tag{4.68}$$

For constant limits a and b, the quantity Z must be a random variable. We can follow exactly the same approach that we used for derivatives and define a Riemann sum that approximates Z, then say that the moments and/or other properties of Z are the limits of the corresponding quantities for the Riemann sum. For example, we can say that

$$Y_n = \Delta t \sum_{j=1}^{n} X(a + j\,\Delta t)$$

with $\Delta t = (b-a)/n$. Then,

$$\mu_Z \equiv E(Z) = \lim_{n\to\infty} E(Y_n) = \lim_{n\to\infty} \Delta t \sum_{j=1}^{n} \mu_X(a + j\,\Delta t) = \int_a^b \mu_X(t)\,dt \tag{4.69}$$

$$E(Z^2) = \lim_{n\to\infty} E(Y_n^2) = \lim_{n\to\infty} (\Delta t)^2 \sum_{j=1}^{n} \sum_{k=1}^{n} \phi_{XX}(a + j\,\Delta t, a + k\Delta t)$$

$$= \int_a^b \int_a^b \phi_{XX}(t,s)\,dt\,ds \tag{4.70}$$

and so forth. Thus, we will say that μ_Z exists if the integral of $\mu_X(t)$ from a to b exists and that $E(Z^2)$ exists if the integral of $\phi_{XX}(t,s)$ over the square domain exists. Clearly, these integrals will exist for finite values of a and b if the first and second moments of $\{X(t)\}$ are finite.

The next integral that we will consider is the stochastic process that can be considered the antiderivative:

$$Z(t) = \int_a^t X(s)\,ds \tag{4.71}$$

with a being any constant. Equations 4.69 and 4.70 now generalize to give the mean value function and autocorrelation function for the new process $\{Z(t)\}$ as

$$\mu_Z(t) \equiv E[Z(t)] = \int_a^t \mu_X(s)\,ds \tag{4.72}$$

$$\phi_{XZ}(t_1,t_2) \equiv E[X(t_1)\,Z(t_2)] = \int_a^{t_2} \phi_{XX}(t_1,s_2)\,ds_2 \tag{4.73}$$

and

$$\phi_{ZZ}(t_1,t_2) \equiv E[Z(t_1)\,Z(t_2)] = \int_a^{t_2} \int_a^{t_1} \phi_{XX}(s_1,s_2)\,ds_1\,ds_2 \tag{4.74}$$

Again, we will generally have no problem with the existence of these moments. Also note that Eqs. 4.72–4.74 are exactly the inverse forms of the relationships in Eqs. 4.55–4.57, confirming that $\{\dot{Z}(t)\} = \{X(t)\}$, which is the desired inverse of Eq. 4.71. The idea of the compatibility of our concepts of stochastic derivative and stochastic integral will be essential in our study of dynamic systems. For example, we will need to be able to use the usual deterministic ideas that velocity is the derivative of displacement, and displacement is the integral of velocity when both displacement and velocity are stochastic processes.

There is no assurance that a stationary process $\{X(t)\}$ will give a stationary integral process $\{Z(t)\}$. For example, $\{X(t)\}$ being mean-value stationary only ensures that μ_X is independent of t; Eq. 4.72 then gives the nonstationary mean value $\mu_Z(t) = \mu_X\, t$ for $\{Z(t)\}$. Similarly, a stationary autocorrelation function for $\{X(t)\}$ may not give a stationary autocorrelation function for $\{Z(t)\}$, as will be illustrated in Example 4.21.

The third type of stochastic integral that we will use is a generalization of Eq. 4.68 in a slightly different way. This time we will keep the limits of integration as constants, but the integral will be a stochastic process because of the presence of a kernel function in the integrand:

$$Z(\eta) = \int_a^b X(t)\, g(t,\eta)\, dt \tag{4.75}$$

For any particular value of the new variable η, this integral is a random variable exactly like Eq. 4.68, and for any reasonably smooth kernel function $g(t,\eta)$, this new family of random variables can be considered a stochastic process $\{Z(\eta)\}$, with η as the index parameter. From Eqs. 4.69 and 4.70, we have

$$\mu_Z(\eta) = \int_a^b \mu_X(t)\, g(t,\eta)\, dt \tag{4.76}$$

and

$$\phi_{ZZ}(\eta_1,\eta_2) = \int_a^b \int_a^b \phi_{XX}(t,s)\, g(t,\eta_1)\, g(s,\eta_2)\, dt\, ds \tag{4.77}$$

Other properties can be derived as needed. This type of stochastic process is needed for the study of dynamics. In the time domain analysis of Chapter 5, η represents another time variable, and η is a frequency parameter in the Fourier analysis of Chapter 6.

****

Example 4.21: Consider the integral process $\{Z(t): t \ge 0\}$ defined by

$$Z(t) = \int_0^t X(s)\, ds$$

for $\{X(t)\}$ being a mean-zero stationary stochastic process with autocorrelation function $R_{XX}(\tau) = e^{-\alpha|\tau|}$, in which α is a positive constant. Find the cross-correlation function of $\{X(t)\}$ and $\{Z(t)\}$ and the autocorrelation function for $\{Z(t)\}$.

From Eq. 4.73 we have

$$\phi_{XZ}(t_1,t_2)=\int_0^{t_2}\phi_{XX}(t_1,s_2)\,ds_2=\int_0^{t_2}R_{XX}(t_1-s_2)\,ds_2=\int_0^{t_2}e^{-\alpha|t_1-s_2|}\,ds_2$$

For $t_2 \le t_1$ this gives

$$\phi_{XZ}(t_1,t_2)=\int_0^{t_2}e^{-\alpha t_1+\alpha s_2}=e^{-\alpha t_1}\frac{e^{\alpha t_2}-1}{\alpha}=\frac{e^{-\alpha(t_1-t_2)}-e^{-\alpha t_1}}{\alpha}$$

and for $t_2 > t_1$ the integral must be split into two parts, giving

$$\phi_{XZ}(t_1,t_2)=\int_0^{t_1}e^{-\alpha t_1+\alpha s_2}\,ds_2+\int_{t_1}^{t_2}e^{\alpha t_1-\alpha s_2}\,ds_2$$

$$=e^{-\alpha t_1}\frac{e^{\alpha t_1}-1}{\alpha}+e^{\alpha t_1}\frac{e^{-\alpha t_2}-e^{-\alpha t_1}}{-\alpha}$$

or

$$\phi_{XZ}(t_1,t_2)=\frac{2-e^{-\alpha t_1}-e^{-\alpha(t_2-t_1)}}{\alpha}$$

The autocorrelation of $\{Z(t)\}$ can now be found as

$$\phi_{ZZ}(t_1,t_2)=\int_0^{t_1}\phi_{XZ}(s_1,t_2)\,ds_1$$

For $t_2 \le t_1$ we find that

$$\phi_{ZZ}(t_1,t_2)=\int_0^{t_2}\frac{2-e^{-\alpha s_1}-e^{-\alpha(t_2-s_1)}}{\alpha}ds_1+\frac{e^{\alpha t_2}-1}{\alpha}\int_{t_2}^{t_1}e^{-\alpha s_1}\,ds_1$$

$$=\frac{1}{\alpha}\left[2t_2-\frac{(e^{-\alpha t_2}+1)(e^{\alpha t_2}-1)}{\alpha}+\frac{(e^{\alpha t_2}-1)(e^{-\alpha t_1}-e^{-\alpha t_2})}{-\alpha}\right]$$

$$=[2\alpha t_2-e^{-\alpha(t_1-t_2)}+e^{-\alpha t_1}+e^{-\alpha t_2}-1]/\alpha^2$$

and for $t_2 > t_1$

$$\phi_{ZZ}(t_1,t_2)=\int_0^{t_1}\frac{2-e^{-\alpha s_1}-e^{-\alpha(t_2-s_1)}}{\alpha}ds_1=\frac{1}{\alpha}\left(2t_1-\frac{e^{-\alpha t_1}-1}{-\alpha}-\frac{e^{-\alpha t_2}(e^{\alpha t_1}-1)}{\alpha}\right)$$

$$=[2\alpha t_1-e^{-\alpha(t_2-t_1)}+e^{-\alpha t_1}+e^{-\alpha t_2}-1]/\alpha^2$$

Note that this function can be combined as

$$\phi_{ZZ}(t_1,t_2)=[2\alpha\min(t_1,t_2)-e^{-\alpha|t_2-t_1|}+e^{-\alpha t_1}+e^{-\alpha t_2}-1]/\alpha^2$$

for any values of t_1 and t_2. In this form it is obvious that the function does have the necessary symmetry, $\phi_{ZZ}(t_1,t_2)=\phi_{ZZ}(t_2,t_1)$. Note that $\{Z(t)\}$ is not second-moment stationary and it is not jointly second-moment stationary with $\{X(t)\}$, even though $\{X(t)\}$ is second-moment stationary.

****

Example 4.22: Consider the integral process $\{Z(t):t\ge 0\}$ defined by

$$Z(t)=\int_0^t X(s)\,ds$$

for $X(t)$ equal to some random variable A_j in each time increment $j\Delta \le t \le (j+1)\Delta$, and with the A_j being identically distributed, mean-zero, and independent. Find the mean value function of $\{Z(t)\}$, the cross-correlation function of $\{X(t)\}$ and $\{Z(t)\}$, and the autocorrelation function for $\{Z(t)\}$.

Because $\mu_X(t)=\mu_A=0$, we have $\mu_Z(t)=\mu_A\, t=0$. In Example 4.9 we found that the autocorrelation function for $\{X(t)\}$ is equal to $E(A^2)$ in squares of dimension Δ along the 45° line of the plane and zero elsewhere. To simplify the presentation of the results, we define a new integer time function as the number of full time increments included in $[0,t]$:

$$k(t)=\text{Int}(t/\Delta)=j \qquad \text{if } \; j\Delta \le t<(j+1)\Delta$$

This allows us to classify more simply any point into one of three sets: points within the squares of dimension Δ along the 45° line of the (t_1,t_2) plane are described by $k(t_1)=k(t_2)$, by $k(t_1)\Delta \le t_2 <[k(t_1)+1]\Delta$, or by $k(t_2)\Delta \le t_1 < [k(t_2)+1]\Delta$; points below the squares are described by $k(t_1)>k(t_2)$, by $t_2<k(t_1)\Delta$, or by $t_1 \ge [k(t_2)+1]\Delta$; and points above the squares are described by $k(t_1)<k(t_2)$, by $t_2\ge[k(t_1)+1]\Delta$, or by $t_1<k(t_2)\Delta$.

Then using

$$\phi_{XZ}(t_1,t_2)=\int_0^{t_2}\phi_{XX}(t_1,s_2)\,ds_2$$

gives

$$\phi_{XZ}(t_1,t_2)=0 \qquad \text{if } k(t_1)>k(t_2)$$

$$\phi_{XZ}(t_1,t_2)=E(A^2)[t_2-k(t_1)\Delta] \qquad \text{if } k(t_1)=k(t_2)$$

$$\phi_{XZ}(t_1,t_2)=E(A^2)\Delta \qquad \text{if } k(t_1)<k(t_2)$$

The function $\phi_{XZ}(t_1,t_2)$ versus t_2 grows linearly from zero to $E(A^2)\Delta$ as t_2 varies from $k(t_1)\Delta$ to $[k(t_1)+1]\Delta$. The same function versus t_1 is a stairstep that is constant at the level $E(A^2)\Delta$ for $t_1<k(t_2)\Delta$, at the level $E(A^2)[t_2-k(t_1)\Delta]$ for $k(t_2)\Delta\le t_1<[k(t_2)+1]\Delta$, and at zero for $t_1\ge[k(t_2)+1]\Delta$.

Now

$$\phi_{ZZ}(t_1,t_2)=\int_0^{t_1}\phi_{XZ}(s_1,t_2)\,ds_1$$

gives

$$\phi_{ZZ}(t_1,t_2) = E(A^2)\, t_1\, \Delta \qquad \text{if } k(t_1) < k(t_2)$$

$$\phi_{ZZ}(t_1,t_2) = E(A^2)\Big(k(t_1)\,\Delta^2 + [t_2 - k(t_1)\,\Delta][t_1 - k(t_2)\,\Delta]\Big) \qquad \text{if } k(t_1) = k(t_2)$$

$$\phi_{ZZ}(t_1,t_2) = E(A^2)\, t_2\, \Delta \qquad \text{if } k(t_1) < k(t_2)$$

Note that if Δ is small compared with t_1 and t_2 then one can say that $\phi_{ZZ}(t_1,t_2) \approx E(A^2)\,\Delta \min(t_1,t_2)$. This expression is exact for $k(t_1) \neq k(t_2)$ and has an error of order Δ^2 for $k(t_1) = k(t_2)$.

****

Example 4.23: Consider the integral process $\{Z(t)\}$ defined by

$$Z(t) = \int_{-\infty}^{t} X(s)\, ds$$

for $X(t)$ equal to some random variable A in the time increment $0 \le t \le \Delta$, and equal to zero elsewhere. Find the mean-value function of $\{Z(t)\}$, the cross-correlation function of $\{X(t)\}$ and $\{Z(t)\}$, and the autocorrelation function for $\{Z(t)\}$.

First we note that the $\{X(t)\}$ process has a mean-value function and autocorrelation function given by

$$\mu_X(t) = \mu_A\, U(t)\, U(\Delta - t)$$

and

$$\phi_{XX}(t_1,t_2) = E(A^2)\, U(t_1)\, U(\Delta - t_1)\, U(t_2)\, U(\Delta - t_2)$$

Now

$$\mu_Z(t) = \int_{-\infty}^{t} \mu_X(s)\, ds$$

gives

$$\mu_Z(t) = \mu_A\, t\, U(t)\, U(\Delta - t) + \mu_A\, \Delta\, U(t - \Delta)$$

which grows linearly from zero to $\mu_A\, \Delta$ on the interval $0 \le t \le \Delta$. Similarly,

$$\phi_{XZ}(t_1,t_2) = \int_{-\infty}^{t_2} \phi_{XX}(t_1,s_2)\, ds_2$$

gives

$$\phi_{XZ}(t_1,t_2) = E(A^2)\, U(t_1)\, U(\Delta - t_1)[t_2\, U(t_2)\, U(\Delta - t_2) + \Delta\, U(t_2 - \Delta)]$$

which is $E(A^2)\, t_2$ within the square $0 \le t_1 < \Delta$, $0 \le t_2 < \Delta$, is $E(A^2)\,\Delta$ for $0 \le t_1 < \Delta$ and $t_2 > \Delta$, and is zero elsewhere. Finally

$$\phi_{ZZ}(t_1,t_2) = \int_{-\infty}^{t_1} \phi_{XZ}(s_1,t_2)\, ds_1$$

gives

$$\phi_{XZ}(t_1,t_2) = E(A^2)[t_1\, U(t_1)U(\Delta - t_1) + \Delta U(t_1 - \Delta)] \times [t_2\, U(t_2)U(\Delta - t_2) + \Delta U(t_2 - \Delta)]$$

which is $E(A^2)\, t_1\, t_2$ within the square $0 \le t_1 < \Delta$ and $0 \le t_2 < \Delta$, is $E(A^2)\, t_1\, \Delta$ for $0 \le t_1 < \Delta$ and $t_2 > \Delta$, is $E(A^2)\, t_2\, \Delta$ for $t_1 > \Delta$ and $0 \le t_2 < \Delta$, is $E(A^2)\, \Delta^2$ for $t_1 > \Delta$, $t_2 > \Delta$, and is zero elsewhere.

****

4.10 Gaussian Stochastic Processes

The definition of a Gaussian stochastic process is extremely simple. In particular, $\{X(t)\}$ is said to be a Gaussian stochastic process if any finite set $\{X(t_1), X(t_2), \cdots, X(t_n)\}$ of random variables from that process is composed of members that are jointly Gaussian. Thus, understanding the special properties of Gaussian processes requires only knowledge of the properties of jointly Gaussian random variables, a number of which are covered in Examples 2.17, 2.19, 2.25, 2.32, 3.10–3.13, 3.16, and 3.23. The jointly Gaussian probability density function for the vector $\vec{V} = [X(t_1), X(t_2), \cdots, X(t_n)]^T$ is

$$p_{\vec{V}}(\vec{u}) = \frac{1}{(2\pi)^{n/2}|\mathbf{K}_{VV}|^{1/2}} \exp\left(-\frac{1}{2}(\vec{u} - \vec{\mu}_V)^T\, \mathbf{K}_{VV}^{-1}\, (\vec{u} - \vec{\mu}_V)\right) \tag{4.78}$$

and the characteristic function is

$$M_{\vec{V}}(\vec{\theta}) = \exp\left(i\,\vec{\theta}^T\, \vec{\mu}_V - \vec{\theta}^T\, \mathbf{K}_{VV}\, \vec{\theta}\right) \tag{4.79}$$

In addition (see Example 2.17), the jointly Gaussian distribution implies that each of the components is a scalar Gaussian random variable with

$$p_{X(t_j)}(u_j) = \frac{1}{(2\pi)^{1/2}\sigma_X(t_j)} \exp\left(-\frac{1}{2}\left[\frac{u_j - \mu_X(t_j)}{\sigma_X(t_j)}\right]^2\right) \tag{4.80}$$

and

$$M_{X(t_j)}(\theta_j) = \exp\left(i\,\theta_j\, \mu_X(t_j) - \theta_j^2\, \sigma_X^2(t_j)\right) \tag{4.81}$$

One of the key features of the Gaussian distribution is the relatively simple form of its probability density functions and characteristic functions, with the components of a jointly Gaussian set $\{X(t_1),X(t_2),\cdots,X(t_n)\}$ depending only on the mean-value vector and the covariance matrix for the set. For a single scalar random variable X, this reduces to dependence on only the mean value and the variance. In addition, the form of the Gaussian probability density function often allows analytical evaluation of quantities of interest. For example, all the moments of a Gaussian random variable X can be described by simple relationships (see Example 3.8). Also important is the fact that the Gaussian probability density function $p_X(u)$ converges to zero sufficiently rapidly for $|u|$ tending to infinity that one is assured that the expectation of many functions of X will exist. For example, $E(e^{aX})$ exists for all values of the parameter a, and that is not true for many other unbounded random variables X.

Particularly important for application to vibration analysis is the fact that any linear combination of jointly Gaussian random variables is itself Gaussian, and jointly Gaussian with other such linear combinations (see Example 2.19). This ensures that linear operations on a Gaussian stochastic process will yield other Gaussian processes and that these new processes are jointly Gaussian with the original process. In particular, the derivative $\dot{X}(t)$ defined in Section 4.8 is a limit of a linear combination of the $X(t)$ and $X(t+h)$ random variables. This results in $\{\dot{X}(t)\}$ being a Gaussian stochastic process. Furthermore, $\{\dot{X}(t)\}$ is jointly Gaussian with $\{X(t)\}$, meaning that any set of random variables from the two processes, such as $\{X(t_1),X(t_2),\cdots,X(t_n),\dot{X}(s_1),\dot{X}(s_2),\cdots,\dot{X}(s_m)\}$, is described by a jointly Gaussian distribution. Similarly, the various integrals defined in Section 4.9 are limits of linear combinations of $X(t_j)$ random variables. Thus, if $\{X(t)\}$ is a Gaussian process, then any integral of $\{X(t)\}$ is either a Gaussian random variable or a Gaussian process and it is jointly Gaussian with the original $\{X(t)\}$ process.

Gaussian stochastic processes play an extremely important role in practical engineering applications. In fact, the majority of applications of stochastic methods to the study of dynamic systems consider only Gaussian processes, even though Gaussian is only one special case out of infinitely many that could be studied. Similarly, the Gaussian special case occupies a very important, although not quite so dominant, position in other engineering application of random variable and stochastic process methods.

There are at least two fundamentally different, important reasons why the Gaussian distribution is used so extensively in applied probability. One is convenience, as illustrated in the preceding paragraphs. The second reason that the

Gaussian distribution is often chosen in modeling real problems is that it often provides quite a good approximation to measured statistical data. Presumably, this is because of the central limit theorem. A very loose statement of this important theorem is that if a random variable X is the sum of a large number of separate components, then X is approximately Gaussian under weak restrictions on the joint distribution of the components. More precisely, if X_j is the sum of the terms $\{Z_1, Z_2, \cdots, Z_j\}$, then the probability density function for the standardized random variable $(X_j - \mu_{X_j})/\sigma_{X_j}$ tends to a Gaussian form as j tends to infinity under weak restrictions on the distribution of the Z_k components. One adequate set of restrictions on the Z_k components is that they all be independent and identically distributed, but neither the condition of identical distribution or that of independence of all the Z_k terms is essential for the proof of the central limit theorem. In practical modeling situations, one generally cannot prove that the central limit theorem is satisfied inasmuch as one rarely has the information that would be needed about the joint probability distribution of the Z_k components contributing to the quantity of interest. On the other hand, the theorem does provide some basis for expecting a random variable X to be approximately Gaussian if many factors contribute to its randomness. Experience seems to confirm that this is a quite good assumption in a great variety of situations in virtually all areas of applied probability, including stochastic structural dynamics.

It should be noted, though, that the Gaussian distribution does have significant limitations that may make it inappropriate for some particular problems. Most obvious is the fact that $p_X(u)$ is symmetric about $u = \mu_X$ for any Gaussian random variable. One consequence of this is that all odd central moments are zero. In particular, the skewness (see Eq. 3.12) is zero. Similarly, the even central moments can all be written in terms of the variance σ_X^2. For example, the fourth central moment is $3\sigma_X^4$, so the kurtosis (see Eq. 3.13) has a value of exactly 3. It is common to use skewness and kurtosis as additional parameters in approximating distributions that cannot be adequately approximated as being symmetric with probability density tails that decay like an exponential of a quadratic. Grigoriu (1995) has a complete book on the use of non-Gaussian processes.

****

Example 4.24: Consider a covariant stationary stochastic process $\{X(t)\}$ for which the joint probability density function of $X(t)$ and the derivative $\dot{X}(t)$ at the same instant of time is given by

$$p_{X(t)\dot{X}(t)}(u,v) = \frac{1}{\pi 2^{1/2} u} \exp\left(-u^2 - \frac{v^2}{2u^2} \right)$$

Find the marginal probability density function $p_{X(t)}(u)$ and the conditional probability density function for the derivative, $p_{\dot{X}(t)}[v \mid X(t) = u]$. Determine whether $X(t)$ is a Gaussian random variable and whether $\{X(t)\}$ is a Gaussian stochastic process.

To find the marginal probability density function we must integrate with respect to v:

$$p_{X(t)}(u) = \frac{e^{-u^2}}{\pi 2^{1/2} u} \int_{-\infty}^{\infty} e^{-v^2/(2u^2)}\, dv = \frac{e^{-u^2}}{\pi^{1/2}} \int_{-\infty}^{\infty} \frac{e^{-v^2/(2u^2)}}{(2\pi)^{1/2} u}\, dv$$

Note that the integrand in the final expression is exactly of the form of a Gaussian distribution with mean zero and variance u^2. Thus, the integral over all values of v must be unity and

$$p_{X(t)}(u) = \frac{e^{-u^2}}{\pi^{1/2}}$$

Comparing this with Eq. 4.80 shows that the random variable $X(t)$ is a Gaussian random variable with mean zero and variance $\sigma_{X(t)}^2 = 1/2$. The conditional distribution can now be found as

$$p_{\dot{X}(t)}[v \mid X(t) = u] = \frac{p_{X(t)\dot{X}(t)}(u,v)}{p_{X(t)}(u)} = \frac{1}{(2\pi)^{1/2} u} \exp\left(-\frac{v^2}{2u^2} \right)$$

This conditional distribution also is seen to be of the Gaussian form, with mean zero and conditional variance u^2. This is not enough information to conclude that the unconditional distribution of $\dot{X}(t)$ is Gaussian, though. We can conclude that $\{X(t)\}$ is not a Gaussian stochastic process, even though $X(t)$ is a Gaussian random variable for every t value. In particular, we note that a Gaussian process and its derivative are jointly Gaussian, but the given $p_{X(t)\dot{X}(t)}(u,v)$ does not have the jointly Gaussian form of Eq. 4.78 (which is also written out more explicitly for two random variables in Example 3.12). Specifically, $p_{X(t)\dot{X}(t)}(u,v)$ is not a constant multiplying an exponential of a quadratic form in u and v. We can also conclude that $\{\dot{X}(t)\}$ is not a Gaussian process by noting that the integral of a Gaussian process is also a Gaussian process. Thus, it is not possible to have $\{\dot{X}(t)\}$ be a Gaussian process since $\{X(t)\}$ is not a Gaussian process.

****

Exercises

****

Time Histories of Stochastic Processes

****

4.1 Let a stochastic process $\{X(t): t \ge 0\}$ be defined by $X(t) = A e^{-Bt}$ in which A and B are independent random variables with μ_A and σ_A^2 known and the probability density function of B being $p_B(u) = b\exp(-bu)U(u)$ in which $b > 0$ is a constant. Find the mean value function $\mu_X(t)$ and the autocorrelation function $\phi_{XX}(t,s)$.

****

4.2 Let a stochastic process $\{X(t): t \ge 0\}$ be defined by $X(t) = A e^{-t} + B e^{-3t}$ in which A and B are random variables with the joint probability density function

$$p_{AB}(u,v) = U(u)U(1-u)U(v)U(1-v)$$

Find the mean value function $\mu_X(t)$ and the autocorrelation function $\phi_{XX}(t,s)$.

****

4.3 Let $\{X(t)\}$ be a stochastic process that depends on three random variables (A,B,C): $X(t) = A + Bt + Ct^2$. To simplify the description of the random variables let them be written as a vector $\vec{V}$ so that the mean values are also a vector $\vec{\mu}_V$ and the mean-squared values and cross-products can be arranged in a matrix $E(\vec{V}\vec{V}^T)$. Specifically, let

$$\vec{V} = \begin{pmatrix} A \\ B \\ C \end{pmatrix}, \qquad \vec{\mu}_V \equiv E(\vec{V}) = \begin{pmatrix} 1 \\ 2 \\ 3 \end{pmatrix}, \qquad E(\vec{V}\vec{V}^T) = \begin{pmatrix} 4 & -1 & 6 \\ -1 & 9 & 0 \\ 6 & 0 & 19 \end{pmatrix}$$

(a) Find the mean value function $\mu_X(t)$.
(b) Find the autocorrelation function $\phi_{XX}(t,s)$.
(c) Find the mean squared value $E[X^2(t)]$.
(d) Find the variance $\sigma_X^2(t)$.

****

4.4 Let the second-moment stationary processes $\{X_1(t)\}$ and $\{X_2(t)\}$ represent the motions at two different points in a complex structural system. Let the correlation matrix for $\{X_1(t)\}$ and $\{X_2(t)\}$ be given by

$$\mathbf{R}_{\vec{X}\vec{X}}(\tau) \equiv E[\vec{X}(t+\tau)\vec{X}^T(t)] = \begin{bmatrix} g(\tau)+g(2\tau) & 2g(\tau)-g(2\tau) \\ 2g(\tau)-g(2\tau) & 4g(\tau)+g(2\tau) \end{bmatrix}$$

in which $g(\tau) = e^{-b|\tau|}[\cos(\omega_0 \tau) + (b/\omega_0)\sin(\omega_0 |\tau|)]$ for constants b and ω_0. Let $\{Z(t)\}$ denote the relative motion between the two points: $Z(t) = X_2(t) - X_1(t)$.
(a) Find the cross-correlation function $R_{X_1Z}(\tau) \equiv E[X_1(t+\tau)Z(t)]$.
(b) Find the autocorrelation function of Z: $R_{ZZ}(\tau) \equiv E[Z(t+\tau)Z(t)]$.

****

Stationarity

****

4.5 For each of the following autocorrelation functions, tell whether the stochastic process $\{X(t)\}$ is second-moment stationary.

(a) $\phi_{XX}(t,s) = \dfrac{1}{1+(t-s)^2}$

(b) $\phi_{XX}(t,s) = \dfrac{1}{1+(t-s)^2} U(1-|t|)\, U(1-|s|)$

(c) $\phi_{XX}(t,s) = 1/(1+t^2+s^2)$

(d) $\phi_{XX}(t,s) = e^{-t^2-s^2}$

****

4.6 For each of the following autocorrelation functions, tell whether the stochastic process $\{X(t)\}$ is second-moment stationary.

(a) $\phi_{XX}(t,s) = \cos(t)\cos(s) + \sin(t)\sin(s)$

(b) $\phi_{XX}(t,s) = \dfrac{\cos(t-s)}{1+(t-s)^2}$

(c) $\phi_{XX}(t,s) = (t-1)(s-1)$

(d) $\phi_{XX}(t,s) = e^{-(t-s)^2}\, U(1-|t|)\, U(1-|s|)$

****

4.7 Consider a stochastic process $\{X(t)\}$ with autocorrelation function

$$\phi_{XX}(t,s) = \frac{t^2 s^2}{1+t^2 s^2 (1+s^2-2ts+t^2)}$$

(a) Find $E[Y^2(t)]$ for $t \ge 0$ and sketch it versus t.
(b) Use the limit as $t \to \infty$ of $\phi_{XX}(t+\tau,t)$ to show that $\{X(t)\}$ tends to become second-moment stationary in that limiting situation.

****

Properties of Mean and Covariance Functions

****

4.8 It is asserted that none of the following $\phi(t,s)$ functions could be the autocorrelation function for any stochastic process. That is, there is no stochastic process $\{X(t)\}$ such that $E[X(t)\,X(s)] = \phi(t,s)$. For each of the ϕ functions, give at least one reason why the assertion must be true.
[Note: The Schwarz inequality is the only condition you should need to check for nonnegative definiteness.]

(a) $\phi(t,s) = \dfrac{1}{(1+s^2)(2+t^2)}$

(b) $\phi(t,s) = \dfrac{(s-t)^2}{1+(s-t)^2}$

(c) $\phi(t,s) = e^{-(s-t)^2}\, U(t-s) + e^{-2(s-t)^2}\, U(s-t)$

(d) $\phi(t,s) = e^{-(s-t)^2}\, U(t+s) + e^{-2(s-t)^2}\, U(-t-s)$

(e) $\phi(t,s) = (s-t)^2 e^{-(s-t)^2}$

****

4.9 It is asserted that none of the following $\phi(t,s)$ functions could be the autocorrelation function for any stochastic process. That is, there is no stochastic process $\{X(t)\}$ such that $E[X(t)\,X(s)] = \phi(t,s)$. For each of the ϕ functions, give at least one reason why the assertion must be true.
[Note: The Schwarz inequality is the only condition you should need to check for nonnegative definiteness.]

(a) $\phi(t,s) = \dfrac{s}{(1+s^2)(1+t^2)}$

(b) $\phi(t,s) = 1 - e^{-(s-t)^2}$

(c) $\phi(t,s) = U(1-t^2-s^2) + e^{-(s-t)^2}\, U(t^2+s^2-1)$

(d) $\phi(t,s) = 1/(1+t^2 s^4)$

(e) $\phi(t,s) = (s-t)^2 \cos(s-t)$

****

4.10 It is asserted that none of the following $R(\tau)$ functions could be the autocorrelation function for any stochastic process. That is, there is no stochastic process $\{X(t)\}$ such that $E[X(t)\,X(s)] = R(t-s)$. For each of the R functions, give at least one reason why the assertion must be true.
[Note: The Schwarz inequality is the only condition you should need to check for nonnegative definiteness.]

(a) $R(\tau) = e^{-\tau^2}\, U(\tau)$

(b) $R(\tau) = \tau^2 / (1+\tau^4)$

(c) $R(\tau) = e^{-2\tau^2}\, U(1-\tau^2)$

(d) $R(\tau) = e^{-2\tau^2} \sin(5\tau)$

****

4.11 Assume that for a stationary process $\{X(t)\}$ you know a conditional probability density function of the form $p_{X(t+\tau)}[v \mid X(t)=u]$. Furthermore, assume that

$$\lim_{\tau\to\infty} p_{X(t+\tau)}[v \mid X(t)=u] = p_{X(t)}(v) \text{ for any } u \text{ and } t \text{ values.}$$

Give an integral expression for the $R_{XX}(\tau)$ autocorrelation function in terms of the conditional probability density function.

(It will be shown in Chapter 9 that a conditional probability density function of this type can sometimes be derived from a Fokker-Planck equation.)

****

Continuity

****

4.12 For each of the following $\phi_{XX}(t,s)$ autocorrelation functions, identify any t values at which $\{X(t)\}$ is not mean-square continuous.

(a) $\phi_{XX}(t,s) = \dfrac{1}{1+(t-s)^2} U(1-|t|)U(1-|s|)$

(b) $\phi_{XX}(t,s) = \dfrac{1}{1+t^2+s^2}$

(c) $\phi_{XX}(t,s) = \cos(t)\cos(s) U\left(\dfrac{\pi}{2}-|t|\right) U\left(\dfrac{\pi}{2}-|s|\right)$

****

4.13 For each of the following $\phi_{XX}(t,s)$ autocorrelation functions, identify any t values at which $\{X(t)\}$ is not mean-square continuous.

(a) $\phi_{XX}(t,s) = (1-t^2)(1-s^2)U(1-t^2)U(1-s^2)$

(b) $\phi_{XX}(t,s) = \cos(t)\cos(s) + \sin(t)\sin(s)$

(c) $\phi_{XX}(t,s) = \cos(t)\cos(s) U(\pi-|t|) U(\pi-|s|)$

****

Derivatives

****

4.14 For each of the following $\phi_{XX}(t,s)$ autocorrelation functions, determine whether there exist $\phi_{X\dot{X}}(t,s)$ and/or $\phi_{\dot{X}\dot{X}}(t,s)$ functions that are finite for all (t,s) values. Identify the (t,s) values for which a finite value of either of the functions does not exist. Also give the value of $E[\dot{X}^2(t)]$ for all t values for which it is finite.

(a) $\phi_{XX}(t,s) = e^{-|t-s|}$

(b) $\phi_{XX}(t,s) = e^{-(t-s)^2} U(ts)$

(c) $\phi_{XX}(t,s) = (1-t^2)(1-s^2)U(1-t^2)U(1-s^2)$

****

4.15 For each of the following $\phi_{XX}(t,s)$ autocorrelation functions, determine whether there exist $\phi_{X\dot{X}}(t,s)$ and/or $\phi_{\dot{X}\dot{X}}(t,s)$ functions that are finite for all (t,s) values. Identify the (t,s) values for which a finite value of either of the functions does not exist. Also give the value of $E[\dot{X}^2(t)]$ for all t values for which it is finite.

(a) $\phi_{XX}(t,s) = e^{-(t-s)^2}\, U(t)\, U(s) + \dfrac{[1-U(t)][1-U(s)]}{1+(t-s)^2}$

(b) $\phi_{XX}(t,s) = t\,s\,U(t)\,U(1-t)\,U(s)\,U(1-s) + e^{-|t-s|}[1-U(1-t)][1-U(1-s)]$

****

4.16 Consider the derivative of the $\{X(t)\}$ process of Exercise 4.7 with

$$\phi_{XX}(t,s) = \frac{t^2 s^2}{1+t^2 s^2 (1+s^2-2ts+t^2)}$$

Find the limit as $t \to \infty$ of $E[\dot{X}^2(t)]$, the mean-square value of the derivative.

****

4.17 For each of the following $R_{XX}(\tau)$ stationary autocorrelation functions, determine whether there exist $R_{X\dot{X}}(\tau)$, $R_{\dot{X}\dot{X}}(\tau)$, $R_{\dot{X}\ddot{X}}(\tau)$, and $R_{\ddot{X}\ddot{X}}(\tau)$ functions that are finite for all τ values. Identify the τ values for which a finite value of any of the R functions does not exist. Also give the values of $E[\dot{X}^2(t)]$ and $E[\ddot{X}^2(t)]$ if they are finite. In each case, a and b are real constants and $a > 0$.

(a) $R_{XX}(\tau) = e^{-a\tau^2}$

(b) $R_{XX}(\tau) = e^{-a|\tau|}\cos(b\tau)$

(c) $R_{XX}(\tau) = (1 + a|\tau|)\, e^{-a|\tau|}$

****

4.18 For each of the following $R_{XX}(\tau)$ stationary autocorrelation functions, determine whether there exist $R_{X\dot{X}}(\tau)$, $R_{\dot{X}\dot{X}}(\tau)$, $R_{\dot{X}\ddot{X}}(\tau)$, and $R_{\ddot{X}\ddot{X}}(\tau)$ functions that are finite for all τ values. Identify the τ values for which a finite value of any of the R functions does not exist. Also give the values of $E[\dot{X}^2(t)]$ and $E[\ddot{X}^2(t)]$ if they are finite. In each case, a and b are real constants and $a > 0$.

(a) $R_{XX}(\tau) = b^2/(b^2+\tau^2)$

(b) $R_{XX}(\tau) = \left(1 - \dfrac{|\tau|}{2}\right) e^{-a|\tau|}$

(c) $R_{XX}(\tau) = \sin(a\tau)/\tau$

****

4.19 Consider the second-order stochastic process $\{X(t)\}$ of Exercise 4.13(a) with $\mu_X(t) \equiv 0$, and autocorrelation function

$$\phi_{XX}(t,s) = (1-t^2)(1-s^2)\,U(1-t^2)\,U(1-s^2)$$

(a) Find $\phi_{\dot{X}\dot{X}}(t,s)$ for all (t,s) values for which it exists and identify any (t,s) values for which it does not exist.

(b) Find $\phi_{\ddot{X}\ddot{X}}(t,s)$ for all (t,s) values for which it exists and identify any (t,s) values for which it does not exist

****

4.20 Consider the stochastic process $\{X(t)\}$ of Exercise 4.1 with $X(t) = Ae^{-Bt}$, in which A and B are independent random variables with μ_A and σ_A^2 known and $p_B(u) = b\exp(-bu)U(u)$ with $b > 0$.

(a) Find the $\mu_{\dot{X}}(t)$ mean-value function of the derivative process $\{\dot{X}(t)\}$.

(b) Find the $\phi_{X\dot{X}}(t,s)$ cross-correlation function between $\{X(t)\}$ and the derivative process.

(c) Find the $\phi_{\dot{X}\dot{X}}(t,s)$ autocorrelation function of the derivative process.

(d) Confirm that $\mu_{\dot{X}}(t)$, $\phi_{X\dot{X}}(t,s)$, and $\phi_{\dot{X}\dot{X}}(t,s)$ are the same whether determined from derivatives of the moment functions for $\{X(t)\}$ or by analyzing the time histories of the $\{\dot{X}(t)\}$ derivative process.

****

4.21 Consider the stochastic process $\{X(t)\}$ of Exercise 4.2 with $X(t) = Ae^{-t} + Be^{-3t}$, in which A and B are random variables with the joint probability density function $p_{AB}(u,v) = U(u)U(1-u)U(v)U(1-v)$.

(a) Find the $\mu_{\dot{X}}(t)$ mean-value function of the derivative process $\{\dot{X}(t)\}$.

(b) Find the $\phi_{X\dot{X}}(t,s)$ cross-correlation function between $\{X(t)\}$ and the derivative process.

(c) Find the $\phi_{\dot{X}\dot{X}}(t,s)$ autocorrelation function of the derivative process.

(d) Confirm that $\mu_{\dot{X}}(t)$, $\phi_{X\dot{X}}(t,s)$, and $\phi_{\dot{X}\dot{X}}(t,s)$ are the same whether determined from derivatives of the moment functions for $\{X(t)\}$ or by analyzing the time histories of the $\{\dot{X}(t)\}$ derivative process.

****

4.22 Consider the stochastic process $\{X(t)\}$ of Exercise 4.3 with $X(t) = A + Bt + Ct^2$, in which the random variables (A,B,C) are arranged in a vector with

$$\vec{V} = \begin{pmatrix} A \\ B \\ C \end{pmatrix}, \qquad \vec{\mu}_V \equiv E(\vec{V}) = \begin{pmatrix} 1 \\ 2 \\ 3 \end{pmatrix}, \qquad E(\vec{V}\vec{V}^T) = \begin{pmatrix} 4 & -1 & 6 \\ -1 & 9 & 0 \\ 6 & 0 & 19 \end{pmatrix}$$

(a) Find the $\mu_{\dot{X}}(t)$ mean-value function of the derivative process $\{\dot{X}(t)\}$.

(b) Find the $\phi_{X\dot{X}}(t,s)$ cross-correlation function between $\{X(t)\}$ and the derivative process.

(c) Find the $\phi_{\dot{X}\dot{X}}(t,s)$ autocorrelation function of the derivative process.

(d) Confirm that $\mu_{\dot{X}}(t)$, $\phi_{X\dot{X}}(t,s)$, and $\phi_{\dot{X}\dot{X}}(t,s)$ are the same whether determined from derivatives of the moment functions for $\{X(t)\}$ or by analyzing the time histories of the $\{\dot{X}(t)\}$ derivative process.

****

4.23 Consider a stochastic process $\{Y(t)\}$ used to model the ground acceleration during an earthquake. This $\{Y(t)\}$ is formed by multiplying a stationary process $\{X(t)\}$ by a specified time function. In particular, $Y(t) = X(t)\, a\, t\, e^{-bt}\, U(t)$ in which a and b are positive constants. Presume that the constant standard deviations σ_X and $\sigma_{\dot{X}}$ are known and that $\mu_X = 0$.
(a) Find $E[Y^2(t)]$.
(b) Find $E[\dot{Y}^2(t)]$.
(c) Find $E[Y(t)\,\dot{Y}(t)]$.
****

4.24 Consider a stochastic process $\{X(t)\}$ with mean $\mu_X(t)$ and variance $\sigma_X^2(t)$ at any time t. Let the conditional distribution of the derivative of the process be given by

$$p_{\dot{X}(t)}[v \mid X(t) = u] = \delta(v - b\,u)$$

in which b is some positive constant and $\delta(\cdot)$ is the Dirac delta function.
(a) Find $E[\dot{X}(t) \mid X(t) = u]$, $E[X(t)\,\dot{X}(t) \mid X(t) = u]$, and $E[\dot{X}^2(t) \mid X(t) = u]$.
(b) Find $E[\dot{X}(t)]$, $E[X(t)\,\dot{X}(t)]$, and $E[\dot{X}^2(t)]$ as functions of $\mu_X(t)$ and $\sigma_X^2(t)$.
(c) Is it possible for $\{X(t)\}$ to be mean-value stationary? That is, is the given information consistent with $\mu_X(t)$ being constant and $\mu_{\dot{X}}(t)$ being zero?
(d) Show that it is not possible for $\{X(t)\}$ to be second-moment stationary by showing that the information found in solving part (c) is inconsistent with $E[X^2(t)]$ and $E[\dot{X}^2(t)]$ being constant.
****

4.25 Let $\{X(t)\}$ be a stationary stochastic process with known values of $E[X^2(t)]$, $E[\dot{X}^2(t)]$, and $E[\ddot{X}^2(t)]$. Find $E[\ddot{X}(t)\,\dddot{X}(t)]$, $E[\dot{X}(t)\,\ddot{X}(t)]$, and $E[X(t)\,\dddot{X}(t)]$ in terms of the known quantities. (Each dot denotes a derivative with respect to t.)
****

4.26 Let $\{X(t)\}$ be a stationary stochastic process with known values of $E[X^2(t)]$, $E[\dot{X}^2(t)]$, $E[\ddot{X}^2(t)]$, and $E[\dddot{X}^2(t)]$. Find $E[\dddot{X}(t)\,\ddddot{X}(t)]$, $E[\ddot{X}(t)\,\ddddot{X}(t)]$, $E[\dot{X}(t)\,\ddddot{X}(t)]$, and $E[X(t)\,\ddddot{X}(t)]$ in terms of the known quantities. (Each dot denotes a derivative with respect to t.)
****

Integrals
****

4.27 Let $\{X(t): t \ge 0\}$ have the autocorrelation function $\phi_{XX}(t_1,t_2) = b\,e^{-c(t_1+t_2)}$ with b and c being positive constants, and let $\{Z(t): t \ge 0\}$ be defined by

$$Z(t) = \int_0^t X(s)\, ds$$

Find the $\phi_{XZ}(t_1,t_2)$ and $\phi_{ZZ}(t_1,t_2)$ functions.
****

4.28 Let $\{X(t): t \ge 0\}$ have $\phi_{XX}(t_1,t_2) = b\, t_1\, t_2\, U(t_1) U(c - t_1) U(t_2) U(c - t_2)$ with b and c being positive constants, and let $\{Z(t): t \ge 0\}$ be defined by

$$Z(t) = \int_0^t X(s)\, ds$$

Find the $\phi_{XZ}(t_1,t_2)$ and $\phi_{ZZ}(t_1,t_2)$ functions.

****

4.29 Let $\{X(t): t \ge 0\}$ have $\phi_{XX}(t_1,t_2) = b(t_1 + c)^{-2}(t_2 + c)^{-2}$ with b and c being positive constants, and let $\{Z(t): t \ge 0\}$ be defined by

$$Z(t) = \int_0^t X(s)\, ds$$

Find the $\phi_{XZ}(t_1,t_2)$ and $\phi_{ZZ}(t_1,t_2)$ functions.

****

4.30 Let $\{X(t)\}$ have the autocorrelation function $\phi_{XX}(t_1,t_2) = b e^{-c(|t_1|+|t_2|)}$ with b and c being positive constants, and let $\{Z(t)\}$ be defined by

$$Z(t) = \int_{-\infty}^t X(s)\, ds$$

Find the $\phi_{XZ}(t_1,t_2)$ and $\phi_{ZZ}(t_1,t_2)$ functions.

****

4.31 Let $\{X(t)\}$ have $\phi_{XX}(t_1,t_2) = b(c - |t_1|)(c - |t_2|) U(c - |t_1|) U(c - |t_2|)$ with b and c being positive constants, and let $\{Z(t)\}$ be defined by

$$Z(t) = \int_{-\infty}^t X(s)\, ds$$

Find the $\phi_{XZ}(t_1,t_2)$ and $\phi_{ZZ}(t_1,t_2)$ functions.

****

4.32 Let $\{X(t)\}$ have the autocorrelation function

$$\phi_{XX}(t_1,t_2) = b e^{-c(|t_1|+|t_2|) - a|t_1 - t_2|}$$

with a, b and c being positive constants, and let $\{Z(t)\}$ be defined by

$$Z(t) = \int_{-\infty}^t X(s)\, ds$$

Find the $\phi_{XZ}(t_1,t_2)$ function.

****

Gaussian Processes

****

4.33 Consider a covariant stationary stochastic process $\{X(t)\}$ for which the joint probability density function of the process and its derivative is given by

$$p_{X(t)\dot{X}(t)}(u,v) = \frac{|u|}{2\pi} \exp\left(-\frac{u^2(1+v^2)}{2} \right)$$

(a) Find the marginal probability density function $p_{X(t)}(u)$. Is it Gaussian?
(b) Find the conditional probability density function for the derivative, $p_{\dot{X}(t)}[v \mid X(t)=u]$. Is it Gaussian?
(c) Find the conditional mean and variance of $\dot{X}(t)$, given the event $X(t)=u$.
(d) Is either $\{X(t)\}$ or $\{\dot{X}(t)\}$ a Gaussian process? Briefly explain your answer.
[Hint: See Example 4.24.]
****

4.34 Consider a covariant stationary stochastic process $\{X(t)\}$ for which the joint probability density function of the process and its derivative is given by

$$p_{X(t)\dot{X}(t)}(u,v)=\frac{2|u|^{1/2}}{\pi}e^{-2u^2-2|u|v^2}$$

(a) Find the marginal probability density function $p_{X(t)}(u)$. Is it Gaussian?
(b) Find the conditional probability density function for the derivative, $p_{\dot{X}(t)}[v \mid X(t)=u]$. Is it Gaussian?
(c) Find the conditional mean and variance of $\dot{X}(t)$, given the event $X(t)=u$.
(d) Is either $\{X(t)\}$ or $\{\dot{X}(t)\}$ a Gaussian process? Briefly explain your answer.
[Hint: See Example 4.24.]
****

Chapter 5
Time Domain Linear Vibration Analysis

5.1 Deterministic Dynamics

Before beginning our study of stochastic dynamics, we will review some of the fundamental ideas of deterministic linear time domain dynamic analysis. For the moment, we will consider our linear time-invariant system to have one scalar excitation $f(t)$ and one scalar response $x(t)$. In Chapter 8, we will use the idea of superposition to extend this analysis to any number of excitation and response components. Treating the excitation as the input to our linear system and the response as its output gives a situation that can be represented by the schematic diagrams in Fig. 5.1.

The function $h_x(t)$ in Fig. 5.1 is defined as the response of the system to one particular excitation. In particular, if $f(t) = \delta(t)$ then $x(t) = h_x(t)$, in which $\delta(\cdot)$ is the Dirac delta function, as introduced in Section 2.4 and explained more fully in Appendix A. We will investigate how to find the $h_x(t)$ function in the following section, but first we want to emphasize that this single function is adequate to characterize the response of the system to any excitation $f(t)$. The steps shown in Fig. 5.1 illustrate this idea. We derive the result by superposition, first writing $f(t)$ as a superposition of Dirac delta functions (see Appendix A). In particular, we write $f(t)$ as the convolution integral:

$$f(t) = \int_{-\infty}^{\infty} f(s)\,\delta(t-s)\,ds \equiv \int_{-\infty}^{\infty} f(t-r)\,\delta(r)\,dr \tag{5.1}$$

in which the first integral is illustrated in Fig. 5.1(c), and the final form is obtained by the change of variables $r = t - s$. (Both of the two equivalent forms are given because either may be more convenient for use in a particular situation.) Now we note that our definition of the $h_x(t)$ function ensures that the excitation component $\delta(t-s)$ will induce a response $h_x(t-s)$ or, equivalently, that $\delta(r)$ will induce $h_x(r)$. Multiplying these delta function pulses of excitation by their amplitudes from Eq. 5.1 and superimposing the responses gives

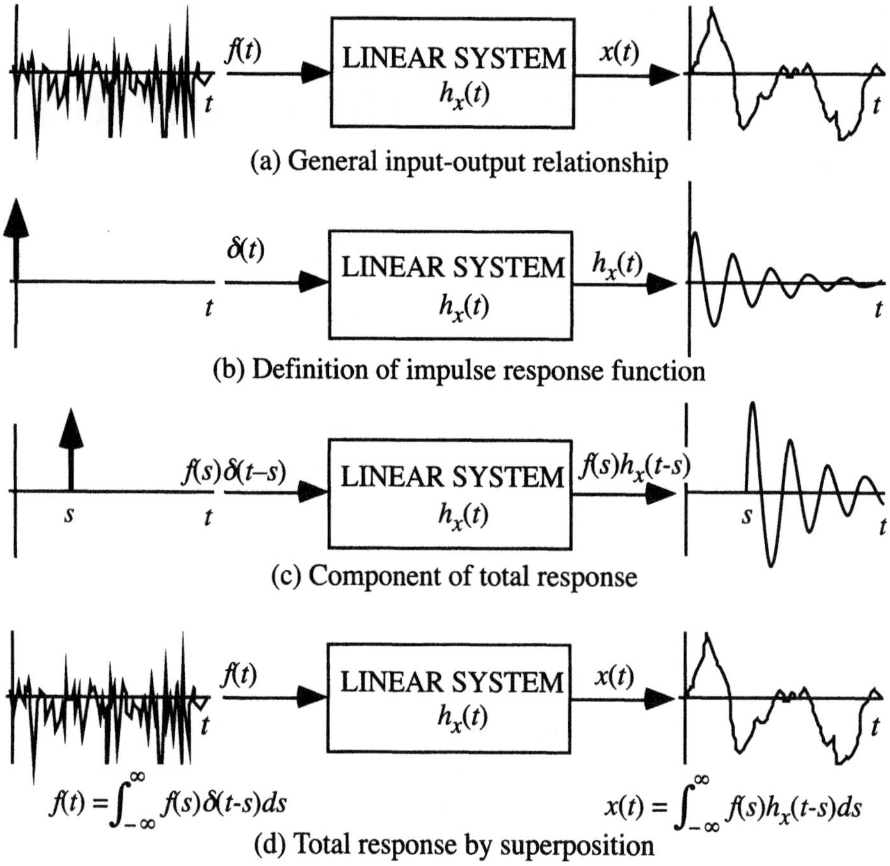

Figure 5.1 Schematic of general linear system.

$$x(t) = \int_{-\infty}^{\infty} f(s)\, h_x(t-s)\, ds \equiv \int_{-\infty}^{\infty} f(t-r)\, h_x(r)\, dr \tag{5.2}$$

Either of the integrals in Eq. 5.2 is called the *Duhamel convolution integral* for the linear system, and the function $h_x(t)$ is called the impulse response function for response x of the system.

It should be noted that the definition of $h_x(t)$ as the $x(t)$ response to $f(t) = \delta(t)$ includes the condition that $h_x(t)$ is the response when there has never been any other excitation of the system except the pulse at time $t = 0$. In particular, $h_x(t)$ is the response when the initial conditions on the system are such that $x(t)$ and all of its derivatives are zero at time $t \to -\infty$. A system is said to be causal if $h_x(t) \equiv 0$ for $t < 0$. This simply means that the response to an excitation $f(t) = \delta(t)$ does not begin to appear until time $t = 0$, which we would certainly expect to be true for a physical system. We do not expect the response to precede the excitation, but one should keep in mind that this is a condition of physics and cannot be proved mathematically. For a causal system, note that the limits of integration in Eq. 5.2 can be modified as

$$x(t) = \int_{-\infty}^{t} f(s)\, h_x(t-s)\, ds \equiv \int_{0}^{\infty} f(t-r)\, h_x(r)\, dr \tag{5.3}$$

The term $f(s)\, h_x(t-s)$ in the first integral of Eq. 5.3 gives the response at time t due to an excitation pulse $f(s)\delta(t-s)$ at time s for $s \le t$. Similarly, $f(t-r)\, h_x(r)$ in the second integral gives the response at time t to an excitation pulse $f(t-r)\delta(r)$ at time $(t-r)$ for $r \ge 0$. For reasons of convenience we will use the $-\infty$ to ∞ limits of Eq. 5.2 in general, with the causal nature of $h_x(t)$ being taken into account as needed in specific examples.

It is also instructive to consider the steady-state response to a static excitation $f(t)$ that has a constant value f_0 for all t. Equation 5.2 gives this steady-state response as $x(t) = f_0\, h_{x,static}$, with

$$h_{x,static} \equiv \int_{-\infty}^{\infty} h_x(r)\, dr \tag{5.4}$$

Thus, the system has a static steady-state x response to this static excitation if and only if $h_{x,static}$ is finite. Of course, the condition for $h_x(r)$ to be integrable so that $|h_{x,static}| < \infty$ is that $h_x(r)$ tends to zero faster than r^{-1} as r goes to infinity. We will find it convenient in the following discussions to classify any linear system as having a bounded or infinite static response based on whether $|h_{x,static}| < \infty$ or $|h_{x,static}| = \infty$, respectively.

In some situations it is more convenient to consider the response of the linear system to be a superposition of the effects of some initial conditions at a fixed time t_0 and the effects of the part of the excitation occurring after time t_0. That is, if a sufficient set of initial conditions is known at time t_0, then we do not need to know the excitation prior to time t_0 in order to find the dynamic response after time t_0. The response of the system caused by excitation after time t_0 is found by limiting the range of integration in the convolution integrals of Eqs. 5.2 or 5.3 to include only $s \ge t_0$ or $r \le t - t_0$, and the additional term giving the response due to the initial conditions is found by considering the homogeneous system with $f(t) = 0$.

Note that the equations presented so far have been restricted to linear systems that do not vary with time. It is also possible to use the basic formulation presented here for linear systems with time-varying properties. The basic change is to replace $h_x(t-s)$ with an $h_{xf}(t,s)$ function of two time parameters. The fundamental convolution integral of Eq. 5.2 then becomes

$$x(t) = \int_{-\infty}^{\infty} f(s)\, h_{xf}(t,s)\, ds \tag{5.5}$$

Substituting $f(s) = \delta(s - t_0)$ into this equation gives $x(t) = h_{xf}(t,t_0)$, which reveals the meaning of $h_{xf}(t,s)$. Precisely, $h_{xf}(t,s)$ is the $x(t)$ response at time t due to an excitation that consists only of a Dirac delta function at time s. For the time-invariant situation this, of course, reduces to the $h_x(t-s)$ function that we have been using.

The definition in Eq. 5.5 can also be extended to give

$$\dot{x}(t) = \int_{-\infty}^{\infty} f(s)\, h_{\dot{x}f}(t,s)\, ds$$

for the derivative of $x(t)$, but this quantity can also be obtained by differentiating Eq. 5.5, giving

$$\dot{x}(t) = \int_{-\infty}^{\infty} f(s) \frac{\partial}{\partial t} h_{xf}(t,s)\, ds$$

Thus,

$$h_{\dot{x}f}(t,s) = \frac{\partial}{\partial t} h_{xf}(t,s) \tag{5.6}$$

which includes the special case of $h_{\dot{x}}(t) = h'_x(t)$ for the time-invariant system.

In this book we will generally restrict our attention to time-invariant systems, but there will be some situations in which it will be useful to consider the time-varying formulation of Eq. 5.5.

5.2 Evaluation of Impulse Response Functions

We will be primarily concerned with analysis of the dynamics of systems governed by differential equations. A general form of such an nth-order linear constant-coefficient differential equation can be written as

$$\sum_{j=0}^{n} a_j \frac{d^j x(t)}{dt^j} = f(t) \tag{5.7}$$

By its definition, the impulse response function for such a system must satisfy the corresponding differential equation

$$\sum_{j=0}^{n} a_j \frac{d^j h_x(t)}{dt^j} = \delta(t) \tag{5.8}$$

subject to the boundary condition that $h_x(t)$ and all of its derivatives tend to zero as $t \to -\infty$. An obvious difficulty with Eq. 5.8 is the fact that the right-hand side is not truly a function. The nature of the Dirac delta function, though, turns out to be a definite advantage. In particular, we know that $h_x(t)$ is identically zero for $t < 0$, so we need to be concerned only with the differential equation for values of t that are greater than or equal to zero. Clearly, we must be very careful at $t = 0$, but for $t > 0$ the differential equation is homogeneous:

$$\sum_{j=0}^{n} a_j \frac{d^j h_x(t)}{dt^j} = 0 \qquad \text{for } t > 0 \tag{5.9}$$

To have a unique solution to this nth-order differential equation, we must have n initial conditions or boundary conditions. We will use initial values of $h_x(t)$ and its first $(n-1)$ derivatives immediately after time zero. In particular, we will write $h_x(0^+)$ for the limit of $h_x(t)$ as t tends to zero from the positive side. Thus, the initial information we need can be described as the values of $h_x(t)$ and its first $(n-1)$ derivatives at time $t = 0^+$. We must derive these initial conditions from the behavior of Eq. 5.8 in the neighborhood of $t = 0$.

Inasmuch as $\delta(t)$ is infinite for $t = 0$, Eq. 5.8 tells us that at least one of the terms on the left-hand side of the equation must also be infinite at $t = 0$. Let us presume for the moment that it is the jth-order term that is infinite. In particular, we will assume that

$$\frac{d^j h_x(t)}{dt^j} = b\,\delta(t)$$

for some b value. This causes difficulties, though, unless $j = n$. In particular, we now have

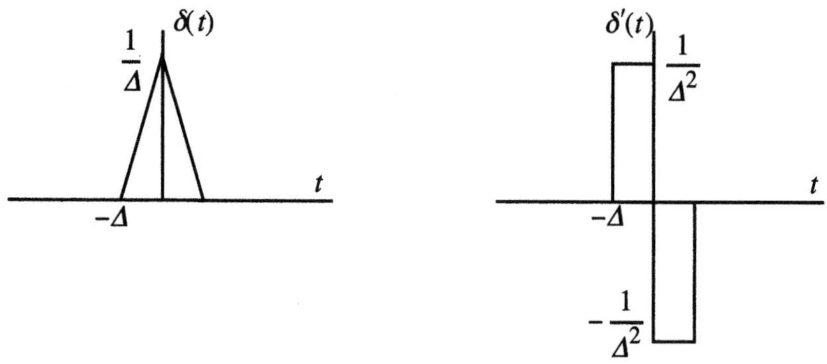

Figure 5.2 Possible approximations for the Dirac delta function and its derivative.

$$\frac{d^{j+1} h_x(t)}{dt^{j+1}} = b \frac{d\delta(t)}{dt} \equiv b\,\delta'(t)$$

As explained in Appendix A, the precise definition of $\delta(t)$ is in terms of the limit of a sequence involving bounded functions. For example, if we consider $\delta(t)$ to be the limit as $\Delta \to 0$ of $[(1-|t|/\Delta)/\Delta]U(\Delta-|t|)$, as shown in Fig. 5.2, then we have $\delta'(t)$ as a limit of $[-\text{sgn}(t)/\Delta^2]U(\Delta-|t|)$ so that $|\delta'(t)|/\delta(t) \to \infty$ as $\Delta \to 0$. This same result holds true for any sequence that we might consider as tending to $\delta'(t)$. Thus, we must consider the magnitude of $\delta'(t)$ to be infinitely larger than that of $\delta(t)$, so we cannot satisfy Eq. 5.8 in the neighborhood of the origin if the jth derivative of $h_x(t)$ is like $\delta(t)$ and the $(j+1)$st derivative also appears in the equation. On the other hand, if we say that the nth derivative of $h_x(t)$ in the neighborhood of the origin is $b\,\delta(t)$, then all the other terms on the left-hand side of Eq. 5.8 are finite, so the equation can be considered satisfied at the origin if $b = a_n^{-1}$. Using

$$\frac{d^n h_x(t)}{dt^n} = a_n^{-1}\,\delta(t) \quad \text{for very small } |t| \tag{5.10}$$

now gives us the initial conditions for Eq. 5.9 at time $t = 0^+$. In particular, the integral of Eq. 5.10 gives

$$\frac{d^{n-1} h_x(t)}{dt^{n-1}} = a_n^{-1}\,U(t) \quad \text{for very small } |t|$$

in which $U(t)$ is the unit step function (see Eqs. 2.5 and 2.14). Thus,

$$\left(\frac{d^{n-1}h_x(t)}{dt^{n-1}}\right)_{t=0^+}=a_n^{-1} \tag{5.11}$$

Further integrations give continuous functions, so

$$\left(\frac{d^j h_x(t)}{dt^j}\right)_{t=0^+}=0 \quad \text{for } j\le n-2 \tag{5.12}$$

This now provides adequate conditions to ensure that we can find $h_x(t)$ as a unique solution of Eq. 5.9.

Inasmuch as determination of the impulse response function involves finding the solution to an initial condition problem, it is closely related to the general problem mentioned at the end of Section 5.1 regarding finding the response of a system with given initial conditions. In general, we expect initial conditions to be given as values of $x(t_0)$ and the derivatives $x^{(j)}(t_0)$ for $j=1,\cdots,n-1$. We have already found that $h_x(t)$ is exactly the solution to the problem with the initial conditions given in Eqs. 5.11 and 5.12. Thus, the response resulting from having the $(n-1)$st-order derivative be $x^{(n-1)}(t_0)$ at time t_0, instead of a_n^{-1} at time 0^+, is simply $[x^{(n-1)}(0^+)/a_n^{-1}]h_x(t-t_0)$. The responses to other initial conditions at time t_0 can also be found by first placing the initial value at time 0^+, then shifting the time axis of the response by the amount t_0.

We will now derive the impulse response functions for some example problems, each of which is described by a constant-coefficient differential equation. We will also list one physical system that is described by each of the differential equations. It should be kept in mind, though, that the impulse response function characterizes the dynamic behavior of any system governed by the given differential equation and not only the particular physical example that is given. Inasmuch as the class of systems governed by any given differential equation is much broader than any particular example that we might give, it is useful to consider the examples to be defined by the differential equations, not by the physical examples cited.

****

Example 5.1: Consider a linear system governed by the first-order differential equation

$$c\,\dot{x}(t)+k\,x(t)=f(t)$$

The accompanying sketch shows one physical system that is governed by this differential equation, with k being a spring stiffness, c a dashpot value, and $f(t)$ an applied force. Find the impulse response function $h_x(t)$ such that Eq. 5.2 describes the solution of the problem.

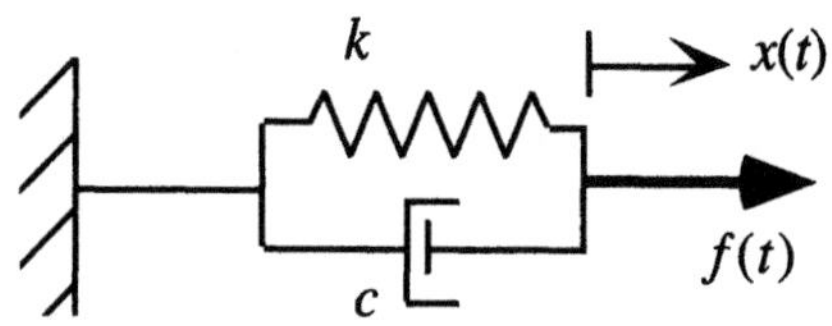

Rewriting Eqs. 5.9 and 5.11 gives

$$c\frac{dh_x(t)}{dt}+k\,h_x(t)=0 \qquad \text{for } t>0$$

with $h_x(0^+)=c^{-1}$. Taking the usual approach of trying an exponential form for the solution of a linear differential equation, we try $h_x(t)=a\,e^{bt}$. This has the proper initial condition if $a=c^{-1}$, and it satisfies the equation if $c\,b+k=0$, or $b=-k/c$, so the answer is

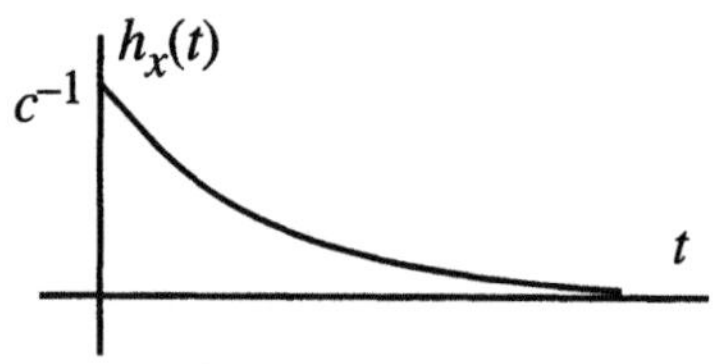

$$h_x(t)=c^{-1}e^{-kt/c}\,U(t)$$

Taking the derivative of this function gives

$$h'_x(t)=-k\,c^{-2}e^{-kt/c}\,U(t)+c^{-1}e^{-kt/c}\,\delta(t)=-k\,c^{-2}e^{-kt/c}\,U(t)+c^{-1}\,\delta(t)$$

in which the last term has been simplified by taking advantage of the fact that $e^{-kt/c}=1$ at the one value for which $\delta(t)$ is not zero. Substitution of $h_x(t)$ and $h'_x(t)$ confirms that we have, indeed, found the solution of the differential equation $c\,h'_x(t)+k\,h_x(t)=\delta(t)$.

Note also that

$$h_{x,static}\equiv\int_{-\infty}^{\infty}h_x(r)\,dr=k^{-1}<\infty$$

so the response for this system has a bounded static value. For the physical model shown, it is fairly obvious that the static response to force f_0 should be f_0/k, so it should be no surprise that $h_{x,static}=k^{-1}$

****

Example 5.2: Consider a linear system governed by the differential equation

$$m\,\ddot{x}(t)+c\,\dot{x}(t)=f(t)$$

The accompanying sketch shows one physical system that is governed by this differential equation, with c being a dashpot value and $f(t)$ the force applied to the mass m. Find the impulse function

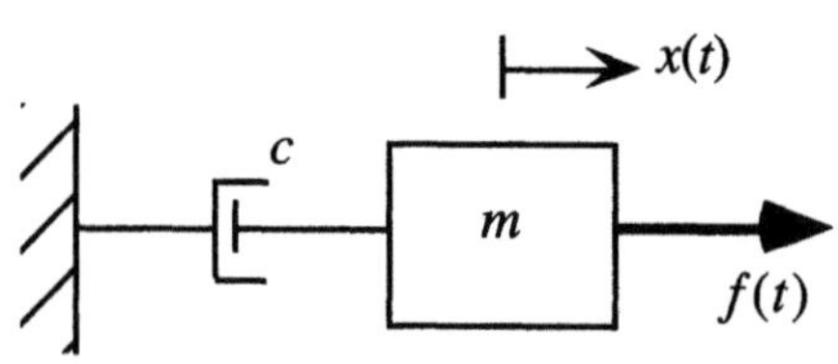

$h_x(t)$ such that Eq. 5.2 describes the solution of the problem.

This time we obtain

$$m\frac{d^2h_x(t)}{dt^2}+c\frac{dh_x(t)}{dt}=0 \quad \text{for } t>0$$

with $h'_x(0^+)=m^{-1}$ and $h_x(0^+)=0$. Proceeding as before, we find that the homogeneous equation is satisfied by $a\,e^{bt}$ with $mb^2+cb=0$. Thus, $b=0$ and $b=-c/m$ are both possible, and we must consider the general solution to be the linear combination of these two terms: $h_x(t)=a_1+a_2\,e^{-ct/m}$. Meeting the initial conditions at $t=0^+$ then gives $-a_2 c/m=m^{-1}$ and $a_1+a_2=0$, so $a_1=-a_2=c^{-1}$ and the impulse response function is

$$h_x(t)=c^{-1}(1-e^{-ct/m})U(t)$$

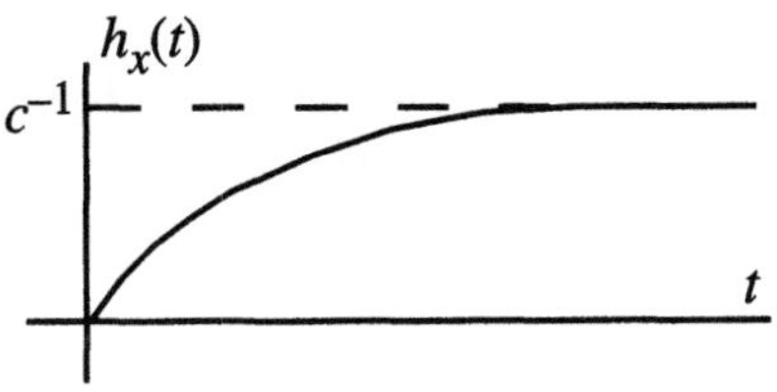

Note that $h_{x,static}=\infty$ for this system, so the static response of $x(t)$ is infinite, which agrees with what one would anticipate for the physical model shown.

One may also note a strong similarity between Examples 5.1 and 5.2. In particular, if we rewrite Example 5.2 in terms of $y(t)=\dot{x}(t)$, then the equation is

$$m\,\dot{y}(t)+c\,y(t)=f(t)$$

which is identical to the equation for $x(t)$ in Example 5.1 if c is replaced by m and k is replaced by c, showing that $h_y(t)=m^{-1}e^{-ct/m}\,U(t)$. However, the impulse response for $y(t)$ should be exactly the derivative of that for $x(t)$, so the solution to Example 5.1 gives us

$$h'_x(t)=h_y(t)=m^{-1}\,e^{-ct/m}\,U(t)$$

which does agree with the $h_x(t)$ that we derived directly for this problem. Also, this shows that the static response of $\dot{x}(t)$ for this system is finite, even though that of $x(t)$ is infinite.

****

Example 5.3: Consider a linear system governed by the differential equation

$$m\,\ddot{x}(t)+c\,\dot{x}(t)+k\,x(t)=f(t)$$

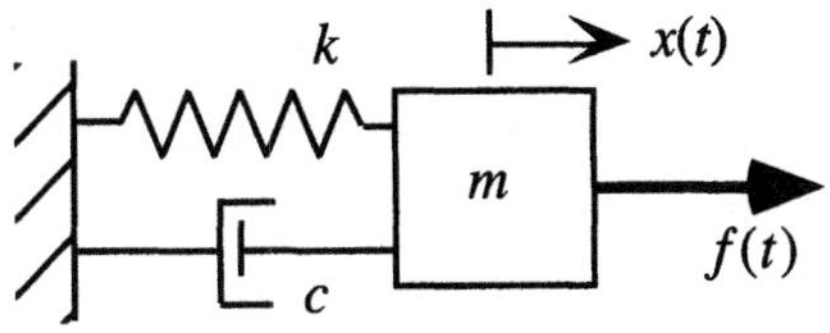

The accompanying sketch shows one physical system that is governed by this differential equation, with m being a mass, k being a spring stiffness, and c being a dashpot value. This system is called the single-degree-of-freedom system and will play a key role in our study of stochastic dynamics, but at this point it is simply one more example of a relatively simple dynamical system. Find the impulse function $h_x(t)$ such that Eq. 5.2 describes the solution of the problem.

Provided that $m \neq 0$, we can divide the equation by m to obtain another common form of the governing equation as

$$\ddot{x} + 2\zeta\,\omega_0\,\dot{x} + \omega_0^2\,x = f(t)/m$$

in which $\omega_0 \equiv (k/m)^{1/2}$ and $\zeta = c/[2(k\,m)^{1/2}]$ are called the undamped natural circular frequency and the fraction of critical damping, respectively, of the system. In the same way as we analyzed Examples 5.1 and 5.2, we find that the response $h_x(t)$ to $f(t) = \delta(t)$ must satisfy

$$\frac{d^2 h_x(t)}{dt^2} + 2\zeta\,\omega_0 \frac{dh_x(t)}{dt} + \omega_0^2\,h_x(t) = 0 \quad \text{for } t > 0$$

with $h_x'(0^+) = m^{-1}$ and $h_x(0^+) = 0$. We will present results only for the situation with $|\zeta| < 1$, because that is the situation that is usually of most practical interest.[1] For $|\zeta| < 1$ the general solution to the homogeneous equation can be written as $h_x(t) = e^{-\zeta\omega_0 t}[A\cos(\omega_d\,t) + B\sin(\omega_d\,t)]$, in which $\omega_d = \omega_0(1-\zeta^2)^{1/2}$ is called the damped natural circular frequency. Applying the initial conditions at time $t = 0^+$ then gives

$$h_x(t) = \frac{e^{-\zeta\,\omega_0\,t}}{m\,\omega_d} \sin(\omega_d\,t)\,U(t)$$

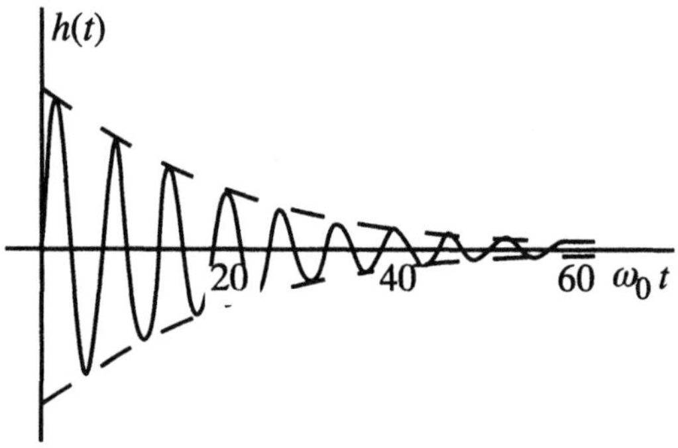

The shape of this function is shown in the accompanying sketch for the special case of $\zeta = 0.05$. The exponential envelope shown dashed in the figure simply omits the sinusoidal term of $h_x(t)$.

Based on the physical model shown, we can predict that the static response of this equation will be bounded with $h_{x,static} = k^{-1}$, as in Example 5.1, and integration of $h_x(t)$ confirms that this is true.

****

Note that the examples included here have been restricted to time-invariant systems. In general, it is more difficult to find the $h_{xf}(t,s)$ function applying to a time-varying system. If the linear system is governed by a differential equation with time-varying coefficients, then the principle for finding $h_{xf}(t,s)$ is the same as for the time-invariant situation. In particular, $h_{xf}(t,s)$ is the solution of the homogeneous equation with zero response prior to time s and initial conditions at time s resulting from $f(t) = \delta(t-s)$. The difficulty in practice, though, is in finding a general analytical homogeneous solution for a variable-coefficient

---

[1]The reader is asked to investigate situations with $|\zeta| > 1$ in Exercises 5.1 and 5.2.

equation. Handbooks of mathematical functions (e.g., Abramowitz and Stegun, 1965) do provide a number of special cases for which solutions are defined as special functions (e.g., Bessel functions, Legendre functions), but each such time-varying equation is a special case that requires careful study. Stated another way, the time-invariant situation, for which the solution is composed of exponential terms, is itself a special case, albeit a common one. The diversity of special cases with time-varying parameters makes it infeasible to study them here.

5.3 Stochastic Dynamics

The simplest formulation of stochastic linear dynamics is simply to replace the deterministic functions $f(t)$ and $x(t)$ in Eq. 5.2 by stochastic processes $\{F(t)\}$ and $\{X(t)\}$, giving

$$X(t) = \int_{-\infty}^{\infty} F(s)\, h_x(t-s)\, ds \equiv \int_{-\infty}^{\infty} F(t-r)\, h_x(r)\, dr \tag{5.13}$$

which can be considered as a special case of the stochastic convolution integral defined in Eq. 4.75. Note, in particular, that Eq. 5.13 must be true for every time history of excitation and response. We can now learn much about the characteristics of the $\{X(t)\}$ process by studying expectations obtained from Eq. 5.13. The first such result is that

$$\mu_X(t) \equiv E[X(t)] = \int_{-\infty}^{\infty} \mu_F(s)\, h_x(t-s)\, ds \equiv \int_{-\infty}^{\infty} \mu_F(t-r)\, h_x(r)\, dr \tag{5.14}$$

which follows directly from reversing the order of integration and expectation, as in Chapter 4. Note that Eq. 5.14 is of exactly the same form as Eq. 5.2, which describes deterministic dynamics. This means that finding the mean value of the stochastic response is always just the same as solving a deterministic problem with excitation $\mu_F(t)$.

In the same way we can write the autocorrelation function for the response as

$$\begin{aligned}\phi_{XX}(t_1,t_2) &\equiv E[X(t_1)\, X(t_2)] \\ &= E\left(\int_{-\infty}^{\infty}\int_{-\infty}^{\infty} F(s_1)\, h_x(t_1-s_1)\, F(s_2)\, h_x(t_2-s_2)\, ds_1\, ds_2\right)\end{aligned}$$

and taking the expectation inside the integrals gives

$$\phi_{XX}(t_1,t_2)=\int_{-\infty}^{\infty}\int_{-\infty}^{\infty}\phi_{FF}(s_1,s_2)\,h_x(t_1-s_1)\,h_x(t_2-s_2)\,ds_1\,ds_2 \tag{5.15}$$

or

$$\phi_{XX}(t_1,t_2)=\int_{-\infty}^{\infty}\int_{-\infty}^{\infty}\phi_{FF}(t_1-r_1,t_2-r_2)\,h_x(r_1)\,h_x(r_2)\,dr_1\,dr_2 \tag{5.16}$$

Clearly, this idea can be extended to any moment function. The general jth moment function of the response can be written as

$$E[X(t_1)\,X(t_2)\cdots X(t_j)]=\int_{-\infty}^{\infty}\cdots\int_{-\infty}^{\infty}E[F(s_1)\cdots F(s_j)]\,h_x(t_1-s_1)\cdots h_x(t_j-s_j)\,ds_1\cdots ds_j \tag{5.17}$$

Again, all of these results follow directly from using the Duhamel convolution integral and reversing the order of integration and expectation.

Similarly, one can consider more than one dynamic response. Thus, if some other response of interest is given by

$$Y(t)=\int_{-\infty}^{\infty}F(s)\,h_y(t-s)\,ds$$

then one can derive cross-product terms such as the cross-correlation function

$$\phi_{XY}(t_1,t_2)\equiv E[X(t_1)\,Y(t_2)]=\int_{-\infty}^{\infty}\int_{-\infty}^{\infty}\phi_{FF}(s_1,s_2)\,h_x(t_1-s_1)\,h_y(t_2-s_2)\,ds_1\,ds_2 \tag{5.18}$$

for the stochastic responses $\{X(t)\}$ and $\{Y(t)\}$ caused by a stochastic excitation $\{F(t)\}$.

Note that the presentation to this point has been very general, with no limitation on the type of response that the stochastic processes $\{X(t)\}$ and/or $\{Y(t)\}$ represent. Usually we will use $\{X(t)\}$ to represent a displacement, but the preceding mathematical expressions certainly do not require that. If $\{X(t)\}$ is a displacement, then $\{\dot X(t)\}$ and $\{\ddot X(t)\}$, respectively, will denote the velocity and acceleration of response, and we will often be interested in these quantities as well.

Recall our definition of $h_x(t)$ as the response $x(t)=h_x(t)$ when $f(t)=\delta(t)$. For $x(t)$ being response displacement, this gives us $\dot{x}(t)=h'_x(t)$ and $\ddot{x}(t)=h''_x(t)$ when $f(t)=\delta(t)$. Thus, the impulse response functions giving the velocity and acceleration responses, respectively, to the Dirac delta function pulse of excitation are $h_{\dot{x}}(t)=h'_x(t)$ (as in Eq. 5.6) and $h_{\ddot{x}}(t)=h''_x(t)$. These expressions can then be used to obtain moments of $\{\dot{X}(t)\}$ and $\{\ddot{X}(t)\}$ in forms that are exactly parallel to Eqs. 5.14–5.17. Two of the more important of these expressions are

$$\phi_{X\dot{X}}(t_1,t_2)=\frac{\partial}{\partial t_2}\phi_{XX}(t_1,t_2)=\int_{-\infty}^{\infty}\int_{-\infty}^{\infty}\phi_{FF}(s_1,s_2)\,h_x(t_1-s_1)\,h'_x(t_2-s_2)\,ds_1\,ds_2 \tag{5.19}$$

and

$$\phi_{\dot{X}\dot{X}}(t_1,t_2)=\frac{\partial^2\phi_{XX}(t_1,t_2)}{\partial t_1\partial t_2}=\int_{-\infty}^{\infty}\int_{-\infty}^{\infty}\phi_{FF}(s_1,s_2)\,h'_x(t_1-s_1)\,h'_x(t_2-s_2)\,ds_1\,ds_2 \tag{5.20}$$

In addition, one can obtain an integral expression for any desired higher-order cross-product term, such as

$$E[X(t_1)\dot{X}(t_2)\ddot{X}(t_3)]=\int_{-\infty}^{\infty}\int_{-\infty}^{\infty}\int_{-\infty}^{\infty}E[F(s_1)F(s_2)F(s_3)]\times$$
$$h_x(t_1-s_1)\,h'_x(t_2-s_2)\,h''_x(t_3-s_3)\,ds_1\,ds_2\,ds_3$$

Other forms of moment information, such as central moments or cumulants, can also be obtained from the moments given here. For example, using $K_{XX}(t_1,t_2)=\phi_{XX}(t_1,t_2)-\mu_X(t_1)\,\mu_X(t_2)$ gives the autocovariance function for $\{X(t)\}$ as

$$K_{XX}(t_1,t_2)=\int_{-\infty}^{\infty}\int_{-\infty}^{\infty}\phi_{FF}(s_1,s_2)\,h_x(t_1-s_1)\,h_x(t_2-s_2)\,ds_1\,ds_2-$$
$$\int_{-\infty}^{\infty}\mu_F(t_1)\,h_x(t_1-s_1)\,ds_1\int_{-\infty}^{\infty}\mu_F(t_2)\,h_x(t_2-s_2)\,ds_2$$

or

$$K_{XX}(t_1,t_2)=\int_{-\infty}^{\infty}\int_{-\infty}^{\infty}K_{FF}(s_1,s_2)\,h_x(t_1-s_1)\,h_x(t_2-s_2)\,ds_1\,ds_2 \tag{5.21}$$

One important aspect of the dynamics of linear systems is the "uncoupling" of the moments of response. That is, the jth moment function of the response depends only on the jth moment function of the excitation. Thus, one can compute the mean-value function of the stochastic response even if the only information given about the excitation is its mean value function. Similarly, knowledge of only the autocorrelation function for the excitation is sufficient to allow calculation of the autocorrelation function for the response, or calculation of the cross-product of two response measures such as $\{X(t)\}$ and $\{\dot{X}(t)\}$. Also, covariance functions for the response depend only on the autocovariance function for the excitation.

If we are given a complete set of initial conditions[2] at some time t_0, then we will find it convenient to modify the convolution integral of Eq. 5.13 to include only excitation after time t_0 and to add terms representing the response to the initial conditions. Because these initial conditions may be random, we will write the results as

$$X(t) = \sum_j Y_j\, g_j(t-t_0) + \int_{t_0}^{\infty} F(s)\, h_x(t-s)\, ds$$

in which the Y_j random variable denotes the jth initial condition value and the $g_j(t-t_0)$ term represents the response due to a unit value of that one initial condition at time $t=t_0$. The expressions for the mean-value and autocovariance functions of the response can then be written as

$$\mu_X(t) = \sum_j \mu_{Y_j}\, g_j(t-t_0) + \int_{t_0}^{\infty} \mu_F(s)\, h_x(t-s)\, ds \tag{5.22}$$

and

$$\begin{aligned} K_{XX}(t_1,t_2) = &\sum_{j_1}\sum_{j_2} K_{Y_{j_1}Y_{j_2}}\, g_{j_1}(t_1-t_0)\, g_{j_2}(t_2-t_0) + \\ &\sum_j \int_{t_0}^{\infty} K_{Y_j F(s)}\, g_j(t_1-t_0)[h_x(t_1-s)+h_x(t_2-s)]\, ds + \\ &\int_{t_0}^{\infty}\int_{t_0}^{\infty} K_{FF}(s_1,s_2) h_x(t_1-s_1)\, h_x(t_2-s_2)\, ds_1\, ds_2 \end{aligned} \tag{5.23}$$

A similar expression can be written out for the response autocorrelation function. In many situations, of course, the excitation after time t_0 may be independent of the initial conditions at time t_0. If this is true, then the $K_{XX}(t_1,t_2)$ result is simplified because the $K_{Y_j F(s)}$ cross-covariance in the second expression is

---

[2]By "a complete set of initial conditions" we mean that enough conditions are given at time t_0 to ensure that the solution for $t > t_0$ is unique.

zero, eliminating this term. Another simplified situation that may arise is when there is no stochastic excitation, in which case the dynamic response is due only to the random initial conditions. Finally, in the special case in which the initial conditions are deterministic, all the covariance terms involving the initial conditions are zero, but the mean value and autocorrelation functions for the response are still affected by the initial conditions.

One can also use a conditional probability distribution for the $\{X(t)\}$ process in Eq. 5.13 to obtain conditional versions of the mean value, autocovariance function, and autocorrelation function. This is possible for any type of conditioning event, but we will be particularly interested in the situation in which the given event, which will be denoted by A, includes a complete set of deterministic initial conditions at time t_0. We can then write the conditional mean and covariance in the same form as Eqs. 5.22 and 5.23 with the random initial condition Y_j replaced by a deterministic y_j whose value is known when A is known. The conditional mean value of the response then depends on the initial conditions and the conditional mean of the excitation as

$$E[X(t)\,|\,A]=\sum_j y_j\, g_j(t-t_0)+\int_{t_0}^{\infty} E[F(s)\,|\,A]\, h_x(t-s)\, ds \tag{5.24}$$

The conditional covariance of the response is

$$\text{Cov}[X(t_1),X(t_2)\,|\,A]=\int_{t_0}^{\infty}\int_{t_0}^{\infty}\text{Cov}[F(s_1),F(s_2)\,|\,A]\, h_x(t_1-s_1)\, h_x(t_2-s_2)\, ds_1\, ds_2 \tag{5.25}$$

which depends only on the conditional covariance of the excitation.

5.4 Response to Stationary Excitation

Consider now the special case in which the excitation process $\{F(t)\}$ has existed since time $t=-\infty$ and is stationary. First, if $\{F(t)\}$ is mean-value stationary, then $\mu_F(t-r)=\mu_F$ is a constant in Eq. 5.14, so

$$\mu_X(t)\equiv E[X(t)]=\mu_F\int_{-\infty}^{\infty} h_x(r)\, dr=\mu_F\, h_{x,static} \tag{5.26}$$

using $h_{x,static}$ defined in Eq. 5.4. This shows that $\mu_X(t)$ is not a function of t in this case. There are, however, two distinct possibilities. Either $h_{x,static}$ is infinite, so $\mu_X(t)$ does not exist, or $\mu_X(t)=\mu_X$ is a constant. In particular, if a system has a finite static response and the excitation is mean-value stationary, then the stochastic response is mean-value stationary. If the system has an infinite static response and $\mu_F \ne 0$, then the stochastic response has an infinite mean value.

Similarly, if the excitation is second-moment stationary then $\phi_{FF}(t_1 - r_1, t_2 - r_2) = R_{FF}(t_1 - r_1 - t_2 + r_2)$ in Eq. 5.16 and this gives

$$\phi_{XX}(t_1,t_2) = \int_{-\infty}^{\infty}\int_{-\infty}^{\infty} R_{FF}(t_1 - t_2 - r_1 + r_2)\, h_x(r_1)\, h_x(r_2)\, dr_1\, dr_2$$

If this second moment of the response $\phi_{XX}(t_1,t_2)$ is finite, then it is a function only of the difference between the two time arguments $(t_1 - t_2)$, showing that $\{X(t)\}$ is second-moment stationary. Thus, we can write $\tau = t_1 - t_2$ to get the stationary autocorrelation function as

$$R_{XX}(\tau) = \int_{-\infty}^{\infty}\int_{-\infty}^{\infty} R_{FF}(\tau - r_1 + r_2)\, h_x(r_1)\, h_x(r_2)\, dr_1\, dr_2 \qquad (5.27)$$

Second-moment stationarity of $\{F(t)\}$ is also sufficient to ensure that the cross-correlation of two responses $\{X(t)\}$ and $\{Y(t)\}$ has the stationary form of

$$R_{XY}(\tau) = \int_{-\infty}^{\infty}\int_{-\infty}^{\infty} R_{FF}(\tau - r_1 + r_2)\, h_x(r_1)\, h_y(r_2)\, dr_1\, dr_2 \qquad (5.28)$$

including as a special case

$$R_{X\dot{X}}(\tau) = \int_{-\infty}^{\infty}\int_{-\infty}^{\infty} R_{FF}(\tau - r_1 + r_2)\, h_x(r_1)\, h_x'(r_2)\, dr_1\, dr_2 \qquad (5.29)$$

Also, Eqs. 5.27–5.29 are equally valid if the stationary R functions are all replaced by stationary autocovariance and cross-covariance G functions.

Let us now consider the issue of whether $R_{XX}(\tau)$ and $G_{XX}(\tau)$ will be bounded for a system with a stationary excitation. From the Schwarz inequality we know that $R_{XX}(\tau)$ is bounded by the mean-squared value of X, and $G_{XX}(\tau)$ is bounded by the variance of X: $|R_{XX}(\tau)| \le E(X^2)$ and $|G_{XX}(\tau)| \le \sigma_X^2$. Thus, $R_{XX}(\tau)$ and $G_{XX}(\tau)$ are bounded for all τ if and only if $E(X^2)$ and σ_X^2, respectively, are bounded. As already noted, μ_X will be infinite if $\mu_F \ne 0$ and the system is such that it does not have a finite static response (i.e., if $|h_{x,static}| = \infty$). The fact that $E(X^2) = \mu_X^2 + \sigma_X^2$ tells us, then, that $E(X^2)$ is also infinite in that situation, even if the variance is finite.

It is possible for the response variance to be infinite even if $h_x(t)$ or μ_F is such that μ_X is finite, and it is also possible for the variance to be finite when the mean is infinite. In order to investigate the boundedness of the variance, we will rewrite Eq. 5.27 for the stationary covariance of the response and set $\tau = 0$ to obtain the response variance as

$$\sigma_X^2 = \int_{-\infty}^{\infty}\int_{-\infty}^{\infty} G_{FF}(r_2 - r_1)\, h_x(r_1)\, h_x(r_2)\, dr_1\, dr_2 = \int_{-\infty}^{\infty} G_{FF}(r_3)\, h_{x,var}(r_3)\, dr_3 \tag{5.30}$$

in which

$$h_{x,var}(\tau) \equiv \int_{-\infty}^{\infty} h_x(s-\tau)\, h_x(s)\, ds \tag{5.31}$$

is a characteristic of the linear system, and we have used the change of variables $r_3 = r_2 - r_1$. The absolute value of $h_{x,var}(\tau)$ can be bounded as

$$h_{x,var}(\tau) \le h_{x,var}(0) = \int_{-\infty}^{\infty} h_x^2(s)\, ds$$

Thus, we can bound the variance of the response as

$$\sigma_X^2 \le h_{x,var}(0) \int_{-\infty}^{\infty} \left| G_{FF}(r_3) \right| dr_3 \tag{5.32}$$

Note, also, that $h_{x,var}(0)$ will be finite if $h_x(t)$ is bounded and the system has a finite static response ($h_{x,static} < \infty$). In particular, $h_{x,var}(0) \le h_{x,max}\, h_{x,static}$ with $h_{x,max} \equiv \max |h_x(t)|$. Thus, we see that the system must have a finite variance of stationary response if it has a finite static response and the excitation is such that its autocorrelation function is absolute value integrable. It should be noted that, although these conditions are sufficient to ensure a finite stationary response covariance, they do not preclude the possibility of covariant stationary response under different conditions.

****

Example 5.4: Investigate the boundedness of the first and second moments of stationary stochastic response of the system $m\,\ddot{X}(t) + c\,\dot{X}(t) = F(t)$, for which the impulse response function was derived in Example 5.2 as

$$h_x(t) = c^{-1}(1 - e^{-ct/m})\, U(t)$$

Because the static response value for this system is infinite ($h_{x,static} = \infty$), we know that the mean value and autocorrelation function of the $\{X(t)\}$ response will generally not exist for an $\{F(t)\}$ process that is mean-value and second-moment stationary. In particular, $\mu_X = \mu_F\, h_{x,static}$ is finite only if $\mu_F = 0$, in which case $\mu_X = 0$ also. Because μ_X is infinite if $\mu_F \neq 0$, we also know that $\phi_{XX}(t_1,t_2)$ is infinite for this situation. In addition, though, $\phi_{XX}(t_1,t_2)$ may be infinite even if $\mu_F = 0$. In particular, the response variance may be infinite because $h_{x,var}(0)$ is infinite. This does not prove that $\sigma_X^2 = \infty$, and we will later see (Example 6.4) that σ_X^2 is infinite for some excitations and finite for others. Recall that the derivative response for this problem does have a finite static value, because its impulse response function of $h_{\dot{x}}(t) = h'_x(t) = m^{-1} e^{-ct/m}\, U(t)$ gives $h_{\dot{x},static} = c^{-1}$. Thus, the mean value and autocorrelation of stationary $\{\dot{X}(t)\}$ response generally do exist for this system, even though those for $\{X(t)\}$ may not.

****

Example 5.5: Investigate the behavior of the first and second moments of stochastic response of the system $c\,\dot{X}(t) + k\,X(t) = 0$, for which the impulse response function was derived in Example 5.1 as $h_x(t) = c^{-1} e^{-kt/c}\, U(t)$. In particular, investigate $\mu_X(t)$, $K_{XX}(t_1,t_2)$, and $\phi_{XX}(t_1,t_2)$ for this situation with no excitation process, but with a random variable initial condition of $X(t_0) = Y$ with μ_Y and σ_Y known.

We can use Eqs. 5.22 and 5.23 if we know the value of the $g(t)$ function. Note that one needs only one initial condition for this first-order system, so the given value of $X(t_0)$ is sufficient to describe a unique solution. Furthermore, Example 5.1 shows that the impulse response function $h_x(t) = c^{-1} e^{-kt/c}\, U(t)$ is the response of the system to an initial condition of $X(0) = c^{-1}$, so we have $g(t) = c\, h_x(t - t_0) = e^{-k(t-t_0)/c}\, U(t - t_0)$. Thus, for $t > t_0$ we have

$$\mu_X(t) = \mu_Y\, g(t - t_0) = \mu_Y\, e^{-k(t-t_0)/c}$$

and

$$K_{XX}(t_1,t_2) = \sigma_Y^2\, g(t_1 - t_0)\, g(t_2 - t_0) = \sigma_Y^2\, e^{-k(t_1+t_2-2t_0)/c}$$

Putting these two together gives the autocorrelation function of the response as

$$\phi_{XX}(t_1,t_2) = E(Y^2)\, g(t_1 - t_0)\, g(t_2 - t_0) = E(Y^2)\, e^{-k(t_1+t_2-2t_0)/c}$$

By letting $t_1 = t_2 = t$, one can also find the response variance as $\sigma_X^2(t) = \sigma_Y^2\, e^{-2k(t-t_0)/c}$, and the mean-squared value is given by a similar expression.

****

All the results and examples in Sections 5.3 and 5.4 to this point have been for time-invariant systems. Most of the general results, such as Eqs. 5.14, 5.15, 5.17, 5.18, and 5.21, can be converted for the time-varying situation by replacing $h_x(t - s)$ with $h_{xf}(t,s)$. The situation treated in the current section, with $\{F(t)\}$ being a stationary process, is particularly significant in practice. The response of a time-varying system to a stationary excitation is a particular type of nonstationary process. Such processes are called "modulated" or "evolutionary" processes, and their properties have been studied in some detail. One particular way in which modulated processes are used is as models of nonstationary excitation processes. For example, one might model an $\{F(t)\}$ process as the modulated process with

$$F(t) = \int_{-\infty}^{\infty} F_S(s)\, h_{fS}(t,s)\, ds \tag{5.33}$$

in which $\{F_S(t)\}$ is a stationary process. If $\{X(t)\}$ represents the response of a linear system excited by $\{F(t)\}$, then we can write

$$X(t) = \int_{-\infty}^{\infty} F(u)\, h_{xf}(t,u)\, du = \int_{-\infty}^{\infty} F_S(s)\, h_{xS}(t,s)\, ds$$

in which

$$h_{xS}(t,s) = \int_{-\infty}^{\infty} h_{xf}(t,u)\, h_{fS}(u,s)\, du \tag{5.34}$$

so $\{X(t)\}$ is also a modulated process for the excitation $\{F_S(t)\}$ with a combined impulse response function $h_{xS}(t,s)$ that is a convolution of h_{FS} and h_{xf}. A very special case of the modulated process is that which arises when $h_{FS}(t,s)$ has the form $h_{fS}(t,s) = \hat{h}(t)\,\delta(t-s)$, so $F(t) = \hat{h}(t)\, F_S(t)$. In this special case, $\{F(t)\}$ is called a uniformly modulated process. The combined impulse response function is

$$h_{xS}(t,s) = \int_{-\infty}^{\infty} h_{xf}(t,u)\, \hat{h}(u)\, \delta(u-s)\, du = h_{xf}(t,s)\, \hat{h}(s)$$

and the response $\{X(t)\}$ to the excitation $\{F(t)\}$ then has

$$X(t) = \int_{-\infty}^{\infty} F_S(s)\, \hat{h}(s)\, h_{xf}(t,s)\, ds$$

so it is the same as for an excitation $\{F_S(t)\, \hat{h}(t)\}$.

Consider now a response problem in which the excitation is modeled as $F(t) = W(t)U(t)$, in which $\{W(t)\}$ is a stationary process and $U(t)$ is the unit step function. Clearly this may be viewed as a modulated excitation, with $\hat{h}(t) = U(t)$. On the other hand, it may also be viewed as an expression of the conditional properties of the response for $t > 0$ given that the system is at rest at time $t = 0$. Thus, no special techniques are necessary to analyze the response of a time-invariant system to such a uniformly modulated excitation, and the variance of the response can be written from Eq. 5.23 as

$$\sigma_X^2(t) = \int_0^\infty \int_0^\infty G_{FF}(s_1 - s_2)\, h_x(t - s_1)\, h_x(t - s_1)\, ds_1\, ds_2$$

or

$$\sigma_X^2(t) = \int_{-\infty}^{t} \int_{-\infty}^{t} G_{FF}(r_2 - r_1)\, h_x(r_1)\, h_x(r_2)\, dr_1\, dr_2$$

which is identical to the first form in Eq. 5.30 except for the time-dependent limits of integration. Using the change of variables of $r_3 = r_2 - r_1$ allows this to rewritten as

$$\sigma_X^2(t) = 2\int_0^\infty G_{FF}(r_3)\, h_{x,var}(r_3,t)\, dr_3 \tag{5.35}$$

in which

$$h_{x,var}(\tau,t) = \int_{-\infty}^{t} h_x(s)\, h_x(s - \tau)\, ds \tag{5.36}$$

Clearly the final form of Eq. 5.30 and Eq. 5.31 are the special cases of Eqs. 5.35 and 5.36 for $t \to \infty$.

Note that using $\hat{h}(t) = U(t)$ gives a particularly simple uniformly modulated process. Analysis of more complicated modulated processes is commonly done in the frequency domain, so this issue will be reconsidered in Chapter 6. The meanings of the terms "modulated" and "uniformly modulated" will also be clearer in the frequency domain.

5.5 Delta-Correlated Excitations

There are a number of important physical problems in which the excitation process $\{F(t)\}$ is so erratic that $F(t)$ and $F(s)$ are almost independent unless t and s are almost equal. For a physical process there generally is some dependence between $F(t)$ and $F(s)$ for t and s sufficiently close, but this dependence may decay very rapidly as the separation between t and s grows. To illustrate this idea, let T_c denote the time difference $t - s$ over which $F(t)$ and $F(s)$ are significantly dependent, making $F(t)$ and $F(s)$ essentially independent if $|t - s| > T_c$. Then if T_c is sufficiently small compared with other characteristic time values for the problem being considered, it may be possible to approximate $\{F(t)\}$ by the limiting process for which $F(t)$ and $F(s)$ are independent for $t \neq s$. This limiting process is called a *delta-correlated process*. The motivation for using a delta-correlated process is strictly convenience. Computing response statistics for a delta-correlated excitation is often much easier than for a nearly delta-correlated one.

We now wish to investigate the details of how a delta-correlated $\{F(t)\}$ excitation process must be defined in order that it will cause a response $\{X(t)\}$ that approximates that for a physical problem with a nearly delta-correlated excitation. We will do this by focusing on the covariance functions of $\{F(t)\}$ and $\{X(t)\}$, starting with the covariant stationary situation. Independence of $F(t)$ and $F(s)$, of course, implies that the covariance $K_{FF}(t,s)$ is zero, as well as corresponding relationships involving higher moment functions of the process.[3] Thus, a covariant stationary delta-correlated $\{F(t)\}$ would have $G_{FF}(\tau)=0$ for $\tau \neq 0$. Recall that Eq. 5.32 gave a bound on the response variance as $h_{x,var}(0)$ multiplied by the integral of the absolute value of $G_{FF}(\tau)$, with $h_{x,var}(0)$ being a characteristic of the linear system. The integral of the absolute value of $G_{FF}(\tau)$, though, will be zero if we choose $G_{FF}(\tau)$ for our delta-correlated process to have a finite value at $\tau = 0$ and to be zero for $\tau \neq 0$. Thus, our delta-correlated excitation process will not give any response unless we allow $G_{FF}(\tau)$ to be infinite for $\tau = 0$. This leads us to the standard form for the covariance function of a stationary delta-correlated process

$$G_{FF}(\tau) = G_0\, \delta(\tau) \tag{5.37}$$

which is a special case of the covariance function for a nonstationary delta-correlated process

$$K_{FF}(t,s) = G_0(t)\, \delta(t-s) \tag{5.38}$$

One should note that Eqs. 5.37 and 5.38 ensure only that $F(t)$ and $F(s)$ are uncorrelated for $t \neq s$, whereas we defined the delta-correlated property to mean that $F(t)$ and $F(s)$ are independent for $t \neq s$. That is, the lack of correlation given in Eqs. 5.37 and 5.38 is necessary for a delta-correlated process, but it is not the definition of the process.[4]

---

[3]Precisely, one can say that $F(t)$ and $F(s)$ are independent if and only if all the cross-cumulant functions are zero, whereas covariance is only the second cross-cumulant function. Some discussion of cumulants is given in Section 3.7.

[4]The term *white noise* is commonly used to refer to processes of the general type of our delta-correlated process. Sometimes, though, *white noise* implies only Eq. 5.38 rather than the more general independence property of a delta-correlated process.

The convenience of using a delta-correlated excitation in studies of dynamic response can be illustrated by substituting Eq. 5.37 for the stationary covariance function of the excitation into Eq. 5.27, giving the response covariance as

$$G_{XX}(\tau) = \int_{-\infty}^{\infty}\int_{-\infty}^{\infty} G_0\, \delta(\tau - r_1 + r_2)\, h_x(r_1)\, h_x(r_2)\, dr_1\, dr_2$$

The presence of the Dirac delta function now allows easy evaluation of one of the integrals, with the result that the response covariance $G_{XX}(\tau)$ is given by a single (rather than a double) integral

$$G_{XX}(\tau) = G_0 \int_{-\infty}^{\infty} h_x(\tau + r)\, h_x(r)\, dr \tag{5.39}$$

A special case of this formula is the variance of the response given by

$$\sigma_X^2 = G_{XX}(0) = G_0 \int_{-\infty}^{\infty} h_x^2(r)\, dr \equiv G_0\, h_{x,var}(0) \tag{5.40}$$

This shows that when the excitation is delta-correlated, the variance actually equals the bound given in Eq. 5.32. This also shows that the response to delta-correlated excitation is always unbounded if the $h_{x,var}(0)$ integral of the square of the impulse response function is infinite.

In the same way, we can substitute Eq. 5.38 into Eq. 5.21 to obtain the covariance of response to a nonstationary delta-correlated excitation as

$$K_{XX}(t_1,t_2) = \int_{-\infty}^{\infty}\int_{-\infty}^{\infty} G_0(s_1)\, \delta(s_1 - s_2)\, h_x(t_1 - s_1)\, h_x(t_2 - s_2)\, ds_1\, ds_2$$

which simplifies to

$$K_{XX}(t_1,t_2) = \int_{-\infty}^{\infty} G_0(s_1)\, h_x(t_1 - s_1)\, h_x(t_2 - s_1)\, ds_1 \tag{5.41}$$

The variance in this situation is given by

$$\sigma_X^2(t) = K_{XX}(t,t) = \int_{-\infty}^{\infty} G_0(s)\, h_x^2(t - s)\, ds = \int_{-\infty}^{\infty} G_0(t - r)\, h_x^2(r)\, dr \tag{5.42}$$

which is seen to be exactly in the form of a convolution of the function giving the intensity of the excitation and the square of the impulse response function for the dynamic system.

Delta-correlated excitations have another convenient feature when one considers the conditional distribution of the response, as in Eqs. 5.24 and 5.25. For a causal system, one can say that the response at any time t_0 is a function of the excitation $F(t)$ only for $t \le t_0$. For a delta-correlated excitation we can say that $F(t)$ for $t > t_0$ is independent of $F(t)$ for $t \le t_0$, and this implies that $F(t)$ for $t > t_0$ is independent of the response of the system at time t_0. Thus, if the conditioning event A involves only the value of the excitation and the response at time t_0, then we can say that $F(t)$ for $t > t_0$ is independent of A. This makes Eqs. 5.24 and 5.25 depend on the unconditional mean and covariance of the excitation, and gives

$$E[X(t)\,|\,A] = \sum_j y_j\, g_j(t-t_0) + \int_{t_0}^{\infty} \mu_F(s)\, h_x(t-s)\, ds \tag{5.43}$$

and

$$\text{Cov}[X(t_1),X(t_2)\,|\,A] = \int_{t_0}^{\infty} G_0(s)\, h_x(t_1-s)\, h_x(t_2-s)\, ds \tag{5.44}$$

Note that the final equation is of exactly the same form as the unconditional covariance in Eq. 5.41, except that it has a different limit of integration.

A disadvantage of using a delta-correlated excitation should also be noted. In particular, modeling the excitation as having an infinite variance always results in some response quantity also having an infinite variance. For example, the general nth order system of Eq. 5.7 will have an infinite variance for the nth order derivative when the excitation is a delta-correlated $\{F(t)\}$. This follows from Eq. 5.40 and the fact that the impulse response function for $d^n x(t)/(d\,t^n)$ behaves like $\delta(t)$ for t near the origin.

****

Example 5.6: Show that one particular delta-correlated process is the stationary "shot noise" defined by

$$F(t) = \sum F_j\, \delta(t - T_j)$$

in which $\{T_1, T_2, \cdots, T_j, \cdots\}$ with $0 \le T_1 \le T_2 \le \cdots \le T_j$ are the random arrival times for a Poisson process $\{Z(t)\}$ (see Example 4.13), and $\{F_1, F_2, \cdots, F_j, \cdots\}$ is a sequence of identically distributed random variables that are independent of each other and of the arrival times. Also show that the process is not Gaussian and evaluate the cumulant functions for $\{F(t)\}$.

The nature of the Poisson process is such that knowledge of past arrival times gives no information about future arrival times. Specifically, the interarrival time $T_{j+1} - T_j$ is independent of $\{T_1, \cdots, T_j\}$. Combined with the independence of the F_j pulse magnitudes, this ensures that $F(t)$ must be independent of $F(s)$ for $t \ne s$. For $s < t$, knowledge of $F(s)$ may give some information about past arrival times and pulse magnitudes, but it gives no information about the likelihood that a pulse will arrive at future time t or about the likely magnitude of such a pulse if it does arrive at time t. Thus, $\{F(t)\}$ must be delta-correlated.
In order to investigate the cumulants, we first define a process $\{Q(t)\}$ that is the integral of $\{F(t)\}$:

$$Q(t) = \int_0^t F(s)\, ds = \sum F_j\, U(t - T_j)$$

then we will use these results in describing $\{F(t)\} = \{\dot{Q}(t)\}$. The cumulant functions are an ideal tool for determining whether a stochastic process is Gaussian, because the cumulants of order higher than two are identically zero for a Gaussian process. Thus, we will proceed directly to seek the cumulant functions of $\{Q(t)\}$ and $\{F(t)\}$. It turns out that this is fairly easily done by consideration of the log-characteristic function of $\{Q(t)\}$, as in Section 3.7.
The nth-order joint characteristic function of $\{Q(t)\}$ for times $\{t_1, \cdots, t_n\}$ can be written as

$$M_{Q(t_1)\cdots Q(t_n)}(\theta_1, \cdots, \theta_n) \equiv E\left(\exp\left[i \sum_{j=1}^{n} \theta_j\, Q(t_j) \right] \right)$$

We can take advantage of the delta-correlated properties of shot noise by rewriting $Q(t_j)$ as

$$Q(t_j) = \sum_{l=1}^{j} \Delta Q_l, \qquad \text{with} \qquad \Delta Q_l \equiv Q(t_l) - Q(t_{l-1})$$

in which $t_0 < t_1 < t_2 < \cdots < t_n$ and $t_0 = 0$, so $Q(t_0) = 0$. Now we can rewrite the characteristic function as

$$M_{Q(t_1)\cdots Q(t_n)}(\theta_1, \cdots, \theta_n) = E\left(\exp\left[i \sum_{j=1}^{n} \sum_{l=1}^{j} \theta_j\, \Delta Q_l \right] \right) = E\left(\exp\left[i \sum_{l=1}^{n} \sum_{j=l}^{n} \theta_j\, \Delta Q_l \right] \right)$$

The fact that $F(t)$ and $F(s)$ are independent random variables for $t \ne s$ allows us to say that $\{\Delta Q_1, \Delta Q_2, \cdots, \Delta Q_n\}$ is a sequence of independent random variables. This allows us to write the characteristic function as

$$M_{Q(t_1)\cdots Q(t_n)}(\theta_1,\cdots,\theta_n)=\prod_{l=1}^{n}E\left(\exp\left[i\,\Delta Q_l\sum_{j=l}^{n}\theta_j\right]\right)=\prod_{l=1}^{n}M_{\Delta Q_l}(\theta_l+\cdots+\theta_n)$$

Thus, we can find the desired joint characteristic function from the characteristic function of ΔQ_l.

For simplicity, we will derive the characteristic function for the increment from time zero to time t_1, that is, for the random variable $\Delta Q_1 \equiv Q(t_1)$. This can most easily be evaluated in two steps. First, we evaluate the conditional expected value

$$E\big(\exp[i\theta Q(t_1)]\,|\,Z(t)=r\big)=E\left(\exp\left[i\theta\sum_{j=1}^{r}F_j\right]\right)=E\left(e^{i\theta F_1}\cdots e^{i\theta F_r}\right)=M_F^r(\theta)$$

in which the final form follows from the fact that $\{F_1,F_2,\cdots,F_j,\cdots\}$ is a sequence of independent, identically distributed random variables, and $M_F(\theta)$ is used to designate the characteristic function of any member of the sequence. Using the probability values for the Poisson process from Example 4.13 then gives the characteristic function of interest as

$$M_{\Delta Q_1}(\theta)=\sum_{r=1}^{\infty}E\big(\exp[i\theta Q(t_1)]\,|\,Z(t_1)=r\big)P[Z(t_1)=r]=\sum_{r=1}^{\infty}M_F^r(\theta)\frac{\mu_Z^r(t_1)}{r!}e^{-\mu_Z(t_1)}$$

The result can be simplified as

$$M_{\Delta Q_1}(\theta)=e^{-\mu_Z(t_1)}\sum_{r=1}^{\infty}\frac{1}{r!}[M_F(\theta)\,\mu_Z(t_1)]^r=\exp[-\mu_Z(t_1)+M_F(\theta)\,\mu_Z(t_1)]$$

because the summation is exactly the power series expansion for the exponential function. The log-characteristic function of ΔQ_1 is then

$$\log[M_{\Delta Q_1}(\theta)]\equiv\log[M_{Q(t_1)}(\theta)]=\mu_Z(t_1)[M_F(\theta)-1]$$

The corresponding function for any arbitrary increment ΔQ_l is obtained by replacing $\mu_Z(t_1)$ with $\mu_Z(t_l)-\mu_Z(t_{l-1})$.

The log-characteristic function for $\Delta Q_1 \equiv Q(t_1)$ provides sufficient evidence to prove that $\{F(t)\}$ is not Gaussian. In particular, we can find the cumulants of $\{Q(t)\}$ by taking derivatives of $\log[M_{Q(t)}(\theta)]$ and evaluating them at $\theta=0$. Inasmuch as

$$\frac{d^j}{d\theta^j}\log[M_{Q(t)}(\theta)]=\mu_Z(t)\frac{d^j}{d\theta^j}M_F(\theta)$$

though, this relates the cumulants of $Q(t)$ to the moments of F, because the derivatives of a characteristic function evaluated at $\theta=0$ always give moment values (see Section 3.7). It is not possible for the higher-order even-order moment values of F, such as $E(F^4)$, to be zero, so we can see that the higher-order even cumulants of $Q(t)$ are not zero. This proves that $Q(t)$ is not a

Gaussian random variable, which then implies that $\{Q(t)\}$ cannot be a Gaussian stochastic process. If $\{Q(t)\}$ is not a Gaussian process, though, then its derivative, $\{F(t)\}$, is also not a Gaussian process.

Using the preceding results, we can evaluate the joint log-characteristic function for $\{Q(t)\}$ as

$$\log[M_{Q(t_1)\cdots Q(t_n)}(\theta_1,\cdots,\theta_n)] = \sum_{l=1}^{n}[\mu_Z(t_l)-\mu_Z(t_{l-1})][M_F(\theta_l+\cdots+\theta_n)-1]$$

The joint cumulant function for $\{Q(t_1),\cdots,Q(t_n)\}$ is then

$$\kappa_n[Q(t_1),\cdots,Q(t_n)] = i^{-n}\left(\frac{\partial^n}{\partial\theta_1\cdots\partial\theta_n}\log[M_{Q(t_1)\cdots Q(t_n)}(\theta_1,\cdots,\theta_n)]\right)_{\theta_1=\cdots=\theta_n=0}$$

However, there is only one term in our expression for the log-characteristic function that gives a nonzero value for the mixed partial derivative. In particular, only the first term , with $l=1$, contains all the arguments $(\theta_1,\cdots,\theta_n)$. Thus, only it contributes to the cumulant function, and we have

$$\kappa_n[Q(t_1),\cdots,Q(t_n)] = i^{-n}\mu_Z(t_1)\left(\frac{\partial^n}{\partial\theta_1\cdots\partial\theta_n}M_F(\theta_1+\cdots+\theta_n)\right)_{\theta_1=\cdots=\theta_n=0}$$

$$= \mu_Z(t_1)\,E[F^n]$$

Recall that the only way in which t_1 was distinctive in the set $(t_1,\cdots,t_n)$ was that it was the minimum of the set. Thus, the general result for $\{Q(t)\}$ is

$$\kappa_n[Q(t_1),\cdots,Q(t_n)] = \mu_Z(\min[t_1,\cdots,t_n])\,E[F^n]$$

The linearity property of cumulants (see Eq. 3.44) then allows us to write the corresponding cumulant function for the $\{F(t)\}$ process as

$$\kappa_n[F(t_1),\cdots,F(t_n)] = E[F^n]\frac{\partial^n}{\partial t_1\cdots\partial t_n}\mu_Z(\min[t_1,\cdots,t_n])$$

$$= E[F^n]\frac{\partial^n}{\partial t_1\cdots\partial t_n}\sum_{j=1}^{n}\mu_Z(t_j)\prod_{\substack{l=1\\ l\neq j}}^{n}U(t_l-t_j)$$

After performing the differentiation and eliminating a number of terms that cancel, this expression can be simplified to

$$\kappa_n[F(t_1),\cdots,F(t_n)] = E[F^n]\,\dot{\mu}_Z(t_n)\prod_{l=1}^{n-1}\delta(t_l-t_n)$$

which confirms the fact that $\{F(t)\}$ is a delta-correlated process. One may also note that choosing $n=2$ in this expression gives the covariance function for shot noise as

$$K_{FF}(t,s) \equiv \kappa_2[F(t),F(s)] = E[F^2]\,\dot{\mu}_Z(t)\,\delta(t-s)$$

One could have derived this covariance function without consideration of characteristic functions, but it would be very difficult to do that for all the higher-order cumulants.

****

Example 5.7: Determine the covariance function for the response $\{X(t)\}$ of a linear system described by $c\,\dot{X}(t)+k\,X(t)=F(t)$ with $X(0)=0$. For $t>0$, $\{F(t)\}$ is a mean-zero, nonstationary delta-correlated process with $\phi_{FF}(t_1,t_2)=G_0\,(1-e^{-\alpha t_1})\,\delta(t_1-t_2)$.

From Example 5.1 we know that $h_x(t)=c^{-1}e^{-kt/c}\,U(t)$, and the fact that $\mu_F(t)=0$ gives $K_{FF}(t_1,t_2)=\phi_{FF}(t_1,t_2)$. Using this information in Eq. 5.21 gives

$$K_{XX}(t_1,t_2)=\int_{-\infty}^{\infty}\int_{-\infty}^{\infty}K_{FF}(s_1,s_2)\,h_x(t_1-s_1)\,h_x(t_2-s_2)\,ds_1\,ds_2$$

$$=\frac{G_0}{c^2}\int_0^{t_2}\int_0^{t_1}(1-e^{-\alpha s_1})\,\delta(s_1-s_2)\,e^{-k(t_1+t_2-s_1-s_2)/c}\,ds_1\,ds_2$$

which can be integrated to give

$$K_{XX}(t_1,t_2)=\frac{G_0\,e^{-k(t_1+t_2)/c}}{c^2}\int_0^{\min(t_1,t_2)}(1-e^{-\alpha s_2})\,e^{2ks_2/c}\,ds_2$$

$$=\frac{G_0\,e^{-k(t_1+t_2)/c}}{c^2}\left(\frac{c[e^{2k\min(t_1,t_2)/c}-1]}{2k}+\frac{c[1-e^{-(\alpha-2k/c)\min(t_1,t_2)}]}{2k-\alpha c}\right)$$

Using the fact that $t_1+t_2-2\min(t_1,t_2)=|t_1-t_2|$ allows this to be simplified somewhat to give

$$K_{XX}(t_1,t_2)=\frac{G_0\,e^{-k|t_1-t_2|/c}}{c}\left(\frac{1}{2k}-\frac{e^{-\alpha\min(t_1,t_2)}}{2k-\alpha c}\right)+\frac{\alpha\,G_0\,e^{-k(t_1+t_2)/c}}{2k(2k-\alpha c)}$$

As a special case one can let $t_1=t_2=t$ to obtain the response variance as

$$\sigma_X^2(t)=\frac{G_0}{c}\left(\frac{1}{2k}-\frac{e^{-\alpha t}}{2k-\alpha c}\right)+\frac{\alpha\,G_0\,e^{-2kt/c}}{2k(2k-\alpha c)}$$

and the stationary limit as t tends to infinity is $\sigma_X^2(t)=G_0/(2kc)$.

****

5.6 Response of Linear Single-Degree-of-Freedom Oscillator

Most of the remainder of this chapter will be devoted to the study of one particular constant-coefficient linear system subjected to stationary delta-correlated excitation. In a sense this constitutes one extended example problem, but it warrants special attention because it will form the basis for much of what we will do in analyzing the stochastic response of other dynamical systems. The system to be studied is the "single-degree-of-freedom" (SDF) oscillator introduced in Example 5.3.

The SDF oscillator may be regarded as the prototypical vibratory system, and it has probably been studied more thoroughly than any other dynamical system. Essentially, it is the simplest differential equation that exhibits fundamentally oscillatory behavior. In particular, as we saw in Example 5.3, its impulse response function is oscillatory if damping is small, and this is a basic characteristic of vibratory systems. For both deterministic and stochastic situations, the dynamic behavior of the linear SDF system can be regarded as providing the fundamental basis for much of both the analysis and the interpretation of results for more complicated vibratory systems. In particular, the response of more complicated (multi-degree-of-freedom) linear systems is typically found by superimposing modal responses, each of which represents the response of an SDF system, and the dynamic behavior of a vibratory nonlinear system is usually interpreted in terms of how it resembles and how it differs from that of a linear system.

In a somewhat similar way, the delta-correlated excitation can be considered prototypical, inasmuch as it is simple and the nature of any other excitation process is often interpreted in terms of how it differs from a delta-correlated process. This idea will be investigated in more detail in Chapter 6.

The differential equation for stochastic motion of the SDF system is commonly written as either

$$m\ddot{X}(t)+c\,\dot{X}(t)+k\,X(t)=F(t) \tag{5.45}$$

or

$$\ddot{X}(t)+2\zeta\,\omega_0\,\dot{X}(t)+\omega_0^2\,X(t)=F(t)/m=-\ddot{Z}(t) \tag{5.46}$$

in which $\omega_0 \equiv (k/m)^{1/2}$ and $\zeta = c/[2(k\,m)^{1/2}]$ are called the undamped natural circular frequency and the fraction of critical damping, respectively, of the system. The usual interpretation of this equation for mechanical vibration problems is that m, k, and c denote the magnitudes of a mass, spring, and dashpot, respectively, attached as shown in Fig. 5.3. For the fixed-base system of Fig. 5.3(a), the term $F(t)$ is the force externally applied to the mass, and $X(t)$ is the location of the mass relative to its position when the system is at rest. When the system has an imposed motion $\{Z(t)\}$ at the base, as in Fig. 5.3(b), $X(t)=Y(t)-Z(t)$ is the change in the location of the mass relative to the moving

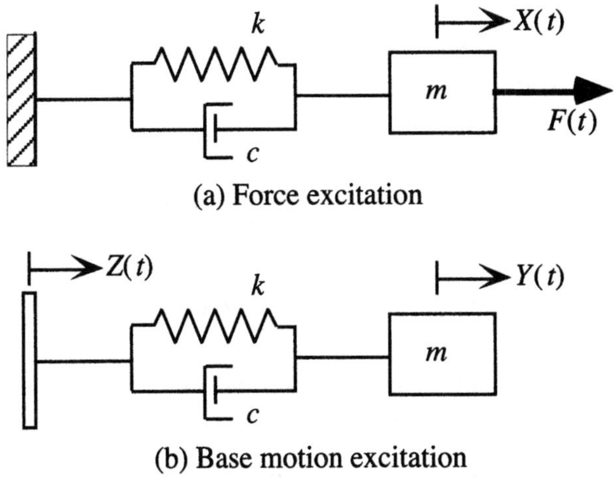

Figure 5.3 Single-degree-of-freedom oscillator.

base and $F(t)$ is the product of the mass and the negative of the base acceleration.[5] In this physical model we expect m, k, and c all to be nonnegative.

First we will consider the nonstationary stochastic response that results when the SDF oscillator is initially at rest and the delta-correlated excitation is suddenly applied to the system at time $t=0$ and then is stationary for all later time. Mathematically this condition can be described by treating the $\{F(t)\}$ excitation process as a uniformly modulated delta-correlated process

$$F(t) = W(t)\,U(t)$$

in which $\{W(t)\}$ is a stationary delta-correlated process and the unit step $U(t)$ is included to eliminate excitation prior to $t=0$, and thereby ensure that the system is at rest at time $t=0$. Using $K_{WW}(t,s) = G_0\,\delta(t-s)$ to denote the covariance function of $\{W(t)\}$ gives

$$K_{FF}(t,s) = G_0\,\delta(t-s)\,U(t)$$

with G_0 being a constant. We will first investigate the general nonstationary response of this oscillator, then consider the limiting behavior for t tending to infinity. Also, Example 5.8 will investigate the other limiting situation of t nearly zero.

---

[5]Alternatively, one can apply the term *SDF* to any system that is governed by a single second-order differential equation like Eq. 5.45.

Recall that the impulse function of Eq. 5.45 was found in Example 5.3 to be $h_x(t)=(m\,\omega_d)^{-1}e^{-\zeta\omega_0 t}\sin(\omega_d\,t)U(t)$ with $\omega_d=\omega_0(1-\zeta^2)^{1/2}$. We will find the nonstationary covariance of the response of Eq. 5.45 by substitution of this $h_x(t)$ and $K_{FF}(t,s)$ into Eq. 5.21, giving

$$K_{XX}(t_1,t_2)=\frac{G_0}{m^2\omega_d^2}\int_{-\infty}^{\infty}\int_{-\infty}^{\infty}\delta(s_1-s_2)U(s_1)e^{-\zeta\omega_0(t_1-s_1)}\sin[\omega_d(t_1-s_1)]\times$$
$$U(t_1-s_1)e^{-\zeta\omega_0(t_2-s_2)}\sin[\omega_d(t_2-s_2)]U(t_2-s_2)\,ds_1\,ds_2$$

Taking advantage of the unit step functions and the Dirac delta function in the integrand, one can rewrite this expression as

$$K_{XX}(t_1,t_2)=\frac{G_0}{m^2\omega_d^2}\int_0^{\min(t_1,t_2)}e^{-\zeta\omega_0(t_1+t_2-2s)}\sin[\omega_d(t_1-s)]\sin[\omega_d(t_2-s)]\,ds\ \times$$
$$U(t_1)U(t_2)$$

One simple way to evaluate this integral is to use the identity $\sin(\alpha)=(e^{i\alpha}-e^{-i\alpha})/(2i)$ for each of the sine terms, giving the integrand as the sum of four different exponential functions of s, each of which can be easily integrated. After substituting the fact that $t_1+t_2-2\min(t_1,t_2)=|t_1-t_2|$ and performing considerable algebraic simplification, the result can be written as

$$K_{XX}(t_1,t_2)=\frac{G_0}{4m^2\zeta\,\omega_0^3}U(t_1)U(t_2)\Bigg(e^{-\zeta\omega_0|t_1-t_2|}\bigg[\cos[\omega_d(t_1-t_2)]+$$
$$\frac{\zeta\omega_0}{\omega_d}\sin[\omega_d\,|t_1-t_2|]\bigg]-e^{-\zeta\omega_0(t_1+t_2)}\bigg[\frac{\omega_0^2}{\omega_d^2}\cos[\omega_d(t_1-t_2)]+$$
$$\frac{\zeta\omega_0}{\omega_d}\sin[\omega_d(t_1+t_2)]-\frac{\zeta^2\omega_0^2}{\omega_d^2}\cos[\omega_d(t_1+t_2)]\bigg]\Bigg)$$

(5.47)

The rather lengthy expression in Eq. 5.47 contains a great deal of information about the stochastic dynamic behavior of the SDF system. We will use it to investigate a number of matters of interest.

In many situations it is also important to study the response levels for derivatives of $X(t)$. The covariance functions for such derivatives could be obtained in exactly the same way as Eq. 5.47 was obtained. For example, we could find $h_{\dot{x}}(t) = h'_x(t)$ and substitute it instead of $h_x(t)$ into Eq. 5.21 to obtain a covariance function for the $\{\dot{X}(t)\}$ process. Rather than performing more integration, though, we can use expressions from Chapter 4 to derive variance properties of derivatives of $\{X(t)\}$ directly from Eq. 5.47. For example, Eq. 4.60 gives the cross-covariance of $\{X(t)\}$ and $\{\dot{X}(t)\}$ as

$$K_{X\dot{X}}(t_1,t_2) = \frac{\partial K_{XX}(t_1,t_2)}{\partial t_2}$$

and Eq. 4.61 gives the covariance function for $\{\dot{X}(t)\}$ as

$$K_{\dot{X}\dot{X}}(t_1,t_2) = \frac{\partial^2 K_{XX}(t_1,t_2)}{\partial t_1 \partial t_2}$$

Substituting from Eq. 5.47 into these expressions gives

$$K_{X\dot{X}}(t_1,t_2) = \frac{G_0}{4m^2 \zeta \omega_0 \omega_d} U(t_1) U(t_2) \Bigg(e^{-\zeta \omega_0 |t_1 - t_2|} \sin[\omega_d (t_1 - t_2)] -$$
$$e^{-\zeta \omega_0 (t_1 + t_2)} \left[\sin[\omega_d (t_1 - t_2)] - \frac{\zeta \omega_0}{\omega_d} \cos[\omega_d (t_1 - t_2)] + \frac{\zeta \omega_0}{\omega_d} \cos[\omega_d (t_1 + t_2)] \right] \Bigg) \tag{5.48}$$

and

$$K_{\dot{X}\dot{X}}(t_1,t_2) = \frac{G_0}{4m^2 \zeta \omega_0} U(t_1) U(t_2) \Bigg(e^{-\zeta \omega_0 |t_1 - t_2|} \left[\cos[\omega_d (t_1 - t_2)] - \frac{\zeta \omega_0}{\omega_d} \sin[\omega_d |t_1 - t_2|] \right] -$$
$$e^{-\zeta \omega_0 (t_1 + t_2)} \left[\frac{\omega_0^2}{\omega_d^2} \cos[\omega_d (t_1 - t_2)] - \frac{\zeta \omega_0}{\omega_d} \sin[\omega_d (t_1 + t_2)] - \frac{\zeta^2 \omega_0^2}{\omega_d^2} \cos[\omega_d (t_1 + t_2)] \right] \Bigg) \tag{5.49}$$

Next let us consider the variance of the $\{X(t)\}$ response by letting $t_1 = t_2 = t$ in Eq. 5.47, giving

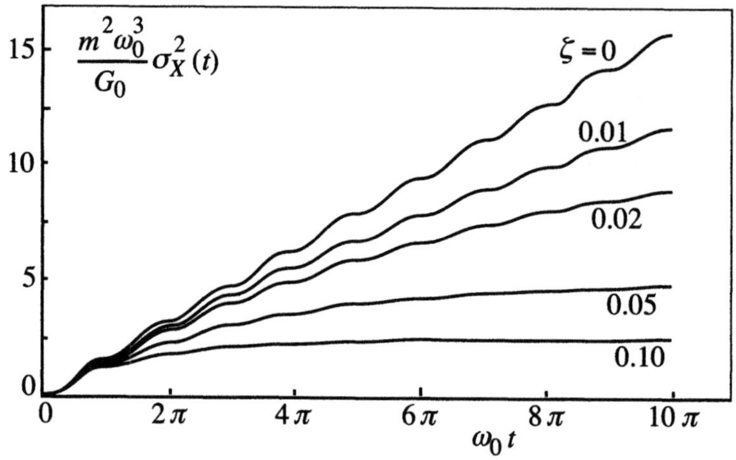

Figure 5.4 Nonstationary growth of variance.

$$\sigma_X^2(t) \equiv K_{XX}(t,t) = \frac{G_0}{4m^2\zeta\omega_0^3}\left(1 - e^{-2\zeta\omega_0 t}\left[\frac{\omega_0^2}{\omega_d^2} + \frac{\zeta\omega_0}{\omega_d}\sin(2\omega_d t) - \frac{\zeta^2\omega_0^2}{\omega_d^2}\cos(2\omega_d t)\right]\right)U(t) \tag{5.50}$$

This important expression was published by Caughey and Stumpf in 1961. It is plotted in a nondimensional form in Fig. 5.4 for several values of the damping parameter ζ. Figure 5.4 also includes a curve for $\zeta = 0$, which cannot be obtained directly from Eq. 5.50. Rather, one must take the limit of the equation as ζ approaches zero, obtaining

$$\sigma_X^2(t) = \frac{G_0}{4m^2\omega_0^3}\left[2\omega_0 t - \sin(2\omega_0 t)\right]U(t) \quad \text{for } \zeta = 0 \tag{5.51}$$

Note that for any damping value other than zero, $\sigma_X^2(t)$ tends to an asymptote as t tends to infinity. In fact, Eq. 5.50 gives $\sigma_X^2(t) \to G_0/(4m^2\zeta\omega_0^3)$ for $t \to \infty$, which agrees exactly with the bound of Eq. 5.32 for any excitation of this system. This confirms the general result derived in Eq. 5.40, that the variance bound of Eq. 5.32 is identical to the true response variance for a stationary delta-correlated excitation process.

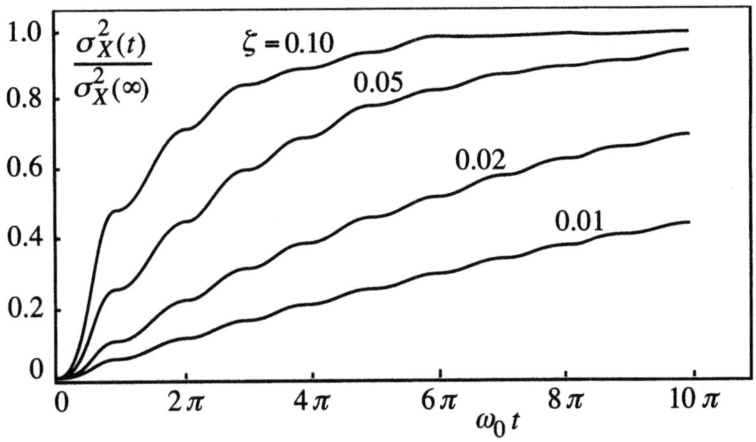

Figure 5.5 Approach of nonstationary variance to asymptote.

The form of Fig. 5.4 emphasizes the important fact that the response variance grows more rapidly when damping is small. On the other hand, it appears that the initial rate of growth is approximately independent of damping. We will examine this aspect in more detail in Example 5.8, when we consider the limiting behavior for t near zero. The form of normalization used in Fig. 5.4 gives the asymptote for each curve as $(4\zeta)^{-1}$ for $t \to \infty$. The fact that the asymptote is different for each curve rather obscures the information about the rate at which each curve approaches its asymptote and how this rate depends on damping. In order to illustrate this aspect more clearly, Fig. 5.5 shows plots of the response variance normalized by the asymptote for several nonzero values of damping. In this form of the plots, each curve tends to unity as t becomes large, and it is clear that the asymptotic level is reached much more quickly in a system with larger damping.

Because response variance is a very important quantity in practical applications of stochastic dynamics, we will also consider some simple approximations and bounds of Eq. 5.50. One possible approximation is to note that the oscillatory terms in Figs. 5.4 or 5.5 seem to be relatively unimportant compared with the exponential approach of the variance to its asymptotic value. This leads to the idea of replacing the oscillatory sine and cosine terms by their average value of zero in Eq. 5.50 and obtaining the averaged expression of

$$\sigma_X^2(t) \approx \frac{G_0}{4 m^2 \zeta \omega_0^3} \left(1 - \frac{\omega_0^2}{\omega_d^2} e^{-2\zeta\omega_0 t} \right) U(t) = \frac{G_0}{4 m^2 \zeta \omega_0^3} \left(1 - \frac{e^{-2\zeta\omega_0 t}}{1 - \zeta^2} \right) U(t)$$

The actual variance value oscillates about this smooth curve, with the amplitude of the oscillation decaying exponentially. Alternatively, one can obtain a simpler result that is almost identical to this for small damping values by neglecting the ζ^2 term and writing

$$\sigma_X^2(t) \approx \frac{G_0}{4\, m^2 \zeta\, \omega_0^3} \left(1 - e^{-2\zeta\omega_0 t}\right) U(t) \tag{5.52}$$

In some situations one may wish to obtain a rigorous upper bound on the nonstationary variance expression of Eq. 5.50. This can be obtained by using the amplitude of the sum of the two oscillatory terms as

$$\text{Ampl}\left(\frac{\zeta\,\omega_0}{\omega_d}\sin(2\omega_d t) - \frac{\zeta^2 \omega_0^2}{\omega_d^2}\cos(2\omega_d t)\right) = \frac{\zeta\,\omega_0}{\omega_d}\left(1 + \frac{\zeta^2\omega_0^2}{\omega_d^2}\right)^{1/2} = \frac{\zeta\,\omega_0^2}{\omega_d^2}$$

Replacing the oscillatory terms by the negative of the amplitude gives the bound

$$\sigma_X^2(t) \le \frac{G_0}{4\, m^2 \zeta\, \omega_0^3} \left(1 - \frac{e^{-2\zeta\omega_0 t}}{1+\zeta}\right) U(t) \tag{5.53}$$

Figure 5.6 compares the exact response variance of Eq. 5.50 with the smooth approximation of Eq. 5.52 and the bound of Eq. 5.53 for the special case of $\zeta = 0.10$. Even for this case with rather large damping it is seen that the curves are all relatively near each other, and for smaller damping they are much closer. It should be noted, though, that the smooth exponential approximation never adequately describes the response growth for very small values of time, such as $\omega_0\, t < \pi/4$. As previously indicated, this small time situation will be examined in more detail in Example 5.8.

The exponential forms of Eqs. 5.52 and 5.53 are valuable if one needs an approximation of the time that it takes for the response variance to reach some specified level. Solving Eq. 5.52 for t shows that $\sigma_X^2(t)$ reaches a specified level B when

$$t \approx \frac{-1}{2\zeta\,\omega_0} \ln\left(1 - \frac{4 m^2 \zeta\, \omega_0^3}{G_0} B\right)$$

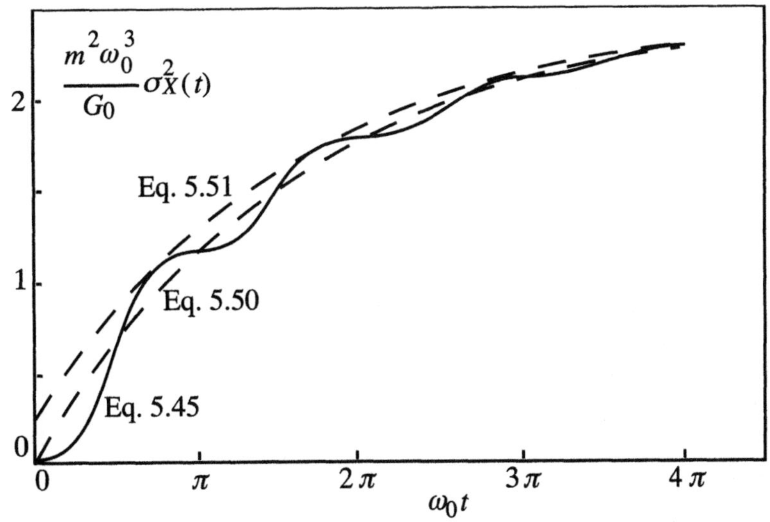

Figure 5.6 Approximation and bound of nonstationary variance for $\zeta = 0.10$.

Similarly, the bound gives $\sigma_X^2(t) \le B$ for

$$t \le \frac{-1}{2\zeta\omega_0} \ln\left(1 - (1+\zeta)\frac{4m^2\zeta\omega_0^3}{G_0} B\right)$$

Such an approximation or bound of the response variance is an important part of the estimation or bounding of the reliability for structural systems that are susceptible to failure due to excessive deformation or stress.

In the same way, we can investigate the variance of $\{\dot{X}(t)\}$ by taking $t_1 = t_2 = t$ in Eq. 5.49. The result can be written as

$$\sigma_{\dot{X}}^2(t) = \frac{G_0}{4m^2\zeta\omega_0}\left(1 - e^{-2\zeta\omega_0 t}\left[\frac{\omega_0^2}{\omega_d^2} - \frac{\zeta\omega_0}{\omega_d}\sin(2\omega_d t) - \frac{\zeta^2\omega_0^2}{\omega_d^2}\cos(2\omega_d t)\right]\right)U(t) \tag{5.54}$$

The similarity between this equation and that for $\sigma_X^2(t)$ in Eq. 5.50 is striking. The only differences are a multiplication of Eq. 5.50 by ω_0^2 and a change of sign of one term. Given this similarity in the equations, it is not surprising to find that $\sigma_X^2(t)$ and $\sigma_{\dot{X}}^2(t)$ grow in a very similar manner. This is shown in Fig. 5.7, wherein the result of Eq. 5.54 is normalized and superimposed upon the results

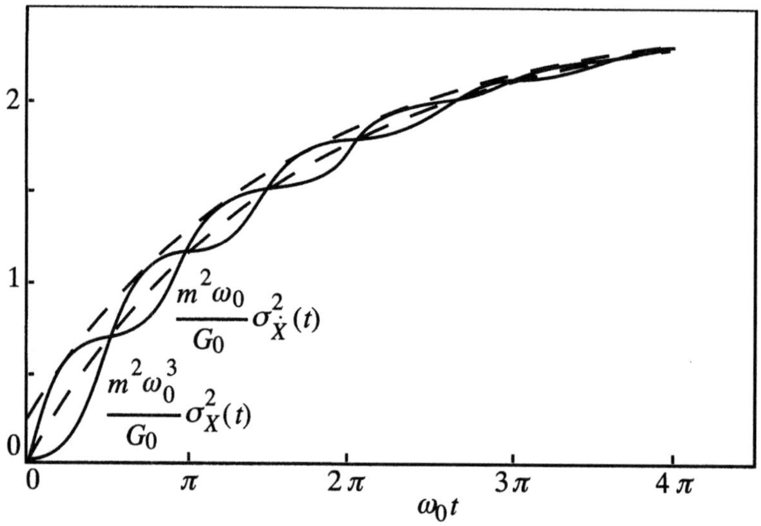

Figure 5.7 Variance of $\{\dot X(t)\}$ compared with results of Fig. 5.6 for $\zeta = 0.10$.

from Fig. 5.6. This figure shows that the primary difference between $\sigma_{\dot X}^2(t)$ and $\sigma_X^2(t)$ is the phase difference caused by the change of sign on the $\sin(2\omega_d t)$ term. This difference has little significance except for small values of t. Figure 5.7 also shows that Eqs. 5.52 and 5.53, when multiplied by ω_0^2, provide a smooth approximation and a bound, respectively, on $\sigma_{\dot X}^2(t)$. Note that for smaller values of damping the growth of $\sigma_X^2(t)$ and $\sigma_{\dot X}^2(t)$ will be even more similar than for the case shown in Fig. 5.7 with $\zeta = 0.10$.

Finally, we note that by taking $t_1 = t_2 = t$ in Eq. 5.48 we can find the cross-covariance of the two random variables $X(t)$ and $\dot X(t)$ denoting the response and its derivative at the same instant of time. The result can be written as

$$
\begin{aligned}
K_{X\dot X}(t,t) &= \frac{G_0}{4m^2\omega_d^2} e^{-2\zeta\omega_0 t}\left[1 - \cos(2\omega_d t)\right] U(t) \\
&= \frac{G_0}{2m^2\omega_d^2} e^{-2\zeta\omega_0 t} \sin^2(\omega_d t)\, U(t)
\end{aligned}
\tag{5.55}
$$

Figure 5.8 shows a plot of this function for $\zeta = 0.10$. Two key features of this cross-covariance are that it tends to zero as t becomes large and it is exactly zero at times t that are an integer multiple of $\pi/(2\omega_d)$. Thus, $X(t)$ and $\dot X(t)$ are uncorrelated twice during each period of response of the oscillator. Furthermore, $K_{X\dot X}(t,t)$ is always relatively small. For a lightly damped system (i.e., for $\zeta << 1$) we can say that the maximum cross-covariance is approximately equal to the

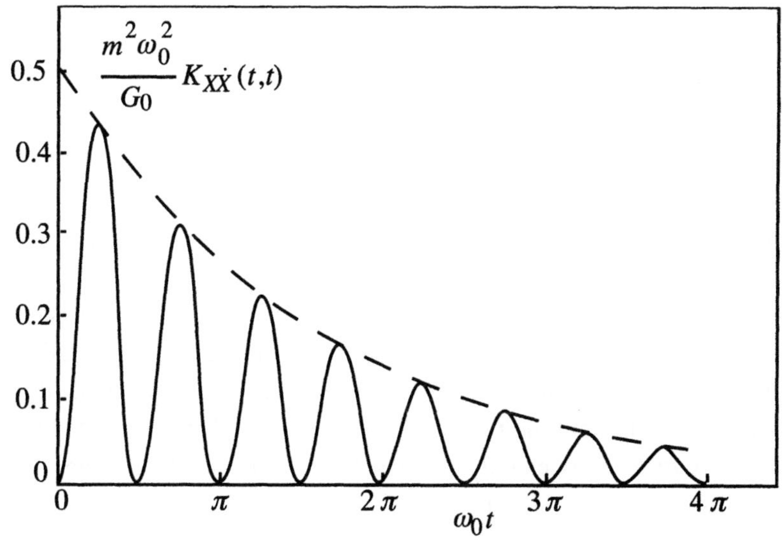

Figure 5.8 Covariance of $\{X(t)\}$ and $\{\dot{X}(t)\}$ for $\zeta = 0.10$.

bound of $G_0 /(2m^2 \omega_d^2)$. Note that this bound does not depend on ζ, whereas the covariances of $\{X(t)\}$ and $\{\dot{X}(t)\}$ both vary like ζ^{-1}, and therefore become very large when $\zeta << 1$. This, then, implies that the correlation coefficient

$$\rho_{X\dot{X}}(t,t) \equiv \frac{K_{X\dot{X}}(t,t)}{\sigma_X(t)\sigma_{\dot{X}}(t)}$$

is generally small for the response of the SDF system. The exception is when t is small, because in that situation $\sigma_X(t)$ and $\sigma_{\dot{X}}(t)$ in the denominator of the correlation coefficient are also small.

It may be instructive to note that one can also obtain Eq. 5.55 from Eq. 5.50, rather than using Eq. 5.47. In particular, Eq. 4.64 states that

$$\frac{d}{dt} E[X^2(t)] = E[2X(t)\dot{X}(t)]$$

and this gives

$$\phi_{X\dot{X}}(t,t) = \frac{1}{2}\frac{d\phi_{XX}(t,t)}{dt}$$

Using this along with the fact that $E[\dot{X}(t)]$ is the derivative of $E[X(t)]$ gives the covariance relationship as

$$K_{X\dot{X}}(t,t)=\frac{1}{2}\frac{d}{dt}\sigma_X^2(t) \tag{5.56}$$

Substituting from Eq. 5.50 into this expression gives exactly Eq. 5.55.

It should be noted that the equations developed in this section may also be used to describe the conditional covariance of the response of the SDF system when one is given the values of $X(t_0)$ and $\dot{X}(t_0)$ at some specified time t_0. In particular, the response at time t due to the portion of a delta-correlated excitation occurring after time t_0 is exactly the same as the response at time $t-t_0$ of a system at rest at time $t=0$. Thus, Eqs. 5.44 and 5.47 give the conditional covariance of the response as

$$\begin{aligned}\mathrm{Cov}[X(t_1),X(t_2)\,|\,X(t_0)=u,\dot{X}(t_0)=v]=\\ \frac{G_0}{4m^2\zeta\omega_0^3}\Bigg(e^{-\zeta\omega_0|t_1-t_2|}\left[\cos[\omega_d(t_1-t_2)]+\frac{\zeta\omega_0}{\omega_d}\sin[\omega_d\,|t_1-t_2\,|]\right]-\\ e^{-\zeta\omega_0(t_1+t_2-2t_0)}\Bigg[\frac{\omega_0^2}{\omega_d^2}\cos[\omega_d(t_1-t_2)]+\frac{\zeta\omega_0}{\omega_d}\sin[\omega_d(t_1+t_2-2t_0)]-\\ \frac{\zeta^2\omega_0^2}{\omega_d^2}\cos[\omega_d(t_1+t_2-2t_0)]\Bigg]\Bigg)\end{aligned} \tag{5.57}$$

provided that t_1 and t_2 are both greater than t_0. From this relationship one can find the conditional variance of $X(t)$ and $\dot{X}(t)$, their cross-covariance, and so forth. Similarly, the conditional mean of Eq. 5.43 can be written as

$$E[X(t)\,|\,X(t_0)=u,\dot{X}(t_0)=v]=u\,g(t-t_0)+v\,m\,h_x(t-t_0)+\int_{t_0}^{\infty}\mu_F(s)\,h_x(t-s)\,ds \tag{5.58}$$

n which

$$g(t)=e^{-\zeta\omega_0 t}\left(\cos(\omega_d t)+\frac{\zeta\omega_0}{\omega_d}\sin(\omega_d t)\right) \tag{5.59}$$

is the deterministic response to a unit displacement initial condition at time zero, and $m\,h_x(t)$ is the response to a unit velocity initial condition at time zero.

We have not commented on the behavior of the response acceleration of the SDF system with delta-correlated excitation, but that can also be investigated by the same procedures. Taking more derivatives of Eq. 5.49 shows that $K_{\dot{X}\ddot{X}}(t_1,t_2)$ is discontinuous at $t_1 = t_2$, so $K_{\ddot{X}\ddot{X}}(t_1,t_2)$ is infinite along that line. Thus, $\sigma_{\ddot{X}}^2(t) = K_{\ddot{X}\ddot{X}}(t,t) = \infty$. This is consistent with the comment in Section 5.5 that a delta-correlated excitation always gives an infinite response variance for some response quantity. For the SDF system with delta-correlated excitation, the response acceleration always has infinite variance.

5.7 Stationary SDF Response to Delta-Correlated Excitation

In the preceding section we analyzed the response of the SDF system to a delta-correlated excitation that was identically zero for $t < 0$ and was the same as a stationary process for $t \geq 0$. Note that a simple shift along the time axis will give us corresponding results for a problem in which the excitation begins to become effective at some time t_0 instead of at zero. That is, replacing t with $t - t_0$ in any of the response equations in the preceding section gives the corresponding response measure for this new loading situation with $F(t) = W(t)\,U(t - t_0)$. One particular situation that we might consider is the one in which $t_0 \to -\infty$ so that the excitation has always been effective. In this situation we expect the response to be stationary, and the fact that $t - t_0 \to \infty$ shows that the stationary response levels are the same as asymptotes for $t \to \infty$ of the nonstationary response from the previous section. Stated differently, the nonstationary response from the previous section tends to stationary response at times long after the instant when the excitation first became active. This simple result will hold true for any linear system with a stationary excitation, if a stationary response exists for the problem.

If we had not already investigated nonstationary levels of $K_{XX}(t_1,t_2)$ for the SDF oscillator, then the easiest way to find the covariance function for stationary response would be to use Eq. 5.27 written for covariance with $G_{FF}(\tau) = G_0\,\delta(\tau)$ to obtain

$$G_{XX}(\tau) = \int_{-\infty}^{\infty}\int_{-\infty}^{\infty} G_{FF}(\tau - r_1 + r_2)\,h_x(r_1)\,h_x(r_2)\,dr_1\,dr_2$$

$$= G_0 \int_{-\infty}^{\infty} h_x(r)\,h_x(r - \tau)\,dr = G_0\;h_{x,var}(\tau)$$

Evaluation of the $h_{x,var}(\tau)$ integral (introduced in Eq. 5.31) is somewhat simpler than the determination of Eq. 5.47, because the limits of integration are simpler. However, inasmuch as we already have Eq. 5.47 we can obtain exactly the same stationary response covariance by letting t_1 and t_2 both go to infinity in that equation while holding $\tau = t_1 - t_2$ to a finite level. The result is

$$G_{XX}(\tau) = G_0 \, h_{x,var}(\tau) = \frac{G_0}{4m^2 \zeta \omega_0^3} e^{-\zeta \omega_0 |\tau|} \left(\cos(\omega_d \tau) + \frac{\zeta \omega_0}{\omega_d} \sin(\omega_d |\tau|) \right) \tag{5.60}$$

Similarly, the cross-covariance of $\{X(t)\}$ and $\{\dot{X}(t)\}$ and the covariance of $\{\dot{X}(t)\}$ can be obtained either from covariance equations of the form of Eqs. 5.27 and 5.28 with $G_{FF}(\tau) = G_0 \, \delta(\tau)$, or from the large t limits of Eqs. 5.48 and 5.49. The results are

$$G_{X\dot{X}}(\tau) = \frac{G_0}{4m^2 \zeta \omega_0 \omega_d} e^{-\zeta \omega_0 |\tau|} \sin(\omega_d \tau) \tag{5.61}$$

and

$$G_{\dot{X}\dot{X}}(\tau) = \frac{G_0}{4m^2 \zeta \omega_0} e^{-\zeta \omega_0 |\tau|} \left(\cos(\omega_d \tau) - \frac{\zeta \omega_0}{\omega_d} \sin(\omega_d |\tau|) \right) \tag{5.62}$$

Figure 5.9 shows normalized versions of these three stationary covariance results for the situation with $\zeta = 0.10$. Note that Eqs. 5.60 and 5.62 are even functions of τ, thereby satisfying the necessary symmetry condition for an autocovariance function. The cross-covariance of Eq. 5.61, however, is an odd function of τ. Comparing Eqs. 5.60 and 5.62 shows that the covariance of $\{\dot{X}(t)\}$ differs from that of $\{X(t)\}$ only by the inclusion of an ω_0^2 factor in the multiplier in front and by the sign of the $\sin(\omega_d |\tau|)$ term. Because this latter term is multiplied by the small parameter ζ, it is not surprising that the normalized plots of $G_{XX}(\tau)$ and $G_{\dot{X}\dot{X}}(\tau)$ in parts (a) and (c) of Fig. 5.9 are almost the same. The difference between the shapes of the $G_{XX}(\tau)$ and $G_{\dot{X}\dot{X}}(\tau)$ plots is significant only when τ is very small. At $\tau = 0$ the slope of the $G_{\dot{X}\dot{X}}(\tau)$ plot is discontinuous while $G_{XX}(\tau)$ is smooth.

The stationary response variance values can now be obtained either by choosing $\tau = 0$ in Eqs. 5.60 and 5.62 or by letting t tend to infinity in the nonstationary response expressions in Eqs. 5.52 and 5.54, giving

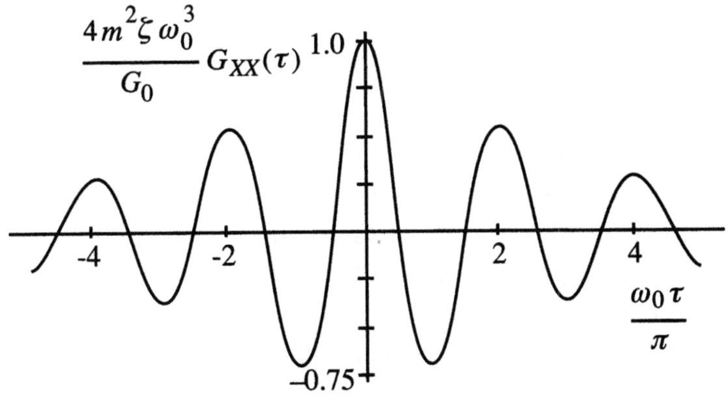

(a) Covariance of $\{X(t)\}$

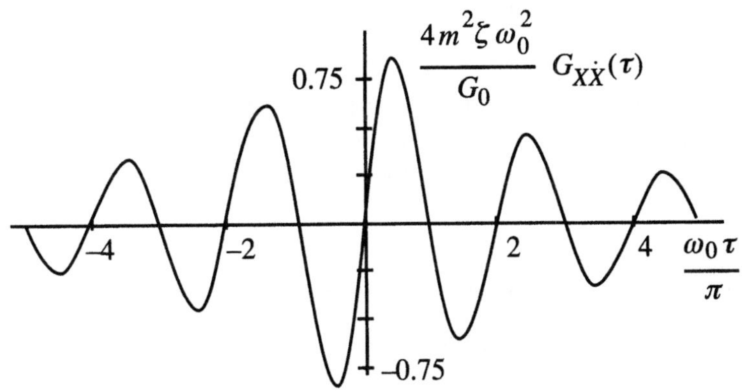

(b) Cross-covariance of $\{X(t)\}$ and $\{\dot{X}(t)\}$

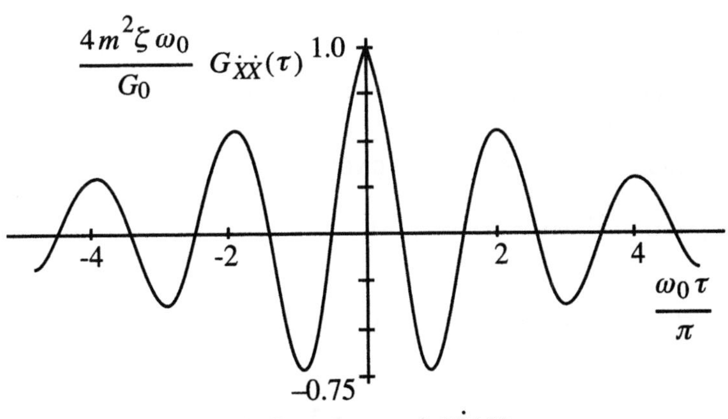

(c) Covariance of $\{\dot{X}(t)\}$

Figure 5.9 Stationary covariance for SDF with $\zeta = 0.10$.

$$\sigma_X^2 = \frac{G_0}{4m^2\zeta\omega_0^3} \quad \text{and} \quad \sigma_{\dot{X}}^2 = \frac{G_0}{4m^2\zeta\omega_0} \tag{5.63}$$

Similarly, either Eq. 5.61 or 5.55 shows that the covariance of the two random variables $X(t)$ and $\dot{X}(t)$ is zero at any instant of time during stationary response. Note that the two stationary variance values given in Eq. 5.63 are exactly the multipliers that have appeared on all the nonstationary and stationary variance values derived for the $\{X(t)\}$ and $\{\dot{X}(t)\}$ processes. One can also rewrite these important values in terms of the original parameters of the physical model shown in Fig. 5.3, obtaining

$$\sigma_X^2 = \frac{G_0}{2kc} \quad \text{and} \quad \sigma_{\dot{X}}^2 = \frac{G_0}{2mc} \tag{5.64}$$

These expressions reveal the possibly surprising facts that the stationary variance of the displacement $\{X(t)\}$ is unaffected by the mass of the system and that the stationary variance of the velocity $\{\dot{X}(t)\}$ is unaffected by the stiffness of the system.

****

Example 5.8: Analyze the response covariance for the SDF of Eq. 5.45 with $c=0$ and $k=0$.

Rather than looking for limits of the general expressions as c and k tend to zero, let us directly analyze the response of the simplified equation $m\ddot{X}(t)=F(t)$. Inasmuch as $\ddot{X}(t)=F(t)/m$, we can directly write the expression for the covariance function of the acceleration as

$$K_{\ddot{X}\ddot{X}}(t,s) = \frac{1}{m^2} K_{FF}(t,s) = \frac{G_0}{m^2}\delta(t-s)U(t)U(s)$$

Repeatedly integrating this expression gives

$$K_{\dot{X}\ddot{X}}(t,s) = \frac{G_0}{m^2} U(t-s)U(t)U(s)$$

$$K_{\dot{X}\dot{X}}(t,s) = \frac{G_0}{m^2}\min(t,s)U(t)U(s) = \frac{G_0}{m^2}\big[t+(s-t)U(t-s)\big]U(t)U(s)$$

$$K_{X\dot{X}}(t,s) = \frac{G_0}{2m^2}\Big[t^2-(t-s)^2U(t-s)\Big]U(t)U(s)$$

and

$$K_{XX}(t,s) = \frac{G_0}{6m^2}\left[s^2(3t-s)U(t-s) + t^2(3s-t)[1-U(t-s)]\right]U(t)U(s)$$

$$= \frac{G_0}{6m^2}[\max(t,s)]^2[3\max(t,s) - \min(t,s)]U(t)U(s)$$

By letting $s=t$ in the $K_{\dot{X}\dot{X}}(t,s)$ and $K_{XX}(t,s)$ expressions we can find the response variances as

$$\sigma_{\dot{X}}^2(t) = \frac{G_0}{m^2}tU(t) \qquad \text{and} \qquad \sigma_X^2(t) = \frac{G_0}{3m^2}t^3 U(t)$$

showing that application of the delta-correlated excitation to a mass with no restoring force gives a velocity variance that grows linearly with time and a displacement variance that grows cubicly. Similarly, the cross-covariance of displacement and velocity at time t is

$$K_{X\dot{X}}(t,t) = \frac{G_0}{2m^2}t^2 U(t)$$

which gives their correlation coefficient as $\rho_{X\dot{X}} \equiv K_{X\dot{X}}(t,t)/(\sigma_X \sigma_{\dot{X}}) = 3^{1/2}/2$. One can go one step further with this type of analysis and notice that the cross-covariance function for velocity and acceleration is finite but has a discontinuity at $s=t$. We have $K_{\dot{X}\ddot{X}}(t,s) = G_0/m^2$ for $s<t$, and this covariance is zero for $s>t$. Although the mass with no restoring force may seem to be a very impractical system, its response actually does describe an important situation. In particular, it is the same as the limit of the response of the general SDF system of Eq. 5.45 immediately after the application of the delta-correlated excitation. The key reason for this is that at $t=0$, both $X(t)$ and $\dot{X}(t)$ in Eq. 5.45 are zero. Thus, we know that the restoring force $c\dot{X}(t)+kX(t)$ starts at zero when $t=0$. Furthermore, we have found that the variance of $X(t)$ and $\dot{X}(t)$ grow continuously, so there must be a small range of t values for which the restoring force remains small. During this range of t values, one can approximate Eq. 5.45 by the simplified expression of this example. In fact, it can be shown that all the variance and covariance functions we have derived here agree exactly with the $t \rightarrow 0$ limits of power series expansions of the corresponding expressions for the SDF system.

One implication of these results is that the variance of velocity and displacement of the SDF system initially grow like $G_0 t/m^2$ and $G_0 t^3/(3m^2)$, respectively. These rates of growth ensure that for small t, the values of displacement are relatively much smaller than those of velocity, and one can see that this initial behavior shows quite clearly for the SDF response in Fig. 5.7. For the simple mass, the linear and cubic growth, respectively, of velocity and displacement variance will eventually cause displacement levels to be relatively large compared with velocity, as both grow toward infinity. For the SDF system the large time behavior is quite different, though, because the spring and dashpot become effective. In particular, both displacement and velocity variance grow in the same manner toward stationary values, as illustrated in Fig. 5.7. For both the simple mass and the SDF system, the variance of acceleration is infinite for all time during which a delta-correlated excitation is applied.

****

5.8 Nearly Delta-Correlated Processes

As stated previously, no physical process is truly delta-correlated, even though delta-correlated processes are often used in practical problems to approximate excitations that are nearly independent of themselves at distinct times t and s. Equations 5.37 and 5.38 illustrate that one must somehow choose G_0 (either as a constant or a function of time) in order to define the covariance function of the approximating delta-correlated process. For a covariant stationary process, one logical way to do this is on the basis of the variance bound in Eq. 5.32. In particular, if one chooses G_0 for the delta-correlated process to be the same as the integral of the absolute value of $G_{FF}(\tau)$ for the physical process

$$G_0 = \int_{-\infty}^{\infty} |G_{FF}(r)| dr \tag{5.65}$$

then Eq. 5.32 will give the same bound on the response variance for the idealized approximation of the excitation as for the physical process excitation.

In order to generalize Eq. 5.65 to a situation with a nonstationary $K_{FF}(s_1,s_2)$, let us introduce new time variables as $\tau = s_1 - s_2$ and $t = (s_1 + s_2)/2$. The first of these variables is the time difference, and the second is a symmetric function of s_1 and s_2 that becomes time s in the limit of $s_1 = s_2 = s$. We then have

$$K_{FF}(s_1,s_2) = K_{FF}\left(t + \frac{\tau}{2}, t - \frac{\tau}{2}\right)$$

For a nearly delta-correlated process it is clear that this covariance function must decay rapidly as τ increases. The generalization of Eq. 5.65 is then to choose

$$G_0(t) = \int_{-\infty}^{\infty} K_{FF}\left(t + \frac{\tau}{2}, t - \frac{\tau}{2}\right) d\tau \tag{5.66}$$

One may note that Eq. 5.65 is the stationary special case of Eq. 5.66.

The choice of G_0 will be given more consideration in Chapter 6, when we consider frequency decompositions of processes. It will be found that, in some applications, we can identify choices for G_0 that are better than Eq. 5.66.

****

Example 5.9: Choose G_0 for a delta-correlated approximation of a stochastic process with autocovariance function

$$G_{FF}(\tau) = A e^{-c|\tau|}$$

From Eqs. 5.65 and 5.40 we see that choosing $G_0 = 2A/c$ will give the response variance of any linear oscillator with the delta-correlated excitation to be the same as the bound of Eq. 5.32 for the response variance for $\{F(t)\}$ excitation. Of course $\{F(t)\}$ is nearly delta-correlated only if c is large, and we can also note that $G_{FF}(\tau)$ can be considered to tend to $(2A/c)\delta(\tau)$ in the limit as $c \to \infty$.

****

5.9 Response to Gaussian Excitation

All the results in this chapter have involved calculating the first and second moments of response of a linear system by writing that response as a convolution integral of the excitation and an impulse response function. As noted in Section 4.10, any integral of a Gaussian process is also a Gaussian process. Thus, any linear system with a Gaussian excitation also has Gaussian responses. Furthermore, different responses, including $\{X(t)\}$ and $\{\dot{X}(t)\}$, are jointly Gaussian.

Knowledge that a response is a Gaussian process is very useful information in practice, because the complete probability density function for any Gaussian random variable or vector is determined by the values of the appropriate first and second moments. For example, knowledge of the $\mu_X(t)$ and $\sigma_X(t)$ functions allows one to compute probabilities such as $P[X(t) > a]$ or $P[a < X(t) < b]$ for any t value. Similarly, knowledge of the $\mu_X(t)$ and $K_{XX}(t_1,t_2)$ functions allows one to compute probabilities of events involving two or more times such as $P[X(t_1) < a, X(t_2) < b, X(t_3) < c]$. Furthermore, knowledge that $X(t_1)$ and $\dot{X}(t_2)$ are jointly Gaussian allows one to write their joint probability density function in terms of the functions $\mu_X(t)$, $\mu_{\dot{X}}(t)$, and $K_{X\dot{X}}(t_1,t_2)$, which can, in turn, be written in terms of $\mu_X(t)$ and $K_{XX}(t_1,t_2)$.

In summary, the calculation of the first and second moments of response is performed in exactly the same way whether the excitation is Gaussian or non-Gaussian. The only difference between the two situations is that knowledge of these first two moments gives complete probability information about a Gaussian response but only partial information about a non-Gaussian response, such as would result from a non-Gaussian excitation.

Exercises

****

Impulse Response Functions

****

5.1 Consider a linear system for which the response $x(t)$ to an excitation $f(t)$ is governed by the differential equation

$$\ddot{x}(t)+5\,\dot{x}(t)+6\,x(t)=f(t)$$

Find the impulse response function $h_x(t)$ by considering $f(t)=\delta(t)$. Sketch your answer.

****

5.2 Consider a linear system for which the response $x(t)$ to an excitation $f(t)$ is governed by the differential equation

$$\ddot{x}(t)+7\,\dot{x}(t)+10\,x(t)=f(t)$$

Find the impulse response function $h_x(t)$ by considering $f(t)=\delta(t)$. Sketch your answer.

****

5.3 Consider a linear system for which the response $x(t)$ to an excitation $f(t)$ is governed by the third-order differential equation

$$\frac{d^3}{dt^3}x(t)+2\zeta\,\omega_0\frac{d^2}{dt^2}x(t)+\omega_0^2\frac{d}{dt}x(t)=f(t)$$

with $0<\zeta<1$. Find the impulse response function $h_x(t)$ by considering $f(t)=\delta(t)$. Specifically:

(a) Find three different functions that each solve the homogeneous third-order equation.

(b) Identify the three necessary initial conditions on $h_x(t)$ at $t=0^+$.

(c) Solve for the constants in the homogeneous solution.

****

5.4 Consider a linear system for which the response $x(t)$ to an excitation $f(t)$ is governed by the third-order differential equation

$$\frac{d^3}{dt^3}x(t)+c\frac{d^2}{dt^2}x(t)=f(t)$$

with $c > 0$. Find the impulse response function $h_x(t)$ by considering $f(t) = \delta(t)$. Specifically:

(a) Find three different functions that each solve the homogeneous third-order equation.

(b) Identify the three necessary initial conditions on $h_x(t)$ at $t = 0^+$.

(c) Solve for the constants in the homogeneous solution.

****

First-Order System

****

5.5 Consider the response $\{X(t)\}$ of a linear system described by

$$c\,\dot{X}(t) + k\,X(t) = F(t)$$

for which the impulse response function was derived in Example 5.1 as $h_x(t) = c^{-1} e^{-kt/c}\, U(t)$.

(a) Find expressions for $\mu_X(t)$ and $\mu_{\dot{X}}(t)$ for response to a general $\{F(t)\}$ excitation.

(b) Find expressions for $\phi_{XX}(t_1,t_2)$, $\phi_{X\dot{X}}(t_1,t_2)$, and $\phi_{\dot{X}\dot{X}}(t_1,t_2)$ for a general $\{F(t)\}$ excitation.

(c) Find simplified expressions for μ_X and $R_{FF}(\tau)$ for an $\{F(t)\}$ process that is mean-value and second-moment stationary.

****

5.6 Let A be a random variable with $\mu_A = 0$ and $\sigma_A^2 = 1$. Let this random variable be taken as the initial condition at time zero $[X(0) = A]$ of a linear system governed by $\dot{X}(t) = b\,X(t)$, in which b is a constant.

(a) Find expressions for $\phi_{XX}(t_1,t_2)$, $\phi_{X\dot{X}}(t_1,t_2)$, and $\phi_{\dot{X}\dot{X}}(t_1,t_2)$.

(b) Are $E[X^2(t)]$ and $E[\dot{X}^2(t)]$ finite for all $t > 0$? What, if any, restrictions on the value of b are required?

(c) Can $\{X(t)\}$ tend to a covariant stationary process as t goes to infinity?

****

5.7 Consider the response $\{X(t)\}$ of the linear system of Example 5.2 described by $m\,\ddot{X}(t) + c\,\dot{X}(t) = F(t)$ in which m and c are positive constants and $\{F(t)\}$ is a stationary stochastic process. Under what conditions will either of the mean-squared responses $E[X^2(t)]$ and $E[\dot{X}^2(t)]$ be stationary and finite?

****

5.8 Consider the response $\{X(t)\}$ of the linear system of Exercise 5.4 described by

$$\frac{d^3}{dt^3} X(t) + c \frac{d^2}{dt^2} X(t) = F(t)$$

in which c is a positive constant and $\{F(t)\}$ is a stationary stochastic process. Under what conditions will any of the mean-squared responses $E[X^2(t)]$, $E[\dot{X}^2(t)]$, and $E[\ddot{X}^2(t)]$ be stationary and finite?

****

5.9 Consider the response $\{X(t)\}$ of a linear system described by

$$\dot{X}(t) + a\,X(t) = W(t)\,U(t)$$

in which $\{W(t)\}$ is delta-correlated process with $\mu_W(t) = 0$ and $\phi_{WW}(t_1,t_2) = G_0\,\delta(t_1 - t_2)$. Note that the unit step causes $\{W(t)\}$ to be applied to the oscillator only for $t \ge 0$. The system is initially at rest: $X(0) = 0$.

(a) Find $\mu_X(t)$ and $\phi_{XX}(t_1,t_2)$ for all time values.

(b) Find $E[X^2(t)]$ for all $t > 0$.

(c) Find the stationary autocorrelation function: $R_{XX}(\tau) = \lim_{t\to\infty} \phi_{XX}(t+\tau,t)$.

****

5.10 Consider the response $\{X(t)\}$ of a linear system described by

$$c\,\dot{X}(t) + k\,X(t) = F(t)$$

with the deterministic initial condition $X(0) = x_0$. For $t > 0$, $\{F(t)\}$ is a mean-zero, nonstationary delta-correlated process with $\mu_F(t) = 0$ and $\phi_{FF}(t_1,t_2) = G_0\,(1 - e^{-\alpha t_1})\,\delta(t_1 - t_2)$.

(a) Find the $\mu_X(t)$ mean-value function of $\{X(t)\}$.

(b) Find the $K_{XX}(t_1,t_2)$ covariance of $\{X(t)\}$.

(c) Find $E[X^2(t)]$.

****

5.11 Consider a linear system whose response $\{X(t)\}$ to an excitation $\{F(t)\}$ is governed by $\ddot{X}(t) + c\,\dot{X}(t) = F(t)$, in which $c > 0$ is a constant. The excitation $\{F(t)\}$ is a delta-correlated process modulated by the unit step function. Specifically,

$$K_{FF}(t_1,t_2) = G_0\,\delta(t_1 - t_2)\,U(t_1)\,U(t_2)$$

(a) Find the variance of $\{X(t)\}$.

(b) Find the variance of $\{\dot{X}(t)\}$.

(c) Discuss the behavior of the response variances as t tends to $+\infty$.

****

5.12 Consider the response $\{X(t)\}$ of a linear system described by

$$\dot{X}(t) + a\,X(t) = W(t)\,U(t)\,U(T - t)$$

in which $T > 0$ and $\{W(t)\}$ is a stationary delta-correlated process with $\mu_W(t) = 0$ and $\phi_{WW}(t_1,t_2) = G_0\,\delta(t_1 - t_2)$. Note that the unit step causes $\{W(t)\}$ to be applied to the oscillator only for $0 \le t \le T$. The system is initially at rest: $X(0) = 0$.

(a) Find $\mu_X(t)$ and $\phi_{XX}(t_1,t_2)$ for all time values.

(b) Sketch $E[X^2(t)]$ versus t (for all $t > 0$).

****

5.13 Consider a linear system governed by the first-order differential equation

$$\dot{X}(t) + a\,X(t) = F(t) \qquad \text{for } t \ge 0$$

in which a is a constant, with $0<a<1$. Let $\{X(t)\}$ have the random initial condition $X(0)=Y$ and let the forcing function consist of one random impulse at time $t=(4a)^{-1}$: $F(t)=Z\delta[t-1/(4a)]$, in which Y and Z are random variables with $\mu_Y=1$, $E(Y^2)=2$, $\mu_Z=0$, $E(Z^2)=5$, $E(YZ)=1.5$.

(a) Find $\mu_X(t)$ for all $t\geq 0$.

(b) Find $E[X^2(t)]$ for all $t\geq 0$.

(c) Sketch your answers to (a) and (b) versus t.

****

5.14 Consider a linear system governed by the first-order differential equation

$$\dot{X}(t)+a\,X(t)=F(t)$$

The excitation $\{F(t)\}$ is a stationary random process with a nonzero mean value, $\mu_F=b$, and an autocovariance function $K_{FF}(t+\tau,t)=G_0\,e^{-c|\tau|}$ for all t and τ. The terms a, b, and c are constants with $a>0$ and $c>0$.

(a) Find the μ_X mean value of the $\{X(t)\}$ response.

(b) Find the $E(X^2)$ mean-squared value of the $\{X(t)\}$ response.

****

5.15 Let the random process $\{Z(t)\}$ denote the ground acceleration during an earthquake. One possible model is a segment from a stationary process described by

$$\dot{Z}(t)+b\,Z(t)=b\,W(t)$$

in which $\{W(t)\}$ is a mean-zero delta-correlated process with $K_{WW}(t,s)=G_0\,\delta(t-s)$.

(a) Find the stationary value of $E(Z^2)$.

(b) Data for North American earthquakes indicate that $b=6\pi$ rad/s (i.e., 3 Hz) is a reasonable choice. Using this b value, find the value of G_0 such that the stationary standard deviation of $\{Z(t)\}$ is $\sigma_Z=0.981$ m/s^2 (i.e., 0.1 g). Also give your units for G_0.

****

5.16 Consider a causal linear system for which the $h_x(t)$ impulse response function is known. That is, if the input is $F(t)$ then the response is

$$X(t)=\int_{-\infty}^{\infty}F(s)\,h_x(t-s)\,ds$$

(a) Find the general integral expression relating the $E[X(t_1)X(t_2)X(t_3)]$ third moment of response to the corresponding $E[F(t_1)F(t_2)F(t_3)]$ input expression.

(b) Simplify the integral in part (a) for the special case of a stationary delta-correlated excitation with $E[F(t_1)F(t_2)F(t_3)]=Q_0\,\delta(t_1-t_3)\,\delta(t_2-t_3)$.

(c) For the special case of $h_x(t)=e^{-at}\,U(t)$ and the delta-correlated excitation of part (b), find the stationary value of $E(X^3)$.

****

SDF System

****

5.17 Consider the response $\{X(t)\}$ of a linear SDF system described by

$$\ddot{X}(t)+2\zeta\,\omega_0\,\dot{X}(t)+\omega_0^2\,X(t)=F(t)$$

with the deterministic initial conditions $X(0)=0$, $\dot{X}(0)=v_0$. For $t>0$, $\{F(t)\}$ is a stationary delta-correlated process with $\mu_F(t)=0$ and $\phi_{FF}(t_1,t_2)=G_0\,\delta(t_1-t_2)$. G_0 and ω_0 are positive constants and $0<\zeta<1$.

(a) Find the $\mu_X(t)$ mean-value function of the $\{X(t)\}$ response for $t\ge 0$.

(b) Find the variance of $\{X(t)\}$ for $t\ge 0$.

(c) Find $E[X^2(t)]$ for $t\ge 0$.

Give all answers in terms of the parameters G_0, ω_0, ζ, and $\omega_d=\omega_0\,(1-\zeta_0^2)^{1/2}$.

****

5.18 Consider the response $\{X(t)\}$ of a linear SDF system described by

$$\ddot{X}(t)+2\zeta\,\omega_0\,\dot{X}(t)+\omega_0^2\,X(t)=F(t)$$

with random initial conditions of $X(0)=Y$, $\dot{X}(0)=0$. The forcing function consists of one random impulse at time T: $F(t)=Z\,\delta(t-T)$ in which $T>0$ is a constant and Y and Z are random variables with

$$E(Y)=1,\ E(Y^2)=2,\ E(Z)=0,\ E(Z^2)=5\,\omega_0^2,\ E(YZ)=1.5\,\omega_0$$

(a) Find $\mu_X(t)$ for all $t\ge 0$.

(b) Find $E[X^2(t)]$ for all $t\ge 0$.

(c) Sketch your answers to (a) and (b) versus $\omega_0 t$ for $\omega_0 T=10$.

****

5.19 Assume that you are concerned about the possibility of collision between two buildings during wind-excited vibration. The east-west clearance between the buildings is 35 mm. Model each building as an SDF system:

$$\ddot{X}_j(t)+2\zeta_j\,\omega_j\,\dot{X}_j(t)+\omega_j^2\,X_j(t)=F_j(t)$$

in which $X_j=$ displacement (to the east) of the top of building j. The building parameters are

Building A: $\omega_A=2\pi$ rad/s, $\zeta_A=0.01$
Building B: $\omega_B=\pi$ rad/s, $\zeta_B=0.01$

The critical condition is with the wind blowing from the north, so you can neglect any static deflection of the buildings. Model the east-west excitation of each building (due to vortex shedding, etc.) as a delta-correlated process: $K_{F_jF_j}(t,s)=G_j\,\delta(t-s)$. Noting that $F_j(t)$ is the force per unit mass in each SDF system, consider the excitation intensities to be $G_A=40$ (mm^2/s^3) and $G_B=80$ (mm^2/s^3).

(a) Find the standard deviation σ_{X_j} for the stationary displacement response at the top of each building.

(b) Let $Y_B = X_B/2$ denote the displacement of the taller building B at the level of the top of building A and $Z(t) = X_A(t) - Y_B(t)$ be the relative displacement at the top of building A. Estimate the value of the stationary standard deviation σ_Z by assuming that $\{X_A(t)\}$ and $\{X_B(t)\}$ are independent random processes.

(c) Presume that you wish to reduce the probability of collision by adding damping to one of the structures. To which structure would you add the damping? Briefly explain your answer.

****

5.20 Assume that you are concerned about the possibility of collision between two buildings during an earthquake. The two buildings are of the same height and north-south clearance between them is 20 cm. Model each building as an SDF system:

$$\ddot{X}_j(t) + 2\zeta_j \omega_j \dot{X}_j(t) + \omega_j^2 X_j(t) = -\ddot{Z}(t)$$

in which $X_j(t)$ is the north-south displacement at the top of building j and $\{\ddot{Z}(t)\}$ is the north-south ground acceleration. The building parameters are

Building A: $\omega_A = 2\pi$ rad/s, $\zeta_A = 0.01$

Building B: $\omega_B = 3\pi$ rad/s, $\zeta_B = 0.01$

Presume that the earthquake has such a long duration that the critical portion of the $\{\ddot{Z}(t)\}$ excitation and the $\{X_j(t)\}$ responses may all be modeled as stationary processes. Let $\{\ddot{Z}(t)\}$ be modeled as a delta-correlated process with $K_{\ddot{Z}\ddot{Z}}(t,s) = (0.1 \text{ m}^2/\text{s}^3)\,\delta(t-s)$.

(a) Find the standard deviation σ_{X_j} for the displacement response at the top of each building.

(b) Let $Z(t) = X_A(t) - X_B(t)$ be the relative displacement at the top of the buildings. Estimate the value of the stationary standard deviation σ_Z by assuming that $\{X_A(t)\}$ and $\{X_B(t)\}$ are independent random processes.

(c) Presume that you wish to reduce the probability of collision by adding damping to one of the structures. To which structure would you add the damping? Briefly explain your answer.

****

5.21 Consider approximating an earthquake ground acceleration of 30-second duration by using a mean-zero, delta-correlated $\{\ddot{Y}(t)\}$ process with

$$K_{\ddot{Y}\ddot{Y}}(t,s) = (1.5 \text{ m}^2/\text{s}^3)\,\delta(t-s)\,U(t)\,U(30-t)$$

in which t is time in seconds. Let the ground velocity have an initial value of zero: $\dot{Y}(0) = 0$. Evaluate $E[\dot{Y}^2(t)]$ and sketch this mean-squared value versus t over the range of $0 \le t \le 50$.

****

5.22 Consider approximating an earthquake ground acceleration by using a nonstationary mean-zero, delta-correlated $\{\ddot{Y}(t)\}$ process with

$$K_{\ddot{Y}\ddot{Y}}(t,s) = (6\ \mathrm{m}^2/\mathrm{s}^3)(e^{-t/10} - e^{-t/5})\delta(t-s)U(t)$$

in which t is time in seconds. Let the ground velocity have an initial value of zero: $\dot{Y}(0) = 0$. Evaluate $E[\dot{Y}^2(t)]$ and sketch this mean-squared value versus t.

****

5.23 Let the random process $\{Z(t)\}$ denote the earthquake ground acceleration model proposed in Exercise 5.15 with

$$\dot{Z}(t) + b\,Z(t) = b\,W(t)$$

in which $\{W(t)\}$ is a delta-correlated process. Let $\{X(t)\}$ represent the response of an SDF structural model excited by this ground motion

$$m\,\ddot{X}(t) + c\,\dot{X}(t) + k\,X(t) = -m\,Z(t)$$

Substitute $Z(t)$ from the second equation into the first equation to obtain one ordinary differential equation relating $\{X(t)\}$ and $\{W(t)\}$.

****

Response to Gaussian Excitation

****

5.24 Let the stationary, mean-zero process $\{X(t)\}$ denote the stress in a particular member of a linear system responding to a Gaussian stochastic excitation. Let the standard deviation of $X(t)$ be $\sigma_X = 75$ MPa. Find $P[X(t) \ge 350\ \text{MPa}]$ for any particular time t.

****

5.25 Let the stationary process $\{X(t)\}$ denote the deflection at the top of a linear offshore structure responding to a Gaussian stochastic sea. Let the mean and standard deviation of $X(t)$ be $\mu_X = 3$ m and $\sigma_X = 2$ m. Find $P[X(t) \ge 10\ \text{m}]$ for any particular time t.

****

Chapter 6
Frequency Domain Analysis

6.1 Frequency Content of a Stochastic Process

The Fourier transform provides the classical method for decomposing a time history into its frequency components (see Appendix B for brief background information on Fourier analysis). For any time history $f(t)$, we will denote the Fourier transform by $\tilde{f}(\omega)$ and define it as

$$\tilde{f}(\omega) = \frac{1}{2\pi} \int_{-\infty}^{\infty} f(t) e^{-i\omega t} \, dt \tag{6.1}$$

The inverse relationship is then

$$f(t) = \int_{-\infty}^{\infty} \tilde{f}(\omega) e^{i\omega t} \, d\omega \tag{6.2}$$

The interpretation of the Fourier transform as a frequency decomposition is based on Eq. 6.2. Inasmuch as an integral may be viewed as the limit of a summation, Eq. 6.2 shows that the original time history $f(t)$ is essentially a summation of harmonic terms, with $\tilde{f}(\omega) d\omega$ being the amplitude of the $e^{i\omega t}$ component having frequency ω. This amplitude is generally complex, as shown by Eq. 6.1, so one must consider its absolute value to determine how much of the total $f(t)$ time history is contributed by frequency ω. A condition ensuring the existence of the Fourier transform $\tilde{f}(\omega)$ is that $f(t)$ be absolute-value integrable:

$$\int_{-\infty}^{\infty} | f(t) | dt < \infty \tag{6.3}$$

Equation 6.2 then produces exactly the original $f(t)$ function at all points of continuity of $f(t)$. At a point of finite discontinuity of $f(t)$, Eq. 6.2 gives a value midway between the left-hand and right-hand limits at the point.

We now wish to apply the Fourier transform idea to a stochastic process $\{X(t)\}$. Simply using a stochastic integrand in Eq. 6.1 gives

$$\tilde{X}(\omega) = \frac{1}{2\pi} \int_{-\infty}^{\infty} X(t) e^{-i\omega t} dt \tag{6.4}$$

which is a stochastic integral of the form of Eq. 4.75. This defines a new stochastic process $\{\tilde{X}(\omega)\}$ on the set of all possible ω values. Taking the expectation of Eq. 6.4 shows that the mean-value function for $\{\tilde{X}(\omega)\}$, if it exists, can be written as

$$\mu_{\tilde{X}}(\omega) = \frac{1}{2\pi} \int_{-\infty}^{\infty} \mu_X(t) e^{-i\omega t} dt = \tilde{\mu}_X(\omega) \tag{6.5}$$

The purpose of the last term in Eq. 6.5 is to emphasize that the mean-value function for the Fourier transform of $\{X(t)\}$ is exactly the Fourier transform of the mean-value function for $\{X(t)\}$.

Next, we consider the second moments of the $\{\tilde{X}(\omega)\}$ process, but we slightly modify the definition of the second-moment function because $\{\tilde{X}(\omega)\}$ is not real. For a complex stochastic process, we will use the standard procedure of defining the second-moment function with a complex conjugate (denoted by the superscript *) on the second term:

$$\phi_{\tilde{X}\tilde{X}}(\omega_1, \omega_2) = E[\tilde{X}(\omega_1) \tilde{X}^*(\omega_2)] \tag{6.6}$$

Note that this is identical to Eq. 4.2 for a real process. One reason for the inclusion of the complex conjugate is so that $\phi_{\tilde{X}\tilde{X}}(\omega_1, \omega_2)$ will be real along the line $\omega_2 = \omega_1 = \omega$. In particular, we have $\phi_{\tilde{X}\tilde{X}}(\omega,\omega) = E[|\tilde{X}^2(\omega)|]$. Using this definition and Eq. 6.4 gives

$$\phi_{\tilde{X}\tilde{X}}(\omega_1, \omega_2) = \frac{1}{(2\pi)^2} \int_{-\infty}^{\infty} \int_{-\infty}^{\infty} \phi_{XX}(t_1,t_2) e^{-i(\omega_1 t_1 - \omega_2 t_2)} dt_1\, dt_2 \tag{6.7}$$

The corresponding definition of the autocovariance function is

$$K_{\tilde{X}\tilde{X}}(\omega_1, \omega_2) = E\Big([\tilde{X}(\omega_1) - \mu_{\tilde{X}}(\omega_1)][\tilde{X}(\omega_2) - \mu_{\tilde{X}}(\omega_2)]^*\Big)$$

and this is easily shown to be

$$K_{\tilde{X}\tilde{X}}(\omega_1, \omega_2) = \frac{1}{(2\pi)^2} \int_{-\infty}^{\infty} \int_{-\infty}^{\infty} K_{XX}(t_1,t_2) e^{-i(\omega_1 t_1 - \omega_2 t_2)} dt_1\, dt_2 \tag{6.8}$$

Expressions for higher moment functions can, obviously, also be derived.

One major difficulty in applying the Fourier transform procedure to many problems of interest is the fact that the expressions just written will not exist if $\{X(t)\}$ is a stationary stochastic process. In particular, Eqs. 6.3 and 6.5 show that $\mu_{\tilde{X}}(\omega)$ may not exist unless $\mu_X(t)$ is absolute value integrable, and this requires that $\mu_X(t)$ tend to zero as $|t|$ tends to infinity. In general, $\mu_{\tilde{X}}(\omega)$ exists for a mean-value stationary process only if the process is mean-zero, in which case $\mu_{\tilde{X}}(\omega)$ is also zero. Similarly, Eqs. 6.7 and 6.8 show that $\phi_{\tilde{X}\tilde{X}}(\omega_1,\omega_2)$ and $K_{\tilde{X}\tilde{X}}(\omega_1,\omega_2)$ may not exist unless $\phi_{XX}(t_1,t_2)$ and $K_{XX}(t_1,t_2)$, respectively, tend to zero as t_1 and t_2 tend to infinity. These conditions cannot be uniformly met for a stochastic process that is second-moment stationary or covariant stationary, respectively. In particular, we have $\phi_{XX}(t_1,t_2)=R_{XX}(t_1-t_2)$ and $K_{XX}(t_1,t_2)= G_{XX}(t_1-t_2)$ for a stationary process, so t_1 and t_2 tending to infinity with a finite value of (t_1-t_2) would not give $\phi_{XX}(t_1,t_2)$ or $K_{XX}(t_1,t_2)$ tending to zero and the integrals might not exist. Thus, we will modify the Fourier transform procedure to obtain a form that will apply to stationary processes.

6.2 Spectral Density Functions for Stationary Processes

To avoid the problem of existence of the Fourier transforms, consider a new stochastic process $\{X_T(\omega)\}$ that is a truncated version of our original stationary process $\{X(\omega)\}$:

$$X_T(t)=X(t)U(t+T/2)U(T/2-t) \tag{6.9}$$

in which $U(\cdot)$ denotes the unit step function (see Eq. 2.5). The Fourier transform of $X_T(\omega)$ is sure to exist because the process is defined to be zero if $|t|>T/2$, as illustrated in Fig. 6.1. In particular, the moment integrals in Eqs. 6.5–6.8 will exist, because the limits of integration will now be bounded. Of course, we must consider the behavior as T goes to infinity or we will not have a description of the complete $\{X(t)\}$ process, and we know that the integrals in Eqs. 6.5–6.8 may diverge in that situation. Thus, we must find a way of normalizing our results in such a way that a limit exists as T goes to infinity.

We really need not give any more attention to the mean-value function of the Fourier transform for a stationary process. If $\{X(t)\}$ is mean-value stationary and $\mu_X \neq 0$, then the Fourier transform of $\mu_X(t)$ contains only an infinite spike at the origin: $\mu_{\tilde{X}}(\omega)=\mu_X\,\delta(\omega)$. The more interesting situation involves the

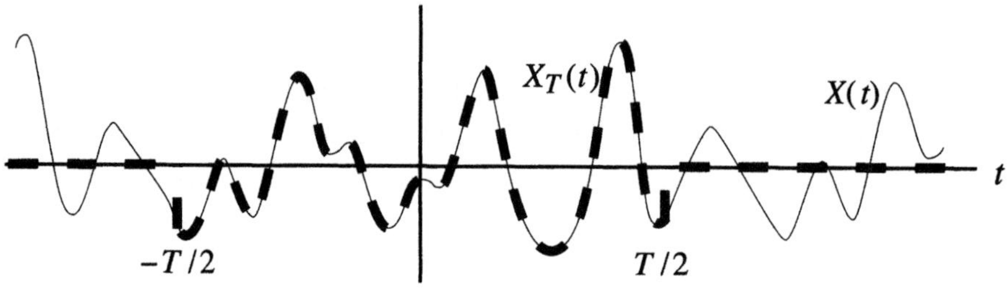

Figure 6.1 Truncated time history of $X(t)$.

second moments of the process. Specifically, we will consider $\{X(t)\}$ to be stationary, so

$$\phi_{\tilde{X}_T \tilde{X}_T}(\omega_1,\omega_2)=\frac{1}{(2\pi)^2}\int_{-T/2}^{T/2}\int_{-T/2}^{T/2} R_{XX}(t_1-t_2)e^{-i(\omega_1 t_1-\omega_2 t_2)}\,dt_1\,dt_2 \tag{6.10}$$

and

$$K_{\tilde{X}_T \tilde{X}_T}(\omega_1,\omega_2)=\frac{1}{(2\pi)^2}\int_{-T/2}^{T/2}\int_{-T/2}^{T/2} G_{XX}(t_1-t_2)e^{-i(\omega_1 t_1-\omega_2 t_2)}\,dt_1\,dt_2 \tag{6.11}$$

In most applications, we will find that two of the random variables that make up a stochastic process are independent if they correspond to observation times that are infinitely far apart. That is, $X(t+\tau)$ becomes uncorrelated with $X(t)$ as $|\tau|$ becomes very large. This causes $G_{XX}(\tau)$ to tend to zero and $R_{XX}(\tau)$ to tend to $(\mu_X)^2$ as $|\tau|$ tends to infinity. One consequence of this behavior is the fact that as T tends to infinity the integral in Eq. 6.10 may grow much more rapidly than the one in Eq. 6.11. Thus, we will focus on the case with better behavior and restrict our attention to covariance only. If the stationary mean value is zero, then this is the same as the second moment of the process. If $\mu_X(t)$ does not equal zero, then analysis of covariance gives us information about how $X(t)$ varies from this deterministic mean value.

We will now analyze the covariance function of the Fourier transform of the truncated stochastic process $\{X_T(\omega)\}$ when $\{X(t)\}$ is a covariant stationary process. The integral in Eq. 6.11 can be rearranged by using exactly the same changes of variables and other manipulations as were used in deriving the ergodicity condition in Eq. 4.50, although the presence of the exponential term makes things a little more complicated this time. The result can be written as

$$K_{\tilde{X}_T\tilde{X}_T}(\omega_1,\omega_2)=\frac{1}{(2\pi)^2}\left(\frac{2\sin[(\omega_1-\omega_2)T/2]}{(\omega_1-\omega_2)}\int_{-T}^{T}G_{XX}(\tau)e^{-i\omega_1\tau}\,d\tau-\right.$$
$$2\cos[(\omega_1-\omega_2)T/2]\int_0^T G_{XX}(\tau)\left[\frac{\sin(\omega_1\tau)-\sin(\omega_2\tau)}{(\omega_1-\omega_2)}\right]d\tau+$$
$$\left.\frac{2\sin[(\omega_1-\omega_2)T/2]}{(\omega_1-\omega_2)}\int_0^T G_{XX}(\tau)\big[\cos(\omega_1\tau)-\cos(\omega_2\tau)\big]\,d\tau\right) \tag{6.12}$$

Note that this expression is quite well behaved if ω_1 and ω_2 are not nearly equal. In particular, if we impose the condition that the autocovariance function is absolutely integrable

$$\int_{-\infty}^{\infty}|G_{XX}(\tau)|\,d\tau<\infty \tag{6.13}$$

then it is clear that we have $K_{\tilde{X}_T\tilde{X}_T}(\omega_1,\omega_2)$ bounded for any T value, including $T\to\infty$, for $\omega_1\neq\omega_2$. The situation is not as good, though, when ω_1 and ω_2 are nearly equal. Letting $\omega_2=\omega$ and $\omega_1=\omega+\Delta\omega$, then taking the limit as $\Delta\omega\to 0$ so that ω_1 and ω_2 approach a common value, gives $\sin(\omega_1\tau)\approx\sin(\omega\tau)+\Delta\omega\,\tau\cos(\omega\tau)$ and $\cos(\omega_1\tau)\approx\cos(\omega\tau)-\Delta\omega\,\tau\sin(\omega\tau)$. Substituting these expressions into Eq. 6.12, along with $\sin(\Delta\omega\,\tau)\approx\Delta\omega\,\tau$, shows that the final term in Eq. 6.12 drops out and the other two terms give

$$K_{\tilde{X}_T\tilde{X}_T}(\omega,\omega)=\frac{1}{(2\pi)^2}\left(T\int_{-T}^{T}G_{XX}(\tau)e^{-i\omega\tau}\,d\tau-2\int_0^T G_{XX}(\tau)\,\tau\,\cos(\omega\tau)\,d\tau\right)$$

Clearly, the first term in this expression will grow proportionally with T as T goes to infinity, under the condition of Eq. 6.13. It is not immediately obvious how the second term will behave, but it can be shown that it grows less rapidly than the first, so the first term always dominates in the limit.[1] Thus, we can say that

---

[1]We will not include a proof of the general case, but it is very easy to verify the special case in which $|G_{XX}(\tau)|\le c\,\tau^{-1-\varepsilon}$ for some positive constants c and ε. Then, the absolute value of the second integral in the equation is bounded by $c\,T^{1-\varepsilon}$ as T goes to infinity, so the second term becomes less significant than the first term.

$$K_{\tilde{X}_T\tilde{X}_T}(\omega,\omega) \rightarrow \frac{T}{(2\pi)^2}\int_{-\infty}^{\infty} G_{XX}(\tau)e^{-i\omega\tau}\,d\tau \quad \text{as } T \rightarrow \infty \tag{6.14}$$

Because Eq. 6.14 grows proportionally with T for large T values, we can define a normalized form that will exist for $T \rightarrow \infty$ by dividing $K_{\tilde{X}_T\tilde{X}_T}(\omega_1,\omega_2)$ by T. Note, though, that this normalized autocovariance of the Fourier transform will go to zero as T goes to infinity if $\omega_1 \neq \omega_2$ because the unnormalized form was finite in that situation. Thus, all the useful information that this normalized form can give us about the stationary process is included in the special case $K_{\tilde{X}_T\tilde{X}_T}(\omega,\omega)$, corresponding to $\omega_1 = \omega_2 = \omega$. Therefore, we define the new function of frequency as

$$\begin{aligned} S_{XX}(\omega) &= \lim_{T\rightarrow\infty} \frac{2\pi}{T} K_{\tilde{X}_T\tilde{X}_T}(\omega,\omega) \\ &\equiv \lim_{T\rightarrow\infty} \frac{2\pi}{T} E\Big([\tilde{X}_T(\omega) - \mu_{\tilde{X}_T}(\omega)][\tilde{X}_T(\omega) - \mu_{\tilde{X}_T}(\omega)]^*\Big) \end{aligned} \tag{6.15}$$

in which the inclusion of the factor of 2π is arbitrary, but traditional. An equivalent form is

$$S_{XX}(\omega) = \lim_{T\rightarrow\infty} \frac{2\pi}{T} E[|\tilde{X}_T(\omega) - \tilde{\mu}_{X_T}(\omega)|^2] \tag{6.16}$$

Noting the fact that $\tilde{X}_T(-\omega) = \tilde{X}_T^*(\omega)$ for a real $X(t)$ allows us to rewrite this expression as

$$S_{XX}(\omega) = \lim_{T\rightarrow\infty} \frac{2\pi}{T} E\Big([\tilde{X}_T(\omega) - \tilde{\mu}_{X_T}(\omega)][\tilde{X}_T(-\omega) - \tilde{\mu}_{X_T}(-\omega)]\Big) \tag{6.17}$$

We will call $S_{XX}(\omega)$ the *autospectral density function* of $\{X(t)\}$. Other terms that are commonly used for it are *power spectral density* or simply *spectral density*. Many authors define the spectral density slightly differently, using the autocorrelation function rather than the autocovariance function. The difficulty with that approach is that it gives a spectral density that generally does not exist if the Fourier transform of $\mu_X(t)$ does not exist, as when $\{X(t)\}$ is mean-value stationary and $\mu_X \neq 0$. This problem is then circumvented by looking at the spectral density of a mean-zero process defined as $Y(t) = X(t) - \mu_X(t)$. The definition used here gives exactly the same result in a more direct manner.

Comparing Eqs. 6.14 and 6.15 shows a very important result:

$$S_{XX}(\omega) = \frac{1}{2\pi} \int_{-\infty}^{\infty} G_{XX}(\tau) e^{-i\omega\tau} \, d\tau \tag{6.18}$$

The $S_{XX}(\omega)$ autospectral density function is exactly the Fourier transform of the $G_{XX}(\tau)$ autocovariance function. Furthermore, this implies that the inverse must also be true:

$$G_{XX}(\tau) = \int_{-\infty}^{\infty} S_{XX}(\omega) e^{i\omega\tau} \, d\omega \tag{6.19}$$

We know that this Fourier transform pair will exist for a process $\{X(t)\}$ satisfying Eq. 6.13. Note also, the similarity of Eqs. 6.18 and 6.8. Equation 6.8 defines $K_{\tilde{X}\tilde{X}}(\omega_1, \omega_2)$ as the double Fourier transform of the general $K_{XX}(t_1,t_2)$ covariance function, one transform with respect to t_1 and one with respect to t_2. For the covariant stationary process we know that $K_{XX}(t_1,t_2)$ can be replaced by a function $G_{XX}(\tau)$ of the one time argument $\tau = t_1 - t_2$, and Eq. 6.18 shows that the autospectral density function is the Fourier transform of this function with respect to its single time argument.

Similar to Eq. 6.15, one can define a cross-spectral density function for two jointly covariant stationary stochastic processes $\{X(t)\}$ and $\{Y(t)\}$ as

$$\begin{aligned} S_{XY}(\omega) &= \lim_{T\to\infty} \frac{2\pi}{T} K_{\tilde{X}_T \tilde{Y}_T}(\omega,\omega) \\ &= \lim_{T\to\infty} \frac{2\pi}{T} E\Big([\tilde{X}_T(\omega) - \tilde{\mu}_{X_T}(\omega)][\tilde{Y}_T(-\omega) - \tilde{\mu}_{Y_T}(-\omega)]\Big) \end{aligned} \tag{6.20}$$

and show that it is the Fourier transform of the cross-covariance function of the processes

$$S_{XY}(\omega) = \frac{1}{2\pi} \int_{-\infty}^{\infty} G_{XY}(\tau) e^{-i\omega\tau} \, d\tau \tag{6.21}$$

There is another alternative to the truncated Fourier transforms used here to characterize the frequency decomposition of a stationary stochastic process. In this alternative form one uses the Lebesque integral to write

$$X(t) = \int_{-\infty}^{\infty} e^{i\omega t} \, d\Xi(\omega)$$

Note that the increment of the new complex, frequency-domain stochastic process $\{\Xi(\omega)\}$ replaces $\tilde{X}(\omega)\,d\omega$ in the Fourier transform formulation. The spectral density relationship is then written in terms of this increment of $\{\Xi(\omega)\}$ as $E[d\Xi(\omega_1)\,d\Xi(\omega_2)] = S_{XX}(\omega_1)\delta(\omega_1+\omega_2)\,d\omega_1\,d\omega_2$. In this book we have chosen to use the truncated Fourier transform approach because it seems easier to comprehend and to relate to physical observations.

6.3 Properties of Spectral Density Functions

First we note three physical features of the autospectral density function. Directly from the definition in Eq. 6.15 or 6.16, we see that for any stochastic process $\{X(t)\}$ we must have all of the following:

$S_{XX}(\omega)$ is always real for all values of ω

$S_{XX}(\omega) \ge 0$ for all values of ω

$S_{XX}(-\omega) = S_{XX}(\omega)$ for all values of ω

The fact that $S_{XX}(\omega)$ is real and symmetric can also be immediately verified from Eq. 6.18, keeping in mind that $G_{XX}(\tau)$ is real and symmetric. Confirming that Eq. 6.18 also implies that $S_{XX}(\omega)$ is nonnegative requires use of the fact that $G_{XX}(\tau)$ is a nonnegative definite function (see Section 3.3).

Possibly the most important feature of the various spectral density functions is that each of them gives all the information about a corresponding covariance (or cumulant) function. In particular, the usual autospectral density function of Eqs. 6.15–6.19 gives all the information about the autocovariance function of the $\{X(t)\}$ process, and the cross-spectral density function of Eqs. 6.20 and 6.21 gives all the information about the cross-covariance of $\{X(t)\}$ and $\{Y(t)\}$. We know that this must be true because the inverse Fourier transform formula (as in Eq. 6.19) allows us to compute the covariance function from knowledge of the spectral density function. Combined with the fundamental idea of the Fourier transform, this shows that a spectral density function is a complete frequency decomposition of a stationary covariance function. For example, $S_{XX}(\omega)$ can definitely be considered as a single-frequency component inasmuch as its definition depends only on the frequency ω part of the Fourier transform (i.e., the frequency decomposition) of $\{X(t)\}$, and knowledge of all the $S_{XX}(\omega)$ frequency components is sufficient information to recreate the covariance function. These are precisely the characteristics of a frequency decomposition of any quantity.

Another important property of autospectral density functions is obtained from the special case corresponding to setting $\tau = 0$ in Eq. 6.19, giving

$$\sigma_X^2 = G_{XX}(0) = \int_{-\infty}^{\infty} S_{XX}(\omega)\, d\omega \tag{6.22}$$

This shows that the variance of a covariant stationary process can always be found from the area under its autospectral density function. This property is frequently used for calculation of variance values in stochastic dynamic analysis.

The symmetry of autospectral density has also led to the use of an alternative form called the single-sided autospectral density. One argument for using this representation is that all the information is contained within the half of the function corresponding to $0 \le \omega \le \infty$, so why bother with the other half. To retain the property that variance is given by an integral of autospectral density, as in Eq. 6.22, the single-sided autospectral density is typically taken to be $2 S_{XX}(\omega)$. It should also be noted that there are other normalizations of the autospectral density in common usage, so one must be careful in interpreting the meaning of any quoted value for the quantity. In particular, some prefer to replace Eq. 6.22 with an integral over frequency in Hz (cycles per second) rather than radians per second, and this requires a modification of autospectral density by a factor of 2π. Furthermore, this variation may be found in conjunction with either the two-sided $S_{XX}(\omega)$ that we use or the single-sided version, giving four different possibilities. Throughout this book we will use only the two-sided autospectral density function defined by Eq. 6.15, because use of various forms seems to add unnecessary possibilities for confusion.

Next we will use Eq. 6.19 to evaluate the derivatives of the autocovariance function, as

$$\frac{d^j G_{XX}(\tau)}{d\tau^j} = (i)^j \int_{-\infty}^{\infty} \omega^j\, S_{XX}(\omega)\, e^{i\omega\tau}\, d\omega$$

and substitute these expressions into relationships between derivatives of autocovariance functions and covariance functions involving the stochastic derivative of a process, as given in Eqs. 4.60 and 4.61. Thus, the cross-covariance of $\{X(t)\}$ and $\{\dot{X}(t)\}$ is given by

$$G_{X\dot{X}}(\tau) = -\frac{d\,G_{XX}(\tau)}{d\tau} = -i\int_{-\infty}^{\infty} \omega\, S_{XX}(\omega)\, e^{i\omega\tau}\, d\omega$$

and the autocovariance of $\{\dot{X}(t)\}$ is

$$G_{\dot{X}\dot{X}}(\tau) = -\frac{d^2 G_{XX}(\tau)}{d\tau^2} = \int_{-\infty}^{\infty} \omega^2\, S_{XX}(\omega)\, e^{i\omega\tau}\, d\omega$$

Comparing these two results with Eq. 6.19 shows that the spectral densities involving $\{\dot{X}(t)\}$ are simply related to those for $\{X(t)\}$. For example, the expression for $G_{X\dot{X}}(\tau)$ is precisely in the form of an inverse Fourier transform of $(-i\,\omega)\, S_{XX}(\omega)$. The uniqueness of the Fourier transform then tells us that

$$S_{X\dot{X}}(\omega) = -i\,\omega\, S_{XX}(\omega) \tag{6.23}$$

Similarly, the expression for $G_{\dot{X}\dot{X}}(\tau)$ tells us that

$$S_{\dot{X}\dot{X}}(\omega) = \omega^2\, S_{XX}(\omega) \tag{6.24}$$

In addition, if we write $\{X^{(j)}(t)\}$ for the jth-order stochastic derivative with respect to t of $\{X(t)\}$, then we can obtain the general result that

$$S_{X^{(j)}X^{(k)}}(\omega) = (-1)^k (i)^{j+k} \omega^{j+k}\, S_{XX}(\omega) \tag{6.25}$$

which includes the special case of

$$S_{X^{(j)}X^{(j)}}(\omega) = \omega^{2j}\, S_{XX}(\omega)$$

for the autospectral density function of the jth derivative process.

6.4 Narrowband Processes

We will now investigate the relationship between the shape of an autospectral density function and the nature of the possible time histories for the stochastic process that it describes. We will begin this investigation with consideration of so-called narrowband processes. A process $\{X(t)\}$ is said to be narrowband if the autospectral density $S_{XX}(\omega)$ is very small except within a narrow band of frequencies. Because $S_{XX}(\omega)$ is an even function, this really means that the band of significant frequencies appears both for positive and negative ω values. We

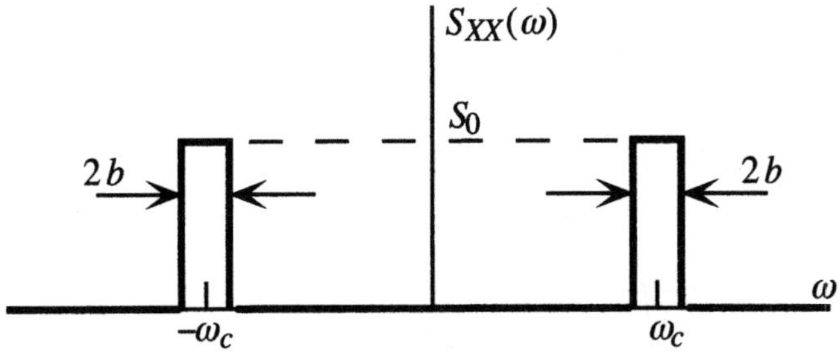

Figure 6.2 Ideal narrowband process.

can state this narrowband condition as $S_{XX}(\omega) \approx 0$ unless $|\omega| \approx \omega_c$ for some given characteristic frequency ω_c. To illustrate this idea more clearly, let us choose the example with $S_{XX}(\omega)$ equal to a constant value S_0 if ω is within a distance b of either $+\omega_c$ or $-\omega_c$, and identically zero for all other frequencies, as shown in Fig. 6.2. Mathematically, this can be written as

$$S_{XX}(\omega) = S_0\, U[|\omega| - (\omega_c - b)]\, U[(\omega_c + b) - |\omega|]$$

We can now use the inverse Fourier transform of Eq. 6.19 to find the corresponding autocovariance function

$$G_{XX}(\tau) = S_0 \int_{-\omega_c - b}^{-\omega_c + b} e^{i\omega\tau}\, d\omega + S_0 \int_{\omega_c - b}^{\omega_c + b} e^{i\omega\tau}\, d\omega = 2 S_0 \int_{\omega_c - b}^{\omega_c + b} \cos(\omega\tau)\, d\omega$$

or

$$G_{XX}(\tau) = 2 S_0 \frac{\sin[(\omega_c + b)\tau] - \sin[(\omega_c - b)\tau]}{\tau} = 4 S_0 \frac{\cos(\omega_c \tau)\sin(b\tau)}{\tau}$$

Letting $\tau \to 0$ in this expression gives the variance of the process as $\sigma_X^2 = 4 b S_0$, which could also have been found from the area under the autospectral density curve, as given in Eq. 6.22. We can use this variance as a normalization factor and write the autocovariance function in a convenient form as

$$G_{XX}(\tau) = \sigma_X^2 \cos(\omega_c \tau) \frac{\sin(b\tau)}{b\tau}$$

For this process to be classified as narrowband, it is necessary that the parameter b be small in some sense. In particular, if $b << \omega_c$, then the oscillations of the $\sin(b\tau)$ term are very slow compared with those at frequency ω_c. In this case, one can consider $\sin(b\tau)/(b\tau)$ to provide a slowly varying envelope for the $\sigma_X^2 \cos(\omega_c\tau)$ term. Figure 6.3 shows a sketch of this behavior for $b = \omega_c/10$. It is

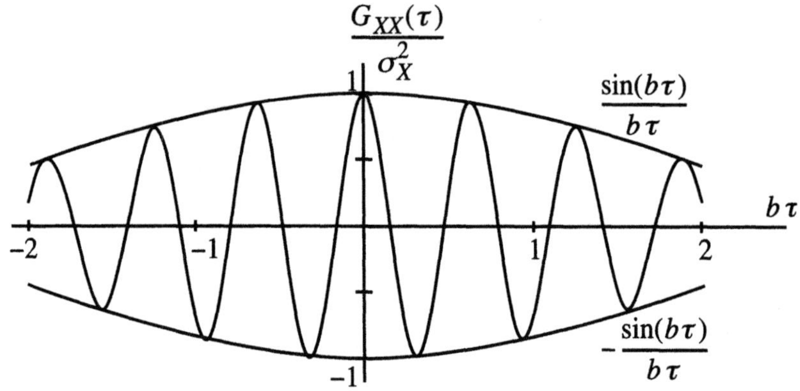

Figure 6.3 Autocovariance function for a narrowband process.

seen that $G_{XX}(\tau)$ is well approximated by $\sigma_X^2 \cos(\omega_c \tau)$ for small values of τ but the amplitude of this cosine function decays as τ increases. This is a fundamental characteristic of any narrowband process and is the feature by which a process can be identified as being narrowbanded based only on knowledge of the autocovariance function.

The ultimate narrowband process is the process considered in Example 4.3, for which the autocovariance function was exactly given as $G_{XX}(\tau) = \sigma_X^2 \cos(\omega_c \tau)$, with no decay of the amplitude of $G_{XX}(\tau)$. The autospectral density for this example is $S_{XX}(\omega) = \sigma_X^2 [\delta(\omega + \omega_c) + \delta(\omega - \omega_c)]$, showing that there are variance contributions only from components at the discrete frequencies $\pm\omega_c$. This is not too surprising, though, inasmuch as the time histories of this process are exactly cosine waves with frequency ω_c and amplitude and phase that are random variables. That is, any particular time history is a simple harmonic function with frequency ω_c and fixed amplitude and phase, although different time histories generally have different values for amplitude and phase. Thus, we see that a more general narrowband process, with an autocovariance function similar to that shown in Fig. 6.3, must be somewhat similar to the process having harmonic time histories. We will examine this idea more carefully because it provides a way to relate the frequency domain concept of a narrowband process directly to the characteristics of the time histories of the process.

The key idea that we want to emphasize is that if any process has an autocovariance function that approximates $\sigma_X^2 \cos(\omega_c \tau)$, then the quantity $Y(t) \equiv [X(t) - \mu_X(t)]$ for almost all of its time histories must closely approximate a harmonic function with frequency ω_c. That is, one can write

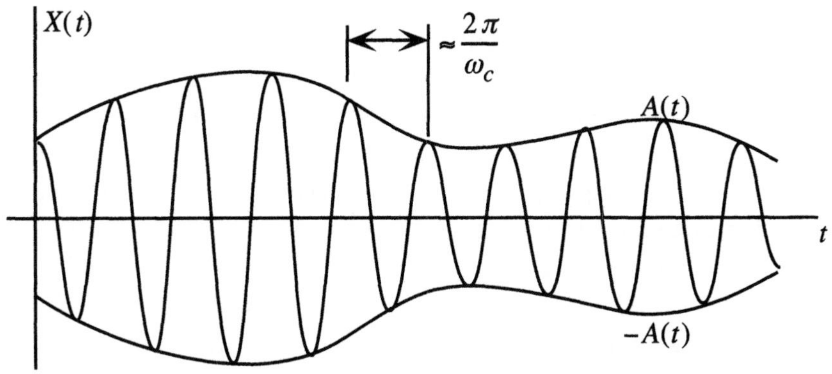

Figure 6.4 Typical time history of a narrowband process.

$$X(t) = \mu_X(t) + A(t)\cos[\omega_c t + \theta(t)] \tag{6.26}$$

in which the amplitude $A(t)$ and phase $\theta(t)$ are slowly varying. More precisely, if $G_{XX}(\tau) \approx \sigma_X^2 \cos(\omega_c \tau)$ over some range $|\tau| \le T_0$, then a segment of length T_0 of a time history of $Y(t)$ will almost surely be well approximated by the harmonic function with frequency ω_c. The reasoning is simply that if $E[Y(t+\tau)Y(t)] \approx E(Y^2)\cos(\omega_c \tau)$, then one can say with probability one that $Y(t+\tau) \approx Y(t)\cos(\omega_c \tau)$. But this condition is true for a range of t and τ values only if $Y(t)$ has the prescribed harmonic form. Thus, we see that for any narrowband process we should expect a typical segment of a time history of $Y(t)$ to approximate a segment from a harmonic function, as illustrated in Fig. 6.4. Conversely, if the time histories of $Y(t) \equiv [X(t) - \mu_X(t)]$ do have the form of harmonic functions with a fixed frequency ω_c and slowly varying amplitude and phase, then we know that $\{X(t)\}$ will have a narrowband autospectral density function. Furthermore, $\sigma_X^2 \approx E[A^2(t)]/2$, and neglecting $\dot{A}(t)$ and $\dot{\theta}(t)$ for this nearly harmonic time history gives $\sigma_{\dot{X}}^2 \approx \omega_c^2\, E[A^2(t)]/2$, which implies that the characteristic frequency ω_c for any narrowband process may be approximated by

$$\omega_{c2} \equiv \frac{\sigma_{\dot{X}}}{\sigma_X} \tag{6.27}$$

The "2" in the subscript of ω_{c2} is to distinguish this definition of characteristic frequency from alternate forms that will be introduced in Chapter 7. The ideas of amplitude and phase of a stochastic process are also investigated in much more detail in Chapter 7, but without the requirement that $\{X(t)\}$ be narrowbanded in nature.

6.5 Broadband Processes and White Noise

As noted in the preceding section, a narrowband process is one for which the variance is all contributed by components in the vicinity of a single frequency ω_c and its reflection $-\omega_c$. At the opposite extreme is a process for which all components contribute equally, such that the spectral density is the same for all ω values. Let us investigate a process $\{X(t)\}$ of this type with $S_{XX}(\omega)=S_0$. The first thing that one may notice about this autospectral density function is that it is not integrable. Thus, Eq. 6.22 gives $\sigma_X^2=\infty$, showing that no physically meaningful process has precisely this autospectral density. Nonetheless, let us investigate the process further, because we will find that it can be used to approximate meaningful processes. Because $S_{XX}(\omega)$ is not integrable, the inverse Fourier transform takes the degenerate form of a Dirac delta function, giving the autocovariance function as $G_{XX}(\tau)=2\pi S_0\,\delta(\tau)$, which does satisfy the forward Fourier transform of Eq. 6.18. Thus, we see that the autocovariance function for this process is the same as for the stationary delta-correlated process introduced in Section 5.5. Conversely, we can say that a delta-correlated process will always give an autospectral density that is a constant. Clearly, the relationship between the autospectral density level S_0 and the covariance parameter G_0 used in Eq. 5.37 is simply $G_0=2\pi S_0$.

A process with a constant value of $S_{XX}(\omega)$ is commonly referred to as *white noise*, by analogy with white light, which supposedly contains equal contributions from all visible frequency components. The basic difference between the terms *delta-correlated* and *white noise* in defining a stochastic process is that the former term focuses attention on the time-domain characterization of the process, whereas the latter focuses attention on the frequency domain. A process $\{X(t)\}$ that is so erratic that $X(t_1)$ and $X(t_2)$ are independent of each other for any two distinct times ($t_1\neq t_2$) also contains all frequencies equally in the frequency decomposition of its covariance.[2] The time histories of any delta-correlated process, of course, must be extremely erratic. In fact, if the time histories had any finite probability of being continuous at some value of t_1, then $X(t_1)$ and $X(t_2)$ could not be independent if t_2 were near t_1. Thus, a truly delta-correlated process is so erratic as to preclude the drawing of sample time histories. Because a truly constant autospectral density gives the unbounded Dirac delta function as the autocovariance function, it is useful also to

---

[2]The frequency decompositions of higher-order cumulants, as represented by bispectra, trispectra, and so forth, are also made up of equal contributions from all frequencies.

examine physically realizable processes that approach this delta-correlated process in the limit. Examples 6.1 and 6.2 present two such situations.

****

Example 6.1: Let $\{X(t)\}$ be a covariant stationary process with autospectral density function given by

$$S_{XX}(\omega) = S_0\, e^{-\gamma|\omega|}$$

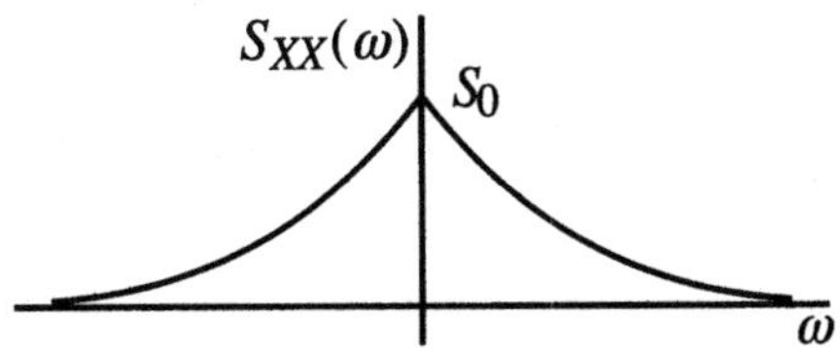

Find the autocovariance function for the process, and verify that it tends to that of a delta-correlated process in the limit as $\gamma \to 0$.

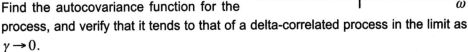

The inverse Fourier transform of $S_{XX}(\omega)$, as in Eq. 6.19, gives

$$G_{XX}(\tau) = S_0 \int_{-\infty}^{\infty} e^{-\gamma|\omega|} e^{i\omega\tau}\, d\omega$$

$$= S_0 \int_{-\infty}^{0} e^{(\gamma+i\tau)\omega}\, d\omega + S_0 \int_{0}^{\infty} e^{(-\gamma+i\tau)\omega}\, d\omega$$

or

$$G_{XX}(\tau) = S_0 \left(\frac{1}{\gamma + i\,\tau} + \frac{1}{\gamma - i\,\tau} \right) = \frac{2\gamma S_0}{\gamma^2 + \tau^2}$$

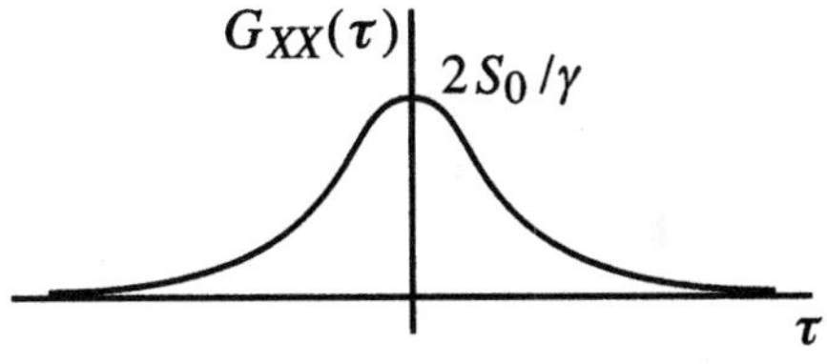

Note that setting $\tau = 0$ gives $\sigma_X^2 = G_{XX}(0) = 2S_0/\gamma$. Now if we let γ tend to zero for any $\tau \neq 0$, we see that $G_{XX}(\tau)$ tends to zero, as it should for a delta-correlated process. On the other hand, for $\tau = 0$, we have $G_{XX}(0) \to \infty$ as γ tends to zero, which again is characteristic of a delta-correlated process. The only remaining issue to check in verifying that this autocovariance is that of a delta-correlated process regards the question of whether its integral with respect to τ exists and is equal to a constant G_0. However, we know without any further explicit integration that the Fourier transform of this autocovariance function is $S_0\, e^{-\gamma|\omega|}$. Setting $\omega = 0$ in this Fourier transform shows that the integral of the autocovariance function exists and has a value of $2\pi S_0$. This confirms that setting $G_0 = 2\pi S_0$ does make $G_{XX}(\tau)$ tend to the form $G_0\, \delta(\tau)$ of Eq. 5.37 as $\gamma \to 0$.

****

Example 6.2: Let $\{X(t)\}$ be a covariant stationary process with autospectral density function given by

$$S_{XX}(\omega) = S_0 \frac{\alpha^2}{\alpha^2 + \omega^2}$$

Find the autocovariance function for the process, and verify that it tends to that of a delta-correlated process in the limit as $\alpha \to \infty$.

One must be slightly more creative in finding the inverse Fourier transform of $S_{XX}(\omega)$ in this example

$$G_{XX}(\tau) = S_0 \alpha^2 \int_{-\infty}^{\infty} \frac{e^{i\omega\tau}}{\alpha^2 + \omega^2} d\omega$$

For someone familiar with the theory of complex variables, the most straightforward approach is to note that the integrand has poles at $\omega = \pm i\alpha$ and use the calculus of residues to evaluate the integral. Alternatively, one might look in a table of Fourier transforms to find the same result. In the present situation, we can avoid doing either of these things by comparing this integral with the results in Example 6.1. In particular, the Fourier transform giving the autospectral density as a function of the autocovariance in that example is

$$S_{XX}(\omega) = S_0 e^{-\gamma|\omega|} = \frac{1}{2\pi} \int_{-\infty}^{\infty} G_{XX}(\tau) e^{-i\omega\tau} d\tau = \frac{1}{2\pi} \int_{-\infty}^{\infty} \frac{2\gamma S_0}{\gamma^2 + \tau^2} e^{-i\omega\tau} d\tau$$

which gives

$$\int_{-\infty}^{\infty} \frac{e^{-i\omega\tau}}{\gamma^2 + \tau^2} d\tau = \frac{\pi}{\gamma} e^{-\gamma|\omega|}$$

A change of variables then gives

$$\int_{-\infty}^{\infty} \frac{e^{i\omega\tau}}{\alpha^2 + \omega^2} d\omega = \frac{\pi}{\alpha} e^{-\alpha|\tau|}$$

Applying this result to the current problem gives $G_{XX}(\tau) = S_0 \pi \alpha e^{-\alpha|\tau|}$. Clearly, this satisfies the conditions of $G_{XX}(\tau) \to 0$ as $\alpha \to \infty$ for $\tau \neq 0$ and $G_{XX}(0) \to \infty$ as $\alpha \to \infty$. Furthermore, the integral of $e^{-\alpha|\tau|}$ from $-\infty$ to ∞ is $2/\alpha$, confirming that the limit of $G_{XX}(\tau)$ can be written as $2\pi S_0 \delta(\tau)$.

****

6.6 Linear Dynamics and Harmonic Transfer Functions

In Chapter 5, we formulated time-domain expressions for stochastic dynamics of linear systems directly from the deterministic Duhamel convolution integral. Now we will pursue a parallel development, using the common deterministic frequency domain formulation of linear dynamics to give us information about stochastic response. Using the Fourier transform, one can describe the input and output, respectively, of a time-invariant linear system either as $f(t)$ and $x(t)$ or as $\tilde{f}(t)$ and $\tilde{x}(t)$, as shown in Fig. 6.5.

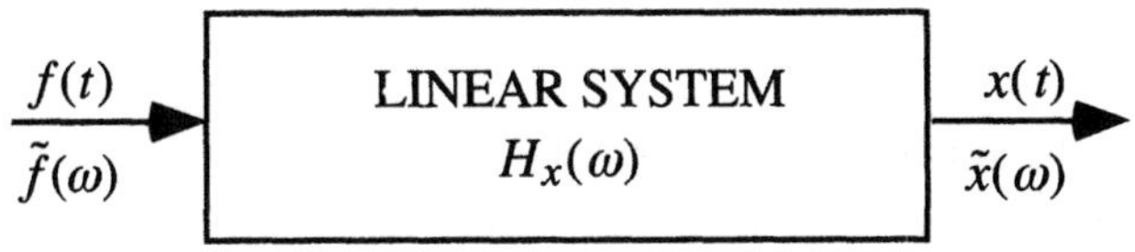

Figure 6.5 Schematic of general linear system.

The function $H_x(\omega)$ in Fig. 6.5 is called the *harmonic transfer function*, and it is defined to be the $x(t)/f(t)$ ratio when $f(t)$ is the pure harmonic $e^{i\omega t}$. That is, if $f(t) = e^{i\omega t}$, then $x(t) = H_x(\omega) e^{i\omega t}$. Using superposition, then, one can say that an excitation $\tilde{f}(\omega) e^{i\omega t}$ induces a response of $H_x(\omega)\tilde{f}(\omega) e^{i\omega t}$, so a time history of input of

$$f(t) = \int_{-\infty}^{\infty} \tilde{f}(\omega) e^{i\omega t}\, d\omega$$

causes a time history of response of

$$x(t) = \int_{-\infty}^{\infty} H_x(\omega)\, \tilde{f}(\omega) e^{i\omega t}\, d\omega \tag{6.28}$$

Comparing this equation with the standard inverse Fourier transform shows that

$$\tilde{x}(\omega) = H_x(\omega)\, \tilde{f}(\omega) \tag{6.29}$$

Equation 6.29 is generally regarded as the standard form of the frequency-domain input-output relationship for linear dynamics, but it should be remembered that it also implies the time-domain relationship of Eq. 6.28.

Before proceeding to stochastic structural dynamics, it is useful to note the relationship between the harmonic transfer function $H_x(\omega)$ and the impulse response function $h_x(t)$ used in the time-domain formulation. One easy way of identifying this relationship is to use the harmonic function $f(t) = e^{i\omega t}$ as the input in the convolution integral of Eq. 5.2. This gives

$$x(t) = \int_{-\infty}^{\infty} f(t-r)\, h_x(r)\, dr = \int_{-\infty}^{\infty} e^{i\omega(t-r)} h_x(r)\, dr = e^{i\omega t} \int_{-\infty}^{\infty} e^{-i\omega r}\, h_x(r)\, dr$$

The fact that this response must be $x(t) = H_x(\omega) e^{i\omega t}$, from the definition of the harmonic transfer function, shows that $H_x(\omega)$ is exactly 2π times the Fourier transform of $h_x(t)$

$$H_x(\omega) = 2\pi\, h_x(\omega) = \int_{-\infty} e^{-i\omega r}\, h_x(r)\, dr \quad (6.30)$$

Switching to stochastic processes $\{F(t)\}$ and $\{X(t)\}$ for input and output of the linear system now gives us the relationship of $\tilde{X}(\omega) = H_x(\omega)\,\tilde{F}(\omega)$ for the stochastic Fourier transform processes $\{\tilde{F}(\omega)\}$ and $\{\tilde{X}(\omega)\}$, just as Eq. 6.29 described the deterministic situation. Taking the expectation of this relationship shows that the Fourier transform of the mean-value function of the output can be written as $\tilde{\mu}_X(\omega) = H_x(\omega)\,\tilde{\mu}_F(\omega)$. Furthermore, for the special case of covariant stationary processes, the definition of autospectral density in Eqs. 6.15–6.17 gives

$$S_{XX}(\omega) = H_x(\omega) H_x(-\omega)\, S_{FF}(\omega) = |H_x(\omega)|^2\, S_{FF}(\omega) \qquad (6.31)$$

This is the fundamental frequency-domain relationship regarding the autocovariance of the response of a linear dynamic system. It can be viewed as a frequency-domain form of Eq. 5.25, which gave the second-moment function of the stationary response. A major difference is that finding the response second-moment or covariance function from Eq. 5.25 involves a double integral in the time domain, whereas Eq. 6.31 gives us the response autospectral density from simply a multiplication of functions in the frequency domain.

After the response autospectral density is found, one can obtain the response variance from a single frequency-domain integration, as in Eq. 6.22, as an alternative to the double convolution integral of Eq. 5.25. Thus, it is sometimes easier to use frequency-domain manipulations to find the response variance rather than using the time-domain relationships. There is one complication with this approach, though. For many problems, the form of the spectral densities is such that one must use the theory of residues from complex variable analysis in order to evaluate conveniently the integral over ω values. The reader who is not proficient with complex analysis should keep in mind that the time-domain integration will always give the same results and may actually be easier to perform. The choice between time-domain and frequency-domain calculations of variance is often primarily a matter of personal preference of the analyst.

One can also use Eq. 6.30 to give frequency-domain interpretations of the bounding quantities $h_{x,static}$ and $h_{x,var}(0)$ introduced in Chapter 5. In particular, one can see directly from Eq. 6.30 that

$$h_{x,static} \equiv \int_{-\infty}^{\infty} h_x(t)\, dt = H_x(0)$$

To find the other quantity of interest we can substitute Eq. 6.30 into an integral to give

$$\int_{-\infty}^{\infty} H_x(\omega)\, H_x(-\omega)\, d\omega = \int_{-\infty}^{\infty}\int_{-\infty}^{\infty}\int_{-\infty}^{\infty} h_x(r_1)\, h_x(r_2)\, e^{-i\omega(r_1-r_2)}\, d\omega\, dr_1\, dr_2$$

and evaluate the integral with respect to ω as $2\pi\delta(r_1 - r_2)$, so

$$\int_{-\infty}^{\infty} H_x(\omega)\, H_x(-\omega)\, d\omega = 2\pi \int_{-\infty}^{\infty} h_x^2(r_1)\, dr_1$$

or

$$h_{x,var}(0) = \frac{1}{2\pi}\int_{-\infty}^{\infty} H_x(\omega)\, H_x(-\omega)\, d\omega$$

Note that the situation with $h_{x,var}(0) = \infty$ is characterized by neither $h_x(t)$ nor $|H_x(\omega)|$ being squared-integrable.

Similar to Eq. 6.31, one can write the cross-spectral density for two response processes $\{X(t)\}$ and $\{Y(t)\}$ as

$$S_{XY}(\omega) = H_x(\omega)\, H_y(-\omega)\, S_{FF}(\omega) = H_x(\omega)\, H_y^*(\omega)\, S_{FF}(\omega) \qquad (6.32)$$

and the cross-spectral density between the excitation and the response is

$$S_{XF}(\omega) = H_x(\omega)\, S_{FF}(\omega)$$

We can also confirm the corresponding relationships involving derivatives, as given in Eqs. 6.23–6.25, by noting that if $x(t) = H_x(\omega)\, e^{i\omega t}$ then $\dot{x}(t) = i\omega H_x(\omega)\, e^{i\omega t}$, so the harmonic transfer function for the derivative response is simply $H_{\dot{x}}(\omega) = i\omega H_x(\omega)$. Because an integral of a cross-spectral density function always gives a cross-covariance term, these cross-covariances can also be evaluated from the frequency-domain analysis.

One other advantage of frequency-domain analysis involves the ease with which the harmonic transfer functions can be evaluated for a system governed by a linear differential equation. For example, if we consider the general nth-order system (as in Eq. 5.7)

$$\sum_{j=0}^{n} a_j \frac{d^j x(t)}{d t^j} = f(t)$$

then substitution of $f(t) = e^{i\omega t}$ and $x(t) = H_x(\omega)\, e^{i\omega t}$ gives

$$H_x(\omega) = \left(\sum_{j=0}^{n} a_j (i\,\omega)^j \right)^{-1} \tag{6.33}$$

Thus, one needs use only algebra to find the harmonic transfer function for the system, whereas finding the corresponding time-domain impulse response function involves solution of an initial value problem for the differential equation.

The input-output autospectral density relationship in Eq. 6.31 shows that any linear system has the effect of shaping autospectral density, in the sense that the autospectral density of the output is $|H_x(\omega)|^2$ times that of the input. Systems specifically designed to produce a particular shape of output autospectral density are commonly called filters, and Eq. 6.31 shows that any linear system can be regarded as a linear filter. Three common categories of such filters are called low-pass, high-pass, and band-pass filters, depending on whether the output autospectral density is dominated by low-frequency components, high-frequency components, or some band of frequency components. It should be noted, though, that not all linear systems fall into these three categories because one can design a linear system to approximate any desired harmonic transfer function.

The frequency-domain analysis procedures for dynamic response are applied to two simple nonoscillatory systems in the following two examples. The much more important situation with oscillatory response is investigated in Section 6.8.

****

Example 6.3: Consider a linear system governed by the differential equation $c\,\dot{x}(t) + k\,x(t) = f(t)$, for which the impulse response function was found in Example 5.1 as $h_x(t) = c^{-1} e^{-kt/c}\, U(t)$. Find the harmonic transfer function from the Fourier transform of $h_x(t)$, as in Eq. 6.30, and compare with the result from Eq. 6.33. Also find the autospectral density and the variance of the response

$\{X(t)\}$ for the situation when $f(t)$ is replaced by a white noise (or delta-correlated) process $\{F(t)\}$ with $S_{FF}(\omega) = S_0$.

From Eq. 6.30 we have

$$H_x(\omega) = \int_{-\infty}^{\infty} e^{-i\omega r} h_x(r)\, dr = c^{-1} \int_0^{\infty} e^{-i\omega r} e^{-kr/c}\, dr = c^{-1}(i\omega + k/c)^{-1}$$
$$= (k + i\omega c)^{-1}$$

which is identical to what Eq. 6.33 gives.

The autospectral density of the response is now given by Eq. 6.31 as

$$S_{XX}(\omega) = |H_x(\omega)|^2 S_{FF}(\omega) = \frac{S_0}{k^2 + \omega^2 c^2}$$

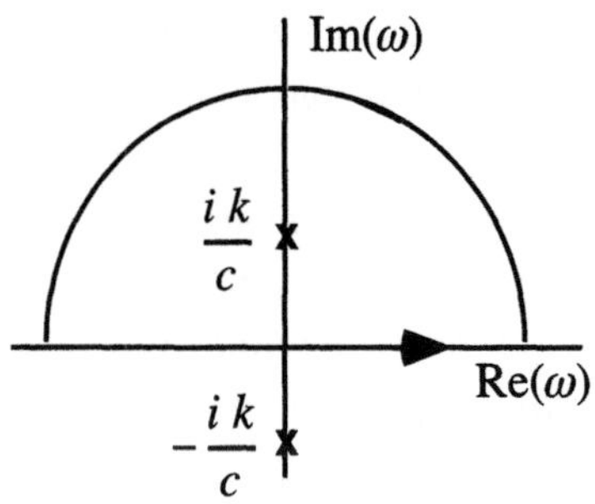

The most direct method to evaluate the response variance from this autospectral density is to note that $S_{XX}(\omega)$ has poles at $\omega = \pm i\, k/c$ and use the calculus of residues.[3] Thus, if we evaluate the integral along the real axis of the complex ω space by closing the contour in the upper half-space, we get the integral as being equal to $2\pi i$ times the residue at $\omega = +i\, k/c$

$$\sigma_X^2 = \int_{-\infty}^{\infty} \frac{S_0}{k^2 + \omega^2 c^2}\, d\omega = (2\pi i) \lim_{\omega \to ik/c} \frac{S_0(\omega - ik/c)}{c^2(\omega - ik/c)(\omega + ik/c)} = \frac{\pi S_0}{c\, k}$$

Having a delta-correlated excitation also simplifies the time-domain evaluation of the response variance from Eq. 5.30 or 5.40. The result, of course, is identical to that obtained here.

****

Example 6.4: Find the harmonic transfer function for the linear system governed by the differential equation $m\, \ddot{x}(t) + c\, \dot{x}(t) = f(t)$, for which the impulse response function was found in Example 5.2 as $h_x(t) = c^{-1}(1 - e^{-ct/m})U(t)$. Also consider the autospectral density and variance of the stochastic response of this system for a stochastic excitation with autospectral density $S_{FF}(\omega)$.

Strictly speaking, the Fourier transform of $h_x(t)$ does not exist for this problem because $h_x(t)$ is not integrable, which is the same condition as $h_{x,static} = \infty$ (as was pointed out in Example 5.2). Thus, one cannot directly apply Eq. 6.30. There is no difficulty, though, in using Eq. 6.33 to write

$$H_x(\omega) = \frac{1}{i\omega c - \omega^2 m}$$

---

[3]The details of this technique are included in any introductory textbook on complex analysis.

Thus, the autospectral density of covariant stationary stochastic response will be

$$S_{XX}(\omega) = \frac{S_{FF}(\omega)}{\omega^2 c^2 + \omega^4 m^2}$$

Note that the denominator of this expression tends to zero as ω tends to zero. In particular, it behaves like ω^2 for small values of ω, and this ensures that the integral of the autospectral density will not exist if $S_{FF}(0) \neq 0$. Thus, this system has infinite response covariance for most stationary stochastic excitations, which agrees with our observation in Example 5.4 that $h_{x,var}(0) = \infty$. We do see, though, that if $S_{FF}(\omega)$ is zero and behaves like $|\omega|^b$ with $b > 1$ at $\omega \approx 0$, then the response variance will be finite. This shows that, for some excitations, a system can have a finite response variance even though it has $h_{x,var}(0) = \infty$.

****

Example 6.5: Estimate the autospectral density $S_{XX}(\omega_c)$ based only on a single long time history of the process $\{X(t)\}$ and the results obtained from an ideal bandpass filter that transmits only those Fourier components with $\omega \approx \pm\omega_c$. In particular, the harmonic transfer function can be written as $H(\omega) = U(|\omega| - \omega_c + \varepsilon)\, U(\omega_c + \varepsilon - |\omega|)$.

Let $\{Z(t)\}$ be the output from the filter when $\{X(t)\}$ is the input. This then gives us $S_{ZZ}(\omega) = S_{XX}(\omega)$ within the pass band and $S_{ZZ}(\omega) = 0$ otherwise, as shown in the accompanying sketch. The $\{Z(t)\}$ process will be mean-zero because the stationary mean value of $\{X(t)\}$ is a component at frequency zero and this component is blocked by the filter. Thus, the variance and mean-squared value of $\{Z(t)\}$ are both given by

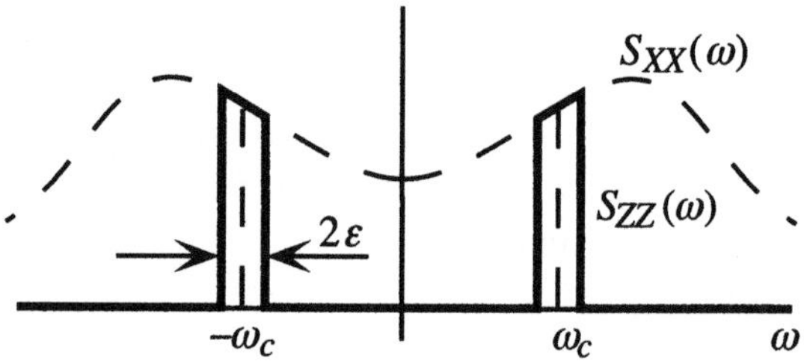

$$E[Z^2(t)] = \int_{-\infty}^{\infty} S_{ZZ}(\omega)\, d\omega = 2\int_0^{\infty} S_{ZZ}(\omega)\, d\omega = 2\int_{\omega_c - \varepsilon}^{\omega_c + \varepsilon} S_{XX}(\omega)\, d\omega$$

and if ε is small enough that $S_{XX}(\omega)$ is nearly linear across the range of integration, then we have

$$E[Z^2(t)] \approx 4\varepsilon\, S_{XX}(\omega_c)$$

Thus, $E[Z^2(t)]/(4\varepsilon)$ can be used as an estimate of the unknown $S_{XX}(\omega_c)$. Assuming covariance ergodicity of $\{Z(t)\}$, the value of $E[Z^2(t)]$ can be estimated from a time average, giving

$$S_{XX}(\omega_c) \approx \frac{1}{4\varepsilon T}\int_0^T Z^2(s)\, ds$$

for a large value of T. Note that the finite bandwidth 2ε tends to smooth the measured autospectral density, which results in a peak in $S_{XX}(\omega)$ being almost lost if it is not wider than 2ε. Based on this fact, it may appear that one would want to choose ε as small as possible to minimize this error due to variation of $S_{XX}(\omega)$ across the pass band. On the other hand, it can be shown that the convergence of the time average of $Z^2(t)$ to its expected value is slower when the bandwidth is smaller. Thus, if one uses a very small ε value, then it is necessary also to have a very large T value, which may not be feasible in some situations.

To estimate the entire $S_{XX}(\omega)$ autospectral density function in this way, one must have a filter with a variable center frequency ω_c (and possibly a variable ε), and repeat the procedure for different ω_c values until the function is defined with sufficient resolution. This technique for estimating $S_{XX}(\omega)$ has been quite important in the past, but it would not be commonly used now, in an age of inexpensive digital computation. Section 6.9 discusses the approach that is usually used now to estimate $S_{XX}(\omega)$ from recorded data.

****

6.7 Evolutionary Spectral Density

The results in this chapter have, up until now, been limited to time-invariant linear systems. If the system is time-varying, then one must expect the harmonic transfer function to vary with time. This function will be written as $H_{xf}(t,\omega)$ and will be defined such that Eq. 6.28 can be generalized to the form

$$x(t) = \int_{-\infty}^{\infty} H_{xf}(t,\omega)\, \tilde{f}(\omega)\, e^{i\omega t}\, d\omega \tag{6.34}$$

An interpretation of $H_{xf}(t,\omega)$ can be found by considering the special case with $f(t) = e^{i\omega_0 t}$, for which $\tilde{f}(\omega) = \delta(\omega - \omega_0)$. Performing the integration then shows that an excitation of $f(t) = e^{i\omega_0 t}$ gives a response of $H_{xf}(t,\omega_0)\, e^{i\omega_0 t}$. Thus, $H_{xf}(t,\omega)$ can be given the same interpretation as the $H_x(\omega)$ defined in the preceding section—it is the amplitude of the $e^{i\omega t}$ harmonic response when the $f(t)$ excitation is a pure harmonic $e^{i\omega t}$. One must be careful with this idea, though, because the $H_{xf}(t,\omega)\, e^{i\omega t}$ response is not a pure harmonic, due to $H_{xf}(t,\omega)$ varying with time t.

One should also recall that $\tilde{f}(\omega)$ exists only if $f(t)$ is absolute-value integrable. Thus, Eq. 6.34 does not apply to all situations, and it is advantageous to give a more general definition of $H_{xf}(t,\omega)$. This may be done by relating it to the time-varying impulse response function introduced in Section 5.1. In particular, letting $f(t)=e^{i\omega t}$ in the time-domain integral of Eq. 5.5, gives

$$x(t)=\int_{-\infty}^{\infty} e^{i\omega s}\, h_{xf}(t,s)\, ds$$

We have already shown, though, that this excitation gives a response of $x(t)=H_{xf}(t,\omega)\, e^{i\omega t}$. Thus, we can say that

$$H_{xf}(t,\omega)=e^{-i\omega t}\int_{-\infty}^{\infty} e^{i\omega s}\, h_{xf}(t,s)\, ds=\int_{-\infty}^{\infty} e^{-i\omega r}\, h_{xf}(t,t-r)\, dr \qquad (6.35)$$

Equation 6.35 can be considered the general definition of $H_{xf}(t,\omega)$, even though Eq. 6.34 gives a more intuitive interpretation of the term. For the special case of a time-invariant system, $h_{xf}(t,t-r)=h_x(r)$, so Eq. 6.35 becomes identical to Eq. 6.30.

Let us now evaluate the time-varying harmonic response function for the $\dot{x}(t)$ derivative. We begin by differentiating Eq. 6.35 to give

$$\frac{\partial}{\partial t} H_{xf}(t,\omega)=-i\,\omega H_{xf}(t,\omega)+e^{-i\omega t}\int_{-\infty}^{\infty} e^{i\omega s}\, h_{\dot{x}f}(t,s)\, ds$$

in which the final term has been simplified by noting that Eq. 5.6 gives $h_{\dot{x}f}(t,s)=\partial\, h_{xf}(t,s)/\partial t$. Comparing the final term of this equation with Eq. 6.35 shows that it is exactly $H_{\dot{x}f}(t,\omega)$. Thus, we can rearrange terms to obtain

$$H_{\dot{x}f}(t,\omega)=\frac{\partial}{\partial t} H_{xf}(t,\omega)+i\,\omega H_{xf}(t,\omega) \qquad (6.36)$$

Consider now a case in which the $\{\tilde{f}(\omega)\}$ Fourier transform exists for the $f(t)$ input to the time-varying system. Equation 6.34 then gives the response as

$$x(t)=\int_{-\infty}^{\infty} H_{xf}(t,\omega)\, \tilde{f}(\omega)\, e^{i\omega t}\, d\omega$$

or

$$\tilde{x}(\omega)=H_{xf}(t,\omega)\, \tilde{f}(\omega) \qquad (6.37)$$

Thus, $H_{xf}(t,\omega)$ "modulates" $\{F(t)\}$ by giving a time-varying alteration of the magnitude of each harmonic component. Recall the special case of uniform modulation introduced in Section 5.4, for which $h_{xF}(t,s)=\hat{h}(t)\,\delta(t-s)$. For this situation, Eq. 6.35 gives $H_{xf}(t,\omega) = h(t)$, so every frequency component in Eq. 6.37 is modulated by the same $\hat{h}(t)$ function. This explains the choice of the term *uniformly modulated* for this special case.

For a *modulated* (also called *evolutionary*) stochastic process, the excitation is a stationary $\{F_S(t)\}$ process, for which the Fourier transform does not exist. The time-varying spectral density of this process is called an evolutionary spectral density and is defined by a modification of Eq. 6.31 (Priestly, 1988):

$$S_{XX}(t,\omega) = H_{xS}(t,\omega)\, H_{xS}(t,-\omega)\, S_{F_S F_S}(\omega) = |H_{xS}(t,\omega)|^2\, S_{F_S F_S}(\omega) \qquad (6.38)$$

The notation $H_{xS}(t,\omega)$ is introduced to indicate that this function relates $\{X(t)\}$ to the stationary $\{F_S(t)\}$ process, and not to some other $\{F(t)\}$ excitation that may itself be a modulated process. Similar to Eq. 6.38, an evolutionary cross-spectral density for two modulated processes depending on the same stationary $\{F_S(t)\}$ can be defined by a modification of Eq. 6.32 as

$$S_{XY}(t,\omega) = H_{xS}(t,\omega)\, H_{yS}(t,-\omega)\, S_{F_S F_S}(\omega) = H_{xS}(t,\omega)\, H^*_{yS}(t,\omega)\, S_{F_S F_S}(\omega) \qquad (6.39)$$

For example, using Eq. 6.36 gives

$$\begin{aligned} S_{X\dot{X}}(t,\omega) &= H_{xS}(t,\omega)\, H_{\dot{x}S}(t,-\omega)\, S_{F_S F_S}(\omega) \\ &= H_{xS}(t,\omega)\left(\frac{\partial}{\partial t} H_{xS}(t,-\omega) - i\,\omega\, H_{xS}(t,-\omega)\right) S_{F_S F_S}(\omega) \qquad (6.40) \\ &= H_{xS}(t,\omega)\frac{\partial}{\partial t} H_{xS}(t,-\omega)\, S_{F_S F_S}(\omega) - i\,\omega\, S_{XX}(\omega) \end{aligned}$$

The final term of the equation agrees with Eq. 6.23 for a time-invariant system, whereas the preceding term reflects an effect of the time variation of the frequency content of the $\{X(t)\}$ process.

One can now use Eq. 5.5 to write the auto-covariance or cross-covariance of the response to a stationary stochastic excitation $\{F_S(t)\}$. In particular,

$$K_{XY}(t_1,t_2) = \int_{-\infty}^{\infty}\int_{-\infty}^{\infty} K_{F_S F_S}(s_1,s_2)\, h_{xS}(t_1,s_1)\, h_{yS}(t_2,s_2)\, ds_1\, ds_2 \qquad (6.41)$$

Because $\{F_S(t)\}$ is covariant stationary, though, one can say that

$$K_{F_SF_S}(s_1,s_2) = G_{F_SF_S}(s_1 - s_2) = \int_{-\infty}^{\infty} S_{F_SF_S}(\omega)\, e^{i\omega(s_1-s_2)}\, d\omega$$

Substituting this relationship into Eq. 6.41 then gives

$$K_{XY}(t_1,t_2) = \int_{-\infty}^{\infty} S_{F_SF_S}(\omega)\left(\int_{-\infty}^{\infty} h_{xS}(t_1,s_1)\, e^{i\omega s_1}\, ds_1\right)\left(\int_{-\infty}^{\infty} h_{yS}(t_2,s_2)\, e^{-i\omega s_2}\, ds_2\right) d\omega$$

and using Eq. 6.35 allows this to be rewritten as

$$K_{XY}(t_1,t_2) = \int_{-\infty}^{\infty} S_{FF}(\omega)\, H_{xS}(t_1,\omega)\, H_{yS}(t_2,-\omega)\, e^{i\omega(t_1-t_2)}\, d\omega$$

An important special case is when $t_1 = t_2 = t$, for which the left-hand side becomes $K_{XY}(t,t)$ and the integrand becomes $S_{FF}(\omega)\, H_{xS}(t,\omega)\, H_{yS}(t,-\omega) \equiv S_{XY}(t,\omega)$. Thus,

$$K_{XY}(t,t) = \int_{-\infty}^{\infty} S_{XY}(t,\omega)\, d\omega$$

which demonstrates that the definition of the evolutionary spectral density preserves the usual property that the single-time covariance of the response can be found from an integral of the appropriate spectral density. Letting $Y = X$ gives variance of the response as

$$\sigma_X^2(t) = \int_{-\infty}^{\infty} S_{XX}(t,\omega)\, d\omega$$

A special case of some interest is that of *modulated white noise*, in which $\{F_S(t)\}$ is white, and therefore has a constant spectral density. From Eq. 6.37 one can then see that the resulting modulated process is quite general inasmuch as it can have a frequency content that varies with time, as well as a magnitude that varies with time. As a result, modulated white noise can be used to approximate many physical processes.

6.8 Response of Linear SDF Oscillator

We will now present an extended example, applying our frequency-domain analysis procedures to investigate the response of the linear single-degree-of-freedom (SDF) system described by Eqs. 5.45 and 5.46

$$m\ddot{X}(t) + c\,\dot{X}(t) + k\,X(t) = F(t)$$

or

$$\ddot{X}(t) + 2\zeta\,\omega_0\,\dot{X}(t) + \omega_0^2\,X(t) = F(t)/m$$

As pointed out in Section 5.6, this system plays a key role in our analysis of oscillatory systems, as distinguished from the simpler systems of Examples 6.3 and 6.4 that have oscillatory response only if they have oscillatory excitation.

First we can find the harmonic transfer function from Eq. 6.33 as

$$H_x(\omega) = \frac{1}{(k + i\,\omega c - \omega^2 m)} = \frac{1}{m(\omega_0^2 + 2i\,\zeta\,\omega_0\,\omega - \omega^2)} \tag{6.42}$$

We could also check this expression by finding the Fourier transform of the impulse response function, which was derived in Example 5.3 as $h_x(t) = (m\,\omega_d)^{-1} e^{-\zeta\omega_0 t}\sin(\omega_d\,t)$, in which $\omega_d = \omega_0(1-\zeta^2)^{1/2}$.

For $\{F(t)\}$ being a covariant stationary process, we can use Eq. 6.42 to write the autospectral density of the response of the SDF system as

$$S_{XX}(\omega) = S_{FF}(\omega)\,|\,H_x(\omega)\,|^2 = \frac{S_{FF}(\omega)}{m^2[(\omega_0^2 - \omega^2)^2 + (2\zeta\,\omega_0\,\omega)^2]} \tag{6.43}$$

Figure 6.6 shows a plot of $m^2\omega_0^4\,|\,H_x(\omega)\,|^2 \equiv k^2\,|\,H_x(\omega)\,|^2$ versus the normalized frequency ω/ω_0 for several values of the damping ratio ζ. Most structural and mechanical vibration problems involve systems with damping values even smaller than the smallest value of $\zeta = 0.05$ shown in Fig. 6.6. In fact, values of ζ of 0.01 or even smaller are not uncommon. The larger damping values in Fig. 6.6 are included to illustrate the effect of damping, rather than because they are typical. If damping is increased still further, the peak of $|\,H_x(\omega)\,|$ eventually completely disappears. In particular, if $\zeta \ge 2^{-1/2}$, then $S_{XX}(\omega)$ simply decays monotonically from its value at $\omega = 0$.

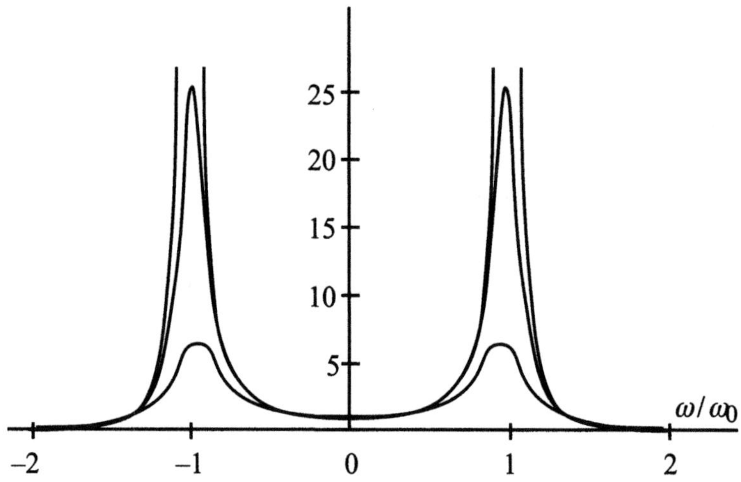

Figure 6.6 $|H_x(\omega)|^2$ for $\zeta = 0.05, 0.10$, and 0.20.

From Eq. 6.43 and Fig. 6.6, one can identify three key frequency ranges in the response of the SDF system. For $\omega \approx 0$, the response autospectral density $S_{XX}(\omega)$ is closely approximated by $S_{FF}(\omega)$ at that frequency divided by $m^2\omega_0^4 \equiv k^2$. Thus, the low-frequency response of the system is dependent only on the stiffness of the oscillator, behaving in the same way as if the response were pseudostatic. These low-frequency components are almost the same as they would be for a system without mass or damping, being governed by $k\,X(t) = F(t)$. At the other extreme, one finds that for $|\omega| >> \omega_0$, $S_{XX}(\omega)$ is approximated by $S_{FF}(\omega)/(m^2\omega^4)$. This implies that, at these frequencies, the autospectral density of the response acceleration has $S_{\ddot{X}\ddot{X}}(\omega) \approx S_{FF}(\omega)/m^2$. Thus, the high-frequency response components are almost the same as they would be for a system without a spring or damping, being governed by $m\,\ddot{X}(t) = F(t)$. More important in most applications is the intermediate situation. If ζ is small, then the "resonant" response components for $\omega \approx \pm\omega_0$ are greatly amplified, so $S_{XX}(\omega)$ is much greater than $S_{FF}(\omega)$. In this situation, the contributions to Eq. 6.43 depending on m and k nearly cancel out, making the response have $S_{XX}(\omega) \approx S_{FF}(\omega)/(2m\zeta\omega_0\,\omega)^2 = S_{FF}(\omega)/(c^2\omega^2)$ or $S_{\dot{X}\dot{X}}(\omega) \approx S_{FF}(\omega)/c^2$, which corresponds to a governing equation of $c\,\dot{X}(t) = F(t)$. Thus, each of the terms in the governing differential equation plays a key role. The $k\,X(t)$ term dominates the low-frequency response, the $m\,\ddot{X}(t)$ term dominates the high-frequency response, and the $c\,\dot{X}(t)$ term dominates the resonant-frequency response, giving the height of the resonant peak of $S_{XX}(\omega)$. It is important to remember that the stochastic response process $\{X(t)\}$ generally consists of a superposition of all frequency components from zero to infinity, all

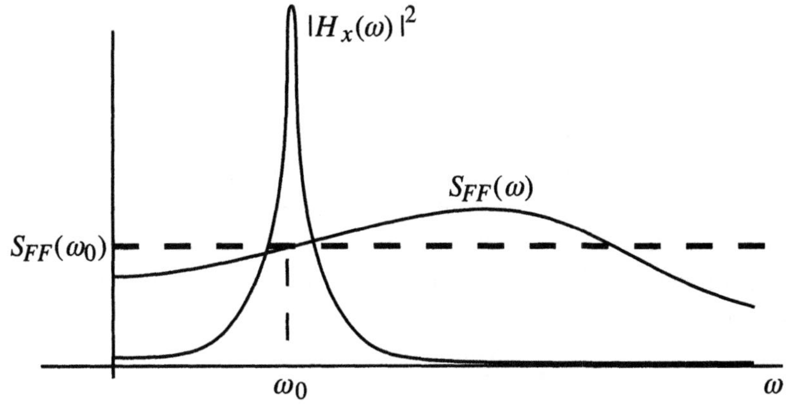

Figure 6.7 Equivalent white noise excitation.

occurring simultaneously. The comments in this paragraph are merely interpretations of the magnitudes of those various components, not discussions of different dynamic problems.

The plot in Fig. 6.6 shows that for small values of ζ the SDF system acts as a bandpass filter, giving substantial amplification only to components of $\{F(t)\}$ that are near $\pm\omega_0$. In fact, unless $S_{FF}(\omega)$ is much larger for some other frequencies than it is for $\omega = \omega_0$, it is clear that the $S_{XX}(\omega)$ response autospectral density will be dominated by the frequencies near $\pm\omega_0$. This leads to the common situation in which the stochastic response of the SDF system can be considered a narrowband process, which sometimes results in significant analytical simplifications. The excitation $\{F(t)\}$, on the other hand, can often be considered a broadband process or even approximated by an "equivalent" white noise, as explained in the following paragraph.

The justification for the approximation of $\{F(t)\}$ by an "equivalent white noise" can be seen in Fig. 6.7. Because $S_{XX}(\omega)$ is the product of $S_{FF}(\omega)$ and $|H_x(\omega)|^2$ and the most significant portion of $S_{XX}(\omega)$ comes from the near-resonant frequencies with $\omega \approx \pm\omega_0$, we can see that $S_{XX}(\omega)$ would not be changed very significantly if the actual $S_{FF}(\omega)$ were replaced by some other broadband function that agreed with $S_{FF}(\omega)$ for $\omega \approx \pm\omega_0$. The simplest such approximation is the white noise with constant autospectral density $S_0 = S_{FF}(\omega_0)$. The most common usage for this approximation is in the computation of the response variance, for which the approximation gives

$$\sigma_X^2 = \int_{-\infty}^{\infty} S_{XX}(\omega)\, d\omega = \int_{-\infty}^{\infty} S_{FF}(\omega) |H_x(\omega)|^2\, d\omega \approx S_{FF}(\omega_0) \int_{-\infty}^{\infty} |H_x(\omega)|^2\, d\omega \tag{6.44}$$

One can compute the value of this response variance by performing the frequency-domain integration in the last term of Eq. 6.44, but the result, of course, must be the same as was found from time-domain integration in Eq. 5.63 for the response to a delta-correlated excitation.[4] Taking the G_0 for the delta-correlated process to be $2\pi S_{FF}(\omega_0)$ gives the approximation of the SDF response variance as

$$\sigma_X^2 \approx \frac{\pi S_{FF}(\omega_0)}{2 m^2 \zeta \omega_0^3} = \frac{\pi S_{FF}(\omega_0)}{c\, k} \tag{6.45}$$

This approximation will generally be quite adequate as long as $S_{XX}(\omega)$ is sharply peaked so that its magnitude is much smaller for other frequencies than it is for $\omega \approx \pm\omega_0$. That is, the approximation can generally be used if $\{X(t)\}$ is narrowband. Although this is a common situation, it should be kept in mind that this condition is not always met. Figure 6.8 illustrates one such situation in which the approximation is not very good for $\zeta = 0.10$, although it does appear to be acceptable for $\zeta = 0.02$. Note that the excitation in this example is not very broadband, having a clear peak near $\omega = 0.8$.

One can also use the concept of a modulated process to consider nonstationary response problems, including the one treated in Section 5.6. In particular, taking $F(t) = W(t)U(t)$, with $\{W(t)\}$ being a stationary process makes $\{F(t)\}$ a uniformly modulated process with a modulation function of $\hat{h}(t) = U(t)$. The response $\{X(t)\}$ is then also a modulated process with a combined modulation function obtained from Eq. 5.33 as

$$h_{xS}(t,s) = \int_{-\infty}^{\infty} h_{xf}(t,u)\, U(u)\, \delta(u-s)\, du = h_x(t-s)\, U(s)$$

in which the $h_{xf}(t,s)$ term has been rewritten as $h_x(t-s)$ to reflect the fact that the SDF system is time invariant.

We can now find the combined time-varying harmonic response function from Eq. 6.35 as

---

[4]Alternatively, one can perform the frequency domain integration by noting that $|H_x(\omega)|^2$ has poles at $\omega = \pm\omega_d \pm i\zeta\omega_0$ and using the calculus of residues.

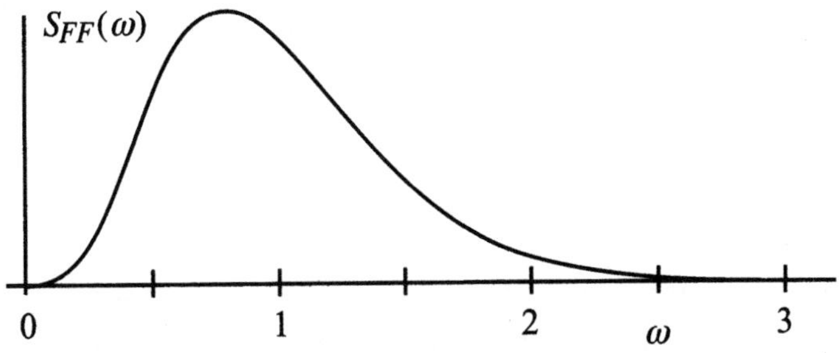

(a) Autospectral density of the excitation

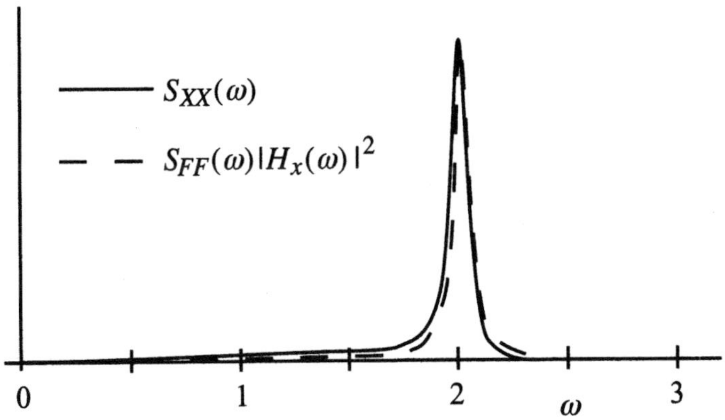

(b) Autospectral density of the response, $\omega_0 = 2$, $\zeta = 0.02$; adequate approximation

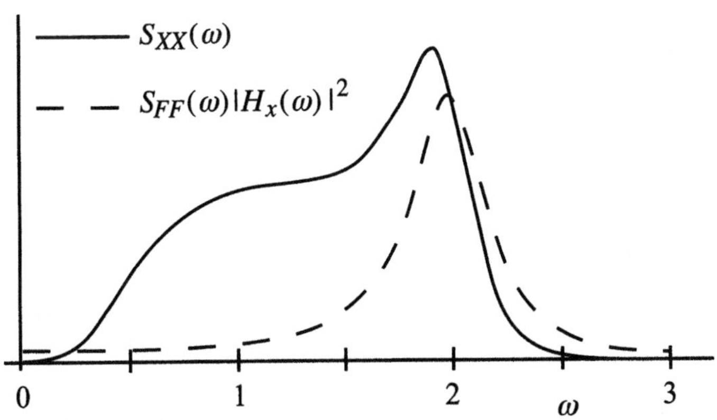

(c) Autospectral density of the response, $\omega_0 = 2$, $\zeta = 0.10$; inadequate approximation

Figure 6.8 Satisfactory and unsatisfactory approximation by equivalent white noise.

$$H_{xS}(t,\omega)=\int_{-\infty}^{\infty}e^{-i\omega r}\,h_x(r)\,U(t-r)\,dr=\int_{-\infty}^{t}e^{-i\omega r}\,h_x(r)\,dr$$

and using $h_x(t)=(m\,\omega_d)^{-1}e^{-\zeta\,\omega_0\,t}\sin(\omega_d\,t)U(t)$, as derived in Example 5.3, allows $H_{xS}(t,\omega)$ to be calculated. The result can be written as

$$H_{xS}(t,\omega)=H_x(\omega)\Big(1-e^{-i\omega t}[g_x(t)+i\,\omega\,m\,h_x(t)]\Big) \tag{6.46}$$

in which $H_x(\omega)$ is the same as in Eq. 6.42 and $g_x(t)=e^{-\zeta\,\omega_0 t}[\cos(\omega_d t)+(\zeta\,\omega_0\,/\,\omega_d)\sin(\omega_d t)]$ is the displacement response due to a unit displacement initial condition, as in Eq. 5.61. Using $S_{FF}(\omega)$ for the spectral density of the stationary excitation now allows the evolutionary spectral density for $\{X(t)\}$ from Eq. 6.38 to be written as

$$\begin{aligned}S_{XX}(t,\omega)&=S_{FF}(\omega)\,|H_{xS}(t,\omega)|^2\\&=S_{FF}(\omega)\,|H_x(\omega)|^2\Big(1-2\,g_x(t)\cos(\omega\,t)-\\&\qquad 2\,\omega\,m\,h_x(t)\sin(\omega\,t)+g_x^2(t)+\omega^2 m^2 h_x^2(t)\Big)\end{aligned} \tag{6.47}$$

Similarly, one can use Eqs. 6.36 and 6.46 to evaluate the time-varying harmonic response function for the velocity response as

$$H_{\dot{x}S}(t,\omega)\;=H_x(\omega)\Big(i\,\omega-e^{-i\omega t}[g_x'(t)+i\,\omega\,m\,h_x'(t)]\Big) \tag{6.48}$$

so that one can calculate the spectral densities

$$S_{X\dot{X}}(t,\omega)=S_{FF}(\omega)\,H_{xS}(t,\omega)\,H_{\dot{x}S}(t,-\omega) \tag{6.49}$$

and

$$S_{\dot{X}\dot{X}}(t,\omega)=S_{FF}(\omega)\,|H_{\dot{x}S}(t,\omega)|^2 \tag{6.50}$$

The time-varying response quantities $\sigma_X(t)$, $K_{X\dot{X}}(t,t)$, and $\sigma_{\dot{X}}(t)$ can then be found by integrating Eqs. 6.47, 6.49, and 6.50, respectively, with respect to ω. The result for variance of response from integration of Eq. 6.47 can be shown to be identical with the result of the time-domain integration in Eq. 5.35. In some situations, though, the frequency-domain integration will be easier to perform accurately. In particular, if the excitation is narrowband, then its covariance function will be oscillatory, giving the integrand in Eq. 5.35 as the product of two oscillatory functions, which can cause difficulties in numerical integration. The spectral density in Eq. 6.47, however, is generally not oscillatory, thereby avoiding this difficulty. For the special case of white noise excitation, the frequency-domain integrals can be evaluated by the calculus of residues, but that will not be done here. We already know the answers for that special case from the time-domain analysis in Section 5.6.

For $S_{FF}(\omega)=S_0$, one may note that the evolutionary spectral densities in Eqs. 6.47, 6.49, and 6.50 all behave as ω^{-2} for $|\omega|\rightarrow\infty$. This is in distinction to the stationary situation in which $S_{XX}(\omega)\sim\omega^{-4}$ and $S_{X\dot{X}}(\omega)\sim|\omega|^{-3}$, and it results in some integrals not existing for the evolutionary spectral densities, even though they are finite for the stationary situation.

6.9 Calculating Autospectral Density from a Sample Time History

In Example 6.5, we presented one method of calculating the autospectral density of a process by filtering a time history but noted that it is not the approach that is now commonly used. We will now present the usual method, which is based directly on the definition of spectral density in terms of a Fourier transform.

The concept of ergodicity discussed in Section 4.7 is very important when one needs to derive properties of a process by using measured data. In particular, if the process has the appropriate type of ergodicity, then one can use a time average over a single long time history to approximate an expected value, rather than needing an ensemble average over a large number of time histories. This property is usually assumed to hold unless there is evidence to the contrary. For autospectral density, however, we can show that ergodicity never holds in as strong a form as we might wish. Note that the definition of autospectral density in Eq. 6.15 involves both an expectation and a limit as the time interval T tends to infinity. It would be very desirable if ergodicity held in such a form that the expectation was unnecessary, and the calculation using a single time history of length T converged to the expected value as $T\rightarrow\infty$. Unfortunately, this is not true. We will now demonstrate that fact and indicate how the matter can be remedied so that a useful estimate of the autospectral density can be obtained from a single time history.

The relationships appearing in Eq. 6.15 can be rewritten in a compact form by defining a new frequency-domain stochastic process $\{Q_T(\omega)\}$ by

$$Q_T(\omega)=\frac{2\pi}{T}\tilde{Y}_T(\omega)\tilde{Y}_T(-\omega)\equiv\frac{1}{2\pi T}\int_{-T/2}^{T/2}\int_{-T/2}^{T/2}Y(t_1)\,Y(t_2)\,e^{-i\omega(t_1-t_2)}\,dt_1\,dt_2$$

in which $\tilde{Y}_T(\omega)$ is the truncated Fourier transform of $Y(t) \equiv X(t) - \mu_X(t)$. The definition of autospectral density is then simply $S_{XX}(\omega) = E[Q_T(\omega)]$. If the variance of this $Q_T(\omega)$ random variable went to zero as $T \to \infty$, then we could be confident that $Q_T(\omega)$ calculated from a long time history would have a high probability of being a good approximation of $S_{XX}(\omega)$. Thus, we wish to investigate the variance of $Q_T(\omega)$, which can be found from its first and second moments. The second moment is given by

$$E[Q_T^2(\omega)] = \frac{1}{(2\pi T)^2} \int_{-T/2}^{T/2} \int_{-T/2}^{T/2} \int_{-T/2}^{T/2} \int_{-T/2}^{T/2} E[Y(t_1) Y(t_2) Y(t_3) Y(t_4)] \times e^{-i\omega(t_1 - t_2 + t_3 - t_4)} \, dt_1 \, dt_2 dt_3 \, dt_4 \tag{6.51}$$

We can use Eq. 3.40 to evaluate the fourth moment term in the integrand. In particular, the fact that $\{Y(t)\}$ is mean-zero reduces this to

$$E[Y(t_1) Y(t_2) Y(t_3) Y(t_4)] = G_{XX}(t_1 - t_2) G_{XX}(t_3 - t_4) + G_{XX}(t_1 - t_4) \times G_{XX}(t_3 - t_2) + G_{XX}(t_1 - t_3) G_{XX}(t_2 - t_4) + \kappa_4[X(t_1), X(t_2), X(t_3), X(t_4)] \tag{6.52}$$

in which κ_4 denotes the fourth cumulant. One can now show that the contribution to $E[Q_T^2(\omega)]$ of the $G_{XX}(t_1 - t_3) G_{XX}(t_2 - t_4)$ term on the right-hand side of Eq. 6.52 tends to zero as $T \to \infty$ because of the signs in the exponential of Eq. 6.51. Also, the contribution of the final term tends to zero provided that the fourth cumulant function of $\{X(t)\}$ meets the sort of restrictions required for second-moment ergodicity (see Section 4.7). Thus, we need be concerned only with the behavior of the first two terms on the right-hand side of Eq. 6.52. The quadruple integral of each of these two terms can be separated into the product of two double integrals: grouped as (t_1, t_2) and (t_3, t_4) for the first term and as (t_1, t_4) and (t_2, t_3) for the second term. Furthermore, each of these double integrals is exactly of the form of $E[Q_T(\omega)]$. Substituting this result into the expression for $E[Q_T^2(\omega)]$ and taking the limit as T tends to infinity gives

$$\lim_{T \to \infty} E[Q_T^2(\omega)] = 2 S_{XX}^2(\omega)$$

from which we find that the variance of $Q_T(\omega)$ tends to $S_{XX}^2(\omega)$. Thus, the standard deviation of $Q_T(\omega)$ tends to the same limit as the mean value, namely $S_{XX}(\omega)$. Clearly, this gives no basis for expecting a sample value of $Q_T(\omega)$ to have a high probability of being near the desired, but unknown, $S_{XX}(\omega)$.

The approach that is typically used to avoid the sampling difficulty is to replace $Q_T(\omega)$ by a smoothed version. We will illustrate this idea by using the simplest sort of smoothing, which is a simple average over an increment of frequency

$$\overline{Q}_T(\omega) = \frac{2\pi}{T}\frac{1}{2\varepsilon}\int_{\omega-\varepsilon}^{\omega+\varepsilon} \tilde{Y}_T(\omega_1)\,\tilde{Y}_T(-\omega_1)\,d\omega_1 \tag{6.53}$$

or

$$\overline{Q}_T(\omega) = \frac{1}{4\pi T\varepsilon}\int_{-T/2}^{T/2}\int_{-T/2}^{T/2} Y(t_1)\,Y(t_2)\int_{\omega-\varepsilon}^{\omega+\varepsilon} e^{-i\omega_1(t_1-t_2)}\,d\omega_1\,dt_1\,dt_2$$

The expected value of $\overline{Q}_T(\omega)$, of course, is not exactly $S_{XX}(\omega)$ but rather an average of this autospectral density over the range $[\omega-\varepsilon,\omega+\varepsilon]$. This is generally not a significant problem if ε is chosen to be sufficiently small. The inclusion of this averaging, though, substantially changes the behavior of the mean-squared value, as can be verified by writing out $E([\overline{Q}_T(\omega)]^2)$ in the same way as Eq. 6.51

$$E[\overline{Q}_T^2(\omega)] = \frac{1}{(4\pi T\varepsilon)^2}\int_{-T/2}^{T/2}\int_{-T/2}^{T/2}\int_{-T/2}^{T/2}\int_{-T/2}^{T/2}\Bigg(E[Y(t_1)\,Y(t_2)Y(t_3)\,Y(t_4)]$$
$$\int_{\omega-\varepsilon}^{\omega+\varepsilon} e^{-i\omega_1(t_1-t_2)}\,d\omega_1\int_{\omega-\varepsilon}^{\omega+\varepsilon} e^{-i\omega_2(t_3-t_4)}\,d\omega_2\Bigg)dt_1\,dt_2 dt_3\,dt_4$$

The change comes from the fact that the second term of the expansion in Eq. 6.52 behaves very differently for $E[\overline{Q}_T^2(\omega)]$ than for $E[Q_T^2(\omega)]$. After performing the integration with respect to frequency on this second term, one obtains an integrand for the fourfold time integration of $G_{XX}(t_1-t_4)\times G_{XX}(t_3-t_2)/[(t_1-t_2)(t_3-t_4)]$ multiplied by a sum of four harmonic exponential terms. One can easily verify that this integrand goes to zero as any of the time arguments goes to infinity except in the special case when they all tend to infinity with $t_1 \approx t_2 \approx t_3 \approx t_4$. This behavior causes the fourfold time integral to give a term that grows like T as $T\to\infty$, but the division by T^2 then ensures that this term contributes nothing to $E[\overline{Q}_T^2(\omega)]$ in the limit.

The first term of the expansion of Eq. 6.52, however, behaves basically as it did for $E[Q_T^2(\omega)]$. In particular, it gives the integrand of the time integration as $G_{XX}(t_1-t_2)G_{XX}(t_3-t_4)/[(t_1-t_2)(t_3-t_4)]$, and this term does not decay as time arguments go to infinity on the two-dimensional set $t_1\approx t_2$, $t_3\approx t_4$, with no restriction that the first pair of time arguments be close to the second pair. Thus, the integral of this term grows like T^2. In fact, it can be shown that its contribution to $E[\overline{Q}_T^2(\omega)]$ is exactly $(E[\overline{Q}_T(\omega)])^2$, so the variance of $\overline{Q}_T(\omega)$

does tend to zero as $T \to \infty$. This guarantees that the $\overline{Q}_T(\omega)$ value obtained from a single sample time history has a high probability of being close to $E[\overline{Q}_T(\omega)]$ if T is sufficiently large. It is interesting that this limiting result holds no matter how small ε is chosen in the frequency averaging. Of course, the rate at which the variance of $\overline{Q}_T(\omega)$ tends to zero does depend on the value of ε.

The usual approach to estimation of $S_{XX}(\omega)$ from recorded data is, then, represented by Eq. 6.53, although a more sophisticated form of frequency averaging may be used in place of that shown. By taking the Fourier transform of a single time history of long duration and performing this averaging over a band of frequencies one can obtain an estimate of $S_{XX}(\omega)$. The bandwidth for the frequency averaging (2ε in Eq. 6.53) is often adjusted empirically to obtain an appropriately smooth estimate of $S_{XX}(\omega)$. This presumes, though, that one has some idea of the form of $S_{XX}(\omega)$ before the estimate is obtained. It must always be kept in mind that if the true $S_{XX}(\omega)$ function has a narrow peak that has a width not significantly exceeding 2ε, then the averaging in Eq. 6.53 will introduce significant error by greatly reducing that peak. Thus, considerable judgment should be exercised in estimating spectral densities.

It should also be noted that records of measured data are always of finite length. Furthermore, they represent observations only at discrete values of time rather than giving continuous time histories. This mandates that the integral Fourier transform considered here be replaced by the discrete Fourier transform (DFT), which is essentially a Fourier series in exponential form (see Appendix B). Efficient numerical schemes, including use of the fast Fourier transform (FFT) algorithm, have been developed to expedite implementation of the basic ideas presented. More detail on the problems of spectral estimation, including extensions to more complicated situations, are given in books on signal processing, such as Bendat and Piersol (1966), Marple (1987), and Priestly (1988).

6.10 Higher-Order Spectral Density Functions

For a non-Gaussian stochastic process, the spectral density analysis presented so far is important but gives only a partial description of the process. For example, one can, in principle, take the inverse Fourier transform of the autospectral density of any covariant stationary process $\{X(t)\}$ and thereby find the autocovariance function of the process. If the process is Gaussian, then this autocovariance function along with the mean-value function gives complete information about the probability density functions of any random variables $\{X(t_1),\cdots X(t_n)\}$ from the process. For example, any higher-order moment functions of $\{X(t)\}$ have simple relationships to the mean and autocovariance functions. If $\{X(t)\}$ is non-Gaussian, then the second-moment information given by the autospectral density is still equally valid, but it gives much less information about the probability distributions of the process because the process is no longer completely determined by its mean and autocovariance functions. Higher-order spectral density functions provide a way to analyze this higher-moment (or cumulant) information in the frequency domain.

The higher-order spectral density functions are most conveniently defined in terms of Fourier transforms of higher-order cumulant functions, either for a single stochastic process or for several processes. We will write out some explicit results for the so-called bispectrum, which relates to the third cumulant function. (See Section 3.7 for more general information on cumulant functions.) We will write the result for three jointly stationary processes $\{X(t)\}$, $\{Y(t)\}$, and $\{Z(t)\}$, because that includes the simpler cases with two or all of the processes being the same. The definition of the bispectrum is related to the Fourier transforms of the processes at three different frequency values

$$S_{XYZ}(\omega_1,\omega_2) = \lim_{T\to\infty} \frac{2\pi}{T} E\Big([\tilde{X}_T(\omega_1)-\tilde{\mu}_{X_T}(\omega_1)][\tilde{Y}_T(\omega_2)-\tilde{\mu}_{Y_T}(\omega_2)] \\ [\tilde{Z}_T(-\omega_1-\omega_2)-\tilde{\mu}_{Z_T}(-\omega_1-\omega_2)]\Big) \tag{6.54}$$

Note that the sum of the three frequency arguments in the right-hand side of Eq. 6.54 is zero. It can be shown that the corresponding expectation of the product of j Fourier transforms is generally bounded as T goes to infinity if the sum of the frequency arguments is not zero but unbounded when the sum of the frequency arguments is zero. Thus, the useful information contained within the jth cumulant function normalized by T is given only by the special case with the sum of the frequency arguments equal zero. Equations 6.17 and 6.20 are also special cases of this general result. Another higher-order spectrum that is sometimes encountered in practical analysis of non-Gaussian processes is called the trispectrum and is related to the Fourier transforms of processes at four frequency values that sum to zero.

From Eqs. 6.54 and 3.39 one can show that

$$S_{XYZ}(\omega_1,\omega_2)=\frac{1}{4\pi^2}\int_{-\infty}^{\infty}\int_{-\infty}^{\infty}\kappa_3[X(t+\tau_1),Y(t+\tau_2),Z(t)]e^{-i\omega_1\tau_1-i\omega_2\tau_2}\,d\tau_1\,d\tau_2$$

This demonstrates that the bispectrum is exactly the second-order Fourier transform of the stationary third cumulant function, just as the ordinary autospectral density function is the Fourier transform of the autocovariance function. This same relationship also extends to higher-order spectral densities. In particular, the trispectrum is the third-order Fourier transform of the stationary fourth-order cumulant function.

****

Example 6.6: Find the higher-order autospectral density function for a general stationary delta-correlated process $\{F(t)\}$ and for the $\{F(t)\}$ shot noise of Example 5.6 when the mean arrival rate is a constant b.

We will say that the general stationary delta-correlated process $\{F(t)\}$ has higher-order cumulants given by

$$\kappa_n[F(t_1),\cdots,F(t_n)]=G_n\,\delta(t_1-t_n)\delta(t_2-t_n)\cdots\delta(t_{n-1}-t_n)$$

This includes Eq. 5.37 as the special case with $n=2$, except that we used the notation G_0 rather than G_2 in that equation. We now write the nth-order autospectral density function as the $(n-1)$th-order Fourier transform of this cumulant. Thus, the general relationship that

$$S_n(\omega_1,\cdots,\omega_{n-1})=\frac{1}{(2\pi)^{n-1}}\int_{-\infty}^{\infty}\cdots\int_{-\infty}^{\infty}\kappa_n[F(t+\tau_1),\cdots,F(t+\tau_{n-1}),F(t)]\times$$

$$\exp[-i\,(\omega_1\tau_1+\cdots+\omega_{n-1}\tau_{n-1})]\,d\tau_1\cdots d\tau_{n-1}$$

gives $S_n(\omega_1,\cdots,\omega_{n-1})=G_n/(2\pi)^{n-1}$. Autospectral density functions of all orders are constants for a stationary delta-correlated process.

For the shot noise of Example 5.6 we found that the nth-order cumulant was

$$\kappa_n[F(t_1),\cdots,F(t_n)]=E[F^n]\,b\,\delta(t_1-t_n)\delta(t_2-t_n)\cdots\delta(t_{n-1}-t_n)$$

in which we have now replaced the nonstationary arrival rate $\dot{\mu}_Z(t)$ with the constant b. Thus, we see that $G_n=E[F^n]b$ and the spectral density is $S_n(\omega_1,\cdots,\omega_{n-1})=E[F^n]b/(2\pi)^{n-1}$.

****

Exercises

****

Spectral Density and Variance

****

6.1 Let $\{X(t)\}$ be a covariant stationary stochastic process with autospectral density function $S_{XX}(\omega) = |\omega| \, e^{-\omega^2}$.

(a) Find the variance of the $\{X(t)\}$ process.

(b) Find the variance of the $\{\dot{X}(t)\}$ derivative process.

****

6.2 Consider a covariant stationary stochastic process $\{X(t)\}$ with autospectral density function $S_{XX}(\omega) = e^{-\omega^2/2}$.

(a) Find the variance of the $\{X(t)\}$ process.

(b) Find the variance of the $\{\dot{X}(t)\}$ derivative process.

****

6.3 Let $\{X(t)\}$ be a covariant stationary stochastic process with autospectral density function

$$S_{XX}(\omega) = S_0 \left| \frac{\omega}{\omega_0} \right|^c U(\omega_0 - |\omega|) + S_0 \left| \frac{\omega_0}{\omega} \right|^c U(|\omega| - \omega_0)$$

in which S_0, ω_0, and c are positive constants.

(a) Find the variance of the $\{X(t)\}$ process.

(b) Find the variance of the $\{\dot{X}(t)\}$ derivative process.

(c) Note any restrictions on c required for your answers to parts (a) and (b).

****

6.4 Let $\{X(t)\}$ be a mean-zero covariant stationary stochastic process with autospectral density function $S_{XX}(\omega) = S_0 \, e^{-|\omega|}$.

(a) Find $E[X^2(t)]$.

(b) Find $E[\dot{X}^2(t)]$.

(c) Find $E[X(t)\,\dot{X}(s)]$.

****

6.5 Consider a covariant stationary process $\{X(t)\}$ with autospectral density

$$S_{XX}(\omega) = S_0 \, U(\omega_0 - |\omega|) + S_0 \left| \frac{\omega_0}{\omega} \right|^4 U(-|\omega| - \omega_0) \qquad \text{with } \omega_0 > 0$$

(a) Find the variance of the $\{X(t)\}$ process.

(b) Find the variance of the $\{\dot{X}(t)\}$ derivative process.

****

6.6 As a first step in modeling an earthquake ground motion, you wish to find a stationary stochastic process $\{Z(t)\}$ simultaneously satisfying the following three conditions:

$$\text{I: } E(\ddot{Z}^2) = 1.0 \ (\text{m/sec}^2)^2, \text{ II: } E(\dot{Z}^2) = 1.0 \ (\text{m/sec})^2, \text{ and III: } E(Z^2) < \infty$$

Consider each of the three following autospectral density curves as candidates for this modeling. For each autospectral density, state whether it is possible to choose real, positive constants a and b such as to satisfy conditions I, II, and III. If it is not possible to satisfy any particular condition, then explain the difficulty encountered. [Note: Evaluation of a or b is not required, only determination of whether solutions exist.]

(a) $S_{\dddot{Z}\dddot{Z}}(\omega) = a\,U(b - |\omega|)$

(b) $S_{\ddot{Z}\ddot{Z}}(\omega) = \omega^4 / (a + b\omega^8)$

(c) $S_{ZZ}(\omega) = (a + b\omega^4)^{-1}$

****

Narrowband and Broadband Processes

****

6.7 Let $\{X(t)\}$ be a covariant stationary stochastic process with autospectral density function

$$S_{XX}(\omega) = S_0 \exp(-\gamma\,|\,\omega + \omega_0\,|) + S_0 \exp(-\gamma\,|\,\omega - \omega_0\,|) \qquad \text{with } \omega_0 > 0$$

(a) Find the $G_{XX}(\tau)$ autocovariance function.

(b) Show that $\gamma >> 1$ gives a narrowband process with

$$\rho_{XX}(\tau) \equiv G_{XX}(\tau)/\sigma_X^2 \approx \cos(\omega_0 \tau)$$

(c) Show that $\gamma \to 0$ gives $G_{XX}(\tau) \to 0$ for $\tau \neq 0$, and $G_{XX}(0) \to \infty$ so that the autocovariance of $\{X(t)\}$ tends to that for a delta-correlated process.

****

6.8 Let $\{X(t)\}$ be a covariant stationary stochastic process with autospectral density function

$$S_{XX}(\omega) = S_0\, e^{-\gamma(\omega+\omega_0)^2} + S_0\, e^{-\gamma(\omega-\omega_0)^2} \qquad \text{with } \gamma > 0$$

(a) Find the $G_{XX}(\tau)$ autocovariance function.

(b) Show that $\gamma >> 1$ gives a narrowband process with

$$\rho_{XX}(\tau) \equiv G_{XX}(\tau)/\sigma_X^2 \approx \cos(\omega_0 \tau)$$

(c) Show that $\gamma \to 0$ gives $G_{XX}(\tau) \to 0$ for $\tau \neq 0$, and $G_{XX}(0) \to \infty$ so that the autocovariance of $\{X(t)\}$ tends to that for a delta-correlated process.

[Hint: Example 3.22 involves evaluation of a similar integral.]

****

Dynamic Response

****

6.9 Consider a linear system whose response $\{X(t)\}$ to an excitation $\{F(t)\}$ is governed by the differential equation $\ddot{X}(t) + c\,\dot{X}(t) = F(t)$, in which $c > 0$ is a constant. Let $\{F(t)\}$ be covariant stationary with autospectral density

$$S_{FF}(\omega) = S_0\, U(10c - |\omega|)$$

(a) Find the autospectral densities of $\{X(t)\}$ and $\{\dot{X}(t)\}$.

(b) Are the variances of $\{X(t)\}$ and $\{\dot{X}(t)\}$ finite? Briefly explain your answer.

****

6.10 Consider a linear system governed by the first-order differential equation

$$\dot{X}(t) + a\,X(t) = F(t) \qquad \text{with } 0 \le a \le 1$$

The excitation $\{F(t)\}$ is covariant stationary with nonzero mean $\mu_F = b$ and autocovariance function $K_{FF}(t+\tau,t) = e^{-|\tau|}$ for all t and τ.

(a) Find the autospectral density of $\{F(t)\}$.
(b) Find the autospectral density of $\{X(t)\}$.

****

6.11 Consider a building subjected to a wind force $\{F(t)\}$. The building is modeled as a linear SDF system, $m\,\ddot{X}(t) + c\,\dot{X}(t) + k\,X(t) = F(t)$, with $m = 200{,}000$ kg, $c = 8.0$ kN·s/m, and $k = 3{,}200$ kN/m. The force $\{F(t)\}$ has a mean of $\mu_F = 20$ kN and an autospectral density of

$$S_{FF}(\omega) = 500\,|\omega|e^{-\omega^2/2}\ (\text{kN})^2/(\text{rad/s}) \qquad \text{for all } \omega$$

(a) Find $E(F^2)$, the mean-squared force on the building.
(b) Find $\mu_X \equiv E(X)$.
(c) Estimate the standard deviation of $\{X(t)\}$ by replacing $\{F(t)\}$ by a constant force of 20 kN plus an "equivalent" white noise excitation.
(d) A delicate instrument is to be mounted in the building. Find the autospectral density of its base acceleration $S_{\ddot{X}\ddot{X}}(\omega)$.

****

6.12 Consider a linear system whose response $\{X(t)\}$ is governed by the differential equation $\ddot{X}(t) + 5\,\dot{X}(t) + 6\,X(t) = F(t)$. The excitation $\{F(t)\}$ is a mean-zero stationary white noise with autospectral density S_0.

(a) Find the $H_x(\omega)$ harmonic transfer function.
(b) Find the $S_{XX}(\omega)$ autospectral density of the response. Sketch your answer.

****

6.13 Consider the response of two structures to an anticipated earthquake. The strong motion portion of the ground acceleration will be modeled as a mean-zero stationary process with autospectral density

$$S_{FF}(\omega) = \frac{6}{400+\omega^2}\ \frac{(\text{m/s}^2)^2}{\text{rad/s}}$$

Each structure will be modeled as an SDF system

$$\ddot{X}_j(t) + 2\zeta_j\,\omega_j\,\dot{X}(t) + \omega_j^2\,X(t) = F(t)$$

with X_j representing the displacement at the top of structure j.

Structure number 1 has $\omega_1 = 10$ rad/s and $\zeta_1 = 0.01$. Structure number 2 is to be built near the first structure and is expected to have $\omega_2 = 15$ rad/s and $\zeta_2 = 0.005$.

(a) Use the concept of an equivalent white noise to estimate the stationary standard deviation of response of each structure.
(b) Let b = static clearance between the two structures. This gives the clearance between the two structures during the earthquake as $Y = b - X_1 + X_2$ (assuming that both are of the same height). Find stationary values of μ_Y and σ_Y by assuming that X_1 and X_2 are independent.

****

Chapter 7
Frequency, Bandwidth, and Amplitude

7.1 General Concepts

Several concepts introduced in Chapters 5 and 6 warrant more detailed analysis. In particular, characteristic frequency, amplitude, and phase of a stochastic process were used for narrowband processes, but no precise definitions were given that would allow the concepts to be extended to broadband situations. Furthermore, no measure of bandwidth was given to allow us to establish standards for judging whether or not a process is narrowband. One reason that we need such a concept is that some approximate methods of analysis of stochastic vibration and stochastic fatigue are based on the assumption that some process is narrowband, and a bandwidth parameter can help define what that means. It may be noted that dynamic analysis is not needed for the current chapter, so this material could be considered an extension of Chapter 4 on the general characteristics of stochastic processes. All of these concepts arose in the consideration of dynamic response in Chapters 5 and 6, however.

We will begin with time-domain analysis that specifically relates to the ideas of characteristic frequency and bandwidth; then we will consider common frequency-domain characterization of these quantities. Finally, we will study ways in which the concepts of amplitude and phase can be applied to broadband and narrowband processes and how they relate to the ideas of characteristic frequency and bandwidth.

7.2 Characteristic Frequency and Bandwidth from Rates of Occurrence

If one is studying a time history from a stochastic process, it is natural to consider the characteristic frequency to be related to the frequency of occurrences of certain events. Perhaps most obvious would be to look at the occurrences of local maxima, which we shall simply call *peaks*, or local mimima, which we shall call *valleys*. Not quite so obvious, but often more useful, is a study of the occurrences of crossings of the level $\mu_X(t)$ by the $X(t)$ time history. In particular, this

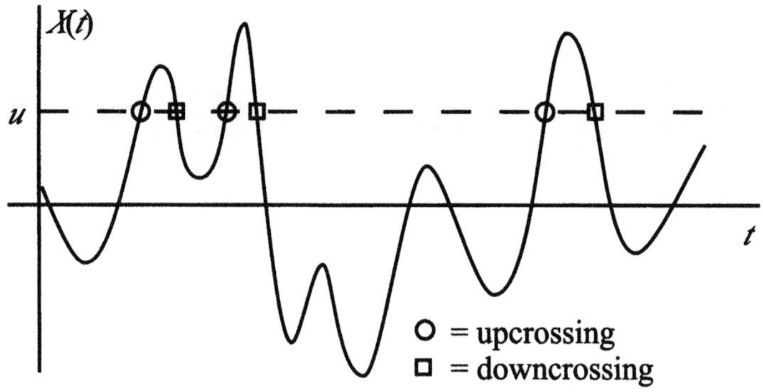

Figure 7.1 Crossings of the level $X(t) = u$.

crossing rate is easier to determine because it is less affected by the possible presence of small "wiggles" due to high-frequency components in the time history. It turns out that it is also slightly easier to give a mathematical description of the rate of crossings. Thus, we begin with the general topic of the rate of occurrence of crossings.

Let $\nu_X^+(u,t)$ denote the expected rate of occurrence of the event $X(t) = u$ with $\dot{X}(t) > 0$, and let $\nu_X^-(u,t)$ denote the expected rate for the event $X(t) = u$ with $\dot{X}(t) < 0$. Commonly these are called the *rate of upcrossings* and the *rate of downcrossings*, respectively, of the level $X = u$, as shown in Fig. 7.1. The expected number of upcrossings during any time interval of finite length, for example, is then the integral of $\nu_X^+(u,t)$ over the interval. Particularly for a nonstationary process, the rate of upcrossings can probably be more clearly understood by relating it to the probability of occurrence of an upcrossing during a small time increment. The basic definition of the expected rate of occurrence of any event can be written as

$$\lim_{\Delta t \to 0} \frac{E(\text{number of occurrences in } [t, t + \Delta t])}{\Delta t}$$

During any infinitesimal time interval $[t, t + \Delta t]$, though, we expect there to be either one or zero occurrences so that the expected number of occurrences is the same as the probability of an occurrence. Thus, we can write the expected rate of upcrossings as

$$\nu_X^+(u,t) = \lim_{\Delta t \to 0} \frac{P(\text{an upcrossing of } u \text{ in } [t, t + \Delta t])}{\Delta t} \tag{7.1}$$

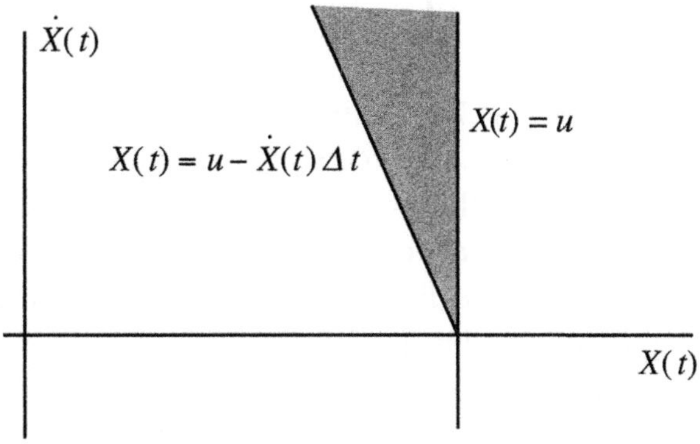

Figure 7.2 Phase diagram showing the event of upcrossing.

The probability of an upcrossing during an infinitesimal interval can be investigated intuitively by considering the phase diagram shown in Fig. 7.2. In particular, we can argue that there can only be an upcrossing of the level u within the interval $[t, t+\Delta t]$ if $X(t)$ at the beginning of the interval is less than u, but close to u, and has a positive derivative. In particular, inasmuch as Δt is infinitesimal, we might consider the derivative to be constant at the value $\dot{X}(t)$ throughout the time interval and conclude that there will be an upcrossing within the interval only if $0 < u - X(t) < \dot{X}(t)\Delta t$, which translates into $u - \dot{X}(t)\Delta t < X(t) < u$. This event is shown shaded on the space of possible values of $X(t)$ and $\dot{X}(t)$ in Fig. 7.2. The probability of this event can now be found by integrating the joint probability density of $X(t)$ and $\dot{X}(t)$ over the shaded region:

$$P(\text{an upcrossing of } u \text{ in } [t, t+\Delta t]) \approx \int_0^\infty \int_{u-v\Delta t}^{u} p_{X(t),\dot{X}(t)}(w,v)\, dw\, dv$$

We now once again use the fact that Δt is infinitesimal to argue that the w variable of integration is always almost the same as u. Replacing $p_{X(t)\dot{X}(t)}(w,v)$ with $p_{X(t)\dot{X}(t)}(u,v)$ in the integrand allows easy evaluation of the integral with respect to w, and gives

$$P(\text{an upcrossing of } u \text{ in } [t, t+\Delta t]) \approx \int_0^\infty (v\,\Delta t)\, p_{X(t)\dot{X}(t)}(u,v)\, dv$$

Substituting this expression into Eq. 7.1 now gives the expected rate of upcrossing as

$$\nu_X^+(u,t) = \int_0^\infty v\, p_{X(t)\dot{X}(t)}(u,v)\, dv \tag{7.2}$$

This derivation of Eq. 7.2 illustrates the ideas involved but is not very rigorous, particularly if the $\dot{X}(t)$ derivative is unbounded. That is, if the probability density function allows the possibility that $\dot{X}(t)$ has arbitrarily large values, then one cannot precisely claim that the range of the w integration from $u - \dot{X}(t)\,\Delta t$ to u is small, as we did in replacing w with u in the integrand. In fact, for any given value of Δt, this distance of $\dot{X}(t)\,\Delta t$ tends to infinity as $\dot{X}(t)$ tends to infinity. It is possible to prove that the limiting process is legitimate despite this difficulty, but rather than pursue that approach we will give an alternative derivation of Eq. 7.2 that relies more on algebra and less on geometry.

Because $\nu_X^+(u,t)$ is the expected rate of upcrossings, we will take the point of view that it is the mean value of the derivative of a counting process $N_X^+(u,t)$ that gives the number of upcrossings since time zero.[1] Figure 7.3 illustrates this idea. First we define a process $Z(t) \equiv U[X(t)-u]$ that steps back and forth between the level zero and unity, depending on whether $X(t)$ is less than or greater than u, respectively, as shown in part (b) of Fig. 7.3. The derivative of this process is always zero or infinity, but it can be written formally as

$$\dot{Z}(t) \equiv \delta[X(t)-u]\,\dot{X}(t)$$

Each of the Dirac delta function pulses in $\dot{Z}(t)$ is shown in Fig. 7.3(c) as an arrow, pointing toward either plus infinity or minus infinity, depending on the sign of $\dot{X}(t)$. Of course, the integral across any one of these delta functions is a step of unit magnitude, because the integral of $\dot{Z}(t)$ is $Z(t)$. By eliminating the delta functions with negative multipliers and then integrating, we can obtain a process that counts the number of upcrossings. This elimination of negative pulses is easily done by multiplying by $U[\dot{X}(t)]$, as shown in Fig. 7.3(d). The resulting counting process, as shown in part (e) of the figure, is $N_X^+(u,t)$, which starts at zero and proceeds to increase by unit step values, because it contains all the positive steps and none of the negative steps of $Z(t)$.

We can now say that $\nu_X^+(u,t)$ is the expected value of the derivative of $N_X^+(u,t)$. Thus, it is the expected value of the $\dot{Z}(t)\,U[\dot{X}(t)]$ process illustrated in part (d) of the figure. Thus, we merely need to substitute for $\dot{Z}(t)$ to obtain

---

[1]We can expect that this derivative of a counting process may have a finite mean value, but an infinite mean-squared value, as was true for the Poisson process in Example 4.1.

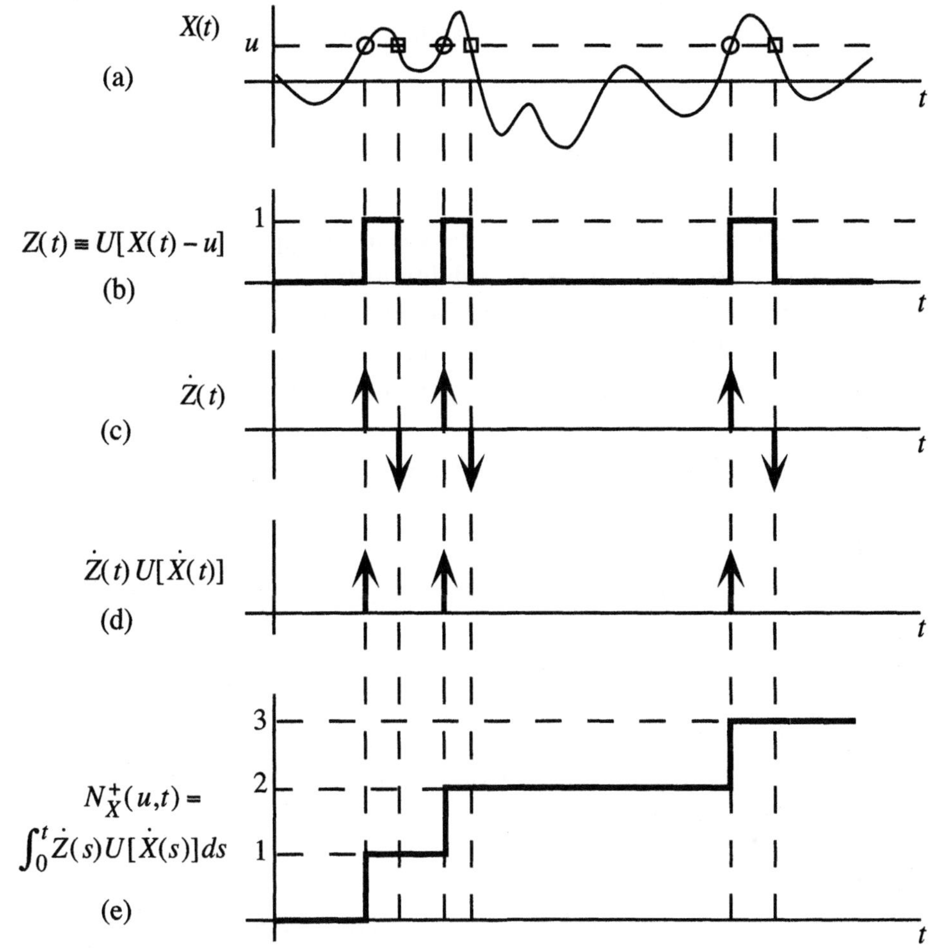

Figure 7.3 Counting the upcrossings.

$$\nu_X^+(u,t) = E\big(\dot Z(t)\, U[\dot X(t)]\big) = E\big(\delta[X(t)-u]\dot X(t)\, U[\dot X(t)]\big)$$

or

$$\nu_X^+(u,t) = \int_{-\infty}^{\infty}\int_{-\infty}^{\infty} p_{X(t)\dot X(t)}(w,v)\,\delta(w-u)\, v\, U(v)\, dw\, dv$$

Performing the integration with respect to w gives

$$\nu_X^+(u,t) = \int_{-\infty}^{\infty} p_{X(t)\dot X(t)}(u,v)\, v\, U(v)\, dv = \int_0^{\infty} v\, p_{X(t)\dot X(t)}(u,v)\, dv$$

which is identical to Eq. 7.2. Alternatively, we can factor the joint probability density function as a product of a marginal and conditional density function and write

$$\nu_X^+(u,t) = p_{X(t)}(u)\int_0^\infty v\, p_{\dot X(t)}[v \mid X(t)=u]\,dv \tag{7.3}$$

The relationships in Eqs. 7.2 and 7.3 are very general, applying to any stationary or nonstationary process with any probability distribution. It should be noted, though, that the value obtained may be infinite if the conditional probability density function of $\dot X(t)$ does not decay sufficiently rapidly. In particular, if $E[\dot X(t)|(t)|X(t)=u]$ does not exist then $\nu_X^+(u,t)$ may be infinite.

To obtain the downcrossing rate, we can essentially reverse the sign of $\dot X(t)$ to obtain

$$\nu_X^-(u,t) = -\int_{-\infty}^0 v\, p_{X(t)\dot X(t)}(u,v)\,dv = p_{X(t)}(u)\int_{-\infty}^0 |v|\, p_{\dot X(t)}[v \mid X(t)=u]\,dv \tag{7.4}$$

From Eqs. 7.2 and 7.4, we can also see that if the $\{X(t)\}$ process is second-order stationary then the crossing rates will be independent of time t. In this special case, we can drop the t argument and write $\nu_X^+(u)$ and $\nu_X^-(u)$ for the upcrossing and downcrossing rates.

****

Example 7.1: Find the expected rate of upcrossings for a covariant stationary Gaussian process $\{X(t)\}$.

For a covariant stationary process, we know that the random variables $X(t)$ and $\dot X(t)$ are uncorrelated. In addition, we know that Gaussian random variables are uncorrelated if and only if they are independent. Thus, we know that $X(t)$ and $\dot X(t)$ are independent for the present situation. This allows us to use the unconditional distribution of $\dot X(t)$ in place of the conditional distribution in Eq. 7.3, giving

$$\nu_X^+(u,t) = p_{X(t)}(u)\int_0^\infty \frac{v}{(2\pi)^{1/2}\sigma_{\dot X}} \exp\left(-\frac{1}{2}\left[\frac{v-\mu_{\dot X}(t)}{\sigma_{\dot X}}\right]^2\right)dv$$

which can be integrated to give

$$\nu_X^+(u,t) = p_{X(t)}(u)\left(\mu_{\dot X}(t)\Phi\left[\frac{\mu_{\dot X}(t)}{\sigma_{\dot X}}\right] + \frac{\sigma_{\dot X}}{(2\pi)^{1/2}}\exp\left[\frac{-\mu_{\dot X}^2(t)}{2\sigma_{\dot X}^2}\right]\right)$$

or

$$\nu_X^+(u,t) = \frac{1}{(2\pi)^{1/2}\sigma_X} \exp\left(-\frac{1}{2}\left[\frac{u-\mu_X(t)}{\sigma_X}\right]^2 \right) \times$$

$$\left(\mu_{\dot X}(t)\Phi\left[\frac{\mu_{\dot X}(t)}{\sigma_{\dot X}}\right] + \frac{\sigma_{\dot X}}{(2\pi)^{1/2}} \exp\left[\frac{-\mu_{\dot X}^2(t)}{2\sigma_{\dot X}^2}\right] \right)$$

in which $\Phi(\cdot)$ denotes the cumulative Gaussian distribution function. Note that μ_X and $\mu_{\dot X}$ are written as functions of time, because covariant stationarity does not imply mean-value stationarity.

Next consider the special case in which $\{X(t)\}$ is also mean-value stationary and thereby strictly stationary. In this case we know that μ_X is a constant and $\mu_{\dot X} = 0$, so our expression can be simplified to

$$\nu_X^+(u) = \frac{\sigma_{\dot X}}{2\pi\sigma_X} \exp\left(-\frac{1}{2}\left[\frac{u-\mu_X}{\sigma_X}\right]^2 \right) \quad \text{(for mean and covariant stationary)}$$

Recall that one definition of characteristic frequency of a narrowband process was given in Eq. 6.27 as $\omega_{c2} = \sigma_{\dot X} / \sigma_X$. Thus, we can extend this definition of ω_{c2} to apply to any stochastic process and say that

$$\nu_X^+(u) = \frac{\omega_{c2}}{2\pi} \exp\left(-\frac{1}{2}\left[\frac{u-\mu_X}{\sigma_X}\right]^2 \right)$$

The maximum value of this crossing rate occurs when $u = \mu_X$ and it is simply $\nu_X^+(\mu_X) = \omega_{c2} / (2\pi)$. Thus, we see that the rate of upcrossings of $u = \mu_X$ by a stationary Gaussian process is simply the energy-based average frequency divided by 2π. The factor of 2π, of course, comes from the fact that ω_{c2} represents a frequency in radians per second, while the rate of mean-upcrossings represents a frequency in cycles per second, or Hz. The interesting thing is that these two quite distinct definitions of process frequency are exactly equivalent for a stationary Gaussian process.

****

Example 7.2: Find the expected rate of upcrossings for a Gaussian process $\{X(t)\}$ that is mean-zero and has a nonstationary covariance.

The conditional Gaussian probability density function of $\dot X(t)$ can be written in the usual Gaussian form (see Example 2.25). In particular, we can write

$$p_{\dot X(t)}[v \mid X(t) = u] = \frac{1}{(2\pi)^{1/2}\sigma_*(t)} \exp\left(-\frac{1}{2}\left[\frac{v-\mu_*(t)}{\sigma_*(t)}\right]^2 \right)$$

in which $\mu_*(t)$ and $\sigma_*(t)$ are the conditional mean and standard deviation of $\dot{X}(t)$ and are given by

$$\mu_*(t) = \frac{K_{X\dot{X}}(t,t)}{K_{XX}(t,t)} u = \rho_{X\dot{X}}(t,t) \frac{\sigma_{\dot{X}}(t)}{\sigma_X(t)} u, \qquad \sigma_*(t) = \sigma_{\dot{X}}(t)[1-\rho_{X\dot{X}}^2(t,t)]^{1/2}$$

Because the conditional distribution of $\dot{X}(t)$ is the usual Gaussian form, the integral in Eq. 7.3 is basically the same as in Example 7.1:

$$\nu_X^+(u,t) = \frac{1}{(2\pi)^{1/2}\sigma_X(t)} \exp\left(\frac{-u^2}{2\sigma_X^2(t)}\right)\left(\mu_*(t)\Phi\left[\frac{\mu_*(t)}{\sigma_*(t)}\right] + \frac{\sigma_*(t)}{(2\pi)^{1/2}} \exp\left[\frac{-\mu_*^2(t)}{2\sigma_*^2(t)}\right]\right)$$

which can be rewritten as

$$\nu_X^+(u,t) = \frac{\sigma_{\dot{X}}(t)}{\sigma_X(t)} \exp\left(\frac{-u^2}{2\sigma_X^2(t)}\right)\left(\frac{\rho_{X\dot{X}}(t,t)\,u}{(2\pi)^{1/2}\sigma_X(t)}\Phi\left[\frac{\rho_{X\dot{X}}(t,t)\,u}{[1-\rho_{X\dot{X}}^2(t,t)]^{1/2}\sigma_X(t)}\right] + \right.$$
$$\left. \frac{[1-\rho_{X\dot{X}}^2(t,t)]^{1/2}}{2\pi} \exp\left[\frac{-\rho_{X\dot{X}}^2(t,t)\,u^2}{2[1-\rho_{X\dot{X}}^2(t,t)]\sigma_X^2(t)}\right]\right)$$

Note that setting the correlation coefficient equal to zero in this expression gives the same result as in Example 7.1. We may also note that a symmetric probability distribution for $X(t)$ and $\dot{X}(t)$, as in this example, gives the rate of downcrossings to be the same as the rate of upcrossings.

Setting $u = 0$ in the previous expression gives the rate of zero-upcrossings or mean-upcrossings as

$$\nu_X^+(0,t) = \nu_X^+(\mu_X,t) = \frac{[1-\rho_{X\dot{X}}^2(t,t)]^{1/2}}{2\pi} \frac{\sigma_{\dot{X}}(t)}{\sigma_X(t)} = \frac{[1-\rho_{X\dot{X}}^2(t,t)]^{1/2}}{2\pi} \omega_{c2}$$

showing that a correlation between $X(t)$ and $\dot{X}(t)$ results in a lowering of the process frequency, as measured by this upcrossing rate.

****

Example 7.3: Find the rate of upcrossings for the stationary dynamic response of a linear SDF oscillator excited by mean-zero stationary Gaussian white noise.

This is a special case of Example 7.1, because the $\{X(t)\}$ response process is Gaussian and covariant stationary. Furthermore, the response is also mean-value stationary. Thus, the key parameters needed are the stationary standard deviations of $X(t)$ and $\dot{X}(t)$. From the results in Sections 5.7 and 6.8, we can write these as

$$\sigma_X = \left(\frac{\pi S_0}{2\zeta\omega_0^3}\right)^{1/2}, \qquad \sigma_{\dot{X}} = \left(\frac{\pi S_0}{2\zeta\omega_0}\right)^{1/2}$$

Thus, we obtain

$$\nu_X^+(u) = \frac{\omega_0}{2\pi} \exp\left(-\frac{1}{2}\left[\frac{u-\mu_X}{\sigma_X}\right]^2 \right)$$

with a maximum value of $\omega_0/(2\pi)$. The expected frequency of mean-upcrossings for this process is exactly the nominal undamped frequency of the oscillator in Hz. If the excitation is not white noise or is not stationary, then this very convenient result will no longer be strictly true.

****

Example 7.4: Find the approximate expected rate of crossings of the mean-value level of a narrowband process $\{X(t)\}$ for which the joint probability density of $X(t)$ and $\dot{X}(t)$ is unknown.

For a narrowband process we can approximate $\nu_X^+[\mu_X(t),t]$ without knowledge of $p_{X(t)\dot{X}(t)}(u,v)$, even though our general formula for calculating the rate of upcrossings depends on the term. This follows directly from the nearly harmonic time history behavior of a narrowband process, as discussed in Section 6.4. For $X(t) \approx \mu_X(t) + A(t)\cos[\omega_c t + \theta(t)]$ with slowly varying amplitude and phase, the crossings of the level $\mu_X(t)$ must occur with an approximate period of $2\pi/\omega_c$, as illustrated in Fig. 6.3. Thus, we must have an approximate expected rate of upcrossings of $\nu_X^+[\mu_X(t),t] \approx \omega_c/2\pi$.

****

The examples illustrate the usefulness of $\nu_X^+(\mu_X,t)$ as a measure of the characteristic frequency (in Hz) of a stochastic process. For a narrowband process it is consistent with the physical notion of the frequency of a nearly harmonic time history, and for a Gaussian process it is exactly in agreement with the ω_{c2} defined in Eq. 6.27 for a stationary Gaussian process.

Because a peak of $X(t)$ occurs whenever $\dot{X}(t)=0$ and $\ddot{X}(t)<0$, we can say that the rate of occurrence of peaks of $\{X(t)\}$ is exactly the rate of downcrossings of the level zero by $\{\dot{X}(t)\}$:

$$\begin{aligned}\nu_P(t) \equiv \nu_{\dot{X}}^-(0,t) &= \int_{-\infty}^0 |w|\, p_{\dot{X}(t),\ddot{X}(t)}(0,w)\, dw \\ &= p_{\dot{X}(t)}(0) \int_{-\infty}^0 |w|\, p_{\ddot{X}(t)}(w \mid \dot{X}(t)=0)\, dw\end{aligned} \tag{7.5}$$

Similarly, the rate of occurrence of valleys of $\{X(t)\}$ is $\nu_V(t) \equiv \nu_{\dot{X}}^+(0,t)$.

Next we note that knowledge of the rate of occurrence of peaks and crossing rates can be used to define a measure of bandwidth of a stochastic process. One can easily verify that any sufficiently long continuous time history of a process must have at least as many peaks as it has upcrossings of any level. Specifically, at least one peak (and one valley) must occur between any two upcrossings of the same level u. From this fact, we can conclude that $\nu_P(t) \ge \nu_X^+(u,t)$ for any u for any process with continuous time histories. The rate of occurrence of peaks of a narrowband process, however, is expected to be only slightly larger than the rate of upcrossings of the mean, based on the similarity of a narrowband time history to a harmonic function with slowly varying amplitude and phase (as discussed in Chapter 6). This property is commonly used to provide a measure of bandwidth that can readily be estimated from a time history. It is called the *irregularity factor* and is defined as

$$IF = \nu_X^+(\mu_X,t)/\nu_P(t) \tag{7.6}$$

The range of possible values of IF is from zero to unity, with IF tending to unity for a narrowband process. We will also use this normalization for other bandwidth measures introduced in the following sections.

****

Example 7.5: Find the rate of peak occurrences and the irregularity factor for a Gaussian process $\{X(t)\}$ that is mean-value stationary and covariant stationary.

We can obtain the rate of occurrences of peaks by rewriting the results in Example 7.1 so that they apply to the $\{\dot{X}(t)\}$ process. In particular, we know that $\mu_{\dot{X}} = 0$ because μ_X is stationary, and we know that the rate of upcrossings is the same as the rate of downcrossings. Thus, we deduce from Example 7.1 that the rate of peak occurrences is $\nu_P = \sigma_{\ddot{X}} /(2\pi\sigma_{\dot{X}})$. The irregularity factor is then easily written as

$$IF = \frac{N_X^+(\mu_X)}{\nu_P} = \frac{\sigma_{\dot{X}}^2}{\sigma_X \, \sigma_{\ddot{X}}}$$

One needs to know only the standard deviations of $X(t)$, $\dot{X}(t)$, and $\ddot{X}(t)$ to evaluate IF for any stationary Gaussian stochastic process.

****

Example 7.6: Find the rate of peak occurrences and the irregularity factor for the $\{X(t)\}$ response of a linear SDF oscillator excited by Gaussian white noise.

As in the previous example, we can rewrite the results in Examples 7.1 and 7.2 so that they apply to the $\{\dot{X}(t)\}$ process. When we do this, we find that $\nu_{\dot{X}}^-(u,t)$

for a Gaussian process depends on $\sigma_{\ddot{X}}(t)$, the standard deviation of the acceleration. We know, though, that $\sigma_{\ddot{X}}(t)$ is infinite for the response of the SDF oscillator to white noise excitation. Thus, we find that the $\nu_P(t)$ rate of peak occurrences is infinite for the response of any linear SDF oscillator excited by Gaussian white noise. Correspondingly, the IF irregularity factor is zero for the response process.

****

The result in Example 7.5 illustrates an important feature of the occurrence rate for peaks. In particular, it was found that the peak occurrence rate is finite for a Gaussian process if and only if $\sigma_{\ddot{X}}$ is finite. We do not have a correspondingly simple rigorous result for non-Gaussian processes, but it is clear from Eq. 7.5 that the existence of a finite peak occurrence rate is dependent on the manner in which $p_{\dot{X}\ddot{X}}(0,w)$ converges to zero as w tends to infinity. In any problem in which $\sigma_{\ddot{X}}$ is infinite, we should anticipate the possibility that the peak occurrence rate may also be infinite. Similarly, if $\sigma_{\dot{X}}$ is infinite then there is the possibility that the crossing rates will be infinite, and this is easily shown to be a rigorous relationship for a Gaussian process.

The response of an SDF oscillator to white noise excitation is one important example in which $\sigma_{\ddot{X}}$ is not finite, so the peak occurrence rate is infinite if the excitation and response are Gaussian. This finding of infinite occurrence rates of peaks in a model of an important physical problem may seem somewhat surprising. That is, we certainly expect the number of peaks in a time history of a physical phenomenon to be finite, and we believe that many physical phenomena are approximated by the response of linear SDF oscillators excited by broadband stochastic excitations. This reveals a shortcoming in our modeling of the excitation as white noise. If we replaced the white noise excitation with an $\{F(t)\}$ that had a finite variance, then we would find that $\sigma_{\ddot{X}}$ would be finite and the peak occurrence rate would be finite. As explained in Sections 5.5 and 6.5, the use of a delta-correlated or white noise excitation always involves an approximation of the physical problem of interest. This approximation gives us many useful results about the response of an oscillator, but it fails to give us the rate of occurrence of peaks. The rate of occurrence of peaks depends quite heavily on the behavior of the high-frequency portion of the autospectral density, and for the SDF oscillator this is approximated by $S_{XX}(\omega) \approx S_{FF}(\omega)/(m^2\omega^4)$ or $S_{\ddot{X}\ddot{X}}(\omega) \approx S_{FF}(\omega)/m^2$, as discussed in Section 6.3. To find the peak occurrence rate for the system, one must know how $S_{FF}(\omega)$ decays as $|\omega|$ becomes large. One can show that the same conclusion applies to any problem in which a broadband force is applied to a finite mass within a system made up of masses, springs, and dashpots, because such a system will always give a linear relationship between $S_{\ddot{X}\ddot{X}}(\omega)$ and $S_{FF}(\omega)$ in the high-frequency region.

7.3 Frequency-Domain Analysis

Probably the most direct frequency-domain method of defining a bandwidth parameter was introduced by Vanmarcke (1972), who noted that the time-invariant $S_{XX}(\omega)$ autospectral density from $\omega = 0$ to $\omega = \infty$ can be likened to a probability density function. In particular, it is a nonnegative function and it has a bounded integral if the $\{X(t)\}$ process has bounded variance. A probability density function might be considered narrow if the associated random variable has a small variance. More precisely, the dimensionless coefficient of variation, giving the ratio of the standard deviation to the mean of the random variable, gives a measure of how much the random variable is likely to differ from its mean. Similarly, one can define moments of the autospectral density function and, from these, calculate a parameter that gives the relative width of the autospectral density in the same way that the coefficient of variation of a random variable gives the relative width of its probability density function. We will define these spectral moments as

$$\lambda_j = \int_{-\infty}^{\infty} |\omega|^j S_{XX}(\omega)\, d\omega = 2\int_0^{\infty} \omega^j S_{XX}(\omega)\, d\omega \tag{7.7}$$

For a very narrowband process concentrated at frequencies $\pm\omega_c$, we see that $\lambda_j \approx (\omega_c)^j \lambda_0$.

Note that in order to have a good analogy with a probability density function, we need to consider a form of autospectral density that has a unit integral, and $S_{XX}(\omega)/\lambda_0$ has this property. Furthermore, the jth moment of $S_{XX}(\omega)/\lambda_0$ is simply λ_j/λ_0. Note that the first spectral moment of $S_{XX}(\omega)/\lambda_0$, corresponding to a mean value in the probability density analogy, is a reasonable description of the characteristic frequency of the process. We will write this as

$$\omega_{c1} \equiv \frac{\lambda_1}{\lambda_0} \tag{7.8}$$

in which the subscript "1" is introduced to distinguish this parameter from the characteristic frequency introduced in Eq. 6.27, which can be rewritten as

$$\omega_{c2} \equiv \frac{\sigma_{\dot{X}}}{\sigma_X} = \left(\frac{\lambda_2}{\lambda_0}\right)^{1/2} \tag{7.9}$$

The bandwidth parameter directly corresponding to coefficient of variation is thus

$$s = \frac{[(\lambda_2 / \lambda_0) - (\lambda_1 / \lambda_0)^2]^{1/2}}{\lambda_1 / \lambda_0} = \left(\frac{\lambda_0 \lambda_2}{\lambda_1^2} - 1\right)^{1/2}$$

The parameter s tends to zero for the limiting narrowband process, and it is always greater than zero for any other situation. It has no general upper limit. Rather than using this parameter directly, though, it is conventional to convert it to a parameter that is always in the range of zero to unity and that tends to unity for a narrowband process. The commonly-used parameter having this normalization is

$$\alpha_1 = \frac{\lambda_1}{(\lambda_0 \lambda_2)^{1/2}} \tag{7.10}$$

which is easily shown to be $\alpha_1 = (s^2 + 1)^{-1/2}$. The fact that s^2 is nonnegative demonstrates that $0 \le \alpha_1 \le 1$.

A slight variation on Eq. 7.10 can be used to define a more general family of bandwidth parameters, each of which has properties similar to α_1, but with some differences. In particular, let

$$\alpha_j = \frac{\lambda_j}{(\lambda_0 \lambda_{2j})^{1/2}} \tag{7.11}$$

The normalization of α_j is the same as that of α_1 inasmuch as α_j is always in the unit interval. That it is nonnegative follows directly from its definition in terms of spectral moments, and one can show that $\alpha_j \le 1$ by demonstrating that $(\lambda_j)^2 \le \lambda_0 \lambda_{2j}$. This latter bound follows from the fact that

$$\int_{-\infty}^{\infty} \left((\lambda_{2j} / \lambda_0)^{1/2} - \omega^j\right)^2 S_{XX}(\omega)\, d\omega = \lambda_{2j} - 2\lambda_j \left(\frac{\lambda_{2j}}{\lambda_0}\right)^{1/2} + \lambda_{2j}$$

and this quantity must be greater than or equal to zero because the integrand is nonnegative. Furthermore, consideration of $\lambda_j \approx (\omega_c)^j \lambda_0$ shows that α_j also tends to unity for a narrowband process. Note that one can also introduce a new definition of characteristic frequency that depends only on λ_j and λ_0:

$$\omega_{cj} \equiv \left(\frac{\lambda_j}{\lambda_0} \right)^{1/j}$$

with ω_{c1} and ω_{c2} of Eqs. 7.8 and 7.9 being special cases. For a narrowband process, all these definitions of characteristic frequency will be nearly the same, but they will give differing values for a broadband process. In a loose sense one can say that the ω_{cj} characteristic frequency is consistent with the α_j bandwidth parameter.

Note that there is little restriction on the quantity j in the definition of the α_j bandwidth parameter or the ω_{cj} characteristic frequency. In fact, it need not even be an integer. One effect of the choice of j is the determination of the types of autospectral density functions for which $\alpha_j = 0$. In particular, $\alpha_j = 0$ if the high-frequency autospectral density decays sufficiently slowly that $\lambda_{2j} = \infty$. For example, if $S_{XX}(\omega)$ decays like $|\omega|^{-6}$, then $\lambda_{2j} = \infty$ for $j \ge 2.5$. One advantage of the α_1 parameter is the fact that it goes to zero only if the autospectral density does not decay more rapidly than $|\omega|^{-3}$, which covers quite a wide variety of practical situations. Out of the entire group of α_j parameters, the one in addition to α_1 that is widely used in practice corresponds to choosing $j = 2$. One disadvantage of α_2, as compared with α_1, is its increased sensitivity to high-frequency components of the autospectral density, resulting in its being zero if the spectral density does not decay more rapidly than $|\omega|^{-5}$. In Example 7.9 we will consider a problem of practical interest that fails to meet this criterion.

Probably the primary reason that investigators often choose to use the α_2 parameter has to do with certain special properties of the even-integer spectral moments. In particular, it is easy to verify that the three spectral moments used in defining α_2 each corresponds to a variance term: $\lambda_0 = \sigma_X^2$, $\lambda_2 = \sigma_{\dot{X}}^2$, and $\lambda_4 = \sigma_{\ddot{X}}^2$. Thus, one can write

$$\alpha_2 = \frac{\lambda_2}{(\lambda_0 \lambda_4)^{1/2}} = \frac{\sigma_{\dot{X}}^2}{\sigma_X \sigma_{\ddot{X}}} \qquad (7.12)$$

Similarly, $\omega_{c2} = \sigma_{\dot{X}} / \sigma_X$, as first defined in Eq. 6.27. Note that for the special case of a stationary Gaussian process, the α_2 result in Eq. 7.12 is identical to the *IF* that was derived in Example 7.5.

Of course, the sequence of even-order spectral moment relationships can also be extended, showing that λ_{2j} corresponds to the variance of the jth derivative of $\{X(t)\}$ whenever j is a nonnegative integer. When j is not an even integer, though, there is not such an easy physical interpretation of λ_j. In particular, the λ_1 spectral moment entering into the definition of α_1 is not easily identified with the variance of any physical quantity. If one knows the autospectral density function, then there is generally no difficulty in evaluating any of the bandwidth measures, but one can also find the value of α_2 and ω_{c2} without knowledge of the complete autospectral density function if the appropriate variances are known.

The α_2 bandwidth parameter also has another interpretation that may help illuminate its significance as a meaningful parameter of the stochastic process. In particular, note that the cross-covariance function for displacement and velocity at time t can be written as $G_{X\dot{X}}(0)$. The derivative of this expression with respect to time t is $G_{X\ddot{X}}(0) + \sigma_{\dot{X}}^2$. For a covariant stationary process, though, this derivative must be zero, because all the covariance terms must be constant. Thus, we have

$$G_{X\ddot{X}}(0) + \sigma_{\dot{X}}^2 = 0 \qquad \text{or} \qquad K_{X\ddot{X}}(t,t) = -\sigma_{\dot{X}}^2$$

From this we find that the correlation coefficient of $X(t)$ and $\ddot{X}(t)$ is exactly

$$\rho_{X\ddot{X}} \equiv \frac{K_{X\ddot{X}}(t,t)}{\sigma_X \sigma_{\ddot{X}}} = -\frac{\sigma_{\dot{X}}^2}{\sigma_X \sigma_{\ddot{X}}} = -\alpha_2 \tag{7.13}$$

The α_2 bandwidth parameter is precisely the negative of the correlation coefficient between the process and its second derivative. Inasmuch as the time histories of a narrowband process are nearly harmonic and the second derivative of any harmonic function is proportional to the negative of the original function, we should certainly expect $X(t)$ and $\ddot{X}(t)$ to have almost perfect negative correlation in the narrowband situation. It seems somewhat fortuitous that the magnitude of the correlation coefficient is exactly α_2, but this does provide one more confirmation that having $\alpha_2 \approx 1$ always implies that the time histories are nearly harmonic.

****

Example 7.7: Find the values of the ω_{c1} and ω_{c2} characteristic frequency parameters and the α_1 and α_2 bandwidth parameters for the covariant stationary process $\{X(t)\}$ of Section 6.4, with autospectral density function given by

$$S_{XX}(\omega) = S_0\, U[|\omega| - (\omega_c - b)]\, U[(\omega_c + b) - |\omega|]$$

The jth spectral density for this $S_{XX}(\omega)$ is easily evaluated as

$$\lambda_j = 2 S_0 \int_{\omega_c - b}^{\omega_c + b} \omega^j \, d\omega = 2 S_0 \frac{(\omega_c + b)^{j+1} - (\omega_c - b)^{j+1}}{j+1}$$

so $\lambda_0 = 4 S_0 b$, $\lambda_1 = 4 S_0 \omega_c b$, $\lambda_2 = 4 S_0 b[(\omega_c)^2 + b^2/3]$, and $\lambda_4 = 4 S_0 b[(\omega_c)^4 + 2(\omega_c)^2 b^2 + b^4/5]$. Thus,

$$\omega_{c1} = \omega_c$$

$$\omega_{c2} = (\omega_c^2 + b^2/3)^{1/2}$$

$$\alpha_1 = \frac{\lambda_1}{(\lambda_0 \lambda_2)^{1/2}} = \frac{\omega_c}{(\omega_c^2 + b^2/3)^{1/2}} = \left(1 + \frac{b^2}{3\omega_c^2}\right)^{-1/2}$$

and

$$\alpha_2 = \frac{\lambda_2}{(\lambda_0 \lambda_4)^{1/2}} = \frac{\omega_c^2 + b^2/3}{(\omega_c^4 + 2\omega_c^2 b^2 + b^4/5)^{1/2}} = \left(1 + \frac{b^2}{3\omega_c^2}\right)\left(1 + 2\frac{b^2}{\omega_c^2} + \frac{b^4}{5\omega_c^4}\right)^{-1/2}$$

For a narrowband situation with $b << \omega_c$, one can use power series expansions to simplify these relationships to

$$\omega_{c1} = \omega_c, \qquad \omega_{c2} \approx \omega_c + \frac{b^2}{6\omega_c}, \qquad \alpha_1 \approx 1 - \frac{b^2}{6\omega_c^2}, \qquad \alpha_2 \approx 1 - \frac{2b^2}{3\omega_c^2}$$

Thus, we see that ω_{c2} tends to $\omega_{c1} = \omega_c$ and both α_1 and α_2 tend to unity as b tends to zero. We also see, though, that α_1 and α_2 approach unity at quite different rates, with $(1 - \alpha_2)$ being four times as large as $(1 - \alpha_1)$.

The most broadband situation we can investigate with this particular autospectral density function is the one with $b = \omega_c$, so $S_{XX}(\omega)$ is flat from $-2\omega_c$ to $+2\omega_c$. In this situation, $\omega_{c2} = (7/6)\omega_c$ and the bandwidth parameters are $\alpha_1 = (3/4)^{1/2} \approx 0.866$ and $\alpha_2 = 5^{1/2}/3 \approx 0.745$. One notes that these bandwidth parameters are not very near zero, even though this would usually be considered to be quite a broadband autospectral density. In general, the bandwidth parameters approach zero only for autospectral densities with slowly decaying tails, and the values obtained here are representative of apparently broadband spectral densities on a bounded set of frequencies.

****

Example 7.8: Find the values of the ω_{c1} and ω_{c2} characteristic frequency parameters and the α_1 and α_2 bandwidth parameters for a covariant stationary process $\{X(t)\}$ with autospectral density function given by $S_{XX}(\omega) = S_0\, e^{-\gamma|\omega|}$, as in Example 6.1.

The *j*th spectral moment can be evaluated in terms of the gamma function (see Example 3.7 for definition), or the factorial if *j* is an integer:

$$\lambda_j = 2S_0 \int_0^\infty \omega^j\, e^{-\gamma\omega}\, d\omega = \frac{2S_0}{\gamma^{j+1}} \int_0^\infty u^j\, e^{-u}\, du = \frac{2S_0}{\gamma^{j+1}} \Gamma(j+1) = \frac{2S_0\, j!}{\gamma^{j+1}}$$

Thus, $\lambda_0 = 2S_0\gamma^{-1}$, $\lambda_1 = 2S_0\gamma^{-2}$, $\lambda_2 = 4S_0\gamma^{-3}$, and $\lambda_4 = 48S_0\gamma^{-5}$. Substituting these expressions into the definitions for ω_{c1}, ω_{c2}, α_1, and α_2 gives

$$\omega_{c1} = \gamma^{-1}, \quad \omega_{c2} \approx 0.707\gamma^{-1}, \quad \alpha_1 \approx 0.707, \quad \alpha_2 \approx 0.408$$

Note that the discrepancy between ω_{c1} and ω_{c2} is unaffected by the value of the parameter γ. Also the values of α_1 and α_2 are unaffected by the value of the γ, which seems somewhat surprising. For γ tending to zero we see that the stochastic process tends to white noise, whereas for large values of γ it has significant autospectral density only near frequency zero. This latter situation might be considered narrowband in a sense, but it is a different sense than we have been discussing. In particular, this process does not have nearly harmonic time histories because it does not have spectral density concentrated at some nonzero frequency ω_c, as in the other narrowband processes we have considered.

Because neither the α_1 nor the α_2 bandwidth parameter can give us information about the effect of the γ parameter, this example illustrates an important point. Namely, it is very difficult to make definite statements about the behavior of time histories based only on a bandwidth parameter. Knowledge of the complete autospectral density function will always give much more comprehensive information than will a bandwidth parameter.

****

Example 7.9: Find the values of the ω_{c1} and ω_{c2} characteristic frequency parameters and the α_1 and α_2 bandwidth parameters for the covariant stationary response $\{X(t)\}$ of an SDF oscillator excited by white noise.

Two of the four spectral moments of interest can be obtained almost trivially. In particular, λ_0 is the response variance, so Eq. 6.45 gives

$$\lambda_0 = \sigma_X^2 = \frac{\pi S_0}{2m^2\zeta\omega_0^3} = \frac{\pi S_0}{c\,k}$$

in which S_0 is the autospectral density of the excitation. Similarly, λ_2 is the variance of the derivative of the response, and this can be found from Eq. 5.63 as

$$\lambda_2 = \sigma_{\dot{X}}^2 = \frac{\pi S_0}{2m^2 \zeta \omega_0} = \frac{\pi S_0}{mc}$$

The λ_1 spectral moment is not a variance quantity, but it can be evaluated as

$$\lambda_1 = 2\int_0^\infty \omega S_{XX}(\omega)\, d\omega = 2S_0 \int_0^\infty \omega |H_x(\omega)|^2\, d\omega$$
$$= \frac{2S_0}{m^2}\int_0^\infty \frac{\omega\, d\omega}{(\omega^2 - \omega_0^2)^2 + (2\zeta\omega_0\omega)^2}$$

This integral over ω from zero to infinity is not appropriate for evaluation by the calculus of residues, which is useful for most of our frequency-domain integrals, but it can be shown that the integrand is the differential of an arctangent function. The result can be written as

$$\lambda_1 = \frac{S_0}{2m^2\zeta\omega_0\omega_d}\left(\frac{\pi}{2} + \tan^{-1}\left[\frac{1-2\zeta^2}{2\zeta(1-\zeta^2)^{1/2}}\right]\right)$$
$$= \frac{S_0}{2m^2\zeta\omega_0\omega_d}\left(\pi - 2\tan^{-1}\left[\frac{\zeta}{(1-\zeta^2)^{1/2}}\right]\right)$$

in which the slightly simpler final form has been derived by use of trigonometric identities. The λ_4 spectral moment is somewhat more of a problem. In particular, $|H_x(\omega)|^2$ behaves like ω^{-4} for $|\omega|$ tending to infinity, and this shows that $\omega^4 |H_x(\omega)|^2$ is not integrable. Actually, this is not surprising inasmuch as $\lambda_4 = \sigma_{\ddot{X}}^2$, and we noted in Section 5.6 that the variance of $\ddot{X}(t)$ is infinite for the SDF system excited by a delta-correlated process. Thus, we have $\lambda_4 = \infty$. From these spectral moments, we can write the characteristic frequency parameters as

$$\omega_{c1} = \frac{\omega_0}{(1-\zeta^2)^{1/2}}\left(1 - \frac{2}{\pi}\tan^{-1}\left[\frac{\zeta}{(1-\zeta^2)^{1/2}}\right]\right), \qquad \omega_{c2} = \omega_0$$

One may note that, in general, ω_{c1} is less than the resonant frequency ω_0 for the oscillator, and for $\zeta << 1$ it tends to ω_0 like $\omega_{c1} \approx \omega_0(1 - 2\zeta/\pi)$, while ω_{c2} is identical to the resonant frequency. Similarly, the bandwidth parameters are

$$\alpha_1 = \frac{1}{(1-\zeta^2)^{1/2}}\left(1 - \frac{2}{\pi}\tan^{-1}\left[\frac{\zeta}{(1-\zeta^2)^{1/2}}\right]\right), \qquad \alpha_2 = 0$$

Clearly, this is a problem for which the two bandwidth parameters give very different results. In fact, the α_2 parameter fails to give any useful information, because it gives $\alpha_2 = 0$, suggesting a broadband $\{X(t)\}$ process regardless of the value of ζ. We anticipate, though, that $\zeta << 1$ will be a narrowband situation, based on the shape of the $S_{XX}(\omega) = S_0 |H_x(\omega)|^2$ function. The α_1 parameter

does behave as we might expect. In particular, it tends to unity as ζ tends to zero, indicating that the response process is narrowband in this situation. For $\zeta << 1$, the limiting behavior is linear in ζ with $\alpha_1 \approx 1 - 2\zeta / \pi$.
According to Eq. 7.13, α_2 is also the negative of the correlation coefficient $\rho_{X\ddot{X}}$. The fact that $\alpha_2 = 0$ for the SDF response to white noise then means that $X(t)$ and $\ddot{X}(t)$, at the same instant of time t, are uncorrelated for this system. This follows directly, though, from the fact that the cross-covariance $K_{X\ddot{X}}$ is finite while the variance of $\ddot{X}(t)$ is infinite.
****

Another way of interpreting the spectral moment functions is in terms of integrals of cross-spectral density functions of the type given by Eqs. 6.23–6.25. For example, one can use the knowledge that $S_{X\dot{X}}(\omega) = -i\,\omega\, S_{XX}(\omega)$ and $S_{\dot{X}\dot{X}}(\omega) = \omega^2 S_{XX}(\omega)$ to write

$$\lambda_1 = i \int_{-\infty}^{\infty} \text{sgn}(\omega)\, S_{X\dot{X}}(\omega)\, d\omega$$

and

$$\lambda_2 = \int_{-\infty}^{\infty} S_{\dot{X}\dot{X}}(\omega)\, d\omega$$

It should be noted that these relationships are not unique. For example, one could use $-S_{\dot{X}X}(\omega)$ in place of $S_{X\dot{X}}(\omega)$ in the expression for λ_1. There is even more ambiguity about which spectral density to use for other spectral moments. For example, $S_{\ddot{X}X}(\omega) = -S_{\dot{X}\dot{X}}(\omega)$, so λ_2 can be rewritten as

$$\lambda_2 = -\int_{-\infty}^{\infty} S_{\ddot{X}X}(\omega)\, d\omega$$

For λ_4 there are three distinct options:

$$\lambda_4 = \int_{-\infty}^{\infty} S_{\ddot{X}\ddot{X}}(\omega)\, d\omega = -\int_{-\infty}^{\infty} S_{\dddot{X}\dot{X}}(\omega)\, d\omega = \int_{-\infty}^{\infty} S_{\ddddot{X}X}(\omega)\, d\omega$$

For a stationary $\{X(t)\}$ process, all of these forms are equivalent, so it makes no difference whether one uses Eq. 7.7 or some of the alternative forms given here to evaluate the spectral moments. If $\{X(t)\}$ is a modulated process, though, we have defined the various evolutionary spectral density functions according to Eqs. 6.38 and 6.39, so it is possible to extend the idea of spectral moments to this situation also. Furthermore, the results do depend on the choice of which formula to extend. For example, using

$$\lambda_1(t) = i\int_{-\infty}^{\infty} \text{sgn}(\omega)\, S_{X\dot{X}}(t,\omega)\, d\omega \quad (7.14)$$

$$\lambda_2(t) = \int_{-\infty}^{\infty} S_{\dot{X}\dot{X}}(t,\omega)\, d\omega = \sigma_{\dot{X}}^2(t) \quad (7.15)$$

and

$$\lambda_4(t) = \int_{-\infty}^{\infty} S_{\ddot{X}\ddot{X}}(t,\omega)\, d\omega = \sigma_{\ddot{X}}^2(t) \quad (7.16)$$

does not give the same values as using

$$\hat{\lambda}_1(t) = \int_{-\infty}^{\infty} |\omega| S_{XX}(t,\omega)\, d\omega \quad (7.17)$$

$$\hat{\lambda}_2(t) = \int_{-\infty}^{\infty} \omega^2 S_{XX}(t,\omega)\, d\omega \quad (7.18)$$

and

$$\hat{\lambda}_4(t) = \int_{-\infty}^{\infty} \omega^4 S_{XX}(t,\omega)\, d\omega$$

In fact, there are situations in which $\hat{\lambda}_1(t)$ and $\hat{\lambda}_2(t)$ have been shown to give infinite values (Corotis et al., 1972), whereas Eqs. 7.14 and 7.15 give finite values for $\lambda_1(t)$ and $\lambda_2(t)$ (Michaelov et al., 1999b). Note that there does seem to be only one logical definition for the nonstationary $\lambda_0(t)$ spectral moment:

$$\lambda_0(t) = \int_{-\infty}^{\infty} S_{XX}(t,\omega)\, d\omega = \sigma_X^2(t) \quad (7.19)$$

Using the time-varying spectral moments of Eqs. 7.14, 7.15, 7.16, and 7.19 allows one to extend the characteristic frequency and bandwidth parameters defined in Eqs. 7.8, 7.9, 7.10, and 7.12 to be functions of time.

The method of introducing the time-varying spectral moments presented here is that used by Michaelov et al. (1999a), but the moments are identical to those introduced earlier by Di Paola (1985) using the so-called pre-envelope process.

****

Example 7.10: Find the time-varying values of the $\omega_{c1}(t)$ and $\omega_{c2}(t)$ characteristic frequency parameters and the $\alpha_1(t)$ and $\alpha_2(t)$ bandwidth parameters for the response $\{X(t)\}$ of an SDF oscillator excited by uniformly modulated white noise: $F(t) = W(t)U(t)$, in which $\{W(t)\}$ is stationary white noise.

Using Eqs. 7.15, 7.16, and 7.19, the even-order spectral moments are given as response variances: $\lambda_0(t)=[\sigma_X(t)]^2$, $\lambda_2(t)=[\sigma_{\dot X}(t)]^2$, and $\lambda_4(t)=[\sigma_{\ddot X}(t)]^2$. The values of the variances have already been investigated in Section 5.6. One of the simplest observations is that $\lambda_4(t)=\infty$ for all values of t whereas $\lambda_0(t)$ and $\lambda_2(t)$ are finite. Thus, $\alpha_2(t)\equiv\lambda_2(t)/[\lambda_0(t)\,\lambda_4(t)]^{1/2}=0$ for all time. Clearly, the $\alpha_2(t)$ bandwidth parameter is of no more use for this modulated response than it was for the stationary response in Example 7.9.

To evaluate the $\alpha_1(t)$ parameter, it is necessary to obtain $\lambda_1(t)$ by integrating $S_{X\dot X}(t,\omega)$. This evolutionary cross-spectral density can be written from Eqs. 6.46, 6.48, and 6.49 as

$$S_{X\dot X}(t,\omega)=S_0\,H_{xS}(t,\omega)\,H_{\dot x S}(t,-\omega)$$
$$=S_0\,|H_x(\omega)|^2\left(1-e^{-i\omega t}[g_x(t)+i\,\omega m\,h_x(t)]\right)\times$$
$$\left(-i\,\omega-e^{i\omega t}[g_x'(t)-i\,\omega m\,h_x'(t)]\right)$$

Thus,

$$\lambda_1(t)=i\,S_0\int_{-\infty}^{\infty}|H_x(\omega)|^2\,\mathrm{sgn}(\omega)\times\left(1-e^{-i\omega t}[g_x(t)+i\,\omega m\,h_x(t)]\right)\times$$
$$\left(-i\,\omega-e^{i\omega t}[g_x'(t)-i\,\omega m\,h_x'(t)]\right)d\omega$$

Expanding the multiplication, substituting $e^{\pm i\omega t}=\cos(\omega t)\pm i\sin(\omega t)$, using symmetry and antisymmetry, and simplifying gives

$$\lambda_1(t)=2S_0\Bigg([1+e^{-2\zeta\omega_0 t}]\int_0^{\infty}|H_x(\omega)|^2\,\omega\,d\omega\;+\;\omega_d^{-1}e^{-\omega_0 t}\sin(\omega_d t)\times$$
$$\left[\omega_0^2\int_0^{\infty}|H_x(\omega)|^2\sin(\omega t)\,d\omega+\int_0^{\infty}|H_x(\omega)|^2\,\omega^2\sin(\omega t)\,d\omega\right]-$$
$$2e^{-\zeta\omega_0 t}\cos(\omega_d t)\int_0^{\infty}|H_x(\omega)|^2\,\omega\cos(\omega t)\,d\omega\Bigg)$$

Note that the first integral in this expression is the one evaluated in Example 7.9 for the stationary response. It is also possible to evaluate the other three integrals analytically, but the results are rather cumbersome. The accompanying sketch shows numerical values for $m^2(\omega_0)^2\lambda_1(t)/S_0$ plotted versus $\omega_0 t$ for $\zeta=0.1$.

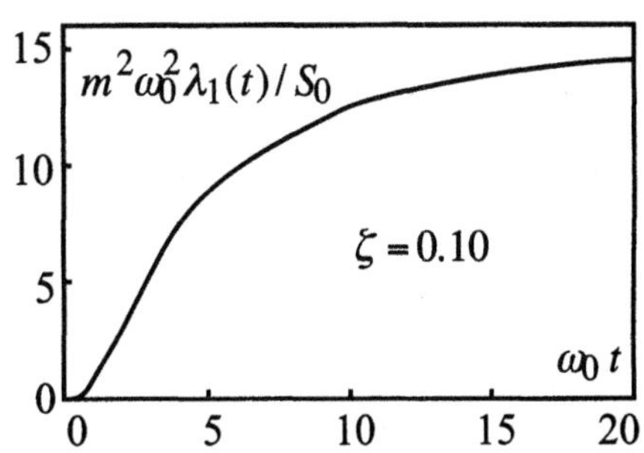

The $\lambda_0(t)$ and $\lambda_2(t)$ spectral moments can be taken directly from Eqs. 5.50 and 5.54 as

$$\lambda_0(t)=\sigma_X^2(t)=\frac{\pi S_0}{2m^2\zeta\,\omega_0^3}\left(1-e^{-2\zeta\omega_0 t}\left[\frac{\omega_0^2}{\omega_d^2}+\frac{\zeta\,\omega_0}{\omega_d}\sin(2\omega_d t)-\frac{\zeta^2\omega_0^2}{\omega_d^2}\cos(2\omega_d t)\right]\right)$$

and

$$\lambda_2(t)=\sigma_{\dot X}^2(t)=\frac{G_0}{4\,m^2\zeta\,\omega_0}\left(1-e^{-2\zeta\omega_0 t}\left[\frac{\omega_0^2}{\omega_d^2}-\frac{\zeta\,\omega_0}{\omega_d}\sin(2\omega_d t)-\frac{\zeta^2\omega_0^2}{\omega_d^2}\cos(2\omega_d t)\right]\right)$$

The two characteristic frequency parameters are now evaluated as $\omega_{c1}(t)=\lambda_1(t)/\lambda_0(t)$ and $\omega_{c2}(t)=[\lambda_2(t)/\lambda_0(t)]^{1/2}$. The accompanying sketches show numerical values if these quantities and of the bandwidth parameter $\alpha_1(t)\equiv\lambda_1(t)/[\lambda_0(t)\,\lambda_2(t)\,]^{1/2}$ for $\zeta=0.1$.

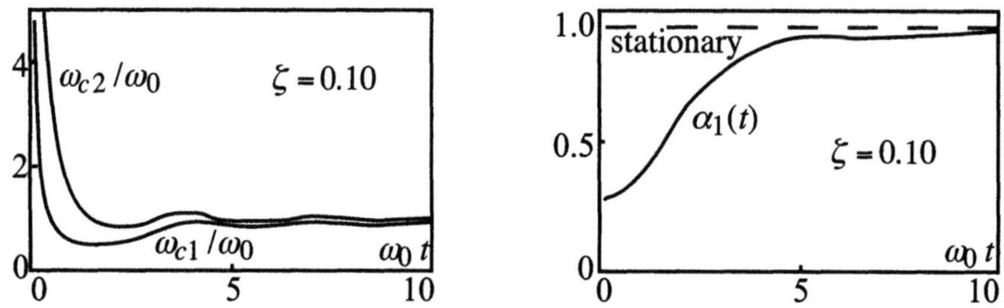

One can observe that $\omega_{c1}(t)$ and $\omega_{c2}(t)$ behave in a similar manner, with both beginning at infinity, then converging to ω_0 as time becomes large. The $\alpha_1(t)$ bandwidth parameter begins quite small, indicating that the response is initially quite broadband, then it converges to the stationary value of 0.94 as time grows. For these results with $\zeta=0.10$, the convergence to stationarity is reasonably close within one period of dynamic response—$\omega_0 t=2\pi$.

Note that Eq. 6.47 gives the time-varying $S_{XX}(t,\omega)$ for the response of this oscillator and shows that this spectral density decays like ω^{-2} as $|\omega|$ tends to infinity. Thus, the alternate time-varying spectral moments, $\hat{\lambda}_1(t)$ and $\hat{\lambda}_2(t)$, from Eqs. 7.17 and 7.18 do not exist, as noted by Corotis et al. (1972).

****

7.4 Amplitude and Phase of a Stationary Stochastic Process

In our discussion of narrowband processes, we used the concepts of amplitude and phase of a stochastic process without giving any precise definition of the terms. If a process does not have time histories that closely resemble harmonic functions, then such an intuitive idea of amplitude and phase is of little use. Furthermore, it is helpful to have a precise definition of the terms even for narrowband situations. This can be done in more than one way. We will begin with what we consider the most obvious approach, then look at alternative definitions.

Let $\{X(t)\}$ be any differentiable, covariant stationary stochastic process. The first requirement for the two new stochastic processes $\{A(t)\}$ and $\{\theta(t)\}$ is that they give

$$Y(t) = A(t)\cos[\omega_a t + \theta(t)] \tag{7.20}$$

in which $\{Y(t)\}$ is a mean-zero process defined as $Y(t) = X(t) - \mu_X(t)$ and ω_a is an average frequency. In particular, the definition of ω_a should ideally be chosen such that $E[\dot{\theta}(t)] = 0$, because a nonzero value for $E[\dot{\theta}(t)]$ should, in itself, be a contribution to the average frequency of the $\{Y(t)\}$ process. Taking the derivative of Eq. 7.20, then, gives

$$\dot{Y}(t) = -A(t)[\omega_a + \dot{\theta}(t)]\sin[\omega_a t + \theta(t)] + \dot{A}(t)\cos[\omega_a t + \theta(t)] \tag{7.21}$$

However, we now choose to impose the restriction that

$$\dot{Y}(t) = -\omega_a A(t)\sin[\omega_a t + \theta(t)] \tag{7.22}$$

which can be accomplished by satisfying the equation

$$\dot{A}(t)\cos[\omega_a t + \theta(t)] = A(t)\dot{\theta}(t)\sin[\omega_a t + \theta(t)] \tag{7.23}$$

It is important to note that Eqs. 7.20 and 7.22 may be viewed as two simultaneous equations that define the new stochastic processes $\{A(t)\}$ and $\{\theta(t)\}$ in terms of the original stochastic processes $\{Y(t)\}$ and $\{\dot{Y}(t)\}$. Specifically, Eqs. 7.20 and 7.22 give the explicit expressions for the amplitude and phase processes as

$$A(t) = \left(Y^2(t) + Z^2(t)\right)^{1/2} \tag{7.24}$$

and

$$\theta(t) = -\tan^{-1}\left(\frac{Z(t)}{Y(t)}\right) - \omega_a t \tag{7.25}$$

in which $Z(t) = \dot{Y}(t)/\omega_a = -A(t)\sin[\omega_a t + \theta(t)]$. Note, in particular, that $A(t) \ge 0$.

Equations 7.24 and 7.25 give definitions of amplitude and phase for any differentiable stochastic process $\{X(t)\}$, whether or not it is narrowband. This is in sharp contrast to the development in Section 6.4 in which we argued that Eq. 6.26, which is equivalent to Eq. 7.20, should hold for a narrowband process as sketched in Fig. 6.4. In the particular case with $\{X(t)\}$ being narrowband, we do anticipate that $\{A(t)\}$ and $\{\theta(t)\}$ in Eqs. 7.24 and 7.25 will be slowly varying so that the time histories of $\{Y(t)\}$ will have the almost harmonic behavior that we identified in Section 6.4 as being appropriate for a narrowband process. Variations in $A(t)$, of course, will be evident in a time history plot such as Fig. 6.4 as changes in the magnitude of the excursions of $Y(t)$ or $X(t)$. Variations in $\theta(t)$ are not quite so readily apparent in a time history, but they are related to changes in the time interval between zero crossings. That is, because $Y(t)=0$ whenever $\omega_a t+\theta(t)$ is an odd multiple of $\pi/2$, the zero crossings will be exactly uniformly spaced at intervals of π/ω_a if $\theta(t)$ has no variation. Any observed variations in the spacing will provide evidence that $\dot\theta(t)\neq 0$, so $\theta(t)$ is not constant. Of course, $\{A(t)\}$ and $\{\theta(t)\}$ can be considered slowly varying processes if $\dot A(t)$ and $\dot\theta(t)$ are small, so we will expect narrowband processes to have small values of these derivatives.

Before we explicitly investigate the derivatives of $\{A(t)\}$ and $\{\theta(t)\}$, we want to generalize their definitions. Even though the amplitude, or envelope, defined in Eqs. 7.24 and 7.25 has the desired characteristics, it turns out that it is only one of the possible choices. That is, there are other possible definitions of the amplitude of a process that will also closely agree with the intuitive concept of amplitude in the narrowband situation, and at least one of these has certain mathematical advantages.[2] We will begin investigating this general idea of amplitude of a stochastic process by looking at the crucial features of $\{A(t)\}$ for a narrowband process. Switching to truncated time histories allows us to use Fourier transforms, as we did in Section 6.2. In particular, the inverse Fourier transform relationship for $Z_T(t)$ is

$$Z_T(t)=\int_{-\infty}^{\infty}\tilde Z_T(\omega)\,e^{i\omega t}\,d\omega=\int_{-\infty}^{\infty} i\frac{\omega}{\omega_a}\,\tilde Y_T(\omega)\,e^{i\omega t}\,d\omega$$

---

[2]The idea of the existence of many definitions of amplitude, each with certain bandwidth implications, was investigated by Winterstein and Cornell in 1985, although this was certainly not the first study of alternative definitions of amplitude [e.g., see Cramer and Leadbetter (1967)].

For a $\{Y(t)\}$ process that is narrowband with average frequency ω_a, the $\tilde{Y}_T(\omega)$ function will be nearly zero except in the neighborhood of the two points $\pm\omega_a$. At $\omega \approx +\omega_a$ it will give $\tilde{Z}_T(\omega) \approx i\,\tilde{Y}_T(\omega)$, and at $\omega \approx -\omega_a$ the relationship will be $\tilde{Z}_T(\omega) \approx -i\,\tilde{Y}_T(\omega)$. These two relationships, then, are the key features of the $\{Z(t)\}$ process because other frequencies contribute very little to its behavior. Thus, if we choose some different "auxiliary" $\{Z(t)\}$ process for which the Fourier transform has this same behavior in the neighborhood of $\omega = \pm\omega_a$, then we can be assured that $A(t) = [Y^2(t) + Z^2(t)]^{1/2}$ will still behave like the amplitude of a narrowband $Y(t)$ process. There are many $\{Z(t)\}$ processes that meet this general condition. In particular, using

$$\tilde{Z}_T(\omega) = i\,g(\omega)\,\tilde{Y}_T(\omega) \tag{7.26}$$

for any odd $g(\cdot)$ function with $g(\omega_a) = 1$ makes

$$Z_T(t) = \int_{-\infty}^{\infty} i\,g(\omega)\,\tilde{Y}_T(\omega)\,e^{i\omega t}\,d\omega$$

be a real function and $\tilde{Z}_T(\omega)$ have the desired behavior in the neighborhood of $\omega = \pm\omega_a$. Using $g(\omega) = \omega/\omega_a$, as in Eqs. 7.20 and 7.22, is only one of infinitely many choices.

Based on Eq. 7.26 we can now investigate the spectral density functions and covariance values related to the new auxiliary process $\{Z(t)\}$. In particular, the autospectral density definition in Eq. 6.15 gives

$$S_{ZZ}(\omega) = g^2(\omega)\,S_{YY}(\omega) \equiv g^2(\omega)\,S_{XX}(\omega)$$

so

$$G_{ZZ}(\tau) = \int_{-\infty}^{\infty} g^2(\omega)\,S_{XX}(\omega)\,e^{i\omega\tau}\,d\omega$$

Similarly,

$$S_{XZ}(\omega) = -i\,g(\omega)\,S_{XX}(\omega) \quad \text{and} \quad G_{XZ}(\tau) = -i\int_{-\infty}^{\infty} g(\omega)\,S_{XX}(\omega)\,e^{i\omega\tau}\,d\omega$$

Taking derivatives of the $G_{ZZ}(\tau)$ and $G_{XZ}(\tau)$ functions then gives the covariance properties of the derivatives, and setting $\tau = 0$ gives various covariance relationships that will be useful in analyzing the behavior of the amplitude. Specifically,

$$\sigma_Z^2 = G_{ZZ}(0) = \int_{-\infty}^{\infty} g^2(\omega)\, S_{XX}(\omega)\, d\omega \tag{7.27}$$

$$K_{XZ} = G_{XZ}(0) = -i \int_{-\infty}^{\infty} g(\omega)\, S_{XX}(\omega)\, d\omega = 0 \tag{7.28}$$

$$K_{X\dot{Z}} = -K_{\dot{X}Z} \equiv G_{X\dot{Z}}(0) = -\int_{-\infty}^{\infty} \omega\, g(\omega)\, S_{XX}(\omega)\, d\omega \tag{7.29}$$

$$K_{\dot{X}\dot{Z}} \equiv G_{\dot{X}\dot{Z}}(0) = -i \int_{-\infty}^{\infty} \omega^2 g(\omega)\, S_{XX}(\omega)\, d\omega = 0 \tag{7.30}$$

$$K_{Z\dot{Z}} \equiv G_{Z\dot{Z}}(0) = -i \int_{-\infty}^{\infty} \omega\, g^2(\omega)\, S_{XX}(\omega)\, d\omega = 0 \tag{7.31}$$

and

$$\sigma_{\dot{Z}}^2 = G_{\dot{Z}\dot{Z}}(0) = \int_{-\infty}^{\infty} \omega^2 g^2(\omega)\, S_{XX}(\omega)\, d\omega \tag{7.32}$$

in which the various covariance values all refer to responses at the same instant of time. In addition, of course, we know that σ_X^2 is the integral of $S_{XX}(\omega)$, $\sigma_{\dot{X}}^2$ is the integral of $\omega^2 S_{XX}(\omega)$, and $K_{X\dot{X}} = 0$. Note that for any odd $g(\omega)$ function we have (X,Z) and $(\dot{X},\dot{Z})$ being uncorrelated pairs, in addition to $(X,\dot{X})$ and $(Z,\dot{Z})$ being uncorrelated pairs. Within the set of four random variables $\{X(t),\dot{X}(t),Z(t),\dot{Z}(t)\}$, the only nonzero correlations are for the pairs $(X,\dot{Z})$ and $(\dot{X},Z)$ and these two covariances are of equal absolute value.

Note that we define $A(t)$ and $\theta(t)$ according to Eqs. 7.24 and 7.25 for any choice of $g(\omega)$ in the definition of $Z(t)$. Furthermore, this implies that Eq. 7.20 holds and

$$Z(t) = -A(t)\sin[(\omega_a t) + \theta(t)] \tag{7.33}$$

for any choice of the $\{Z(t)\}$ process. Equations 7.22 and 7.23, however, hold only for the special case of $Z(t) = \dot{Y}(t)/\omega_a$. For the more general case one must use Eq. 7.21 for $\dot{X}(t)$, and the corresponding relationship for $\dot{Z}(t)$ is

$$\dot{Z}(t) = -A(t)[\omega_a + \dot{\theta}(t)]\cos[(\omega_a t) + \theta(t)] - \dot{A}(t)\sin[(\omega_a t) + \theta(t)] \tag{7.34}$$

We will use the notation $A_2(t)$ for the special case of $Z(t) = \dot{Y}(t)/\omega_a$ to distinguish this amplitude from other possibilities. This $A_2(t)$ is commonly called the *energy-based amplitude*, because it involves a sum of $X^2(t)$ and $\dot{X}^2(t)$, which may be related to the potential energy and the kinetic energy in an oscillator.

We now wish to investigate the properties of the newly defined quantities $A(t)$ and $\theta(t)$ and their derivatives, $\dot{A}(t)$ and $\dot{\theta}(t)$. We wish to show that $\dot{A}(t)$ and $\dot{\theta}(t)$ will be small for a narrowband $\{X(t)\}$ process, but we also wish to investigate the probability distributions of the various quantities. First we note that Eqs. 7.24 and 7.25 give

$$\dot{A}(t) = \frac{Y(t)\dot{Y}(t) + Z(t)\dot{Z}(t)}{A(t)} = \frac{Y(t)\dot{Y}(t) + Z(t)\dot{Z}(t)}{[Y^2(t) + Z^2(t)]^{1/2}} \tag{7.35}$$

and

$$\dot{\theta}(t) = \frac{Z(t)\dot{Y}(t) - Y(t)\dot{Z}(t)}{A^2(t)} - \omega_a = \frac{Z(t)\dot{Y}(t) - Y(t)\dot{Z}(t)}{Y^2(t) + Z^2(t)} - \omega_a \tag{7.36}$$

but these expressions are rather cumbersome for direct investigation of the distributions of $\dot{A}(t)$ and $\dot{\theta}(t)$. Rather, we will look at the probability distribution of the set $(A,\dot{A},\theta,\dot{\theta}) \equiv \{A(t),\dot{A}(t),\theta(t),\dot{\theta}(t)\}$. For a special case we will show that we can write this probability distribution explicitly, then we will proceed to give approximate results for the more general case.

Note that Eqs. 7.20, 7.21, 7.33, and 7.34 are four simultaneous equations giving a unique relationship between the vector $\vec{V} \equiv [Y,\dot{Y},Z,\dot{Z}]^T$ and the vector $\vec{U} \equiv [A,\dot{A},\theta,\dot{\theta}]^T$, in which all components of $\vec{U}$ and $\vec{V}$ are evaluated at the same instant of time. In particular, if we say that $\vec{q}(\cdot)$ is a nonlinear vector function such that $\vec{V} = \vec{q}(A,\dot{A},\theta,\dot{\theta})$, then Eq. 2.41 gives

$$p_{A\dot{A}\theta\dot{\theta}}(u,v,\xi,\psi) = |J|_{\vec{V}=\vec{q}(u,v,\xi,\psi)}\, p_{\vec{V}}[\vec{q}(u,v,\xi,\psi)]$$

in which the Jacobian J is the determinant of the matrix giving the partial derivatives of all components of $\vec{V}$ with respect to each component of $\vec{U}$. These derivatives are found from Eqs. 7.20, 7.21, 7.33, and 7.34, and substitution and simplification gives $J = A^2$. Thus, $p_{A\dot{A}\theta\dot{\theta}}(u,\xi,v,\psi)$ is found by evaluating the function $A^2\, p_{Y\dot{Y}Z\dot{Z}}(y,\dot{y},z,\dot{z})$ at $y = u\cos(\eta)$, $\dot{y} = -u(\omega_a + \psi)\sin(\eta) + v\cos(\eta)$, $z = -u\sin(\eta)$, $\dot{z} = -u(\omega_a + \psi)\cos(\eta) - v\sin(\eta)$, in which $\eta = \omega_a t + \xi$. This expression applies in general, but it is most useful when $\{Y(t)\}$ and $\{Z(t)\}$ are such that $p_{Y\dot{Y}Z\dot{Z}}(y,\dot{y},z,\dot{z})$ has a relatively simple form. In particular, it is very useful when the joint distribution is Gaussian.

Note that if $\{X(t)\}$ is a Gaussian process, then $\{Y(t)\}$, $\{\tilde{Y}_T(\omega)\}$, $\{\tilde{Z}_T(\omega)\}$, and $\{Z(t)\}$ are all jointly Gaussian processes, because they are obtained from linear operations on $\{X(t)\}$. For $\{Y(t)\}$ and $\{Z(t)\}$ being jointly Gaussian processes, though, we know that $(Y,\dot{Y},Z,\dot{Z})$ are jointly Gaussian random variables. As shown in Example 3.13, uncorrelated jointly Gaussian random variables are also independent, so the $(Y,\dot{Z})$ pair is independent of $(\dot{Y},Z)$: $p_{Y\dot{Y}Z\dot{Z}}(y,\dot{y},z,\dot{z}) = p_{Y\dot{Z}}(y,\dot{z})\, p_{\dot{Y}Z}(\dot{y},z)$. Using the Gaussian probability density function from Example 3.10 or 3.12 then gives

$$
\begin{aligned}
p_{Y\dot{Y}Z\dot{Z}}(y,\dot{y},z,\dot{z}) &= p_{Y\dot{Z}}(y,\dot{z})\, p_{\dot{Y}Z}(\dot{y},z) \\
&= \frac{1}{2\pi\sigma_X\sigma_{\dot{X}}\sigma_Z\sigma_{\dot{Z}}(1-\rho_{X\dot{Z}}^2)^{1/2}(1-\rho_{\dot{X}Z}^2)^{1/2}} \times \\
&\quad \exp\!\left(\frac{-1}{2(1-\rho_{X\dot{Z}}^2)}\left[\frac{y^2}{\sigma_X^2} - 2\rho_{X\dot{Z}}\frac{y\,\dot{z}}{\sigma_X\sigma_{\dot{Z}}} + \frac{\dot{z}^2}{\sigma_{\dot{Z}}^2}\right] - \right. \\
&\quad \left. \frac{1}{2(1-\rho_{\dot{X}Z}^2)}\left[\frac{\dot{y}^2}{\sigma_{\dot{X}}^2} - 2\rho_{\dot{X}Z}\frac{\dot{y}\,z}{\sigma_{\dot{X}}\sigma_Z} + \frac{z^2}{\sigma_Z^2}\right]\right)
\end{aligned}
\tag{7.37}
$$

in which $\rho_{X\dot{Z}} = K_{X\dot{Z}}/(\sigma_X\sigma_{\dot{Z}})$ and $\rho_{\dot{X}Z} = K_{\dot{X}Z}/(\sigma_{\dot{X}}\sigma_Z) = -K_{X\dot{Z}}/(\sigma_{\dot{X}}\sigma_Z)$. The desired probability density is then found by evaluating this at $y = u\cos(\eta)$, $\dot{y} = -u(\omega_a+\psi)\sin(\eta) + v\cos(\eta)$, $z = -u\sin(\eta)$, and $\dot{z} = -u(\omega_a+\psi)\cos(\eta) - v\sin(\eta)$.

Before proceeding to a specific choice of the $g(\omega)$ function, which gives a precise definition of $A(t)$ and $\theta(t)$, let us note a fairly general property of the distribution of $A(t)$. From Example 2.32 we know that A defined according to Eq. 7.24 will have the Rayleigh distribution if Y and Z are independent and have identical mean-zero Gaussian distributions. For the current situation we have established that they are independent, mean-zero, and Gaussian, but they have identical distributions only if $\sigma_Z = \sigma_X$. For a narrowband process, Eq. 7.27 does indicate that this condition is approximately satisfied, because $g^2(\pm\omega_a) = 1$ and the variances will come almost entirely from these frequencies. Thus, $A(t)$ will be approximately Rayleigh distributed for a narrowband Gaussian $\{X(t)\}$ process. For a more broadband $\{X(t)\}$ process, though, we will have the Rayleigh distribution only for certain $g(\omega)$ functions that give $\sigma_Z = \sigma_X$. We will investigate two specific examples that do meet this condition.

First let us use $g(\omega)=\omega/\omega_a$ so that the auxiliary process is $Z=\dot{Y}/\omega_a$ and A is the energy-based amplitude A_2. We will denote the corresponding phase angle as θ_2. As noted previously, the ω_a parameter should be chosen such that $E[\dot{\theta}_2]=0$. In order to accomplish this it would be reasonable to leave ω_a unspecified until we have an expression for $E[\dot{\theta}_2]$, but doing this gives rather cumbersome expressions for the joint probability density. Rather than do this, we will guess that $\omega_a=\omega_{c2}=\sigma_{\dot{X}}/\sigma_X$, then later verify that this does give $E[\dot{\theta}_2]=0$. Note that Eq. 7.27 gives $\sigma_Z=\sigma_X=(\lambda_0)^{1/2}$, so we do know that A_2 has the Rayleigh distribution. In addition, Eqs. 7.28–7.32 give $\sigma_{\dot{X}}=(\lambda_2)^{1/2}$ and $\sigma_{\dot{Z}}=(\lambda_0\lambda_4/\lambda_2)^{1/2}$. Also, we note that $\dot{Z}=\ddot{Y}/\omega_a$, so $\rho_{X\dot{Z}}=\rho_{X\ddot{X}}$, which was found in Eq. 7.13 to be $-\alpha_2=-\lambda_2/(\lambda_0\lambda_4)^{1/2}$. The additional term in the probability density, though, is $\rho_{\dot{X}Z}=1$, which causes some difficulty. In particular, having $Z=\dot{Y}/\omega_a$ gives a degenerate probability distribution that has all the probability assigned to a three-dimensional subset of the four-dimensional space. This can be written as

$$p_{Y\dot{Y}Z\dot{Z}}(y,\dot{y},z,\dot{z})=p_{Y\dot{Z}}(y,\dot{z})\,p_{\dot{Y}}(\dot{y})\,\delta(z-\dot{y}/\omega_a)$$

or

$$p_{Y\dot{Y}Z\dot{Z}}(y,\dot{y},z,\dot{z})=\frac{\delta(z-\dot{y}/\omega_a)}{2\pi\sigma_X\,\sigma_{\dot{X}}\,\sigma_{\dot{Z}}\,(1-\rho_{X\dot{Z}}^2)^{1/2}}\times$$
$$\exp\left(\frac{-1}{2(1-\rho_{X\dot{Z}}^2)}\left[\frac{y^2}{\sigma_X^2}-2\rho_{X\dot{Z}}\frac{y\,\dot{z}}{\sigma_X\,\sigma_{\dot{Z}}}+\frac{\dot{z}^2}{\sigma_{\dot{Z}}^2}\right]-\frac{1}{2}\frac{\dot{y}^2}{\sigma_{\dot{X}}^2}\right)$$

Substituting for z and $\dot{y}$ in $\delta(z-\dot{y}/\omega_a)$ gives

$$\delta(z-\dot{y}/\omega_a)=\delta\left(\frac{\psi\,u\sin(\eta)-v\cos(\eta)}{\omega_a}\right)=\frac{\omega_a}{u\,|\sin(\eta)|}\delta\left(\psi-\frac{v\cos(\eta)}{u\sin(\eta)}\right)$$

in which the final form has been obtained by using Eq. 2.49. Thus, the probability density is nonzero only when $\psi=v\cos(\eta)/[u\sin(\eta)]$ in the exponential of the joint probability density function. Making this substitution and using the expressions for y, $\dot{y}$, $\dot{z}$, σ_X, $\sigma_{\dot{X}}$, $\sigma_{\dot{Z}}$, $\rho_{X\dot{Z}}$, and ω_a gives, after considerable simplification,

$$p_{A_2\dot{A}_2\theta_2\dot{\theta}_2}(u,v,\xi,\psi)=u^2\,p_{Y\dot{Y}Z\dot{Z}}[q(u,v,\xi,\psi)]$$

or

$$p_{A_2\dot{A}_2\theta_2\dot{\theta}_2}(u,v,\xi,\psi) =$$

$$\frac{\lambda_2^{1/2}\, u\,\delta\left(\psi - \dfrac{v\cos(\eta)}{u\sin(\eta)}\right)\exp\left(-\dfrac{u^2}{2\lambda_0} - \dfrac{\lambda_2 v^2}{2(\lambda_0\lambda_4 - \lambda_2^2)\sin^2(\eta)}\right)}{(2\pi)^{3/2}\lambda_0(\lambda_0\lambda_4 - \lambda_2^2)^{1/2}\,|\sin(\eta)|} \tag{7.38}$$

The particular form of Eq. 7.38 allows some simplification. In particular, it can be factored into

$$p_{A_2\dot{A}_2\theta_2\dot{\theta}_2}(u,v,\xi,\psi) = p_{A_2}(u)\, p_{\dot{A}_2\theta_2}(v,\xi)\, p_{\dot{\theta}_2}(\psi\,|\,A_2 = u, \dot{A}_2 = v, \theta_2 = \xi)$$

in which

$$p_{A_2}(u) = \frac{u\, e^{-u^2/\lambda_0}}{\lambda_0} \tag{7.39}$$

$$p_{\dot{A}_2\theta_2}(v,\xi) = \frac{\lambda_2^{1/2}\exp\left(-\dfrac{\lambda_2 v^2}{2(\lambda_0\lambda_4 - \lambda_2^2)\sin^2(\eta)}\right)}{(2\pi)^{3/2}(\lambda_0\lambda_4 - \lambda_2^2)^{1/2}\,|\sin(\eta)|} \tag{7.40}$$

and

$$p_{\dot{\theta}_2}(\psi\,|\,A_2 = u, \dot{A}_2 = v, \theta_2 = \xi) = \delta\left(\psi - \frac{v\cos(\eta)}{u\sin(\eta)}\right) \tag{7.41}$$

First we note that Eq. 7.39 confirms that $A_2(t)$ has the Rayleigh distribution, as already noted. In addition, $A_2(t)$ is independent of $\dot{A}_2(t)$ and $\theta_2(t)$, because $p_{A_2\dot{A}_2\theta_2}(u,v,\xi)$ can be factored into $p_{A_2}(u)\, p_{\dot{A}_2\theta_2}(v,\xi)$. Also, Eq. 7.41 tells us that we can consider $\dot{\theta}_2(t)$ to be a function of the other three random variables:

$$\dot{\theta}_2(t) = \frac{\dot{A}_2(t)\cos[\omega_a t + \theta_2(t)]}{A_2(t)\sin[\omega_a t + \theta_2(t)]}$$

Next we note that the Gaussian form of Eq. 7.40 allows us easily to integrate it with respect to v to give

$$p_{\theta_2}(\xi) = (2\pi)^{-1} \quad \text{for } -\pi \le \xi \le \pi$$

so θ_2 has a uniform distribution on its set of possible values. The conditional distribution of $\dot{A}_2$ given θ_2 is now

$$p_{\dot{A}_2}(v\,|\,\theta_2=\xi)=\frac{\lambda_2^{1/2}\exp\left(-\dfrac{\lambda_2 v^2}{2(\lambda_0\lambda_4-\lambda_2^2)\sin^2(\eta)}\right)}{(2\pi)^{1/2}(\lambda_0\lambda_4-\lambda_2^2)^{1/2}\,|\sin(\eta)|}$$

which has a Gaussian form. Thus, we find that

$$E(\dot{A}_2\,|\,\theta_2=\xi)=0,\quad E(\dot{A}_2^2\,|\,\theta_2=\xi)=\frac{(\lambda_0\lambda_4-\lambda_2^2)\sin^2(\eta)}{\lambda_2}$$

and multiplying by $p_{\theta_2}(\xi)=(2\pi)^{-1}$ and integrating with respect to ξ or η gives

$$E(\dot{A}_2)=0,\quad E(\dot{A}_2^2)=\frac{\lambda_0\lambda_4-\lambda_2^2}{2\pi\lambda_2}\int_{-\pi}^{\pi}\sin^2(\eta)\,d\eta=\frac{\lambda_0\lambda_4-\lambda_2^2}{2\lambda_2}$$

The zero mean value of $\dot{A}_2$ is no surprise and follows directly from the fact that $\{A_2(t)\}$ is a stationary process. The variance is more interesting though, inasmuch as it relates to the bandwidth of the process. The result can be rewritten as

$$E(\dot{A}_2^2)=\frac{\lambda_0\lambda_4}{2\lambda_2}(1-\alpha_2^2)=\frac{\sigma_{\dot{X}}^2}{2}\left(\frac{1}{\alpha_2^2}-1\right) \tag{7.42}$$

showing that $E(\dot{A}_2^2)\to 0$ as $\alpha_2\to 1$. This does confirm that $\{A_2(t)\}$ is a slowly varying process if it is narrowband on the basis of $\alpha_2\approx 1$.

Next we consider the properties of $\dot{\theta}_2$. Rewriting the Dirac delta function in Eq. 7.38 according to Eq. 2.49 as

$$\delta\left(\psi-\frac{v\cos(\eta)}{u\sin(\eta)}\right)=\frac{u^2\sin(\eta)}{v\cos(\eta)}\,\delta\left(u-\frac{v\cos(\eta)}{\psi\sin(\eta)}\right)=\frac{u}{\psi}\,\delta\left(u-\frac{v\cos(\eta)}{\psi\sin(\eta)}\right)$$

allows one to write $p_{\dot{\theta}_2}(\psi)$ as a triple integral of this joint probability density function, then perform the integrations with respect to u and v, yielding

$$p_{\dot\theta_2}(\psi) = \frac{\lambda_0^{3/2}\lambda_2^{1/2}\lambda_4(1-\alpha_2^2)}{2\pi}\int_0^{2\pi}\frac{\cos^2(\eta)}{[\lambda_0\lambda_4(1-\alpha_2^2)\cos^2(\eta)+\lambda_0\lambda_2\psi^2]^{3/2}}\,d\eta$$

Note that the symmetry of this expression with respect to ψ gives $E(\dot\theta_2)=0$. This confirms that our choice of $\omega_a=\omega_{c2}$ was correct for the A_2 definition of amplitude. We also note that

$$E(\dot\theta_2^2) = \int_{-\infty}^{\infty}\psi^2\, p_{\dot\theta_2}(\psi)\,d\psi = \infty$$

because $p_{\dot\theta_2}(\psi)$ only decays like $|\psi|^{-3}$ as $|\psi|\mapsto\infty$, and this is true for any nonzero value of α_2. Thus, we cannot say that $\{\theta(t)\}$ is a slowly varying process in the sense of having a small value of $E(\dot\theta_2^2)$. We can note, though, that $p_{\dot\theta_2}(\psi)\to 0$ as $\alpha_2\to 1$ for all $\psi\neq 0$, and $p_{\dot\theta_2}(0)=\infty$ for any value of α_2 (including $\alpha_2\approx 1$). This may be considered a weaker definition of the condition that $\dot\theta_2$ is small for a narrowband process according to the α_2 bandwidth parameter. That is, it implies that $p_{\dot\theta_2}(\psi)\to\delta(\psi)$ for $\alpha_2\to 1$.

Next we turn our attention to a different $g(\omega)$ function, giving us an alternative specific definition of $A(t)$ and $\theta(t)$. In particular, let $g(\omega)=\text{sgn}(\omega)\equiv U(\omega)-U(-\omega)$. In fact, the $Z(t)$ auxiliary process defined in this way has a special name, being called the *Hilbert transform* of $Y(t)$.[3] One of the special features of the $\{Z(t)\}$ Hilbert transform process is that it has the same autospectral density as $\{Y(t)\}$, because it has $g^2(\omega)=1$ everywhere. Among the implications of this are the fact that $\sigma_Z=\sigma_X=(\lambda_0)^{1/2}$ and $\sigma_{\dot Z}=\sigma_{\dot X}=(\lambda_2)^{1/2}$. In addition, Eq. 7.29 gives the nonzero cross-covariance term of interest as

$$K_{X\dot Z} = -\int_{-\infty}^{\infty}\omega\,\text{sgn}(\omega)\,S_{XX}(\omega)\,d\omega \equiv -\lambda_1 \tag{7.43}$$

so

$$\rho_{X\dot Z} = -\rho_{\dot X Z} = -\lambda_1/(\lambda_0\lambda_2)^{1/2} = -\alpha_1$$

The amplitude, or envelope, defined using the Hilbert transform of $Y(t)$ was apparently first introduced by Cramer and Leadbetter in 1967 and is sometimes called the *Cramer and Leadbetter amplitude*. We will denote it by the symbol $A_1(t)$ and use $\theta_1(t)$ for the corresponding phase angle. We will try using $\omega_a=\omega_{c1}=\lambda_1/\lambda_0$, then verify that this does give $E(\dot\theta_1)=0$. Note that we again have $\sigma_Z=\sigma_X$, so we know that $A_1(t)$ also has the Rayleigh distribution.

---

[3]A time domain representation of the Hilbert transform is given by the rather ill-behaved integral $Z(t)=\pi^{-1}\int_{-\infty}^{\infty}[Y(s)/(s-t)]\,ds$.

As before, we use Eq. 7.37 to obtain the joint distribution of A_1, $\dot{A}_1$, θ_1, and $\dot{\theta}_1$ for Gaussian $\{X(t)\}$ and $\{Z(t)\}$ processes. The result can be written as

$$p_{A_1\theta_1\dot{A}_1\dot{\theta}_1}(u,\xi,v,\psi) = p_{A_1\dot{\theta}_1}(u,\psi)\, p_{\theta_1}(\xi)\, p_{\dot{A}_1}(v)$$

in which

$$p_{\theta_1}(\xi) = \frac{1}{2\pi} \quad \text{for } -\pi \le \xi \le \pi$$

$$p_{\dot{A}_1}(v) = \frac{1}{(2\pi)^{1/2}(1-\alpha_1^2)^{1/2}\lambda_2^{1/2}} \exp\left(\frac{-v^2}{2(1-\alpha_1^2)\lambda_2}\right) \tag{7.44}$$

and

$$p_{A_1\dot{\theta}_1}(u,\psi) = \frac{u^2}{(2\pi)^{1/2}\lambda_0^{1/2}(\lambda_0\lambda_2-\lambda_1^2)^{1/2}} \exp\left(-\frac{u^2}{2}\left[\frac{1}{\lambda_0} + \frac{\lambda_0\psi^2}{\lambda_0\lambda_2-\lambda_1^2}\right]\right) \tag{7.45}$$

Thus, $\theta_1(t)$ is independent of $A_1(t)$, $\dot{A}_1(t)$, and $\dot{\theta}_1(t)$, and it is uniformly distributed on $[-\pi,\pi]$. Similarly, $\dot{A}_1(t)$ is independent of $A_1(t)$, $\theta_1(t)$, and $\dot{\theta}_1(t)$, and it has a Gaussian distribution with

$$E[\dot{A}_1(t)] = 0, \quad E[\dot{A}_1^2(t)] = (1-\alpha_1^2)\lambda_2 = (1-\alpha_1^2)\sigma_{\dot{X}}^2(t) \tag{7.46}$$

Clearly $E[\dot{A}_1^2(t)] \to 0$ as $\alpha_1 \to 1$, confirming that $\{A_1(t)\}$ is a slowly varying process if it is narrowband as measured by the value of α_1.

The $A_1(t)$ amplitude and the rate of change of the $\theta_1(t)$ phase are dependent. Integrating Eq. 7.45 with respect to ψ gives

$$p_{A_1}(u) = \frac{u}{\lambda_0} e^{-u^2/(2\lambda_0)}$$

confirming that $A_1(t)$ has the same Rayleigh distribution as $A_2(t)$. Similarly, integrating Eq. 7.45 with respect to u gives

$$p_{\dot{\theta}_1}(\psi) = \frac{\lambda_0(\lambda_0\lambda_2-\lambda_1^2)}{(\lambda_0\lambda_2-\lambda_1^2+\lambda_0^2\psi^2)^{3/2}} = \frac{\lambda_0^{1/2}\lambda_2(1-\alpha_1^2)}{[\lambda_2(1-\alpha_1^2)+\lambda_0\psi^2]^{3/2}}$$

As for $\dot\theta_2$, the symmetry of the probability distribution gives $E(\dot\theta_1)=0$, confirming that ω_{c1} was the proper choice for ω_a for this definition of amplitude and phase. As before, $p_{\dot\theta}(\psi)$ only decays like $|\psi|^{-3}$ as $|\psi|\rightarrow\infty$, so $E[(\dot\theta_1)^2]=\infty$. Again, though, $p_{\dot\theta_1}(\psi)\rightarrow\delta(\psi)$ as $\alpha_1\rightarrow 1$, confirming that $\dot\theta_1$ is small for a narrowband process according to the α_1 bandwidth parameter.

From Eqs. 7.42 and 7.46 we can see a significant difference between the two definitions of amplitude for a broadband process. In particular, Eq. 7.46 shows that the mean-squared value of $\dot A_1$ is always less than the variance of $\dot X$, because $0\le\alpha_1\le 1$. Thus, we can say that the Cramer and Leadbetter amplitude $A_1(t)$ varies less rapidly than $X(t)$, even for a broadband process. On the other hand, Eq. 7.42 shows that the $A_2(t)$ energy-based amplitude varies less rapidly than $X(t)$ only if $(\alpha_2)^2>1/3$, which is not satisfied for some broadband processes.

The Cramer and Leadbetter amplitude also provides an example of an important distinction between Gaussian processes and more general processes also consisting of Gaussian random variables. In particular, recall that Eq. 7.44 shows that $\dot A_1(t)$ is a Gaussian random variable at every time t. On the other hand, we know that $\{\dot A_1(t)\}$ is not a Gaussian process, because the integral of a Gaussian process is also a Gaussian process, and we know that $A_1(t)$ does not have the Gaussian distribution. The explanation of this apparent dilemma is that a set of random variables $\{\dot A_1(t_1),\dot A_1(t_2),\cdots,\dot A_1(t_n)\}$ is not generally jointly Gaussian, even though each of the random variables has a Gaussian marginal distribution. This rather confusing type of stochastic process consisting of a family of marginally Gaussian, but not jointly Gaussian, random variables arises in various areas of application. We shall encounter it again in Chapter 10 when we study the dynamics of nonlinear systems.

We will now turn our attention to the more general situation in which $\{X(t)\}$ is not a Gaussian process. Without a specific probability distribution for $\{Y,\dot Y,Z,\dot Z\}\equiv\{Y(t),\dot Y(t),Z(t),\dot Z(t)\}$, of course, we will not be able to give such a detailed analysis of $\{A,\dot A,\theta,\dot\theta\}\equiv\{A(t),\dot A(t),\theta(t),\dot\theta(t)\}$ as before, but we will be able to make some limited observations.

First we note that we can multiply Eq. 7.35 by A to obtain

$$E(A^2\dot A^2)=E[(Y\dot Y+Z\dot Z)^2]=E(Y^2\dot Y^2)+2E(Y\dot Y Z\dot Z)+E(Z^2\dot Z^2)$$

Similarly multiplying Eq. 7.36 by A^2 gives

$$E(A^4\dot{\theta}^2) = E[(Z\dot{Y} - Y\dot{Z})^2] - 2\omega_a E[A^2(Z\dot{Y} - Y\dot{Z})] + \omega_a^2 E(A^4)$$

or

$$E(A^4\dot{\theta}^2) = E(\dot{Y}^2 Z^2) - 2E(Y\dot{Y}Z\dot{Z}) + E(Y^2\dot{Z}^2) - 2\omega_a E(Y^2\dot{Y}Z) + 2\omega_a E(Y^3\dot{Z}) - 2\omega_a E(\dot{Y}Z^3) + 2\omega_a E(YZ^2\dot{Z}) + \omega_a^2 E(Y^4) + \omega_a^2 E(Y^2 Z^2) + \omega_a^2 E(Z^4)$$

Within the fourth-order cross-product terms that appear in these expressions, we also know that Y and $\dot{Y}$ are uncorrelated, that Z and $\dot{Z}$ are uncorrelated, that X and Z are uncorrelated, and that $\dot{X}$ and $\dot{Z}$ are uncorrelated. Without the Gaussian condition we cannot say that these zero correlations imply independence, but we can anticipate that this condition may be approximately satisfied in many situations. In particular, we can write each of the fourth-order cross-products using Eq. 3.40, which for mean-zero random variables gives

$$E(R_1 R_2 R_3 R_4) = K_{R_1R_2}K_{R_3R_4} + K_{R_1R_3}K_{R_2R_4} + K_{R_1R_4}K_{R_2R_3} + \kappa_4(R_1,R_2,R_3,R_4)$$

in which the various κ_4 fourth cumulants in the equations are exactly zero for jointly Gaussian processes and will be smaller than the other terms in most practical non-Gaussian situations. Substituting for all the fourth-moment terms and simplifying gives

$$E(A^2\dot{A}^2) = \sigma_X^2\sigma_{\dot{X}}^2 - 2K_{X\dot{Z}}^2 + \sigma_Z^2\sigma_{\dot{Z}}^2 + f_1(\kappa_4) \tag{7.47}$$

and

$$E(A^4\dot{\theta}^2) = \sigma_Z^2\sigma_{\dot{X}}^2 + 6K_{X\dot{Z}}^2 + \sigma_X^2\sigma_{\dot{Z}}^2 + 8\omega_a K_{X\dot{Z}}(\sigma_X^2 + \sigma_Z^2) + \omega_a^2(3\sigma_X^4 + 2\sigma_X^2\sigma_Z^2 + 3\sigma_Z^4) + f_2(\kappa_4) \tag{7.48}$$

in which $f_1(\kappa_4)$ and $f_2(\kappa_4)$ denote the combined effects of all fourth cumulant contributions to the two equations.

For the $(A_2,\theta_2,\omega_{c2})$ definition of (A,θ,ω_a) introduced earlier we know that $\sigma_X = \sigma_Z = (\lambda_0)^{1/2}$, $\sigma_{\dot{X}} = (\lambda_2)^{1/2}$, $\sigma_{\dot{Z}} = (\lambda_0\lambda_4/\lambda_2)^{1/2}$, and $K_{X\dot{Z}} = -(\lambda_0\lambda_2)^{1/2}$. Substituting and simplifying gives

$$E(A_2^2\dot{A}_2^2) = (\lambda_0^2\lambda_4/\lambda_2)(1-\alpha_2^2) + f_1(\kappa_4) \tag{7.49}$$

and

$$E(A_2^4\dot{\theta}_2^2) = (\lambda_0^2\lambda_4/\lambda_2)(1-\alpha_2^2) + f_2(\kappa_4) \tag{7.50}$$

Note that Eq. 7.49 is completely in agreement with Eq. 7.42 that we derived for the Gaussian special case, in which A_2 and $\dot{A}_2$ are independent with $E(A_2^2) = 2\lambda_0$ and $\kappa_4 = 0$. Because we anticipate that the κ_4 term will be much smaller than $(\lambda_0)^2\lambda_4/\lambda_2$ in most practical non-Gaussian situations, we see that Eq. 7.49 does indicate that $E(A_2^2\dot{A}_2^2)$ is small if $\alpha_2 \approx 1$. Furthermore, because A_2^2 is not generally small, we can take this as an indication that $\dot{A}_2^2$ is small for a process that is narrowband based on the value of α_2. The same logic can be used to argue that Eq. 7.50 indicates that $\dot{\theta}_2^2$ is also small for a process that is narrowband based on the value of α_2.

Similarly, the $(A_1,\theta_1,\omega_{c1})$ definition of (A,θ,ω_a) gives $\sigma_Z = \sigma_X = (\lambda_0)^{1/2}$, $\sigma_{\dot{X}} = \sigma_{\dot{Z}} = (\lambda_2)^{1/2}$, and $K_{X\dot{Z}} = -\lambda_1$, which yields

$$E(A_1^2\dot{A}_1^2) = 2(\lambda_0\lambda_2 - \lambda_1) = 2\lambda_0\lambda_2(1-\alpha_1^2) + f_1(\kappa_4)$$

and

$$E(A_2^4\dot{\theta}_2^2) = 2\lambda_0\lambda_2(1-\alpha_1^2) + f_2(\kappa_4)$$

Again, we can argue that this indicates that $\dot{A}_1^2$ and $\dot{\theta}_1^2$ are small for a process that is narrowband based on the value of α_1.

Note that we found that $\rho_{X\dot{Z}_1} = -\alpha_1$ for the Cramer and Leadbetter amplitude, just as $\rho_{X\dot{Z}_2} = -\alpha_2$ for the energy-based amplitude. This suggests that it may be reasonable to use $-\rho_{X\dot{Z}}$ as the bandwidth parameter for any choice of the auxiliary function in the definition of the amplitude.

****

Example 7.11: Investigate the rates of change of the amplitude and phase of the covariant stationary $\{X(t)\}$ response of an SDF oscillator excited by white noise using both the energy-based and the Cramer and Leadbetter definitions of the terms.

Because the quantities of interest are directly related to spectral moments and bandwidth parameters, we can simply use the results obtained in Example 7.9. Inasmuch as the rates of change of the energy-based amplitude and phase are directly related to the value $[1-(\alpha_2)^2]$, the fact that $\alpha_2 = 0$ for the SDF response to white noise indicates that even when damping is very small the

$A_2(t)$ amplitude and the $\theta_2(t)$ phase are not slowly varying. This seems surprising, inasmuch as sample time histories of $\{X(t)\}$ for this small damping situation surely do resemble slowly varying harmonic functions of time. Nonetheless, the energy-based amplitude and phase are defined in such a way that they are affected too much by the high-frequency components of $\{X(t)\}$, and they do not accurately reflect the narrowband nature of the autospectral density in the vicinity of the resonant peak for this particular system. The $A_1(t)$ amplitude and the $\theta_1(t)$ phase according to the Cramer and Leadbetter definitions, however, do vary slowly for the small ζ situations for which sample time histories resemble slowly varying harmonic functions, because $[1-(\alpha_1)^2]$ goes to zero in this situation.

****

Example 7.12: Find the expected rate of upcrossings for the Cramer and Leadbetter amplitude of a stationary Gaussian process that is mean-zero.

From Eq. 7.3, the rate of upcrossings for $A_1(t)$ is

$$\nu_{A_1}^+(u) = p_{A_1(t)}(u)\int_0^\infty v\, p_{\dot{A}_1(t)}[v \mid A_1(t)=u]\,dv$$

However, we know that $A_1(t)$ and $\dot{A}_1(t)$ are independent, so the conditioning in the integrand can be ignored. Furthermore, we know that $A_1(t)$ has the Rayleigh distribution of Eq. 7.39 and that $\dot{A}_1(t)$ is Gaussian with mean-zero and variance $\sigma_{\dot{A}} = [1-(\alpha_1)^2]^{1/2}\sigma_{\dot{X}}$. Thus, we can write

$$\nu_{A_1}^+(u) = \left(\frac{u}{\sigma_X^2}\exp\left[\frac{-u^2}{2\sigma_X^2}\right]\right)\int_0^\infty \frac{v\, e^{-v^2/(2\sigma_{\dot{A}}^2)}}{(2\pi)^{1/2}\sigma_{\dot{A}}}\,dv = \left(\frac{u}{\sigma_X^2}\exp\left[\frac{-u^2}{2\sigma_X^2}\right]\right)\frac{\sigma_{\dot{A}}}{(2\pi)^{1/2}}$$

or

$$\nu_{A_1}^+(u) = \left(\frac{u}{\sigma_X}\exp\left[\frac{-u^2}{2\sigma_X^2}\right]\right)\left(\frac{\sigma_{\dot{X}}}{\sigma_X}\right)\left(\frac{1-\alpha_1^2}{2\pi}\right)^{1/2} U(u)$$

Note that the first expression on the right-hand side is a dimensionless quantity giving the form of the dependence on u. The upcrossing rate tends to zero as u approaches zero, which is consistent with the fact that $A_1(t)$ is almost always above such a small u value.

****

Example 7.13: Find the expected rate of upcrossings for the energy-based amplitude of a Gaussian process that is mean-zero.

At first glance this calculation seems more difficult than the one in Example 7.12, because we do not have a closed-form solution for the probability density function for the derivative of the $A_2(t)$ energy-based amplitude. We do, however, know that $A_2(t)$ and $\dot{A}_2(t)$ are independent and that the probability density of the latter quantity is obtained from Eq. 7.40 as the integral

$$p_{\dot{A}_2}(v) = \int_0^{2\pi} \frac{\lambda_2^{1/2}}{(2\pi)^{3/2}(\lambda_0\lambda_4 - \lambda_2^2)^{1/2}\,|\sin(\eta)|} \exp\left(-\frac{\lambda_2 v^2}{2(\lambda_0\lambda_4 - \lambda_2^2)\sin^2(\eta)}\right) d\eta$$

By reversing the order of integration we can integrate first with respect to v, then with respect to η, to obtain

$$\int_0^\infty v\, p_{\dot{A}_2}(v)\, dv = \frac{(\lambda_0\lambda_4 - \lambda_2^2)^{1/2}}{(2\pi)^{3/2}\lambda_2^{1/2}} \int_0^{2\pi} |\sin(\eta)|\, d\eta = \frac{4(\lambda_0\lambda_4 - \lambda_2^2)^{1/2}}{(2\pi)^{3/2}\lambda_2^{1/2}}$$

$$= \frac{4\sigma_X\sigma_{\ddot{X}}(1-\alpha_2^2)^{1/2}}{(2\pi)^{3/2}\sigma_{\dot{X}}}$$

so

$$\nu_{A_2}^+(u,t) = p_{A_2}(u)\frac{4\sigma_X\sigma_{\ddot{X}}(1-\alpha_2^2)^{1/2}}{(2\pi)^{3/2}\sigma_{\dot{X}}} = \left(\frac{u}{\sigma_X}\exp\left[\frac{-u^2}{2\sigma_X^2}\right]\right)\left(\frac{\sigma_{\ddot{X}}}{\sigma_{\dot{X}}}\right)\left(\frac{4(1-\alpha_2^2)^{1/2}}{(2\pi)^{3/2}}\right)$$

for $u > 0$. Note that the u dependence is the same as in Example 7.4 for $A_1(t)$, but the basic frequency of occurrence is now proportional to $(\sigma_{\ddot{X}}/\sigma_{\dot{X}})(1-\alpha_2^2)^{1/2}$, as compared with $(\sigma_{\dot{X}}/\sigma_X)(1-\alpha_1^2)^{1/2}$ for $A_1(t)$.

****

7.5 Amplitude of a Modulated Stochastic Process

In some situations it is also desirable to define an amplitude and phase for a modulated process, with an evolutionary spectral density. Based on Eqs. 5.33 and 6.34 we can write a modulated process $\{X(t)\}$ as

$$X(t) = \int_{-\infty}^{\infty} F_S(s)\, h_{xS}(t,s)\, ds = \lim_{T\to\infty} \int_{-\infty}^{\infty} H_{xF_S}(t,\omega)\, \tilde{F}_{S,T}(\omega)\, e^{i\omega t}\, d\omega \qquad (7.51)$$

in which $\{F_S(t)\}$ is a stationary process and

$$\tilde{F}_{S,T}(\omega) = \frac{1}{2\pi}\int_{-T/2}^{T/2} e^{-i\omega t}\, F_S(t)\, dt$$

If $\{Z_S(t)\}$ is now a stationary auxiliary process for $\{F_S(t)\}$ defined according to Eq. 7.26 so that

$$\tilde{Z}_{S,T}(\omega) = i\, g(\omega)\, \tilde{F}_{S,T}(\omega)$$

then an appropriate nonstationary auxiliary process for $\{X(t)\}$ can be written as the modulated form of $\{Z_S(t)\}$

$$Z(t) = \int_{-\infty}^{\infty} Z_S(s)\, h_{xS}(t,s)\, ds = \lim_{T\to\infty} \int_{-\infty}^{\infty} H_{xS}(t,\omega)\, \tilde{Z}_{S,T}(\omega)\, e^{i\omega t}\, d\omega$$

or

$$Z(t) = \lim_{T\to\infty} \int_{-\infty}^{\infty} i\, g(\omega)\, H_{xS}(t,\omega)\, \tilde{F}_{S,T}(\omega)\, e^{i\omega t}\, d\omega \qquad (7.52)$$

and an appropriate amplitude for the modulated $\{X(t)\}$ process is again given by Eq. 7.24.

Comparing Eqs. 7.51 and 7.52 shows that the time-varying harmonic response function for the modulated auxiliary process is

$$H_{zS}(t,\omega) = i\, g(\omega) H_{xS}(t,\omega)$$

Similarly,

$$H_{\dot{z}S}(t,\omega) = i\, g(\omega) H_{\dot{x}S}(t,\omega)$$

and $H_{\dot{x}S}(t,\omega)$ is given by Eq. 6.36 as

$$H_{\dot{x}S}(t,\omega) = \frac{\partial}{\partial t} H_{xS}(t,\omega) + i\,\omega\, H_{xS}(t,\omega)$$

Expressions for the various evolutionary spectral densities relating $\{X(t)\}$, $\{\dot{X}(t)\}$, $\{Z(t)\}$, and $\{\dot{Z}(t)\}$ can now be found from Eqs. 6.38 and 6.39. In fact, all the evolutionary spectral density relationships for $(X,\dot{X},Z,\dot{Z})$ can be shown to agree in form with the corresponding stationary relationships involving $S_{XX}(\omega)$, $S_{X\dot{X}}(\omega)$, and $S_{\dot{X}\dot{X}}(\omega)$.

Consider now the special case of $\{X(t)\}$ being a uniformly modulated process for which $h_{xS}(t,s) = \hat{h}(t)\delta(t-s)$. This gives $X(t) = \hat{h}(t) F_S(t)$ and $Z(t) = \hat{h}(t) Z_S(t)$, so Eq. 7.24 gives $A(t) = \hat{h}(t) A_S(t)$, with $A_S(t)$ denoting the amplitude of $\{F_S(t)\}$. Thus, for the particular case of uniform modulation, the amplitude of the modulated process is exactly the modulation of the amplitude of the stationary process. This is not true in general, though. In particular, substitution of $Y(t)$ and $Z(t)$ for the general situation gives

$$A(t) = \left(\int_{-\infty}^{\infty}\int_{-\infty}^{\infty} \left[Y_S(s_1) Y_S(s_2) + Z_S(s_1) Z_S(s_2) \right] h_{xS}(t,s_1)\, h_{xS}(t,s_2)\, ds_1\, ds_2 \right)^{1/2}$$

in which $Y_S(t) = F_S(t) - \mu_{F_S}(t)$. This expression is significantly more complicated than a simple modulation of $A_S(t)$. This form of *evolutionary* amplitude was used by Yang in 1972 for the Cramer and Leadbetter definition of amplitude, and it has been used in various studies since that time.

Michaelov et al. (1999b) have investigated the probability distributions of $A_1(t)$, $\dot{A}_1(t)$, $\theta_1(t)$, and $\dot{\theta}_1(t)$ for the special case with $\{X(t)\}$ being a Gaussian process. The details will not be repeated here, but the general procedure is the same as was used for the stationary process in Section 7.4. The results also are similar to the stationary case. In particular, $A_1(t)$ has a Rayleigh distribution and $\dot{A}_1(t)$ has a conditional distribution that is Gaussian:

$$p_{A_1(t)}(u) = \frac{u}{\sigma^2_{X_1}(t)} \exp\left(-\frac{u^2}{2\sigma^2_{X_1}(t)} \right) \tag{7.53}$$

and

$$p_{\dot{A}_1(t)}[v \,|\, A_1(t) = u] = \frac{1}{(2\pi)^{1/2} \sigma^*} \exp\left(-\frac{1}{2}\left[\frac{v-\mu^*}{\sigma^*}\right]^2 \right) \tag{7.54}$$

with

$$\mu^* \equiv E[\dot{A}_1(t) \,|\, A_1(t) = u] = \rho_{X\dot{X}}(t) \left(\frac{\sigma_{\dot{X}}(t)}{\sigma_X(t)} \right) u \tag{7.55}$$

and

$$\sigma^* \equiv \text{Var}[\dot{A}_1(t) \,|\, A_1(t) = u] = \sigma_{\dot{X}}(t) \left(1 - \rho^2_{X\dot{X}}(t) - \alpha_1^2(t) \right)^{1/2} \tag{7.56}$$

It has also been shown that obtaining $E[\dot{\theta}_1(t)] = 0$ requires that

$$\begin{aligned} \omega_{c1}(t) &= -\frac{K_{X\dot{Z}}(t,t)}{\sigma^2_X(t)} = -\frac{1}{\sigma^2_X(t)} \int_{-\infty}^{\infty} S_{X\dot{Z}}(t,\omega)\, d\omega \\ &= \frac{i}{\sigma^2_X(t)} \int_{-\infty}^{\infty} g(\omega)\, S_{X\dot{X}}(t,\omega)\, d\omega = \frac{\lambda_1(t)}{\lambda_0(t)} \end{aligned} \tag{7.57}$$

in which the final form has been written by using Eq. 7.14. Similarly, it has been shown that an appropriate bandwidth parameter is $-\rho_{X\dot{Z}}(t,t)$. All of these results are in complete agreement with those in Section 7.4 for the stationary situation. It should also be noted that letting the $-\rho_{X\dot{Z}}(t,t)$ bandwidth parameter tend to unity does not imply that $\dot{A}_1(t)$ becomes zero, because the variation of $A_1(t)$ also depends on the time-variation of $H_{xS}(t,\omega)$.

****

Example 7.14: Find the expected rate of upcrossings for the Cramer and Leadbetter amplitude of a modulated Gaussian process that is mean-zero.

As in the stationary situation of Example 7.12, the rate of upcrossings for $A_1(t)$ is

$$\nu_{A_1}^+(u,t) = p_{A_1(t)}(u)\int_0^\infty v\, p_{\dot A_1(t)}[v \mid A_1(t)=u]\,dv$$

and the necessary probability distributions are given in Eqs. 7.53–7.56. The result is slightly more complicated than in the stationary situation because the conditional distribution of $\dot A_1(t)$ is not mean-zero in the current situation. Using Eq. 7.54 gives the integral over v in exactly the same form as in Example 7.1:

$$\int_0^\infty v\, p_{\dot A_1(t)}[v \mid A_1(t)=u]\,dv = \mu^*\Phi\left(\frac{\mu^*}{\sigma^*}\right) + \frac{\sigma^*}{(2\pi)^{1/2}}\exp\left(-\frac{(\mu^*)^2}{2(\sigma^*)^2}\right)$$

so

$$\nu_{A_1}^+(u,t) = \frac{u}{\sigma_X(t)}\exp\left(-\frac{u^2}{2\sigma_X^2(t)}\right)U(u)\frac{\sigma_{\dot X}(t)}{\sigma_X(t)}\times$$

$$\left(\frac{\rho_{X\dot X}(t)\,u}{\sigma_X(t)}\Phi\left[\frac{\rho_{X\dot X}(t)\,u}{\sigma_X(t)[1-\rho_{X\dot X}^2(t)-\alpha_1^2(t)]^{1/2}}\right]+\right.$$

$$\left.\left(\frac{1-\rho_{X\dot X}^2(t)-\alpha_1^2(t)}{2\pi}\right)^{1/2}\exp\left[-\frac{1}{2}\left(\frac{\rho_{X\dot X}^2(t)\,u^2}{\sigma_X^2(t)[1-\rho_{X\dot X}^2(t)-\alpha_1^2(t)]}\right)\right]\right)$$

Note that this result is consistent with Example 7.12 in that letting $\rho_{X\dot X}(t)$ tend to zero does give exactly the stationary result.

****

Exercises

****

General Upcrossing Rates

****

7.1 Each of the formulas gives the joint probability density of $X(t)$ and $\dot X(t)$ for a particular $\{X(t)\}$ process at a particular time t. For each process, find the expected rate of upcrossings $\nu_X^+(u,t)$ for all u values.

(a) $p_{X(t)\dot X(t)}(u,v) = 9u^2v^2\,U(u)U(1-u)U(v)U(1-v)$

(b) $p_{X(t)\dot{X}(t)}(u,v) = \dfrac{2u}{(2\pi)^{1/2} b^2 \sigma_2} \exp\left(-\dfrac{v^2}{2\sigma_2^2}\right) U(u) U(b-u)$

(c) $p_{X(t)\dot{X}(t)}(u,v) = \dfrac{3(1-u^2)}{2(2\pi)^{1/2} \sigma_2} \exp\left(-\dfrac{v^2}{2\sigma_2^2}\right) U(1-|u|)$

(d) $p_{X(t)\dot{X}(t)}(u,v) = \dfrac{\lambda}{(2\pi)^{1/2} \sigma_2} \exp\left(-\lambda u - \dfrac{v^2}{2\sigma_2^2}\right) U(u)$

****

7.2 Each of the formulas gives the joint probability density of $X(t)$ and $\dot{X}(t)$ for a particular covariant stationary stochastic process $\{X(t)\}$. For each process, find the expected rate of upcrossings $\nu_X^+(u)$ for all u values.

(a) $p_{X(t)\dot{X}(t)}(u,v) = \dfrac{1}{\pi a^2} U(a^2 - u^2 - v^2)$

(b) $p_{X(t)\dot{X}(t)}(u,v) = \dfrac{1}{2(2\pi)^{1/2} b \sigma_2} \exp\left(-\dfrac{v^2}{2\sigma_2^2}\right) U(b-|u|)$

(c) $p_{X(t)\dot{X}(t)}(u,v) = \dfrac{1}{2\pi \sigma_1 \sigma_2} \exp\left(-\dfrac{u^2}{2\sigma_1^2} - \dfrac{v^2}{2\sigma_2^2}\right)$

(d) $p_{X(t)\dot{X}(t)}(u,v) = \dfrac{3\alpha^{1/3}}{2\Gamma(1/3)} \left(\dfrac{\gamma}{\pi}\right)^{1/2} e^{-\alpha|u^3| - \gamma v^2}$

****

7.3 The stationary response $\{X(t)\}$ for a certain nonlinear oscillator is characterized by

$$p_{X(t)\dot{X}(t)}(u,v) = A \exp\left(-\lambda |u| - \frac{v^2}{2\sigma_2^2}\right)$$

(a) Find the value of the constant A, in terms of λ and σ_2.

(b) Find $\nu_X^+(u)$ for all u values.

(c) Compare $\nu_X^+(u)$ with $\nu_Y^+(u)$ for a stationary Gaussian process $\{Y(t)\}$ with the same mean and autocovariance function as $\{X(t)\}$.

****

7.4 The stationary response $\{X(t)\}$ for a certain nonlinear oscillator is characterized by

$$p_{X(t)\dot{X}(t)}(u,v) = A \exp\left(-\alpha u^4 - \frac{v^2}{2\sigma_2^2}\right)$$

(a) Find the value of the constant A, in terms of α and σ_2.

(b) Find $\nu_X^+(u)$ for all u values.

(c) Compare $\nu_X^+(u)$ with $\nu_Y^+(u)$ for a stationary Gaussian process $\{Y(t)\}$ with the same mean and autocovariance function as $\{X(t)\}$.

[Hint: $\Gamma(1/4) \approx 3.6256$ and $\Gamma(3/4) \approx 1.2254$.]

****

Upcrossing Rates for Narrowband Processes

****

7.5 Let $\{X(t)\}$ be a mean-zero covariant stationary stochastic process with an autospectral density function of

$$S_{XX}(\omega) = S_0\,[\exp(-\gamma\,|\,\omega+\omega_0\,|) + \exp(-\gamma\,|\,\omega-\omega_0\,|)]$$

in which S_0, ω_0, and γ are positive constants. Under what limitations on S_0, ω_0, and/or γ can you use this information to approximate $\nu_X^+(0)$ without further information about the probability distribution of $\{X(t)\}$? Find this approximation, and explain your answer.

****

7.6 Let $\{X(t)\}$ be a mean-zero covariant stationary stochastic process with an autospectral density function of $S_{XX}(\omega) = A\,|\,\omega\,|^b\, e^{-c|\omega|}$, in which A, b, and c are positive constants. Under what limitations on A, b, and/or c can you use this information to approximate $\nu_X^+(0)$ without further information about the probability distribution of $\{X(t)\}$? Find this approximation, and explain your answer.

****

7.7 Let $\{X(t)\}$ have a mean-value function and the covariant stationary autospectral density function introduced in Section 6.4:

$$S_{XX}(\omega) = S_0\, U[(\,|\,\omega\,| - (\omega_c - b)]\, U[(\omega_c + b) - |\,\omega\,|]$$

in which S_0, ω_c, and b are positive constants. Under what limitations on S_0, ω_c, and/or b can you use this information to approximate $\nu_X^+[\mu_X(t)]$ without further information about the probability distribution of $\{X(t)\}$? Find this approximation, and explain your answer.

****

Upcrossing Rates for Gaussian Processes

****

7.8 Let $\{X(t)\}$ be the mean-zero, covariant stationary stochastic process of Exercise 7.5, but with the additional stipulation that it is Gaussian.

(a) Find $\nu_X^+(u)$ for all u values.

(b) Find the value of the irregularity factor *IF*.

****

7.9 Let $\{X(t)\}$ be the mean-zero, covariant stationary stochastic process of Exercise 7.6, but with the additional stipulation that it is Gaussian.

(a) Find $\nu_X^+(u)$ for all u values.

(b) Find the value of the irregularity factor *IF*.

****

7.10 Let $\{X(t)\}$ be the mean-zero, covariant stationary stochastic process of Exercise 7.7, but with the additional stipulation that it is Gaussian.

(a) Find $\nu_X^+(u)$ for all u values.

(b) Find the value of the irregularity factor *IF*.

****

7.11 Let $\{X(t)\}$ be a covariant stationary Gaussian stochastic process with a mean-value function $\mu_X(t)$ and an autospectral density of

$$S_{XX}(\omega) = S_0 \frac{|\omega|}{\omega_0} U(\omega_0 - |\omega|)$$

(a) Find $\nu_X^+(u)$ for all u values.

(b) Find the value of the irregularity factor *IF*.

****

7.12 Let $\{X(t)\}$ be a covariant stationary Gaussian stochastic process with a constant mean value μ_X and an autospectral density of

$$S_{XX}(\omega) = |\omega|^5 e^{-\omega^2}$$

(a) Find $\nu_X^+(u)$ for all u values.

(b) Find the value of the irregularity factor *IF*.

****

7.13 Let $\{X(t)\}$ be a covariant stationary Gaussian stochastic process with a constant mean value μ_X and an autospectral density of

$$S_{XX}(\omega) = |\omega| e^{-\omega^2}$$

(a) Find $\nu_X^+(u)$ for all u values.

(b) Find the value of the irregularity factor *IF*.

****

7.14 Let $\{X(t)\}$ be a mean-zero covariant stationary Gaussian stochastic process with an autocovariance function of

$$G_{XX}(\tau) = e^{-c\tau^2} \cos(a\tau) \qquad \text{with } c > 0$$

(a) Find $\nu_X^+(u)$ for all u values.

(b) Find the value of the irregularity factor *IF*.

****

7.15 Let $\{X(t)\}$ be a mean-zero covariant stationary Gaussian stochastic process with an autocovariance function of

$$G_{XX}(\tau) = \cos(a\tau)/(1 + c\tau^2) \qquad \text{with } c > 0$$

(a) Find $\nu_X^+(u)$ for all u values.

(b) Find the value of the irregularity factor *IF*.

****

Frequency-Domain Analysis

****

7.16 Let $\{X(t)\}$ be the covariant stationary stochastic process of Exercise 7.5 with

$$S_{XX}(\omega) = S_0 \exp\left(-\gamma\,|\,\omega + \omega_0\,|\right) + S_0 \exp\left(-\gamma\,|\,\omega - \omega_0\,|\right)$$

in which S_0, ω_0, and γ are positive constants.

(a) Find the λ_0, λ_1, λ_2, and λ_4 spectral moments.

(b) Find the ω_{c1} and ω_{c2} characteristic frequencies.

(c) Find the α_1 and α_2 bandwidth parameters.

(d) Discuss the behavior of the autospectral density, the bandwidth parameters, and the characteristic frequencies both for $\gamma \to \infty$ and for $\gamma \to 0$.

****

7.17 Let $\{X(t)\}$ be the covariant stationary stochastic process of Exercise 6.3 with

$$S_{XX}(\omega) = S_0 \left|\frac{\omega}{\omega_0}\right|^c U(\omega_0 - |\,\omega\,|) + S_0 \left|\frac{\omega_0}{\omega}\right|^c U(|\,\omega\,| - \omega_0)$$

in which S_0, ω_0, and c are positive constants.

(a) Find the λ_0, λ_1, λ_2, and λ_4 spectral moments.

(b) Find the ω_{c1} and ω_{c2} characteristic frequencies.

(c) Find the α_1 and α_2 bandwidth parameters.

(d) Discuss the behavior of the autospectral density, the bandwidth parameters, and the characteristic frequencies both for $c \to \infty$ and for $c \to 0$.

****

7.18 Let $\{X(t)\}$ be a covariant stationary stochastic process with autospectral density function $S_{XX}(\omega) = A\,|\,\omega\,|^b\, e^{-c|\omega|}$ in which A, b, and c are positive constants.

(a) Find the λ_0, λ_1, λ_2, and λ_4 spectral moments.

(b) Find the ω_{c1} and ω_{c2} characteristic frequencies.

(c) Find the α_1 and α_2 bandwidth parameters.

(d) For what values of b and/or c is $\{X(t)\}$ a narrowband process? What is the dominant frequency of this narrowband process?

****

7.19 Consider the covariant stationary process $\{X(t)\}$ of Exercise 6.5 with

$$S_{XX}(\omega) = S_0\, U(\omega_0 - |\,\omega\,|) + S_0 \left|\frac{\omega_0}{\omega}\right|^4 U(-|\,\omega\,| - \omega_0) \qquad \text{with } \omega_0 > 0$$

(a) Find the λ_0, λ_1, λ_2, and λ_4 spectral moments.
(b) Find the ω_{c1} and ω_{c2} characteristic frequencies.
(c) Find the α_1 and α_2 bandwidth parameters.

****

7.20 Let $\{X(t)\}$ be a covariant stationary process with autocovariance function

$$G_{XX}(\tau) = e^{-c\tau^2}\cos(a\tau) \qquad \text{with } c > 0$$

(a) Find the ω_{c2} characteristic frequency.
(b) Find the α_2 bandwidth parameter.
(c) Under what conditions on the parameters a and c might $\{X(t)\}$ be considered a narrowband process?

****

7.21 Let $\{X(t)\}$ be a covariant stationary process with autocovariance function

$$G_{XX}(\tau) = \cos(a\tau)/(1 + c\tau^2) \qquad \text{with } c > 0$$

(a) Find the ω_{c2} characteristic frequency.
(b) Find the α_2 bandwidth parameter.
(c) Under what conditions on the parameters a and c might $\{X(t)\}$ be considered a narrowband process?

****

Chapter 8
Matrix Analysis of Linear Systems

8.1 Generalization of Scalar Formulation

We have used matrices, to this point, to organize arrays of information, such as all the possible covariances within a set of scalar random variables. We will now do the same thing for the properties of dynamic systems, leading to matrix equations of dynamic equilibrium. Such matrix analysis is necessary for most practical problems, and there are several methods for using matrix formulations within stochastic analysis. We will look at some of the characteristics of each.

Recall that Chapters 5 and 6 began with very general deterministic descriptions of the response of a linear system. In Chapter 5 this gave the time history of a single response as a Duhamel convolution integral of an impulse response function and the time history of a single excitation, and in Chapter 6 the Fourier transform of the response time history was found to be a simple product of a harmonic transfer function and the Fourier transform of the excitation. We now wish to extend those presentations to include consideration of systems with multiple inputs and multiple outputs.

Let n_X be the number of response components in which we are interested, and let n_F be the number of different excitations of the system. We will denote the responses as $\{X_j(t)\}$ for $j=1,\cdots,n_X$ and the excitations as $\{F_l(t)\}$ for $l=1,\cdots,n_F$, and begin by considering evaluation of the jth component of the response. In many situations the excitations may be components of force and the responses may be components of displacement, as illustrated in Fig. 8.1, but this is not necessary. The formulation is general enough to include any definitions of excitation and response, except that we will consider the system to be causal in the sense that the responses at time t are caused only by the excitations at prior times.

For the time domain formulation we use Eq. 5.2 to give the response due to the lth component of the excitation, and using superposition we can say that the total response is the sum over l of these components. Thus, we have

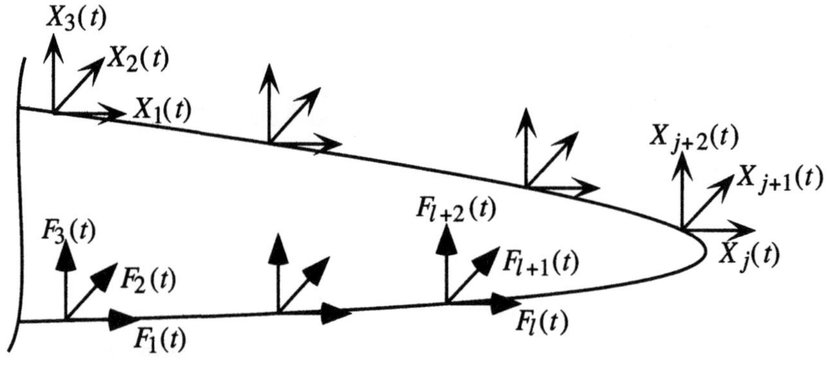

Figure 8.1 Selected components of force and displacement for an airplane wing.

$$X_j(t) = \sum_{l=1}^{n_F} \int_{-\infty}^{\infty} h_{jl}(t-s) F_l(s) ds \tag{8.1}$$

in which the impulse response function $h_{jl}(t)$ is defined to be the response component $X_j(t)$ due to a Dirac delta function excitation with $F_l(t) = \delta(t)$ and $F_r(t) = 0$ for $r \neq l$. This equation, though, is simply the jth row of the matrix equation

$$\vec{X}(t) = \int_{-\infty}^{\infty} \mathbf{h}(t-s) \vec{F}(s) ds \tag{8.2}$$

Thus, Eq. 8.2 is a very convenient representation of the set of n_X equations that describe the n_X responses in terms of the n_F excitations. Each component of the rectangular matrix of dimension $n_X \times n_F$ is defined as stated. Alternatively, this information can be organized so that an entire column of the $\mathbf{h}(t)$ matrix is defined in one equation. Namely,

$$\vec{X}(t) = [h_{1l}(t), \cdots, h_{n_X l}(t)]^T$$

is the response to

$$\vec{F}(t) = [\underbrace{0,\cdots,0}_{l-1}, \delta(t), \underbrace{0,\cdots,0}_{n_F - l}]^T$$

and we define the entire matrix by considering different l values.

In a similar way we can use Eq. 6.29 to give the Fourier transform of the jth response due to the lth excitation component, and use superposition to obtain

$$\tilde{X}_j(\omega) = \sum_{l=1}^{n_F} H_{jl}(\omega)\,\tilde{F}_l(\omega) \tag{8.3}$$

or

$$\tilde{\vec{X}}(\omega) = \mathbf{H}(\omega)\,\tilde{\vec{F}}(\omega) \tag{8.4}$$

in which the lth column of the harmonic transfer matrix is defined such that

$$\vec{X}(t) = [H_{1l}(\omega),\cdots,H_{n_X l}(\omega)]^T\, e^{i\omega t}$$

is the response to

$$\vec{F}(t) = [\underbrace{0,\cdots,0}_{l-1},e^{i\omega t},\underbrace{0,\cdots,0}_{n_F-l}]^T$$

Of course, we can show that any component of the impulse response function matrix and the corresponding component of the harmonic transfer function matrix satisfy a Fourier transform relationship like Eq. 6.30 for the scalar case. The collection of all these relationships can be written in matrix form as

$$\mathbf{H}(\omega) = \int_{-\infty}^{\infty} \mathbf{h}(t)\, e^{-i\omega t}\, dt$$

meaning that the scalar Fourier transform integral is applied to each component of the matrix in the integrand.

We can now use Eqs. 8.2 and 8.4 in writing matrix versions of various expressions in Chapters 5 and 6 that describe the stochastic response in terms of the excitation. The mean-value relationships are obtained by simply taking the expectations of the two equations; they take the form

$$\vec{\mu}_X(t) = \int_{-\infty}^{\infty} \mathbf{h}(t-s)\,\vec{\mu}_F(s)\,ds \tag{8.5}$$

and

$$\tilde{\vec{\mu}}_X(\omega) = \mathbf{H}(\omega)\,\tilde{\vec{\mu}}_F(\omega) \tag{8.6}$$

with the jth components of these equations being exactly equivalent to summations over l of Eqs. 5.14 and 6.29, respectively, written to give the jth component of response due to the lth component of excitation. Similarly, we can write the autocorrelation function for the response as

$$\boldsymbol{\phi}_{XX}(t,s) \equiv E[\vec{X}(t)\vec{X}^T(s)] = \int_{-\infty}^{\infty}\int_{-\infty}^{\infty} \mathbf{h}(t-u)\,\boldsymbol{\phi}_{FF}(u,v)\,\mathbf{h}^T(s-v)\,du\,dv \tag{8.7}$$

and the autocovariance as

$$\begin{aligned}\mathbf{K}_{XX}(t,s) &= \boldsymbol{\phi}_{XX}(t,s) - \vec{\mu}_X(t)\vec{\mu}_X^T(s)\\ &= \int_{-\infty}^{\infty}\int_{-\infty}^{\infty} \mathbf{h}(t-u)\,\mathbf{K}_{FF}(u,v)\,\mathbf{h}^T(s-v)\,du\,dv\end{aligned} \tag{8.8}$$

For the special case of stationarity, these relationships can be rewritten in any of several slightly simpler forms, including

$$\mathbf{R}_{XX}(\tau) = \int_{-\infty}^{\infty}\int_{-\infty}^{\infty} \mathbf{h}(u)\,\mathbf{R}_{FF}(\tau-u+v)\,\mathbf{h}^T(v)\,du\,dv \tag{8.9}$$

and

$$\mathbf{G}_{XX}(\tau) = \int_{-\infty}^{\infty}\int_{-\infty}^{\infty} \mathbf{h}(u)\,\mathbf{G}_{FF}(\tau-u+v)\,\mathbf{h}^T(v)\,du\,dv \tag{8.10}$$

Similarly, the autospectral density for the covariant stationary situation is described by

$$\mathbf{S}_{XX}(\omega) = \mathbf{H}(\omega)\,\mathbf{S}_{FF}(\omega)\,\mathbf{H}^{*T}(\omega) \tag{8.11}$$

The jth diagonal components of Eqs. 8.7–8.11 are exactly equivalent to summations over l of the expressions in Chapters 5 and 6 giving autocorrelation, autocovariance, and autospectral density of the jth response due to the lth component of excitation. Similarly, the off-diagonal components of these matrix equations give the cross-correlation, cross-covariance, and cross-spectral density functions for different components of response.

Note that if the $\vec{X}(t_0) = \vec{Y}$ value of the response at time t_0 is known, then one can write Eq. 8.2 in the alternative form of

$$\vec{X}(t) = \mathbf{g}(t-t_0)\,\vec{Y} + \int_{t_0}^{t} \mathbf{h}(t-s)\,\vec{F}(s)\,ds \tag{8.12}$$

in which the vector $\vec{Y}$ contains a complete set of initial conditions at time t_0, and $\mathbf{g}(t)$ is a matrix that gives the time histories of free vibration (i.e., the homogeneous solution) to unit values of the possible initial conditions. Using this form of the time history solution makes it possible to write vector analogies of Eqs. 5.22 and 5.23, which describe the input and output in the corresponding scalar situation. Vector versions of Eqs. 5.24 and 5.25 giving the conditional mean and conditional covariance of the response can also be obtained.

8.2 Multi-Degree-of-Freedom Systems

One of the most commonly encountered situations involving a linear system with multiple inputs and multiple outputs is the so-called *multi-degree-of-freedom* (MDF) system. As a simple generalization of Eq. 5.45, the MDF equation of motion can be written as

$$\mathbf{m}\ddot{\vec{X}}(t) + \mathbf{c}\dot{\vec{X}}(t) + \mathbf{k}\vec{X}(t) = \vec{F}(t) \tag{8.13}$$

in which $\mathbf{m}$, $\mathbf{c}$, and $\mathbf{k}$ are square matrices of dimension $n \times n$ and the location and orientation of the components of $\vec{F}(t)$ are identical to those of $\vec{X}(t)$.[1] Often we have problems of this type in which the elements of the $\mathbf{m}$, $\mathbf{c}$, and $\mathbf{k}$ matrices represent physical masses, dashpots, and springs, respectively, but this is not always the case. In our examples, it will always be true that $\mathbf{m}$, $\mathbf{c}$, and $\mathbf{k}$ will represent inertia terms, energy dissipation terms, and restoring force terms, but the exact meaning of any component will depend on the set of coordinates used in describing the problem. One can write certain energy expressions, however, that are true for any choice of coordinates. In particular, the kinetic energy, the potential energy, and the rate of energy dissipation (power dissipation), respectively, are given by

$$KE(t) = \frac{1}{2}\dot{\vec{X}}^T(t)\,\mathbf{m}\,\dot{\vec{X}}(t) \tag{8.14}$$

$$PE(t) = \frac{1}{2}\vec{X}^T(t)\,\mathbf{k}\,\vec{X}(t) \tag{8.15}$$

[1]For example, if $X_j(t)$ is the absolute displacement of node j of a system, then $F_j(t)$ is the external force at that node, but if $X_j(t)$ is a relative displacement between two nodes, then $F_j(t)$ is the difference between the two applied nodal forces.

and

$$PD(t) = \dot{\vec{X}}^T(t)\,\mathbf{c}\,\dot{\vec{X}}(t) \qquad (8.16)$$

Also, it is always true that $\mathbf{m}$, $\mathbf{c}$, and $\mathbf{k}$ are symmetric matrices. Note that the nonnegativity of the kinetic and potential energy in Eqs. 8.14 and 8.15 make it mandatory that $\mathbf{m}$ and $\mathbf{k}$, respectively, not have any complex or negative eigenvalues. For example, if $\mathbf{m}$ had a negative eigenvalue, then Eq. 8.14 would give a negative value of kinetic energy if $\vec{X}(t)$ happened to be parallel to the corresponding eigenvector. Also, Eq. 8.16 requires that $\mathbf{c}$ have no negative eigenvalues if this matrix represents the effect of components like dashpots that can never, even for an instant, add energy to the system.

The meaning of the $\mathbf{m}$, $\mathbf{c}$, and $\mathbf{k}$ matrices is particularly simple when all the mass in the system is in n discrete masses, and the coordinates, denoted by the components of $\vec{X}(t)$, each represent one component of displacement of a mass relative to a fixed frame of reference. (If multidimensional motion of any mass is possible, then we will presume that orthogonal coordinates are used to describe the total motion.) In this case the $\mathbf{m}$ matrix is diagonal, because the total kinetic energy of Eq. 8.14 is simply a sum of quadratic terms, each depending on the magnitude of one mass and the square of a component of velocity of that mass. The terms of $\mathbf{k}$ and $\mathbf{c}$ can then be described as follows. The lth column of $\mathbf{k}$, which can be written as $[k_{1l},\cdots,k_{nl}]^T$, gives the vector of forces in the n coordinates that will result in a unit static displacement in coordinate $X_l(t)$ along with zero displacement in every other coordinate. Similarly, a set of static forces $[c_{1l},\cdots,c_{nl}]^T$ would give a unit velocity in coordinate $X_l(t)$ along with zero velocity in every other coordinate, if $\mathbf{k}$ were set equal to zero. In this situation with the coordinates representing displacement relative to a fixed frame of reference, $-k_{jl}$ usually represents a physical spring connecting the masses at nodes j and l, and k_{jj} represents the sum of all springs attached to the mass at node j. In a corresponding way, the elements of $\mathbf{c}$ correspond to dashpots connected to the masses. These statements are strictly true only if the springs and dashpots are aligned with the directions of the orthogonal coordinates.

****

Example 8.1: Find the $\mathbf{m}$, $\mathbf{k}$, and $\mathbf{c}$ matrices for the two-degree-of-freedom (2DF) system shown in the accompanying sketch.

The equations of motion for the two masses are

$$m_1\ddot{X}_1(t) + c_1\dot{X}_1(t) + c_2[\dot{X}_1(t) - \dot{X}_2(t)] + k_1 X_1(t) + k_2[X_1(t) - X_2(t)] = F_1(t)$$

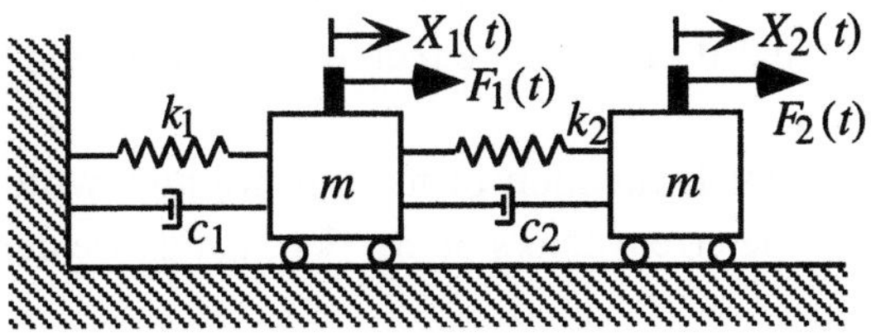

and

$$m_2\ddot{X}_2(t)+c_2[\dot{X}_2(t)-\dot{X}_1(t)]+k_2[X_2(t)-X_1(t)]=F_2(t)$$

These equations exactly agree with Eq. 8.13 with

$$\mathbf{m}=\begin{pmatrix} m_1 & 0 \\ 0 & m_2 \end{pmatrix},\quad \mathbf{k}=\begin{pmatrix} k_1+k_2 & -k_2 \\ -k_2 & k_2 \end{pmatrix},\quad \mathbf{c}=\begin{pmatrix} c_1+c_2 & -c_2 \\ -c_2 & c_2 \end{pmatrix}$$

****

8.3 Uncoupled Modes of MDF Systems

The most common solution of Eq. 8.13 uses the eigenvectors and eigenvalues of the system. In particular, we will presume that $\mathbf{m}$ is positive definite (i.e., has no zero eigenvalues), which ensures that its inverse exists. We will then let the matrix $\boldsymbol{\theta}$, of dimension $n\times n$, have columns that are the eigenvectors of $\mathbf{m}^{-1}\mathbf{k}$. This means that

$$\mathbf{m}^{-1}\mathbf{k}\boldsymbol{\theta}=\boldsymbol{\theta}\boldsymbol{\lambda} \tag{8.17}$$

in which $\boldsymbol{\lambda}$ is a diagonal matrix of dimension $n\times n$, with the (j,j) element being the eigenvalue corresponding to the eigenvector located in column j of $\boldsymbol{\theta}$. Let us now define two new symmetric matrices as

$$\hat{\mathbf{m}}\equiv\boldsymbol{\theta}^T\mathbf{m}\boldsymbol{\theta} \tag{8.18}$$

$$\hat{\mathbf{k}}\equiv\boldsymbol{\theta}^T\mathbf{k}\boldsymbol{\theta} \tag{8.19}$$

An important property of $\boldsymbol{\theta}$ is easily obtained by rewriting Eq. 8.17 as $\mathbf{k}\boldsymbol{\theta}=\mathbf{m}\boldsymbol{\theta}\boldsymbol{\lambda}$ and multiplying this equation by $\boldsymbol{\theta}^T$ on the left to give $\hat{\mathbf{k}}=\hat{\mathbf{m}}\boldsymbol{\lambda}$. The transpose of this equation is $\hat{\mathbf{k}}=\boldsymbol{\lambda}\hat{\mathbf{m}}$, because $\hat{\mathbf{k}}$, $\hat{\mathbf{m}}$, and the diagonal matrix $\boldsymbol{\lambda}$ are all symmetric. From these two equations we see that $\hat{\mathbf{m}}\boldsymbol{\lambda}=\boldsymbol{\lambda}\hat{\mathbf{m}}$, showing that the matrices $\hat{\mathbf{m}}$ and $\boldsymbol{\lambda}$ commute. Provided that $\hat{\mathbf{m}}\neq\mathbf{0}$ and the diagonal elements of $\boldsymbol{\lambda}$ are distinct, this condition can be satisfied only if $\hat{\mathbf{m}}$ is diagonal. If $\mathbf{m}^{-1}\mathbf{k}$ has one or more repeated eigenvalues (i.e., the elements of $\boldsymbol{\lambda}$ are not all distinct), then it is not required that $\hat{\mathbf{m}}$ be exactly diagonal, but it is possible to choose the eigenvectors such that $\hat{\mathbf{m}}$ is diagonal. Thus, we will presume that $\hat{\mathbf{m}}$ is diagonal.

Furthermore, the fact that $\hat{\mathbf{k}} = \hat{\mathbf{m}}\boldsymbol{\lambda}$ shows that $\hat{\mathbf{k}}$ is also diagonal. It is also significant that the $\boldsymbol{\lambda}$ and $\boldsymbol{\theta}$ matrices are real for feasible $\mathbf{m}$ and $\mathbf{k}$ matrices.

To simplify the MDF equation of motion, we now write $\vec{X}(t)$ as a linear expansion in terms of the eigenvectors of $\mathbf{m}^{-1}\mathbf{k}$. That is, we define a vector $\vec{Z}(t)$ such that

$$\vec{X}(t) = \boldsymbol{\theta}\vec{Z}(t) \tag{8.20}$$

giving the jth component of $\vec{Z}(t)$ as being the projection of $\vec{X}(t)$ on the jth eigenvector. Note that the component θ_{jl} now gives the magnitude of the $X_j(t)$ response component resulting from a unit magnitude of $Z_l(t)$. In the usual terminology, the lth eigenvector (i.e., column l of $\boldsymbol{\theta}$) is the lth mode shape and $Z_l(t)$ is the lth modal amplitude. Substituting Eq. 8.20 into Eq. 8.13 and multiplying the equation on the left by $\boldsymbol{\theta}^T$ gives

$$\boldsymbol{\theta}^T \mathbf{m}\boldsymbol{\theta}\ddot{\vec{Z}}(t) + \boldsymbol{\theta}^T \mathbf{c}\boldsymbol{\theta}\dot{\vec{Z}}(t) + \boldsymbol{\theta}^T \mathbf{k}\boldsymbol{\theta}\vec{Z}(t) = \boldsymbol{\theta}^T \vec{F}(t)$$

which can be rewritten according to Eqs. 8.18 and 8.19 as

$$\ddot{\vec{Z}}(t) + \boldsymbol{\beta}\dot{\vec{Z}}(t) + \boldsymbol{\lambda}\vec{Z}(t) = \hat{\mathbf{m}}^{-1}\boldsymbol{\theta}^T \vec{F}(t) \tag{8.21}$$

in which the new matrix $\boldsymbol{\beta}$ is defined such that

$$\hat{\mathbf{c}} \equiv \boldsymbol{\theta}^T \mathbf{c}\boldsymbol{\theta} = \hat{\mathbf{m}}\boldsymbol{\beta} \tag{8.22}$$

Although one can always rewrite the MDF equation of motion in the form of Eq. 8.21, this formulation is not particularly useful unless some limitations are placed on the $\mathbf{c}$ damping matrix. In particular, the formulation is very useful when $\boldsymbol{\beta}$ is diagonal, in which case the jth row of the equation has the simple form

$$\ddot{Z}_j(t) + \beta_{jj}\dot{Z}_j(t) + \lambda_{jj}Z_j(t) = \frac{1}{\hat{m}_{jj}} \sum_{l=1}^{n} \theta_{lj}\, F_l(t) \tag{8.23}$$

Various terms are used to designate this special case in which $\boldsymbol{\beta}$ is diagonal. Caughey (1960a) called it the situation with *classical normal modes*, and other terms meaning the same thing include *classical damping* and *uncoupled modes*. The last term emphasizes the key fact shown in Eq. 8.23—that one can solve for

$Z_j(t)$ from an equation that is completely uncoupled from the similar equations governing the behavior of $Z_l(t)$ for $l \neq j$.

For the situation with uncoupled modes, we see from Eq. 8.23 that the behavior of any $Z_j(t)$ modal amplitude is governed by a scalar, second-order, differential equation that is essentially the same as the SDF equation we have previously considered. In fact, if we define the modal frequency ω_j and the modal damping ζ_j such that $\omega_j = (\lambda_{jj})^{1/2}$ and $2\zeta_j \omega_j = \beta_{jj}$, then Eq. 8.23 takes exactly the form of Eq. 5.46:

$$\ddot{Z}_j(t) + 2\zeta_j \omega_j \dot{Z}_j(t) + \omega_j^2 Z_j(t) = \frac{1}{\hat{m}_{jj}} \sum_{l=1}^{n} \theta_{lj} F_l(t) \tag{8.24}$$

Thus, modal analysis reduces the MDF problem to the solution of a set of SDF problems plus matrix algebra, provided that the system does have uncoupled modes.

We will now look at the conditions under which the modal equations do uncouple, giving Eq. 8.24 as the modal equation of motion. First we note that if the matrix $\boldsymbol{\beta}$ is diagonal, then it commutes with the diagonal matrix $\boldsymbol{\lambda}$; that is $\boldsymbol{\lambda}\boldsymbol{\beta} = \boldsymbol{\beta}\boldsymbol{\lambda}$. Furthermore, if all the elements of $\boldsymbol{\lambda}$ are distinct, then $\boldsymbol{\lambda}\boldsymbol{\beta} = \boldsymbol{\beta}\boldsymbol{\lambda}$ only if $\boldsymbol{\beta}$ is diagonal. Now we investigate the restrictions on $\mathbf{m}$, $\mathbf{c}$, and $\mathbf{k}$ that will result in $\boldsymbol{\lambda}\boldsymbol{\beta} = \boldsymbol{\beta}\boldsymbol{\lambda}$. Solving $\hat{\mathbf{k}} = \hat{\mathbf{m}}\boldsymbol{\lambda} = \boldsymbol{\theta}^T \mathbf{k} \boldsymbol{\theta}$ for $\boldsymbol{\lambda}$ and Eq. 8.22 for $\boldsymbol{\beta}$ gives their products as $\boldsymbol{\lambda}\boldsymbol{\beta} = \hat{\mathbf{m}}^{-1}\boldsymbol{\theta}^T \mathbf{k} \boldsymbol{\theta} \hat{\mathbf{m}}^{-1} \boldsymbol{\theta}^T \mathbf{c} \boldsymbol{\theta}$ and $\boldsymbol{\beta}\boldsymbol{\lambda} = \hat{\mathbf{m}}^{-1}\boldsymbol{\theta}^T \mathbf{c} \boldsymbol{\theta} \hat{\mathbf{m}}^{-1} \boldsymbol{\theta}^T \mathbf{k} \boldsymbol{\theta}$. From Eq. 8.18 we can note that $\hat{\mathbf{m}}^{-1}\boldsymbol{\theta}^T = \boldsymbol{\theta}^{-1}\mathbf{m}^{-1}$, and substitution of this relationship gives $\boldsymbol{\lambda}\boldsymbol{\beta} = \boldsymbol{\beta}\boldsymbol{\lambda}$ if and only if $\mathbf{k}\mathbf{m}^{-1}\mathbf{c} = \mathbf{c}\mathbf{m}^{-1}\mathbf{k}$. Thus, we can be assured of having uncoupled modal equations if the coefficient matrices in the original equation of motion satisfy this condition. Alternatively, it can be stated that the modal equations uncouple if and only if $\mathbf{m}^{-1}\mathbf{c}$ and $\mathbf{m}^{-1}\mathbf{k}$ commute. If this condition is not met, then the $\boldsymbol{\beta}$ matrix will not be diagonal and different modal equations will be coupled by the damping terms, requiring simultaneous solution of the equations. It is possible to solve the problem when the uncoupling condition is not met, but it is necessary to use some different method. Two methods that do not require any restriction on the $\mathbf{m}$, $\mathbf{c}$, and $\mathbf{k}$ matrices are presented in Sections 8.5 and 8.6. Nonetheless, the most common method of solving the MDF problem is by uncoupling the equations, as presented here, or some equivalent variation of this method.

The general condition of $\mathbf{k}\mathbf{m}^{-1}\mathbf{c} = \mathbf{c}\mathbf{m}^{-1}\mathbf{k}$ for uncoupled modal equations was presented by Caughey and O'Kelley in 1965. A much less general condition that is sufficient to ensure the existence of uncoupled modes, but is not necessary for their existence, is called the *Rayleigh condition*, which is given by $\mathbf{c} = a_1\,\mathbf{m} + a_2\,\mathbf{k}$ for some scalar constants a_1 and a_2. It is easy to verify that the Rayleigh condition is a special case of $\mathbf{k}\,\mathbf{m}^{-1}\mathbf{c} = \mathbf{c}\,\mathbf{m}^{-1}\mathbf{k}$. One can also show (Caughey, 1960a) that other sufficient conditions can be written as

$$\mathbf{c} = \mathbf{m} \sum_{j=0}^{J} a_j (\mathbf{m}^{-1}\mathbf{k})^j \tag{8.25}$$

which is sometimes called a generalized Rayleigh condition. It can be shown that if the upper limit J of the summation is chosen to be greater than or equal to $n-1$, then this condition is equivalent to the general condition of $\mathbf{k}\,\mathbf{m}^{-1}\mathbf{c} = \mathbf{c}\,\mathbf{m}^{-1}\mathbf{k}$, but Eq. 8.25 is a more restrictive condition for any smaller value of J.

It seems somewhat surprising that it has been possible to model a great variety of physical systems by using the rather restrictive version of the MDF equations with uncoupled modes. One reason that this is true is that we often have little information about the precise form for the **c** damping matrix, so we are free to choose it in a way that simplifies the analysis. This is in contrast to the **m** and **k** matrices, whose elements typically are well approximated by calculations of physical mass and stiffness terms. In most cases the energy dissipation in real structures occurs in complicated, usually nonlinear, and poorly understood phenomena such as friction and local yielding. What is best known about the energy dissipation is its overall level; the details of precisely where and how it occurs are ill defined. One of the most common ways of modeling the energy dissipation does not even begin with a choice of a **c** matrix, but rather with the choice of the ζ_j modal damping values. Knowing the λ_j values from analysis of **m** and **k**, one can then find a diagonal $\boldsymbol{\beta}$ matrix from $\beta_j = 2\zeta_j \omega_j$. If desired, one can also solve Eq. 8.22 for $\mathbf{c} = (\boldsymbol{\theta}^T)^{-1}\hat{\mathbf{m}}\boldsymbol{\beta}\boldsymbol{\theta}^{-1}$, which can be rewritten using Eq. 8.18 as $\mathbf{c} = \mathbf{m}\boldsymbol{\theta}\boldsymbol{\beta}\,\hat{\mathbf{m}}^{-1}\boldsymbol{\theta}^T\mathbf{m}$. Often, though, there is no need to know the values of the terms of **c**, so this calculation is not performed.

****

Example 8.2: Consider the system of Example 8.1 with: $m_1 = 2m$, $m_2 = m$, $k_1 = k_2 = k$, and $c_1 = c_2 = c$, in which *m*, *k*, and *c* are positive scalar constants. Show that the system does have uncoupled modes, and find the $\boldsymbol{\theta}$, $\boldsymbol{\lambda}$, $\hat{\mathbf{m}}$, and $\boldsymbol{\beta}$ matrices and the values of the modal frequencies and damping values.

Substituting the parameter values into $\mathbf{m}$, $\mathbf{k}$, and $\mathbf{c}$ from Example 8.1 gives

$$\mathbf{m} = m\begin{pmatrix} 2 & 0 \\ 0 & 1 \end{pmatrix}, \quad \mathbf{k} = k\begin{pmatrix} 2 & -1 \\ -1 & 1 \end{pmatrix}, \quad \mathbf{c} = c\begin{pmatrix} 2 & -1 \\ -1 & 1 \end{pmatrix}$$

Inverting $\mathbf{m}$ is almost trivial, and matrix multiplication then gives

$$\mathbf{k}\,\mathbf{m}^{-1}\mathbf{c} = \mathbf{c}\,\mathbf{m}^{-1}\mathbf{k} = \frac{kc}{m}\begin{pmatrix} 3 & -2 \\ -2 & 1.5 \end{pmatrix}$$

The equality of $\mathbf{k}\,\mathbf{m}^{-1}\mathbf{c}$ and $\mathbf{c}\,\mathbf{m}^{-1}\mathbf{k}$ assures us that uncoupled modes do exist. Eigenanalysis of $\mathbf{m}^{-1}\mathbf{k}$ gives

$$\boldsymbol{\theta} = \begin{pmatrix} 1 & 1 \\ 1.414 & -1.414 \end{pmatrix}, \quad \boldsymbol{\lambda} = \frac{k}{m}\begin{pmatrix} 0.2929 & 0 \\ 0 & 1.707 \end{pmatrix}$$

Note that the columns of $\boldsymbol{\theta}$ are the eigenvectors of $\mathbf{m}^{-1}\mathbf{k}$ and their ordering is arbitrary. We have followed the common convention of putting the lower frequency mode in the first column. Note also that although the "directions" of the eigenvectors are unique, their "lengths" are arbitrary. A different scaling of these eigenvectors will affect the values of the components of $\hat{\mathbf{m}}$, but not of $\boldsymbol{\beta}$. Using $\boldsymbol{\theta}$ as written gives

$$\hat{\mathbf{m}} = m\begin{pmatrix} 4 & 0 \\ 0 & 4 \end{pmatrix}, \quad \boldsymbol{\beta} = \frac{c}{m}\begin{pmatrix} 0.2929 & 0 \\ 0 & 1.707 \end{pmatrix}$$

Taking the square root of the elements of $\boldsymbol{\lambda}$ gives the modal frequencies as $\omega_1 = 0.5412(k/m)^{1/2}$ and $\omega_2 = 1.307(k/m)^{1/2}$. Dividing the elements of $\boldsymbol{\beta}$ by $2\omega_j$ gives the modal damping values as $\zeta_1 = 0.2706\, c/(km)^{1/2}$ and $\zeta_2 = 0.6533\, c/(km)^{1/2}$.

It may also be noted that one could have performed this latter analysis without checking to see if $\mathbf{k}\,\mathbf{m}^{-1}\mathbf{c}$ and $\mathbf{c}\,\mathbf{m}^{-1}\mathbf{k}$ were equal. That is, one can perform the eigenanalysis to find $\boldsymbol{\theta}$, then evaluate $\boldsymbol{\beta}$ regardless of whether the system has uncoupled modes. The form of $\boldsymbol{\beta}$, in fact, then provides the information about the existence of uncoupled modes. A diagonal $\boldsymbol{\beta}$ matrix is sufficient evidence of uncoupled modes. Checking to see whether $\mathbf{k}\,\mathbf{m}^{-1}\mathbf{c} = \mathbf{c}\,\mathbf{m}^{-1}\mathbf{k}$ allows one to avoid doing the eigenanalysis if the equations will not uncouple anyway.

****

Example 8.3: Show that the system of Example 8.1 with $m_1 = 1000$ kg, $m_2 = 500$ kg, $k_1 = k_2 = 500$ kN/m, $c_1 = 0.5$ kN/(m/s), and $c_2 = 2.0$ kN/(m/s) does not have uncoupled modes.

Substituting the parameter values into $\mathbf{m}$, $\mathbf{k}$, and $\mathbf{c}$ from Example 8.1 gives

$$\mathbf{m} = \begin{pmatrix} 1000 & 0 \\ 0 & 500 \end{pmatrix}\text{kg}, \quad \mathbf{k} = \begin{pmatrix} 10^6 & -5\times 10^5 \\ -5\times 10^5 & 5\times 10^5 \end{pmatrix}\frac{\text{N}}{\text{m}}, \quad \mathbf{c} = \begin{pmatrix} 2500 & -2000 \\ -2000 & 2000 \end{pmatrix}\frac{\text{N}\cdot\text{s}}{\text{m}}$$

From these matrices, we find that

$$\mathbf{c}\,\mathbf{m}^{-1}\mathbf{k} = \begin{pmatrix} 4.5 & -3.25 \\ -4.0 & 3 \end{pmatrix} \times 10^6 \frac{\mathrm{N}^2}{\mathrm{m}\cdot\mathrm{s}}, \qquad \mathbf{k}\,\mathbf{m}^{-1}\mathbf{c} = \begin{pmatrix} 4.5 & -4.0 \\ -3.25 & 3 \end{pmatrix} \times 10^6 \frac{\mathrm{N}^2}{\mathrm{m}\cdot\mathrm{s}}$$

This shows that the system does not meet the condition necessary for uncoupled modes. If we had not performed this check and had proceeded to find $\boldsymbol{\theta}$, which is the same as in Example 8.2, then we would find that

$$\boldsymbol{\beta} = \begin{pmatrix} 0.4216 & -0.7071 \\ -0.7955 & 6.078 \end{pmatrix} \mathrm{rad/s}$$

The fact that this matrix is not diagonal also shows that uncoupled modes do not exist.

****

Example 8.4: Consider the system of Example 8.1 with $m_1 = m_2 = 100$ kg, $k_1 = k_2 = 500$ N/m. Find the values of dashpots in the system that will give uncoupled modes with 10% damping in the lower-frequency mode and 1% damping in the higher-frequency mode.

Using

$$\mathbf{m} = \begin{pmatrix} 100 & 0 \\ 0 & 100 \end{pmatrix} \mathrm{kg}, \qquad \mathbf{k} = \begin{pmatrix} 200 & -100 \\ -100 & 100 \end{pmatrix} \frac{\mathrm{N}}{\mathrm{m}}$$

gives

$$\boldsymbol{\theta} = \begin{pmatrix} 1 & 1 \\ 1.618 & -0.618 \end{pmatrix}, \quad \boldsymbol{\lambda} = \begin{pmatrix} 0.3820 & 0 \\ 0 & 2.618 \end{pmatrix} (\mathrm{rad/s})^2, \quad \hat{\mathbf{m}} = \begin{pmatrix} 361.8 & 0 \\ 0 & 138.2 \end{pmatrix} \mathrm{kg}$$

Setting $\omega_j = (\lambda_j)^{1/2}$ and $\beta_j = 2\zeta_j \omega_j$, with $\zeta_1 = 0.10$ and $\zeta_2 = 0.01$, gives

$$\boldsymbol{\beta} = \begin{pmatrix} 0.1236 & 0 \\ 0 & 0.0324 \end{pmatrix} \mathrm{rad/s}$$

Solving Eq. 8.22 for $\mathbf{c}$ gives $\mathbf{c} = (\boldsymbol{\theta}^{-1})^T \hat{\mathbf{m}} \boldsymbol{\beta} \boldsymbol{\theta}^{-1}$. Rather than invert a matrix that is not diagonal, we can now solve Eq. 8.18 for the inverse as $\boldsymbol{\theta}^{-1} = \hat{\mathbf{m}}^{-1} \boldsymbol{\theta}^T \mathbf{m}$ and use this to obtain

$$\mathbf{c} = \mathbf{m}\boldsymbol{\theta}\boldsymbol{\beta}\,\hat{\mathbf{m}}^{-1}\boldsymbol{\theta}^T\mathbf{m} = \begin{pmatrix} 5.758 & 4.081 \\ 4.081 & 9.839 \end{pmatrix}$$

(This technique for finding $\boldsymbol{\theta}^{-1}$ is not particularly important for a 2DF but can save significant computational time for large matrices.) Using $\mathbf{m}$, $\mathbf{k}$, and $\mathbf{c}$ now gives the scalar equations of motion as

$$100\,\ddot{X}_1(t) + 5.758\,\dot{X}_1(t) + 4.081\,\dot{X}_2(t) + 200\,X_1(t) - 100\,X_2(t) = F_1(t)$$

and

$$100\,\ddot{X}_2(t) + 4.081\,\dot{X}_1(t) + 9.839\,\dot{X}_2(t) - 100\,X_1(t) + 100\,X_2(t) = F_2(t)$$

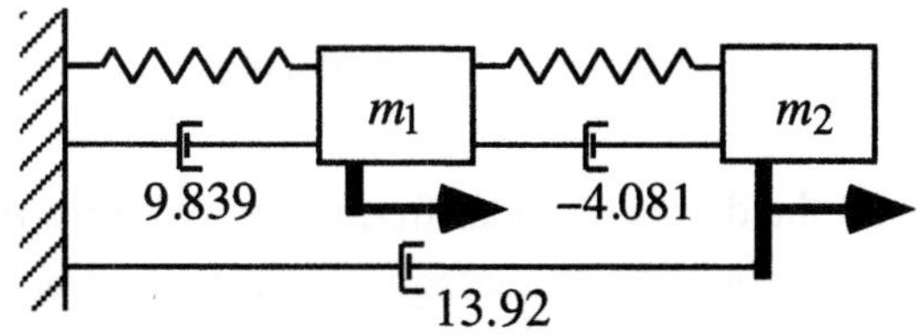

The accompanying sketch shows an arrangement of dashpots that will give these equations of motion. Note, in particular, that the value of the dashpot connecting the two masses is negative. Obviously, a negative dashpot is not a physical element, but that is the result obtained any time that an off-diagonal element of **c** is found to be positive. Furthermore, it is not unusual for such a procedure of assigning values of modal damping to result in a model that does not correspond to physical dashpots. This anomaly does not cause any difficulty in the mathematical analysis or any pathological behavior in the dynamic response, despite its apparent lack of logical explanation. One should always keep in mind, though, that even when it corresponds to positive dashpots, the linear damping matrix **c** usually represents no more than a crude approximation of the actual energy losses that limit the dynamic response of a real system.

****

8.4 Time-Domain Stochastic Analysis of Uncoupled MDF Systems

There are two different approaches that one can now use in performing a time-domain analysis of the MDF system described by Eq. 8.13, with uncoupled modal responses described by Eq. 8.24. The results, of course, are the same, but the manipulations are somewhat different. The first approach is to use the results of Section 8.3 strictly for the deterministic analysis that gives the $\mathbf{h}(t)$ matrix of impulse response functions needed in order to use the stochastic analysis of Section 8.1. The second approach is to consider the equations of Section 8.3 to be stochastic and to use them directly in finding such things as the mean and covariance of the $\vec{X}(t)$ response process. We will look briefly at each approach.

To find the $\mathbf{h}(t)$ matrix, we must consider the response of the MDF system of Eq. 8.13 to excitations that consist solely of a single Dirac delta function pulse. As noted in Section 8.1 the lth column of $\mathbf{h}(t)$, written as $[h_{1l}(t),\cdots,h_{nl}(t)]^T$, is the $\vec{X}(t)$ response vector when the only excitation is $\vec{F}_l(t)=\delta(t)$. We can now use Eq. 8.23 or 8.24 to obtain the excitation of the jth modal equation as $\theta_{lj}\,\delta(t)/\gamma_{jj}$. Thus, one finds that the modal responses to this Dirac delta function excitation at node l are $Z_j(t)=\theta_{lj}\,\hat{h}_{jj}(t)$ in which $\hat{h}_{jj}(t)$ represents the impulse response function of the jth modal equation, and from our study of the SDF system this is known to be

$$\hat{h}_{jj}(t) = \frac{1}{\hat{m}_{jj}\,\omega_{dj}} e^{-\zeta_j \omega_j t} \sin(\omega_{dj} t)\, U(t) \tag{8.26}$$

with the damped modal frequency given by $\omega_{dj} \equiv \omega_j [1-(\zeta_j)^2]^{1/2}$. Note that $\hat{m}_{jj}$ in this equation is analogous to the mass m in the SDF formula of Example 5.3. Thus, $\hat{m}_{jj}$ can be considered the modal mass. Using these modal impulse responses in Eq. 8.20 gives any element of column l of $\mathbf{h}(t)$ as

$$h_{rl}(t) = X_r(t) = \sum_{j=1}^{n} \theta_{rj}\, Z_j(t) = \sum_{j=1}^{n} \theta_{rj} \theta_{lj}\, \hat{h}_{jj}(t) \tag{8.27}$$

From this equation for a typical element of the $\mathbf{h}(t)$ matrix, we can write the entire relationship in matrix form as

$$\mathbf{h}(t) = \boldsymbol{\theta}\, \hat{\mathbf{h}}(t)\, \boldsymbol{\theta}^T \tag{8.28}$$

in which $\hat{\mathbf{h}}(t)$ denotes a diagonal matrix of the modal impulse response functions given in Eq. 8.26.

These equations show that it is quite straightforward to obtain the MDF impulse response function matrix from the analysis of uncoupled equations. The largest computational effort in the procedure is the eigenanalysis to find $\boldsymbol{\theta}$ and $\boldsymbol{\lambda}$. The other operations involve only arithmetic. Note that our original MDF equation allowed the possibility of an exciting force at each of the n coordinate points, and we have chosen to include all of the coordinates in our definition of $\mathbf{h}(t)$. Obviously this gives a square $\mathbf{h}(t)$ matrix. If we wish to perform our stochastic analysis only for a subset of the coordinate points, then we can consider r in Eq. 8.27 to vary only over that subset, thereby obtaining a rectangular $\mathbf{h}(t)$ matrix appropriate for those points. Once the $\mathbf{h}(t)$ matrix is determined, the mean-value vector for the stochastic response can be found from Eq. 8.5, the autocorrelation matrix is given by Eq. 8.7 or 8.9, and the autocovariance matrix is given by Eq. 8.8 or 8.10.

Now we will consider the alternative of doing stochastic analysis directly on the equations of Section 8.3, rather than using those equations to obtain the $\mathbf{h}(t)$ matrix. From Eq. 8.20 we can write the mean-value vector of the response as $\vec{\mu}_X(t) = \boldsymbol{\theta}\,\vec{\mu}_Z(t)$, the autocorrelation matrix as $\boldsymbol{\phi}_{XX}(t,s) = \boldsymbol{\theta}\,\boldsymbol{\phi}_{ZZ}(t,s)\,\boldsymbol{\theta}^T$, and the autocovariance matrix as $\mathbf{K}_{XX}(t,s) = \boldsymbol{\theta}\,\mathbf{K}_{ZZ}(t,s)\,\boldsymbol{\theta}^T$. The time histories of the components of the stochastic $\vec{Z}(t)$ vector of modal responses are obtained from Eq. 8.23 as

$$Z_j(t) = \sum_{k=1}^{n} \theta_{kj} \int_{-\infty}^{\infty} \hat{h}_{jj}(t-s)\, F_k(s)\, ds \tag{8.29}$$

or in matrix form as

$$\vec{Z}(t) = \int_{-\infty}^{\infty} \hat{\mathbf{h}}(t-u)\, \boldsymbol{\theta}^T \vec{F}(u)\, du \tag{8.30}$$

Thus, the mean-value vector for the response is

$$\vec{\mu}_X(t) = \boldsymbol{\theta} \vec{\mu}_Z(t) = \int_{-\infty}^{\infty} \boldsymbol{\theta} \hat{\mathbf{h}}(t-u)\, \boldsymbol{\theta}^T \vec{\mu}_F(u)\, du \tag{8.31}$$

and the autocorrelation matrix for $\{\vec{Z}(t)\}$ is

$$\boldsymbol{\phi}_{ZZ}(t,s) = \int_{-\infty}^{\infty}\int_{-\infty}^{\infty} \hat{\mathbf{h}}(t-u)\, \boldsymbol{\theta}^T \boldsymbol{\phi}_{FF}(u,v)\, \boldsymbol{\theta} \hat{\mathbf{h}}^T(s-v)\, du\, dv \tag{8.32}$$

which gives the corresponding result for $\{\vec{X}(t)\}$ as

$$\boldsymbol{\phi}_{XX}(t,s) = \boldsymbol{\theta} \int_{-\infty}^{\infty}\int_{-\infty}^{\infty} \hat{\mathbf{h}}(t-u)\, \boldsymbol{\theta}^T \boldsymbol{\phi}_{FF}(u,v)\, \boldsymbol{\theta} \hat{\mathbf{h}}^T(s-v)\, du\, dv\, \boldsymbol{\theta}^T \tag{8.33}$$

Similarly the autocovariance matrices are

$$\mathbf{K}_{ZZ}(t,s) = \int_{-\infty}^{\infty}\int_{-\infty}^{\infty} \hat{\mathbf{h}}(t-u)\, \boldsymbol{\theta}^T \mathbf{K}_{FF}(u,v)\, \boldsymbol{\theta} \hat{\mathbf{h}}^T(s-v)\, du\, dv \tag{8.34}$$

and

$$\mathbf{K}_{XX}(t,s) = \boldsymbol{\theta} \int_{-\infty}^{\infty}\int_{-\infty}^{\infty} \hat{\mathbf{h}}(t-u)\, \boldsymbol{\theta}^T \mathbf{K}_{FF}(u,v)\, \boldsymbol{\theta} \hat{\mathbf{h}}^T(s-v)\, du\, dv\, \boldsymbol{\theta}^T \tag{8.35}$$

It must be remembered that the expressions in Eqs. 8.32–8.35 are called autocorrelation and autocovariance matrices because each refers to only one response vector process, either $\{\vec{Z}(t)\}$ or $\{\vec{X}(t)\}$. At the same time, most of the elements of these $\boldsymbol{\phi}$ and $\mathbf{K}$ matrices are cross-correlation and cross-covariance terms of the scalar components of the response vectors. Only the diagonal terms of the matrices are autocorrelation and autocovariance terms of the scalar components. For example, the typical term of Eq. 8.34 can be written as $[K_{ZZ}(t,s)]_{jl} \equiv K_{Z_j(t)Z_l(s)}$.

One can learn much about the relative importance of the various modal contributions to the response by considering the special case of response at time t due to a white noise excitation process $\{\vec{F}(t)\}$. In particular, we will consider the case with $\vec{F}(t)$ and $\vec{F}(s)$ being uncorrelated vectors for $t \neq s$, so that $\mathbf{K}_{FF}(t,s) = 2\pi \mathbf{S}_0\, \delta(t-s)$, in which the autospectral density matrix $\mathbf{S}_0$ is generally a full square matrix. This allows the various components of $\vec{F}(t)$ to be correlated at any one instant of time t, even though they are uncorrelated at distinct times. Substituting this relationship into Eq. 8.35 for the case of $t=s$ gives the matrix of response variance and covariance values as

$$\mathbf{K}_{XX}(t,t) = 2\pi \boldsymbol{\theta} \int_{-\infty}^{\infty} \hat{\mathbf{h}}(t-u)\, \boldsymbol{\theta}^T \mathbf{S}_0\, \boldsymbol{\theta}\, \hat{\mathbf{h}}^T(t-u)\, du\, \boldsymbol{\theta}^T$$

so the typical component can be written as

$$K_{X_j(t)X_l(t)} = 2\pi \sum_{r_1=1}^{n} \sum_{r_2=1}^{n} \sum_{r_3=1}^{n} \sum_{r_4=1}^{n} \theta_{jr_1} \theta_{r_2 r_1} [S_0]_{r_2 r_3} \theta_{r_3 r_4} \theta_{l r_4}\, q_{r_1 r_4} \tag{8.36}$$

in which

$$q_{r_1 r_4} \equiv \int_{-\infty}^{\infty} \hat{h}_{r_1 r_1}(t-u)\, \hat{h}_{r_4 r_4}(t-u)\, du = \int_0^{\infty} \hat{h}_{r_1 r_1}(v)\, \hat{h}_{r_4 r_4}(v)\, dv$$

One can now substitute Eq. 8.26 into this integrand and perform the integration to obtain

$$q_{r_1 r_4} = \frac{2(\zeta_{r_1}\omega_{r_1} + \zeta_{r_4}\omega_{r_4})}{\hat{m}_{r_1 r_1} \hat{m}_{r_4 r_4} [(\omega_{r_1}^2 - \omega_{r_4}^2)^2 + 4\omega_{r_1}\omega_{r_4}(\zeta_{r_1}\omega_{r_1} + \zeta_{r_4}\omega_{r_4})(\zeta_{r_1}\omega_{r_4} + \zeta_{r_4}\omega_{r_1})]} \tag{8.37}$$

Setting $r_4 = r_1$ in this expression gives the result for the special case as

$$q_{r_1 r_1} = \frac{1}{4\, \hat{m}_{r_1 r_1}^2\, \zeta_{r_1}\, \omega_{r_1}^3} \tag{8.38}$$

which agrees exactly with the SDF result in Eq. 5.63.

Note that Eq. 8.38 is proportional to modal damping raised to the –1 power, so it is quite large for the commonly considered systems with small damping values. Furthermore, Eq. 8.37 is of the order of modal damping to the +1 power, provided that ω_{r_1} and ω_{r_4} are well separated. Thus, if all modal frequencies are well separated, then the "off-diagonal" terms with $r_4 \neq r_1$ in Eq. 8.36 are much smaller than are the "on-diagonal" terms with $r_4 = r_1$. In particular, the ratio between the off-diagonal and on-diagonal terms is of order damping squared. For modal damping values of less than 10%, for instance, one can expect the off-diagonal terms to contribute relatively little to the covariance of the $\vec{X}(t)$ responses. Thus, in many situations it is possible to neglect the terms from Eq. 8.36 with $r_4 \neq r_1$ and approximate Eq. 8.35 by considering only the $r_4 = r_1$ terms, giving

$$K_{X_j(t)X_l(s)} \approx \sum_{r_1=1}^{n} \sum_{r_2=1}^{n} \sum_{r_3=1}^{n} \theta_{jr_1}\theta_{r_2r_1}[\mathbf{K}_{FF}(u,v)]_{r_2r_3}\theta_{r_3r_1}\theta_{lr_1}\, q_{r_1r_1} \qquad (8.39)$$

Furthermore, one notes from Eq. 8.38 that the $q_{r_1r_1}$ modal contributions are proportional to the modal frequency raised to the –3 power, so high-frequency modes will generally contribute much less to the response than will low-frequency modes. This leads to the possibility of further approximating Eq. 8.39 by limiting the range of r_1 to include only the lower-frequency modes.

Although the approximations in the preceding paragraph have been justified by consideration of the response to white noise, they can also be considered appropriate for other broadband excitations. It is important to notice that if two modes have nearly equal frequencies, then the approximation in Eq. 8.39 is not valid. For example, if $\omega_l - \omega_j$ is of the order of the damping, then the denominator in Eq. 8.37 is of the order of damping squared, so q_{jl} may be equally as significant as q_{jj} and q_{ll}. Thus, one must exercise some caution in deciding whether to use the simplification of Eq. 8.39. For situations with closely spaced modal frequencies, it is sometimes convenient to rewrite Eqs. 8.37 and 8.38 in the form of the correlation coefficient for the modal responses: $\rho_{r_1r_4} = q_{r_1r_4}(q_{r_1r_1}\, q_{r_4r_4})^{-1/2}$. Numerical investigations (Der Kiureghian, 1980) have shown that this correlation coefficient is approximately correct for nonwhite broadband excitations, and this has formed the basis for a *complete-quadratic-combination* (CQC) method for accurately computing the response of MDF systems, even if they have closely spaced modal frequencies (Wilson et al., 1981). It should also be noted that some continuous structures are very likely to have closely spaced modal frequencies. Elishakoff (1995) has particularly noted this fact for shells. The cross-correlation of modal responses should generally be expected to be important in these situations.

****

Example 8.5: Find the $\mathbf{h}(t)$ impulse response function matrix for the 2DF model of Example 8.2, with $m = 500$ kg, $k = 500$ kN/m, and $c = 500$ N/(m/s), for which

$$\mathbf{m} = \begin{pmatrix} 1000 & 0 \\ 0 & 500 \end{pmatrix} \text{kg}, \quad \mathbf{k} = \begin{pmatrix} 10^6 & -5\times 10^5 \\ -5\times 10^5 & 5\times 10^5 \end{pmatrix} \frac{\text{N}}{\text{m}}, \quad \mathbf{c} = \begin{pmatrix} 1000 & -500 \\ -500 & 500 \end{pmatrix} \frac{\text{N}\cdot\text{s}}{\text{m}}$$

It was found in Example 8.2 that this system has uncoupled modes, and the results there give $\hat{m}_{11} = \hat{m}_{22} = 2000$ kg, $\omega_1 = 17.11$ rad/s, $\omega_2 = 41.32$ rad/s, $\zeta_1 = 0.00856$, and $\zeta_2 = 0.02066$. Thus, Eq. 8.26 gives

$$\hat{\mathbf{h}}(t) = \frac{1}{2000} \begin{pmatrix} \dfrac{e^{-0.1464t}\sin(17.11t)}{17.11} & 0 \\ 0 & \dfrac{e^{-0.8536t}\sin(41.31t)}{41.31} \end{pmatrix} U(t)$$

Using Eq. 8.28 with

$$\mathbf{\theta} = \begin{pmatrix} 1 & 1 \\ 1.414 & -1.414 \end{pmatrix}$$

then gives the elements of $\mathbf{h}(t)$ as

$$h_{11}(t) = \theta_{11}^2 \hat{h}_{11}(t) + \theta_{12}^2 \hat{h}_{22}(t)$$
$$= [2.922\times 10^{-5} e^{-0.1464t}\sin(17.11t) + 1.210\times 10^{-5} e^{-0.8536t}\sin(41.31t)]U(t)$$
$$h_{12}(t) = h_{21}(t) = \theta_{11}\theta_{21} \hat{h}_{11}(t) + \theta_{12}\theta_{22} \hat{h}_{22}(t)$$
$$= [4.132\times 10^{-5} e^{-0.1464t}\sin(17.11t) - 1.712\times 10^{-5} e^{-0.8536t}\sin(41.31t)]U(t)$$
$$h_{22}(t) = h_{21}(t) = \theta_{21}^2 \hat{h}_{11}(t) + \theta_{22}^2 \hat{h}_{22}(t)$$
$$= [5.843\times 10^{-5} e^{-0.1464t}\sin(17.11t) + 2.421\times 10^{-5} e^{-0.8536t}\sin(41.31t)]U(t)$$

****

Example 8.6: Find $E(X_1^2)$ for the stationary response of the oscillator shown here when the base acceleration $\{\ddot{Y}(t)\}$ is mean-zero white noise with autospectral density $S_0 = 0.1$ (m/s²)/(rad/s).

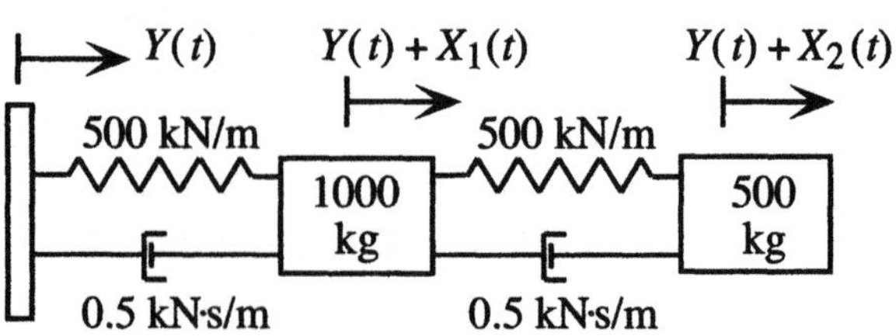

The system is described by Eq. 8.13 with $\mathbf{m}$, $\mathbf{k}$, and $\mathbf{c}$ as given in Example 8.5, $F_1(t) = -m_{11}\ddot{Y}(t)$, and $F_2(t) = -m_{22}\ddot{Y}(t)$. One approach is to write

$$X_1(t) = -\int_0^\infty [m_{11}h_{11}(\tau) + m_{22}h_{12}(\tau)]\ddot{Y}(t-\tau)\,d\tau$$

Using $E[\ddot{Y}(u)\ddot{Y}(v)] = 2\pi S_0\,\delta(u-v)$ and substituting the $h_{11}(s)$ and $h_{12}(s)$ impulse response functions found in Example 8.5 then gives

$$E(X_1^2) = 2\pi S_0 \int_0^\infty [m_{11}h_{11}(s) + m_{22}h_{12}(s)]^2\,ds$$

$$= 2\pi S_0 \left(2.488\times 10^{-3} \int_0^\infty e^{-0.2929 s} \sin^2(17.11 s)\,ds + \right.$$

$$3.536\times 10^{-4} \int_0^\infty e^{-1.0 s} \sin(17.11 s)\sin(41.31 s)\,ds +$$

$$\left. 1.257\times 10^{-5} \int_0^\infty e^{-1.707 s} \sin^2(41.31 s)\,ds \right)$$

Performing the integration then reduces this to

$$E(X_1^2) = 2\pi S_0 (4.246\times 10^{-3} + 2.498\times 10^{-7} + 3.680\times 10^{-6})$$

$$= 2.671\times 10^{-3}\ \text{m}^2$$

Note that the first term in this form is due to the first mode of the system, the second term is cross-modal, and the final term is due to the second mode. Clearly the cross-modal term and the second mode term are both much smaller than the first mode term, as discussed in conjunction with Eq. 8.39. The insignificance of the second mode is exaggerated in this example by the fact that the excitation has $F_1(t)$ in phase with $F_2(t)$, and this distribution of forces is not effective in exciting the second mode, which has $X_1(t)$ 180° out of phase with $X_2(t)$.

Alternatively, one can investigate the mean-squared modal responses, then use these in finding $E(X_1^2)$. For the excitation given, the modal equations of Eq. 8.21 can be written as

$$\hat{\mathbf{m}}\ddot{\vec{Z}}(t) + \hat{\mathbf{m}}\boldsymbol{\beta}\dot{\vec{Z}}(t) + \hat{\mathbf{m}}\boldsymbol{\lambda}\vec{Z}(t) = -\boldsymbol{\theta}^T \begin{pmatrix} m_{11} \\ m_{22} \end{pmatrix} \ddot{Y}(t) = -\begin{pmatrix} 1707 \\ 292.9 \end{pmatrix} \ddot{Y}(t)$$

and this gives the modal responses as

$$E(Z_1^2) = \frac{\pi S_0}{2\hat{m}_{11}^2 \zeta_1 \omega_1^3}(1707)^2 = (2\pi S_0)(4.246\times 10^{-3})$$

and

$$E(Z_2^2) = \frac{\pi S_0}{2\hat{m}_{22}^2 \zeta_2 \omega_2^3}(292.9)^2 = (2\pi S_0)(3.680\times 10^{-6})$$

The cross-modal response from Eq. 8.37 is

$$E(Z_1 Z_2) = \frac{4\pi S_0(\zeta_1\omega_1 + \zeta_2\omega_2)(1707)(292.9)}{\hat{m}_{11}\hat{m}_{22}[(\omega_1^2 - \omega_2^2)^2 + 4\omega_1\omega_2(\zeta_1\omega_1 + \zeta_2\omega_2)(\zeta_1\omega_2 + \zeta_2\omega_1)]}$$
$$= (2\pi S_0)(1.249\times 10^{-7})$$

Using the fact that $X_1(t) = \theta_{11} Z_1(t) + \theta_{12} Z_2(t)$ then gives

$$E(X_1^2) = \theta_{11}^2 E(Z_1^2) + 2\theta_{11}\theta_{12} E(Z_1 Z_2) + \theta_{12}^2 E(Z_2^2)$$
$$= (2\pi S_0)(4.246\times 10^{-3} + 2.498\times 10^{-7} + 3.680\times 10^{-6})$$
$$= 2.671\times 10^{-3}\ \mathrm{m}^2$$

Note that exactly the same individual terms are summed in the two approaches to solving the problem. The difference is simply in the order of performance of the operations.

****

Example 8.7: Consider the system of Example 8.2 for the special case in which $F_2(t) \equiv 0$ and $F_1(t)$ is mean-zero white noise with an autospectral density of S_0. Find the autocorrelation function of the $\{X_1(t)\}$ response component. That is, find $R_{X_1X_1}(\tau) \equiv E[X_1(t+\tau)X_1(t)]$.

Using modal superposition, we note that $\mathbf{R}_{XX}(\tau) = \boldsymbol{\theta}\mathbf{R}_{ZZ}(\tau)\boldsymbol{\theta}^T$, so

$$R_{X_1X_1}(\tau) = \theta_{11}^2 R_{Z_1Z_1}(\tau) + \theta_{11}\theta_{12}[R_{Z_1Z_2}(\tau) + R_{Z_2Z_1}(\tau)] + \theta_{12}^2 R_{Z_2Z_2}(\tau)$$

The uncoupled modal equations of motion can be written as

$$\ddot{\vec{Z}}(t) + \boldsymbol{\beta}\dot{\vec{Z}}(t) + \boldsymbol{\lambda}\vec{Z}(t) = \hat{\mathbf{m}}^{-1}\boldsymbol{\theta}^T \vec{F}(t)$$

or

$$\ddot{Z}_1(t) + \beta_{11}\dot{Z}_1(t) + \lambda_{11} Z_1(t) = \theta_{11} F_1(t)/\gamma_{11}$$
$$\ddot{Z}_2(t) + \beta_{22}\dot{Z}_2(t) + \lambda_{22} Z_2(t) = \theta_{12} F_1(t)/\gamma_{22}$$

The responses of these equations can be written by Duhamel integrals as

$$Z_1(t) = \frac{\theta_{11}}{\hat{m}_{11}}\int_0^\infty F_1(t-s)\frac{e^{-\zeta_1\omega_1 s}}{\omega_{d1}}\sin(\omega_{d1} s)\,ds$$

$$Z_2(t) = \frac{\theta_{12}}{\hat{m}_{22}}\int_0^\infty F_1(t-s)\frac{e^{-\zeta_2\omega_2 s}}{\omega_{d2}}\sin(\omega_{d2} s)\,ds$$

The $R_{Z_1Z_1}(\tau)$ and $R_{Z_2Z_2}(\tau)$ terms are easily found, because they are essentially the same as the autocorrelation function of the response of an SDF system. The cross-modal terms, though, are new and must be evaluated. Multiplying appropriate versions of the two modal response integrals and taking the expected value gives

$$R_{Z_1Z_2}(\tau) \equiv E[Z_1(t+\tau)\,Z_2(t)] = \frac{\theta_{11}\,\theta_{12}}{\hat{m}_{11}\,\hat{m}_{22}} \int_0^\infty \int_0^\infty R_{F_1F_1}(t+\tau-s_1, t-s_2) \times$$
$$\frac{e^{-\zeta_1\omega_1 s_1} e^{-\zeta_2\omega_2 s_2}}{\omega_{d1}\,\omega_{d2}} \sin(\omega_{d1} s_1) \sin(\omega_{d2} s_2)\, ds_1\, ds_2$$

Substituting the relationship $R_{F_1F_1}(t+\tau-s_1, t-s_2) = 2\pi S_0\, \delta(\tau - s_1 + s_2)$ for white noise now reduces this to a single integral:

$$R_{Z_1Z_2}(\tau) = \frac{\theta_{11}\,\theta_{12}}{\hat{m}_{11}\,\hat{m}_{12}} \frac{2\pi S_0}{\omega_{d1}\,\omega_{d2}} e^{-\zeta_1\omega_1\tau} \int_{\max(0,-\tau)}^\infty e^{-(\zeta_1\omega_1+\zeta_2\omega_2)s_2} \times$$
$$\sin[\omega_{d1}(\tau+s_2)] \sin(\omega_{d2} s_2)\, ds_2$$

For $\tau > 0$, the result of integration and simplification can be written as

$$R_{Z_1Z_2}(\tau) = \frac{\theta_{11}\,\theta_{12}}{\hat{m}_{11}\,\hat{m}_{22}} \frac{2\pi S_0}{\omega_{d1}} e^{-\zeta_1\omega_1\tau} \times$$
$$\frac{2\omega_{d1}(\zeta_1\,\omega_1+\zeta_2\,\omega_2)\cos(\omega_{d1}\tau) + [(\zeta_1\,\omega_1+\zeta_2\,\omega_2)^2 + \omega_{d2}^2 - \omega_{d1}^2]\sin(\omega_{d1}\tau)}{(\omega_1^2-\omega_2^2)^2 + 4\omega_1\omega_2(\zeta_1\,\omega_1+\zeta_2\,\omega_2)(\zeta_1\,\omega_2+\zeta_2\,\omega_1)}$$

while $\tau < 0$ gives

$$R_{Z_1Z_2}(\tau) = \frac{\theta_{11}\,\theta_{12}}{\hat{m}_{11}\,\hat{m}_{22}} \frac{2\pi S_0}{\omega_{d2}} e^{\zeta_2\omega_2\tau} \times$$
$$\frac{2\omega_{d2}(\zeta_1\,\omega_1+\zeta_2\,\omega_2)\cos(\omega_{d2}\tau) - [(\zeta_1\,\omega_1+\zeta_2\,\omega_2)^2 + \omega_{d1}^2 - \omega_{d2}^2]\sin(\omega_{d2}\tau)}{(\omega_1^2-\omega_2^2)^2 + 4\omega_1\omega_2(\zeta_1\,\omega_1+\zeta_2\,\omega_2)(\zeta_1\,\omega_2+\zeta_2\,\omega_1)}$$

It should now be noted that stationarity allows us to write

$$R_{Z_2Z_1}(\tau) \equiv E[Z_2(t+\tau)Z_1(t)] = E[Z_2(t)Z_1(t-\tau)] = R_{Z_1Z_2}(-\tau)$$

so $R_{Z_2Z_1}(\tau)$ can be found from the expressions already derived.

By modifying Eq. 5.60 to apply to our modal equations, we find the other two necessary terms as

$$R_{Z_1Z_1}(\tau) = \frac{\theta_{11}^2}{\hat{m}_{11}^2} \frac{\pi S_0}{2\zeta_1\,\omega_1^3} e^{-\zeta_1\omega_1|\tau|} \left(\cos(\omega_{d1}\tau) + \frac{\zeta_1\,\omega_1}{\omega_{d1}} \sin(\omega_{d1}\,|\tau|) \right)$$

and

$$R_{Z_2Z_2}(\tau) = \frac{\theta_{12}^2}{\hat{m}_{22}^2} \frac{\pi S_0}{2\zeta_2\,\omega_2^3} e^{-\zeta_2\omega_2|\tau|} \left(\cos(\omega_{d2}\tau) + \frac{\zeta_2\,\omega_2}{\omega_{d2}} \sin(\omega_{d2}\,|\tau|) \right)$$

so $R_{X_1X_1}(\tau)$ can be found by superposition.

If damping values are small and the modal frequencies are well separated, it is unlikely that $R_{Z_1Z_2}(\tau)$ and $R_{Z_2Z_1}(\tau)$ will contribute significantly to the value of $R_{X_1X_1}(\tau)$. The argument is the same as was given in regard to Eqs. 8.37 and 8.38. Namely, the $R_{Z_1Z_2}(\tau)$ and $R_{Z_2Z_1}(\tau)$ terms are of the order of damping, whereas $R_{Z_1Z_1}(\tau)$ and $R_{Z_2Z_2}(\tau)$ are much larger because they are of the order of damping to the –1 power. Thus, one may often neglect the cross-modal terms, which is equivalent to assuming that the modal responses are independent of each other. Furthermore, if one modal frequency is much larger than the other then the presence of the frequency cubed term in the denominators of $R_{Z_1Z_1}(\tau)$ and $R_{Z_2Z_2}(\tau)$ may cause the higher-frequency mode to contribute very little to the autocorrelation function, just as was previously noted for the covariance matrix for any single value of time.

****

8.5 Frequency-Domain Analysis of MDF Systems

As with time-domain analysis, there are two ways to proceed with the frequency-domain analysis of MDF systems. In the first approach one uses deterministic analysis of the MDF equation of motion to obtain the $\mathbf{H}(\omega)$ harmonic transfer matrix and then uses the equations of Section 8.1 for the stochastic analysis. The second approach consists of direct stochastic analysis of the MDF equations. Within the first approach, though, we will consider two different possible techniques for finding $\mathbf{H}(\omega)$.

If the MDF system has uncoupled modes, one can first use Eq. 8.24 to find an $\hat{\mathbf{H}}(\omega)$ harmonic transfer matrix describing the modal responses to a harmonic excitation and then use that in finding an $\mathbf{H}(\omega)$ that describes the $\vec{X}(t)$ response vector. In particular, using an excitation that consists of only one harmonic component $F_l(t) = e^{i\omega t}$ in Eq. 8.24 gives

$$\ddot{Z}_j(t) + 2\zeta_j \omega_j \dot{Z}_j(t) + \omega_j^2 Z_j(t) = \frac{\theta_{lj}}{\hat{m}_{jj}} e^{i\omega t}$$

and the fact that the response is defined to be $Z_j(t) = \hat{H}_{jl}(\omega)\, e^{i\omega t}$ gives the (j,l) element of the harmonic transfer matrix as

$$\hat{H}_{jl}(\omega) = \frac{\theta_{lj}}{\hat{m}_{jj}(\omega_j^2 - \omega^2 + 2i\zeta_j \omega_j \omega)}$$

Alternatively, this can be written in matrix form as

$$\hat{\mathbf{H}}(\omega) = \hat{\mathbf{m}}^{-1}[\boldsymbol{\lambda} - \omega^2\,\mathbf{I} + i\,\omega\boldsymbol{\beta}]^{-1}\boldsymbol{\theta}^T \tag{8.40}$$

in which the inverse operations are essentially trivial, because they are for diagonal matrices. One can now use Eq. 8.20 to find the $\vec{X}(t)$ responses to the single harmonic excitation component, thereby obtaining the $\mathbf{H}(\omega)$ matrix as

$$\mathbf{H}(\omega) = \boldsymbol{\theta}\hat{\mathbf{H}}(\omega)\boldsymbol{\theta}^T = \boldsymbol{\theta}\hat{\mathbf{m}}^{-1}[\boldsymbol{\lambda} - \omega^2\,\mathbf{I} + i\,\omega\boldsymbol{\beta}]^{-1}\boldsymbol{\theta}^T \tag{8.41}$$

One can also derive the $\mathbf{H}(\omega)$ matrix without using the modal equations. In particular, one can take the Fourier transform of Eq. 8.13, giving

$$[\mathbf{k} - \omega^2\mathbf{m} + i\,\omega\,\mathbf{c}]\,\tilde{\vec{X}}(\omega) = \tilde{\vec{F}}(\omega)$$

then solve this equation for

$$\tilde{\vec{X}}(\omega) = [\mathbf{k} - \omega^2\mathbf{m} + i\,\omega\,\mathbf{c}]^{-1}\,\tilde{\vec{F}}(\omega)$$

Comparing this with Eq. 8.4 shows that

$$\mathbf{H}(\omega) = [\mathbf{k} - \omega^2\mathbf{m} + i\,\omega\,\mathbf{c}]^{-1} \tag{8.42}$$

A little matrix manipulation shows that Eqs. 8.41 and 8.42 are exactly equivalent. In fact, Eq. 8.41 can be considered the version of Eq. 8.42 that results from using eigenanalysis to simplify the problem of inverting the matrix. In Eq. 8.41 the inversion is an almost trivial operation inasmuch as the matrix is diagonal, whereas the inversion in Eq. 8.42 is of a general square matrix and will generally be done numerically unless the dimension n is quite small. Use of Eq. 8.41, however, involves significant computation in the determination of $\boldsymbol{\theta}$ and $\boldsymbol{\lambda}$ from eigenanalysis. If one wants $\mathbf{H}(\omega)$ for only a single frequency, there is no particular advantage of one approach over the other. In general, though, one wants to know $\mathbf{H}(\omega)$ for many frequencies, and in this situation Eq. 8.41 is more efficient because the eigenanalysis needs to be performed only once, whereas the matrix to be inverted in Eq. 8.42 is different for each frequency value. If n is sufficiently small, one can perform the inversion in Eq. 8.42 analytically as a function of ω, and this is a very practical approach to solving the problem.

There is one major advantage to Eq. 8.42, as compared with Eq. 8.41; namely, Eq. 8.42 does not require the existence of uncoupled modes. Thus, one can use this approach for almost any $\mathbf{m}$, $\mathbf{c}$, and $\mathbf{k}$ matrices. Exceptions do exist, such as when ω is equal to a modal frequency of $\mathbf{m}^{-1}\mathbf{k}$ and $\mathbf{c}$ is such that it gives no damping in that particular mode; but these pathological cases are not of great practical interest.

One should note that Eq. 8.42 also gives us an alternative method for finding the matrix $\mathbf{h}(t)$ of impulse response functions. In particular, the inverse Fourier transform of Eq. 8.42 gives

$$\mathbf{h}(t) = \frac{1}{2\pi}\int_{-\infty}^{\infty} [\mathbf{k} - \omega^2 \mathbf{m} + i\,\omega\,\mathbf{c}]^{-1} e^{i\omega t} d\omega \tag{8.43}$$

If the system of interest has uncoupled modes, then Eq. 8.43 will generally not be as efficient as the procedure in Section 8.4; however, Eq. 8.43 does provide a possible method for finding $\mathbf{h}(t)$ for a system that does not have uncoupled modes. A more commonly used alternative method will be presented in Section 8.6.

In Section 8.4, we use the properties of the impulse response functions to show that terms involving the interaction of modes often contribute much less to response covariance than do terms involving only a single mode. Consideration of the harmonic transfer functions confirms this result and illustrates it in a rather graphic way. As in deriving Eqs. 8.36–8.38, we limit our attention to the special case of a stationary white noise excitation with $\mathbf{K}_{FF}(t,s) = 2\pi\mathbf{S}_0\,\delta(t-s)$. The autospectral density of the response from Eq. 8.11 is then given by

$$\begin{aligned}\mathbf{S}_{XX}(\omega) &= \mathbf{H}(\omega)\mathbf{S}_0\,\mathbf{H}^{T*}(\omega)\\ &= \boldsymbol{\theta}\,\hat{\mathbf{m}}^{-1}[\boldsymbol{\lambda} - \omega^2\,\mathbf{I} + i\,\omega\boldsymbol{\beta}]^{-1}\boldsymbol{\theta}^T\mathbf{S}_0\,\boldsymbol{\theta}[\boldsymbol{\lambda} - \omega^2\,\mathbf{I} - i\,\omega\boldsymbol{\beta}]^{-1}\hat{\mathbf{m}}^{-1}\boldsymbol{\theta}^T\end{aligned}$$

in which the final form has been obtained by use of Eq. 8.41. Using the notation $\lambda_{jj} = (\omega_j)^2$ and $\beta_{jj} = 2\zeta_j\omega_j$ then allows us to write the frequency domain equivalent of Eq. 8.36 as

$$S_{X_jX_l}(\omega) = \sum_{r_1=1}^{n}\sum_{r_2=1}^{n}\sum_{r_3=1}^{n}\sum_{r_4=1}^{n}\theta_{jr_1}\theta_{r_2r_1}[S_0]_{r_2r_3}\theta_{r_3r_4}\theta_{lr_4}\frac{w_{r_1}(\omega)\,w_{r_4}(-\omega)}{\hat{m}_{r_1r_1}\hat{m}_{r_4r_4}\lambda_{r_1r_1}\lambda_{r_4r_4}}$$

in which

$$w_{r_1}(\omega) = \frac{\lambda_{r_1r_1}}{\lambda_{r_1r_1} - \omega^2 + i\,\omega\,\beta_{r_1r_1}} = \left(1 - \frac{\omega^2}{\omega_{r_1}^2} + 2i\,\zeta_{r_1}\frac{\omega}{\omega_{r_1}}\right)^{-1}$$

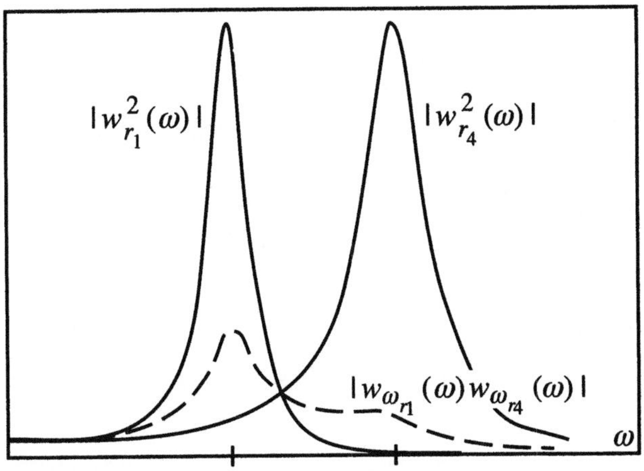

Figure 8.2 Cross-modal contributions to autospectral density.

Note that the $w_r(\omega)$ functions have the form of the harmonic transfer function for an SDF system, because they result from the modal harmonic responses. When damping is small, the absolute values of these functions have very narrow peaks, as was shown in Figure 6.6. When ω_{r_1} and ω_{r_4} are not nearly equal, as illustrated in Fig. 8.2, the product of $w_{r1}(\omega)$ and $w_{r4}(-\omega)$ must generally be very small in comparison with $|w_{r_1}(\omega)|^2$ and $|w_{r_4}(\omega)|^2$. Only if $\omega_{r_1}-\omega_{r_4}$ is of the order of the modal damping will the peaks overlap to such an extent that the contribution from $w_{r_1}(\omega)\,w_{r_4}(-\omega)$ will be significant. This confirms our finding in Section 8.4 that if the damping in the system is small and the modal frequencies are well separated, then the contributions of "cross-modal" terms are much smaller than contributions from single modes.

****

Example 8.8: Find the $\mathbf{H}(\omega)$ harmonic transfer matrix for the 2DF system of Example 8.5 with

$$\mathbf{m}=\begin{pmatrix}1000 & 0\\ 0 & 500\end{pmatrix}\text{kg},\quad \mathbf{k}=\begin{pmatrix}10^6 & -5\times10^5\\ -5\times10^5 & 5\times10^5\end{pmatrix}\frac{\text{N}}{\text{m}},\quad \mathbf{c}=\begin{pmatrix}1000 & -500\\ -500 & 500\end{pmatrix}\frac{\text{N}\cdot\text{s}}{\text{m}}$$

and investigate the pole locations for $\mathbf{H}(\omega)$.

From the results in Examples 8.2 and 8.5, we have

$$\boldsymbol{\theta}=\begin{pmatrix}1 & 1\\ 1.414 & -1.414\end{pmatrix},\quad \boldsymbol{\lambda}=\begin{pmatrix}292.9 & 0\\ 0 & 1707.\end{pmatrix}(\text{rad/s})^2$$

and this choice of $\boldsymbol{\theta}$ gives

$$\hat{\mathbf{m}} = \begin{pmatrix} 2000 & 0 \\ 0 & 2000 \end{pmatrix} \text{kg}, \qquad \boldsymbol{\beta} = \begin{pmatrix} 0.2929 & 0 \\ 0 & 1.707 \end{pmatrix} \text{rad/s}$$

Thus, we can obtain the harmonic transfer matrix from Eq. 8.41 as

$$\mathbf{H}(\omega) = \boldsymbol{\theta}\, \hat{\mathbf{m}}^{-1} \begin{pmatrix} 292.9 - \omega^2 + 0.2929\, i\, \omega & 0 \\ 0 & 1707 - \omega^2 + 1.707\, i\, \omega \end{pmatrix}^{-1} \boldsymbol{\theta}^T$$

The components are

$$H_{11}(\omega) = \frac{H_{22}(\omega)}{2} = \frac{5\times 10^{-4}}{292.9 - \omega^2 + 0.2929\, i\, \omega} + \frac{5\times 10^{-4}}{1707 - \omega^2 + 1.707\, i\, \omega}$$

and

$$H_{12}(\omega) = H_{21}(\omega) = \frac{7.071\times 10^{-4}}{292.9 - \omega^2 + 0.2929\, i\, \omega} - \frac{7.071\times 10^{-4}}{1707. - \omega^2 + 1.707\, i\, \omega}$$

Alternatively, one can use Eq. 8.42 without eigenanalysis to find

$$\mathbf{H}(\omega) = \begin{pmatrix} 10^6 - 1000\omega^2 + 1000 i\omega & -0.5(10^6 + 1000 i\omega) \\ -0.5(10^6 + 1000 i\omega) & 0.5(10^6 - 1000\omega^2 + 1000 i\omega) \end{pmatrix}^{-1}$$

$$= \frac{1}{D} \begin{pmatrix} 1000 - \omega^2 + i\omega & 1000 + i\omega \\ 1000 + i\omega & 2(1000 - \omega^2 + i\omega) \end{pmatrix}$$

in which the denominator is $D = 1000(\omega^4 - 2i\,\omega^3 - 2000.5\,\omega^2 + 1000 i\,\omega + 5\times 10^5)$, or $D = 1000(292.9 - \omega^2 + 0.2929 i\,\omega)(1707 - \omega^2 + 1.707 i\omega)$. The reader can confirm that the two versions of $\mathbf{H}(\omega)$ are identical.

Recall that one could choose to evaluate the $\mathbf{h}(t)$ impulse response function matrix by using the inverse Fourier transform of $\mathbf{H}(\omega)$. To do this, one would probably use the calculus of residues, giving the integral as a sum of terms coming from the poles of $\mathbf{H}(\omega)$; that is, from the values of complex ω at which $\mathbf{H}(\omega)$ is infinite. Obviously these pole locations are the solutions of $D = 0$. The four pole locations can be found by solving the two complex quadratic equations $\omega^2 - 0.2929 i\,\omega - 292.9 = 0$ and $\omega^2 - 1.707 i\,\omega - 1707 = 0$. The solutions are $\pm 17.11 + 0.1465 i$ and $\pm 41.31 + 0.8535 i$. These values are a particular example of the general relationship that the poles of $\mathbf{H}(\omega)$ are located at $\pm\omega_j [1 - (\zeta_j)^2]^{1/2} + i\,\zeta_j \omega_j$. It is observed that the real values of the pole locations are the damped natural frequencies of the system, and the imaginary values are one-half the β_j modal damping values. The fact that the imaginary parts are always positive is necessary in order that the system satisfy the causality condition of $\mathbf{h}(t) = \mathbf{0}$ for $t < 0$.

****

Example 8.9: Find the autospectral density matrix for the stationary response of the system of Examples 8.5, 8.6, and 8.8 with stationary, mean-zero, white noise base acceleration with $S_0 = 0.1\,(\text{m/s}^2)/(\text{rad/s})$.

The autospectral density matrix of the $\vec{F}(t)$ excitation process is

$$\mathbf{S}_{FF}(\omega) = S_0 \begin{pmatrix} m_{11}^2 & m_{11}m_{22} \\ m_{11}m_{22} & m_{22}^2 \end{pmatrix} = S_0 \begin{pmatrix} 10^6 & 5\times 10^5 \\ 5\times 10^5 & 2.5\times 10^5 \end{pmatrix}$$

so using $\mathbf{H}(\omega)$ from Example 8.8 gives the $S_{XX}(\omega) = \mathbf{H}(\omega)\, S_{FF}(\omega)\, \mathbf{H}^{*T}(\omega)$ response autospectral density from Eq. 8.11 as

$$\frac{S_0}{D^2}(10^6)\begin{pmatrix} \omega^4 - 2998\omega^2 + 2.25\times 10^6 & \omega^4 + 0.5i\omega^3 - 3497\omega^2 + 3\times 10^6 \\ \omega^4 + 0.5i\omega^3 - 3497\omega^2 + 3\times 10^6 & \omega^4 - 3996\omega^2 + 4\times 10^6 \end{pmatrix}$$

in which $D = 1000(\omega^4 - 2i\,\omega^3 - 2000.5\,\omega^2 + 1000i\,\omega + 5\times 10^5)$, as given in Example 8.8. Note that the value of $E(X_1^2)$ found in Example 8.6 could also be evaluated from the integral of $\mathbf{S}_{XX}(\omega)$ from $-\infty$ to ∞.

****

Example 8.10: Find the $\mathbf{H}(\omega)$ harmonic transfer matrix for the 2DF system of Example 8.3 with

$$\mathbf{m} = \begin{pmatrix} 1000 & 0 \\ 0 & 500 \end{pmatrix}\text{kg},\quad \mathbf{k} = \begin{pmatrix} 1000 & -500 \\ -500 & 500 \end{pmatrix}\frac{\text{kN}}{\text{m}},\quad \mathbf{c} = \begin{pmatrix} 2.5 & -2.0 \\ -2.0 & 2.0 \end{pmatrix}\frac{\text{kN}\cdot\text{s}}{\text{m}}$$

We know that this system does not have uncoupled modes, so we use Eq. 8.42 to obtain

$$\mathbf{H}(\omega) = [\mathbf{k} - \omega^2\mathbf{m} + i\,\omega\,\mathbf{c}]^{-1}$$

$$= 1000\begin{pmatrix} 1000 + 2.5\,i\,\omega - \omega^2 & -500 - 2\,i\,\omega \\ -500 - 2\,i\,\omega & 500 + 2.0\,i\,\omega - 0.5\,\omega^2 \end{pmatrix}^{-1}$$

$$= \frac{1}{D}\begin{pmatrix} 1000 + 4\,i\,\omega - \omega^2 & 1000 + 4\,i\,\omega \\ 1000 + 4\,i\,\omega & 2000 + 5\,i\,\omega - 2\omega^2 \end{pmatrix}$$

in which $D = 1000(5\times 10^5 + 2500\,i\,\omega - 2002\,\omega^2 - 6.5\,i\,\omega^3 + \omega^4)$.

Note that the $\mathbf{h}(t)$ impulse response function matrix can now be found as the inverse Fourier transform of $\mathbf{H}(\omega)$. Solving $D = 0$ gives the location of the four poles of $\mathbf{H}(\omega)$ as $\omega = \pm 17.12 + 0.2106i$ and $\omega = \pm 41.20 + 3.039i$, and the inverse Fourier transform integral can be evaluated by the calculus of residues. Rather than performing this operation, we will derive the impulse response functions for this system by an alternative method in Example 8.14.

****

8.6 State-Space Formulation of Equations of Motion

We will now introduce a formulation that can be used for the MDF system but that can also be used for more general problems. In particular, it is possible to write the equations of motion for any linear system of order n_Y as a set of n_Y first-order differential equations, or a single first-order differential equation for a vector with n_Y components:

$$\mathbf{A}\dot{\vec{Y}}(t)+\mathbf{B}\vec{Y}(t)=\vec{Q}(t) \tag{8.44}$$

in which $\vec{Y}(t)$ contains only expressions describing the response, **A** and **B** are matrices determined from the coefficients in the original equations of motion, and the $\vec{Q}(t)$ vector involves only excitation terms in the original equations of motion. The $\vec{Y}(t)$ vector is called the *state vector*, and its components are called *state variables*. Knowledge of $\vec{Y}(t_0)$ for some particular time t_0 gives a complete set of initial conditions for finding a unique solution for $\vec{Y}(t)$ for $t>t_0$.

For a problem in which we begin with a coupled set of equations of motion that involve J variables $\{X_1(t),\cdots,X_J(t)\}$ with derivatives up to order n_j in the variable $X_j(t)$, the order of the system, and the dimension of the arrays in Eq. 8.46, will be

$$n_Y=\sum_{j=1}^{J} n_j \tag{8.45}$$

In this case, the components of the state vector $\vec{Y}(t)$ will generally be taken as $X_j(t)$ and its first n_j-1 derivatives, for $j=1,\cdots,J$. For example, the MDF system of Eq. 8.13 with matrices of dimension $n\times n$ is of order $n_Y=2n$, and the state variables are usually taken as the components of $\vec{X}(t)$ and $\dot{\vec{X}}(t)$: $\vec{Y}^T(t)=[\vec{X}^T(t),\dot{\vec{X}}^T(t)]$. One way of converting Eq. 8.13 into the form of Eq. 8.44 is to rewrite it as $\ddot{\vec{X}}(t)+\mathbf{m}^{-1}\mathbf{c}\,\dot{\vec{X}}(t)+\mathbf{m}^{-1}\mathbf{k}\,\vec{X}(t)=\mathbf{m}^{-1}\vec{F}(t)$ and put this into the second "row" of Eq. 8.44, along with the trivial equation $\dot{\vec{X}}(t)-\dot{\vec{X}}(t)=\vec{0}$ in the top row, giving

$$\mathbf{A}=\mathbf{I}_{2n}=\begin{pmatrix}\mathbf{I}_n & \mathbf{0}\\ \mathbf{0} & \mathbf{I}_n\end{pmatrix},\qquad \mathbf{B}=\begin{pmatrix}\mathbf{0} & -\mathbf{I}_n\\ \mathbf{m}^{-1}\mathbf{k} & \mathbf{m}^{-1}\mathbf{c}\end{pmatrix},\qquad \vec{Q}(t)=\begin{pmatrix}\vec{0}\\ \mathbf{m}^{-1}\vec{F}(t)\end{pmatrix}$$

in which $\mathbf{I}_j$ denotes the $j\times j$ identity matrix. Note that each "element" shown in **A** and **B** is a submatrix of dimension $n\times n$ and each "element" of $\vec{Q}(t)$ is a vector of dimension n. In general, though, the choice of **A**, **B**, and $\vec{Q}(t)$ is not unique, and there is some advantage in using the alternative form of

$$\mathbf{A}=\begin{pmatrix} -\mathbf{k} & \mathbf{0} \\ \mathbf{0} & \mathbf{m} \end{pmatrix}, \quad \mathbf{B}=\begin{pmatrix} \mathbf{0} & \mathbf{k} \\ \mathbf{k} & \mathbf{c} \end{pmatrix}, \quad \vec{Q}(t)=\begin{pmatrix} \vec{0} \\ \vec{F}(t) \end{pmatrix} \quad (8.46)$$

in which both **A** and **B** are symmetric. The mathematical advantages of symmetry will be pursued in solving problems. The state-space formulation of the MDF problem is commonly called the *Foss method.*

****

Example 8.11: Find definitions of $\vec{Y}(t)$, $\mathbf{A}$, $\mathbf{B}$, and $\vec{Q}(t)$ such that Eq. 8.44 describes the system with the equation of motion

$$\sum_{j=0}^{n} a_j \frac{d^j X(t)}{dt^j} = F(t)$$

We note that the order of the system is n, and as suggested, we take the state variables to be $X(t)$ and its first $n-1$ derivatives, so

$$\vec{Y}(t)=\left(X(t), \dot{X}(t), \ddot{X}(t), \cdots, \frac{d^{n-1}X(t)}{dt^{n-1}} \right)^T$$

The original equation of motion then relates $\dot{Y}_n(t)$ to the components of $\vec{Y}(t)$ and the excitation $F(t)$. The other $n-1$ scalar equations come from the fact that the derivative of each of the other state variables is itself a state variable: $\dot{Y}_j(t) = Y_{j+1}(t)$ for $j=1,\cdots,n-1$. Thus, the most obvious way to obtain Eq. 8.44 for the equation of motion is to choose

$$\mathbf{A}=\mathbf{I}_n, \qquad \mathbf{B}=\begin{pmatrix} 0 & -1 & 0 & \cdots & 0 \\ 0 & 0 & -1 & \ddots & \vdots \\ \vdots & \ddots & \ddots & \ddots & 0 \\ 0 & \cdots & 0 & 0 & -1 \\ \frac{a_0}{a_n} & \frac{a_1}{a_n} & \frac{a_2}{a_n} & \cdots & \frac{a_{n-1}}{a_n} \end{pmatrix}, \qquad \vec{Q}(t)=\frac{1}{a_n}\begin{pmatrix} 0 \\ 0 \\ \vdots \\ 0 \\ F(t) \end{pmatrix}$$

An alternative choice with symmetric **A** and **B** matrices is

$$\mathbf{A}=\begin{pmatrix} 0 & \cdots & 0 & -a_0 & 0 \\ 0 & \cdots & -a_0 & -a_1 & 0 \\ \vdots & \ddots & \vdots & \vdots & \vdots \\ -a_0 & \cdots & -a_{n-3} & -a_{n-2} & 0 \\ 0 & 0 & \cdots & 0 & a_n \end{pmatrix}, \quad \mathbf{B}=\begin{pmatrix} 0 & 0 & \cdots & 0 & a_0 \\ 0 & \ddots & 0 & a_0 & a_1 \\ \vdots & 0 & \ddots & \vdots & \vdots \\ 0 & a_0 & \cdots & a_{n-3} & a_{n-2} \\ a_0 & a_1 & \cdots & a_{n-2} & a_{n-1} \end{pmatrix}$$

with $\vec{Q}(t)=[0,\cdots,0,F(t)]^T$.

****

Example 8.12: Define terms such that the oscillator shown here is described by Eq. 8.44.

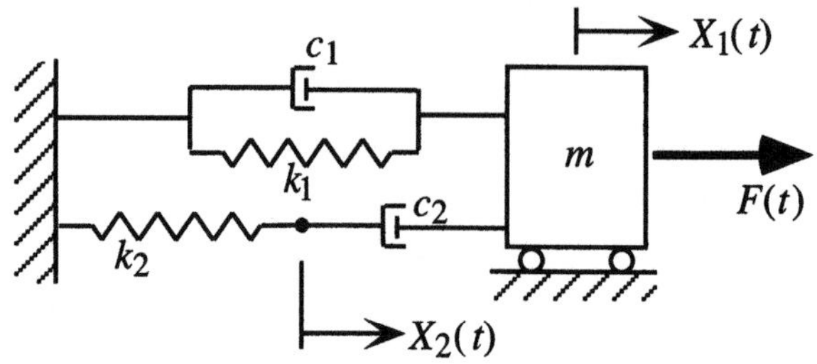

First we write the differential equations of motion in terms of the variables $X_1(t)$ and $X_2(t)$ shown in the accompanying sketch. One such form is

$$m\ddot{X}_1(t)+c_1\dot{X}_1(t)+k_1X_1(t)+k_2X_2(t)=F(t)$$

and

$$k_2X_2(t)=c_2[\dot{X}_1(t)-\dot{X}_2(t)]$$

Because these equations involve up to the second derivative of $X_1(t)$ but only the first derivative of $X_2(t)$, we can define the state vector as $\vec{Y}(t)=[X_1(t),X_2(t),\dot{X}_1(t)]^T$. The first equation can then be written as $m\dot{Y}_3(t)+c_1Y_3(t)+k_1Y_1(t)+k_2Y_2(t)=F(t)$ and the second equation as $k_2Y_2(t)=c_2[Y_3(t)-\dot{Y}_2(t)]$. The third equation that will be needed is the general relationship that $\dot{Y}_1(t)=Y_3(t)$ for this state vector. Putting these three equations directly into the standard form of Eq. 8.44 gives

$$\mathbf{A}=\mathbf{I}_3,\qquad \mathbf{B}=\begin{pmatrix}0&0&-1\\0&k_2/c_2&-1\\k_1/m&k_2/m&c_1/m\end{pmatrix},\qquad \vec{Q}(t)=\frac{1}{m}\begin{pmatrix}0\\0\\F(t)\end{pmatrix}$$

A symmetric alternative form is

$$\mathbf{A}=\begin{pmatrix}-k_1&0&0\\0&-k_2&0\\0&0&m\end{pmatrix},\qquad \mathbf{B}=\begin{pmatrix}0&0&k_1\\0&-k_2^2/c_2&k_2\\k_1&k_2&c_1\end{pmatrix},\qquad \vec{Q}(t)=\begin{pmatrix}0\\0\\F(t)\end{pmatrix}$$

****

The fact that Eq. 8.44 involves only first-order derivatives allows its solution to be easily written in terms of the matrix exponential function. In particular, the homogeneous solution of Eq. 8.44 can be written as $\vec{X}(t)=\exp[-t\,\mathbf{A}^{-1}\mathbf{B}]$ in which the matrix exponential is defined as

$$\exp(\mathbf{A})=\sum_{j=0}^{\infty}\frac{1}{j!}\mathbf{A}^j \qquad (8.47)$$

for any square matrix $\mathbf{A}$, and with $\mathbf{A}^0$ defined to be the identity matrix of the same dimension as $\mathbf{A}$, and $\mathbf{A}^j=\mathbf{A}\,\mathbf{A}^{j-1}$ for $j\ge 1$. This relationship gives the derivative with respect to t of $\exp[-t\mathbf{A}^{-1}\mathbf{B}]$ as $-\mathbf{A}^{-1}\mathbf{B}\exp[-t\mathbf{A}^{-1}\mathbf{B}]$, so the solution of Eq. 8.44 has the same form as for a scalar equation. Similarly, the general inhomogeneous solution can be written as the convolution integral

$$\vec{Y}(t) = \int_{-\infty}^{t} \exp[-(t-s)\,\mathbf{A}^{-1}\mathbf{B}]\;\mathbf{A}^{-1}\vec{Q}(s)\,ds \tag{8.48}$$

Although the solution in Eq. 8.48 is mathematically correct, it is not very convenient for numerical computations. Simplified numerical procedures result from diagonalizing $\mathbf{A}^{-1}\mathbf{B}$ by the use of eigenanalysis. We will use the notation of $\boldsymbol{\lambda}$ and $\boldsymbol{\theta}$ for the matrices of eigenvalues and eigenvectors of $\mathbf{A}^{-1}\mathbf{B}$, just as we did for $\mathbf{m}^{-1}\mathbf{k}$ in Section 8.3. Thus, we have $\mathbf{A}^{-1}\mathbf{B}\boldsymbol{\theta} = \boldsymbol{\theta}\boldsymbol{\lambda}$ or $\mathbf{A}^{-1}\mathbf{B} = \boldsymbol{\theta}\boldsymbol{\lambda}\boldsymbol{\theta}^{-1}$. We can readily verify that $(\mathbf{A}^{-1}\mathbf{B})^j = \boldsymbol{\theta}\boldsymbol{\lambda}^j\boldsymbol{\theta}^{-1}$ for any jth power of the matrix, and this establishes that $\exp[-t\mathbf{A}^{-1}\mathbf{B}] = \boldsymbol{\theta}\exp[-t\boldsymbol{\lambda}]\boldsymbol{\theta}^{-1}$. Thus, the general solution in Eq. 8.48 can be written as

$$\vec{Y}(t) = \int_{-\infty}^{t} \boldsymbol{\theta}\exp[-(t-s)\boldsymbol{\lambda}]\boldsymbol{\theta}^{-1}\mathbf{A}^{-1}\,\vec{Q}(s)\,ds \tag{8.49}$$

in which $\exp[-(t-s)\boldsymbol{\lambda}]$ is a diagonal matrix involving only the scalar exponential function. In particular, $\exp[-(t-s)\lambda_{jj}]$ is the (j,j) element of this diagonal matrix exponential. Unfortunately, the eigenvalues and eigenvectors of $\mathbf{A}^{-1}\mathbf{B}$ are generally not real, so one must perform complex mathematics.

The form in Eq. 8.49 is much simpler than that in Eq. 8.48, inasmuch as the exponential function is now evaluated only for scalar quantities. It involves the inverse of the $\boldsymbol{\theta}$ matrix, however, and calculating inverses is not easy for large matrices, especially when they are complex. One solution to this problem is to perform the eigenanalysis for $(\mathbf{A}^{-1}\mathbf{B})^T = \mathbf{B}^T(\mathbf{A}^{-1})^T$ finding the matrix $\boldsymbol{\eta}$ such that $(\mathbf{A}^{-1}\mathbf{B})^T\boldsymbol{\eta} = \boldsymbol{\eta}\boldsymbol{\lambda}$, because $(\mathbf{A}^{-1}\mathbf{B})^T$ and $\mathbf{A}^{-1}\mathbf{B}$ have the same eigenvalues. One then finds that $\boldsymbol{\eta}^T\boldsymbol{\theta}$ is diagonal, so it is easy to calculate the inverse as $\boldsymbol{\theta}^{-1} = (\boldsymbol{\eta}^T\boldsymbol{\theta})^{-1}\boldsymbol{\eta}^T$. The penalty in using this technique is the cost of performing a second eigenanalysis. For the special case in which $\mathbf{A}$ and $\mathbf{B}$ are symmetric, there exists a much simpler approach. In particular, we follow exactly the same approach as we used for the eigenanalysis of $\mathbf{m}^{-1}\mathbf{k}$ to show that $\hat{\mathbf{A}} = \boldsymbol{\theta}^T\mathbf{A}\boldsymbol{\theta}$ is a diagonal matrix. Simple algebra then shows that $\boldsymbol{\theta}^{-1} = \hat{\mathbf{A}}^{-1}\boldsymbol{\theta}^T\mathbf{A}$. Thus, in this situation one needs only to evaluate the diagonal matrix $\hat{\mathbf{A}} = \boldsymbol{\theta}^T\mathbf{A}\boldsymbol{\theta}$ by matrix multiplication, invert that diagonal matrix, and use matrix multiplication to evaluate $\boldsymbol{\theta}^{-1} = \hat{\mathbf{A}}^{-1}\boldsymbol{\theta}^T\mathbf{A}$. Using this symmetric form in Eq. 8.49 gives

$$\vec{Y}(t) = \int_{-\infty}^{t} \boldsymbol{\theta} \exp[-(t-s)\boldsymbol{\lambda}]\, \hat{\mathbf{A}}^{-1} \boldsymbol{\theta}^T\, \vec{Q}(s)\, ds \tag{8.50}$$

The ensuing equations will be written using the $\boldsymbol{\theta}^{-1}$ notation, but the simplified ways of calculating this matrix can be very useful in solving problems. In particular, it should be kept in mind that using the symmetric form of **A** and **B**, as shown in Eq. 8.46, allows simple evaluation of $\boldsymbol{\theta}^{-1}$ for any MDF system.

There are several possible approaches to performing the stochastic analysis of Eq. 8.44, just as there were in Sections 8.4 and 8.5 for the equations considered there. One approach is to use Eq. 8.49 only in finding the $\mathbf{h}(t)$ and $\mathbf{H}(\omega)$ matrices, for use in the equations of Section 8.1. For example, we can say that $\vec{Y}(t) = [h_{1l}(t), \cdots, h_{n_Y l}(t)]^T$ when the excitation is a single Dirac delta function pulse in the $Q_l(t)$ component of excitation, and Eqs. 8.48 and 8.49 then give

$$\mathbf{h}(t) = \exp[-t\,\mathbf{A}^{-1}\mathbf{B}]\,\mathbf{A}^{-1}\, U(t) = \boldsymbol{\theta} \exp[-t\boldsymbol{\lambda}]\boldsymbol{\theta}^{-1}\mathbf{A}^{-1}\, U(t) \tag{8.51}$$

Similarly, $\vec{Y}(t) = [H_{1l}(\omega), \cdots, H_{n_Y l}(\omega)]^T e^{i\omega t}$ when the excitation is a single harmonic term in the $Q_l(t)$ component of excitation, and Eq. 8.44 directly gives $[i\,\omega\,\mathbf{A} + \mathbf{B}]\mathbf{H}(\omega) = \mathbf{I}_{n_Y}$, so

$$\mathbf{H}(\omega) = [i\,\omega\,\mathbf{A} + \mathbf{B}]^{-1} = [i\,\omega\,\mathbf{I} + \mathbf{A}^{-1}\mathbf{B}]^{-1}\mathbf{A}^{-1} = \boldsymbol{\theta}[i\,\omega\,\mathbf{I} + \boldsymbol{\lambda}]^{-1}\boldsymbol{\theta}^{-1}\mathbf{A}^{-1} \tag{8.52}$$

The alternative of direct stochastic analysis of Eq. 8.48 or 8.49 gives the mean-value vector as

$$\begin{aligned} \vec{\mu}_Y(t) &= \int_{-\infty}^{t} \exp[-(t-s)\,\mathbf{A}^{-1}\mathbf{B}]\ \mathbf{A}^{-1}\, \vec{\mu}_Q(s)\, ds \\ &= \int_{-\infty}^{t} \boldsymbol{\theta} \exp[-(t-s)\boldsymbol{\lambda}]\boldsymbol{\theta}^{-1}\mathbf{A}^{-1}\, \vec{\mu}_Q(s)\, ds \end{aligned} \tag{8.53}$$

and the autocorrelation matrix as

$$\begin{aligned} \boldsymbol{\phi}_{YY}(t,s) = \int_{-\infty}^{s}\int_{-\infty}^{t} &\exp[-(t-u)\mathbf{A}^{-1}\mathbf{B}]\,\mathbf{A}^{-1}\,\boldsymbol{\phi}_{QQ}(u,v)\,(\mathbf{A}^{-1})^T \times \\ &\exp[-(s-v)\,\mathbf{B}^T (\mathbf{A}^{-1})^T]\, du\, dv \end{aligned} \tag{8.54}$$

or

$$\boldsymbol{\phi}_{YY}(t,s) = \int_{-\infty}^{s}\int_{-\infty}^{t} \boldsymbol{\theta} \exp[-(t-u)\boldsymbol{\lambda}]\boldsymbol{\theta}^{-1}\mathbf{A}^{-1}\boldsymbol{\phi}_{QQ}(u,v)\,(\mathbf{A}^{-1})^T \times (\boldsymbol{\theta}^{-1})^T \exp[-(s-v)\boldsymbol{\lambda}]\boldsymbol{\theta}^T\,du\,dv \tag{8.55}$$

Similarly, the autocovariance is

$$\mathbf{K}_{YY}(t,s) = \int_{-\infty}^{s}\int_{-\infty}^{t} \boldsymbol{\theta} \exp[-(t-u)\boldsymbol{\lambda}]\boldsymbol{\theta}^{-1}\mathbf{A}^{-1}\mathbf{K}_{QQ}(u,v)\,(\mathbf{A}^{-1})^T \times (\boldsymbol{\theta}^{-1})^T \exp[-(s-v)\boldsymbol{\lambda}]\boldsymbol{\theta}^T\,du\,dv \tag{8.56}$$

These expressions all agree with what one would obtain from Eqs. 8.6–8.8 in Section 8.1. The autospectral density matrix for covariant stationary response is most easily obtained from Eq. 8.11 as

$$\begin{aligned}\mathbf{S}_{YY}(\omega) &= [i\omega\,\mathbf{A}+\mathbf{B}]^{-1}\mathbf{S}_{QQ}(\omega)[-i\omega\,\mathbf{A}^T+\mathbf{B}^T]^{-1} \\ &= \boldsymbol{\theta}[i\omega\,\mathbf{I}+\boldsymbol{\lambda}]^{-1}\boldsymbol{\theta}^{-1}\mathbf{A}^{-1}\mathbf{S}_{QQ}(\omega)(\mathbf{A}^T)^{-1}(\boldsymbol{\theta}^{T*})^{-1}[-i\omega\,\mathbf{I}+\boldsymbol{\lambda}^*]^{-1}\boldsymbol{\theta}^{T*}\end{aligned} \tag{8.57}$$

Let us now consider the response after given initial conditions. As previously noted, $\vec{Y}(0)$ gives the complete initial condition vector for the system. In addition, the initial value response matrix $\mathbf{g}(t)$ of Eq. 8.12 is $\mathbf{g}(t) \equiv \mathbf{h}(t)\,\mathbf{A}$. Thus, $\mathbf{A}$ and the $\mathbf{h}(t)$ matrix described by Eq. 8.51 give all the system properties involved in using Eq. 8.12:

$$\vec{Y}(t) = \exp[-(t-t_0)\,\mathbf{A}^{-1}\mathbf{B}]\,\vec{Y}(t_0) + \int_{t_0}^{t} \exp[-(t-s)\mathbf{A}^{-1}\mathbf{B}]\,\mathbf{A}^{-1}\vec{Q}(s)\,ds \tag{8.58}$$

Direct stochastic analysis of this equation gives

$$\vec{\mu}_Y(t) = \exp[-(t-t_0)\,\mathbf{A}^{-1}\mathbf{B}]\,\vec{\mu}_Y(t_0) + \int_{t_0}^{t} \exp[-(t-s)\mathbf{A}^{-1}\mathbf{B}]\,\mathbf{A}^{-1}\vec{\mu}_Q(s)\,ds \tag{8.59}$$

and if the $\vec{Q}(s)$ excitation for $s > t_0$ is independent of the $Y(t_0)$ initial condition, then the covariance reduces to

$$\mathbf{K}_{YY}(t,s) = \exp[-(t-t_0)\mathbf{A}^{-1}\mathbf{B}]\,\mathbf{K}_{YY}(t_0,t_0)\exp[-(s-t_0)\mathbf{B}^T(\mathbf{A}^{-1})^T]$$
$$+\int_{t_0}^{s}\int_{t_0}^{t}\exp[-(t-u)\mathbf{A}^{-1}\mathbf{B}]\,\mathbf{A}^{-1}\,\mathbf{K}_{QQ}(u,v)(\mathbf{A}^{-1})^T\exp[-(s-v)\mathbf{B}^T(\mathbf{A}^{-1})^T]\,du\,dv \tag{8.60}$$

The corresponding expressions for the conditional mean and covariance, analogous to Eqs. 5.57 and 5.58 for the scalar problem, are

$$E[\vec{Y}(t)\,|\,\vec{Y}(t_0)=\vec{w}\,] = \exp[-(t-t_0)\mathbf{A}^{-1}\mathbf{B}]\vec{w}\, +$$
$$\int_{t_0}^{t}\exp[-(t-s)\mathbf{A}^{-1}\mathbf{B}]\,\mathbf{A}^{-1}E[\vec{Q}(s)\,|\,\vec{Y}(t_0)=\vec{w}\,]\,ds \tag{8.61}$$

and

$$\mathbf{K}[\vec{Y}(t),\vec{Y}(s)\,|\,\vec{Y}(t_0)=\vec{w}\,] = \int_{t_0}^{s}\int_{t_0}^{t}\exp[-(t-u)\mathbf{A}^{-1}\mathbf{B}]\,\mathbf{A}^{-1}\,\times$$
$$\mathbf{K}[\vec{Q}(u),\vec{Q}(v)\,|\,\vec{Y}(t_0)=\vec{w}\,](\mathbf{A}^{-1})^T\exp[-(s-v)\mathbf{B}^T(\mathbf{A}^{-1})^T]\,du\,dv \tag{8.62}$$

For the special case of a delta-correlated excitation, of course, many of these relationships are somewhat simplified. For example, if we use the notation that $\mathbf{K}_{QQ}(u,v) = 2\pi\,\mathbf{S}_0(u)\,\delta(u-v)$, then the autocovariance of the response can be written as a single, rather than double, integral as

$$\mathbf{K}_{YY}(t,s) = 2\pi\int_{-\infty}^{\min(t,s)}\exp[-(t-u)\mathbf{A}^{-1}\mathbf{B}]\,\mathbf{A}^{-1}\,\mathbf{S}_0(u)(\mathbf{A}^{-1})^T\,\times$$
$$\exp[-(s-u)\mathbf{B}^T(\mathbf{A}^T)^{-1}]\,du \tag{8.63}$$

or

$$\mathbf{K}_{YY}(t,s) = \exp[-(t-t_0)\mathbf{A}^{-1}\mathbf{B}]\,\mathbf{K}_{YY}(t_0,t_0)\exp[-(s-t_0)\mathbf{B}^T(\mathbf{A}^{-1})^T]$$
$$+2\pi\int_{t_0}^{\min(t,s)}\exp[-(t-u)\mathbf{A}^{-1}\mathbf{B}]\,\mathbf{A}^{-1}\,\mathbf{S}_0(u)(\mathbf{A}^{-1})^T\exp[-(s-u)\mathbf{B}^T(\mathbf{A}^{-1})^T]\,du \tag{8.64}$$

and the conditional mean and conditional variance expressions are

$$E[\vec{Y}(t)\,|\,\vec{Y}(t_0)=\vec{w}\,] = \exp[-(t-t_0)\mathbf{A}^{-1}\mathbf{B}]\vec{w}\, +$$
$$\int_{t_0}^{t}\exp[-(t-s)\mathbf{A}^{-1}\mathbf{B}]\,\mathbf{A}^{-1}\,\vec{\mu}_Q(s)\,ds \tag{8.65}$$

and

$$\mathrm{Var}[\vec{Y}(t),\vec{Y}(s)\,|\,\vec{Y}(t_0)=\vec{w}]=2\pi\int_{t_0}^{t}\exp[-(t-u)\mathbf{A}^{-1}\mathbf{B}]\,\mathbf{A}^{-1}\times \mathbf{S}_0(u)\,(\mathbf{A}^{-1})^T\exp[-(s-u)\mathbf{B}^T(\mathbf{A}^{-1})^T]\,du \tag{8.66}$$

Each of these expressions can be diagonalized by use of $\boldsymbol{\theta}$ and $\boldsymbol{\theta}^{-1}$.

****

Example 8.13: Find the state-space formulations of the $\mathbf{h}(t)$ impulse response matrix and the $\mathbf{H}(\omega)$ harmonic transfer matrix for the 2DF system of Examples 8.5 and 8.8 using the state-space vector $\vec{Y}(t)=[X_1(t),X_2(t),\dot{X}_1(t),\dot{X}_2(t)]^T$.

Using Eq. 8.46 and the $\mathbf{m}$, $\mathbf{c}$, and $\mathbf{k}$ matrices from Example 8.5, we find that

$$\mathbf{A}=\begin{pmatrix} -10^6 & 5\times10^5 & 0 & 0\\ 5\times10^5 & -5\times10^5 & 0 & 0\\ 0 & 0 & 1000 & 0\\ 0 & 0 & 0 & 500\end{pmatrix}$$

and

$$\mathbf{B}=\begin{pmatrix} 0 & 0 & 10^6 & -5\times10^5\\ 0 & 0 & -5\times10^5 & 5\times10^5\\ 10^6 & -5\times10^5 & 1000 & -500\\ -5\times10^5 & 5\times10^5 & -500 & 500\end{pmatrix}$$

Eigenanalysis of $\mathbf{A}^{-1}\mathbf{B}$ yields

$$\boldsymbol{\theta}=\begin{pmatrix} -0.3536-41.32i & -0.3536+41.32i & -0.3536+17.11i & -0.3536-17.11i\\ -0.5-58.43i & -0.5+58.43i & -0.5-24.20i & -0.5+24.20i\\ 707.1 & 707.1 & -707.1 & -707.1\\ 1000 & 1000 & 1000 & 1000\end{pmatrix}$$

and the nonzero elements of $\boldsymbol{\lambda}$ and $\hat{\mathbf{A}}$ are $\lambda_{11}=0.1464-17.11i$, $\lambda_{22}=\lambda_{11}^*$, $\lambda_{33}=0.8536-41.31i$, $\lambda_{44}=\lambda_{33}^*$, $\hat{A}_{11}=2.000\times10^9-1.711\times10^7 i$, $\hat{A}_{22}=\hat{A}_{11}^*$, $\hat{A}_{33}=1.999\times10^9-4.131\times10^7 i$, and $\hat{A}_{44}=\hat{A}_{33}^*$. Note that the eigenvalues of $\mathbf{A}^{-1}\mathbf{B}$ are $\pm i\,\omega_j[1-(\zeta_j)^2]^{1/2}+\zeta_j\omega_j$. That is, the imaginary parts of the eigenvalues are the same as the damped frequencies of the uncoupled modes, and the real parts are the same as the corresponding elements of the $\boldsymbol{\beta}$ modal damping matrix. This, then, shows that the absolute values of the eigenvalues are like undamped natural frequencies. These general relationships will be found to be true for any MDF system with uncoupled modes.

The impulse response matrix can now be obtained from $\mathbf{h}(t) = \boldsymbol{\theta} e^{-t\boldsymbol{\lambda}} \boldsymbol{\theta}^{-1} \mathbf{A}^{-1} U(t) = \boldsymbol{\theta} e^{-t\boldsymbol{\lambda}} \hat{\mathbf{A}}^{-1} \boldsymbol{\theta}^T U(t)$, in which the nonzero elements of $e^{-t\boldsymbol{\lambda}}$ are given by $e^{-t\lambda_{jj}}$ on the diagonal. The harmonic transfer matrix can be obtained from $\mathbf{H}(\omega) = \boldsymbol{\theta}[i\omega\mathbf{I} + \boldsymbol{\lambda}]^{-1} \boldsymbol{\theta}^{-1} \mathbf{A}^{-1} = \boldsymbol{\theta}[i\omega\mathbf{I} + \boldsymbol{\lambda}]^{-1} \hat{\mathbf{A}}^{-1} \boldsymbol{\theta}^T$, in which the nonzero elements of $[i\omega\mathbf{I} + \boldsymbol{\lambda}]^{-1}$ are given by $1/(i\,\omega + \lambda_{jj})$ on the diagonal.
For example, the $X_1(t)$ response to a Dirac delta function in $Q_3(t) = F_1(t)$ is

$$h_{13}(t) = [14.61i\, e^{(-0.1464-17.11i)t} - 14.61\times 10^{-5} i\, e^{(-0.1464+17.11i)t}$$
$$-6.052i\, e^{(-0.8536+41.31i)t} + 6.052i\, e^{(-0.8536-41.31i)t}] \times 10^{-6}\, U(t)$$

or

$$h_{13}(t) = [29.22 e^{-0.1464t} \sin(17.11t) + 12.10 e^{-0.8536t} \sin(41.31t)] \times 10^{-6}\, U(t)$$

The corresponding harmonic transfer term giving the magnitude of the $X_1(t)$ response to a unit amplitude harmonic function in $Q_3(t) = F_1(t)$ is

$$H_{13}(\omega) = \frac{1 + 0.001 i\,\omega - 0.001\omega^2}{5\times 10^5 + 1000 i\,\omega - 2000.5\omega^2 - 2i\,\omega^3 + \omega^4}$$

These expressions are in perfect agreement with the results in Examples 8.5 and 8.8 for the $X_1(t)$ response to given Dirac delta and unit harmonic functions for $F_1(t)$. For this matrix of dimension 4×4, it is also feasible to obtain the expressions for $\mathbf{H}(\omega)$ by analytically inverting $[i\omega\mathbf{A} + \mathbf{B}]$ without using eigenanalysis. This approach is generally not feasible, though, if the dimension of the problem is not quite small.

****

Example 8.14: Find the state-space formulations of the $\mathbf{h}(t)$ impulse response matrix and the $\mathbf{H}(\omega)$ harmonic transfer matrix for the 2DF system of Examples 8.3 and 8.10 using the state-space vector $\vec{Y}(t) = [X_1(t), X_2(t), \dot{X}_1(t), \dot{X}_2(t)]^T$.

The approach is the same as in Example 8.13, but some of the results are different because this system does not have uncoupled modes. From Eq. 8.46 and the $\mathbf{m}$, $\mathbf{c}$, and $\mathbf{k}$ matrices from Exercise 8.3, we find that

$$\mathbf{A} = \begin{pmatrix} -10^6 & 5\times 10^5 & 0 & 0 \\ 5\times 10^5 & -5\times 10^5 & 0 & 0 \\ 0 & 0 & 1000 & 0 \\ 0 & 0 & 0 & 500 \end{pmatrix}$$

and

$$\mathbf{B} = \begin{pmatrix} 0 & 0 & 10^6 & -5\times 10^5 \\ 0 & 0 & -5\times 10^5 & 5\times 10^5 \\ 10^6 & -5\times 10^5 & 2500 & -2000 \\ -5\times 10^5 & 5\times 10^5 & -2000 & 2000 \end{pmatrix}$$

Eigenanalysis of $\mathbf{A}^{-1}\mathbf{B}$ yields

$$\boldsymbol{\theta}=\begin{pmatrix} 1.994-16.932i & 1.994+16.932i & 0.2396-41.35i & 0.2396+41.35i \\ -1.781+24.14i & -1.781-24.14i & -0.7189-58.42i & -0.7189+58.42i \\ -703.6-30.68i & -703.6+30.68i & 707.7+12.81i & 707.7-12.81i \\ 1000 & 1000 & 1000 & 1000 \end{pmatrix}$$

and the nonzero elements of $\boldsymbol{\lambda}$ and $\hat{\mathbf{A}}$ are $\lambda_{11}=3.039+41.20i$, $\lambda_{22}=\lambda_{11}^*$, $\lambda_{33}=0.2106-17.12i$, $\lambda_{44}=\lambda_{33}^*$, $\hat{A}_{11}=1.972\times 10^9+2.320\times 10^8 i$, $\hat{A}_{22}=\hat{A}_{11}^*$, $\hat{A}_{33}=2.001\times 10^9+1.169\times 10^7 i$, and $\hat{A}_{44}=\hat{A}_{33}^*$.

Now the impulse response function is given by $\mathbf{h}(t)=\boldsymbol{\theta}e^{-t\boldsymbol{\lambda}}\boldsymbol{\theta}^{-1}\mathbf{A}^{-1}U(t)=\boldsymbol{\theta}e^{-t\boldsymbol{\lambda}}\hat{\mathbf{A}}^{-1}\boldsymbol{\theta}^T U(t)$. For example, the $X_1(t)$ response to a Dirac delta function in $Q_3(t)=F_1(t)$ is

$$h_{13}(t)=[-(2.641+60.42i)e^{(-3.039+41.20i)t}-(2.641-60.42i)e^{(-3.039-41.20i)t}+(2.641+146.3i)e^{(-0.2106-17.12i)t}+(2.641-146.3i)e^{(-0.2106+17.12i)t}]\times 10^{-7}\,U(t)$$

or

$$h_{13}(t)=5.282\times 10^{-7}[e^{-0.2106t}\cos(17.12t)-e^{-3.039t}\cos(41.20t)]+2.925\times 10^{-5}e^{-0.2106t}\sin(17.12t)+1.208\times 10^{-5}e^{-3.039t}\sin(41.20t)]U(t)$$

Note the presence of cosine terms in these impulse response functions. Such terms never appear in the impulse response functions for a system with uncoupled modes, because they do not appear in the impulse response function for an SDF system. The primary effect of a damping matrix that does not allow the modes to be uncoupled is a change in phase of the motions. Note that the imaginary parts of the eigenvalues play the role in the impulse response function of the damped natural frequency ω_d in the SDF system and the real parts of the eigenvalues are like $\zeta\omega_0$ values. As when the system does have uncoupled modes, the absolute values of the eigenvalues are like undamped natural frequencies. In fact, one can show that they are precisely the square roots of the eigenvalues of $\mathbf{m}^{-1}\mathbf{k}$ for this problem.

The corresponding harmonic transfer function is given by $\mathbf{H}(\omega)=\boldsymbol{\theta}[i\omega\mathbf{I}+\boldsymbol{\lambda}]^{-1}\boldsymbol{\theta}^{-1}\mathbf{A}^{-1}=\boldsymbol{\theta}[i\omega\mathbf{I}+\boldsymbol{\lambda}]^{-1}\hat{\mathbf{A}}^{-1}\boldsymbol{\theta}^T$. The term giving the magnitude of the $X_1(t)$ response to a unit amplitude harmonic function in $Q_3(t)=F_1(t)$ is

$$H_{13}(\omega)=\frac{1+0.004i\omega-0.001\omega^2}{5\times 10^5+2500i\omega-2002\omega^2-6.5i\omega^3+\omega^4}$$

and

$$H_{14}(\omega)=\frac{1+0.004i\omega}{5\times 10^5+2500i\omega-2002\omega^2-6.5i\omega^3+\omega^4}$$

gives the magnitude of the $X_1(t)$ response to a unit amplitude harmonic in $Q_4(t) = F_2(t)$. Again, these results are in perfect agreement with the results in Example 8.10. One may also note that the poles of the harmonic response function that were found in Example 8.10 are the same as i times the eigenvalues of $\mathbf{A}^{-1}\mathbf{B}$, emphasizing the similarity between the two approaches.
Note that there is considerable redundancy in the state-space formulation of the eigenanalysis for an MDF system. In particular, the complex conjugates of the eigenvalues and eigenvectors are also eigenvalues and eigenvectors, respectively (unless the damping values are very large). This fact can be used in reducing the amount of computation required (Igusa et al., 1984; Veletsos and Ventura, 1986).
****

Example 8.15: Find the autospectral density function and the mean-squared value of the stationary $\{X_1(t)\}$ response of the system of Examples 8.3, 8.10, and 8.14 when it is subjected to the mean-zero white noise base acceleration excitation of Examples 8.6 and 8.9 with $S_0 = 0.1\ (\text{m/s}^2)/(\text{rad/s})$.

From Eq. 8.46 we see that the excitation vector in the state-space formulation is $\vec{Q}(t) = -\ddot{Y}(t)[0,0,m_1,m_2]^T$. The autospectral density matrix for this excitation is then

$$\mathbf{S}_{QQ} = S_0 \begin{pmatrix} 0 & 0 & 0 & 0 \\ 0 & 0 & 0 & 0 \\ 0 & 0 & m_1^2 & m_1 m_2 \\ 0 & 0 & m_1 m_2 & m_2^2 \end{pmatrix} = S_0 \begin{pmatrix} 0 & 0 & 0 & 0 \\ 0 & 0 & 0 & 0 \\ 0 & 0 & 10^6 & 5\times 10^5 \\ 0 & 0 & 5\times 10^5 & 2.5\times 10^5 \end{pmatrix}$$

and Eq. 8.11 gives $\mathbf{S}_{YY}(\omega) = \mathbf{H}(\omega)\mathbf{S}_{QQ}(\omega)\,\mathbf{H}^{*T}(\omega)$ for the $\mathbf{H}(\omega)$ determined in Example 8.14. Substituting gives Eq. 8.57, and symmetry allows this also to be rewritten as

$$\mathbf{S}_{YY}(\omega) = \boldsymbol{\theta}[i\omega\,\mathbf{I} + \boldsymbol{\lambda}]^{-1}\hat{\mathbf{A}}^{-1}\boldsymbol{\theta}^T\mathbf{S}_{QQ}(\omega)\boldsymbol{\theta}^*\hat{\mathbf{A}}^{-1*}[-i\omega\,\mathbf{I} + \boldsymbol{\lambda}^*]^{-1}\boldsymbol{\theta}^{T*}$$

The (1,1) component of this expression is

$$[\mathbf{S}_{YY}(\omega)]_{11} = \sum_{j_1=1}^{4}\sum_{j_2=1}^{4}\sum_{j_3=1}^{4}\sum_{j_4=1}^{4} \frac{\theta_{1j_1}\theta_{j_2 j_1}[\mathbf{S}_{QQ}]_{j_2 j_3}\theta^*_{j_3 j_4}\theta^*_{1j_4}}{(i\omega + \lambda_{j_1 j_1})\hat{A}_{j_1 j_1}\hat{A}^*_{j_4 j_4}(-i\omega + \lambda^*_{j_4 j_4})}$$

which can be written as

$$[\mathbf{S}_{YY}(\omega)]_{11} = \frac{(\omega^4 - 2964\omega^2 + 2.25\times 10^6)\,S_0}{(\omega^4 - 585.4\omega^2 + 8.584\times 10^4)(\omega^4 - 3376\omega^2 + 2.912\times 10^6)}$$

Although $[\mathbf{S}_{YY}(\omega)]_{11}$ does have peaks in the vicinities of both natural frequencies of the undamped system, the high-frequency peak is so small that it is not very significant. As in Example 8.6, the insignificance of the higher mode is exaggerated by the phase of the excitation components. The following sketches show the two peaks of the autospectral density for $S_0 = 0.1$. Note, though, that the vertical scale is greatly different in the two sketches.

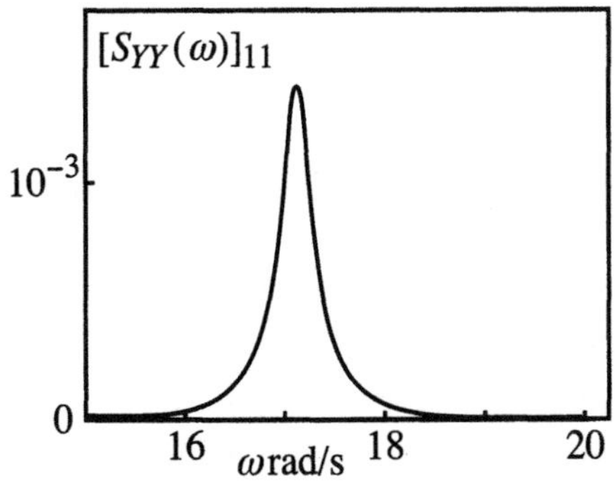

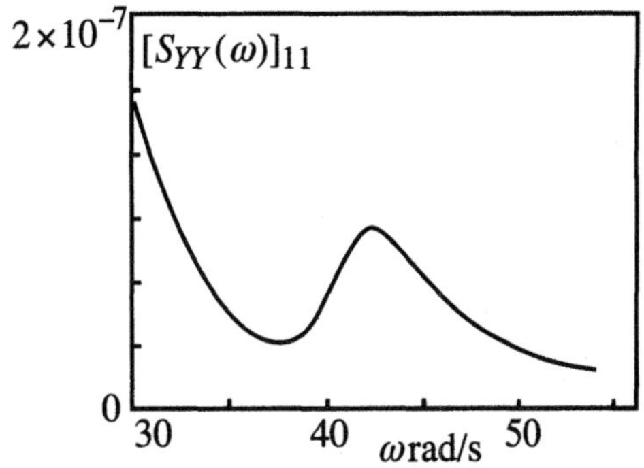

The response covariance for this stationary, mean-zero problem can be found from Eq. 8.63, which can be rewritten as

$$\mathbf{K}_{YY}(t,t) = 2\pi \int_0^\infty \exp[-u\,\mathbf{A}^{-1}\mathbf{B}]\,\mathbf{A}^{-1}\mathbf{S}_{QQ}(\mathbf{A}^{-1})^T \exp[-u\,\mathbf{B}^T(\mathbf{A}^T)^{-1}]\,du$$

or

$$\mathbf{K}_{YY}(t,t) = 2\pi\boldsymbol{\theta}\int_0^\infty \exp[-u\,\boldsymbol{\lambda}]\,\hat{\mathbf{A}}^{-1}\boldsymbol{\theta}^T\,\mathbf{S}_{QQ}\,\boldsymbol{\theta}\,\hat{\mathbf{A}}^{-1}\exp[-u\,\boldsymbol{\lambda}]\,du\,\boldsymbol{\theta}^T$$

The (1,1) component is then

$$E(X_1^2) = [\mathbf{K}_{YY}(t,t)]_{11} =$$

$$2\pi \sum_{j_1=1}^{4}\sum_{j_2=3}^{4}\sum_{j_3=3}^{4}\sum_{j_4=1}^{4} \frac{\theta_{1j_1}\theta_{j_2j_1}(\mathbf{S}_{QQ})_{j_2j_3}\theta_{j_3j_4}\theta_{1j_4}}{\hat{A}_{j_1j_1}\hat{A}_{j_4j_4}} \int_0^\infty e^{-u(\lambda_{j_1j_1}+\lambda_{j_4j_4})}\,du$$

$$= 2\pi \sum_{j_1=1}^{4}\sum_{j_2=3}^{4}\sum_{j_3=3}^{4}\sum_{j_4=1}^{4} \frac{\theta_{1j_1}\theta_{j_2j_1}(\mathbf{S}_{QQ})_{j_2j_3}\theta_{j_3j_4}\theta_{1j_4}}{\hat{A}_{j_1j_1}\hat{A}_{j_4j_4}(\lambda_{j_1j_1}+\lambda_{j_4j_4})} = 1.860\times10^{-3}\ \text{m}^2$$

****

Exercises

****

Response Modes and Impulse Response Functions

****

8.1 Let the vector process $\{\vec{X}(t)\}$ be the response of a system with the equation of motion

$$\ddot{\vec{X}}(t) + \begin{pmatrix} 0.10 & -0.02 \\ -0.02 & 0.10 \end{pmatrix}\dot{\vec{X}}(t) + \begin{pmatrix} 26 & -10 \\ -10 & 26 \end{pmatrix}\vec{X}(t) = \begin{pmatrix} F_1(t) \\ F_2(t) \end{pmatrix}$$

(a) Demonstrate that this system has uncoupled modes.

(b) Find the $\mathbf{h}(t)$ impulse response function matrix.

****

8.2 Consider the 2DF system shown in Example 8.1 with $m_1 = 400\,\text{kg}$, $m_2 = 300\,\text{kg}$, $k_1 = 1200\,\text{N/m}$, $k_2 = 600\,\text{N/m}$, $c_1 = 100\,\text{N}\cdot\text{s/m}$, and $c_2 = 50\,\text{N}\cdot\text{s/m}$.

(a) Demonstrate that this system has uncoupled modes.

(b) Find the $\mathbf{h}(t)$ impulse response function matrix.

****

8.3 Find the $\mathbf{h}(t)$ impulse response function matrix for the 2DF system of Example 8.4.

****

8.4 Consider the 2DF system shown in Example 8.1 with $m_1 = m_2 = m$, $k_1 = 2k$, $k_2 = k$, $c_1 = 3c$, and $c_2 = c$, in which *m*, *k*, and *c* are scalar constants.

(a) Find the **m**, **k**, and **c** matrices such that this system is described by Eq. 8.13.

(b) Show that the system does not have uncoupled modes.

****

8.5 Consider a three-degree-of-freedom system for which the mass matrix is $\mathbf{m} = m\,\mathbf{I}_3$, with *m* being a scalar constant and $\mathbf{I}_3$ being the 3×3 identity matrix. The modes are uncoupled and the mode shapes, natural frequencies, and damping values are given by

$$[3,5,6]^T, \qquad [2,0,-1]^T, \qquad [1,-3,2]^T$$

$$\omega_1 = (k/m)^{1/2}, \qquad \omega_2 = 2(k/m)^{1/2}, \qquad \omega_3 = 3(k/m)^{1/2}$$

$$\zeta_1 = 0.05, \qquad \zeta_2 = 0.05, \qquad \zeta_3 = 0.05$$

in which *k* is another scalar constant.

(a) Find the **k** stiffness matrix.

(b) Find the **c** damping matrix.

(c) On a sketch of the model, such as the one shown, indicate the values of the individual springs and dashpots.

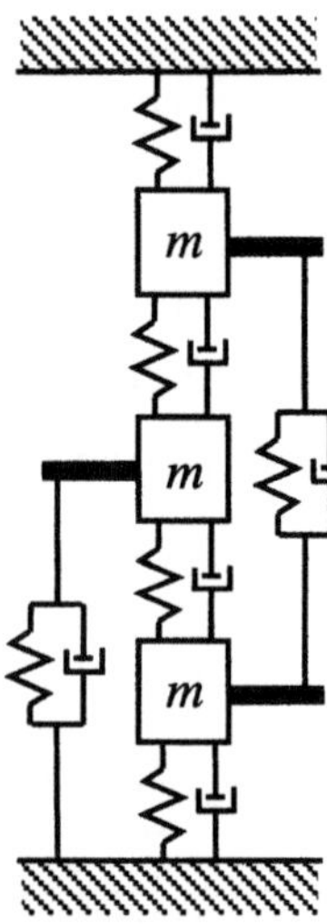

****

8.6 Consider a three-degree-of-freedom system for which the mass matrix is $\mathbf{m} = m\,\mathbf{I}_3$, with *m* being a scalar constant and $\mathbf{I}_3$ being the 3×3 identity matrix. The modes are uncoupled and the mode shapes, natural frequencies, and damping values are given by

$$[2,5,9]^T, \qquad [9,0,-2]^T, \qquad [2,17,9]^T$$

$$\omega_1 = (k/m)^{1/2}, \qquad \omega_2 = (3k/m)^{1/2}, \qquad \omega_3 = (5k/m)^{1/2}$$

$$\zeta_1 = 0.01, \qquad \zeta_2 = 0.01, \qquad \zeta_3 = 0.02$$

in which *k* is another scalar constant.

(a) Find the **k** stiffness matrix.
(b) Find the **c** damping matrix.
(c) On a sketch of the model, such as the one shown in Exercise 8.5, indicate the values of the individual springs and dashpots.

****

8.7 Let the stationary processes $\{X_1(t)\}$ and $\{X_2(t)\}$ represent the motions at two different points in a complicated system. Let the correlation matrix for $\{X_1(t)\}$ and $\{X_2(t)\}$ be given by

$$\mathbf{R}_{XX}(\tau) \equiv E[\vec{X}(t+\tau)\,\vec{X}^T(t)] = \begin{pmatrix} g(\tau)+g(2\tau) & 2g(\tau)-g(2\tau) \\ 2g(\tau)-g(2\tau) & 4g(\tau)+g(2\tau) \end{pmatrix}$$

in which $g(\tau) = e^{-b|\tau|}[\cos(\omega_0 \tau) + (b/\omega_0)\sin(\omega_0 |\tau|)]$ for positive constants b and ω_0. Let $\{Y(t)\}$ denote the relative motion between the two points: $Y(t) = X_2(t) - X_1(t)$.
(a) Find the mean-squared value of $Y(t)$.
(b) Find the $R_{\dot{X}_1 Y}(\tau) \equiv E[\dot{X}_1(t+\tau)\,Y(t)]$ cross-correlation function.

****

Harmonic Transfer Functions

****

8.8 Consider the 2DF system of Exercise 8.1.
(a) Find the $\mathbf{H}(\omega)$ harmonic transfer function matrix.
(b) Show that $\mathbf{H}(\omega)$ has poles at ω values of $\pm\omega_j[1-(\zeta_j)^2]^{1/2} + i\,\zeta_j\omega_j$.

****

8.9 Consider the 2DF system of Exercise 8.2.
(a) Find the $\mathbf{H}(\omega)$ harmonic transfer function matrix.
(b) Show that $\mathbf{H}(\omega)$ has poles at ω values of $\pm\omega_j[1-(\zeta_j)^2]^{1/2} + i\,\zeta_j\omega_j$.

****

8.10 Consider the 2DF system of Example 8.4 and Exercise 8.3.
(a) Find the $\mathbf{H}(\omega)$ harmonic transfer function matrix.
(b) Show that $\mathbf{H}(\omega)$ has poles at ω values of $\pm\omega_j[1-(\zeta_j)^2]^{1/2} + i\,\zeta_j\omega_j$.

****

8.11 Consider the 2DF system of Exercise 8.4 with $m = 1.0$ kg, $k = 50$ kN/m, and $c = 1.0$ N·s/m.
(a) Find the $\mathbf{H}(\omega)$ harmonic transfer function matrix.
(b) Locate the poles of $\mathbf{H}(\omega)$, and use this information to identify appropriate values of the modal parameters ω_j and ζ_j.

****

MDF Response

****

8.12 For the 2DF system of Exercises 8.1 and 8.8, let $F_1(t) \equiv 0$ and $\{F_2(t)\}$ be a stationary, mean-zero white noise with autospectral density of $S_0 = 10.0$.

(a) Find the $S_{X_1X_1}(\omega)$ autospectral density function for the $\{X_1(t)\}$ stationary response component.

(b) Approximate the $R_{X_1X_1}(\tau)$ autocorrelation function for the stationary $\{X_1(t)\}$ response component by treating the modal responses as though they were independent.

(c) Approximate the $E[(X_1)^2]$ mean-squared value for the stationary $\{X_1(t)\}$ response component by using the same simplification as in part (b).

****

8.13 For the 2DF system of Exercise 8.2, let $\{F_1(t)\}$ and $\{F_2(t)\}$ be independent stationary white noise processes with mean-value vector and autospectral density matrix of

$$\vec{\mu}_F = \begin{pmatrix} 4 \\ 6 \end{pmatrix} \text{N}, \qquad \mathbf{S}_{FF}(\omega) = \begin{pmatrix} 1.0 & 0 \\ 0 & 1.5 \end{pmatrix} \frac{\text{N}^2 \cdot \text{s}}{\text{rad}}$$

(a) Find the $\vec{\mu}_X$ mean-value vector for the stationary response.

(b) Find the $S_{X_2X_2}(\omega)$ autospectral density function for the $\{X_2(t)\}$ stationary response component.

(c) Approximate the $G_{X_2X_2}(\tau)$ autocovariance function for the stationary $\{X_2(t)\}$ response component by treating the modal responses as though they were independent.

(d) Approximate the $E(X_2^2)$ mean-squared value for the stationary $\{X_2(t)\}$ response component by using the same simplification as in part (b).

****

8.14 Consider the 2DF system of Examples 8.5 and 8.9, with an excitation having $F_2(t) \equiv 0$ and $\{F_1(t)\}$ being a stationary, mean-zero process with autospectral density of $S_{F_1F_1}(\omega) = 10^{-18}\omega^{20}e^{-0.5|\omega|}$ N·s/rad.
Find the autospectral density of the stationary $\{X_1(t)\}$ response component, and show a sketch of it.

****

8.15 Consider a 2DF linear system that has modes with natural circular frequencies of $\omega_1 = 3$ rad/s and $\omega_2 = 8$ rad/s and damping values of $\zeta_1 = 0.01$ and $\zeta_2 = 0.02$. The excitation is a broadband Gaussian process. Analysis has shown that the stationary modal responses are mean-zero and have standard deviation values of $\sigma_{Z_1} = 20$ mm and $\sigma_{Z_2} = 10$ mm. The response of primary interest is $X(t) = Z_1(t) + Z_2(t)$.

(a) Estimate the σ_X standard deviation of stationary response.

(b) Estimate the $\sigma_{\dot{X}}$ standard deviation of velocity of stationary response.
(c) Approximate the rate of upcrossings of the level zero by the stationary $\{X(t)\}$ process.

[Note: If you find that insufficient information is given, then use reasonable approximations and explain your reasoning.]

****

8.16 A linear structure has been studied by modal analysis. It has been determined that only four modes contribute significantly to the dynamic response. The undamped natural frequencies and damping ratios of these modes are as given:

$$\omega_1 = 3 \text{ rad/s} \quad \omega_2 = 5 \text{ rad/s} \quad \omega_3 = 8 \text{ rad/s} \quad \omega_4 = 12 \text{ rad/s}$$

$$\zeta_1 = 0.01 \quad \zeta_2 = 0.01 \quad \zeta_3 = 0.02 \quad \zeta_4 = 0.03$$

The process $\{X(t)\}$ represents a critical distortion in the structure, and it is written as the sum of the four modal contributions: $X(t) = Y_1(t) + Y_2(t) + Y_3(t) + Y_4(t)$. The excitation is modeled as a broadband, zero-mean, Gaussian process, and analysis of the dynamic response of the structure has given $\sigma_{Y_1} = 20$ mm, $\sigma_{Y_2} = 10$ mm, $\sigma_{Y_3} = 2$ mm, and $\sigma_{Y_4} = 0.5$ mm.

(a) Estimate the σ_X standard deviation of stationary response.
(b) Estimate the $\sigma_{\dot{X}}$ standard deviation of velocity of stationary response.
(c) Approximate the rate of upcrossings of the level zero by the stationary $\{X(t)\}$ process.

[Note: If you find that insufficient information is given, then use reasonable approximations and explain your reasoning.]

****

State-Space Formulation

****

8.17 Consider the 2DF system of Exercises 8.1 and 8.8.

(a) Find symmetric **A** and **B** matrices and the $\vec{Q}(t)$ vector such that this system is described by Eq. 8.44 with $\vec{Y}(t) = [X_1(t), X_2(t), \dot{X}_1(t), \dot{X}_2(t)]^T$.
(b) Verify that the eigenvalues of $\mathbf{A}^{-1}\mathbf{B}$ are $\pm i\,\omega_j[1-(\zeta_j)^2])^{1/2} + \zeta_j\omega_j$. That is, show that the values $\lambda = \pm i\,\omega_j[1-(\zeta_j)^2]^{1/2} + \zeta_j\omega_j$ satisfy the eigenvalue relationship $|\mathbf{A}^{-1}\mathbf{B} - \lambda\,\mathbf{I}_4| = 0$.

****

8.18 Consider the 2DF system of Exercises 8.2 and 8.9.

(a) Find symmetric **A** and **B** matrices and the $\vec{Q}(t)$ vector such that this system is described by Eq. 8.44 with $\vec{Y}(t) = [X_1(t), X_2(t), \dot{X}_1(t), \dot{X}_2(t)]^T$.
(b) Verify that the eigenvalues of $\mathbf{A}^{-1}\mathbf{B}$ are $\pm i\,\omega_j[1-(\zeta_j)^2])^{1/2} + \zeta_j\omega_j$. That is, show that the values $\lambda = \pm i\,\omega_j[1-(\zeta_j)^2]^{1/2} + \zeta_j\omega_j$ satisfy the eigenvalue relationship $|\mathbf{A}^{-1}\mathbf{B} - \lambda\,\mathbf{I}_4| = 0$.

****

8.19 Consider the 2DF system of Example 8.4 and Exercises 8.3 and 8.10.

(a) Find symmetric **A** and **B** matrices and the $\vec{Q}(t)$ vector such that this system is described by Eq. 8.44 with $\vec{Y}(t)=[X_1(t),X_2(t),\dot{X}_1(t),\dot{X}_2(t)]^T$.

(b) Verify that the eigenvalues of $\mathbf{A}^{-1}\mathbf{B}$ are $\pm i\,\omega_j[1-(\zeta_j)^2])^{1/2}+\zeta_j\omega_j$. That is, show that the values $\lambda=\pm i\,\omega_j[1-(\zeta_j)^2]^{1/2}+\zeta_j\omega_j$ satisfy the eigenvalue relationship $|\,\mathbf{A}^{-1}\mathbf{B}-\lambda\,\mathbf{I}_4\,|=0$.

****

8.20 Consider the 2DF system of Exercise 8.11.

(a) Find symmetric **A** and **B** matrices and the corresponding $\vec{Q}(t)$ vector such that this system is described by Eq. 8.44 with a state vector of $\vec{Y}(t)=[X_1(t),X_2(t),\dot{X}_1(t),\dot{X}_2(t)]^T$.

(b) Verify that the eigenvalues of $\mathbf{A}^{-1}\mathbf{B}$ are $\lambda=\omega_p/i$, with ω_p denoting the pole locations found in Exercise 8.11.

****

8.21 Consider the system of Example 8.12 with $k_1=k_2=k$, $c_1=0.01(k\,m)^{1/2}$, and $c_2=50\,c_1$, in which m and k are positive scalar constants.

(a) Find the eigenvalues and eigenvectors of the state-space matrix $\mathbf{A}^{-1}\mathbf{B}$. (Two eigenvalues are complex and one is real.)

(b) Find the $\mathbf{H}(\omega)$ harmonic transfer matrix for the state-space vector $\vec{Y}(t)=[X_1(t),X_2(t),\dot{X}_1(t)]^T$.

****

Chapter 9
Direct Stochastic Analysis of Linear Systems

9.1 Basic Concept

In this chapter we will study linear stochastic vibration problems by using techniques that are fundamentally different from those we have used so far. These "direct" stochastic analysis techniques provide an alternative to the "indirect" time-domain and frequency-domain approaches that were introduced in Chapters 5 and 6. The sense in which the time-domain and frequency-domain analyses are indirect is that both begin with a deterministic solution for the linear problem, then consider the effect of replacing the deterministic excitation with a stochastic process. In the approaches presented in this chapter, the stochastic analysis comes first. In particular, deterministic differential equations are derived in which the unknowns are statistical properties of the stochastic response rather than the time history of response. The derivation of these equations is accomplished directly from the stochastic differential equation of motion governing the system, not from solutions of the equation of motion. Solution of these deterministic differential equations, then, directly gives the evolution over time of one or more particular response properties.

One advantage of the direct methods over the indirect methods is that they have more potential for nonlinear problems, because they do not begin with an initial assumption of linearity. Superposition methods, such as the Duhamel convolution integral and the Fourier transform, may be used in solving the deterministic differential equations when they are linear, but they are not used in deriving the equations. Use of direct methods for nonlinear problems will be explored in Chapter 10. Even for linear systems, though, the direct methods are sometimes more convenient than the indirect methods, and/or they can provide information that is not readily available from the indirect methods.

There are two basic types of response quantities that will be considered here. In the simplest situation one derives ordinary differential equations that govern the evolution of particular moments of the response. These are called *state-space moment equations*, because it is always necessary to consider a coupled set of equations governing the evolution of moments of all the state variables. A slight variation on state-space moment analysis involves using equations that govern the evolution of coupled sets of cumulants of the response state variables. The more complicated situation involves a partial differential equation governing the evolution of the probability density function of the state variables. This is called the Fokker-Planck or Kolmogorov equation. We will consider all these situations, but we will give more emphasis to the simpler equations involving moments or cumulants. In all situations it will be found that the problem is simplified if the quantity being studied is stationary so time derivatives disappear. This is particularly significant for state-space moment or cumulant equations, because this results in the reduction of ordinary differential equations to algebraic equations.

It is usually easiest to apply the methods presented in this chapter by using the state-space formulation of the equations of motion, as introduced in Section 8.6. However, this is definitely not essential, and we illustrate this idea in Example 9.1 by deriving moment equations for the second moments of a single-degree-of-freedom (SDF) oscillator directly from the original form of the equation of motion. The important point to be remembered is that the state-space formulation of the equations of motion and the direct methods of analysis of state variables are separate concepts that can be efficiently used together but that can also be used separately.

****

Example 9.1: Let $\{X(t)\}$ denote the response of the SDF oscillator governed by the equation of motion

$$m\,\ddot{X}(t) + c\,\dot{X}(t) + k\,X(t) = F(t)$$

For such a second-order differential equation it is customary to designate $X(t)$ and $\dot{X}(t)$ as state variables and to seek equations involving the moments of these variables. To do this we begin by multiplying each side of the equation of motion by $X(t)$, then taking the expected value. This gives

$$m\,E[X(t)\,\ddot{X}(t)] + c\,E[X(t)\,\dot{X}(t)] + k\,E[X^2(t)] = E[X(t)\,F(t)]$$

Note that the second and third terms involve only moment functions of the state variables, but the first term involves $\ddot{X}(t)$, which is not a state variable. This first term can be rewritten, though, in terms of moment functions involving $X(t)$ and $\dot{X}(t)$ as

$$E[X(t)\,\ddot{X}(t)] = \frac{d}{dt}E[X(t)\,\dot{X}(t)] - E[\dot{X}^2(t)]$$

Thus, the equation becomes

$$m\frac{d}{dt}E[X(t)\,\dot{X}(t)] - m\,E[\dot{X}^2(t)] + c\,E[X(t)\,\dot{X}(t)] + k\,E[X^2(t)] = E[X(t)\,F(t)]$$

This involves only moment functions of the state variables and the term $E[X(t)\,F(t)]$, which involves excitation as well as a state variable. Similarly, one can multiply the original equation of motion by $\dot{X}(t)$ and take the expected value to obtain

$$m\,E[\dot{X}(t)\,\ddot{X}(t)] + c\,E[\dot{X}^2(t)] + k\,E[X(t)\,\dot{X}(t)] = E[\dot{X}(t)\,F(t)]$$

and this can be rewritten in terms of the three moment functions of the state variables as

$$\frac{m}{2}\frac{d}{dt}E[\dot{X}^2(t)] + c\,E[\dot{X}^2(t)] + k\,E[X(t)\,\dot{X}(t)] = E[\dot{X}(t)\,F(t)]$$

Note also that the general relationship

$$\frac{d}{dt}E[X^2(t)] = 2\,E[X(t)\,\dot{X}(t)]$$

gives a third equation involving only the moment functions of the state variables. If we can somehow find the values of $E[X(t)\,F(t)]$ and $E[\dot{X}(t)\,F(t)]$, then this will give us three simultaneous linear differential equations that can be solved to find the three moment functions $E[X^2(t)]$, $E[X(t)\,\dot{X}(t)]$, and $E[\dot{X}^2(t)]$. In vector form these equations can be written as

$$m\frac{d}{dt}\begin{pmatrix} E[X^2(t)] \\ E[X(t)\dot{X}(t)] \\ E[\dot{X}^2(t)] \end{pmatrix} + \begin{pmatrix} 0 & -2m & 0 \\ k & c & -m \\ 0 & 2k & 2c \end{pmatrix}\begin{pmatrix} E[X^2(t)] \\ E[X(t)\dot{X}(t)] \\ E[\dot{X}^2(t)] \end{pmatrix} = \begin{pmatrix} 0 \\ E[X(t)\,F(t)] \\ 2\,E[\dot{X}(t)\,F(t)] \end{pmatrix}$$

A tremendous simplification in this equation results for the special case of second-moment stationary response of the oscillator. In particular, if $E[X^2(t)]$, $E[X(t)\,\dot{X}(t)]$, and $E[\dot{X}^2(t)]$ are all independent of time, then the derivative terms completely drop out of the equations and $E[X(t)\,\dot{X}(t)] = 0$. In the original scalar form of the equations, this gives $-m\,E[\dot{X}^2(t)] + k\,E[X^2(t)] = E[X(t)\,F(t)]$ and $c\,E[\dot{X}^2(t)] = E[\dot{X}(t)\,F(t)]$ as the two algebraic equations governing the values of $E[X^2(t)]$ and $E[\dot{X}^2(t)]$.

****

Several important general properties of state-space moment analysis are illustrated by Example 9.1. First, we note that no assumptions or approximations were made in deriving the moment equations, other than the assumed existence of the various expected values. Thus, the equations will be true for almost any $\{F(t)\}$ excitation of the oscillator. In addition, we find that we cannot derive one moment equation that governs the behavior of only one moment function. In particular, we found in the example that three simultaneous equations governed the behavior of the three possible second-moment functions of the state variables. This coupling of all moments of the same order is typical of the method. For linear systems, though, it is also typical that the moment equations of one order are uncoupled from those of any other order. For instance, no mean-value terms or third moment terms appear in the second-order moment equations of Example 9.1. Furthermore, the property that one need solve only algebraic equations, rather than differential equations, when evaluating stationary moments of the state variables is also a general feature of the method. Finally, we note that Example 9.1 illustrates the fact that in order to make the moment equations useful, we must find a way to evaluate certain cross-product values involving both the excitation and the response. We will begin to address this problem in Section 9.4, after introducing a systematic procedure for deriving state-space moment equations.

It may be noted that there is no particular reason for the methods of direct stochastic analysis to follow the time-domain and frequency-domain methods of indirect analysis presented in Chapters 5, 6, and 8. The method does typically involve matrix equations, so some of the material in Chapter 8 is relevant here; other than that, however, this chapter could have come directly after Chapter 4. In fact, in a Russian book in 1958 (Lebedev), the author begins with state-space moment equations as the fundamental tool for analyzing stochastic response. At least within the United States, the method has been much less frequently used than the indirect methods, but that is changing somewhat in recent years.

9.2 Derivation of State-Space Moment and Cumulant Equations

Let the $\{X(t)\}$ process be the response of a general linear system excited by a stochastic process $\{F(t)\}$. A state-space moment equation results any time that we multiply a governing differential equation of motion by some power of $X(t)$ and take the expected value of both sides of the resulting equation. For example, if the equation of motion can be written as

$$\sum_{j=0}^{n} a_j \frac{d^j X(t)}{d t^j} = F(t) \tag{9.1}$$

and we multiply both sides of this equation by $X^k(t)$ and then take the expected value, we obtain

$$\sum_{j=0}^{n} a_j E\left(X^k(t) \frac{d^j X(t)}{dt^j} \right) = E[X^k(t) F(t)] \tag{9.2}$$

We call this a moment equation, because each term is of the form of a cross-product of $\{X(t)\}$ and either $\{F(t)\}$ or a derivative of $\{X(t)\}$. Other moment equations can be obtained by multiplying Eq. 9.1 by different terms. For example, a multiplier that is the kth power of the same process at some other time s gives

$$\sum_{j=0}^{n} a_j E\left(X^k(s) \frac{d^j X(t)}{dt^j} \right) = E[X^k(s) F(t)]$$

Similarly, we could multiply by a power of some derivative of $\{X(t)\}$, such as

$$\sum_{j=0}^{n} a_j E\left(\dot{X}^k(t) \frac{d^j X(t)}{dt^j} \right) = E[\dot{X}^k(t) F(t)]$$

or we could multiply by a term involving some different process, including $\{F(t)\}$

$$\sum_{j=0}^{n} a_j E\left(F^k(t) \frac{d^j X(t)}{dt^j} \right) = E[F^{k+1}(t)]$$

Each of these procedures gives a moment equation, as does any combination of the procedures.

We can obtain cumulant equations in the same way. Specifically, recall the linearity property of cumulants given in Eq. 3.44:

$$\kappa_{n+1}\left(W_1,\cdots,W_n,\sum a_j Z_j\right) = \sum a_j \kappa_{n+1}(W_1,\cdots,W_n,Z_j) \tag{9.3}$$

for any set of W_l and Z_j random variables. This shows that the joint cumulant having one argument that is a linear combination of random variables can be written as a linear combination of joint cumulants. In our state-space analysis we are generally dealing with equations of motion that have the form of linear combinations of random variables. The linearity property of cumulants allows us to deal with joint cumulants of such equations of motion and other random variables. For example, if we take the joint cumulant of Eq. 9.1 and $X^k(t)$ we obtain

$$\sum_{j=0}^{n} a_j \kappa_2\left(X^k(t), \frac{d^j X(t)}{dt^j}\right) = \kappa_2\left(X^k(t), F(t)\right) \qquad (9.4)$$

which is the cumulant version of the moment equation in Eq. 9.2. Each of our other moment equations can also be rewritten as a cumulant equation by simply replacing each moment term by the corresponding cumulant term.

Clearly it is not difficult to derive moment or cumulant equations. In fact, we can easily derive any number of such equations. The significant issues that must be addressed relate to the identification of sets of equations that can be solved and to the simplification of the cross-product or cross-cumulant terms involving both the response and the excitation, such as the right-hand side of Eqs. 9.2 and 9.4.

As mentioned in Section 9.1, it is usually easier to implement the state-space analysis techniques by using the state-space formulation of the equations of motion. Thus, we will direct our attention to systems governed by the equation

$$\mathbf{A}\dot{\vec{Y}}(t) + \mathbf{B}\vec{Y}(t) = \vec{Q}(t) \qquad (9.5)$$

in which $\vec{Y}(t)$ is the state vector of dimension n_Y describing the response, **A** and **B** are constant square matrices, and the $\vec{Q}(t)$ vector describes the excitation. As shown in Section 8.6, this form of the equations can be applied to all systems for which the original equations of motion are linear ordinary differential equations. Even though use of this form of the equations is somewhat arbitrary, it is standard for direct stochastic analysis and it emphasizes the similarity of the various problems, regardless of the original form of the equations. As in Section 8.6, mathematical manipulations will sometimes be significantly simplified if **A** and **B** can be chosen to be symmetric.

9.3 Equations for First and Second Moments and Covariance

We now wish to identify the sets of moment equations governing the behavior of the response moments of a particular order. To do this, we must find moment equations including only those particular moments and their derivatives with respect to time. In this section we look only at the mean-value function, the second-moment autocorrelation function, and the second-cumulant covariance function, which is also the second-order central moment function. First- and second-moment information, of course, gives a complete description of a

Gaussian process, so one reason for emphasizing first- and second-moment analysis is the fact that many problems can be modeled by stochastic processes that are Gaussian or nearly Gaussian. Furthermore, as for random variables, first- and second-moment information can be considered the first two steps, and usually the most important steps, in a more accurate description of a non-Gaussian process. In this chapter, as in Chapters 5–8, it is important to remember that the analysis of first and second moments or cumulants is in no way limited to the Gaussian problem. That is, the analysis of mean and covariance, for example, is carried out in exactly the same way for Gaussian and non-Gaussian processes; the difference in the two situations is whether this first- and second-cumulant information gives a complete or a partial description of the process of interest.

The simplest, and almost trivial, case of state-space moment analysis is for the first moment, or mean-value function, which is also the first cumulant. Simply taking the expected value of both sides of Eq. 9.5 gives the first-order moment equation as

$$\mathbf{A}\dot{\vec{\mu}}_Y(t) + \mathbf{B}\vec{\mu}_Y(t) = \vec{\mu}_Q(t) \tag{9.6}$$

in which we have used the standard notation $\vec{\mu}_Y(t) \equiv E[\vec{Y}(t)]$ and $\vec{\mu}_Q(t) \equiv E[\vec{Q}(t)]$. Thus, the mean values of the state variables are governed by a first-order vector differential equation that has the same form as Eq. 9.5. There is a major difference, though, in that Eq. 9.6 is not a stochastic equation. That is, neither $\vec{\mu}_Y(t)$ or $\vec{\mu}_Q(t)$ is stochastic, so this is a set of ordinary deterministic differential equations.

Next we consider the derivation of equations involving the second moments of the state variables. Inasmuch as the state variables are the n_Y components of the state vector $\vec{Y}(t)$, it is convenient to write the second moments as the matrix $\boldsymbol{\phi}_{YY}(t,t) \equiv E[Y(t)\,Y^T(t)]$. This matrix contains all the second-moment terms, with the (j,l) component being $E[Y_j(t)\,Y_l(t)]$. Note that the matrix is symmetric, so that there are $n_Y(n_Y+1)/2$, rather than $(n_Y)^2$ distinct scalar moment functions to be determined in the second-order moment analysis. The state-space analysis of these second moments will involve finding and solving a matrix differential equation governing the behavior of the $\boldsymbol{\phi}_{YY}(t,t)$ matrix. This differential equation must involve the derivative of the matrix with respect to time, and this can be written as

$$\frac{d}{dt}\boldsymbol{\phi}_{YY}(t,t) = E[\dot{\vec{Y}}(t)\,\vec{Y}^T(t)] + E[\vec{Y}(t)\,\dot{\vec{Y}}^T(t)] \tag{9.7}$$

An equation involving the first term on the right-hand side of Eq. 9.7 is easily obtained by multiplying Eq. 9.5 by $\vec{Y}^T(t)$ on the right and $\mathbf{A}^{-1}$ on the left, then taking the expected value, giving

$$E[\dot{\vec{Y}}(t)\vec{Y}^T(t)] + \mathbf{A}^{-1}\mathbf{B}\,\boldsymbol{\phi}_{YY}(t,t) = \mathbf{A}^{-1}E[\vec{Q}(t)\vec{Y}^T(t)] \equiv \mathbf{A}^{-1}\boldsymbol{\phi}_{QY}(t,t)$$

Similarly, the transpose of this relationship describes the behavior of the final term in Eq. 9.7. Using the fact that $\boldsymbol{\phi}_{YY}(t,t)$ is symmetric, this is

$$E[\vec{Y}(t)\dot{\vec{Y}}^T(t)] + \boldsymbol{\phi}_{YY}(t,t)(\mathbf{A}^{-1}\mathbf{B})^T = E[\vec{Y}(t)\vec{Q}^T(t)](\mathbf{A}^{-1})^T \equiv \boldsymbol{\phi}_{YQ}(t,t)(\mathbf{A}^{-1})^T$$

Adding these two equations and using Eq. 9.7, then gives

$$\frac{d}{dt}\boldsymbol{\phi}_{YY}(t,t) + (\mathbf{A}^{-1}\mathbf{B})\boldsymbol{\phi}_{YY}(t,t) + \boldsymbol{\phi}_{YY}(t,t)(\mathbf{A}^{-1}\mathbf{B})^T = \mathbf{A}^{-1}\boldsymbol{\phi}_{QY}(t,t) + \boldsymbol{\phi}_{YQ}(t,t)(\mathbf{A}^{-1})^T \tag{9.8}$$

This is a compact form of the general set of equations for the second moments of the state variables. The left-hand side of the equation involves the $\boldsymbol{\phi}_{YY}(t,t)$ matrix of unknowns, its first derivative, and the system matrices $\mathbf{A}$ and $\mathbf{B}$. The right-hand side involves cross-products of the response variables and the excitation terms. Use of Eq. 9.8 for the study of moments is commonly called Lyapunov analysis.

The matrix differential equation derived for the second moments is, of course, equivalent to a set of simultaneous scalar differential equations. In particular, the (j,l) component of Eq. 9.8 is

$$\frac{d}{dt}E[Y_j(t)\,Y_l(t)] + \sum_{r=1}^{n_Y}\sum_{s=1}^{n_Y} A_{jr}^{-1}B_{rs}E[Y_s(t)\,Y_l(t)] + \sum_{r=1}^{n_Y}\sum_{s=1}^{n_Y} E[Y_j(t)\,Y_r(t)]\,B_{sr}A_{ls}^{-1} = \sum_{r=1}^{n_Y} A_{jr}^{-1}E[Q_r(t)\,Y_l(t)] + \sum_{r=1}^{n_Y} E[Y_j(t)Q_r(t)]\,A_{lr}^{-1}$$

The symmetry of the equation makes the (l,j) component identical to this (j,l) component, so the total number of unique scalar differential equations is $n_Y(n_Y+1)/2$. Thus, the number of equations is equal to the number of unknowns, and a solution is possible if the cross-products on the right-hand side are determined. Uniqueness of the solution will be ensured by knowledge of $n_Y(n_Y+1)/2$ initial conditions, such as the initial values of all components of $\boldsymbol{\phi}_{YY}(t,t)$. Note that Example 9.1 was a special case of second-moment analysis with $n_Y = 2$, giving the number of equations and unknowns as 3.

Equations 9.6 and 9.8 both confirm the fact that if the moments of interest are stationary, then one needs to solve only algebraic equations rather than differential equations. This is quite possibly the most important feature of state-space analysis. Neither the time-domain analysis of Chapter 5 nor the frequency-domain analysis of Chapter 6 gives such a tremendous simplification for the stationary problem.

****

Example 9.2: Find coupled scalar equations for the first- and second-order moments for the third-order system of Example 8.12 governed by the equations

$$m\,\ddot{X}_1(t) + c_1\dot{X}_1(t) + k_1 X_1(t) + k_2 X_2(t) = F(t)$$

and

$$k_2 X_2(t) = c_2[\dot{X}_1(t) - \dot{X}_2(t)]$$

In Example 8.12 we showed that this system is governed by Eq. 9.5 with

$$\vec{Y}(t) = \begin{pmatrix} X_1(t) \\ X_2(t) \\ \dot{X}_1(t) \end{pmatrix},\quad \mathbf{A} = \begin{pmatrix} -k_1 & 0 & 0 \\ 0 & -k_2 & 0 \\ 0 & 0 & m \end{pmatrix},\quad \mathbf{B} = \begin{pmatrix} 0 & 0 & k_1 \\ 0 & -k_2^2/c_2 & k_2 \\ k_1 & k_2 & c_1 \end{pmatrix},\quad \vec{Q}(t) = \begin{pmatrix} 0 \\ 0 \\ F(t) \end{pmatrix}$$

Using Eq. 9.6 for the equations for the first-order moments gives three first-order differential equations as

$$-k_1\,\dot{\mu}_{Y_1}(t) + k_1\,\mu_{Y_3}(t) = 0$$

$$-k_2\,\dot{\mu}_{Y_2}(t) - (k_2^2/c_2)\mu_{Y_2}(t) - k_2\mu_{Y_3}(t) = 0$$

and

$$m\,\dot{\mu}_{Y_3}(t) + k_1\,\mu_{Y_1}(t) + k_2\,\mu_{Y_2}(t) + c_1\,\mu_{Y_3}(t) = \mu_F(t)$$

Alternatively, one can simply take the expected values of the original coupled equations of motion, giving two equations, with one of them involving a second derivative:

$$m\,\ddot{\mu}_{X_1}(t) + c_1\,\dot{\mu}_{X_1}(t) + k_1\,\mu_{X_1}(t) + k_2\,\mu_{X_2}(t) = \mu_F(t)$$

and

$$k_2\,\mu_{X_2}(t) = c_2[\dot{\mu}_{X_1}(t) - \dot{\mu}_{X_2}(t)]$$

Given that $n_Y = 3$, we see that it is possible to use symmetry to reduce the problem to six simultaneous scalar equations governing six second-moment terms. One can use either the upper triangular portion or the lower triangular portion of Eq. 9.8 to give these six equations. Evaluating $\mathbf{A}^{-1}\mathbf{B}$ as

$$\mathbf{A}^{-1}\mathbf{B} = \begin{pmatrix} 0 & 0 & -1 \\ 0 & k_2/c_2 & -1 \\ k_1/m & k_2/m & c_1/m \end{pmatrix}$$

and proceeding through the lower triangular components of Eq. 9.8 in the order (1,1), (2,1), (2,2), (3,1), (3,2), (3,3) gives the equations as

$$\frac{d}{dt}E[Y_1^2(t)] - 2E[Y_1(t)Y_3(t)] = 0$$

$$\frac{d}{dt}E[Y_1(t)Y_2(t)] + \frac{k_2}{c_2}E[Y_1(t)Y_2(t)] - E[Y_1(t)Y_3(t)] - E[Y_2(t)Y_3(t)] = 0$$

$$\frac{d}{dt}E[Y_2^2(t)] + 2\frac{k_2}{c_2}E[Y_2^2(t)] - 2E[Y_2(t)Y_3(t)] = 0$$

$$\frac{d}{dt}E[Y_1(t)Y_3(t)] + \frac{k_1}{m}E[Y_1^2(t)] + \frac{k_2}{m}E[Y_1(t)Y_2(t)] + \frac{c_1}{m}E[Y_1(t)Y_3(t)] - E[Y_3^2(t)] = \frac{1}{m}E[F(t)Y_1(t)]$$

$$\frac{d}{dt}E[Y_2(t)Y_3(t)] + \frac{k_1}{m}E[Y_1(t)Y_2(t)] + \frac{k_2}{m}E[Y_2^2(t)] + \left(\frac{k_2}{c_2} + \frac{c_1}{m}\right)E[Y_2(t)Y_3(t)] - E[Y_3^2(t)] = \frac{1}{m}E[F(t)Y_2(t)]$$

and

$$\frac{d}{dt}E[Y_3^2(t)] + 2\frac{k_1}{m}E[Y_1(t)Y_3(t)] + 2\frac{k_2}{m}E[Y_2(t)Y_3(t)] + 2\frac{c_1}{m}E[Y_3^2(t)] = \frac{2}{m}E[F(t)Y_3(t)]$$

Note that it is also possible to write these relationships as one matrix equation of the form

$$\dot{\vec{V}}(t) + \mathbf{G}\vec{V}(t) = \vec{\psi}(t)$$

with a vector of unknowns of

$$\vec{V}(t) = \left(E[Y_1^2(t)], E[Y_1(t)Y_2(t)], E[Y_2^2(t)], E[Y_1(t)Y_3(t)], E[Y_2(t)Y_3(t)], E[Y_3^2(t)]\right)^T$$

and a 6×6 matrix of coefficients. Clearly, this version of the second-moment equations has the same form as the vector equation of motion in Eq. 9.5, or the vector first-moment equation in Eq. 9.6. Alternative ways of finding the solutions of Equation 9.8 will be discussed in Section 9.5.

****

In many cases, we will find it to be more convenient to use state-space covariance analysis rather than second-moment analysis. This is primarily because of simplifications that result in the cross-product terms involving both the excitation and the response, as on the right-hand side of Eq. 9.8. In addition, if the process of interest is Gaussian, the mean-value and covariance functions give us the elements needed for the standard form of the probability density function. It turns out to be very simple to derive a state-space covariance equation similar to Eq. 9.8. One approach is to substitute Eqs. 9.6 and 9.8 into the general relationship describing the covariance matrix as $\mathbf{K}_{YY}(t,t) = \boldsymbol{\phi}_{YY}(t,t) - \vec{\mu}_Y(t)[\vec{\mu}_Y(t)]^T$, which shows that $\mathbf{K}_{YY}(t,t)$ is governed by an equation that is almost identical in form to Eq. 9.8:

$$\frac{d}{dt}\mathbf{K}_{YY}(t,t) + (\mathbf{A}^{-1}\mathbf{B})\,\mathbf{K}_{YY}(t,t) + \mathbf{K}_{YY}(t,t)(\mathbf{A}^{-1}\mathbf{B})^T = \mathbf{A}^{-1}\mathbf{K}_{QY}(t,t) + \mathbf{K}_{YQ}(t,t)(\mathbf{A}^{-1})^T \tag{9.9}$$

Alternatively, Eq. 9.9 can be derived by using the relationships for cumulants, without consideration of first and second moments. First, we note that the cross-covariance matrix for any two vectors can be viewed as a two-dimensional array of the possible second-order cross-cumulants of the components of the vectors, so we can write $\mathbf{K}_{QY}(t,s) = \kappa_2[Q(t),Y(s)]$, for example. The linearity property of cumulants, as given in Eq. 9.3, can then be used to show that the derivative of $\mathbf{K}_{YY}(t,t)$ can be written as

$$\frac{d}{dt}\mathbf{K}_{YY}(t,t) = \mathbf{K}_{\dot{Y}Y}(t,t) + \mathbf{K}_{Y\dot{Y}}(t,t)$$

which is the cumulant version of Eq. 9.7. Furthermore, Eq. 9.3 allows us to take the joint cumulant of $\vec{Y}(t)$ and Eq. 9.5 and write the result as

$$\mathbf{K}_{\dot{Y}Y}(t,t) + \mathbf{A}^{-1}\mathbf{B}\,\mathbf{K}_{YY}(t,t) = \mathbf{A}^{-1}\mathbf{K}_{QY}(t,t)$$

Adding this equation to its transpose gives exactly Eq. 9.9. The advantage of this second derivation is that it can also be directly generalized to obtain higher-order cumulant equations that will be considered in Section 9.7.

9.4 Simplifications for Delta-Correlated Excitation

We now investigate the right-hand side of our state-space equations, beginning with the simplest situation. In particular, we begin by looking at the state-space covariance equation in which the $\{\vec{Q}(t)\}$ excitation vector stochastic process is delta-correlated. We do not impose any stationarity condition, nor do we assume that different components of $\vec{Q}(t)$ be independent of each other at any time t. We do require, though, that $\vec{Q}(t)$ and $\vec{Q}(s)$ are independent vector random variables for $t \neq s$, meaning that knowledge of the values of $\vec{Q}(t)$ gives no information about the possible values of $\vec{Q}(s)$ or about the probability distribution on those possible values. For this situation, the covariance matrix of the excitation process is given by

$$\mathbf{K}_{QQ}(t,s) = 2\pi \mathbf{S}_0(t)\,\delta(t-s) \tag{9.10}$$

in which $\mathbf{S}_0(t)$ is the nonstationary autospectral density matrix for $\{\vec{Q}(t)\}$.

As the first step in simplifying the right-hand side of Eq. 9.9, we write $\vec{Y}(t)$ in the form

$$\vec{Y}(t) = \vec{Y}(t_0) + \int_{t_0}^{t} \dot{\vec{Y}}(u)\,du$$

in which t_0 is any time prior to t. We can now solve Eq. 9.5 for $\dot{\vec{Y}}(u)$ and substitute into this expression, giving

$$\vec{Y}(t) = \vec{Y}(t_0) + \mathbf{A}^{-1} \int_{t_0}^{t} \vec{Q}(u)\,du - \mathbf{A}^{-1}\mathbf{B} \int_{t_0}^{t} \vec{Y}(u)\,du \tag{9.11}$$

Transposing this equation and substituting it into $\mathbf{K}_{QY}(t,t)$ gives

$$\mathbf{K}_{QY}(t,t) = \mathbf{K}_{QY}(t,t_0) + \int_{t_0}^{t} \mathbf{K}_{QQ}(t,u)\,du\,(\mathbf{A}^{-1})^T - \int_{t_0}^{t} \mathbf{K}_{QY}(t,u)\,du\,(\mathbf{A}^{-1}\mathbf{B})^T \tag{9.12}$$

We will show, though, that only one of the terms on the right-hand side of this equation contributes to $\mathbf{K}_{QY}(t,t)$ when $\{\vec{Q}(t)\}$ is a delta-correlated process.

First, we note that because $\vec{Q}(t)$ is independent of $\vec{Q}(u)$ for $u < t$, we can also argue from the principle of cause and effect that $\vec{Q}(t)$ is independent of $\vec{Y}(t_0)$ for $t > t_0$. That is, the response at time t_0 is due to the portion of the excitation that occurred at times up to time t_0, but $\vec{Q}(t)$ is independent of that portion of the excitation. Thus, $\vec{Q}(t)$ is independent of $\vec{Y}(t_0)$. This independence, of course, implies that $\mathbf{K}_{QY}(t,t_0) = 0$, eliminating the first term on the right-hand side of Eq. 9.12. Furthermore, this same property allows the last term in Eq. 9.12 to be written as

$$\int_{t_0}^{t} \mathbf{K}_{QY}(t,u)\,du = \int_{t-\Delta t}^{t} \mathbf{K}_{QY}(t,u)\,du$$

in which the time increment Δt can be taken to be arbitrarily small. We note, though, that if $\mathbf{K}_{QY}(t,t)$ is finite, then this term is an integral of a finite integrand over an infinitesimal interval, so it also is equal to zero. Thus, if the second term on the right-hand side of Eq. 9.12 is finite, then it is exactly equal to $\mathbf{K}_{QY}(t,t)$, inasmuch as the other terms on the right-hand side of the equation are then zero.

Simply substituting the Dirac delta function autocovariance of Eq. 9.10 into the integral in the second term on the right-hand side of Eq. 9.12 demonstrates that the term is finite, but leaves some ambiguity about its value. In particular, we can write an integral similar to, but more general than, the one of interest as

$$\int_{t_0}^{s} \mathbf{K}_{QQ}(t,u)\,du = 2\pi \int_{t_0}^{s} \mathbf{S}_0(t)\,\delta(t-u)\,du = 2\pi \mathbf{S}_0(t)\,U(s-t)$$

For $s < t$ this integral is zero, and for $s > t$ it is $2\pi \mathbf{S}_0(t)$. We are interested in the situation with $s = t$, though, for which the upper limit of the integral is precisely aligned with the Dirac delta pulse of the integrand. This is the one value for the upper limit for which there is uncertainty about the value of the integral. We resolve this problem in the following paragraph, but for the moment we simply write

$$\mathbf{K}_{QY}(t,t) = \int_{t_0}^{t} \mathbf{K}_{QQ}(t,u)\,du\,(\mathbf{A}^{-1})^T \tag{9.13}$$

Note that the transpose of this term also appears in Eq. 9.9, and we write it in similar fashion as

$$\mathbf{K}_{YQ}(t,t) = \mathbf{A}^{-1} \int_{t_0}^{t} \mathbf{K}_{QQ}(u,t)\,du$$

so Eq. 9.9 can be written as

$$\frac{d}{dt}\mathbf{K}_{YY}(t,t)+(\mathbf{A}^{-1}\mathbf{B})\,\mathbf{K}_{YY}(t,t)+\mathbf{K}_{YY}(t,t)(\mathbf{A}^{-1}\mathbf{B})^T=$$
$$\mathbf{A}^{-1}\left(\int_{t_0}^{t}\mathbf{K}_{QQ}(t,u)\,du+\int_{t_0}^{t}\mathbf{K}_{QQ}(u,t)\,du\right)(\mathbf{A}^{-1})^T \tag{9.14}$$

To resolve the ambiguity about the value of the right-hand side of Eq. 9.14, we note that the Dirac delta form of the autocovariance in Eq. 9.10 gives

$$\int_{t_0}^{t}\int_{t_0}^{t}\mathbf{K}_{QQ}(u,v)\,du\,dv=2\pi\int_{t_0}^{t}\mathbf{S}_0(v)U(t-v)\,dv=2\pi\int_{t_0}^{t}\mathbf{S}_0(v)\,dv \tag{9.15}$$

for any $t_0<t$. Taking the derivative with respect to t of each side of this equation gives

$$\int_{t_0}^{t}\mathbf{K}_{QQ}(t,v)\,dv+\int_{t_0}^{t}\mathbf{K}_{QQ}(u,t)\,du=2\pi\mathbf{S}_0(t) \tag{9.16}$$

Note, though, that the left-hand side of this equation is exactly the same as the term on the right-hand side of Eq. 9.14. Thus, we find that the state-space covariance equation can be written as

$$\frac{d}{dt}\mathbf{K}_{YY}(t,t)+\mathbf{A}^{-1}\mathbf{B}\,\mathbf{K}_{YY}(t,t)+\mathbf{K}_{YY}(t,t)(\mathbf{A}^{-1}\mathbf{B})^T=2\pi\,\mathbf{A}^{-1}\mathbf{S}_0(t)(\mathbf{A}^{-1})^T \tag{9.17}$$

for the special case when the $\{\vec{Q}(t)\}$ excitation is delta-correlated with nonstationary autospectral density matrix $\mathbf{S}_0(t)$.

Similarly, we can simplify the right-hand side of Eq. 9.8 describing the second moments of the state variables by noting that

$$\mathbf{A}^{-1}\boldsymbol{\phi}_{QY}(t,t)+\boldsymbol{\phi}_{YQ}(t,t)(\mathbf{A}^{-1})^T=\mathbf{A}^{-1}\mathbf{K}_{QY}(t,t)+\mathbf{K}_{YQ}(t,t)(\mathbf{A}^{-1})^T+$$
$$\mathbf{A}^{-1}\vec{\mu}_Q(t)\vec{\mu}_Y^T(t)+\vec{\mu}_Y(t)\vec{\mu}_Q^T(t)(\mathbf{A}^{-1})^T$$

so we have

$$\frac{d}{dt}\boldsymbol{\phi}_{YY}(t,t)+\mathbf{A}^{-1}\mathbf{B}\boldsymbol{\phi}_{YY}(t,t)+\boldsymbol{\phi}_{YY}(t,t)(\mathbf{A}^{-1}\mathbf{B})^T = \\ 2\pi\,\mathbf{A}^{-1}\mathbf{S}_0(t)(\mathbf{A}^{-1})^T+\mathbf{A}^{-1}\vec{\mu}_Q(t)\vec{\mu}_Y^T(t)+\vec{\mu}_Y(t)\vec{\mu}_Q^T(t)(\mathbf{A}^{-1})^T \tag{9.18}$$

for the autocorrelation matrix of the response to the nonstationary delta-correlated excitation. This equation is somewhat less convenient than Eq. 9.17, because it involves both the mean-value vector and the autospectral density matrix of the excitation. This is the first demonstration of any significant difference between moment equations and cumulant equations. The difference in complexity between moment and cumulant equations becomes more substantial when the terms considered are of higher order.

9.5 Solution of the State-Space Equations

First we reiterate that the stationary solutions of Eqs. 9.6 and 9.17 are found from algebraic equations. In particular, if $\vec{\mu}_Q$ and $\mathbf{S}_0$ are independent of time and we seek only stationary values of $\vec{\mu}_Y$ and $\mathbf{K}_{YY}$, we have equations of

$$\mathbf{B}\,\vec{\mu}_Y=\vec{\mu}_Q \tag{9.19}$$

and

$$\mathbf{A}^{-1}\mathbf{B}\,\mathbf{K}_{YY}+\mathbf{K}_{YY}(\mathbf{A}^{-1}\mathbf{B})^T=2\pi\,\mathbf{A}^{-1}\mathbf{S}_0(\mathbf{A}^{-1})^T \tag{9.20}$$

Obviously, the solution of Eq. 9.19 for the stationary mean value can be written as

$$\vec{\mu}_Y=\mathbf{B}^{-1}\,\vec{\mu}_Q \tag{9.21}$$

but the solution of Eq. 9.20 for the stationary covariance matrix is not so obvious. We will return to this matter after writing the solutions for the nonstationary response quantities.

The solution of Eq. 9.6 for the nonstationary mean vector is very simple. In fact, it is exactly the same as was given in Section 8.6 for time-domain analysis of the state-space formulation of the equations of motion:

$$\vec{\mu}_Y(t)=\vec{\mu}_Y(t_0)\exp[-(t-t_0)\,\mathbf{A}^{-1}\mathbf{B}]+\int_{t_0}^{t}\exp[-(t-s)\mathbf{A}^{-1}\mathbf{B}]\,\mathbf{A}^{-1}\vec{\mu}_Q(s)\,ds \tag{9.22}$$

in which $\vec{\mu}_Y(t_0)$ is presumed to be a given initial condition. Similarly, the formal solution of Eq. 9.17 for the nonstationary $\mathbf{K}_{YY}(t,t)$ covariance matrix can be written as

$$\mathbf{K}_{YY}(t,t) = \exp[-(t-t_0)(\mathbf{A}^{-1}\mathbf{B})]\,\mathbf{A}^{-1}\mathbf{K}_{YY}(t_0,t_0)(\mathbf{A}^{-1})^T \exp[-(t-t_0)(\mathbf{A}^{-1}\mathbf{B})^T] + 2\pi\int_{t_0}^{t} \exp[-(t-s)(\mathbf{A}^{-1}\mathbf{B})]\,\mathbf{A}^{-1}\mathbf{S}_0(s)(\mathbf{A}^{-1})^T \exp[-(t-s)(\mathbf{A}^{-1}\mathbf{B})^T]\,ds \tag{9.23}$$

Note that taking S_0 to be a constant and ignoring the initial conditions in Eq. 9.23 gives a stationary solution of

$$\mathbf{K}_{YY} = 2\pi\int_0^{\infty} e^{-r\,\mathbf{A}^{-1}\mathbf{B}}\,\mathbf{A}^{-1}\mathbf{S}_0\,(\mathbf{A}^{-1})^T\, e^{-r\,(\mathbf{A}^{-1}\mathbf{B})^T{}^T}\,dr$$

This is a form of solution of Eq. 9.20, but it is rather awkward for a stationary solution. Of course, none of these formal solutions involving matrix exponentials is very useful until a method is used for diagonalizing the $\mathbf{A}^{-1}\mathbf{B}$ matrix.

As outlined in Section 8.6, the $\boldsymbol{\lambda}$ and $\boldsymbol{\theta}$ matrices of eigenvalues and eigenvectors are found from the equation $\mathbf{A}^{-1}\mathbf{B}\boldsymbol{\theta} = \boldsymbol{\theta}\boldsymbol{\lambda}$, in which $\boldsymbol{\lambda}$ is diagonal. Using a change of variables of

$$\vec{Y}(t) = \boldsymbol{\theta}\vec{Z}(t) \tag{9.24}$$

then allows Eq. 9.5 to be rewritten as

$$\dot{\vec{Z}}(t) + \boldsymbol{\lambda}\vec{Z}(t) = \vec{P}(t) \tag{9.25}$$

in which

$$\vec{P}(t) = \boldsymbol{\theta}^{-1}\mathbf{A}^{-1}\vec{Q}(t) \tag{9.26}$$

Consider now the response quantities for the new vector $\vec{Z}(t)$. The stationary mean value, as in Eq. 9.21, is given by

$$\vec{\mu}_Z = \boldsymbol{\lambda}^{-1}\vec{\mu}_P = \boldsymbol{\lambda}^{-1}\boldsymbol{\theta}^{-1}\mathbf{A}^{-1}\vec{\mu}_Q \tag{9.27}$$

Similarly, the equation for the stationary covariance, as in Eq. 9.20, is

$$\boldsymbol{\lambda}\,\mathbf{K}_{ZZ} + \mathbf{K}_{ZZ}\,\boldsymbol{\lambda}^T = 2\pi \mathbf{S}_{PP} = 2\pi \boldsymbol{\theta}^{-1}\mathbf{A}^{-1}\mathbf{S}_0\,(\boldsymbol{\theta}^{-1}\mathbf{A}^{-1})^T$$

This equation can be solved for any scalar component as

$$[\mathbf{K}_{ZZ}]_{jl} = \frac{2\pi [\mathbf{S}_{PP}]_{jl}}{\lambda_j + \lambda_l} \tag{9.28}$$

Using Eq. 9.24, it is possible to use Eqs. 9.27 and 9.28 to write the corresponding quantities for $\{\vec{Y}(t)\}$ as

$$\vec{\mu}_Y = \boldsymbol{\theta}\,\vec{\mu}_Z \equiv \boldsymbol{\theta}\boldsymbol{\lambda}^{-1}\boldsymbol{\theta}^{-1}\mathbf{A}^{-1}\vec{\mu}_Q = \mathbf{B}^{-1}\vec{\mu}_Q, \qquad \mathbf{K}_{YY} = \boldsymbol{\theta}\,\mathbf{K}_{ZZ}\,\boldsymbol{\theta}^T \tag{9.29}$$

The first equation is identical to Eq. 9.21, and the second provides a usable solution of Eq. 9.20.

Similarly, we can obtain expressions corresponding to Eqs. 9.22 and 9.23 for the uncoupled equations for $\{\vec{Z}(t)\}$ for the nonstationary situation

$$\vec{\mu}_Z(t) = \vec{\mu}_Z(t_0)\exp[-(t-t_0)\,\boldsymbol{\lambda}] + \int_{t_0}^{t} \exp[-(t-s)\boldsymbol{\lambda}]\,\vec{\mu}_P(s)\,ds \tag{9.30}$$

and

$$\mathbf{K}_{ZZ}(t,t) = \exp[-(t-t_0)\,\boldsymbol{\lambda}]\;\mathbf{K}_{ZZ}(t_0,t_0)\exp[-(t-t_0)\,\boldsymbol{\lambda}] + 2\pi\int_{t_0}^{t} \exp[-(t-s)\,\boldsymbol{\lambda}]\;\mathbf{S}_{PP}(s)\exp[-(t-s)\,\boldsymbol{\lambda}]\,ds \tag{9.31}$$

As for the stationary situation, Eq. 9.29 is used to convert these expressions into the ones for $\{\vec{Y}(t)\}$. Note that the matrix exponential terms in Eqs. 9.30 and 9.31 are almost trivial, because $\exp[-t\,\boldsymbol{\lambda}]$ is diagonal with $e^{-t\lambda_{jj}}$ being its (j,j) element.

It should also be recalled that if **A** and **B** are both symmetric, then computations are considerably simplified. In particular, one can find $\boldsymbol{\theta}^{-1}$ by using very simple operations, because $\boldsymbol{\theta}^{-1} = \hat{\mathbf{A}}^{-1}\boldsymbol{\theta}^T\mathbf{A}$, in which $\hat{\mathbf{A}} \equiv \boldsymbol{\theta}^T\mathbf{A}\boldsymbol{\theta}$ is almost trivial to invert inasmuch as it is a diagonal matrix.

****

Example 9.3: Using state-space analysis, find the variance of the response of the system, in which $\{F(t)\}$ is stationary with autospectral density S_0.

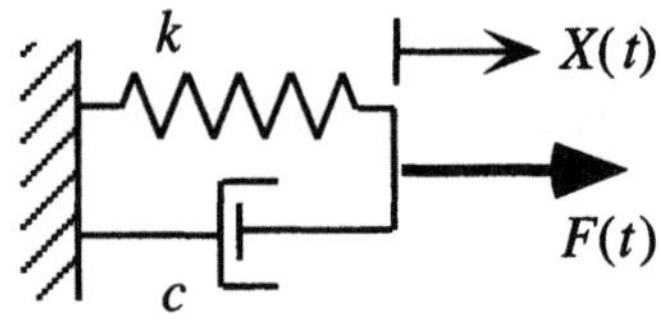

Putting the equation of motion into the form of Eq. 9.5, we have

$$c\,\dot{X}(t)+k\,X(t)=F(t)$$

so $n_Y=1$ and the state "vector" is the scalar $X(t)$. The "matrices" $\mathbf{A}$ and $\mathbf{B}$ have dimension (1,1) and are c and k, respectively, and the "vector" $\vec{Q}(t)$ is the scalar $F(t)$. The covariance "matrix" in Eq. 9.17 is then the scalar variance $K_{XX}(t)\equiv\sigma_X^2(t)$, and the relationship is

$$\frac{d}{dt}\sigma_X^2(t)+2\frac{k}{c}\sigma_X^2(t)=\frac{2\pi S_0}{c^2}$$

For the special case of stationary response, the variance is found by setting the derivative term equal to zero and solving the algebraic equation to obtain

$$\sigma_X^2=\frac{\pi S_0}{kc}$$

If we are given an initial condition of $\sigma_X^2(t_0)$ for some particular time, then the solution of the differential equation can be written as

$$\sigma_X^2(t)=\sigma_X^2(t_0)e^{-2kt/c}+\frac{2\pi S_0}{c^2}\int_{t_0}^{t}e^{-2k(t-s)/c}\,ds$$

$$=\sigma_X^2(t_0)e^{-2kt/c}+\frac{\pi S_0}{kc}[1-e^{-2k(t-t_0)/c}]$$

as in Eq. 9.22. For the common situation in which the system is at rest at time $t=0$, we have

$$\sigma_X^2(t)=\frac{\pi S_0}{kc}[1-e^{-2kt/c}]$$

All of these results, of course, are exactly the same as one would obtain by the time-domain and frequency-domain analysis techniques of Chapters 5 and 6, respectively. Their determination by state-space analysis, though, is somewhat simpler, particularly for the stationary situation.

****

Example 9.4: Find the response variance and covariance values for an SDF oscillator with a mean-zero, stationary, delta-correlated excitation with autospectral density S_0.

The symmetric state-space description of the system is Eq. 9.5 with $n_Y=2$:

$$\vec{Y}(t)=\begin{pmatrix}X(t)\\ \dot{X}(t)\end{pmatrix},\quad \mathbf{A}=\begin{pmatrix}-k & 0\\ 0 & m\end{pmatrix},\quad \mathbf{B}=\begin{pmatrix}0 & k\\ k & c\end{pmatrix},\quad \vec{Q}(t)=\begin{pmatrix}0\\ F(t)\end{pmatrix}$$

Thus, the state-space equation for the covariance, from Eq. 9.17, is

$$\frac{d}{dt}\mathbf{K}_{YY}(t,t)+\begin{pmatrix}0 & -1\\ k/m & c/m\end{pmatrix}\mathbf{K}_{YY}(t,t)+\mathbf{K}_{YY}(t,t)\begin{pmatrix}0 & k/m\\ -1 & c/m\end{pmatrix}=\frac{2\pi S_0}{m^2}\begin{pmatrix}0 & 0\\ 0 & 1\end{pmatrix}$$

Using the change of variables of $k=m\omega_0^2$, $c=2m\zeta\omega_0$, and $\omega_d=\omega_0(1-\zeta^2)^{1/2}$, eigenanalysis of $\mathbf{A}^{-1}\mathbf{B}$ gives eigenvalues of $\lambda_{11}=\zeta\omega_0-i\omega_d$ and $\lambda_{22}=\lambda_{11}^*$, with the matrix of eigenvectors being

$$\boldsymbol{\theta}=\begin{pmatrix}1 & 1\\ -\lambda_{11} & -\lambda_{22}\end{pmatrix}$$

Also, $\hat{\mathbf{A}}\equiv\boldsymbol{\theta}^T\mathbf{A}\boldsymbol{\theta}$ has nonzero elements of $\hat{\mathbf{A}}_{11}=-2im\omega_d\lambda_{11}$ and $\hat{\mathbf{A}}_{22}=2im\omega_d\lambda_{22}$ and

$$\boldsymbol{\theta}^{-1}=\hat{\mathbf{A}}^{-1}\boldsymbol{\theta}^T\mathbf{A}=\frac{i}{2\omega_d}\begin{pmatrix}-\lambda_{22} & -1\\ \lambda_{11} & 1\end{pmatrix}$$

Using the transformation $\vec{Y}(t)=\boldsymbol{\theta}\vec{Z}(t)$ and $\vec{P}(t)=\boldsymbol{\theta}^{-1}\mathbf{A}^{-1}\vec{Q}(t)$ from Eqs. 9.24 and 9.26 allows us to use Eqs. 9.28, 9.29, and 9.31 for the response covariance. The excitation spectral density matrix in the $\{\vec{Z}(t)\}$ equations is

$$\mathbf{S}_{PP}=\frac{S_0}{m^2}\boldsymbol{\theta}^{-1}\begin{pmatrix}0 & 0\\ 0 & 1\end{pmatrix}(\boldsymbol{\theta}^{-1})^T=\frac{S_0}{4m^2\omega_d^2}\begin{pmatrix}-1 & 1\\ 1 & -1\end{pmatrix}$$

For stationary response we use Eq. 9.28 to obtain

$$\mathbf{K}_{ZZ}=\frac{\pi S_0}{4m^2\omega_d^2}\begin{pmatrix}-1/\lambda_{11} & 1/(2\zeta\omega_0)\\ 1/(2\zeta\omega_0) & -1/\lambda_{22}\end{pmatrix}$$

and

$$\mathbf{K}_{YY}=\boldsymbol{\theta}\mathbf{K}_{ZZ}\boldsymbol{\theta}^T=\frac{\pi S_0}{2m^2}\begin{pmatrix}1/(\zeta\omega_0^3) & 0\\ 0 & 1/(\zeta\omega_0)\end{pmatrix}=\pi S_0\begin{pmatrix}1/(ck) & 0\\ 0 & 1/(cm)\end{pmatrix}$$

The (1,1) and (2,2) elements, respectively, confirm the values obtained in Chapters 5 and 6 for σ_X and $\sigma_{\dot{X}}$, and the off-diagonal terms confirm the known fact that $X(t)$ and $\dot{X}(t)$ are uncorrelated random variables in stationary response.

Because the dimension of the system is small, it is also feasible to obtain these results without uncoupling the equations. In particular, the (1,1), (2,1), and (2,2) elements of the equation $\mathbf{A}^{-1}\mathbf{B}\,\mathbf{K}_{YY}(t,t)+\mathbf{K}_{YY}(t,t)(\mathbf{A}^{-1}\mathbf{B})^T=2\pi\mathbf{A}^{-1}S_0$ are

$$-2K_{X\dot{X}}(t,t)=0$$

$$-K_{\dot{X}\dot{X}}(t)+\frac{k}{m}K_{XX}(t)+\frac{c}{m}K_{X\dot{X}}(t)=0$$

$$2\frac{k}{m}K_{X\dot{X}}(t)+2\frac{c}{m}K_{\dot{X}\dot{X}}(t)=\frac{2\pi S_0}{m^2}$$

and solution of these three simultaneous equations gives the results obtained by matrix manipulation.

Let us now presume that we are given an initial condition of $\mathbf{K}_{YY}(t_0,t_0)$ at some specific time t_0. This will result in a nonstationary $\mathbf{K}_{YY}(t,t)$ matrix, and in order to find this nonstationary solution we will need to use the matrix exponential. Using the diagonalized equations for $Z(t)=\boldsymbol{\theta}^{-1}Y(t)$ and the information that $\mathbf{S}_{PP}$ is independent of time, we find from Eq. 9.31 that

$$[\mathbf{K}_{ZZ}(t,t)]_{jl}=\exp[-(t-t_0)(\lambda_j+\lambda_l)]\,[\mathbf{K}_{ZZ}(t_0,t_0)]_{jl}+$$

$$2\pi[\mathbf{S}_{PP}]_{jl}\int_{t_0}^{t}\exp[-(t-s)(\lambda_j+\lambda_l)]\,ds$$

Note that $\mathbf{K}_{ZZ}(t_0,t_0)=\boldsymbol{\theta}^{-1}\mathbf{K}_{YY}(t_0,t_0)(\boldsymbol{\theta}^{-1})^T$ in this expression. The integration may now be performed to obtain

$$[\mathbf{K}_{ZZ}(t,t)]_{jl}=\exp[-(t-t_0)(\lambda_j+\lambda_l)]\,[\mathbf{K}_{ZZ}(t_0,t_0)]_{jl}+$$

$$2\pi[\mathbf{S}_{PP}]_{jl}\frac{1-\exp[-(t-t_0)(\lambda_j+\lambda_l)]}{\lambda_j+\lambda_l}$$

The real parts of both eigenvalues are positive, giving each element of $\mathbf{K}_{ZZ}(t,t)$ an exponential decay of its initial condition, plus an exponential growth toward its previously found stationary value of $2\pi[\mathbf{S}_{PP}]_{jl}/(\lambda_j+\lambda_l)$. Both the decaying and the growing terms, though, also contain trigonometric terms because of the imaginary parts of the eigenvalues.

For the special case in which the initial condition is $\mathbf{K}_{YY}(0,0)=\mathbf{0}$, the results are

$$\sigma_X^2(t)=[K_{YY}(t,t)]_{11}$$

$$=\frac{\pi S_0}{2m^2\zeta\omega_0^3}\left(1-e^{-2\zeta\omega_0 t}\left[\frac{\omega_0^2}{\omega_d^2}+\frac{\zeta\omega_0}{\omega_d}\sin(2\omega_d t)-\frac{\omega_0^2}{\omega_d^2}\cos(2\omega_d t)\right]\right)$$

$$\mathrm{Cov}[X(t),\dot{X}(t)]=[K_{YY}(t,t)]_{12}=\frac{\pi S_0}{2m^2\omega_d^2}e^{-2\zeta\omega_0 t}\left[1-\cos(2\omega_d t)\right]$$

and

$$\sigma_{\dot{X}}^2(t)=[K_{YY}(t,t)]_{22}$$

$$=\frac{\pi S_0}{2m^2\zeta\omega_0}\left(1-e^{-2\zeta\omega_0 t}\left[\frac{\omega_0^2}{\omega_d^2}-\frac{\zeta\omega_0}{\omega_d}\sin(2\omega_d t)-\frac{\omega_0^2}{\omega_d^2}\cos(2\omega_d t)\right]\right)$$

These results are identical to those in Chapter 5, where they were obtained by the indirect method of using the Duhamel integral time-domain solution of the original equation of motion.

****

Example 9.5: Find the mean and variance of the $X(t)$ response for an SDF oscillator with $m=100\,\mathrm{kg}$, $k=10\,\mathrm{kN/m}$, and $c=40\,\mathrm{N\cdot s/m}$. The excitation is a nonstationary delta-correlated process having a mean value of $\mu_F(t)=\mu_0\,t\,e^{-\alpha t}$ with $\mu_0=1\,\mathrm{kN/s}$ and $\alpha=0.8\,\mathrm{s}^{-1}$ and a covariance function of $K_{FF}(t,s)=2\pi S_0\,t\,e^{-\alpha t}\,\delta(t-s)\,U(t)$ with $S_0=1000\,\mathrm{N}^2/\mathrm{rad}$. The system is initially at rest.

The state-space description of the system has the same form as in Example 9.4. Thus, the equation for the mean value of the response is

$$\mathbf{A}\dot{\vec{\mu}}_Y(t)+\mathbf{B}\vec{\mu}_Y(t)=\vec{\mu}_F(t)=\mu_0\, te^{-\alpha t}\,U(t)\begin{bmatrix}0\\1\end{bmatrix}$$

or

$$\begin{pmatrix}-10^4 & 0\\ 0 & 100\end{pmatrix}\dot{\vec{\mu}}_Y(t)+\begin{pmatrix}0 & 10^4\\ 10^4 & 40\end{pmatrix}\vec{\mu}_Y(t)=10\,te^{-0.8t}\,U(t)\begin{bmatrix}0\\1\end{bmatrix}$$

Its solution can be written as

$$\vec{\mu}_Y(t)=\frac{\mu_0}{m}\int_0^t \exp[-(t-s)\mathbf{A}^{-1}\mathbf{B}]\,se^{-\alpha s}\,ds\begin{bmatrix}0\\1\end{bmatrix}$$

$$=10\int_0^t \boldsymbol{\theta}\exp[-(t-s)\boldsymbol{\lambda}]\boldsymbol{\theta}^{-1}se^{-0.8s}\,ds\begin{bmatrix}0\\1\end{bmatrix}$$

in which $\boldsymbol{\theta}$ is the 2×2 matrix with columns that are the eigenvectors of $\mathbf{A}^{-1}\mathbf{B}$, and $\boldsymbol{\lambda}=\boldsymbol{\theta}^{-1}\mathbf{A}^{-1}\mathbf{B}\boldsymbol{\theta}$ is the diagonal matrix of eigenvalues. Eigenanalysis yields $\lambda_{11}=\zeta\omega_0-i\omega_d=0.2-9.998i$, $\lambda_{22}=\zeta\omega_0+i\omega_d=0.2+9.998i$ and

$$\boldsymbol{\theta}=\begin{bmatrix}1 & 1\\ -\lambda_{11} & -\lambda_{22}\end{bmatrix}=\begin{bmatrix}1 & 1\\ -0.2+9.998i & -0.2-9.998i\end{bmatrix}$$

The diagonal matrix $\hat{\mathbf{A}}=\boldsymbol{\theta}^T\mathbf{A}\boldsymbol{\theta}$ has elements $\hat{A}_{11}=-19992-399.9i$ and $\hat{A}_{22}=-19992+399.9i$, and

$$\boldsymbol{\theta}^{-1}=\hat{\mathbf{A}}^{-1}\boldsymbol{\theta}^T\mathbf{A}=\begin{pmatrix}0.5-0.01002i & -0.05001i\\ 0.5+0.01002i & 0.05001i\end{pmatrix}$$

The first term of the $\vec{\mu}_Y(t)$ vector is the mean of $X(t)$, and it is given by

$$\mu_X(t)=10\sum_{r=1}^{2}\theta_{1r}\,\theta_{r2}^{-1}\int_0^t e^{-(t-s)\lambda_{rr}}se^{-\alpha s}\,ds$$

which reduces to

$$\mu_X(t)=(0.001192+0.09968\,t)e^{-0.8t}-$$
$$e^{-0.09899t}[0.001192\cos(9.9980\,t)+0.009899\sin(9.9980\,t)]$$

This result is shown here, along with $\mu_F(t)/k$, which is the static response that one would obtain by neglecting the dynamics of the system.

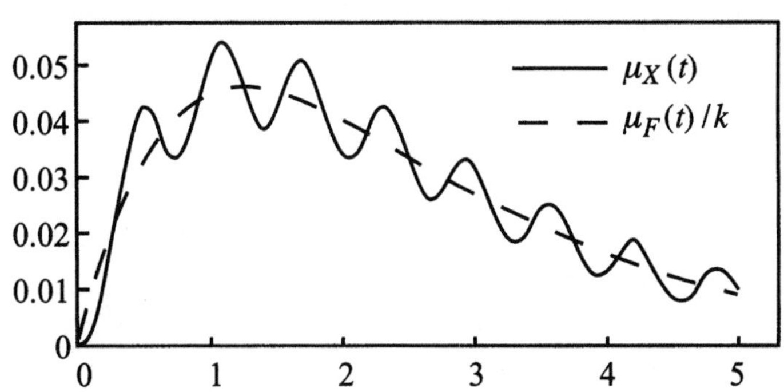

For the covariance analysis we can use the results derived in Example 9.4, for which our parameters give

$$\mathbf{S}_{PP} = \frac{S_0\, t e^{-\alpha t}}{4 m^2 \omega_d^2} \begin{pmatrix} -1 & 1 \\ 1 & -1 \end{pmatrix} = t e^{-0.8t} \begin{pmatrix} -2.501\times 10^{-4} & 2.501\times 10^{-4} \\ 2.501\times 10^{-4} & -2.501\times 10^{-4} \end{pmatrix}$$

Equation 9.31 then gives the components of $\mathbf{K}_{ZZ}(t,t)$, which can be written as

$$[\mathbf{K}_{ZZ}(t,t)]_{jl} = \int_0^t [\mathbf{S}_{PP}(s,s)]_{jl}\, e^{-(\lambda_{jj}+\lambda_{ll})(t-s)}\, ds$$

$$= \pm 2.501\times 10^{-4}\, e^{-0.8t}\, \beta^{-2}(1 - e^{\beta t} + \beta t)$$

in which $\beta = 0.8 - \lambda_{jj} - \lambda_{ll}$. The covariance matrix for $\vec{Y}(t)$ is then found from $\mathbf{K}_{YY}(t,t) = \mathbf{\theta}\,\mathbf{K}_{YY}(t,t)\,\mathbf{\theta}^T$, and the (1,1) component is

$$\sigma_X^2(t) = (-1.965\times 10^{-2} - 7.854\times 10^{-3}\, t)\, e^{-0.8t} +$$
$$e^{-0.4t}[1.964\times 10^{-2} + 7.851\times 10^{-6}\cos(20.00t) -$$
$$3.142\times 10^{-7}\sin(20.00t)]$$

The accompanying sketch compares this result with a "pseudostationary" approximation based only on stationary analysis. This approximation is simply $\pi S_0(t)/(kc)$, in which $K_{FF}(t+\tau,t) = 2\pi S_0(t)\delta(\tau)$ gives $S_0(t) = S_0 t e^{-\alpha t} = 1000\, t e^{-0.8t}$. The pseudostationary approximation is obtained by substituting the nonstationary autospectral density of the excitation into the formula giving the response variance for a stationary excitation. This approximation is good for problems in which $S_0(t)$ varies slowly, but it is not surprising that it fails for the present problem.

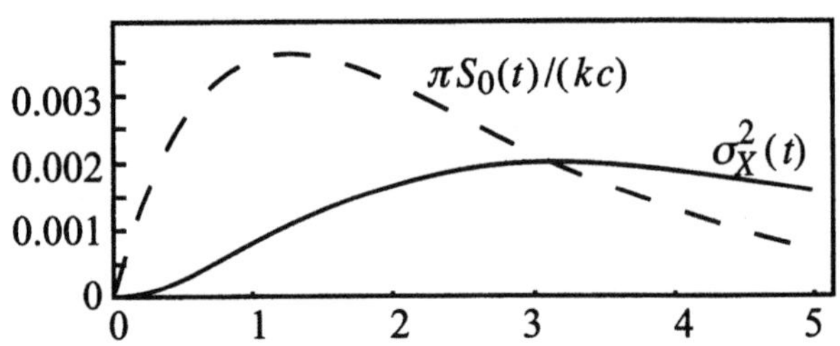

****

Example 9.6: Consider the 2DF system shown with $m_1 = 1000\,\text{kg}$, $m_2 = 500\,\text{kg}$, $k_1 = k_2 = 5\times 10^5\,\text{N/m}$, $c_1 = c_2 = 500\,\text{N}\cdot\text{s/m}$. Find the stationary covariance matrix for the response to a white noise base acceleration $a(t)$ with autospectral density $S_{aa} = 0.1\,(\text{m/s}^2)/(\text{rad/s})$. The $X_j(t)$ coordinates denote motion relative to the base.

The equation governing this system is $\mathbf{A}\dot{\vec{Y}}(t) + \mathbf{B}\vec{Y}(t) = \vec{Q}(t)$ in which $n_Y = 4$ and symmetric $\mathbf{A}$ and $\mathbf{B}$ matrices are given by Eq. 8.46:

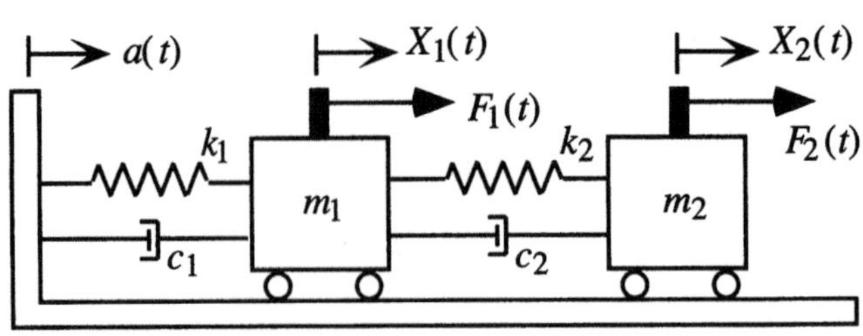

$$\vec{Y}(t)=\begin{pmatrix}X_1(t)\\X_2(t)\\\dot{X}_1(t)\\\dot{X}_2(t)\end{pmatrix},\qquad \mathbf{A}=\begin{pmatrix}-10^6 & 5\times10^5 & 0 & 0\\5\times10^5 & -5\times10^5 & 0 & 0\\0 & 0 & 1000 & 0\\0 & 0 & 0 & 500\end{pmatrix}$$

$$\mathbf{B}=\begin{pmatrix}0 & 0 & 10^6 & -5\times10^5\\0 & 0 & -5\times10^5 & 5\times10^5\\10^6 & -5\times10^5 & 1000 & -500\\-5\times10^5 & 5\times10^5 & -500 & 500\end{pmatrix},\qquad \vec{Q}(t)=\begin{pmatrix}0\\0\\-m_1\,a(t)\\-m_2\,a(t)\end{pmatrix}$$

The spectral density matrix for the $\{\vec{Q}(t)\}$ excitation vector is

$$\mathbf{S}_{QQ}=\begin{pmatrix}0 & 0 & 0 & 0\\0 & 0 & 0 & 0\\0 & 0 & m_1^2\,S_{aa} & m_1 m_2\,S_{aa}\\0 & 0 & m_1 m_2\,S_{aa} & m_2^2\,S_{aa}\end{pmatrix}=\begin{pmatrix}0 & 0 & 0 & 0\\0 & 0 & 0 & 0\\0 & 0 & 10^5 & 5\times10^4\\0 & 0 & 5\times10^4 & 2.5\times10^4\end{pmatrix}\mathrm{N}^2\cdot\mathrm{s/rad}$$

From Eq. 9.17 we see that the stationary covariance matrix is the solution of $\mathbf{A}^{-1}\mathbf{B}\,\mathbf{K}_{YY}(t,t)+\mathbf{K}_{YY}(t,t)(\mathbf{A}^{-1}\mathbf{B})^T=2\pi\,\mathbf{A}^{-1}\mathbf{S}_{QQ}(\mathbf{A}^{-1})^T$.

This equation can now be solved by performing the eigenanalysis of $\mathbf{A}^{-1}\mathbf{B}$. In particular, using $\boldsymbol{\theta}$ as the matrix of eigenvectors of $\mathbf{A}^{-1}\mathbf{B}$ gives $\hat{\mathbf{A}}=\boldsymbol{\theta}^T\mathbf{A}\boldsymbol{\theta}$ as a diagonal matrix and $\boldsymbol{\theta}^{-1}=\hat{\mathbf{A}}^{-1}\boldsymbol{\theta}^T\mathbf{A}$. The diagonal eigenvalue matrix is then given by $\boldsymbol{\lambda}=\boldsymbol{\theta}^{-1}\mathbf{A}^{-1}\mathbf{B}\boldsymbol{\theta}$. Using $\vec{Y}(t)=\boldsymbol{\theta}\,\vec{Z}(t)$ gives

$$\boldsymbol{\lambda}\,\mathbf{K}_{ZZ}(t,t)+\mathbf{K}_{ZZ}(t,t)\,\boldsymbol{\lambda}=2\pi\mathbf{S}_{PP}\equiv2\pi\boldsymbol{\theta}^{-1}\mathbf{A}^{-1}\mathbf{S}_{QQ}(\mathbf{A}^{-1})^T(\boldsymbol{\theta}^{-1})^T$$

so the (j,l) component of $\mathbf{K}_{ZZ}(t,t)$ is given by

$$[\mathbf{K}_{ZZ}(t,t)]_{jl}=2\pi[\mathbf{S}_{PP}]_{jl}/(\lambda_{jj}+\lambda_{ll})$$

After calculating all elements of $\mathbf{K}_{ZZ}(t,t)$, the evaluation of $\mathbf{K}_{YY}(t,t)$ involves only $\mathbf{K}_{ZZ}(t,t)=\boldsymbol{\theta}\,\mathbf{K}_{ZZ}(t,t)\,\boldsymbol{\theta}^T$. The result is

$$\mathbf{K}_{YY}(t,t)=\begin{pmatrix}2.67\times10^{-3} & 3.77\times10^{-3} & 0 & 1.57\times10^{-4}\\3.77\times10^{-3} & 5.34\times10^{-3} & 1.57\times10^{-4} & 0\\0 & 1.57\times10^{-4} & 7.85\times10^{-1} & 1.10\\1.57\times10^{-4} & 0 & 1.10 & 1.57\end{pmatrix}$$

Note that the diagonal elements give the variances of the components of $\vec{Y}(t)$, and the off-diagonal elements give covariances of those components. The (1,1) element gives the variance of $X_1(t)$ as $2.67\times10^{-3}\,\mathrm{m}^2$, which agrees exactly with the result in Example 8.6 for the solution of this same problem by conventional time-domain methods. Note that the (1,3), (3,1) elements of the matrix are zero, confirming the fact that $\mathrm{Cov}[X_1(t),\dot{X}_1(t)]=0$ for stationary response. Similarly, the (2,4), and (4,2) elements are zero because they are $\mathrm{Cov}[X_2(t),\dot{X}_2(t)]=0$.

****

Example 9.7: Consider the stationary response of an SDF oscillator with $m = 100\,\text{kg}$, $k = 10{,}000\,\text{N/m}$, and $c = 40\,\text{N}\cdot\text{s/m}$. The excitation is a stationary delta-correlated process having a mean value of $\mu_F(t) = \mu_0 = 1000\,\text{N}$ and a covariance function of $K_{FF}(t,s) = 2\pi S_0\,\delta(t-s) = 2\pi(1000\,\text{N}^2\cdot\text{s/rad})\,\delta(t-s)$. Find the conditional mean and variance of the $X(t)$ and $\dot{X}(t)$ responses for $t > 5\,\text{s}$ given that $X(5) = 0.05\,\text{m}$ and $\dot{X}(5) = 0$.

This oscillator has the same parameters as the one considered in Example 9.5, so we can use basically the same state-space equations as were developed there—only the excitation and the initial conditions are different. Thus, the conditional mean value of the response is found from the solution of

$$\frac{d}{dt}E\Big(\vec{Y}(t)\,|\,\vec{Y}(5) = [0.05,0]^T\Big) + \mathbf{A}^{-1}\mathbf{B}\,E\Big(\vec{Y}(t)\,|\,\vec{Y}(5) = [0.05,0]^T\Big) = \mathbf{A}^{-1}\begin{bmatrix}0\\ \mu_0\end{bmatrix} = \begin{bmatrix}0\\10\end{bmatrix}$$

The result for $t > 5\,\text{s}$ can be written as

$$E\Big(\vec{Y}(t)\,|\,\vec{Y}(5) = [0.05,0]^T\Big) = \exp[-(t-5)\mathbf{A}^{-1}\mathbf{B}]\begin{bmatrix}0.05\\0\end{bmatrix} + \int_5^t \exp[-(t-s)\mathbf{A}^{-1}\mathbf{B}]\,ds\begin{bmatrix}0\\10\end{bmatrix}$$

or

$$E\Big(\vec{Y}(t)\,|\,\vec{Y}(5) = [0.05,0]^T\Big) = \exp[-(t-5)\mathbf{A}^{-1}\mathbf{B}]\begin{bmatrix}0.05\\0\end{bmatrix} +$$

$$(\mathbf{A}^{-1}\mathbf{B})^{-1}\Big(\mathbf{I}_2 - \exp[-(t-5)\mathbf{A}]\Big)\begin{bmatrix}0\\10\end{bmatrix}$$

$$= \boldsymbol{\theta}\exp[-(t-5)\boldsymbol{\lambda}]\boldsymbol{\theta}^{-1}\begin{bmatrix}0.05\\0\end{bmatrix} + \boldsymbol{\theta}\boldsymbol{\lambda}^{-1}\boldsymbol{\theta}^{-1}\begin{bmatrix}0\\10\end{bmatrix} - \boldsymbol{\theta}\boldsymbol{\lambda}^{-1}\exp[-(t-5)\boldsymbol{\lambda}]\boldsymbol{\theta}^{-1}\begin{bmatrix}0\\10\end{bmatrix}$$

Using the 2×2 $\boldsymbol{\theta}$ and $\boldsymbol{\lambda}$ matrices from Example 9.5 gives

$$E[X(t)\,|\,X(5) = 0.05, \dot{X}(5) = 0]$$
$$= 0.1 - e^{-0.2(t-5)}\Big(0.05\cos[9.998(t-5)] + 10^{-3}\sin[9.998(t-5)]\Big)$$

and

$$E[\dot{X}(t)\,|\,X(5) = 0.05, \dot{X}(5) = 0] = 0.5001\,e^{-0.2(t-5)}\sin[9.998(t-5)]$$

The values are shown in the following sketches.

For the conditional analysis of covariance, we first note that the given deterministic initial conditions at $t = 5$ imply that the covariance matrix is zero at that time. This gives the solution as $\mathbf{K}_{YY}(t,t) = \boldsymbol{\theta}\mathbf{K}_{ZZ}(t,t)\boldsymbol{\theta}^T$ with

$$\mathbf{K}_{ZZ}(t,t) = 2\pi\int_5^t \exp[-(t-s)\,\boldsymbol{\lambda}]\,\mathbf{S}_{PP}\exp[-(t-s)\,\boldsymbol{\lambda}]\,ds$$

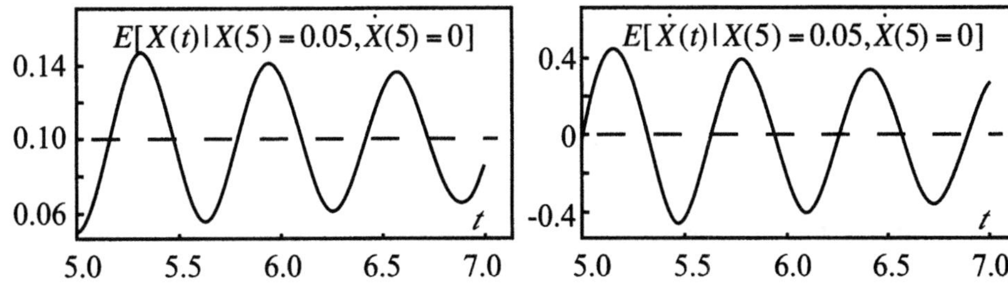

in which

$$\mathbf{S}_{PP}=\boldsymbol{\theta}^{-1}\mathbf{A}^{-1}\mathbf{S}_{QQ}(\boldsymbol{\theta}^{-1}\mathbf{A}^{-1})^T=\boldsymbol{\theta}^{-1}\mathbf{A}^{-1}\begin{pmatrix}0 & 0\\ 0 & S_0\end{pmatrix}(\boldsymbol{\theta}^{-1}\mathbf{A}^{-1})^T$$

Thus,

$$[\mathbf{K}_{ZZ}(t,t]_{jl}=\frac{2\pi[\mathbf{S}_{PP}]_{jl}}{\lambda_j+\lambda_l}\left(1-e^{-(\lambda_j+\lambda_l)(t-5)}\right)$$

The numerical result for the (1,1) and (2,2) elements of $\mathbf{K}_{YY}(t,t)$ give

$$\text{Var}[X(t)|X(5)=0.05,\dot{X}(5)=0]=[\mathbf{K}_{YY}(t,t)]_{1,1}=$$

$$\frac{\pi}{1000}\Big[2.5-e^{-0.4(t-5)}\big(2.501-0.001\cos[20.00(t-5)]+0.05001\sin[20.00(t-5)]\big)\Big]$$

$$\text{Var}[\dot{X}(t)|X(5)=0.05,\dot{X}(5)=0]=[\mathbf{K}_{YY}(t,t)]_{22}$$

$$\frac{\pi}{10}\Big[2.5-e^{-0.4(t-5)}\big(2.501-0.001\cos[20.00(t-5)]-0.05001\sin[20.00(t-5)]\big)\Big]$$

These results are not sketched here because they have exactly the same form as the zero-start variances shown in Fig. 5.7.

****

9.6 Energy Balance and Covariance

Although we have derived the state-space moment equations in Sections 9.3 and 9.4 without explicit consideration of energy, it turns out that in many situations the equations for second moments can also be viewed as energy balance relationships. This is not surprising inasmuch as potential energy and kinetic energy are second-moment properties of displacement and velocity, respectively, in spring-mass systems. Furthermore, a rate of energy addition or dissipation is a cross-product of a force and a velocity. We illustrate this idea by consideration of the SDF oscillator.

First let us consider the equation of motion for the SDF oscillator rewritten as

$$m\,\ddot{X}(t) = F(t) - c\,\dot{X}(t) - k\,X(t)$$

The terms on the right-hand side of this equation are the forces on the mass, which is moving with velocity $\dot{X}(t)$. Thus, if we multiply the equation by $\dot{X}(t)$, we will obtain an equation involving the power into the mass from these three terms:

$$\begin{aligned} m\,\dot{X}(t)\,\ddot{X}(t) &= F(t)\,\dot{X}(t) - [c\,\dot{X}(t)]\,\dot{X}(t) - [k\,X(t)]\,\dot{X}(t) \\ &= PA(t) - PD(t) - k\,X(t)\,\dot{X}(t) \end{aligned} \qquad (9.32)$$

in which we have used the notation $PA(t)$ for the power added by the force $F(t)$ and $PD(t)$ for the power dissipation in the dashpot. The final term in the equation is the rate of energy transfer from the spring to the mass, and not surprisingly it is the same as the rate of decrease of the potential energy $PE(t) = k\,X^2(t)/2$ in the spring

$$k\,X(t)\,\dot{X}(t) = \frac{d}{dt} PE(t)$$

Similarly, the term on the left-hand side of Eq. 9.32 is the rate of change of the kinetic energy $KE(t) = m\,\dot{X}^2(t)/2$ in the mass

$$m\,\dot{X}(t)\,\ddot{X}(t) = \frac{d}{dt} KE(t)$$

Thus, one can rearrange the terms of Eq. 9.32 to give

$$\frac{d}{dt}\big[KE(t) + PE(t)\big] + PD(t) = PA(t)$$

which simply states that the power added is the sum of the power dissipated and the rate of change of the energy in the system. This confirms the well-known fact that Newtonian mechanics (force equals mass times acceleration) results in conservation of energy. The expectation of this energy balance relationship is one of our state-space equations for the SDF system. Precisely this equation appears in our investigation of the SDF system in Example 9.1, and when Eq. 9.18 is used to give the second moments of the system, this energy-balance relationship is the (2,2) component scalar equation.

It is particularly interesting to study the $PA(t)$ term for the special case in which the excitation of an SDF or MDF system is mean-zero, white noise. That is, we want to find terms like $E[F_j(t)\,\dot{X}_j(t)]$ with $F_j(t)$ being a force applied to a mass m_j and $\dot{X}_j(t)$ being the velocity of the mass. Let n be the number of degrees of freedom of the system so that $n_Y = 2n$ is the dimension of the state vector. Using the symmetric state-space formulation of Eq. 8.46 with

$$\vec{Y}(t) = [X_1(t),\cdots,X_n(t),\dot{X}_1(t),\cdots,\dot{X}_n(t)]^T, \quad \vec{Q}(t) = [0,\cdots,0,F_1(t),\cdots,F_n(t)]^T$$

the mean-zero white noise condition can be written as

$$\boldsymbol{\phi}_{QQ}(t,s) = \mathbf{K}_{QQ}(t,s) = 2\pi\mathbf{S}_0(t)\,\delta(t-s)$$

with

$$\mathbf{S}_0(t) = \begin{bmatrix} 0 & \cdots & 0 & 0 & \cdots & 0 \\ \vdots & & \vdots & \vdots & & \vdots \\ 0 & \cdots & 0 & 0 & \cdots & 0 \\ 0 & \cdots & 0 & S_{11}(t) & \cdots & S_{1n}(t) \\ \vdots & & \vdots & \vdots & & \vdots \\ 0 & \cdots & 0 & S_{n1}(t) & \cdots & S_{nn}(t) \end{bmatrix} \tag{9.33}$$

Note that $S_{jl}(t)$ is the nonstationary cross-spectral density of the $F_j(t)$ and $F_l(t)$ force components, and it is the $(n+j,n+l)$ component of $\mathbf{S}_0(t)$.

Our notation now gives the expected rate of energy addition by the jth component of force in the MDF system as

$$E[PA_j(t)] \equiv E[F_j(t)\,\dot{X}_j(t)] = E[Q_{n+j}(t)\,Y_{n+j}(t)]$$

Thus, it is a diagonal term from the matrix $E[Q^T(t)\,Y(t)] = \mathbf{K}_{QY}(t,t)$ or its transpose, $E[Y^T(t)\,Q(t)] = \mathbf{K}_{YQ}(t,t)$. We can find the value of such terms, though, from Eqs. 9.13 and 9.16. In particular, those equations give

$$\mathbf{K}_{QY}(t,t)\,\mathbf{A}^T + \mathbf{A}\,\mathbf{K}_{YQ}(t,t) = \int_{t_0}^{t} \mathbf{K}_{QQ}(u,t)\,du + \int_{t_0}^{t} \mathbf{K}_{QQ}(t,v)\,dv = 2\pi\,\mathbf{S}_0(t) \tag{9.34}$$

The symmetric form of **A** contains the mass matrix **m** in the lower-right portion, and this **m** matrix is diagonal for the coordinates we are using. Thus, the $(n+j,n+j)$ term of $\mathbf{K}_{QY}(t,t)\,\mathbf{A}^T$ and of $\mathbf{A}\,\mathbf{K}_{YQ}(t,t)$ is $m_j\,E[Q_{n+j}(t)\,Y_{n+j}(t)]$, giving

$$m_j\,E[Q_{n+j}(t)\,Y_{n+j}(t)] = \pi[\mathbf{S}_0(t)]_{n+j,n+j} = \pi\,S_{jj}(t)$$

and

$$E[PA_j(t)] = \frac{\pi\,S_{jj}(t)}{m_j} \tag{9.35}$$

From Eq. 9.35 we note the rather remarkable property that the rate of energy addition to the MDF system by a mean-zero, delta-correlated force depends only on the autospectral density of the force and the magnitude of the mass to which it is applied. It is unaffected by the magnitudes of any springs or dashpots in the system and of any other masses or forces in the system. Furthermore, it does not depend on the level of response of the system. For example, if a stationary excitation of this type is applied to a system that is initially at rest, then the rate of energy addition does not change as the response builds up from zero to its stationary level. This property of $E[PA_j(t)]$ also extends to the situation in which the excitation of the system is a mean-zero, delta-correlated base motion, rather than a force. When the base of an MDF system is moved with acceleration $\{a(t)\}$ and $\{\dot{X}(t)\}$ represents motion relative to the base, then we know that the equations of motion have the same form as usual, but with excitation components of $F_j(t) = -m_j\,a(t)$. For this situation, Eq. 9.35 gives the rate of energy addition by $F_j(t)$ as $E[PA_j(t)] = m_j\pi\,S_{aa}(t)$. Again, this rate depends only on the autospectral density of the excitation and the magnitude of the m_j mass and is unaffected by response levels or by other parameters of the system.

It should be noted that Eq. 9.35 is generally not true if the processes involved are not mean-zero. In particular, the rate of energy addition to a mass m_j by a delta-correlated force $\{F_j(t)\}$ with a nonzero mean is

$$E[PA_j(t)] = \frac{\pi\,S_{jj}(t)}{m_j} + \mu_{F_j}(t)\,\mu_{\dot{X}_j}(t)$$

In this situation we see that the rate of energy addition generally does depend on the response and may vary with time even if the excitation is stationary.

In addition to the energy balance, we can note that Eq. 9.34 gives important general relationships regarding the covariance of force and response components of an MDF system. For example, for j and l not exceeding n, the $(n+j,n+l)$ component of Eq. 9.34 gives

$$m_j K_{F_j(t)\dot{X}_l(t)} + m_l K_{\dot{X}_j(t)F_l(t)} = 2\pi S_{jl}(t) \tag{9.36}$$

which includes the special case of

$$K_{F_j(t)\dot{X}_j(t)} = \frac{\pi S_{jj}(t)}{m_j} \tag{9.37}$$

for the covariance, at any instant of time, between a particular component of force and the velocity of the mass to which it is applied. Considering the $(n+j,l)$ component of Eq. 9.34 gives a similar, and even simpler, result regarding the covariance of $F_j(t)$ and any component of response $X_l(t)$ at the same instant of time. Noting that $Q_j(t) \equiv 0$ and that Eq. 9.33 gives $[\mathbf{S}_0(t)]_{n+j,l} = 0$ also, we obtain

$$K_{F_j(t)X_l(t)} \equiv K_{Q_{n+j}(t)Y_l(t)} = 0 \tag{9.38}$$

Thus, we see that $F_j(t)$ is uncorrelated with any component of response $X_l(t)$ at the same instant of time, including the $X_j(t)$ component giving the motion at the point of application of the force.

It should be emphasized that the covariance relationships in Eqs. 9.36–9.38 apply only for a delta-correlated force applied directly to a mass. If a delta-correlated force is applied at some other point in a system or to a system with no mass, as in Example 9.3, then the relationships generally do not hold.

****

Example 9.8: Verify that the rate of energy addition of Eq. 9.35 is consistent with the expressions derived in Section 5.6 for the nonstationary variance values of displacement and velocity for the response of the SDF system with a stationary delta-correlated excitation.

We can write the response variance values of Eqs. 5.50 and 5.54 as

$$\sigma_X^2(t) = \frac{G_0}{2kc}\left\{1 - e^{-2\zeta\omega_0 t}\left[\frac{\omega_0^2}{\omega_d^2} + \frac{\zeta\omega_0}{\omega_d}\sin(2\omega_d t) - \frac{\zeta^2\omega_0^2}{\omega_d^2}\cos(2\omega_d t)\right]\right\}U(t)$$

and

$$\sigma_{\dot X}^2(t) = \frac{G_0}{2mc}\left\{1 - e^{-2\zeta\omega_0 t}\left[\frac{\omega_0^2}{\omega_d^2} - \frac{\zeta\omega_0}{\omega_d}\sin(2\omega_d t) - \frac{\zeta^2\omega_0^2}{\omega_d^2}\cos(2\omega_d t)\right]\right\}U(t)$$

for a system that is at rest at time $t=0$ and has an excitation $\{F(t)\}$ with autocovariance of $K_{FF}(t,s) = G_0\,\delta(t-s)\,U(t)$. Letting the mean values be zero, we use these expressions in calculating the total energy in the oscillator as

$$E[PE(t)+KE(t)] = \frac{k\sigma_X^2(t)}{2} + \frac{m\sigma_{\dot X}^2(t)}{2}$$

$$= \frac{G_0}{2c}\left\{1 - e^{-2\zeta\omega_0 t}\left[\frac{\omega_0^2}{\omega_d^2} - \frac{\zeta^2\omega_0^2}{\omega_d^2}\cos(2\omega_d t)\right]\right\}U(t)$$

The fact that the $(\zeta\omega_0/\omega_d)\sin(2\omega_d t)$ term, which appears in both $\sigma_X^2(t)$ and $\sigma_X^2(t)$, does not appear in the total energy expression shows that this term represents a transfer of energy back and forth between potential energy and kinetic energy, with no net energy addition. Taking the derivative of the expected total energy, and noting that $2\zeta\omega_0 = c/m$, gives the rate of energy growth in the system as

$$\frac{d}{dt}E[PE(t)+KE(t)] = \frac{G_0}{2m}e^{-2\zeta\omega_0 t}\left(\frac{\omega_0^2}{\omega_d^2} - \frac{\zeta\omega_0}{\omega_d}\sin(2\omega_d t) - \frac{\zeta^2\omega_0^2}{\omega_d^2}\cos(2\omega_d t)\right)U(t)$$

Adding this expression to $c\sigma_X^2(t)$, which is the expected rate of energy dissipation in the system, gives the rate of energy addition as $E[PA(t)] = G_0/(2m)\,U(t)$. Recalling that the relationship between G_0 and the autospectral density is $G_0 = 2\pi S_0$ confirms Eq. 9.35.

Similar verification is given by Example 5.8, in which a delta-correlated force is applied to a mass that has no restoring force elements. In this case we found that the variance of $\dot X(t)$ grows linearly with time as $\sigma_X^2(t) = (G_0/m^2)\,t\,U(t)$. When the mean of velocity is zero, this indicates a linear growth of kinetic energy. Inasmuch as this particular system has no potential energy or energy dissipation, the rate of growth of the kinetic energy is the rate of energy addition by the force. The value found is $G_0/(2m)$, which again confirms Eq. 9.35.

****

Example 9.9: Verify that Eq. 9.35 gives the rate of energy addition by a mean-zero shot noise force applied to a mass m that is connected to some springs and dashpots.

As in Example 5.6, we will write the shot noise force as

$$F(t) = \sum F_l\,\delta(t - T_l)$$

in which $\{T_1, T_2, \cdots, T_l, \cdots\}$ is the sequence of arrival times for a Poisson process and $\{F_1, F_2, \cdots, F_l, \cdots\}$ is a sequence of identically distributed random variables that are independent of each other and of the arrival times. We will first consider the energy added to the mass by the single pulse $F_l\,\delta(t - T_l)$. Let the mass have displacement and velocity $X(T_l^-)$ and $\dot{X}(T_l^-)$, respectively, immediately prior to the pulse. Because the forces from attached springs and dashpots will be finite, they will cause no instantaneous changes in $X(t)$ or $\dot{X}(t)$, but the pulse of applied force will cause an instantaneous change in $\dot{X}(t)$. Thus, the state of the mass immediately after the impulse will be $X(T_l^+) = X(T_l^-)$, and $\dot{X}(T_l^+) = \dot{X}(T_l^-) + F_l / m$. Because displacement is continuous, there will be no change in the potential energy of the system during the instant of the excitation pulse, but the kinetic energy will instantaneously change by an amount

$$\Delta KE = \frac{m}{2}[\dot{X}^2(T_l^+) - \dot{X}^2(T_l^-)] = m\,\dot{X}(T_l^-) F_l + \frac{F_l^2}{2m}$$

We know that F_l is independent of the past excitation, and this ensures that it is also independent of $\dot{X}(T_l^-)$. In conjunction with the fact that $E(F_l) = 0$ for a mean-zero process, this gives $E[\dot{X}(T_l^-) F_l] = 0$. Thus, the expected value of the energy added by the pulse $F_l\,\delta(t - T_l)$ can be written as $E(F^2)/(2m)$. Even though the energy added by any one individual pulse does depend on the velocity of the mass at the time of the pulse, this term may be either positive or negative and its expected value is zero. The expected rate of energy addition per unit time by the force on mass m is then $E(F^2)/(2m)$ multiplied by the expected rate of arrival of the pulses. We can find the autospectral density of this shot noise process from the results in Example 5.6, and it again confirms that the expected rate of energy addition is as given in Eq. 9.35.

****

9.7 Higher Moments and Cumulants Using Kronecker Notation

One difficulty in investigating higher moments or cumulants of the state variables is the choice of an appropriate notation. We found that the first moments could be written as the vector $\vec{\mu}_Y(t)$ and the second moments could be written as the matrix $\boldsymbol{\phi}_{YY}(t,t)$. Continuing with this approach gives the third moments as constituting a third-order tensor, the fourth moments as a fourth-order tensor, and so forth. There is no problem with writing expressions for components such as the (j,k,l,m) component of the fourth-order tensor as $E[Y_j(t)Y_k(t)Y_l(t)Y_m(t)]$, but there is some difficulty with the presentation of general relationships for the higher moments, comparable to Eqs. 9.6 and 9.8 for the first and second moments, as well as with the organization of the large amount of information involved. That is, $\vec{\mu}_Y(t)$, $\boldsymbol{\phi}_{YY}(t,t)$, and $\mathbf{K}_{YY}(t,t)$ provide very convenient one- and two-dimensional arrays when that is needed, but we have not defined a notation or appropriate algebra for arrays with dimension higher than two. Because Kronecker notation provides one convenient way to handle this matter, we now present its basic concepts.

The fundamental operation of Kronecker algebra is the product denoted by $\otimes$ and defined as follows: If $\mathbf{A}$ is an $n_A \times r_A$ rectangular matrix and $\mathbf{B}$ is an $n_B \times r_B$ rectangular matrix, then $\mathbf{C} = \mathbf{A} \otimes \mathbf{B}$ is a rectangular matrix of dimension $n_A n_B \times r_A r_B$ with components $C_{(j_A-1)n_B+j_B,(l_A-1)r_B+l_B} = A_{j_A l_A} B_{j_B l_B}$. This relationship can also be written as

$$\mathbf{C} = \begin{bmatrix} A_{11}\mathbf{B} & A_{12}\mathbf{B} & \cdots & A_{1r_A}\mathbf{B} \\ A_{21}\mathbf{B} & A_{22}\mathbf{B} & \cdots & A_{2r_A}\mathbf{B} \\ \vdots & \vdots & & \vdots \\ A_{n_A 1}\mathbf{B} & A_{n_A 2}\mathbf{B} & \cdots & A_{n_A r_A}\mathbf{B} \end{bmatrix} \tag{9.39}$$

Note that the Kronecker product is much more general than matrix multiplication and requires no restrictions on the dimensions of the arrays being multiplied. One can easily verify that the definition gives

$$(\mathbf{A}+\mathbf{B}) \otimes (\mathbf{C}+\mathbf{D}) = \mathbf{A} \otimes \mathbf{C} + \mathbf{A} \otimes \mathbf{D} + \mathbf{B} \otimes \mathbf{C} + \mathbf{B} \otimes \mathbf{D} \tag{9.40}$$

and

$$\mathbf{A} \otimes (\mathbf{B} \otimes \mathbf{C}) = (\mathbf{A} \otimes \mathbf{B}) \otimes \mathbf{C} \tag{9.41}$$

for any arrays $\mathbf{A}$, $\mathbf{B}$, $\mathbf{C}$, and $\mathbf{D}$, showing that the operation is distributive and associative. On the other hand, $\mathbf{B} \otimes \mathbf{A} \neq \mathbf{A} \otimes \mathbf{B}$, so the operation is not commutative. A simplified notation is used for the Kronecker product of a matrix with itself. In particular, we define Kronecker powers as $\mathbf{A}^{[2]} = \mathbf{A} \otimes \mathbf{A}$ and $\mathbf{A}^{[j]} = \mathbf{A} \otimes \mathbf{A}^{[j-1]}$ for $j > 2$. Note that in the particular case in which $\mathbf{A}$ and $\mathbf{B}$ each have only one column, we find that $\mathbf{A} \otimes \mathbf{B}$ also has only one column. That is, the Kronecker product of two vectors is a vector and the Kronecker power of a vector is a vector. This situation will be of particular interest in our state-space analysis. Because our equations of motion contain matrix products, we also need a general relationship involving the combination of matrix products and Kronecker products. This formula is

$$(\mathbf{A}\,\mathbf{B})\otimes(\mathbf{C}\,\mathbf{D})=(\mathbf{A}\otimes\mathbf{C})(\mathbf{B}\otimes\mathbf{D}) \tag{9.42}$$

provided that the dimensions of the arrays are such that the matrix products $\mathbf{A}\,\mathbf{B}$ and $\mathbf{C}\,\mathbf{D}$ exist.

Even though we found the matrix formulation to be very convenient for the equations governing the second moments of the state variables, we now rewrite those expressions using the Kronecker notation to illustrate its application. The array of second-moment functions is given by the vector $E[\vec{Y}^{[2]}(t)]\equiv E[\vec{Y}(t)\otimes\vec{Y}(t)]$. Note that this vector contains every element of the $\boldsymbol{\phi}_{YY}(t,t)$ matrix that we previously considered. The only difference is that the elements are now arranged as a vector of length $(n_Y)^2$ instead of in a matrix of dimension $n_Y\times n_Y$. The derivative of the Kronecker power is given by

$$\frac{d}{dt}E[\vec{Y}^{[2]}(t)]=E[\dot{\vec{Y}}(t)\otimes\vec{Y}(t)]+E[\vec{Y}(t)\otimes\dot{\vec{Y}}(t)]$$

The terms on the right-hand side of this differential equation governing the second moments can be obtained by taking the Kronecker product of $\vec{Y}(t)$ with the equation of motion in Eq. 9.5 rewritten as $\dot{\vec{Y}}(t)+\mathbf{A}^{-1}\mathbf{B}\,\vec{Y}(t)=\mathbf{A}^{-1}\,\vec{Q}(t)$, then taking the expected value. This gives

$$E[\dot{\vec{Y}}(t)\otimes\vec{Y}(t)]+E[\mathbf{A}^{-1}\mathbf{B}\,\vec{Y}(t)\otimes\vec{Y}(t)]=E[\mathbf{A}^{-1}\vec{Q}(t)\otimes\vec{Y}(t)]$$

and

$$E[\vec{Y}(t)\otimes\dot{\vec{Y}}(t)]+E[\vec{Y}(t)\otimes\mathbf{A}^{-1}\mathbf{B}\,\vec{Y}(t)]=E[\vec{Y}(t)\otimes\mathbf{A}^{-1}\vec{Q}(t)]$$

and adding these expressions gives

$$\begin{aligned}\frac{d}{dt}E[\vec{Y}^{[2]}(t)]+E[\mathbf{A}^{-1}\mathbf{B}\,\vec{Y}(t)\otimes\vec{Y}(t)]+E[\vec{Y}(t)\otimes\mathbf{A}^{-1}\mathbf{B}\,\vec{Y}(t)]=\\ E[\mathbf{A}^{-1}\vec{Q}(t)\otimes\vec{Y}(t)]+E[\vec{Y}(t)\otimes\mathbf{A}^{-1}\vec{Q}(t)]\end{aligned} \tag{9.43}$$

We can simplify this relationship by noting that $[\mathbf{A}^{-1}\mathbf{B}\,\vec{Y}(t)]\otimes\vec{Y}(t)\equiv[\mathbf{A}^{-1}\mathbf{B}\,\vec{Y}(t)]\otimes[\mathbf{I}\,\vec{Y}(t)]$, in which $\mathbf{I}\equiv\mathbf{I}_{n_Y}$ is the identity matrix. Based on Eq. 9.40, though, this is the same as $(\mathbf{A}^{-1}\mathbf{B}\otimes\mathbf{I})\,\vec{Y}^{[2]}(t)$. Performing this same operation for $E[\,\vec{Y}(t)\otimes\mathbf{A}^{-1}\mathbf{B}\,\vec{Y}(t)\,]$ $E[\,\vec{Y}\,(t)\otimes\mathbf{A}^{-1}\mathbf{B}\,\vec{Y}\,(t)]$ gives

$$\frac{d}{dt}E[\vec{Y}^{[2]}(t)] + [(\mathbf{A}^{-1}\mathbf{B} \otimes \mathbf{I}) + (\mathbf{I} \otimes \mathbf{A}^{-1}\mathbf{B})]E[\vec{Y}^{[2]}(t)] = E[\mathbf{A}^{-1}\vec{Q}(t) \otimes \vec{Y}(t)] + E[\vec{Y}(t) \otimes \mathbf{A}^{-1}\vec{Q}(t)] \tag{9.44}$$

This equation governing the second moments of the state variables is exactly equivalent to Eq. 9.8. In fact, the (j,l) component of Eq. 9.8 is identical to the $(n_Y - 1)j + l$ component of Eq. 9.44. As noted before, symmetry causes the (l,j) component of Eq. 9.8 to be identical to the (j,l) component. Similarly, the $(n_Y - 1)l + j$ and the $(n_Y - 1)j + l$ components of Eq. 9.44 are identical.

Inasmuch as Eqs. 9.44 and 9.8 have identical components, there is really no advantage in using the Kronecker notation for the second-moment equations. This new notation is easily extended to higher dimensions, however, whereas the matrix notation is not so easily extended. The array of third moments of the state variables, for example, can be written as $E[\vec{Y}^{[3]}(t)]$ and its derivative is

$$\frac{d}{dt}E[\vec{Y}^{[3]}(t)] = E[\vec{Y}(t) \otimes \vec{Y}(t) \otimes \dot{\vec{Y}}(t)] + E[\vec{Y}(t) \otimes \dot{\vec{Y}}(t) \otimes \vec{Y}(t)] + E[\dot{\vec{Y}}(t) \otimes \vec{Y}(t) \otimes \vec{Y}(t)]$$

Substituting for $\dot{\vec{Y}}(t)$ from the equation of motion, arranging terms, and simplifying according to Eq. 9.42 then gives

$$\frac{d}{dt}E[\vec{Y}^{[3]}(t)] + [(\mathbf{I} \otimes \mathbf{I} \otimes \mathbf{A}^{-1}\mathbf{B}) + (\mathbf{I} \otimes \mathbf{A}^{-1}\mathbf{B} \otimes \mathbf{I}) + (\mathbf{A}^{-1}\mathbf{B} \otimes \mathbf{I} \otimes \mathbf{I})]E[\vec{Y}^{[3]}(t)] = E[\vec{Y}(t) \otimes \vec{Y}(t) \otimes \mathbf{A}^{-1}\vec{Q}(t)] + E[\vec{Y}(t) \otimes \mathbf{A}^{-1}\vec{Q}(t) \otimes \vec{Y}(t)] + E[\mathbf{A}^{-1}\vec{Q}(t) \otimes \vec{Y}(t) \otimes \vec{Y}(t)]$$

or

$$\frac{d}{dt}E[\vec{Y}^{[3]}(t)] + [(\mathbf{I}^{[2]} \otimes \mathbf{A}^{-1}\mathbf{B}) + (\mathbf{I} \otimes \mathbf{A}^{-1}\mathbf{B} \otimes \mathbf{I}) + (\mathbf{A}^{-1}\mathbf{B} \otimes \mathbf{I}^{[2]})]E[\vec{Y}^{[3]}(t)] = E[\vec{Y}^{[2]}(t) \otimes \mathbf{A}^{-1}\vec{Q}(t)] + E[\vec{Y}(t) \otimes \mathbf{A}^{-1}\vec{Q}(t) \otimes \vec{Y}(t)] + E[\mathbf{A}^{-1}\vec{Q}(t) \otimes \vec{Y}^{[2]}(t)]$$

In this way one can write the general expression for the jth moments as

$$\frac{d}{dt}E[\vec{Y}^{[j]}(t)] + \mathbf{C}\,E[\vec{Y}^{[j]}(t)] = \vec{\psi}(t) \tag{9.45}$$

in which

$$\mathbf{C}=\sum_{l=1}^{j}\underbrace{\mathbf{I}\otimes\cdots\otimes\mathbf{I}}_{l-1}\otimes\mathbf{A}^{-1}\mathbf{B}\otimes\underbrace{\mathbf{I}\otimes\cdots\otimes\mathbf{I}}_{j-l}=\sum_{l=1}^{j}\mathbf{I}^{[l-1]}\otimes\mathbf{A}^{-1}\mathbf{B}\otimes\mathbf{I}^{[j-l]} \tag{9.46}$$

and

$$\begin{aligned}\vec{\psi}(t)&=\sum_{l=1}^{j}E[\underbrace{\vec{Y}(t)\otimes\cdots\otimes\vec{Y}(t)}_{l-1}\otimes\mathbf{A}^{-1}\vec{Q}(t)\otimes\underbrace{\vec{Y}(t)\otimes\cdots\otimes\vec{Y}(t)}_{j-l}]\\&=\sum_{l=1}^{j}E[\vec{Y}^{[l-1]}(t)\otimes\mathbf{A}^{-1}\vec{Q}(t)\otimes\vec{Y}^{[j-l]}(t)]\end{aligned} \tag{9.47}$$

The equations for higher-order cumulants are essentially identical to those of Eq. 9.45 for higher moments. In fact, the only complication in going from moments to cumulants is in the definition of a cumulant notation corresponding to the Kronecker products and powers. This can be done, though, by analogy with the product terms. In particular, we will define an array of cumulant terms that is organized in exactly the same way as the elements of the corresponding Kronecker product moments. For the general situation in which we have matrices $(\mathbf{M}_1,\mathbf{M}_2,\cdots,\mathbf{M}_j)$ of random variables, we note that any particular component of the jth-order moment matrix $E(\mathbf{M}_1\otimes\mathbf{M}_2\otimes\cdots\otimes\mathbf{M}_j)$ is a product of one element each from the matrices in $(\mathbf{M}_1,\mathbf{M}_2,\cdots,\mathbf{M}_j)$. We now define the corresponding component of the matrix $\kappa_j^{\otimes}(\mathbf{M}_1,\mathbf{M}_2,\cdots,\mathbf{M}_j)$ to be the jth-order joint cumulant of those same j elements, one from each of the matrices in $(\mathbf{M}_1,\mathbf{M}_2,\cdots,\mathbf{M}_j)$. The linearity property of the cumulants, as given in Eq. 9.3, then allows us to rewrite Eq. 9.45 as

$$\begin{aligned}&\frac{d}{dt}\kappa_j^{\otimes}[\vec{Y}(t),\cdots,\vec{Y}(t)]+\mathbf{C}\kappa_j^{\otimes}[\vec{Y}(t),\cdots,\vec{Y}(t)]=\\&\qquad\sum_{l=1}^{j}\kappa_j^{\otimes}[\underbrace{\vec{Y}(t),\cdots,\vec{Y}(t)}_{l-1},\mathbf{A}^{-1}\vec{Q}(t),\underbrace{\vec{Y}(t),\cdots,\vec{Y}(t)}_{j-l}]\end{aligned} \tag{9.48}$$

in which the matrix $\mathbf{C}$ is as defined in Eq. 9.46.

Let us now consider the special case in which the $\{\vec{Q}(t)\}$ excitation vector process is delta-correlated. By this we mean that

$$\kappa_j^{\otimes}[\vec{Q}(t_1),\cdots,\vec{Q}(t_j)]=(2\pi)^{j-1}\,\vec{S}_j(t_j)\,\delta(t_1-t_j)\cdots\delta(t_{j-1}-t_j) \tag{9.49}$$

in which $\vec{S}_j(t)$ is a vector of dimension $(n_Y)^j$. Its components are spectral density terms that give the nonstationary intensity of the jth cumulants of the $\vec{Q}(t)$ excitation. The elements of $\vec{S}_2(t)$ are ordinary autospectral and cross-spectral densities, the elements of $\vec{S}_3(t)$ are bispectral densities, and so forth. Using Eq. 9.49 one can write a generalization of Eq. 9.15 as

$$\int_{t_0}^{t}\cdots\int_{t_0}^{t}\kappa_j^{\otimes}[\vec{Q}(t_1),\cdots,\vec{Q}(t_j)]\,dt_1\cdots dt_j=(2\pi)^{j-1}\int_{t_0}^{t}\vec{S}_j(v)\,dv$$

for any $t_0<t$. The derivative with respect to t of this expression is

$$\sum_{k=1}^{j}\int_{t_0}^{t}\cdots\int_{t_0}^{t}\kappa_j^{\otimes}[\vec{Q}(t_1),\cdots,\vec{Q}(t_{k-1}),\vec{Q}(t),\vec{Q}(t_{k+1}),\cdots,\vec{Q}(t_j)]\times$$
$$dt_1\cdots dt_{k-1}dt_{k+1}\cdots dt_j=(2\pi)^{j-1}\vec{S}_j(t)$$

and substituting the form of Eq. 9.11 for the response terms in the right-hand side of Eq. 9.48 gives exactly the left-hand side of this expression premultiplied by $(\mathbf{A}^{-1})^{[j]}\equiv[\mathbf{A}^{-1}\otimes\cdots\otimes\mathbf{A}^{-1}]$. Thus, Eq. 9.48 becomes

$$\frac{d}{dt}\kappa_j^{\otimes}[\vec{Y}(t),\cdots,\vec{Y}(t)]+\mathbf{C}\kappa_j^{\otimes}[\vec{Y}(t),\cdots,\vec{Y}(t)]=(2\pi)^{j-1}(\mathbf{A}^{-1})^{[j]}\vec{S}_j(t) \qquad (9.50)$$

This generalization of Eq. 9.17 describes all jth-order cumulants of the response process $\{\vec{Y}(t)\}$ resulting from the delta-correlated excitation process $\{\vec{Q}(t)\}$.

The general solution of Eq. 9.50, of course, can be written as

$$\kappa_j^{\otimes}[\vec{Y}(t),\cdots,\vec{Y}(t)]=(2\pi)^{j-1}\int_{-\infty}^{t}e^{-(t-u)\mathbf{C}}(\mathbf{A}^{-1})^{[j]}\vec{S}_j(u)\,du \qquad (9.51)$$

and the corresponding array of conditional cumulants given $\vec{Y}(t_0)=\vec{y}$ at some particular time is

$$\kappa_j^{\otimes}[\vec{Y}(t),\cdots,\vec{Y}(t)\,|\,\vec{Y}(t_0)=\vec{y}]=(2\pi)^{j-1}\int_{t_0}^{t}e^{-(t-u)\mathbf{C}}(\mathbf{A}^{-1})^{[j]}\vec{S}_j(u)\,du \quad \text{for } j>2 \qquad (9.52)$$

Note that Eq. 9.52 gives the conditional cumulants as being zero at time $t=t_0$, in agreement with the fact that all cumulant components of order higher than unity are zero for any deterministic vector, such as $\vec{y}$. Similar solutions can be written for stochastic initial conditions.

Because the matrix exponential in the integrands of Eqs. 9.51 and 9.52 are not convenient for computation, let us consider a form giving uncoupled equations. As before, let $\vec{Y}(t) = \boldsymbol{\theta}\,\vec{Z}(t)$, with $\boldsymbol{\theta}$ being the matrix of eigenvectors giving $\mathbf{A}^{-1}\mathbf{B}\boldsymbol{\theta} = \boldsymbol{\theta}\boldsymbol{\lambda}$, in which $\boldsymbol{\lambda}$ is the diagonal matrix of eigenvalues. Using the $\dot{\vec{Z}}(t) + \boldsymbol{\lambda}\,\vec{Z}(t) = \boldsymbol{\theta}^{-1}\mathbf{A}^{-1}\,\vec{Q}(t)$ equation of motion for $\vec{Z}(t)$, one now finds the equation for cumulants of $\vec{Z}(t)$ as

$$\frac{d}{dt}\kappa_j^{\otimes}[\vec{Z}(t),\cdots,\vec{Z}(t)] + \hat{\mathbf{C}}\kappa_j^{\otimes}[\vec{Z}(t),\cdots,\vec{Z}(t)] = (2\pi)^{j-1}(\boldsymbol{\theta}^{-1}\mathbf{A}^{-1})^{[j]}\,\vec{S}_j(t) \tag{9.53}$$

in which

$$\hat{\mathbf{C}} = \sum_{l=1}^{j}\mathbf{I}^{[l-1]}\otimes\boldsymbol{\lambda}\otimes\mathbf{I}^{[j-l]} \tag{9.54}$$

It is easy to verify that the Kronecker product of two diagonal matrices is itself a diagonal matrix, and this is sufficient to demonstrate that $\hat{\mathbf{C}}$ is diagonal, confirming that the components in Eq. 9.53 are uncoupled. Thus,

$$\kappa_j^{\otimes}[\vec{Z}(t),\cdots,\vec{Z}(t)] = (2\pi)^{j-1}\int_{-\infty}^{t} e^{-(t-u)\hat{\mathbf{C}}}\,(\boldsymbol{\theta}^{-1}\mathbf{A}^{-1})^{[j]}\,\vec{S}_j(u)\,du \tag{9.55}$$

in which $e^{-(t-u)\hat{\mathbf{C}}}$ is a diagonal matrix with $e^{-(t-u)\hat{C}_{ll}}$ as its (l,l) component. After evaluating $\kappa_j^{\otimes}[\vec{Z}(t),\cdots,\vec{Z}(t)]$ from the uncoupled equations, one finds the result for $\vec{Y}(t)$ as

$$\kappa_j^{\otimes}[\vec{Y}(t),\cdots,\vec{Y}(t)] = \boldsymbol{\theta}^{[j]}\,\kappa_j^{\otimes}[\vec{Z}(t),\cdots,\vec{Z}(t)] \tag{9.56}$$

Note that the dimension of the vectors $E[\vec{Y}^{[j]}(t)]$ and $\kappa_j^{\otimes}[\vec{Y}(t),\cdots,\vec{Y}(t)]$ of unknowns in Eqs. 9.45 and 9.50, respectively, is $(n_Y)^j$, but the number of distinct moments or cumulants of order j is much less than this. For example, $\vec{Y}^{[j]}(t)$ includes all $j!$ permutations of the terms in $Y_{l_1}Y_{l_2}\cdots Y_{l_j}$ for distinct values of $(l_1,\cdots,l_j)$, but all of these permutations are identical. Exactly the same symmetry applies to the cumulants. Of course, the $j!$ equations governing these different representations of the same term are also identical. Thus, Eqs. 9.45 and 9.50 contain much redundant information. This is exactly the same situation as was found in Eqs. 9.8 and 9.9 for the state-space equations for the autocorrelation and autocovariance matrices. The symmetry of these matrices of unknowns resulted in all the off-diagonal terms being included twice in Eqs. 9.8 and 9.9. For higher-order moments or cumulants, the amount of redundancy is much greater, though. The actual number of unique moments or cumulants of order j is given by the binomial coefficient $(n_Y + j-1)!/[(n_Y -1)!\, j!]$. One particular way to take account of this symmetry in order to consider only unique cumulants of a given order was presented by Lutes and Chen (1992), but its implementation is somewhat cumbersome.

It should be noted that higher-order moment equations are significantly less convenient than cumulant equations for delta-correlated excitations. Some indication of this fact was given in Eq. 9.18, using the autocorrelation matrix notation for the second moments of response. It can be illustrated more clearly by considering the simple situation of higher moments of a stationary scalar process. In particular, if the vectors $\vec{Y}(t)$ and $\vec{Q}(t)$ have dimension one, then the excitation in Eq. 9.45 becomes

$$\psi(t) = (j/A)\, E[Y^{j-1}(t)\, Q(t)]$$

and rewriting $Y(t) = Y(t_0) + \Delta Y$ with

$$\Delta Y = \int_{t_0}^{t} [Q(u) - (B/A)\, Y(u)]\, du$$

as in Eq. 9.11, one obtains

$$Y^{j-1}(t) = \sum_{l=1}^{j} \frac{(j-1)!}{(l-1)!(j-l)!} Y^{j-l}(t_0)(\Delta Y)^{l-1}$$

and

$$E[Y^{j-l}(t_0)(\Delta Y)^{l-1} Q(t)] = E[Y^{j-l}(t_0)]\, E[(\Delta Y)^{l-1} Q(t)] =$$
$$E[Y^{j-l}(t)] \int_{t_0}^{t} \cdots \int_{t_0}^{t} E[Q(u_1) \cdots Q(u_{l-1}) Q(t)]\, du_1 \cdots du_{l-1}$$

in which $E[Y^{j-l}(t_0)]$ has been replaced with $E[Y^{j-l}(t)]$ because of stationarity and some negligible terms have been omitted, as was done in deriving Eq. 9.13. Next we note that

$$\int_{t_0}^{t}\cdots\int_{t_0}^{t} E[Q(u_1)\cdots Q(u_{l-1})Q(t)]\,du_1\cdots du_{l-1} =$$
$$\frac{1}{l}\frac{d}{dt}\int_{t_0}^{t}\cdots\int_{t_0}^{t} E[Q(u_1)\cdots Q(u_l)]\,du_1\cdots du_l = \frac{(2\pi)^{l-1}}{l} S_l$$

in which S_l is the order-l spectral density of $\{Q(t)\}$. Thus

$$\psi(t) = \frac{1}{A}\sum_{l=1}^{j}\frac{j!}{l!\,(j-l)!}E[Y^{j-l}(t)](2\pi)^{l-1}S_l \tag{9.57}$$

showing that the excitation for the jth moment equation involves all orders of spectral density of $\{Q(t)\}$ up to order j.

An additional difficulty that arises for higher-order moments of a vector $\{\vec{Y}(t)\}$ process can be demonstrated by consideration of the $\vec{\psi}(t)$ vector for $j=3$:

$$\vec{\psi}(t) = E[\mathbf{A}^{-1}\vec{Q}(t)\otimes\vec{Y}(t)\otimes\vec{Y}(t)] + E[\vec{Y}(t)\otimes\mathbf{A}^{-1}\vec{Q}(t)\otimes\vec{Y}(t)] +$$
$$E[\vec{Y}(t)\otimes\vec{Y}(t)\otimes\mathbf{A}^{-1}\vec{Q}(t)]$$

Substituting $\vec{Y}(t) = \vec{Y}(t_0) + \Delta\vec{Y}$ in this expression expands each of these three terms into four terms, giving a total of 12. For example,

$$E[\vec{Y}(t)\otimes\mathbf{A}^{-1}\vec{Q}(t)\otimes\vec{Y}(t)] = E[\vec{Y}(t_0)\otimes\mathbf{A}^{-1}\vec{Q}(t)\otimes\vec{Y}(t_0)] + E[\vec{Y}(t_0)\otimes$$
$$\mathbf{A}^{-1}\vec{Q}(t)\otimes\Delta\vec{Y}] + E[\Delta\vec{Y}\otimes\mathbf{A}^{-1}\vec{Q}(t)\otimes\vec{Y}(t_0)] + E[\Delta\vec{Y}\otimes\mathbf{A}^{-1}\vec{Q}(t)\otimes\Delta\vec{Y}]$$

Not only does each of the terms make a contribution, but some of them are also awkward to handle. For example, consider the first term on the right-hand side of the equation. Independence of $\vec{Q}(t)$ and $\vec{Y}(t_0)$ can be used to give $E[\vec{Y}(t_0)\otimes\mathbf{A}^{-1}\vec{Q}(t)\otimes\vec{Y}(t_0)] = E[\vec{Y}(t_0)\otimes\mathbf{A}^{-1}\vec{\mu}_Q(t)\otimes\vec{Y}(t_0)]$. A component of this array is made up of the product of one component from $\mathbf{A}^{-1}\vec{\mu}_Q(t)$ and one component from $E[\vec{Y}(t_0)\otimes\vec{Y}(t_0)]$, but the ordering of the terms, due to $\mathbf{A}^{-1}\vec{\mu}_Q(t)$ being in the middle of the Kronecker product, precludes direct factoring of the expression. A Kronecker algebra operation to handle such difficulties by using a so-called *permutant matrix* (Di Paola et al., 1992) does exist, but this further complicates the equation. Overall, it seems preferable in many situations to find the response cumulants from solution of Eq. 9.50 and then find any needed response moments from these quantities rather than to work directly with higher-order moment equations.

One may note that the results presented thus far in this chapter have all been limited to situations with delta-correlated excitations. The general form of Eq. 9.45 does apply for any form of stochastic excitation, as do the special cases of Eqs. 9.8, 9.9, and 9.44. However, the right-hand side of each of those equations contains terms that depend on both the unknown dynamic response and the known excitation. It is only for the delta-correlated situation that these equations are reduced to a form that can be solved. This limitation of state-space analysis can be eased if the excitation can be modeled as a filtered delta-correlated process. That is, for many problems it is possible to model the $\{\vec{Q}(t)\}$ process as the output from a linear filter that has a delta-correlated input. If this can be done for the excitation of Eq. 9.5, then it is possible to apply state-space moment or cumulant analysis to the system of interest, as well as to the linear filter. The procedure is to define a composite linear system that includes both the filter and the system of interest. The excitation of this composite system is the input to the filter, and the response includes $\{\vec{Q}(t)\}$ as well as $\{\vec{Y}(t)\}$.

****

Example 9.10: Consider the third-order cumulants of the response of the SDF oscillator with a stationary, delta-correlated excitation. Find expressions for the stationary values of the response cumulants.

The state-space description of the system is as in Example 9.4, with $n_Y = 2$,

$$\vec{Y}(t) = \begin{pmatrix} X(t) \\ \dot{X}(t) \end{pmatrix}, \quad \mathbf{A}^{-1} = \begin{pmatrix} -k^{-1} & 0 \\ 0 & m^{-1} \end{pmatrix}, \quad \mathbf{A}^{-1}\mathbf{B} = \begin{pmatrix} 0 & -1 \\ k/m & c/m \end{pmatrix}, \quad \vec{Q}(t) = \begin{pmatrix} 0 \\ F(t) \end{pmatrix}$$

The vector giving the Kronecker representation of the third-order cumulants is $V(t) \equiv \kappa_3^{\otimes}[Y(t), Y(t), Y(t)]$, which can be written out explicitly as

$$\vec{V}(t) = \Big(\kappa_3[X(t),X(t),X(t)],\, \kappa_3[X(t),X(t),\dot{X}(t)],\, \kappa_3[X(t),\dot{X}(t),X(t)],$$
$$\kappa_3[X(t),\dot{X}(t),\dot{X}(t)],\, \kappa_3[\dot{X}(t),X(t),X(t)],\, \kappa_3[\dot{X}(t),X(t),\dot{X}(t)],$$
$$\kappa_3[\dot{X}(t),\dot{X}(t),X(t)],\, \kappa_3[\dot{X}(t),\dot{X}(t),\dot{X}(t)]\Big)^T$$

The right-hand side of the equation

$$\dot{\vec{V}}(t) + \mathbf{C}\vec{V}(t) = (2\pi)^2 (\mathbf{A}^{-1})^{[3]} \vec{S}_3$$

is very simple. In fact, it is easily shown that it has only one nonzero component—$(2\pi)^2 S_3 / m^3$ in the eighth position, in which the scalar S_3 represents the bispectrum of the $\{F(t)\}$ process:

$$\kappa_3[F(t_1),F(t_2),F(t_3)] = (2\pi)^2 S_3\, \delta(t_1 - t_3)\, \delta(t_2 - t_3)$$

Expansion of the matrix $\mathbf{C}=\mathbf{A}^{-1}\mathbf{B}\otimes\mathbf{I}^{[2]}+\mathbf{I}\otimes\mathbf{A}^{-1}\mathbf{B}\otimes\mathbf{I}+\mathbf{I}^{[2]}\otimes\mathbf{A}^{-1}\mathbf{B}$ gives

$$\mathbf{C}=\begin{pmatrix} 0 & -1 & -1 & 0 & -1 & 0 & 0 & 0 \\ k/m & c/m & 0 & -1 & 0 & -1 & 0 & 0 \\ k/m & 0 & c/m & -1 & 0 & 0 & -1 & 0 \\ 0 & k/m & k/m & 2c/m & 0 & 0 & 0 & -1 \\ k/m & 0 & 0 & 0 & c/m & -1 & -1 & 0 \\ 0 & k/m & 0 & 0 & k/m & 2c/m & 0 & -1 \\ 0 & 0 & k/m & 0 & k/m & 0 & 2c/m & -1 \\ 0 & 0 & 0 & k/m & 0 & k/m & k/m & 3c/m \end{pmatrix}$$

Solving $\mathbf{C}\vec{V}(t)=(2\pi)^{2}(\mathbf{A}^{-1})^{[3]}\vec{S}_3$ then gives the third cumulants of the stationary response as

$$\vec{V}=\frac{(2\pi)^2 S_3}{m^3}\left(\frac{2m^3}{6c^2k+3k^2m},0,0,\frac{m^2}{6c^2+3km},0,\frac{m^2}{6c^2+3km},\right.$$

$$\left.\frac{m^2}{6c^2+3km},\frac{2cm^2}{6c^2+3km}\right)^T$$

Note that the fourth, sixth, and seventh components of $\vec{V}$ are identical. We could tell that this must be true, though, by the fact that the definition of the $\vec{V}$ vector gives each of these components to be the joint cumulant of $X(t)$, $\dot{X}(t)$, and $\dot{X}(t)$. Similarly, the second, third, and fifth components of $\vec{V}$ are all equal to the joint cumulant of $X(t)$, $X(t)$, and $\dot{X}(t)$. These components not only are identical but also are identically zero in the stationary situation. This follows from the fact that

$$\kappa_3[X(t),X(t),\dot{X}(t)]=\frac{1}{3}\frac{d}{dt}\kappa_3[X(t),X(t),X(t)]$$

just as

$$E[X^2(t)\,\dot{X}(t)]=\frac{1}{3}\frac{d}{dt}E[X^3(t)]$$

****

9.8 State-Space Equations for Stationary Autocovariance

We have derived equations governing the behavior of the moments and cumulants of $\vec{Y}(t)$ at any time t by considering joint moments or joint cumulants, respectively, of $\vec{Y}(t)$ and the equation of motion at time t. These equations usually constitute the most useful form of state-space analysis, but Spanos (1983) has demonstrated that the method can also be generalized to give equations governing autocorrelation or autocovariance functions. These more general relationships can be derived by considering joint moments or cumulants of either $\vec{Y}(s)$ or $\vec{Q}(s)$ and the equation of motion at time t, with no restriction of s being equal to t. We illustrate this idea only for one of the simplest situations involving the second-moment autocorrelation and cross-correlation functions. In principle, the method can also be applied to higher-order moments or cumulants, but the calculations become more complex. To obtain an equation for a cross-correlation function we simply multiply Eq. 9.5 on the right by $\vec{Q}^T(s)$ and then take the expected value, namely

$$\mathbf{A}\boldsymbol{\phi}_{\dot{Y}Q}(t,s)+\mathbf{B}\boldsymbol{\phi}_{YQ}(t,s)=\boldsymbol{\phi}_{QQ}(t,s) \tag{9.58}$$

Because the first term in this equation represents a partial derivative of $\boldsymbol{\phi}_{YQ}(t,s)$, this is a partial differential equation governing the behavior of this cross-correlation function between the $\{\vec{Y}(t)\}$ and $\{\vec{Q}(t)\}$ processes. Rather than pursue this general relationship, though, we will simplify the presentation by restricting our attention to the situation with a stationary cross-product $\boldsymbol{\phi}_{YQ}(t,s)=\mathbf{R}_{YQ}(t-s)$. In this case we have the ordinary differential equation

$$\mathbf{A}\,\mathbf{R}'_{YQ}(\tau)+\mathbf{B}\,\mathbf{R}_{YQ}(\tau)=\mathbf{R}_{QQ}(\tau) \tag{9.59}$$

in which the prime denotes a derivative with respect to τ. The solution can be written as

$$\mathbf{R}_{YQ}(\tau)=\int_{-\infty}^{\tau}\exp[-(\tau-u)\mathbf{A}^{-1}\mathbf{B}]\,\mathbf{A}^{-1}\mathbf{R}_{QQ}(u)\,du \tag{9.60}$$

Similarly, when we multiply the equation of motion at time t by $\vec{Y}^T(s)$ on the right, we obtain an equation for the autocorrelation function as

$$\mathbf{A}\,\mathbf{R}'_{YY}(\tau)+\mathbf{B}\,\mathbf{R}_{YY}(\tau)=\mathbf{R}_{QY}(\tau) \tag{9.61}$$

and the solution can be written as

$$\mathbf{R}_{YY}(\tau)=\int_{-\infty}^{\tau}\exp[-(\tau-u)\mathbf{A}^{-1}\mathbf{B}]\,\mathbf{A}^{-1}\mathbf{R}_{QY}(u)\,du \tag{9.62}$$

The "excitation" in this expression, though, is almost the same as the term given in Eq. 9.60. In particular, noting that $\mathbf{R}_{QY}(\tau)=[\mathbf{R}_{YQ}(-\tau)]^T$ and making a change of variables of $v=-u$ in the integral allows us to rewrite Eq. 9.60 as

$$\mathbf{R}_{QY}(\tau) = \int_{\tau}^{\infty} \mathbf{R}_{QQ}(v)(\mathbf{A}^{-1})^T \exp[(\tau - v)(\mathbf{A}^{-1}\mathbf{B})^T]\, dv \tag{9.63}$$

Substitution of Eq. 9.63 into 9.62 then gives

$$\mathbf{R}_{YY}(\tau) = \int_{-\infty}^{\tau}\int_{u}^{\infty} \exp[-(\tau - u)\mathbf{A}^{-1}\mathbf{B}]\,\mathbf{A}^{-1}\mathbf{R}_{QQ}(v)(\mathbf{A}^{-1})^T \times \exp[-(v-u)(\mathbf{A}^{-1}\mathbf{B})^T]\, dv\, du \tag{9.64}$$

Alternatively, one could rewrite Eq. 9.61 to obtain $\mathbf{R}_{YQ}(\tau) = -\mathbf{R}'_{YY}(\tau)\,\mathbf{A}^T + \mathbf{R}_{YY}(\tau)\mathbf{B}^T$ and substitute this into Eq. 9.59 to give a second-order differential equation of

$$\mathbf{A}\,\mathbf{R}''_{YY}(\tau)\mathbf{A}^T + \mathbf{B}\,\mathbf{R}'_{YY}(\tau)\mathbf{A}^T - \mathbf{A}\,\mathbf{R}'_{YY}(\tau)\,\mathbf{B}^T - \mathbf{B}\,\mathbf{R}_{YY}(\tau)\,\mathbf{B}^T = -\mathbf{R}_{QQ}(\tau)$$

then verify that this equation is satisfied by Eq. 9.64.

One can, of course, also rewrite all these expressions in terms of cross-covariance rather than cross-product terms and obtain the results corresponding to Eqs. 9.63 and 9.64 as

$$\mathbf{G}_{QY}(\tau) = \int_{\tau}^{\infty} \mathbf{G}_{QQ}(v)(\mathbf{A}^{-1})^T \exp[(\tau - v)(\mathbf{A}^{-1}\mathbf{B})^T]\, dv \tag{9.65}$$

and

$$\mathbf{G}_{YY}(\tau) = \int_{-\infty}^{\tau}\int_{u}^{\infty} \exp[-(\tau - u)\mathbf{A}^{-1}\mathbf{B}]\,\mathbf{A}^{-1}\mathbf{G}_{QQ}(v)(\mathbf{A}^{-1})^T \times \exp[-(v-u)(\mathbf{A}^{-1}\mathbf{B})^T]\, dv\, du \tag{9.66}$$

Consider now the special case with $\{\vec{Q}(t)\}$ being a stationary white noise process with $\mathbf{G}_{QQ}(\tau) = 2\pi\mathbf{S}_0\,\delta(\tau)$. Substituting this expression into Eqs. 9.65 and 9.66 gives

$$\mathbf{G}_{QY}(\tau) = 2\pi\mathbf{S}_0\,(\mathbf{A}^{-1})^T \exp[\tau\,(\mathbf{A}^{-1}\mathbf{B})^T]\; U(-\tau) \tag{9.67}$$

and

$$\mathbf{G}_{YY}(\tau) = 2\pi \exp[-\tau\,\mathbf{A}^{-1}\mathbf{B}] \int_{-\infty}^{\min(0,\tau)} \exp[u\,\mathbf{A}^{-1}\mathbf{B}]\,\mathbf{A}^{-1}\,\mathbf{S}_0\,(\mathbf{A}^{-1})^T \times \exp(u(\mathbf{A}^{-1}\mathbf{B})^T)\, du \tag{9.68}$$

These results are consistent with those obtained in Section 9.5 for the special case of $\tau = 0$. Although the discontinuity in Eq. 9.67 at $\tau = 0$ leaves some doubt about the value of $\mathbf{K}_{QY}(t,t) \equiv \mathbf{G}_{QY}(0)$, Eq. 9.68 does allow evaluation of $\mathbf{K}_{YY}(t,t) \equiv \mathbf{G}_{YY}(0)$ without ambiguity, and the result is consistent with Eq. 9.23.

Note that the results in this section require knowledge only of the autocorrelation or autocovariance function for the $\{Q(t)\}$ excitation. Thus, for example, the results in Eqs. 9.67 and 9.68 are based only on the white noise property that $Q(t)$ and $Q(s)$ are uncorrelated for $s \neq t$. If we wish to obtain similar results for higher-order cumulant functions, we will need to consider the corresponding higher-order moment or cumulant functions of the excitation. In particular, results comparable to Eqs. 9.67 and 9.68 would require that $\{Q(t)\}$ satisfy other portions of the general delta-correlated relationship. We will not investigate these more general and complicated problems, though. Note that any higher-order cumulant function will depend on more than one time argument, so the formulation would involve partial differential equations. Only the stationary second moment and second cumulant (autocorrelation and autocovariance) involve only a single time argument and are therefore governed by ordinary differential equations.

It may also be noted that the results in this section are more general than those in the earlier portions of the chapter inasmuch as they include solutions for excitations that are not delta-correlated. In particular, Eqs. 9.60 and 9.64 describe characteristics of the response to an excitation that has the autocorrelation function $R_{QQ}(\tau)$, rather than being delta-correlated.

****

Example 9.11: Use Eq. 9.68 to find the autocovariance functions of $\{X(t)\}$ and $\{\dot{X}(t)\}$ for the response of an SDF oscillator excited by a stationary delta-correlated force with autospectral density of S_0.

Using the eigenvectors and eigenvalues of $\mathbf{A}^{-1}\mathbf{B}$, we first rewrite Eq. 9.68 as

$$\mathbf{G}_{YY}(\tau) = 2\pi\boldsymbol{\theta}\exp[-\tau\boldsymbol{\lambda}]\int_{-\infty}^{\min(0,\tau)} \exp[u\boldsymbol{\lambda}]\boldsymbol{\theta}^{-1}\mathbf{A}^{-1}\mathbf{S}_0(\mathbf{A}^{-1})^T(\boldsymbol{\theta}^{-1})^T \times \exp(u\boldsymbol{\lambda})\boldsymbol{\theta}^{\mathrm{T}}\,du$$

and write out the expression for the (j,l) term as

$$[\mathbf{G}_{YY}(\tau)]_{jl} = 2\pi\sum_{r_1}\sum_{r_2}\sum_{r_3}\sum_{r_4}\theta_{jr_1}e^{-\lambda_{r_1r_1}\tau}\theta^{-1}_{r_1r_2}[A^{-1}\mathbf{S}_0(A^{-1})^T]_{r_2r_3}\times \theta^{-1}_{r_4r_3}\theta_{lr_4}\int_{-\infty}^{\min(0,\tau)} e^{(\lambda_{r_1r_1}+\lambda_{r_4r_4})u}\,du$$

The state-space formulation of the SDF oscillator as given in Example 9.4 has $\vec{Q}(t)=[0,F(t)]^T$, $A_{12}=A_{21}=0$, and $A_{22}=m$. This gives

$$\mathbf{S}_0 = 2\pi S_0 \begin{pmatrix} 0 & 0 \\ 0 & 1 \end{pmatrix}, \qquad \mathbf{A}^{-1}\mathbf{S}_0(\mathbf{A}^{-1})^T = \frac{2\pi S_0}{m^2}\begin{pmatrix} 0 & 0 \\ 0 & 1 \end{pmatrix}$$

Thus, the only contributions to $[\mathbf{G}_{YY}(\tau)]_{jl}$ come from terms with $r_2=r_3=2$. Using this fact, performing the integration with respect to u, and simplifying gives

$$[\mathbf{G}_{YY}(\tau)]_{jl} = \frac{2\pi S_0}{m^2}\sum_{r_1=1}^{2}\sum_{r_4=1}^{2}\theta_{jr_1}\theta^{-1}_{r_1 2}\theta^{-1}_{r_4 2}\theta_{lr_4}\frac{e^{\lambda_{r_1,r_1}\min(0,-\tau)}\,e^{\lambda_{r_4 r_4}\min(0,\tau)}}{\lambda_{r_1 r_1}+\lambda_{r_4 r_4}}$$

in which $\lambda_{11}=\zeta\omega_0 - i\,\omega_d = c/(2m) - i\,(k/m)[1-c^2/(4k\,m)]^{1/2}$ and $\lambda_{22}=\lambda_{11}^*$. The eigenvector matrix for $\mathbf{A}^{-1}\mathbf{B}$ is as given in Example 9.5, and its inverse is easily found:

$$\boldsymbol{\theta} = \begin{pmatrix} 1 & 1 \\ -\lambda_{11} & -\lambda_{22} \end{pmatrix} \qquad \boldsymbol{\theta}^{-1} = \frac{-1}{2i\,\omega_d}\begin{pmatrix} -\lambda_{22} & -1 \\ \lambda_{11} & 1 \end{pmatrix}$$

We then find the desired covariance functions by performing the summation over r_1 and r_4. The results from the diagonal elements are

$$G_{XX}(\tau) = [\mathbf{G}_{YY}(\tau)]_{11} = \frac{\pi S_0}{2m^2\zeta\omega_0^3}e^{-\zeta\omega_0|\tau|}\left(\cos(\omega_d\tau)+\frac{\zeta\omega_0}{\omega_d}\sin(\omega_d|\tau|)\right)$$

and

$$G_{\dot{X}\dot{X}}(\tau) = [\mathbf{G}_{YY}(\tau)]_{22} = \frac{\pi S_0}{2m^2\zeta\omega_0}e^{-\zeta\omega_0|\tau|}\left(\cos(\omega_d\tau)-\frac{\zeta\omega_0}{\omega_d}\sin(\omega_d|\tau|)\right)$$

These two expressions are identical to the results in Eqs. 5.60 and 5.62, obtained using the Duhamel convolution integral. Similarly, the $[\mathbf{G}_{YY}(\tau)]_{12}$ element can be shown to be identical to the $G_{X\dot{X}}(\tau)$ function given in Eq. 5.61.

****

9.9 Fokker-Planck Equation

Note the general approach that has been used in deriving the various state-space moment and cumulant equations. We chose a moment or cumulant of interest, took its derivative with respect to time, then used this information in deriving a differential equation that governs the evolution over time of the particular quantity. This same general procedure will now be used in deriving a differential equation that governs the evolution of the probability density of a nonstationary process. We will give the full derivation of the equations for a scalar $\{Y(t)\}$ process, because that significantly simplifies the mathematical expressions, then indicate the generalization to a vector $\{\vec{Y}(t)\}$ process.

Formally we can write the derivative of a nonstationary probability density function as

$$\frac{\partial}{\partial t} p_{Y(t)}(u) = \lim_{\Delta t \to 0} \frac{1}{\Delta t} \left[p_{Y(t+\Delta t)}(u) - p_{Y(t)}(u) \right] \tag{9.69}$$

We now need to rewrite this expression in terms of conditional moments of the increment $\Delta Y \equiv Y(t+\Delta t) - Y(t)$ of $\{Y(t)\}$, because it turns out that in many problems we can find these conditional moments directly from the stochastic equation of motion. There are several ways of proceeding, but we will present only one. Stratonovich (1963) used a conditional characteristic function of ΔY, and the approach used here is similar, but without explicit use of the characteristic function. First, we write the probability density for $Y(t+\Delta t)$ as a marginal probability integral of the joint probability density of $Y(t)$ and $\Delta Y \equiv Y(t+\Delta t) - Y(t)$:

$$p_{Y(t+\Delta t)}(u) = \int_{-\infty}^{\infty} p_{Y(t)Y(t+\Delta t)}(v,u)\, dv = \int_{-\infty}^{\infty} p_{Y(t)}(v)\, p_{\Delta Y}[u - v \mid Y(t) = v]\, dv \tag{9.70}$$

We now rewrite the conditional probability density function for ΔY as

$$p_{\Delta Y}[u - v \mid Y(t) = v] = \int_{-\infty}^{\infty} p_{\Delta Y}[w \mid Y(t) = v]\, \delta(w - u + v)\, dw$$

in order to introduce a range of ΔY values that can be used in obtaining the moments of the increment. The Dirac delta function in the preceding expression is now written as an inverse Fourier transform integral:

$$\delta(w - u + v) = \frac{1}{2\pi} \int_{-\infty}^{\infty} e^{i\theta w} e^{-i\theta(u-v)}\, d\theta = \frac{1}{2\pi} \int_{-\infty}^{\infty} \sum_{j=0}^{\infty} \frac{(i\theta w)^j}{j!} e^{-i\theta(u-v)}\, d\theta \tag{9.71}$$

The power series expansion of the $e^{iw\theta}$ allows us now to perform the integration with respect to w to obtain the desired moments of the increment,[1] giving

---

[1]Performing this integration without the power series expansion would give the characteristic function used by Stratonovich.

$$p_{\Delta Y}[u-v\,|\,Y(t)=v]=\frac{1}{2\pi}\int_{-\infty}^{\infty}e^{-i\theta(u-v)}\sum_{j=0}^{\infty}\frac{(i\theta)^j}{j!}E[(\Delta Y)^j\,|\,Y(t)=v]\,d\theta \tag{9.72}$$

Substituting Eq. 9.72 into Eq. 9.70 allows us to write the unconditional probability density for $Y(t+\Delta t)$ as

$$p_{Y(t+\Delta t)}(u)=p_{Y(t)}(u)+\sum_{j=1}^{\infty}\frac{1}{2\pi}\int_{-\infty}^{\infty}p_{Y(t)}(v)\int_{-\infty}^{\infty}e^{-i\theta(u-v)}\frac{(i\theta)^j}{j!}E[(\Delta Y)^j\,|\,Y(t)=v]\,d\theta\,dv \tag{9.73}$$

in which the term with $j=0$ has been separated from the summation and simplified by again using the integral relationship in Eq. 9.71, but with $v-u$ in place of $w-u+v$. The reason for separating this one term is so that it will cancel the first term in Eq. 9.69, giving

$$\frac{\partial}{\partial t}p_{Y(t)}(u)=\sum_{j=1}^{\infty}\frac{1}{2\pi}\int_{-\infty}^{\infty}p_{Y(t)}(v)\int_{-\infty}^{\infty}e^{-i\theta(u-v)}\frac{(i\theta)^j}{j!}C^{(j)}(v,t)\,d\theta\,dv \tag{9.74}$$

in which we have introduced a new notation of

$$C^{(j)}(v,t)\equiv\lim_{\Delta t\to 0}\frac{1}{\Delta t}E[(\Delta Y)^j\,|\,Y(t)=v] \tag{9.75}$$

presuming that these limits exist. Repeated integration by parts with respect to v now allows us to rewrite Eq. 9.74 as

$$\frac{\partial}{\partial t}p_{Y(t)}(u)=\sum_{j=1}^{\infty}\frac{1}{2\pi}\int_{-\infty}^{\infty}\frac{(-1)^j}{j!}\int_{-\infty}^{\infty}e^{-i\theta(u-v)}\frac{\partial^j}{\partial v^j}[C^{(j)}(v,t)\,p_{Y(t)}(v)]\,d\theta\,dv$$

and integration with respect to θ using Eq. 9.71 gives a Dirac delta function in the argument so that the integration with respect to v is almost trivial and gives

$$\frac{\partial}{\partial t}p_{Y(t)}(u)=\sum_{j=1}^{\infty}\frac{(-1)^j}{j!}\frac{\partial^j}{\partial u^j}[C^{(j)}(u,t)\,p_{Y(t)}(u)] \tag{9.76}$$

This is the Fokker-Planck equation for the scalar process $\{Y(t)\}$. It is also sometimes called the Kolmogorov forward equation. The qualifier "forward" in the latter terminology is to distinguish this equation from a similar partial differential equation called the Kolmogorov backward equation. The backward equation involves derivatives with respect to t_0 and u_0 of a conditional probability density function, $p_{Y(t)}[u \mid Y(t_0)=u_0]$.

Note that Eq. 9.76 is very general, applying to the probability density function of virtually any scalar stochastic process. In particular, we have not used any equation of motion in the derivation of Eq. 9.76. If we can now find the values of the $C^{(j)}(u,t)$ terms that are appropriate for a particular $\{Y(t)\}$ process, then we will have a partial differential equation that must be satisfied by the nonstationary probability density function. For an important class of dynamic problems, it turns out that we can derive the $C^{(j)}(u,t)$ terms from direct use of the stochastic equation of motion. This is quite logical, inasmuch as these terms are moments of the increment of the $\{Y(t)\}$ process, and we can expect that a procedure similar to state space moment analysis should yield values for such moments. One term that is used for the $C^{(j)}(u,t)$ coefficients is *derivate moments*. Lin (1967) attributes this terminology to Moyal in 1949. Stratonovich (1963) calls them *intensity functions*.

For the scalar equation of motion

$$A\dot{Y}(t)+BY(t)=Q(t)$$

we can see that

$$\Delta Y = Y(t+\Delta t)-Y(t)=\int_t^{t+\Delta t}\dot{Y}(s)\,ds=\frac{1}{A}\int_t^{t+\Delta t}Q(s)\,ds-\frac{B}{A}\int_t^{t+\Delta t}Y(s)\,ds$$

so

$$C^{(1)}(u,t)=\lim_{\Delta t\to 0}\frac{1}{\Delta t}\left(\frac{1}{A}\int_t^{t+\Delta t}E[Q(s)\mid Y(t)=u]\,ds-\frac{B}{A}\int_t^{t+\Delta t}E[Y(s)\mid Y(t)=u]\,ds\right)$$

Provided that $Y(t)$ is continuous in the vicinity of time t, we can say that $E[Y(s)\mid Y(t)=u]\approx u$ in the final integral. Similarly, we presume that $E[Q(s)\mid Y(t)=u]$ is continuous in the first integral, giving

$$C^{(1)}(u,t) = A^{-1} E[Q(t) \mid Y(t) = u] - A^{-1} B u \tag{9.77}$$

In a similar fashion

$$C^{(2)}(u,t) = \lim_{\Delta t \to 0} \frac{1}{\Delta t} \left(\frac{1}{A^2} \int_t^{t+\Delta t} \int_t^{t+\Delta t} E[Q(s_1)Q(s_2) \mid Y(t) = u]\, ds_1\, ds_2 - \frac{2B}{A^2} \int_t^{t+\Delta t} \int_t^{t+\Delta t} E[Q(s_1)Y(s_2) \mid Y(t) = u]\, ds_1\, ds_2 + \frac{B^2}{A^2} \int_t^{t+\Delta t} \int_t^{t+\Delta t} E[Y(s_1)Y(s_2) \mid Y(t) = u]\, ds_1\, ds_2 \right)$$

The double integrals in this expression will all be of order $(\Delta t)^2$, so they can contribute nothing to $C^{(2)}(u,t)$ unless their integrands are infinite. Thus, the last term will contribute nothing provided that $E[Y(s_1)Y(s_2) \mid Y(t) = u]$ is finite, and the second term will contribute nothing provided that $E[Q(s_1)Y(s_2) \mid Y(t) = u]$ is finite. We have shown in Section 9.4, though, that $\phi_{YY}(s_1,s_2)$ and $\phi_{QY}(s_1,s_2)$ are finite even when the $\{Q(t)\}$ excitation is delta-correlated. Thus, we find that only the first term may make a contribution. Exactly the same arguments can be made for the higher-order integrals that occur in $C^{(j)}(u,t)$ for $j > 2$, with the result that

$$C^{(j)}(u,t) = \frac{1}{A^j} \lim_{\Delta t \to 0} \frac{1}{\Delta t} \int_t^{t+\Delta t} \cdots \int_t^{t+\Delta t} E[Q(s_1) \cdots Q(s_j) \mid Y(t) = u]\, ds_1 \cdots ds_j \quad \text{for } j \ge 2 \tag{9.78}$$

Note that Eq. 9.76 depends on the A and B coefficients from the equation of motion only through their effect on the $C^{(1)}(u,t)$ coefficient.

The derivation of a Fokker-Planck equation for a vector process $\{\vec{Y}(t)\}$ can be accomplished by the method given for the scalar situation of Eq. 9.76. We will not repeat the details, but the key difference between the scalar and vector situations is that the expansion of the Dirac delta function, as in Eq. 9.71, must be performed for each component of the vector, giving

$$p_{\Delta \vec{Y}}[\vec{u} - \vec{v} \mid \vec{Y}(t) = \vec{v}] = \prod_{l=1}^{n_Y} \frac{1}{2\pi} \int_{-\infty}^{\infty} e^{-i\theta_l (u_l - v_l)} \sum_{j_l=0}^{\infty} \frac{(i\,\theta_l)^{j_l}}{j_l!} E[(\Delta Y_l)^{j_l} \mid \vec{Y}(t) = \vec{v}]\, d\theta_l$$

In the subsequent simplification, similar to Eq. 9.73, it is only the one term with $j_1 = \cdots = j_{n_Y} = 0$ from the multiple summation that is separated out as $p_{\vec{Y}(t)}(u)$. Thus, all the other terms remain in the final equation, which can be written as

$$\frac{\partial}{\partial t} p_{\vec{Y}(t)}(\vec{u}) = \underbrace{\sum_{j_1=0}^{\infty} \cdots \sum_{j_{nY}=0}^{\infty}}_{(\text{except } j_1 = \cdots = j_{nY} = 0)} \frac{(-1)^{j_1 + \cdots + j_{nY}}}{j_1! \cdots j_{n_Y}!} \frac{\partial^{j_1 + \cdots + j_{nY}}}{\partial u_1^{j_1} \cdots \partial u_{n_Y}^{j_{nY}}} \times \left[C^{(j_1,\cdots,j_{nY})}(\vec{u},t) p_{\vec{Y}(t)}(\vec{u}) \right] \tag{9.79}$$

with

$$C^{(j_1,\cdots,j_{nY})}(\vec{u},t) = \lim_{\Delta t \to 0} \frac{1}{\Delta t} E[(\Delta Y_1)^{j_1} \cdots (\Delta Y_{n_Y})^{j_{nY}} \mid \vec{Y}(t) = \vec{u}] \tag{9.80}$$

This is the general form for the Fokker-Planck equation for a vector process. Of course, Eqs. 9.75 and 9.76 are simply the special cases of Eqs. 9.80 and 9.79, respectively, with $n_Y = 1$.

For a system governed by the vector state-space equation of motion

$$\mathbf{A}\dot{\vec{Y}}(t) + \mathbf{B}\vec{Y}(t) = \vec{Q}(t)$$

one can derive the analogs of Eqs. 9.77 and 9.78 as

$$\vec{C}^{(1)}(\vec{u},t) = \mathbf{A}^{-1} E[\vec{Q}(t) \mid \vec{Y}(t) = \vec{u}] - \mathbf{A}^{-1}\mathbf{B}\vec{u} \tag{9.81}$$

and

$$C^{(j_1,\cdots,j_{nY})}(\vec{u},t) = \lim_{\Delta t \to 0} \frac{1}{\Delta t} \int_t^{t+\Delta t} \cdots \int_t^{t+\Delta t} E\Big([\mathbf{A}^{-1}\vec{Q}(s_{11})]_1 \cdots \times [\mathbf{A}^{-1}\vec{Q}(s_{1j_1})]_1 \cdots [\mathbf{A}^{-1}\vec{Q}(s_{n_Y 1})]_{n_Y} \cdots [\mathbf{A}^{-1}\vec{Q}(s_{n_Y j_{nY}})]_{n_Y} \mid \vec{Y}(t) = \vec{u} \Big) \times ds_{11} \cdots ds_{1j_1} \cdots ds_{n_Y 1} \cdots ds_{n_Y j_{nY}} \tag{9.82}$$

The components of $\vec{C}^{(1)}(\vec{u},t)$ in Eq. 9.81 give only the $C^{(j_1,\cdots,j_{nY})}(\vec{u},t)$ terms with $j_1 + \cdots + j_{nY} = 1$. In particular, the lth component of $\vec{C}^{(1)}(\vec{u},t)$ is $C^{(j_1,\cdots,j_{nY})}(\vec{u},t)$ for $j_l = 1$ and all other $j_r = 0$. All other $C^{(j_1,\cdots,j_{nY})}(\vec{u},t)$ terms are evaluated from Eq. 9.82.

The development up to this point has been very general, but practical application of the partial differential equation in Eq. 9.79 generally is limited to the special case in which $\{\vec{Y}(t)\}$ is a so-called Markov process. The defining property of a Markov process can be written as

$$p_{\vec{Y}(t)}[\vec{v}\,|\,\vec{Y}(s_1)=\vec{u}_1,\cdots,\vec{Y}(s_l)=\vec{u}_l]=p_{\vec{Y}(t)}[\vec{v}\,|\,\vec{Y}(s_l)=\vec{u}_l]$$

if $s_1 \le s_2 \le \cdots \le s_l \le t$. That is, knowing several past values $\vec{Y}(s_1)=\vec{u}_1,\cdots$, $\vec{Y}(s_l)=\vec{u}_l$ of the process gives no more information about likely future values than does knowing only the most recent value $\vec{Y}(s_l)=\vec{u}_l$. This property can also be given by the statement that the future of the process is conditionally independent of the past, given the present value. The response of our causal linear state-space equation does have the Markov property in the special case when the excitation is delta-correlated. The easiest way to see this is to consider the Duhamel integral, as in Eq. 8.58, in which $\vec{Y}(t)$ for $t > t_0$ is written as a function of $\vec{Y}(t_0)$ plus an integral involving $\vec{Q}(s)$ for $s \ge t_0$. If $\{\vec{Q}(t)\}$ were not delta-correlated, then knowing additional values of $\vec{Y}(\tau)$ for $\tau < t_0$ might give additional information about the probable values of $\vec{Y}(t)$ by giving information about probable values of $\vec{Q}(s)$ for $s \ge t_0$. When $\{\vec{Q}(t)\}$ is delta-correlated, though, $\vec{Q}(s)$ for $s \ge t_0$ is independent of $\vec{Q}(\tau)$ for $\tau \le t_0$ and therefore it is conditionally independent of $\vec{Y}(\tau)$ for $\tau < t_0$. Thus, we would learn nothing more about the probable values of $\vec{Y}(t)$ by knowing additional values of $\vec{Y}(\tau)$ for $\tau < t_0$. This is the Markov property. For such a Markov process one can eliminate the conditioning in Eq. 9.80, giving the coefficients for Eq. 9.79 as

$$C^{(j_1,\cdots,j_{n_Y})}(\vec{u},t)=\lim_{\Delta t\to 0}\frac{1}{\Delta t}E[(\Delta Y_1)^{j_1}\cdots(\Delta Y_{n_Y})^{j_{n_Y}}] \tag{9.83}$$

because $\Delta\vec{Y}$ is independent of $\vec{Y}(t)$.

Clearly, the Markov property allows us to eliminate the conditioning from Eqs. 9.81 and 9.82 for the response of a linear system to a delta-correlated excitation, giving

$$\vec{C}^{(1)}(\vec{u},t)=\mathbf{A}^{-1}\vec{\mu}_Q(t)-\mathbf{A}^{-1}\mathbf{B}\,\vec{u} \tag{9.84}$$

and

$$C^{(j_1,\cdots,j_{nY})}(\vec{u},t) = \lim_{\Delta t\to 0}\frac{1}{\Delta t}\int_t^{t+\Delta t}\cdots\int_t^{t+\Delta t} E\Big([\mathbf{A}^{-1}\vec{Q}(s_{11})]_1\cdots\times$$
$$[\mathbf{A}^{-1}\vec{Q}(s_{1j_1})]_1\cdots[\mathbf{A}^{-1}\vec{Q}(s_{n_Y 1})]_{n_Y}\cdots[\mathbf{A}^{-1}\vec{Q}(s_{n_Y j_{nY}})]_{n_Y}\Big)\times \qquad (9.85)$$
$$ds_{11}\cdots ds_{1j_1}\cdots ds_{n_Y 1}\cdots ds_{n_Y j_{nY}}$$

In addition, though, the property of delta-correlation also allows us to evaluate the integral in Eq. 9.85 so that we know all the coefficients in Eq. 9.79. In this situation there is the possibility of finding the nonstationary $p_{\vec{Y}(t)}(u)$ probability density function by solving that partial differential equation.

In order to describe precisely the delta-correlated excitation, we will use the notation

$$\kappa_J[Q_1(s_{1,1}),\cdots,Q_1(s_{1,j_1}),\cdots,Q_{n_Y}(s_{n_Y,1}),\cdots,Q_{n_Y}(s_{n_Y,j_{nY}})] =$$
$$(2\pi)^{J-1} S_{j_1,\cdots,j_{nY}}(s_{n_Y,j_{nY}})\,\delta(s_{1,1}-s_{n_Y,j_{nY}})\cdots\times$$
$$\delta(s_{1,j_1}-s_{n_Y,j_{nY}})\cdots\delta(s_{n_Y,1}-s_{n_Y,j_{nY}})\cdots\delta(s_{n_Y,j_{nY}-1}-s_{n_Y,j_{nY}})$$
$$(9.86)$$

for the *J*th-order cross-cumulant, with $J = j_1 +\cdots+ j_{nY}$. This is a component version of the higher-order spectral density representation given in Kronecker notation in Eq. 9.49. Now we must consider the relationship between the *J*th-order moments in Eq. 9.85 and the cumulants in Eq. 9.86. In principle one can always write a *J*th-order moment for any set of random variables as a function of cumulants up to that order, but these general relationships are not simple when J is large. In our situation, though, the delta-correlation property allows us to solve the problem in a simple way. In particular, we note that the integral in Eq. 9.85 is over a volume of size $(\Delta t)^J$, and the only way that an integral over this volume can be linear in Δt is if the integrand behaves like a product of $J-1$ Dirac delta functions. Thus, when we relate the *J*th-order moment in Eq. 9.85 to cumulants we need consider only terms that have a product of $J-1$ Dirac delta functions coming from Eq. 9.86. Fortunately, there is only one such term in Eq. 9.85, and it is exactly the *J*th-order cumulant of the terms in the *J*th-order moment. To illustrate this fact in a relatively simple situation, consider the special case of Eq. 9.85 with $j_1 = J$ and $j_2 =\cdots= j_{n_Y} = 0$. The first few terms in the expansion of the *J*th-order moment in terms of cumulants are

$$
\begin{aligned}
E[Q_1(s_1)\cdots Q_1(s_J)] &= \kappa_J[Q_1(s_1),\cdots,Q_1(s_J)] \\
&+ b_1 \sum_l E[Q_1(s_l)]\kappa_{J-1}[Q_1(s_1),\cdots,Q_1(s_{l-1}),Q_1(s_{l+1}),\cdots,Q_1(s_J)] \\
&+ b_2 \sum_{l_1}\sum_{l_2} \kappa_2[Q_1(s_{l_1}),Q_1(s_{l_2})]\kappa_{J-2}[\text{other } Qs] \\
&+ b_3 \sum_{l_1}\sum_{l_2}\sum_{l_3} \kappa_3[Q_1(s_{l_1}),Q_1(s_{l_2}),Q_1(s_{l_3})]\kappa_{J-3}[\text{other } Qs] + \cdots
\end{aligned}
$$

The first term on the right-hand side of the equation does include a product of $J-1$ Dirac delta functions, but each of the other terms shown includes only $J-2$ Dirac delta functions and therefore contributes only a term of order $(\Delta t)^2$ to the integral of Eq. 9.85. Other terms not written out contribute even less-significant terms. For example, there are terms that correspond to subdividing $\{Q_1(s_1),\cdots,Q_1(s_J)\}$ into three subgroups, then taking the cumulant of each of these subgroups. Each of these terms gives a contribution of order $(\Delta t)^3$ to the integral.

Using the fact that only the Jth-order cumulant term makes an order Δt contribution to Eq. 9.85 and using Eq. 9.86 in evaluating this contribution gives

$$
C^{(j_1,\cdots,j_{nY})}(\vec{u},t) = (2\pi)^{J-1} S_{j_1,\cdots,j_{nY}}(t) \tag{9.87}
$$

for the higher-order coefficients in the Fokker-Planck equation for the response of the linear system to a nonstationary delta-correlated excitation.

The situation is considerably simplified in the special case when the $\{\vec{Q}(t)\}$ excitation is Gaussian. The fact that all cumulants beyond the second order are identically zero for a Gaussian process means that most of the terms given by Eq. 9.87 are also zero. In particular, one needs to use only second-order coefficients in this situation. Furthermore, Eq. 9.87 shows that the second-order coefficients are simply related to the components of the autospectral density matrix for the excitation process as

$$
C^{(\overbrace{0,\cdots,0}^{j-1},1,0,\cdots 0,1,\overbrace{0,\cdots,0}^{nY-l})}(\vec{u},t) = 2\pi[\mathbf{S}_0(t)]_{jl} \qquad \text{for } j \neq l \tag{9.88}
$$

and

$$
C^{(\overbrace{0,\cdots,0}^{j-1},2,\overbrace{0,\cdots,0}^{nY-j})}(\vec{u},t) = 2\pi[\mathbf{S}_0(t)]_{jj} \tag{9.89}
$$

with the usual definition of the autospectral density matrix such that $\mathbf{K}_{QQ}(t,s) = 2\pi \mathbf{S}_0(t)\,\delta(t-s)$ for the delta-correlated process. The Fokker-Planck relationship of Eq. 9.79 can then be written as

$$\frac{\partial}{\partial t} p_{\vec{Y}(t)}(\vec{u}) = -\sum_{j=1}^{n_Y} \frac{\partial}{\partial u_j} [C_j^{(1)}(\vec{u},t)\, p_{\vec{Y}(t)}(\vec{u})] + \pi \sum_{j=1}^{n_Y} \sum_{l=1}^{n_Y} [\mathbf{S}_0(t)]_{jl} \frac{\partial^2}{\partial u_j\, \partial u_l} p_{\vec{Y}(t)}(\vec{u}) \tag{9.90}$$

with $C_j^{(1)}(\vec{u},t)$ given by Eq. 9.84. Of course, finding solutions of the problem also requires consideration of appropriate initial conditions and boundary conditions. It may be noted that Stratonovich (1963) reserved the term *Fokker-Planck equation* for the Gaussian situation of Eq. 9.90 rather than using it for the more general forward Kolmogorov equation.

Using the formulas given, it is relatively easy to find the coefficients in the Fokker-Planck equation for the situations with a delta-correlated excitation. The problem of solving this partial differential equation may be quite difficult, though, particularly if the excitation is not Gaussian so that higher-order coefficients are nonzero. The problem is simpler if the excitation is Gaussian, but there is also less reason to use the Fokker-Planck approach in that situation. We already know that the response is Gaussian for a linear system with a Gaussian excitation, and a Gaussian distribution is completely defined in terms of its first and second moments or cumulants. Thus, for a Gaussian problem it is usually easier to use state-space analysis of mean and variance and then substitute these values into a Gaussian probability density function rather than to solve the Fokker-Planck equation to find that probability density function directly. Examples 9.12 and 9.13 will illustrate the derivation of the Fokker-Planck coefficients in a few situations, and the verification of the Gaussian solution for simple cases with Gaussian excitations.

It should be noted that it is also possible to derive state-space moment or cumulant equations from the Fokker-Planck equation. This will be illustrated in Example 9.14 for the second moment of a very simple system. There seems to be no obvious advantage to derivation of the moment or cumulant equations in this way, though, because they can be derived by the methods of Sections 9.3 and 9.4 without consideration of a partial differential equation.

As with state-space moment or cumulant analysis, practical application of the Fokker-Planck equation is typically limited to problems with delta-correlated excitations. In the case of the Fokker-Planck equation there is no such limitation on the equation itself, but evaluation of the coefficients requires the Markov property. Furthermore, this property is dependent on delta-correlation of the excitation. Just as in moment or cumulant analysis, though, the results can be extended to include any problem in which the excitation can be modeled as a filtered delta-correlated process.

****

Example 9.12: Let $\{X(t)\}$ denote the response of a dashpot with coefficient c subjected to a nonstationary Gaussian white noise force $\{F(t)\}$ with mean $\mu_F(t)$ and autospectral density $S_0(t)$.

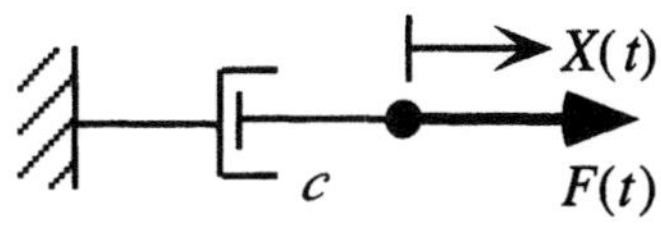

Find the coefficients in the Fokker-Planck equation and find its solution for the situation with $X(0)=0$.

The equation of motion can be written as $c\dot{X}(t)=F(t)$, which is the special case of Eq. 9.5 with $n_Y=1$, $\vec{Y}(t)=X(t)$, $\mathbf{A}=c$, $\mathbf{B}=0$, and $\vec{Q}(t)=F(t)$. Because the excitation is Gaussian, we can use Eq. 9.90 for the Fokker-Planck equation. The first-order coefficient in the Fokker-Planck equation is found from Eq. 9.84 as $C^{(1)}(u,t)=\mu_F(t)/c$, and Eq. 9.89 gives the second-order coefficient as $C^{(2)}(u,t)=2\pi S_0(t)/c^2$. Thus, the Fokker-Planck equation is

$$\frac{\partial}{\partial t}p_{X(t)}(u)=-\frac{\mu_F(t)}{c}\frac{\partial}{\partial u}p_{X(t)}(u)+\frac{\pi S_0(t)}{c^2}\frac{\partial^2}{\partial u^2}p_{X(t)}(u)$$

Note that this equation is particularly simple because $C^{(1)}(u,t)$, as well as $C^{(2)}(u,t)$, is independent of u.

Rather than seeking a solution of this partial differential equation in a direct manner, we will take the simpler approach of using our prior knowledge that the response should be Gaussian and simply verify that the equation is solved by the Gaussian probability density function. That is, we will substitute

$$p_{X(t)}(u)=\frac{1}{(2\pi)^{1/2}\sigma_X(t)}\exp\left(-\frac{1}{2}\left[\frac{u-\mu_X(t)}{\sigma_X(t)}\right]^2\right)$$

into the Fokker-Planck equation and verify that the equation is satisfied if $\mu_X(t)$ and $\sigma_X(t)$ are chosen properly. For the scalar Gaussian probability density function, the three needed derivative terms can be written as

$$\frac{\partial}{\partial u}p_{X(t)}(u)=\frac{-1}{\sigma_X(t)}\left(\frac{u-\mu_X(t)}{\sigma_X(t)}\right)p_{X(t)}(u)$$

$$\frac{\partial^2}{\partial u^2}p_{X(t)}(u)=\frac{1}{\sigma_X^2(t)}\left(\left[\frac{u-\mu_X(t)}{\sigma_X(t)}\right]^2-1\right)p_{X(t)}(u)$$

and

$$\frac{\partial}{\partial t} p_{X(t)}(u) = \left(\frac{\dot{\sigma}_X(t)}{\sigma_X(t)} \left[\left(\frac{u - \mu_X(t)}{\sigma_X(t)} \right)^2 - 1 \right] + \frac{\dot{\mu}_X(t)}{\sigma_X(t)} \left(\frac{u - \mu_X(t)}{\sigma_X(t)} \right) \right) p_{X(t)}(u)$$

$$= \frac{\partial}{\partial u} \left(\sigma_X(t) \dot{\sigma}_X(t) \frac{\partial}{\partial u} p_{X(t)}(u) - \dot{\mu}_X(t)\, p_{X(t)}(u) \right)$$

Substitution of these relationships into the Fokker-Planck equation gives

$$\frac{\partial}{\partial u} \left(\sigma_X(t) \dot{\sigma}_X(t) \frac{\partial}{\partial u} p_{X(t)}(u) - \dot{\mu}_X(t)\, p_{X(t)}(u) + \frac{\mu_F(t)}{c} p_{X(t)}(u) - \frac{\pi S_0(t)}{c^2} \frac{\partial}{\partial u} p_{X(t)}(u) \right) = 0$$

or

$$\frac{\partial}{\partial u} \left(\left[\frac{1}{2} \frac{d}{dt} \sigma_X^2(t) - \frac{\pi S_0(t)}{c^2} \right] \frac{\partial}{\partial u} p_{X(t)}(u) - \left[\dot{\mu}_X(t) - \frac{\mu_F(t)}{c} \right] p_{X(t)}(u) \right) = 0$$

The term in the parentheses is zero for $|u| = \infty$ because of the nature of the probability density function. Thus, it must be zero everywhere because its derivative is zero everywhere. Furthermore, the only way that the term can be zero for all u values is for the coefficients of both $p_{X(t)}(u)$ and its derivative to be zero. Thus, we find that $\mu_X(t)$ and $\sigma_X(t)$ are governed by the equations

$$\frac{d}{dt} \mu_X(t) = \frac{\mu_F(t)}{c}, \qquad \frac{d}{dt} \sigma_X^2(t) = \frac{2\pi S_0(t)}{c^2}$$

The solutions of these ordinary differential equations, of course, depend on the initial conditions on the problem. For our stated initial condition of $X(0) = 0$, we know that $\mu_X(0) = 0$ and $\sigma_X(0) = 0$, so the solutions are

$$\mu_X(t) = \frac{1}{c} \int_0^t \mu_F(s)\, ds, \qquad \sigma_X^2(t) = \frac{2\pi}{c^2} \int_0^t S_0(s)\, ds$$

Note that we have verified only that the Gaussian probability density function with these parameter values is one solution of the Fokker-Planck equation with the given initial conditions. In fact, it is the unique solution of the problem for the given initial conditions. The values of the nonstationary mean and variance of $\{X(t)\}$, of course, can be confirmed through other methods of analysis.

****

Example 9.13: Let $\{X(t)\}$ denote the response of the system shown when subjected to a nonstationary Gaussian white noise force $\{F(t)\}$ with mean $\mu_F(t)$ and autospectral density $S_0(t)$. Find the coefficients in the Fokker-Planck equation, verify that it has a Gaussian solution, and find the

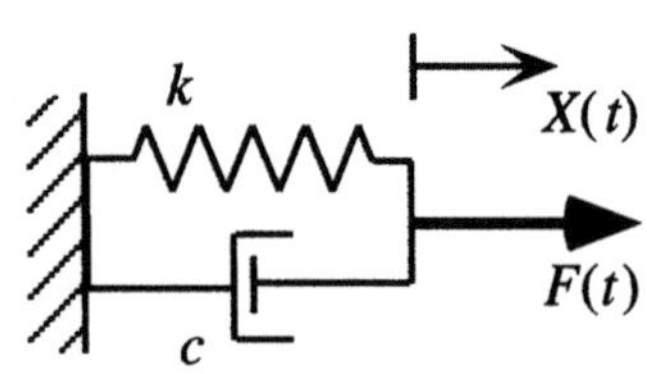

equations governing the evolution of the mean and variance in that Gaussian solution.

Proceeding in the same way as in Example 9.12, the equation of motion is Eq. 9.5 with $n_Y = 1$, $\vec{Y}(t) = X(t)$, $\mathbf{A} = c$, $\mathbf{B} = k$, and $\vec{Q}(t) = F(t)$. The coefficients in the Fokker-Planck equation are found from Eqs. 9.84 and 9.87 as

$$C^{(1)}(u,t) = \frac{\mu_F(t)}{c} - \frac{k}{c}u, \qquad C^{(2)}(u,t) = \frac{2\pi S_0(t)}{c^2}$$

Thus, the Fokker-Planck relationship from Eq. 9.90 is

$$\frac{\partial}{\partial t} p_{X(t)}(u) = -\frac{\partial}{\partial u}\left(\left[\frac{\mu_F(t)}{c} - \frac{k}{c}u\right] p_{X(t)}(u)\right) + \frac{\pi S_0(t)}{c^2}\frac{\partial^2}{\partial u^2} p_{X(t)}(u)$$

Substituting the relationships given in Example 9.12 for the derivatives of a Gaussian probability density function converts this equation into

$$\frac{\partial}{\partial u}\left(\sigma_X(t)\dot{\sigma}_X(t)\frac{\partial}{\partial u} p_{X(t)}(u) - \dot{\mu}_X(t)\, p_{X(t)}(u) + \left[\frac{\mu_F(t)}{c} - \frac{k}{c}u\right] p_{X(t)}(u) - \frac{\pi S_0(t)}{c^2}\frac{\partial}{\partial u} p_{X(t)}(u)\right) = 0$$

This requires that

$$\frac{-1}{\sigma_X(t)}\left(\frac{u - \mu_X(t)}{\sigma_X(t)}\right)\left(\sigma_X(t)\dot{\sigma}_X(t) - \frac{\pi S_0(t)}{c^2}\right) - \dot{\mu}_X(t) + \frac{\mu_F(t)}{c} - \frac{k}{c}u = 0$$

This expression includes only terms that are linear in u and terms that are independent of u. Obviously, both must be zero to ensure that the equation is satisfied for all u values. Thus, we have

$$\frac{-1}{\sigma_X^2(t)}\left(\sigma_X(t)\dot{\sigma}_X(t) - \frac{\pi S_0(t)}{c^2}\right) = \frac{k}{c}$$

and

$$\frac{\mu_X(t)}{\sigma_X^2(t)}\left(\sigma_X(t)\dot{\sigma}_X(t) - \frac{\pi S_0(t)}{c^2}\right) - \dot{\mu}_X(t) = -\frac{\mu_F(t)}{c}$$

Combining these two equations gives

$$\frac{d}{dt}\mu_X(t) + \frac{k}{c}\mu_X(t) = \frac{\mu_F(t)}{c}$$

as the ordinary differential equation governing the evolution of the mean-value function in the Gaussian probability density function. Rearranging the first of the two equations gives

$$\frac{d}{dt}\sigma_X^2(t) + 2\frac{k}{c}\sigma_X^2(t) = \frac{2\pi S_0(t)}{c^2}$$

as the corresponding equation for the variance. Note that the Gaussian assumption used here includes the implicit assumption that the given initial condition on $p_{X(t)}(u)$ can be described by a Gaussian probability density function (possibly with zero initial variance, as in Example 9.12). The Fokker-Planck equation will be the same for other initial conditions as well, but the solution will be different from the one given here.

****

Example 9.14: Give the complete Fokker-Planck equation for the 2DF building model shown when the excitation is an $\{a(t)\}$ base acceleration that is a nonstationary shot noise. The terms k_1 and k_2 represent the total stiffness of all columns in the stories, and $X_1(t)$ and $X_2(t)$ are motions relative to the base of the structure.

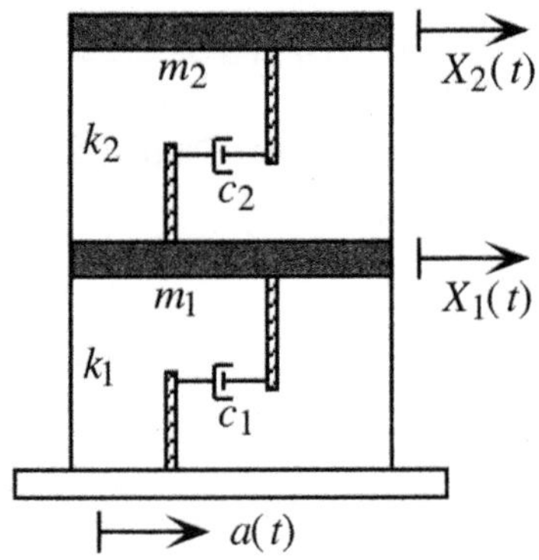

The equation of motion is Eq. 9.5 with $n_Y = 4$, and

$$\vec{Y}(t) = \begin{pmatrix} X_1(t) \\ X_2(t) \\ \dot{X}_1(t) \\ \dot{X}_2(t) \end{pmatrix}, \quad \mathbf{A} = \begin{pmatrix} -k_1 - k_2 & k_2 & 0 & 0 \\ k_2 & -k_2 & 0 & 0 \\ 0 & 0 & \frac{c_1+c_2}{m_1} & \frac{-c_2}{0} \\ 0 & 0 & 0 & m_2 \end{pmatrix}$$

$$\mathbf{B} = \begin{pmatrix} 0 & 0 & k_1+k_2 & -k_2 \\ 0 & 0 & -k_2 & k_2 \\ k_1+k_2 & -k_2 & c_1+c_2 & -c_2 \\ -k_2 & k_2 & -c_2 & c_2 \end{pmatrix} \quad \vec{Q}(t) = \begin{pmatrix} 0 \\ 0 \\ -m_1 a(t) \\ -m_2 a(t) \end{pmatrix}$$

The first-order coefficients in the Fokker-Planck equation are found from Eq. 9.84, which can be written as

$$C_j^{(1)} = [\mathbf{A}^{-1}\vec{\mu}_Q(t)]_j - [\mathbf{A}^{-1}\mathbf{B}\,\vec{u}]_j = \sum_{l=1}^{4} \mathbf{A}_{lj}^{-1}\,\mu_{Q_j}(t) - \sum_{l=1}^{4} [\mathbf{A}^{-1}\mathbf{B}]_{lj}\,u_l$$

giving $C_1^{(1)} = -u_3$, $C_2^{(1)} = -u_4$,

$$C_3^{(1)} = -\mu_a(t) + \frac{k_1+k_2}{m_1}u_1 - \frac{k_2}{m_1}u_2 + \frac{c_1+c_2}{m_1}u_3 - \frac{c_2}{m_1}u_4$$

$$C_4^{(1)} = -\mu_a(t) + \frac{k_2}{m_2}u_1 - \frac{k_2}{m_2}u_2 + \frac{c_2}{m_2}u_3 - \frac{c_2}{m_2}u_4$$

which can be further simplified by noting that $\mu_a(t) = \mu_F\,\dot{\mu}_Z(t)$ for the nonstationary shot noise, in which F denotes a typical pulse of the shot noise, and $\dot{\mu}_Z(t)$ is the expected arrival rate for the pulses (see Example 5.6). Because the excitation is a non-Gaussian delta-correlated process, the nonzero higher-order coefficients are found from Eq. 9.87, with $S_{j_1 j_2 j_3 j_4}(t)$ being the $(j_3 + j_4)$-

order autospectral density of $\{-a(t)\}$ if $j_1 = j_2 = 0$ and being zero otherwise. In Example 6.6 we found that the *J*th-order autospectral density of the shot noise is $E(F^J)\,\dot\mu_Z(t)/(2\pi)^{J-1}$. Thus, Eq. 9.87 gives

$$C^{(j_1,j_2,j_3,j_4)} = E[(-F)^{j_3+j_4}]\,\dot\mu_Z(t) \qquad \text{for } j_1 = j_2 = 0$$
$$= 0 \qquad \text{otherwise}$$

Using the simplified notation of $p(\vec u)$ for $p_{X_1(t)X_2(t)\dot X_1(t)\dot X_2(t)}(u_1,u_2,u_3,u_4)$, the Fokker-Planck relationship from Eq. 9.79 is

$$\frac{\partial}{\partial t}p(\vec u) = -\frac{\partial}{\partial u_1}[-u_3\, p(\vec u)] - \frac{\partial}{\partial u_2}[-u_4\, p(\vec u)] -$$
$$\frac{\partial}{\partial u_3}\left(\left[-\mu_F\dot\mu_Z(t) + \frac{k_1+k_2}{m_1}u_1 - \frac{k_2}{m_1}u_2 + \frac{c_1+c_2}{m_1}u_3 - \frac{c_2}{m_1}u_4\right]p(\vec u)\right) -$$
$$\frac{\partial}{\partial u_4}\left(\left[-\mu_F\dot\mu_Z(t) + \frac{k_2}{m_2}u_1 - \frac{k_2}{m_2}u_2 + \frac{c_2}{m_2}u_3 - \frac{c_2}{m_2}u_4\right]p(\vec u)\right) +$$
$$E(F^2)\dot\mu_Z(t)\left(\frac{1}{2}\frac{\partial^2}{\partial u_3^2}p(\vec u) + \frac{\partial^2}{\partial u_3\partial u_4}p(\vec u) + \frac{1}{2}\frac{\partial^2}{\partial u_4^2}p(\vec u)\right) +$$
$$E(F^3)\dot\mu_Z(t)\left(\frac{1}{6}\frac{\partial^3}{\partial u_3^3}p(\vec u) + \frac{1}{2}\frac{\partial^3}{\partial u_3^2\partial u_4}p(\vec u) + \frac{1}{2}\frac{\partial^3}{\partial u_3\partial u_4^2}p(\vec u) + \frac{1}{6}\frac{\partial^3}{\partial u_4^3}p(\vec u)\right) + \cdots$$

or

$$\frac{\partial}{\partial t}p(\vec u) = u_3\frac{\partial}{\partial u_1}p(\vec u) + u_4\frac{\partial}{\partial u_2}p(\vec u) - \frac{c_1+c_2}{m_1}p(\vec u) + \frac{c_2}{m_2}p(\vec u) -$$
$$\left(-\mu_F\dot\mu_Z(t) + \frac{k_1+k_2}{m_1}u_1 - \frac{k_2}{m_1}u_2 + \frac{c_1+c_2}{m_1}u_3 - \frac{c_2}{m_1}u_4\right)\frac{\partial}{\partial u_3}p(\vec u) -$$
$$\left(-\mu_F\dot\mu_Z(t) + \frac{k_2}{m_2}u_1 - \frac{k_2}{m_2}u_2 + \frac{c_2}{m_2}u_3 - \frac{c_2}{m_2}u_4\right)\frac{\partial}{\partial u_4}p(\vec u) +$$
$$\dot\mu_Z(t)\sum_{J=2}^{\infty}E(F^J)\sum_{l=0}^{J}\frac{J!}{(J-l)!\,l!}\frac{\partial^2}{\partial u_3^l\partial u_4^{J-l}}p(\vec u)$$

****

Example 9.15: For the system of Example 9.13, but with $\mu_F(t) = 0$, derive the state-space equation for the mean-squared value of $X(t)$ by using the Fokker-Planck equation.

Setting $\mu_F(t) = 0$ in the Fokker-Planck equation we found in Example 9.13 gives

$$\frac{\partial}{\partial t}p_{X(t)}(u) = -\frac{\partial}{\partial u}\left(-\frac{k}{c}u\, p_{X(t)}(u)\right) + \frac{\pi S_0(t)}{c^2}\frac{\partial^2}{\partial u^2}p_{X(t)}(u)$$

Multiplying this equation by u^2 and integrating over all possible u values gives

$$\int_{-\infty}^{\infty} u^2 \frac{\partial}{\partial t} p_{X(t)}(u)\,du = -\int_{-\infty}^{\infty} u^2 \frac{\partial}{\partial u}\left(-\frac{k}{c} u\, p_{X(t)}(u)\right) du + \frac{\pi S_0(t)}{c^2} \int_{-\infty}^{\infty} u^2 \frac{\partial^2}{\partial u^2} p_{X(t)}(u)\,du$$

The left-hand side of the equation is easily simplified by reversing the order of integration and differentiation to give an integral that is exactly the derivative with respect to time of $E[X^2(t)]$. The other two integrals can be evaluated by integrating by parts—once in the first integral on the right-hand side of the equation and twice in the other term. This gives

$$\frac{d}{dt} E[X^2(t)] = -2\frac{k}{c} \int_{-\infty}^{\infty} u^2 p_{X(t)}(u)\,du + 2\frac{\pi S_0(t)}{c^2} \int_{-\infty}^{\infty} p_{X(t)}(u)\,du$$

which can be rewritten as

$$\frac{d}{dt} E[X^2(t)] + 2\frac{k}{c} E[X^2(t)] = \frac{2\pi S_0(t)}{c^2}$$

Note that this is identical in form to the variance equation obtained in Example 9.3 directly from the equation of motion.

****

It should be noted that this introduction to Fokker-Planck analysis is quite elementary in nature and omits many advanced topics. More detailed analyses, such as those given by Ibrahim (1985), Soong and Grigoriu (1993), and Lin and Cai (1995), extend its application to problems in which the parameters of the system vary stochastically. In this situation, which is considered to have an internal or multiplicative excitation, one must use care in interpreting the meaning of a stochastic differential equation. There are two common interpretations, called *Itô calculus* and *Stratonovich calculus*, of the differential of the response process. If the parametric excitation is Gaussian white noise, for example, the Itô and Stratonovich results differ by a quantity called the *Wong-Zakai correction term*. More recent results show that both the Itô and Stratonovich results can be considered to be approximations of an "exact" interpretation of the differential (Caddemi and Di Paola, 1997; Lutes, 2002). These issues of interpretation of the differential do not arise for the problems considered here because we consider the stochastic excitation always to be applied externally, which is also referred to as an additive excitation. For the external excitation, all interpretations of the stochastic differential become identical and the simplified procedures presented here are applicable.

Exercises

****

First and Second Moments and Covariance

****

9.1 Consider a building subjected to a wind force $\{F(t)\}$. The building is modeled as a linear SDF system with $m = 200{,}000\,\text{kg}$, $c = 8\,\text{kN}\cdot\text{s/m}$, and $k = 3{,}200\,\text{kN/m}$ in $m\ddot{X}(t) + c\dot{X}(t) + kX(t) = F(t)$. The force $\{F(t)\}$ is modeled as a stationary process having a mean value of $\mu_F = 20\,\text{kN}$ and an autospectral density of $S_{FF}(\omega) = 0.5\,(\text{kN})^2/(\text{rad/s})$ for all ω. Solve the appropriate state-space cumulant equations to find:

(a) The stationary mean value of the displacement response.
(b) The stationary variance of the displacement response.

****

9.2 Perform second-moment state-space analysis of the system shown. The excitation is a mean-zero, stationary white noise with autospectral density of $S_{FF}(\omega) = S_0$ for all ω.

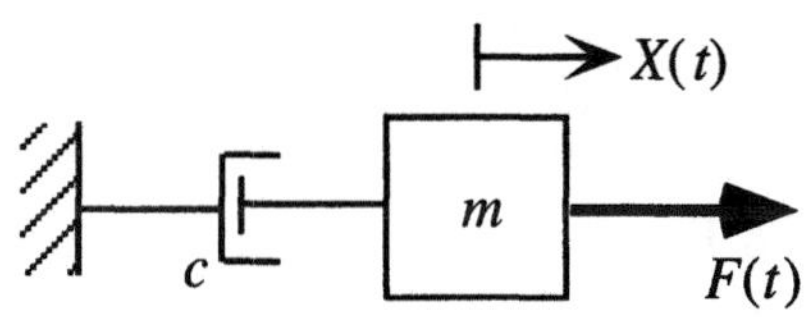

(a) Using $n_Y = 2$ and $\vec{Y}(t) = [X(t), \dot{X}(t)]^T$, solve the state-space equation to find the stationary variance of velocity $\{\dot{X}(t)\}$.
(b) Show that the equation of part (a) fails to give a stationary solution for the variance of $\{X(t)\}$.
(c) Show that one can also solve for the variance of $\{\dot{X}(t)\}$ by using the state-space equation with $n_Y = 1$ and $\vec{Y}(t) = \dot{X}(t)$.

****

9.3 Consider a linear system governed by the following third-order differential equation:

$$\dddot{X}(t) + a\ddot{X}(t) + b\dot{X}(t) + cX(t) = F(t)$$

with $a > 0$, $b > 0$, $c > 0$, and $ab > c$, where each overdot denotes a derivative with respect to time. Let $\{F(t)\}$ be a stationary white noise process with $E[F(t)] = 0$ and $E[F(t)F(s)] = 2\pi S_0\,\delta(t-s)$.

(a) Using a state vector of $\vec{Y}(t) = [X(t), \dot{X}(t), \ddot{X}(t)]^T$, formulate the state-space moment equations for the second moments of the response.
(b) Find the stationary values of $E(X^2)$, $E(\dot{X}^2)$, and $E(\ddot{X}^2)$ by solving the state-space moment equations.

****

9.4 Consider a linear system governed by the following fourth-order differential equation:

$$\ddddot{X}(t) + a_3\dddot{X}(t) + a_2\ddot{X}(t) + a_1\dot{X}(t) + a_0X(t) = F(t)$$

with $a_0 > 0$, $a_1 > 0$, $a_2 > 0$, $a_3 > 0$, and $a_1 a_2 a_3 > a_1^2 + a_0 a_3^2$, where each overdot denotes a derivative with respect to time. Let $\{F(t)\}$ be a stationary white noise process with $E[F(t)] = 0$ and $E[F(t) F(s)] = 2\pi S_0 \delta(t-s)$.

(a) Using a state vector of $\vec{Y}(t) = [X(t), \dot{X}(t), \ddot{X}(t), \dddot{X}(t)]^T$, formulate the state-space moment equations for the second moments of the response.

(b) Find the stationary values of $E(X^2)$, $E(\dot{X}^2)$, $E(\ddot{X}^2)$ and $E(\dddot{X}^2)$ by solving the state-space moment equations.

****

9.5 Consider the state-space cumulant equations derived in Example 9.2 for the third-order system with $m \ddot{X}_1(t) + c_1 \dot{X}_1(t) + k_1 X_1(t) + k_2 X_2(t) = F(t)$ and $k_2 X_2(t) = c_2 [\dot{X}_1(t) - \dot{X}_2(t)]$. Let the excitation be a mean-zero stationary white noise with autocovariance function $K_{FF}(t+\tau, t) = 2\pi S_0(t) \delta(\tau)$. Formulate and solve the stationary state-space equations to find expressions for all the second-moment quantities included in the matrix $\boldsymbol{\phi}_{YY}(t,t)$ for a state vector of $\vec{Y}(t) = [X_1(t), X_2(t), \dot{X}_1(t)]^T$.

****

9.6 For the structure and excitation of Exercise 9.1, solve the appropriate state-space equations in order to find:

(a) The conditional mean of $\{X(t)\}$ given $X(0) = 10$ mm as an initial condition.

(b) The conditional variance of $\{X(t)\}$ given an initial condition of $X(0) = 10$ mm.

****

9.7 Consider a building subjected to an earthquake ground acceleration $\{a(t)\}$. The building is modeled as a linear SDF system:

$$m \ddot{X}(t) + c \dot{X}(t) + k X(t) = -m\, a(t)$$

with $m = 200{,}000\,\text{kg}$, $c = 32\,\text{kN}\cdot\text{s/m}$, and $k = 3{,}200\,\text{kN/m}$. The ground acceleration $\{a(t)\}$ is modeled as a nonstationary, mean-zero, delta-correlated process having an autocovariance function of

$$K_{aa}(t+\tau, t) = 0.04\,\text{m}^2/(\text{rad}\cdot\text{s}^3)(e^{-0.2t} - e^{-0.25t}) U(t) \delta(\tau)$$

(a) Use state-space cumulant equations to find the response standard deviations $\sigma_X(t)$ and $\sigma_{\dot{X}}(t)$.

(b) Compare the exact value of $\sigma_X(t)$ from part (a) with the "pseudostationary" result of $[\pi S_0(t) m^2 / (kc)]^{1/2}$, in which $S_0(t)$ is defined by $K_{FF}(t+\tau, t) = 2\pi S_0(t) \delta(\tau)$ for $F(t) = -m\, a(t)$ (see Example 9.5).

****

Energy Balance

****

9.8 Consider the system of Exercise 9.2.

(a) Use the variance result to find the rate of energy dissipation in the system.

(b) Confirm that this rate of energy dissipation equals the rate of energy addition calculated from Eq. 9.35.

****

9.9 Let the excitation processes $\{F_1(t)\}$ and $\{F_2(t)\}$ in Example 9.6 both be mean-zero.

(a) Use the variance and covariance results to find the rate of energy dissipation in the system.

(b) Confirm that this rate of energy dissipation equals the rate of energy addition calculated from Eq. 9.35.

****

9.10 You wish to approximate the response levels at the top of a multistory building. Past experience shows that this response is usually dominated by the fundamental mode, so you choose to develop an equivalent SDF model based only on that mode. For your structure, the fundamental mode has a natural frequency ω_0 rad/sec, damping ζ, and an approximately linear mode shape: $X(t) = X_T(t)\, y / L$, in which X is the displacement at height y, L is the height of the building, and X_T is the top displacement. Assume that the total mass M is uniformly distributed over the height (M/L per meter).

(a) Find an expression for the kinetic energy (KE) in the structure in terms of the top velocity $\dot{X}_T$.

(b) Assume that an earthquake gives the base of the building a mean-zero white noise acceleration with an autospectral density of S_0. Find the expected value of the rate at which energy is added to the structure: $E(PA)$.

(c) Let PD denote the rate at which energy is dissipated by the structure, and assume that in stationary response $E(PD) = 4\zeta\omega_0 E(KE)$ (as for an SDF system). Using the results of (a) and (b), find the mean-squared values of the top velocity and displacement for stationary response.

****

Kronecker Formulation

****

9.11 Write out all four component equations of expression 9.44 for the second moments of the response of an SDF oscillator using the **A** and **B** matrices given in Example 9.4. Verify that these expressions are identical to those from Eq. 9.8.

****

9.12 Write out all nine component equations of expression 9.44 for second moments using the **A** and **B** matrices for the third-order system of Example 9.2 and Exercise 9.5. Verify that these expressions are identical to those from Eq. 9.8.

****

Fokker-Planck Equation

****

9.13 Let $\{X(t)\}$ be the response of an SDF oscillator with an excitation that is a mean-zero, Gaussian, stationary white noise with an autospectral density of S_0.

(a) Give the complete Fokker-Planck equation for the system.

(b) Verify that this equation is satisfied by a stationary Gaussian probability density function with second moments given by Eq. 5.64 of Chapter 5.

****

9.14 Let $\{X(t)\}$ be the response of the third-order system of Exercise 9.3 with an excitation that is a mean-zero, Gaussian, stationary white noise with an autospectral density value of S_0.

(a) Give the complete Fokker-Planck equation for the system.

(b) Verify that this equation is satisfied by a stationary Gaussian probability density function with the second-moment values found in Exercise 9.3.

****

9.15 Give the complete Fokker-Planck equation for the fourth-order system of Exercise 9.4 with a stationary Gaussian white noise excitation that has a mean value of μ_F and an autospectral density value of S_0.

****

9.16 Give the complete Fokker-Planck equation for the third-order system of Example 9.2 and Exercises 9.5 and 9.12 in the situation in which the stationary, mean-zero, white noise excitation is also Gaussian.

****

Chapter 10
Introduction to Nonlinear Stochastic Vibration

10.1 Approaches to the Problem

Exact analytical solutions have been found for only relatively few problems of nonlinear stochastic dynamics. Thus, much of the analysis of such problems relies on various approximate techniques. The existing analytical solutions are important, though, in at least two ways. Most obvious is the situation in which some important nonlinear system can be adequately approximated by a different nonlinear model for which an analytical solution is known so that the solution can be used directly. Systems with exact analytical solutions can also be used in validating the assumptions of some approximate technique prior to its application to another problem for which no analytical solution is known.

In this brief introduction to nonlinear stochastic vibration, we will focus on two aspects of the problem. We will review some exact analytical formulations of nonlinear problems, along with the corresponding solutions for some significant special cases. In addition, we will present the most commonly used category of approximate methods, called equivalent linearization. Even though it involves the most difficult equations of any of our methods, Fokker-Planck analysis will be presented first. The reason for this seemingly odd choice is that Fokker-Planck analysis provides exact analytical solutions to certain problems. These relatively limited exact results can then be used as a basis for judging the value and the shortcomings of simpler approximate solutions.[1] Before presenting the details of either exact or approximate solutions, though, we will discuss some basic ideas of nonlinear analysis.

Probably the greatest obstacle to analysis of nonlinear problems is the fact that superposition is generally not applicable. Because almost all of our linear

---

[1]There is no difficulty in proceeding directly from Section 10.1 to Section 10.3, in case the reader wishes to bypass the Fokker-Planck analysis of nonlinear systems. This is certainly recommended for anyone who has not become familiar with the material in Section 9.9.

analysis techniques are based on the concept of finding a general solution by superimposing particular solutions, it is necessary that we find alternative formulations for nonlinear problems. The time-domain and frequency-domain analysis methods of Chapters 5, 6, and 8 are particularly limited for nonlinear problems because they both invoke the concept of superposition at the very outset of the analysis procedure. For example, the Duhamel convolution integral of Chapter 5 can be viewed as giving the $X(t)$ response at time t as a superposition of the responses at that time due to $F(u)$ excitation contributions at all earlier times. Similarly, the fundamental idea in Chapter 6 is that for any given frequency ω, the $\tilde{X}(\omega)$ Fourier transform of the response is the same as the response to the $\tilde{F}(\omega)$ Fourier transform of the excitation, and this result is valid only if superposition applies.

The derivation of state-space moment/cumulant equations as introduced in Chapter 9, on the other hand, does not depend on superposition or any other property associated with linearity. Thus, one can also derive these equations for nonlinear systems. In this regard, though, it is important to distinguish between the derivation and the solution of the state-space equations. For example, the state-space moment and cumulant equations that we derived in Chapter 9 were linear, so we used superposition, in the form of convolution integrals, in writing their solutions. For nonlinear systems, we will find that the state-space moment and cumulant equations are also nonlinear. Thus, the form of the solutions of these equations will be different from that in Chapter 9, even though the method of derivation is unchanged. The situation is somewhat different for the Fokker-Planck equation, because it has variable coefficients that may be nonlinear functions of the state variables, even for a linear system. Nonetheless, the increased complexity associated with nonlinear analysis still applies. For example, a nonlinear system with a Gaussian excitation has a non-Gaussian response, and the simple Fokker-Planck solutions demonstrated in Chapter 9 were only for Gaussian processes.

The most commonly used approach for seeking approximate answers to nonlinear dynamic problems is to somehow replace the nonlinear system with a linear system. We will consider the term *equivalent linearization* to apply to all such methods, regardless of the technique used to choose the form or parameters of the linear substitute system. *Statistical linearization* is the most commonly used term to describe techniques for choosing the linearized system on the basis of minimum mean-squared error, but *equivalent linearization* can also be applied to minimization of some other probabilistic measure of discrepancy between the nonlinear and linear systems.

The primary reason for using equivalent linearization techniques, of course, is their simplicity. The contents of this book up until this point are an example of the extensive literature that exists on the stochastic dynamics of linear systems. If a nonlinear system can be replaced by a linear one, then all the techniques for linear systems can be applied to give approximate predictions regarding the response of the nonlinear problem. Interpretation, as well as prediction, though, can be a basis for considering linearization. Even if we have in some way determined the true response levels for a nonlinear system, we may find it useful to seek a linear system with similar response behavior so that we can interpret our results based on our experience with linear dynamics.

It should be emphasized that the concept of "equivalence" between any linear and nonlinear systems is always limited in scope. For example, we can sometimes find a linear system that matches certain mean and variance values of a nonlinear system, but there are always other statistics that are not the same for the two systems. Probably the simplest and most significant difference between linear and nonlinear models has to do with the Gaussian probability distribution. As previously noted, any linear system with a Gaussian excitation also has a Gaussian response. This allows the prediction of various response probability values based only on knowledge of the mean and variance of the response process. For a nonlinear system this relationship no longer holds. That is, a nonlinear system with a Gaussian excitation has a non-Gaussian response. Thus, even if an equivalent linearization scheme gives us good approximations of the mean and variance of the response of a nonlinear system, this will generally not allow us to compute accurately the probability of some rare event. Various techniques do exist for obtaining improved estimates of such probabilities, but this requires knowledge of more than mean and variance. Thus, one must always be careful to remember the limitations of any linearization and not to make erroneous conclusions based on inappropriate analogies with linear response. Despite these limitations, equivalent linearization can often give very useful approximations of the response of a nonlinear system.

The state-space moment/cumulant equations and equivalent linearization methods presented here are probably the most commonly used techniques for nonlinear stochastic analysis. Nonetheless, it should be kept in mind that there are a number of other approaches, and they continue to be developed. For example, perturbation provides a straightforward and general approach for a system that deviates only slightly from linear behavior (for example, see Lin, 1967; Nigam, 1983). We have chosen to emphasize linearization rather than perturbation for several reasons. These include the following facts: First-order perturbation usually agrees with linearization; it is difficult to find higher-order perturbation terms; perturbation sequences do not necessarily converge for finite nonlinearities; and interpretation is generally simpler for linearization than for perturbation models.

Stochastic averaging is another important technique that is not considered here. Although not a fundamentally different approach to the problem, this technique is useful inasmuch as it reduces the order of a problem by using time averages of certain slowly varying quantities (Lin and Cai, 1995; Roberts and Spanos, 1986). Mention was also made at the beginning of this section of the possibility of seeking an equivalence between a given nonlinear system and a substitute nonlinear system for which the solution is known. This method has been used with some success, even when implemented with relatively crude concepts of equivalence (for example, by Chen and Lutes, 1994), and some work has been done on techniques for improving the equivalent nonlinearization procedure by using mean-squared error minimization (Caughey, 1986; Cai and Lin, 1988; Zhu and Yu, 1989; To and Li, 1991). The usefulness of the method is limited in practice, though, by the relatively small number of models for which exact solutions are known [see Lin and Cai (1995) for a summary]. Another approach that has received considerable attention in recent years extends the time- and frequency-domain integrals of Chapters 5 and 6 by using the Volterra series (Schetzen, 1980; Rugh, 1981). Truncating this series has given useful approximate results for some problems, particularly those involving polynomial nonlinearities (Choi et al., 1985; Naess and Ness, 1992; Winterstein et al., 1994). The approach has also been combined with equivalent nonlinearization to give a technique of first replacing the problem of interest with one having a polynomial nonlinearity, then using the approximate solution for the substitute system (Spanos and Donley, 1991; Kareem et al., 1995; Li et al., 1995). Further developments can certainly be expected in the future.

10.2 Fokker-Planck Equation for Nonlinear System

The general Fokker-Planck equation for the probability density of any vector process $\{\vec{Y}(t)\}$ was given in Eq. 9.79 as

$$\frac{\partial}{\partial t} p_{\vec{Y}(t)}(\vec{u}) = \underbrace{\sum_{j_1=0}^{\infty} \cdots \sum_{j_{nY}=0}^{\infty}}_{(\text{except } j_1=\cdots=j_{nY}=0)} \frac{(-1)^{j_1+\cdots+j_{nY}}}{j_1!\cdots j_{nY}!} \frac{\partial^{j_1+\cdots+j_{nY}}}{\partial u_1^{j_1}\cdots\partial u_{n_Y}^{j_{nY}}} \times$$

$$[C^{(j_1,\cdots,j_{nY})}(\vec{u},t)\, p_{\vec{Y}(t)}(\vec{u})] \quad (10.1)$$

This equation applies equally well to the response of a nonlinear system and a linear system. Stated in another way, the equation applies to the probability density function of the vector process independent of whether that process is related to any dynamics problem. Thus, we may also find the equation useful in analyzing nonlinear vibration problems if we can determine the $C^{(j_1,\cdots,j_{n_Y})}(\vec{u},t)$ coefficients for the situation in which $\vec{Y}(t)$ is the vector of state variables for a nonlinear system.

We will now consider a very general formulation of nonlinear dynamics problems with a state vector $\vec{Y}(t)$ described by

$$\dot{\vec{Y}}(t)+\vec{g}[\vec{Y}(t)]=\vec{Q}(t) \tag{10.2}$$

Note that this equation is identical to Eq. 9.5, except that the linear restoring force $\mathbf{A}^{-1}\mathbf{B}\,\vec{Y}(t)$ has now been replaced by the nonlinear vector function $\vec{g}[\,\vec{Y}(t)]$, and the excitation has been simplified by replacing $\mathbf{A}^{-1}\,\vec{Q}(t)$ with $\vec{Q}(t)$. In particular, each component of $\vec{g}[\,\vec{Y}(t)]$ is a scalar nonlinear function of the state variables: $g_j\,[\vec{Y}(t)]\equiv g_j[Y_1(t),\cdots,Y_{n_Y}(t)]$ for $j=1,\cdots,n_Y$.

Using the same procedure as in Section 9.9, we can say that

$$\Delta Y_j=\int_t^{t+\Delta t}\dot{Y}_j(s)\,ds=\int_t^{t+\Delta t}Q_j(s)\,ds-\int_t^{t+\Delta t}g_j[\vec{Y}(s)]\,ds$$

so the vector of first-order coefficients, corresponding to Eq. 9.81, is

$$\vec{C}^{(1)}(u,t)=\lim_{\Delta t\to 0}\frac{1}{\Delta t}E[\Delta\vec{Y}\,|\,\vec{Y}(t)=\vec{u}]=E[\vec{Q}(t)\,|\,\vec{Y}(t)=\vec{u}]-\vec{g}(\vec{u})$$

Provided that $E(Q_j(s_1)g_l[Y(s_2)]\,|\,Y(t)=\vec{u})$ is finite for all choices of j and l, we can say that the coefficients based on higher-order moments of the increments of $\{\vec{Y}(t)\}$ must depend only on terms like the integral of $Q_j(s)$ appearing on the right-hand side of Eq. 10.2. Furthermore, these integrals are identical to those for a linear system, so these higher-order coefficients are identical to those given by Eq. 9.82 for the linear system.

If we now restrict the excitation process $\{\vec{Q}(t)\}$ to be delta-correlated so that $\vec{Q}(t)$ is independent of $\vec{Q}(s)$ for $s\neq t$, we can say that $\{\vec{Y}(t)\}$ has the Markov property and the Fokker-Planck coefficients are given by

$$\vec{C}^{(1)}(\vec{u},t) = \vec{\mu}_Q(t) - \vec{g}(\vec{u}) \tag{10.3}$$

and

$$\begin{aligned} C^{(j_1,\cdots,j_{nY})}(\vec{u},t) &\equiv \lim_{\Delta t \to 0} \frac{1}{\Delta t} E[(\Delta Y_1)^{j_1} \cdots (\Delta Y_{n_Y})^{j_{nY}} \mid \vec{Y}(t) = \vec{u}] \\ &= (2\pi)^{J-1} S_{j_1,\cdots,j_{nY}}(t) \end{aligned} \tag{10.4}$$

in which $J = j_1 + \cdots + j_{n_Y}$ and $S_{j_1,\cdots,j_{nY}}(t)$ is a nonstationary spectral density of order J for the delta-correlated process $\{\vec{Q}(t)\}$. Clearly Eq. 10.3 is the obvious nonlinear generalization of Eq. 9.84, and Eq. 10.4 is actually identical to Eq. 9.87. These two equations then give the coefficients in the Fokker-Planck relationship of Eq. 10.1 for a nonlinear system with a delta-correlated excitation. In the special case in which the excitation is also Gaussian, the higher-order cumulants of the excitation are zero so that $C^{(j_1,\cdots,j_{nY})}(\vec{u},t) = 0$ for $J > 2$ and

$$C^{(\overbrace{0,\cdots,0}^{j-1},1,0,\cdots \overbrace{0,1,0,\cdots,0}^{n_Y - l})}(\vec{u},t) = 2\pi[\mathbf{S}_0(t)]_{jl} \text{ for } j \neq l \tag{10.5}$$

and

$$C^{(\overbrace{0,\cdots,0}^{j-1},2,\overbrace{0,\cdots,0}^{n_Y - j})}(\vec{u},t) = 2\pi[\mathbf{S}_0(t)]_{jj} \tag{10.6}$$

in which $\mathbf{S}_0(t)$ is defined by the covariance relationship $\mathbf{K}_{QQ}(t,s) = 2\pi \mathbf{S}_0(t)\delta(t-s)$.

We now turn our attention to certain special cases for which solutions of Eq. 10.1 have been found for nonlinear systems. We start with the simplest situation of a system with a first-order differential equation of motion:

$$\dot{X}(t) + f'[X(t)] = F(t) \tag{10.7}$$

in which $f'(u)$ is the derivative of a nonlinear function $f(u)$ satisfying $f(u) \to \infty$ as $|u| \to \infty$ and the excitation $\{F(t)\}$ is a mean-zero, stationary Gaussian white noise with autospectral density S_0. This equation could describe the behavior of a linear dashpot and a nonlinear spring attached in parallel or a mass attached to a nonlinear dashpot, although the latter model would require that $\{X(t)\}$ represent the velocity of the system.

From Eqs. 10.3 and 10.6 we have the first- and second-order coefficients in the Fokker-Planck equation for the system of Eq. 10.7 as

$$C^{(1)}(u,t) = -g(u) = -f'(u)$$

and

$$C^{(2)}(u,t) = 2\pi S_0(t)$$

Substituting these expressions into the Fokker-Planck equation gives

$$\frac{\partial}{\partial t} p_{X(t)}(u) = -\frac{\partial}{\partial u}[-f'(u)\, p_{X(t)}(u)] + \pi S_0 \frac{\partial^2}{\partial u^2} p_{X(t)}(u)$$

$$= \frac{\partial}{\partial u}\left[f'(u)\, p_{X(t)}(u) + \pi S_0 \frac{\partial}{\partial u} p_{X(t)}(u) \right]$$

It is easy to verify that a stationary solution of this equation is

$$p_{X(t)}(u) = A \exp\left(\frac{-f(u)}{\pi S_0} \right) \tag{10.8}$$

so

$$\frac{\partial}{\partial u} p_{X(t)}(u) = -\frac{f'(u)}{\pi S_0} p_{X(t)}(u)$$

and the term in the brackets of the Fokker-Planck equation is identically zero, giving

$$\frac{\partial}{\partial t} p_{X(t)}(u) = 0$$

as it must be for a stationary process. The definition of $p_{X(t)}(u)$ given in Eq. 10.8 is completed by evaluating the constant A from

$$A^{-1} = \int_{-\infty}^{\infty} \exp\left(\frac{-f(u)}{\pi S_0} \right) du \tag{10.9}$$

****

Example 10.1: Let $\{X(t)\}$ denote the response of the system with the equation of motion

$$\dot{X}(t) + k_1 X(t) + k_3 X^3(t) = F(t)$$

with $k_1 \ge 0$, $k_3 \ge 0$, and {F(t)} being a mean-zero, stationary Gaussian white noise with autospectral density S_0. Find the probability distribution and the variance of the stationary $\{X(t)\}$ response.

Using $f'(u) = k_1 u + k_3 u^3$ gives $f(u) = k_1 u^2 /2 + k_3 u^4 /4$ so that Eqs. 10.8 and 10.9 give

$$p_{X(t)}(u) = \frac{\exp\left(\frac{-1}{\pi S_0}\left[\frac{k_1 u^2}{2} + \frac{k_3 u^4}{4}\right]\right)}{\int_{-\infty}^{\infty} \exp\left(\frac{-1}{\pi S_0}\left[\frac{k_1 v^2}{2} + \frac{k_3 v^4}{4}\right]\right) dv}$$

It is not convenient to find a closed-form expression for the integral in the denominator of this expression or of the integral of $u^2 p_{X(t)}(u)$ that is needed in the evaluation of the variance σ_X^2. Thus, numerical integration will be used to provide sample results. First, though, it is convenient to introduce a normalizing constant of $\sigma_0^2 = \pi S_0 / k_1$, which is the variance of $X(t)$ for the limiting linear case of $k_3 = 0$. Using a change of variables of $u = w\sigma_0$ then gives the probability density function as

$$p_{X(t)}(w\sigma_0) = \frac{\exp\left(-\frac{w^2}{2} - \alpha \frac{w^4}{4}\right)}{\int_{-\infty}^{\infty} \exp\left(-\frac{v^2}{2} - \alpha \frac{v^4}{4}\right) dv}$$

in which $\alpha = (k_3 \sigma_0^2 / k_1)$. The following sketch shows how the variance of $X(t)$ depends on the dimensionless parameter α. Also, another sketch shows a common measure of the non-Gaussianity of the response of the nonlinear system. In particular, we have numerically evaluated the fourth moment $E(X^4)$ so that we can plot $kurtosis = E(X^4) / \sigma_X^4$. Note that the fourth moment and kurtosis have been used to illustrate non-Gaussianity because the third moment and skewness are zero due to the symmetry of the problem. The kurtosis, of course, is 3.0 for a Gaussian process, and this limit is reached at $\alpha = 0$.

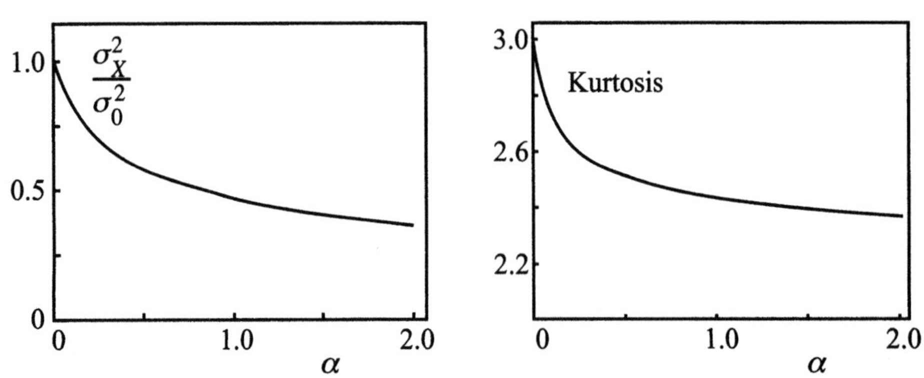

Next we show the probability density function for $X(t)$ for two particular values of the α nonlinearity parameter. In each case a Gaussian probability density function with the same variance as the non-Gaussian one is shown for comparison.

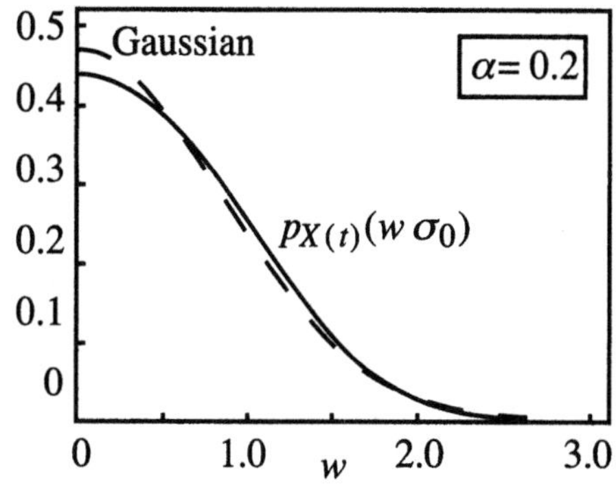

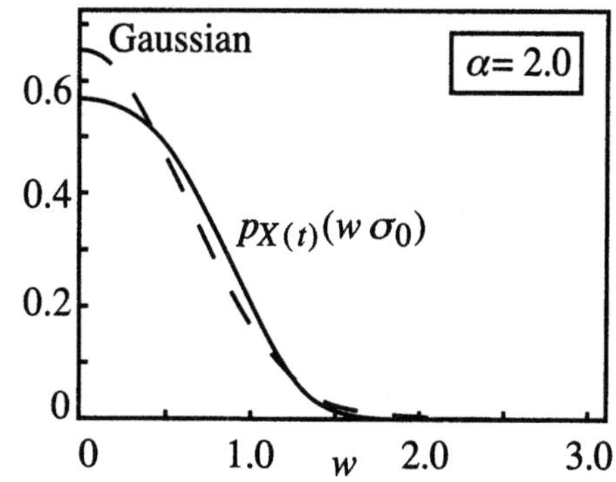

As previously noted, the limiting case of $k_3 = 0$ gives a Gaussian response with variance σ_0^2. For the other limiting case of $k_1 = 0$ it is also possible to evaluate the integrals in closed form. In particular, the integrals of interest can all be written in terms of gamma functions (see Example 3.7) as

$$\int_0^\infty \exp(-a\,u^4)\,du = \frac{\Gamma(0.25)}{4a^{0.25}} \approx 0.9064\,a^{-0.25}$$

$$\int_0^\infty u^2 \exp(-a\,u^4)\,du = \frac{\Gamma(0.75)}{4a^{0.75}} \approx 0.3064\,a^{-0.75}$$

and

$$\int_0^\infty u^4 \exp(-a\,u^4)\,du = \frac{\Gamma(1.25)}{4a^{1.25}} = \frac{\Gamma(0.25)}{16a^{1.25}} \approx 0.2266\,a^{-1.25}$$

Substitution gives $\sigma_X^2 \sim 0.6760(\pi S_0 / k_3)^{1/2}$ and $kurtosis \approx 2.188$.

****

Example 10.2: Let $\{X(t)\}$ denote the response of the system with the equation of motion

$$\dot{X}(t) + k\,|\,X(t)\,|^b \operatorname{sgn}[X(t)] = F(t)$$

with $\{F(t)\}$ being a mean-zero, stationary Gaussian white noise with autospectral density S_0. Find the probability distribution and the variance of the stationary $\{X(t)\}$ response.

Using $f'(u) = k\,|\,u\,|^b \operatorname{sgn}(u) = k\,|\,u\,|^{(b-1)}\,u$ gives $f(u) = k\,|\,u\,|^{b+1}/(b+1)$ so that Eqs. 10.8 and 10.9 give

$$p_{X(t)}(u) = A \exp\left[\frac{-k\,|\,u\,|^{b+1}}{(b+1)\,\pi\,S_0}\right]$$

Using the change of variables $w = k\,u^{n+1}/[\pi(n+1)\,S_0]$ allows A to be evaluated as follows:

$$A^{-1} = 2\int_0^\infty \exp\left(\frac{-k\,u^{b+1}}{(b+1)\pi\,S_0}\right)du = 2\int_0^\infty e^{-w}\left(\frac{w^{-b/(b+1)}}{b+1}\left[\frac{(b+1)\pi\,S_0}{k}\right]^{1/(b+1)}\right)dw$$

and

$$A = \left(\frac{k}{(b+1)\pi\,S_0}\right)^{1/(b+1)}\left(\frac{b+1}{2}\right)\left(\Gamma\left[\frac{1}{b+1}\right]\right)^{-1}$$

The variance of $\{X(t)\}$ is then given by

$$\sigma_X^2 = E(X^2) = 2A\int_0^\infty u^2 \exp\left(\frac{-k\,u^{b+1}}{(b+1)\pi\,S_0}\right)du$$

$$= 2A\int_0^\infty e^{-w}\left(\frac{w^{(j-b)/(b+1)}}{b+1}\left[\frac{(b+1)\pi\,S_0}{k}\right]^{(j+1)/(b+1)}\right)dw$$

or

$$\sigma_X^2 = \left(\frac{\pi\,S_0}{k}\right)^{2/(b+1)}(b+1)^{2/(b+1)}\,\Gamma\left(\frac{3}{b+1}\right)\left(\Gamma\left[\frac{1}{b+1}\right]\right)^{-1}$$

****

One of the more general second-order differential equations for which a solution is known involves a stochastic process $\{X(t)\}$ with an equation of motion of

$$m\,\ddot{X}(t) + f_2'\big(H[X(t),\dot{X}(t)]\big)\dot{X}(t) + f_1'[X(t)] = F(t) \tag{10.10}$$

in which $f_1'(\cdot)$ and $f_2'(\cdot)$ are the derivatives of nonlinear functions satisfying $u\,f_1'(u) \ge 0$ for all u, $f_1(u) \to \infty$ for $|u| \to \infty$, $f_2'(u) \ge 0$ for all $u \ge 0$, $f_2(u) \to \infty$ for $u \to \infty$, and H is a nonnegative term that represents the energy in the system. Note that the $f_2'(H[X(t),X(t)])$ term of Eq. 10.10 is a nonlinear damping term, because it corresponds to a force that opposes the direction of the velocity of the mass. Similarly, $f_1'[X(t)]$ is a nonlinear spring term, giving a restoring force that opposes the displacement of the mass from equilibrium.

The term $f_1(u)$ is exactly the potential energy in the nonlinear system, because it is the integral of the nonlinear restoring force. Thus, the total energy in the system is

$$H[X(t),\dot{X}(t)] = f_1[X(t)] + \frac{m}{2}\dot{X}^2(t)$$

Using $\vec{Y}(t) = [X(t),\dot{X}(t)]^T$ as the state vector gives the terms in Eq. 10.2 as

$$\vec{g}[\vec{Y}(t)] = \begin{pmatrix} -\dot{X}(t) \\ f_1'[X(t)]/m + f_2'\left(H[X(t),\dot{X}(t)]\right)\dot{X}(t)/m \end{pmatrix}$$

and $\vec{Q}(t) = [0, F(t)/m]^T$. We will restrict the excitation $\{F(t)\}$ to be mean-zero, stationary, Gaussian white noise so that it is delta-correlated and the first-order coefficients in the Fokker-Planck equation are

$$\vec{C}^{(1)}(\vec{u},t) = -\vec{g}(\vec{u}) = \begin{pmatrix} u_2 \\ -f_1'(u_1)/m - f_2'[H(u_1,u_2)]\,u_2/m \end{pmatrix} \tag{10.11}$$

Note that $\mathbf{K}_{QQ}(t,s) = 2\pi\mathbf{S}_0\,\delta(t-s)$ gives only one nonzero element in the $\mathbf{S}_0$ matrix. That element is the autospectral density of $\{F(t)/m\}$, which we will denote by the scalar S_0/m^2. This gives the only nonzero higher-order coefficient in the Fokker-Planck equation as

$$C^{(0,2)}(\vec{u},t) = 2\pi[\mathbf{S}_0]_{22} = 2\pi S_0/m^2 \tag{10.12}$$

Substituting Eqs. 10.11 and 10.12 into Eq. 10.1 gives the Fokker-Planck equation for the system of Eq. 10.10 as

$$\frac{\partial}{\partial t}p_{\vec{Y}(t)}(\vec{u}) = -\frac{\partial}{\partial u_1}[u_2\, p_{\vec{Y}(t)}(\vec{u})] -$$
$$\frac{1}{m}\frac{\partial}{\partial u_2}\left[\left(-f_1'(u_1) - f_2'[H(u_1,u_2)]\,u_2\right)p_{\vec{Y}(t)}(\vec{u})\right] + \frac{\pi S_0}{m^2}\frac{\partial^2}{\partial u_2^2}p_{\vec{Y}(t)}(\vec{u}) \tag{10.13}$$

No exact nonstationary solution of this equation has been found, but Caughey (1965) demonstrated that it has a stationary solution of

$$p_{\vec{Y}(t)}(\vec{u}) = A\,\exp\left(\frac{-1}{\pi S_0}f_2[H(u_1,u_2)]\right) \tag{10.14}$$

in which

$$A^{-1} = \int_{-\infty}^{\infty}\int_{-\infty}^{\infty} \exp\left(\frac{-1}{\pi S_0} f_2[H(u_1,u_2)]\right) du_1\, du_2$$

so that the integral of $p_{\vec{Y}(t)}(\vec{u})$ is unity.

To verify that Eq. 10.14 satisfies the Fokker-Planck equation, we can note that

$$\frac{\partial^2}{\partial u_2^2} p_{\vec{Y}(t)}(\vec{u}) = \frac{\partial}{\partial u_2}\left(p_{\vec{Y}(t)}(\vec{u}) \left[\frac{-1}{\pi S_0} f_2'[H(u_1,u_2)] \frac{\partial H(u_1,u_2)}{\partial u_2}\right]\right)$$
$$= \frac{\partial}{\partial u_2}\left(p_{\vec{Y}(t)}(\vec{u}) \left[\frac{-m\, u_2}{\pi S_0} f_2'[H(u_1,u_2)]\right]\right)$$

and substituting this into Eq. 10.13 gives

$$\frac{\partial}{\partial t} p_{\vec{Y}(t)}(\vec{u}) = -\frac{\partial}{\partial u_1}[u_2\, p_{\vec{Y}(t)}(\vec{u})] + \frac{1}{m}\frac{\partial}{\partial u_2}[f_1'(u_1)\, p_{\vec{Y}(t)}(\vec{u})]$$

Evaluation of the two derivatives on the right-hand side of this expression then gives

$$\frac{\partial}{\partial t} p_{\vec{Y}(t)}(\vec{u}) = 0$$

confirming that Eq. 10.14 is a stationary solution of Eq. 10.13.

Note that the $f_1'[X(t)]$ nonlinear spring term in Eq. 10.10 has a very natural form and describes many cases that may be of practical interest, several of which had been investigated prior to Caughey's presentation of the general solution. Simple examples include odd polynomials, such as the cubic function in the common Duffing oscillator. The $f_2'(H[X(t),\dot{X}(t)])\,\dot{X}(t)$ nonlinear damping term, on the other hand, seems rather unnatural. Clearly the form was chosen to give a Fokker-Planck equation that could be solved rather than to provide direct modeling of common physical problems. The model does serve the purpose, though, of giving exact solutions for a class of examples having nonlinear damping. Furthermore, a model with the f_2' damping coefficient varying as a function of the energy in the system may be just as reasonable as either of the more obvious choices of having the damping force (and also the damping coefficient) depend only on the $\dot{X}(t)$ value or having the damping coefficient depend on $|X(t)|$. In many physical problems we have so little information about the true nature of the energy dissipation that we would have difficulty determining which of these models would be more appropriate. For narrowband motion with a given dominant frequency, we could use the model of Eq. 10.10 or either of the other two models suggested to represent any reasonable nonlinear relationship between the energy loss per cycle and the amplitude of the motion.

An important special case of Eq. 10.10 arises when $f_2'(H[X(t),\dot{X}(t)])$ is a constant, which we will denote by c. In this situation, Eq. 10.14 gives

$$p_{\vec{Y}(t)}(\vec{u}) = A \exp\left(\frac{-c\,H(u_1,u_2)}{\pi S_0}\right) = A \exp\left(\frac{-c\,f_1(u_1)}{\pi S_0}\right) \exp\left(\frac{-m\,c\,u_2^2}{2\pi S_0}\right) \quad (10.15)$$

The fact that this stationary probability density function can be factored into the product of a function of u_1 and a function of u_2 shows that $X(t)$ and $\dot{X}(t)$ are independent random variables, for any given t value. Furthermore, the form of the dependence on u_2, which is the dummy variable for $\dot{X}(t)$, shows that $\dot{X}(t)$ is a Gaussian random variable with a variance of

$$\sigma_{\dot{X}}^2(t) = \frac{\pi S_0}{m\,c} \quad (10.16)$$

One can easily verify that this value is identical to that obtained in Chapters 5 and 6 for the response of a linear SDF system to white noise excitation. Of course, this result could have been predicted, inasmuch as Eq. 10.10 with $f_2'[H(u_1,u_2)] = c$ includes the linear SDF system as the special case when $f_1'(u_1) = k$. Nonetheless, it seems somewhat surprising that the probability distribution and the variance of $\dot{X}(t)$ are not affected by the presence of nonlinearity in the spring of the SDF oscillator.

It should be noted that $\{\dot{X}(t)\}$ is not a Gaussian process even though $\dot{X}(t)$ is a Gaussian random variable for every value of t for the system with $f_2' = c$. This is easily proved as follows. If $\{\dot{X}(t)\}$ were a Gaussian process, then its integral $\{X(t)\}$ would also be a Gaussian process, as discussed in Section 4.10. However, Eq. 10.15 shows that $p_{X(t)}(u)$ has the form

$$p_{X(t)}(u) = \hat{A} \exp\left(\frac{-c\,f_1(u_1)}{\pi S_0}\right) \quad (10.17)$$

with $\hat{A} = A(2\pi)^{1/2}\sigma_{\dot{X}}$, and this clearly is not a Gaussian distribution unless $f_1'(u_1) = k\,u_1$. Thus, $\{X(t)\}$ is not a Gaussian process, and this proves that $\{\dot{X}(t)\}$ is not a Gaussian process. Even though $\dot{X}(t)$ is a Gaussian random variable for every value of t in the situation with linear damping, it appears that random variables such as $\dot{X}(t)$ and $\dot{X}(s)$ are not jointly Gaussian, as is required for a Gaussian process.

****

Example 10.3: Let $\{X(t)\}$ denote the response of the nonlinear oscillator with the equation of motion

$$m\,\ddot{X}(t) + c\,\dot{X}(t) + k_1 X(t) + k_3\,X^3(t) = F(t)$$

with $k_1 \ge 0$ and $k_3 \ge 0$ and $\{F(t)\}$ being a mean-zero, stationary Gaussian white noise with autospectral density S_0. Find the probability distribution and the variance of the stationary $\{X(t)\}$ and $\{\dot{X}(t)\}$ responses.

Using $f_2'(u) = c$ and $f_1(u) = k_1 u + k_3\,u^3$ gives

$$H(u_1,u_2) = \frac{k_1 u_1^2}{2} + \frac{k_3 u_1^4}{4} + \frac{m\,u_2^2}{2}$$

and

$$p_{\vec{Y}(t)}(\vec{u}) = A\,\exp\left(\frac{-c}{\pi S_0}\left[\frac{k_1 u_1^2}{2} + \frac{k_3 u_1^4}{4} + \frac{m\,u_2^2}{2}\right]\right)$$

Factoring this expression, as in Eqs. 10.15 and 10.17, gives $\dot{X}(t)$ as being Gaussian with variance $\sigma_{\dot{X}}^2 = \pi S_0 /(c\,m)$ and $X(t)$ as having a non-Gaussian probability density function of

$$p_{X(t)}(u_1) = A(2\pi)^{1/2}\sigma_{\dot{X}}\exp\left(\frac{-c}{\pi S_0}\left[\frac{k_1 u_1^2}{2} + \frac{k_3 u_1^4}{4}\right]\right)$$

One may note that this probability distribution has the same form as that of $X(t)$ in Example 10.1. The only difference in the exponent is the presence of the c term in the present problem. The multiplier in front of the exponential, of course, is simply the value that assures that the integral of the probability density function is unity. If we let σ_0^2 denote the response variance for the limiting linear case of $k_3 = 0$, $\sigma_0^2 = \pi S_0 /(c\,k_1)$, then the change of variables $u = w\,\sigma_0$ gives the probability density function as

$$p_{X(t)}(w\,\sigma_0) = \frac{\exp\left(-\frac{w^2}{2} - \alpha\frac{w^4}{4}\right)}{\int_{-\infty}^{\infty}\exp\left(-\frac{v^2}{2} - \alpha\frac{v^4}{4}\right)dv}$$

with $\alpha = (k_3 \sigma_0^2 / k_1)$. This expression is identical to that found in Example 10.1. Thus, the plots of probability density function, variance, and kurtosis shown in Example 10.1 also apply to the $X(t)$ response in this oscillator problem.

****

Example 10.4: Let $\{X(t)\}$ denote the response of the nonlinear oscillator with the equation of motion

$$m\,\ddot{X}(t) + c_1[\dot{X}^3(t) + (k/m)\,X^2(t)\,\dot{X}(t)] + k\,X(t) = F(t)$$

with $\{F(t)\}$ being a mean-zero, stationary Gaussian white noise with autospectral density S_0. Find the probability distribution and the variance of the stationary $\{X(t)\}$ and $\{\dot{X}(t)\}$ responses.

Because this system has a linear spring the energy term is simply $H[X(t),\dot{X}(t)] = [m\,\dot{X}^2(t) + k\,X^2(t)]/2$. Thus, the equation of motion agrees with Eq. 10.10 if we choose $f_1'(u) = k\,u$ and $f_2'(u) = 2c_1 u/m$. This gives $f_2(u) = c_1 u^2/m$ so that Eq. 10.14 becomes

$$p_{X(t)\dot{X}(t)}(u_1,u_2) = A\exp\left(\frac{-c_1(m\,u_2^2 + k\,u_1^2)^2}{4\pi\,m\,S_0}\right)$$

This joint probability density function does not factor into a function of u_1 multiplied by a function of u_2, which is an indication that $X(t)$ and $\dot{X}(t)$ are not independent random variables for this problem. The most direct way to find the marginal probability density functions for $X(t)$ and $\dot{X}(t)$ is to integrate over all u_2 and u_1 values, respectively. For example

$$p_{X(t)}(u_1) = A\int_{-\infty}^{\infty}\exp\left(\frac{-c_1(m\,u_2^2 + k\,u_1^2)^2}{4\pi\,m\,S_0}\right)du_2$$

with A evaluated from the fact that the integral of the probability density function with respect to u_1 must be unity. Further integration then gives the variance values for $X(t)$ as

$$\sigma_X^2 = A\int_{-\infty}^{\infty} u_1^2\,p_{X(t)}(u_1)\,du_1 = A\int_{-\infty}^{\infty}\int_{-\infty}^{\infty} u_1^2\exp\left(\frac{-c_1(m\,u_2^2 + k\,u_1^2)^2}{4\pi\,m\,S_0}\right)du_2\,du_1$$

However, one can make some conclusions about the relative magnitudes and the probability distributions of $X(t)$ and $\dot{X}(t)$ without any integration. In particular, if a normalized version of $\dot{X}(t)$ is introduced as $Z(t) = \dot{X}(t)/\omega_0$ with $\omega_0 = (k/m)^{1/2}$, then we have

$$p_{X(t)Z(t)}(u_1,u_2) = A\,\omega_0\exp\left(\frac{-c_1 k^2(u_2^2 + u_1^2)^2}{4\pi\,m\,S_0}\right)$$

The symmetry of this relationship shows that $p_{X(t)}(u) = p_{Z(t)}(u) = \omega_0\,p_{\dot{X}(t)}(\omega_0\,u)$. Thus, the probability distribution of $Z(t) = \dot{X}(t)/\omega_0$ is identical to that of $X(t)$. Among other things, this shows that $\sigma_{\dot{X}} = \omega_0 \sigma_X$. It also shows that the kurtosis of $\dot{X}(t)$ is the same as that of $X(t)$.

For this particular problem one can obtain exact results for the value of A, the variance of the response, and so forth, by converting the double integral with respect to u_1 and u_2 into polar coordinates. Specifically, if we let $u_1 = r\cos(\theta)$ and $u_2 = r\sin(\theta)$ then

$$A^{-1} = \omega_0 \int_0^{2\pi} \int_0^{\infty} \exp\left(\frac{-c_1 k^2 r^4}{4\pi m S_0}\right) r\,dr\,d\theta$$

which can be evaluated as

$$A^{-1} = 2\pi\omega_0 \int_0^{\infty} \exp\left(\frac{-c_1 k^2 r^4}{4\pi m S_0}\right) r\,dr = 2\pi\omega_0 \frac{\Gamma(1/2)}{4}\left(\frac{4m\pi S_0}{c_1 k^2}\right)^{1/2} = \pi^2\left(\frac{S_0}{kc_1}\right)^{1/2}$$

because $\Gamma(1/2) = \pi^{1/2}$. Similarly,

$$\sigma_X^2 = A\omega_0 \int_0^{2\pi} \int_0^{\infty} \exp\left(\frac{-c_1 k^2 r^4}{4\pi m S_0}\right) r^3 \cos^2(\theta)\,dr\,d\theta$$

or

$$\sigma_X^2 = \pi A\omega_0 \int_0^{\infty} \exp\left(\frac{-c_1 k^2 r^4}{4\pi m S_0}\right) r^3\,dr$$

$$= \pi A\omega_0 \frac{m\pi S_0}{c_1 k^2} = \frac{1}{k}\left(\frac{m S_0}{c_1}\right)^{1/2}$$

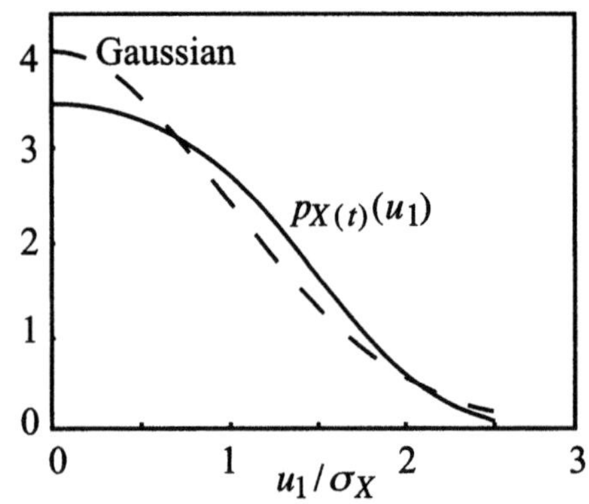

From the fact that $\sigma_{\dot{X}} = \omega_0 \sigma_X$, we can also write $(\sigma_{\dot{X}})^2 = [S_0/(mc_1)]^{1/2}$. The accompanying sketch shows numerical values for the marginal probability density function of $X(t)$, because there is not a simple analytical solution for this function. A Gaussian distribution is also shown for comparison.

****

It should be kept in mind that this discussion of Fokker-Planck analysis of nonlinear systems is very introductory in nature and is included largely for the purpose of presenting exact solutions to a few nonlinear problems. Stochastic averaging is often used to reduce the order of problems, and other special techniques have been developed for investigating the stability and bifurcation of solutions. A summary of some of these results, including references to experimental and numerical simulation studies, has been given by Ibrahim (1995).

10.3 Statistical Linearization

The simplest linearization situation involves finding the coefficients in a linear function such that it approximates a given nonlinear function $g[X(t)]$. The usual approach for doing this in a stochastic problem involves minimizing the mean-squared error. We will write the linear function as $b_0 + b_1[X(t) - \mu_X(t)]$ and choose b_0 and b_1 to minimize $E([\hat{E}^2(t)])$ in which the error $\hat{E}(t)$ is defined as

$$\hat{E}(t) = b_0 + b_1[X(t) - \mu_X(t)] - g[X(t)]$$

Setting the derivatives with respect to b_0 and b_1 equal to zero gives the two equations

$$0 = \frac{d}{d\,b_0} E\left([\hat{E}^2(t)]\right) = 2E\left(\hat{E}(t)\frac{d\hat{E}(t)}{d\,b_0}\right) = 2E[\hat{E}(t)] = 2\left[b_0 - E\left(g[X(t)]\right)\right]$$

and

$$0 = \frac{d}{d\,b_1} E\left([\hat{E}^2(t)]\right) 2E\left(\hat{E}(t)\frac{d\hat{E}(t)}{d\,b_1}\right) = 2E\left(\hat{E}(t)[X(t) - \mu_X(t)]\right)$$
$$= 2b_1\sigma_X^2(t) - 2\text{Cov}\left(X(t), g[X(t)]\right)$$

Determining b_0 and b_1 is precisely a problem in classical linear regression, which is discussed very briefly in Section 3.3. The general solution is

$$b_0 = E\left(g[X(t)]\right) \tag{10.18}$$

and

$$b_1 = \frac{\text{Cov}\left(X(t), g[X(t)]\right)}{\sigma_X^2(t)} \tag{10.19}$$

If $\{X(t)\}$ is mean-zero, which can always be achieved by appropriate definition of the process, then Eq. 10.19 can be rewritten as

$$b_1 = \frac{E\left(X(t)\, g[X(t)]\right)}{E[X^2(t)]} \quad \text{for } \mu_X(t) = 0 \tag{10.20}$$

If one knows the probability distribution of $X(t)$, then it is possible, at least in principle, to solve Eq. 10.18 and either Eq. 10.19 or 10.20 to find the optimal values of b_0 and b_1. One difficulty of using this procedure in a dynamics problem, though, is that one may not know the necessary probability distribution of $X(t)$. For example, if $\{X(t)\}$ is the response of a nonlinear dynamic system, in which $g[X(t)]$ is a nonlinear restoring force, then the probability distribution of $X(t)$ will be unknown. In this situation it is necessary to make some approximation. Another limitation of the procedure regards its accuracy. Even if one can exactly satisfy Eqs. 10.18 and 10.19, that does not ensure that the linear function is a good approximation of $g[X(t)]$. The fitting procedure has only minimized the second moment of the $\hat{E}(t)$ error term and has not even considered other measures of error. Despite these facts, it is often possible to get very useful results from statistical linearization.

The Gaussian distribution is most commonly used to approximate an unknown probability distribution of $\{X(t)\}$. One justification for this approach is based on the common assumption that many naturally occurring excitations are approximately Gaussian, along with the consequences of linearization. Specifically, if the excitation of a nonlinear system is Gaussian, then the dynamic response of a linearized substitute system is also Gaussian, so this is used as the approximation of the distribution of $\{X(t)\}$. Stated another way, if the excitation is Gaussian and the system nonlinearity is not very severe, then linearization should work well and the response of the nonlinear system is expected to be approximately Gaussian. In most applications of linearization, though, it is not feasible to demonstrate that the Gaussian assumption is valid. Fortunately, the choice of the b_0 and b_1 linearization parameters requires only estimates of certain first- and second-moment terms, as shown in Eqs. 10.18 and 10.19. It is quite possible for the Gaussian approximation to give acceptable estimates of these moments even when it does not accurately match the tails of the distribution of $\{X(t)\}$.

Now we generalize the statistical linearization procedure to include the case of a nonlinear function $g[\vec{X}(t)]$ of a number of random variables arranged as the components of the vector process $\{\vec{X}(t)\}$. In this case we write the linear approximation as

$$g[\vec{X}(t)] \approx b_0 + \sum_{j=1}^{n} b_j [X_j(t) - \mu_{X_j}(t)] \tag{10.21}$$

and choose the values of the coefficients $b_0, b_1, \cdots, b_n$ so as to minimize the mean-squared value of the error written as

$$\hat{E}(t) = b_0 + \sum_{j=1}^{n} b_j [X_j(t) - \mu_{X_j}(t)] - g[\vec{X}(t)]$$

The result can be written as

$$b_0 = E\Big(g[\vec{X}(t)]\Big) \tag{10.22}$$

and

$$\vec{b} = \mathbf{K}_{XX}^{-1}(t,t)\,\mathrm{Cov}\Big(\vec{X}(t), g[\vec{X}(t)]\Big) = \mathbf{K}_{XX}^{-1}(t,t)\, E\Big([\vec{X}(t) - \vec{\mu}_X(t)]\, g[\vec{X}(t)]\Big) \tag{10.23}$$

in which $\vec{b} = [b_1,\cdots,b_n]^T$ and $\mathrm{Cov}(\vec{X}(t), g[\vec{X}(t)])$ is a vector containing the covariance of the scalar $g[\vec{X}(t)]$ with each component of $\vec{X}(t)$. These equations, of course, include Eqs. 10.18 and 10.19 as the special case with $n = 1$.

When it is presumed that $\{\vec{X}(t)\}$ has the Gaussian distribution, it is possible to rewrite Eqs. 10.22 and 10.23 in an alternative form that is sometimes more convenient to use. In particular, it was shown in Example 3.11 that the jointly Gaussian distribution gives

$$E\Big([\vec{X}(t) - \vec{\mu}_X(t)]\, g[\vec{X}(t)]\Big) = \mathbf{K}_{XX}\, E\left(\left[\frac{\partial g(\vec{X})}{\partial X_1},\cdots,\frac{\partial g(\vec{X})}{\partial X_n}\right]^T\right)$$

and substitution of this result into Eq. 10.23 gives

$$b_j = E\left(\frac{\partial g[\vec{X}(t)]}{\partial X_j(t)}\right) \tag{10.24}$$

For $n \ge 2$, the use of Eq. 10.24 will usually be somewhat simpler than use of Eq. 10.23. Each formulation requires the evaluation of n expectations of nonlinear functions of $\vec{X}(t)$. However, the expectations in Eq. 10.23 directly give the components in $\mathbf{K}_{XX}\,\vec{b}$, so determination of $\vec{b}$ generally requires the solution of simultaneous equations. On the other hand, the expectations in Eq. 10.24 directly give the components of $\vec{b}$. Equation 10.24, of course, also applies for the special case of $n = 1$. In that case, though, it is not necessarily any easier to use Eq. 10.24 than Eq. 10.19 for finding the value of b_1.

The methods of equivalent linearization have also been extended by Falsone and Rundo Sotera (2003) to a class of non-Gaussian distributions for $X(t)$. In particular, if $X(t)$ has a Type A Gram-Charlier distribution, then one can use Gaussian relationships in evaluating the expectations in Eqs. 10.18 and

10.20. This follows from the fact that the Gram-Charlier probability density function is a Gaussian form multiplied by a polynomial. Falsone and Rundo Sotera have also modified Eq. 10.24 to apply to the Gram-Charlier situation. It should be noted that linearization of a dynamics problem with a non-Gaussian response cannot model any non-Gaussianity induced by the nonlinearity of the original problem. That is, the non-Gaussianity in the response of a linearized problem can come only from the non-Gaussianity of the excitation.

10.4 Linearization of Dynamics Problems

We will now demonstrate the use of statistical linearization for some relatively simple dynamics problems that contain either scalar or vector nonlinear terms $g[X(t)]$, $g[\vec{X}(t)]$, or $\vec{g}[\vec{X}(t)]$, in which $\{X(t)\}$ or $\{\vec{X}(t)\}$ denotes the response process of the system. The basic approach in each situation is quite straightforward. Each nonlinear term is replaced by its linear approximation, according to the formulas in Section 10.3. This gives the linearization parameters as functions of various expectations that depend, in turn, on the statistics of the response process. The response of the linearized system is then found from the methods of linear analysis, using time-domain integration, frequency-domain integration, or state-space moment/cumulant methods, according to the preference of the analyst. This step gives the response parameters as functions of the linearization parameters. Thus, we generally now have simultaneous equations, some giving the linearization parameters as functions of the response levels, and others giving the response levels as functions of the linearization parameters. The final step is to solve these equations to find the response levels as functions of the excitation and the parameters of the nonlinear system. In most situations this solution can be found only by iteration. Using assumed values of the linearization parameters gives initial estimates of the response levels and using the estimated response levels gives improved estimates of the linearization parameters, and so forth. In some situations, though, it is possible to solve the simultaneous equations by analytical methods. We will illustrate a few of these situations, and then we will give an example in which the iterative approach is used. Finally, we will outline the application of the method to general multi-degree-of-freedom problems with nonstationary response, although we will not devote the space necessary to work such advanced examples. Much more extensive investigations of statistical linearization are given by Roberts and Spanos (1990).

We begin with the nonlinear first-order system described by

$$\dot{X}(t) + g[X(t)] = F(t) \qquad (10.25)$$

which is identical to Eq. 10.7 with the nonlinear term rewritten as $g(\cdot)$ instead of $f'(\cdot)$. For simplicity, we take $g(\cdot)$ to be an odd function and the $\{F(t)\}$ excitation process to be a mean-zero, stationary, Gaussian white noise with an autocorrelation function of $R_{FF}(\tau) = 2\pi S_0\,\delta(\tau)$.

Based on the antisymmetry of $g[X(t)]$, we can see that $\mu_X = 0$ and $E\big(g[X(t)]\big) = 0$. Thus, using Eqs. 10.18 and 10.20 for the linearization of $g[X(t)]$ gives Eq. 10.25 as being replaced by

$$\dot{X}(t) + b_1 X(t) = F(t) \qquad (10.26)$$

with the b_1 coefficient given by Eq. 10.20. Thus, Eq. 10.20 gives the dependence of the linearization parameter b_1 on the response levels of the process. That is, if the probability distribution of $\{X(t)\}$ were known, then Eq. 10.20 would give us the value of b_1. The converse relationship giving the dependence of the statistics of $X(t)$ on b_1 is found by linear stochastic analysis of Eq. 10.26. In particular, one can use the methods of Chapters 5, 6, or 9 to show that the stationary solution of Eq. 10.26 has

$$E[X^2(t)] = \frac{\pi S_0}{b_1} \qquad (10.27)$$

Eliminating b_1 between Eqs. 10.20 and 10.27 gives

$$E\big(X(t)\, g[X(t)]\big) = \pi S_0 \qquad (10.28)$$

as a necessary condition of linearization. One can then use this equation to find $\sigma_X^2 \equiv E[X^2(t)]$, provided that one can relate $E(X(t)\, g[X(t)])$ to σ_X^2 for the given $g(\cdot)$ function. As noted in the preceding section, this is usually done by assuming that $X(t)$ is Gaussian. The assumption that $X(t)$ is Gaussian also allows us the alternative of using Eq. 10.24 in place of Eq. 10.20 for relating a_1 to the response values, giving

$$E\left(g'[X(t)]\right) = \frac{\pi S_0}{E[X^2(t)]} \tag{10.29}$$

as being equivalent to Eq. 10.28. In this form it appears that Eq. 10.28 may be slightly simpler than Eq. 10.29.

****

Example 10.5: Let $\{X(t)\}$ denote the response of the system of Example 10.1 with an equation of motion of

$$\dot{X}(t) + k_1 X(t) + k_3 X^3(t) = F(t)$$

in which $\{F(t)\}$ is a mean-zero, stationary, Gaussian, white noise with autospectral density S_0. Using statistical linearization, find the value of the b_1 parameter of Eqs. 10.26 and 10.27, and estimate the value of σ_X^2 for stationary response.

Using the approximation that $X(t)$ is Gaussian and mean-zero gives $E(g[X(t)]) = 0$ and

$$E\left(X(t)\, g[X(t)]\right) = k_1 E[X^2(t)] + k_3\, E[X^4(t)] = k_1\sigma_X^2 + 3k_3\sigma_X^4$$

so that Eq. 10.18 confirms that $b_0 = 0$ and Eq. 10.20 gives

$$b_1 = \frac{E\left(X(t)\, g[X(t)]\right)}{E[X^2(t)]} = k_1 + 3k_3\,\sigma_X^2$$

Equivalently, we could have obtained this latter result by using Eq. 10.24 to write

$$b_1 = E\left(\frac{\partial\, g[X(t)]}{\partial X(t)}\right) = E[k_1 + 3k_3 X^2(t)] = k_1 + 3k_3\,\sigma_X^2$$

Either Eq. 10.28 or 10.29 shows that the result of using this expression in conjunction with Eq. 10.27 gives $k_1\sigma_X^2 + 3k_3\sigma_X^4 = \pi S_0$ which has a solution of

$$\sigma_X^2 = \frac{k_1}{6k_3}\left(\left[1 + \frac{12 k_3 \pi S_0}{k_1^2}\right]^{1/2} - 1\right)$$

Using the notation of $\sigma_0^2 = \pi S_0 / k_1$ for the variance of the system in the limiting case of $k_3 = 0$, as was done in Example 10.1, this can be rewritten as

$$\frac{\sigma_X^2}{\sigma_0^2} = \frac{k_1}{6k_3\sigma_0^2}\left(\left[1 + \frac{12 k_3 \sigma_0^2}{k_1}\right]^{1/2} - 1\right)$$

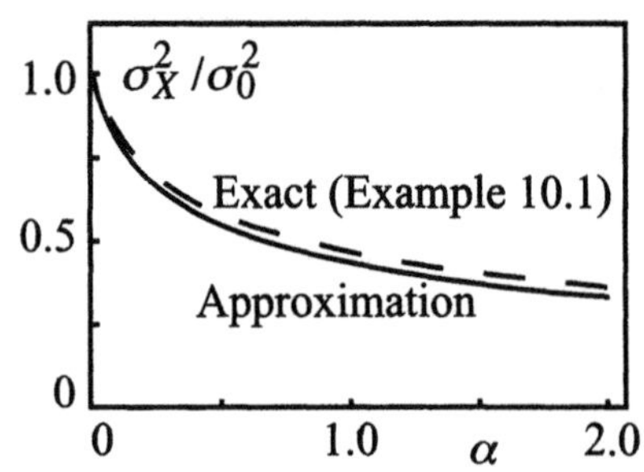

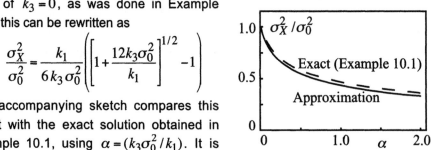

The accompanying sketch compares this result with the exact solution obtained in Example 10.1, using $\alpha = (k_3\sigma_0^2 / k_1)$. It is

obvious that statistical linearization gives a very good approximation of the response variance for this nonlinear problem.
Recall that it was shown in Example 10.1 that the kurtosis of $X(t)$ varies from 3.0 down to less than 2.4 for the range of α values shown in the sketch. Thus, this example shows a situation in which statistical linearization gives a good estimate of variance even though the response is significantly non-Gaussian. It must be kept in mind, though, that this statistical linearization procedure does not give us any estimate of the non-Gaussianity of the response, and using a Gaussian approximation for $X(t)$ could give very poor estimates of the probability of occurrence of rare events. For example, letting $\alpha = 2$ in the exact solution of Example 10.1 gives $P[X(t) > 3\sigma_X(t)] = 3.21 \times 10^{-5}$, whereas a Gaussian distribution gives this probability as 1.35×10^{-3}, or almost two orders of magnitude larger than the exact value. Statistical linearization is surprisingly accurate in predicting the variance, but caution is necessary in using the linearization results for other purposes.
****

Example 10.6: Let $\{X(t)\}$ denote the response of the system of Example 10.2 with the equation of motion

$$\dot{X}(t) + k\,|X(t)|^a \operatorname{sgn}[X(t)] = F(t)$$

with $\{F(t)\}$ being a mean-zero, stationary Gaussian white noise with autospectral density S_0. Using statistical linearization, find the value of the b_1 parameter of Eqs. 10.26 and 10.27, and estimate the value of σ_X^2 for stationary response.

Using the approximation that $\{X(t)\}$ is Gaussian and mean-zero gives $E(g[X(t)]) = 0$ and

$$E\big(X(t)\,g[X(t)]\big) = k\,E\Big(|X(t)|^{a+1}\Big) = 2k\int_0^\infty \frac{u^{a+1}\exp[-u^2/(2\sigma_X^2)]}{(2\pi)^{1/2}\sigma_X}\,du =$$

$$2k\frac{\sigma_X}{(2\pi)^{1/2}}\int_0^\infty (2\sigma_X^2\,w)^{a/2}e^{-w}\,dw = \frac{2^{(a+1)/2}\,k\sigma_X^{a+1}}{\pi^{1/2}}\Gamma\left(\frac{a}{2}+1\right)$$

so that Eq. 10.18 confirms that $b_0 = 0$ and Eq. 10.20 gives

$$b_1 = \frac{E\big(X(t)\,g[X(t)]\big)}{E[X^2(t)]} = k\sigma_X^{a-1}\frac{2^{(a+1)/2}}{\pi^{1/2}}\Gamma\left(\frac{a}{2}+1\right) = k\,a\,\sigma_X^{a-1}\frac{2^{(a-1)/2}}{\pi^{1/2}}\Gamma\left(\frac{a}{2}\right)$$

Equivalently, we could have obtained this latter result by using Eq. 10.24 to write

$$b_1 = E\left(\frac{\partial g[X(t)]}{\partial X(t)}\right) = k\,a\,E[|X(t)|^{a-1}] = 2k\,a\int_0^\infty \frac{u^{a-1}\exp[-u^2/(2\sigma_X^2)]}{(2\pi)^{1/2}\sigma_X}\,du =$$

$$k\,a\,\sigma_X^{a-1}\frac{2^{a/2}}{(2\pi)^{1/2}}\Gamma\left(\frac{a}{2}\right)$$

Either Eq. 10.28 or 10.29 shows that the result of using this expression in conjunction with Eq. 10.27 gives

$$k\,a\frac{2^{a/2}}{(2\pi)^{1/2}}\Gamma\left(\frac{a}{2}\right)\sigma_X^{a+1}=\pi S_0$$

which has a solution of

$$\sigma_X^2=\left(\frac{\pi S_0}{k}\right)^{2/(a+1)}\left(\frac{(2\pi)^{1/2}}{2^{a/2}\,a\,\Gamma(a/2)}\right)^{2/(a+1)}$$

The accompanying sketch compares this result with the exact solution obtained in Example 10.2 by giving values of the ratio

$$R=\frac{[\sigma_X^2]_{equiv.\,lin.}}{[\sigma_X^2]_{exact}}$$

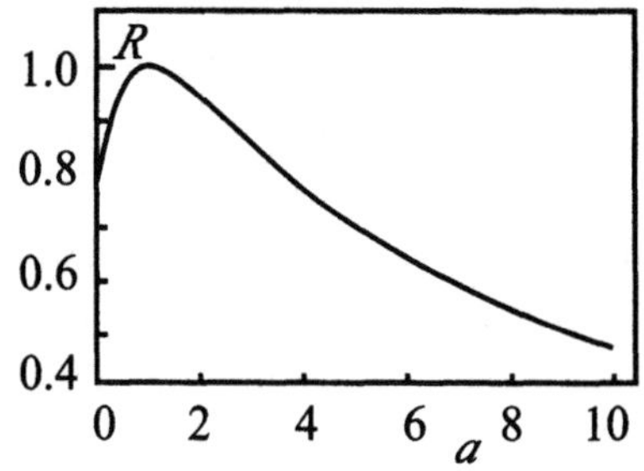

For this system it is clear that statistical linearization gives a good approximation of the response variance only when a is relatively near unity.

****

Next we consider the linearization of an oscillator with a nonlinear restoring force, which may depend on both $X(t)$ and $\dot{X}(t)$:

$$m\,\ddot{X}(t)+f[X(t),\dot{X}(t)]=F(t) \tag{10.30}$$

The nonlinear function is then linearized according to Eq. 10.21 to give

$$m\,\ddot{X}(t)+b_2\dot{X}(t)+b_1X(t)+b_0=F(t) \tag{10.31}$$

with b_0 being found from Eq. 10.22 as

$$b_0=E\Big(f[X(t),\dot{X}(t)]\Big)$$

The b_1 and b_2 coefficients can be found by using Eq. 10.23, which gives two simultaneous equations:

$$b_1\sigma_X^2(t)+b_2\,\rho(t)\,\sigma_X(t)\,\sigma_{\dot{X}}(t)=\mathrm{Cov}\Big(X(t),f[X(t),\dot{X}(t)]\Big) \tag{10.32}$$

and

$$b_1\,\rho(t)\,\sigma_X(t)\,\sigma_{\dot{X}}(t)+b_2\sigma_{\dot{X}}^2(t)=\mathrm{Cov}\Big(\dot{X}(t),f[X(t),\dot{X}(t)]\Big) \tag{10.33}$$

in which $\rho(t)\sigma_X(t)\sigma_{\dot{X}}(t)$ denotes the covariance of $X(t)$ and $\dot{X}(t)$.

If we restrict our attention to the special case in which $\{X(t)\}$ is a stationary process, then we know that $\rho(t)=0$ and $\mu_{\dot{X}}(t)=0$, so Eqs. 10.32 and 10.33 uncouple to give

$$b_1=\frac{\text{Cov}\Big(X(t),f[X(t),\dot{X}(t)]\Big)}{\sigma_X^2(t)} \tag{10.34}$$

and

$$b_2=\frac{\text{Cov}\Big(\dot{X}(t),f[X(t),\dot{X}(t)]\Big)}{\sigma_{\dot{X}}^2(t)}=\frac{E\Big(\dot{X}(t)\,f[X(t),\dot{X}(t)]\Big)}{E[\dot{X}^2(t)]} \tag{10.35}$$

If we make the further restriction that $\mu_X(t)=0$, we can rewrite Eq. 10.34 as

$$b_1=\frac{E\Big(X(t)f[X(t),\dot{X}(t)]\Big)}{E[X^2(t)]}$$

One situation that will lead to this special case of $X(t)$ being mean-zero is for $\{F(t)\}$ to have a probability distribution that is symmetric about zero, and $f(\cdot,\cdot)$ to be odd in the sense that $f(u_1,u_2)=-f(-u_1,-u_2)$.

Let us now consider the special case in which $\{F(t)\}$ is mean-zero, stationary, white noise with autospectral density S_0. For this excitation we know that the response of the linearized SDF oscillator described by Eq. 10.31 is stationary, has a mean value of $\mu_X=-b_0/b_1$, and has variance values of

$$\sigma_X^2=\frac{\pi S_0}{b_1 b_2},\qquad \sigma_{\dot{X}}^2=E[\dot{X}^2]=\frac{\pi S_0}{m\,b_2}$$

One can now use these two equations in conjunction with Eqs. 10.34 and 10.35 and eliminate b_1 and b_2 to obtain necessary conditions for stationary response as

$$\text{Cov}\Big(X(t),f[X(t),\dot{X}(t)]\Big)=m\,\sigma_{\dot{X}}^2(t)=m\,E[\dot{X}^2(t)] \tag{10.36}$$

and

$$\text{Cov}\Big(\dot{X}(t),f[X(t),\dot{X}(t)]\Big)=E\Big(\dot{X}(t)\,f[X(t),\dot{X}(t)]\Big)=\pi S_0/m \tag{10.37}$$

If the $\{F(t)\}$ process is Gaussian, then we also have the choice of using Eq. 10.24 in place of Eq. 10.23. This gives

$$b_1 = E\left(\frac{\partial}{\partial X(t)} f[X(t),\dot{X}(t)]\right), \qquad b_2 = E\left(\frac{\partial}{\partial \dot{X}(t)} f[X(t),\dot{X}(t)]\right)$$

so Eqs. 10.34 and 10.35 give necessary equations for the stationary response as

$$E\left(\frac{\partial}{\partial X(t)} f[X(t),\dot{X}(t)]\right) = \frac{m\sigma_{\dot{X}}^2}{\sigma_X^2}, \qquad E\left(\frac{\partial}{\partial \dot{X}(t)} f[X(t),\dot{X}(t)]\right) = \frac{\pi S_0}{m\sigma_{\dot{X}}^2} \tag{10.38}$$

For the case of Gaussian response, these expressions are equivalent to Eqs. 10.36 and 10.37, but they are not necessarily any simpler than those equations.

It is also instructive to consider the special case in which the nonlinearity of Eq. 10.30 corresponds to a linear damping and a nonlinear spring. Following the notation of Eqs. 10.15–10.17, we write this as

$$f[X(t),\dot{X}(t)] = c\,\dot{X}(t) + f_1'[X(t)]$$

Now we can use the fact that $\text{Cov}[X(t),\dot{X}(t)] = E[X(t)\dot{X}(t)] = 0$ for stationary response to obtain

$$\text{Cov}\Big(X(t), f[X(t),\dot{X}(t)]\Big) = \text{Cov}\Big(X(t), f_1'[X(t)]\Big) \tag{10.39}$$

Similarly, we can use the fact that

$$E\Big(\dot{X}(t)\, f_1'[X(t)]\Big) \equiv \frac{d}{dt} E\Big(f_1[X(t)]\Big) \tag{10.40}$$

must be zero for stationary response to say that

$$E\Big(\dot{X}(t)\, f[X(t),\dot{X}(t)]\Big) = c\, E[\dot{X}^2] = c\,\sigma_{\dot{X}}^2 \tag{10.41}$$

From Eqs. 10.34 and 10.35 we now find that

$$b_1 = \frac{\text{Cov}\left(X(t), f_1'[X(t)]\right)}{\sigma_X^2}, \qquad b_2 = c \tag{10.42}$$

We also know that

$$b_0 = E\left(f_1'[X(t)]\right) \tag{10.43}$$

because $\dot{X}(t)$ is mean-zero for stationary response.

In fact, we could have obtained the results in Eqs. 10.42 and 10.43 by using a slightly simpler approach. Because the $c\,\dot{X}(t)$ term in our $f[X(t),\dot{X}(t)]$ function is linear, there really is no need to include it in our linearization procedure. That is, we can say that our equation of motion for the oscillator is

$$m\,\ddot{X}(t) + c\,\dot{X}(t) + f_1'[X(t)] = F(t) \tag{10.44}$$

and the only linearization required is of the term $f_1'[X(t)] = b_0 + b_1 X(t)$. This one-dimensional linearization then gives exactly Eqs. 10.42 and 10.43, respectively, for b_1 and b_0, and the fact that Eq. 10.42 gives $b_2 = c$ demonstrates that either of these formulations of statistical linearization gives the same linearized system. This illustrates a general property of statistical linearization. If a linear term is somehow included within the nonlinear function that is being linearized, then that linear term will appear unchanged in the results of the linearization.

Substituting Eq. 10.41 into Eq. 10.37 now gives us the linearization estimate of $\sigma_{\dot{X}}^2$ as $\pi S_0/(c\,m)$. This result of the approximate method, though, is identical to the exact result presented in Eq. 10.16 for the special case in which the white noise excitation is also Gaussian. Thus, if the excitation is Gaussian, then statistical linearization gives an exact value for $\sigma_{\dot{X}}^2$ for the system with linear damping and a nonlinear spring. A formula related to the corresponding estimate of σ_X^2 is obtained by substituting Eq. 10.39 and $\sigma_{\dot{X}}^2$ into Eq. 10.36, giving

$$\text{Cov}\left(X(t), f_1'[X(t)]\right) = \pi S_0 / c \tag{10.45}$$

Actual determination of σ_X^2 from this formula requires an assumption about the probability distribution of $X(t)$. If $X(t)$ is assumed to be Gaussian and mean-zero, then one also has the alternative of using Eq. 10.38 to write

$$\sigma_X^2 \, E\big(f_1''[X(t)]\big) = \pi \, S_0 / c \tag{10.46}$$

which is then equivalent to Eq. 10.45.

****

Example 10.7: Let $\{X(t)\}$ denote the response of the oscillator of Example 10.3 with an equation of motion of

$$m\,\ddot{X}(t) + c\,\dot{X}(t) + k_1 X(t) + k_3\,X^3(t) = F(t)$$

in which $\{F(t)\}$ is a mean-zero, stationary Gaussian white noise with autospectral density S_0. Using statistical linearization, find the value of the b_1 and b_2 parameters of Eqs. 10.31, 10.34, and 10.35, and estimate the value of σ_X^2 and $\sigma_{\dot{X}}^2$ for stationary response.

Because this equation of motion is a special case of Eq. 10.44, we know that $b_2 = c$ and $\sigma_{\dot{X}}^2 = \pi \, S_0 / (c\,m)$. Using the notation of Eq. 10.44 we can then write the nonlinear term as $f_1'(u_1) = k_1 u_1 + k_3\, u_1^3$ so Eq. 10.42 gives

$$b_1 = \frac{E\big(X(t)\, f_1'[X(t)]\big)}{\sigma_X^2} = \frac{k_1 E[X^2] + k_3\, E[X^4]}{\sigma_X^2} = k_1 + k_3 E[X^4] / \sigma_X^2$$

Using the assumption that $X(t)$ is Gaussian now gives $b_1 = k_1 + 3k_3\sigma_X^2$. Note that this expression for b_1 is identical to the results of Eq. 10.24, involving the expected value of the partial derivative of $f_1'[X(t)]$.

Using Eq. 10.45 now gives

$$3k_3\,\sigma_X^4 + k_1 \sigma_X^2 = m\,\sigma_{\dot{X}}^2 = \frac{\pi\, S_0}{c}$$

The variance of $X(t)$ is found by solving this quadratic equation to give

$$\sigma_X^2 = \frac{k_1}{6k_3}\left(\left[1 + \frac{12 k_3 \pi\, S_0}{c k_1^2}\right]^{1/2} - 1\right)$$

Using the notation $\sigma_0^2 = \pi\, S_0 / (c\, k_1)$ for the variance in the limiting case of $k_3 = 0$, as was done in Example 10.3, allows the σ_X^2 result to be written in normalized form as

$$\frac{\sigma_X^2}{\sigma_0^2} = \frac{k_1}{6k_3\,\sigma_0^2}\left(\left[1 + \frac{12 k_3 \sigma_0^2}{k_1}\right]^{1/2} - 1\right)$$

Note, though, that this expression is identical to the one obtained and plotted in Example 10.5. Thus, the plot shown in that example also demonstrates that statistical linearization gives good variance approximations for $X(t)$ in the current problem.

****

Example 10.8: Let $\{X(t)\}$ denote the response of the oscillator with an equation of motion of

$$m\ddot{X}(t)+c_1\dot{X}^3(t)+c_2X^2(t)\dot{X}(t)+kX(t)=F(t)$$

in which $\{F(t)\}$ is a mean-zero, stationary Gaussian white noise with autospectral density S_0. Using statistical linearization, find the value of the linearization parameters, and estimate the value of σ_X^2 and $\sigma_{\dot{X}}^2$ for stationary response.

As in Eqs. 10.44–10.46, there is no need to include a linear term in our linearization. Thus, we can write the problem as

$$m\ddot{X}(t)+f[X(t),\dot{X}(t)]+kX(t)=F(t)$$

with the linearization problem being the determination of b_1 and b_2 in

$$f(u_1,u_2)=c_1u_2^3+c_2u_1^2u_2\approx b_1u_1+b_2u_2$$

No b_0 term is included because the symmetry of the problem gives $E(f[X(t),\dot{X}(t)])=0$. Using Eq. 10.23 and noting that $X(t)$ and $\dot{X}(t)$ are uncorrelated for stationary response gives

$$b_1=\frac{E\big(X(t)f[X(t),\dot{X}(t)]\big)}{E[X^2(t)]}=\frac{c_1E[X(t)\dot{X}^3(t)]+c_2E[X^3(t)\dot{X}(t)]}{\sigma_X^2}=\frac{c_1E[X(t)\dot{X}^3(t)]}{\sigma_X^2}$$

and

$$b_2=\frac{E\big(\dot{X}(t)f[X(t),\dot{X}(t)]\big)}{E[\dot{X}^2(t)]}=\frac{c_1E[\dot{X}^4(t)]+c_2E[X^2(t)\dot{X}^2(t)]}{\sigma_{\dot{X}}^2}$$

Note that the expression for b_1 has been simplified by using the information that $E[X^3(t)\dot{X}(t)]=0$ for stationary response. The other expectations in these expressions are unknown, except that we can approximate them by using a simplifying assumption for the probability distribution of $X(t)$ and $\dot{X}(t)$. In particular, if we assume that $X(t)$ and $\dot{X}(t)$ are jointly Gaussian, then they are also independent, which implies that $E[X(t)\dot{X}^3(t)]=0$, $E[\dot{X}^4(t)]=3\sigma_{\dot{X}}^4$, and $E[X^2(t)\dot{X}^2(t)]=\sigma_X^2\sigma_{\dot{X}}^2$. This gives the linearization coefficients as $b_1=0$ and $b_2=3c_1\sigma_{\dot{X}}^2+c_2\sigma_X^2$. Thus, the linearized system is

$$m\ddot{X}(t)+(3c_1\sigma_{\dot{X}}^2+c_2\sigma_X^2)\dot{X}(t)+kX(t)=F(t)$$

We know that the response of this linearized system has

$$\sigma_X^2 = \frac{\pi S_0}{k[3c_1\sigma_{\dot X}^2 + c_2\sigma_X^2]}$$

and $\sigma_{\dot X}^2 = (k/m)\sigma_X^2$. Substituting this latter relationship into the former one and solving for σ_X^2 gives

$$\sigma_X^2 = \frac{(\pi S_0)^{1/2}}{\left(\frac{3c_1k^2}{m} + c_2 k\right)^{1/2}}$$

For the special case in which $c_2 = c_1 k/m$, this nonlinear system is the same as the one studied in Example 10.4. Thus, we have an exact solution for that special case, and we can use it as a basis of comparison to determine the accuracy of the approximate result obtained here. For this particular value of c_2, the approximate result becomes

$$\sigma_X^2 = \frac{1}{2k}\left(\frac{m\pi S_0}{c_1}\right)^{1/2}$$

and this is 11% lower than the exact variance value for this particular problem. This same error also applies to $\sigma_{\dot X}^2$, because the approximate result gives $\sigma_{\dot X}^2/\sigma_X^2 = k/m$, just as was true in the exact solution. It should be kept in mind, as well, that this problem is quite significantly non-Gaussian, as was shown in Example 10.4.

****

Example 10.9: Let $\{X(t)\}$ denote the response of the nonlinear oscillator of Examples 10.3 and 10.7 with the equation of motion of

$$m\ddot X(t) + c\dot X(t) + k_1 X(t) + k_3 X^3(t) = F(t)$$

but with $\{F(t)\}$ being a mean-zero, stationary Gaussian process with a nonwhite autospectral density of

$$S_{FF}(\omega) = S_0\, U(0.7\omega_0 - |\omega|)$$

in which ω_0 is defined as $\omega_0 = (k_1/m)^{1/2}$. That is, $S_{FF}(\omega) = S_0$ for $|\omega| < 0.7\omega_0$ and equals zero otherwise. Using statistical linearization, estimate the variance of the stationary $\{X(t)\}$ and $\{\dot X(t)\}$ responses for the special case of $k_3 = 2ck_1^2/(\pi S_0)$ and $\zeta_0 \equiv c/(2m\omega_0) = 0.05$.

For this excitation we are not able to use simple closed-form analytical solutions of the linearized system. Thus, we use iteration in finding a solution. We can directly use the linearized model found in Example 10.7:

$$m\ddot X(t) + c\dot X(t) + b_1 X(t) = F(t)$$

with $b_1 = k_1 + k_3 E(X^4)/\sigma_X^2$. We also use the assumption that $X(t)$ is approximately Gaussian to convert this into $b_1 = k_1 + 3k_3\sigma_X^2$. We do not have a closed-form solution for the response variance for this linearized problem, so we use the harmonic transfer function of $H_x(\omega) = [b_1 - m\omega^2 + 2ic\omega]^{-1}$ and integrate this to obtain

$$\sigma_X^2 = S_0 \int_{-0.7\omega_0}^{0.7\omega_0} |H_x(\omega)|^2 \, d\omega = 2S_0 \int_0^{0.7\omega_0} \frac{d\omega}{(b_1 - m\omega^2)^2 + (2c\omega)^2}$$

For convenience, we use a nondimensional frequency of $\eta = \omega / \omega_0$ to obtain

$$\sigma_X^2 = \frac{2S_0\,\omega_0}{b_1^2} \int_0^{0.7} \left(\left[1 - \frac{k_1 \eta^2}{b_1} \right]^2 + \left[\frac{2k_1 \xi_0 \eta}{b_1} \right]^2 \right)^{-1} d\eta$$

To begin the iterative process we simply neglect the k_3 term, taking $b_1 = k_1$ and numerically evaluating the integral to obtain $\sigma_X^2 = (2S_0\,\omega_0 / k_1^2)(1.114)$. Using this value along with the given value of k_3 gives an estimate of $b_1 = k_1 + 3k_3\sigma_X^2 = k_1(1.142)$. Using this value in a new evaluation of the frequency integral gives an improved estimate of the response variance as $\sigma_X^2 = (2S_0\,\omega_0 / k_1^2)(1.029)$. This, in turn, gives an improved estimate of the linearization parameter of $b_1 = k_1 + 3k_3\sigma_X^2 = k_1(1.131)$. Carrying out two more steps of the iteration gives $\sigma_X^2 = (2S_0\,\omega_0 / k_1^2)(1.034)$ and $b_1 = k_1 + 3k_3\sigma_X^2 = k_1(1.132)$ then $\sigma_X^2 = (2S_0\,\omega_0 / k_1^2)(1.033)$ and $b_1 = k_1 + 3k_3\sigma_X^2 = k_1(1.132)$. This represents convergence, so our estimate of the response variance is

$$\sigma_X^2 = \frac{2S_0\,\omega_0}{k_1^2}(1.033) = \frac{2.067\,S_0}{(k_1^3 m)^{1/2}}$$

For this linear system we know that $\sigma_{\dot X}^2 = (a_1 / m)\sigma_X^2$, so our estimate of $\sigma_{\dot X}^2$ is

$$\sigma_{\dot X}^2 = \left(\frac{1.132k_1}{m} \right) \left(\frac{2.067\,S_0}{(k_1^3 m)^{1/2}} \right) = \frac{2.340\,S_0}{(k_1 m^3)^{1/2}}$$

****

We can also use statistical linearization for the general formulation of the nonlinear dynamics problem given in Eq. 10.2. In particular, if the equation of motion is

$$\dot{\vec{Y}}(t) + \vec{g}[\vec{Y}(t)] = \vec{Q}(t)$$

then the linearization will consist of using an approximation of the form

$$\vec{g}[\vec{Y}(t)] \approx \vec{b}(t) + \mathbf{B}(t)[\vec{Y}(t) - \vec{\mu}_Y(t)] \tag{10.47}$$

The linearization parameters in the vector $\vec{b}(t)$ and the matrix $\mathbf{B}(t)$ of this relationship are found from Eqs. 10.22 and 10.23 to be given by

$$\vec{b}(t) = E\left(\vec{g}[\vec{Y}(t)]\right) \tag{10.48}$$

and the solution of

$$\mathbf{K}_{YY}(t,t)\,\mathbf{B}^T(t) = \mathbf{K}_{Yg}(t,t) \tag{10.49}$$

in which $\mathbf{K}_{Yg}(t,t)$ denotes the cross-covariance matrix $\text{Cov}(\vec{Y}(t), \vec{g}[\vec{Y}(t)])$. That is,

$$\mathbf{K}_{Yg}(t,t) = E\left(\vec{Y}(t)\,\vec{g}^{\,T}[\vec{Y}(t)]\right) - \vec{\mu}_Y(t)\,E\left(\vec{g}^{\,T}[\vec{Y}(t)]\right) \tag{10.50}$$

Using the linearization of Eq. 10.47 gives the substitute linear equation of motion as

$$\dot{\vec{Y}}(t) + \vec{b}(t) + \mathbf{B}(t)[\vec{Y}(t) - \vec{\mu}_Y(t)] = \vec{Q}(t)$$

This relationship can be put into a more useful form by noting that the expected value of the original nonlinear equation of motion gives

$$\vec{\mu}_{\dot{Y}}(t) + E\left(\vec{g}[\vec{Y}(t)]\right) = \vec{\mu}_Q(t)$$

and combining this with Eq. 10.48 gives

$$\vec{b}(t) = \vec{\mu}_Q(t) - \vec{\mu}_{\dot{Y}}(t) \tag{10.51}$$

This allows the linearized equation of motion to be rewritten as

$$\frac{d}{dt}[\vec{Y}(t) - \vec{\mu}_Y(t)] + \mathbf{B}(t)[\vec{Y}(t) - \vec{\mu}_Y(t)] = [\vec{Q}(t) - \vec{\mu}_Q(t)] \tag{10.52}$$

Note that this equation involves only the deviation of $\vec{Y}(t)$ away from its mean value. Thus, it can tell us about the covariance and higher-order cumulants of the response, but it does not govern the behavior of $\vec{\mu}_Y(t)$. This first-moment behavior is given by Eq. 10.51, which can be rewritten as

$$\vec{\mu}_{\dot{Y}}(t) = \vec{\mu}_Q(t) - \vec{b}(t) \tag{10.53}$$

One can analyze Eq. 10.52 by using any method appropriate for linear systems. In particular, one can use state-space analysis to write cumulant equations as in Chapter 9. The covariance of the response, in particular, is governed by a slightly simplified form of Eq. 9.9:

$$\frac{d}{dt}\mathbf{K}_{YY}(t,t)+\mathbf{B}(t)\,\mathbf{K}_{YY}(t,t)+\mathbf{K}_{YY}(t,t)\,\mathbf{B}^T(t)=\mathbf{K}_{QY}(t,t)+\mathbf{K}_{YQ}(t,t) \quad (10.54)$$

and if the excitation is delta-correlated this can be reduced to

$$\frac{d}{dt}\mathbf{K}_{YY}(t,t)+\mathbf{B}(t)\,\mathbf{K}_{YY}(t,t)+\mathbf{K}_{YY}(t,t)\,\mathbf{B}^T(t)=2\pi\mathbf{S}_0(t) \quad (10.55)$$

which is a special case of Eq. 9.17. It should also be noted that if the $\{\vec{Q}(t)\}$ excitation is a Gaussian process, then the response from Eq. 10.54 will also be Gaussian. Furthermore, if this result is anticipated in the evaluation of the linearization parameters, then one can use Eq. 10.24 in place of Eq. 10.23, giving

$$B_{jl}(t)=E\left(\frac{\partial}{\partial Y_l(t)}g_j[\vec{Y}(t)]\right)$$

as being equivalent to Eq. 10.49. This formulation can sometimes simplify the evaluation of the linearization coefficients.

By switching to Kronecker notation (as in Section 9.7), we could also write the expressions for higher-order cumulants in the same fashion. Note that if the excitation and response processes are stationary, then $\vec{b}$ and $\mathbf{B}$ are independent of t so that our linearized cumulant equations have constant coefficients as in the simpler examples considered earlier in this section.

10.5 Linearization of Hysteretic Systems

Oscillators having a hysteretic restoring force represent a special category of nonlinear systems that are often of practical importance. By a *hysteretic force* we mean that the force at a particular time $t=t_1$ depends not only on the values of state variables such as $X(t_1)$ and $\dot{X}(t_1)$ at that instant of time but also on the time history of $\{X(t)\}$ for $t<t_1$. We use a notation of $g[\{X(t)\}]$ for such a hysteretic term. Some of the simplest hysteretic models are obtained by introducing idealized Coulomb friction in conjunction with spring elements. For example, simply placing the Coulomb slider in series with a linear spring as in

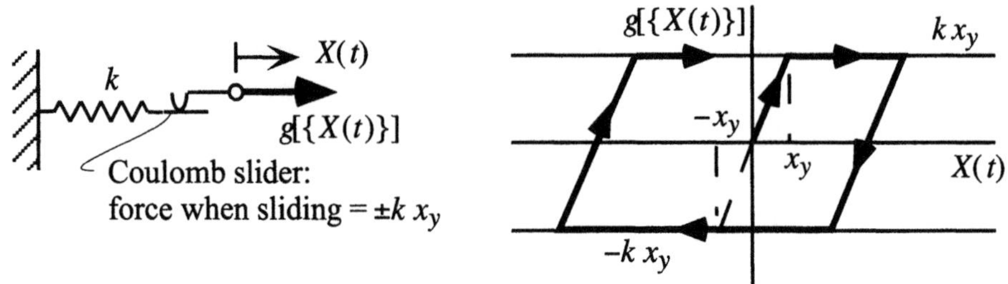

(a) Physical model with hysteresis (b) Hysteretic restoring force

Figure 10.1 System with elastoplastic hysteresis.

Fig. 10.1(a) gives the hysteretic restoring force shown in Fig. 10.1(b). In addition to representing the system with friction, this model with a force that is always bounded by constant values, designated here as $\pm k\,x_y$, approximates the Fig. 10.1(a) gives the hysteretic restoring force shown in Fig. 10.1(b). In addition to representing the system with friction, this model with a force that is always bounded by constant values, designated here as $\pm k\,x_y$, approximates the force-deflection behavior for a uniaxially loaded member made of a material with an elastoplastic stress-strain relationship. Because of this latter fact, the model is commonly called the elastoplastic nonlinearity. The time history behavior is very simple, with the slope of the force-deflection relationship being k whenever $|g[\{X(t)\}]| < k\,x_y$.

By adding another spring in parallel with the elastoplastic element of Fig. 10.1, we can obtain a system with so-called bilinear hysteretic behavior. When we add a mass and a dashpot, we obtain the bilinear hysteretic oscillator shown in Fig. 10.2. There are two important reasons for adding the parallel spring, designated as k_2 in Fig. 10.2, to the bilinear hysteretic oscillator. The first reason is based on the physical fact that most real systems do not lose all their stiffness at the first onset of yielding. That is, the restoring force in a real system usually does continue to increase somewhat after the inception of yielding. The second reason for giving k_2 a nonzero value is that the variance of $X(t)$ grows without bound as t increases for the oscillator with $k_2 = 0$. Thus, the system with $k_2 = 0$ presents some mathematical difficulties inasmuch as it never has a stationary response to a stationary excitation. In a sense, these two justifications for $k_2 \neq 0$ can be related by the argument that at least one reason that designers avoid the possibility of having a system that is completely elastoplastic is its inherent lack of stability, which is also what leads to the absence of a stationary stochastic solution.

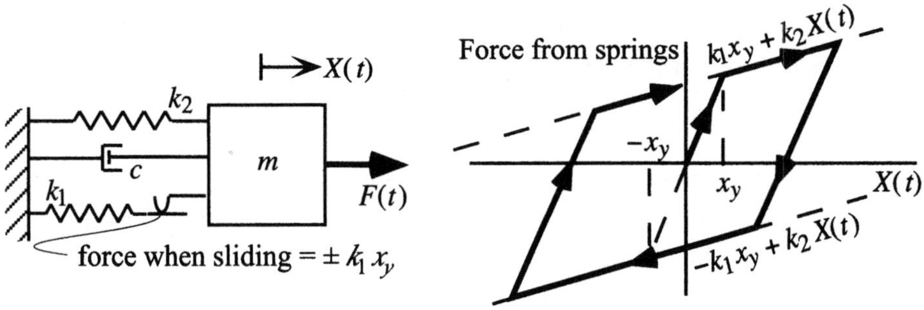

Figure 10.2 Oscillator with bilinear hysteresis.

Real physical systems, of course, often have much more complicated hysteretic nonlinearities than the bilinear behavior represented by Figs. 10.1 and 10.2. The linearization methods that we present here can be applied to a system with any form of hysteresis loop, but we limit our attention to nondeteriorating systems, in the sense that a periodic time history for $X(t)$ leads to a periodic time history for $g[\{X(t)\}]$.

We now consider the dynamics of an oscillator including a hysteretic nonlinear element in series with a linear spring and dashpot. This hysteretic element can be elastoplastic as shown in Fig. 10.1 or of some other form. The equation of motion for the oscillator is

$$m\ddot{X}(t) + c\dot{X}(t) + k_2 X(t) + g[\{X(t)\}] = F(t) \tag{10.56}$$

The first method that we will present follows quite closely the general procedure used in Section 10.3. In particular, we use a linear approximation of

$$g[\{X(t)\}] \approx b_0 + b_1 X(t) + b_2 \dot{X}(t) \tag{10.57}$$

Note that the inclusion of $\dot{X}(t)$ in this linearization may seem somewhat arbitrary, inasmuch as it does not appear explicitly in the nonlinear $g[\{X(t)\}]$ term. There are several ways that one can justify the inclusion of the $\dot{X}(t)$ term in Eq. 10.57, but the consideration of energy dissipation is probably the simplest. We know that the area of the hysteresis loop of a nonlinear spring represents an amount of energy dissipation during one cycle of motion across the spring element. Because energy dissipation has a very important influence on the level of dynamic response in most stochastic vibration problems, it is clear that we should not use a linearized model that lacks the ability to account for this hysteretic energy dissipation. Including the $\dot{X}(t)$ term in Eq. 10.57 amounts to introducing an additional viscous damping term into the equation of motion, which does provide for possible matching of the energy dissipation in the original nonlinear system.

The mean-squared error in the approximation of Eq. 10.57 is minimized by using Eqs. 10.22 and 10.23 to find the linearization coefficients, just as in the nonhysteretic situation. We simplify this, though, by limiting our attention to the situation in which the excitation and the restoring force have symmetry such that $X(t)$ and $g[\{X(t)\}]$ are mean-zero. This then gives $b_0 = 0$, from Eq. 10.22, and the other coefficients are found from Eq. 10.23 as

$$b_1 = \frac{E\big(X(t)\, g[\{X(t)\}]\big)}{E[X^2(t)]} \tag{10.58}$$

and

$$b_2 = \frac{E\big(\dot{X}(t)\, g[\{X(t)\}]\big)}{E[\dot{X}^2(t)]} \tag{10.59}$$

The hysteretic nature of the nonlinearity complicates the evaluation of the terms in the numerators of Eqs. 10.58 and 10.59. We can make an assumption about the probability distribution of $X(t)$, just as we did in Sections 10.3 and 10.4, but that is still not enough information to allow evaluation of $E(X(t)\, g[\{X(t)\}])$ and $E(\dot{X}(t)\, g[\{X(t)\}])$. The difficulty, of course, is that $g[\{X(t)\}]$ also depends on the past time history of $X(t)$, so one also needs some assumption about the nature of the possible time histories. The usual approach is to assume that $\{X(t)\}$ is a narrowband process. In particular, we assume that we can write

$$X(t) = A(t) \cos[\omega_a t + \theta(t)] \tag{10.60}$$

with $\theta(t)$ being uniformly distributed on the interval from zero to 2π and $\{A(t)\}$ and $\{\theta(t)\}$ being independent processes that vary slowly. That is, we assume that $\dot{\theta}(t)$ is generally small compared with the average frequency ω_a, and $\dot{A}(t)$ is generally small compared with $\omega_a A(t)$. With these assumptions we can always say that $A(t)$ and $\theta(t)$ have been at approximately their present values throughout the previous cycle. If this is true, then we know that $g[\{X(t)\}]$ lies on the perimeter of the hysteresis loop with the amplitude $A(t)$. Furthermore, we can find the distribution of $g[\{X(t)\}]$ around that perimeter by using the uniform probability distribution of $\theta(t)$. In particular, we can write the conditional expectation of $X(t)\, g[\{X(t)\}]$ as

$$E\big(X(t) g[\{X(t)\}] \mid A(t)=u\big) = u\, g_c(u)/2$$

with

$$g_c(u) = \frac{1}{\pi}\int_0^{2\pi} \cos(\psi)\, g[u\cos(\psi)]\, d\psi \tag{10.61}$$

in which the integral is evaluated with $g[u\cos(\psi)]$ following the perimeter of the hysteresis loop of amplitude u. Taking the unconditional expectation by using an appropriate probability distribution for $A(t)$ then gives

$$E\big(X(t)\, g[\{X(t)\}]\big) = \frac{E\big(A(t)\, g_c[A(t)]\big)}{2} = \frac{1}{2}\int_0^{\infty} u\, g_c(u)\, p_{A(t)}(u)\, du \tag{10.62}$$

One can also use Eq. 10.60 to rewrite the denominator of Eq. 10.58 as $E[X^2(t)] = E[A^2(t)]/2$, giving

$$b_1 = \frac{E\big(A(t)\, g_c[A(t)]\big)}{E[A^2(t)]} \tag{10.63}$$

The expected value of $\dot{X}(t)\, g[\{X(t)\}]$ is evaluated in a directly parallel fashion. In particular, the derivative of the narrowband process is approximated as

$$\dot{X}(t) = -\omega_a\, A(t) \sin[\omega_a t + \theta(t)] \tag{10.64}$$

which gives $E[\dot{X}^2(t)] = \omega_a^2 E[A^2(t)]/2$. Again presuming that $g[\{X(t)\}]$ is on the perimeter of the hysteresis loop of amplitude $A(t)$ gives

$$E\big(\dot{X}(t)\, g[\{X(t)\}] \mid A(t)=u\big) = -\omega_a\, u\, g_s(u)/2$$

with

$$g_s(u) = \frac{1}{\pi}\int_0^{2\pi} \sin(\psi)\, g[u\cos(\psi)]\, d\psi \tag{10.65}$$

so

$$E\big(\dot{X}(t)\, g[\{X(t)\}]\big) = \frac{-\omega_a}{2} E\big(A(t)\, g_s[A(t)]\big) = \frac{-\omega_a}{2}\int_0^{\infty} u\, g_s(u)\, p_{A(t)}(u)\, du \tag{10.66}$$

and

$$b_2 = -\frac{1}{\omega_a} \frac{E\big(A(t)\, g_s[A(t)]\big)}{E[A^2(t)]} \tag{10.67}$$

One can also simplify this result a little by noting that a change of variables of $w = u\cos(\psi)$ in Eq. 10.65 gives

$$g_s(u) = \frac{-1}{\pi\, u} \int_{loop} g(w)\, dw = \frac{-1}{\pi\, u} \times (\text{Area of Hysteresis Loop with Amplitude } u)$$

It should be noted that the linearization presented here is essentially the same as that given by Caughey (1960b), although his justification of Eqs. 10.61 and 10.65 is slightly different. In particular, he considers these integrals to represent a time average over one cycle of oscillator response, as in the Krylov-Bogoliubov method for linearization of deterministic problems. This interpretation, which can be considered an application of stochastic averaging, makes Eqs. 10.62 and 10.66 become combination averages—expected values of time averages.

It is now necessary to make some assumption about the probability distribution of the $A(t)$ amplitude in order to evaluate the expectations in Eqs. 10.62 and 10.66 and thus obtain the values of the b_1 and b_2 linearization coefficients from Eqs. 10.63 and 10.67. The usual assumption is to say that the narrowband $\{X(t)\}$ process is nearly Gaussian, and this implies that the Rayleigh distribution can be used for $A(t)$. This is essentially the same argument as was used in Section 7.4, where $A(t)$ for a Gaussian process had exactly the Rayleigh probability distribution when it was defined such as to exactly satisfy Eqs. 10.60 and 10.64. Using the Rayleigh probability distribution (or some other known function) for $p_{A(t)}(u)$ then gives b_1 and b_2 as functions of σ_X. Overall, it may be seen that when the narrowband assumption is used for $\{X(t)\}$, the difficulty in linearizing a hysteretic function of $\{X(t)\}$ is only slightly greater than in linearizing a nonhysteretic function of $X(t)$ and $\dot{X}(t)$.

After the linearization coefficients have been evaluated, one can substitute $b_1 X(t) + b_2 \dot{X}(t)$ in place of $g[\{X(t)\}]$ in Eq. 10.56. This gives the approximate equation of motion as

$$m\, \ddot{X}(t) + (c + b_2)\, \dot{X}(t) + (k_2 + b_1)\, X(t) = F(t)$$

from which it is obvious that b_1 represents an additional stiffness term and b_2 represents additional damping. That is, this linearization procedure replaces the elastoplastic element with a spring and a dashpot in parallel. If this linearized system does have a narrowband response, then its average frequency must be approximately

$$\omega_a = \left(\frac{k_2 + b_1}{m} \right)^{1/2} \tag{10.68}$$

because that is the resonant frequency of the new equation of motion. Furthermore, we can use the usual expression for the variance of the response of the linear SDF oscillator to say that

$$\sigma_X^2 = \frac{\pi S_0}{(c + b_2)(k_2 + b_1)} \tag{10.69}$$

The variance of $\dot{X}(t)$, of course, is approximated by $\omega_a^2 \sigma_X^2$. Numerical results are obtained by simultaneously solving Eqs. 10.63, 10.67, 10.68, and 10.69.

Note that it is reasonable to expect the accuracy of this linearization procedure to depend on the adequacy of two particular approximations: the narrowband assumption used in obtaining Eqs. 10.61 and 10.65 and the probability distribution assumed for the $A(t)$ amplitude of the narrowband response. If the system is nearly linear, then the narrowband assumption should be appropriate. Also, if the excitation of the system is Gaussian and the nonlinearity is small, then the response should be nearly Gaussian so that a Rayleigh approximation for the amplitude is justified.

****

Example 10.10: Find the linearization parameters and the mean-squared response levels for the bilinear hysteretic oscillator shown in Fig. 10.2 when the $\{F(t)\}$ excitation is a stationary, mean-zero, Gaussian white noise with autospectral density S_0.

The system is described by Eq. 10.56 with $g[\{X(t)\}]$ as given in Fig. 10.1. Recall that the symmetry of the situation gives $b_0 = 0$. To evaluate b_1 and b_2 we need to evaluate the $g_c(u)$ and $g_s(u)$ functions defined by Eqs. 10.61 and 10.65. First we note that for $u < x_y$, the elastoplastic term is $g[u\cos(\psi)] = k_1 u \cos(\psi)$, so $g_c(u) = k_1 u$, $g_s(u) = 0$ for $u < x_y$. For $u > x_y$, we split the integration with respect to ψ in Eq. 10.61 into four parts, corresponding to the four straight segments of the hysteresis loop. The result is

$$g_c(u)=\frac{1}{\pi}\Bigg(\int_0^{\psi^*}\cos(\psi)k_1[x_y-u(1-\cos\psi)]\,d\psi+\int_{\psi^*}^{\pi}\cos(\psi)(-k_1x_y)\,d\psi+$$
$$\int_{\pi}^{\pi+\psi^*}\cos(\psi)\,k_1[-x_y+u(1-\cos\psi)]\,d\psi+\int_{\pi+\psi^*}^{2\pi}\cos(\psi)(k_1x_y)\,d\psi\Bigg)$$

in which $\psi^*=\cos^{-1}[(u-2x_y)/u]$, as shown in the following sketch. After simplification, this becomes

$$g_c(u)=\frac{k_1u\psi^*}{\pi}-\frac{2k_1}{\pi}\left(\frac{u-2x_y}{u}\right)[x_y(u-x_y)]^{1/2}\qquad\text{for } u>x_y$$

One can obtain $g_s(u)$ either by performing a similar integration around the loop or by finding the area of the hysteresis loop in Fig. 10.1. The result is

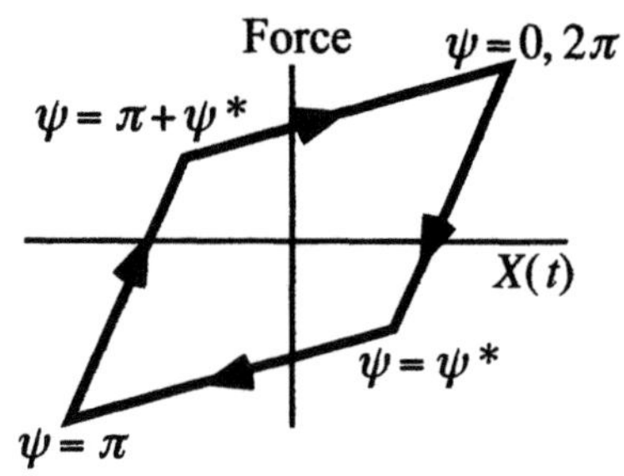

$$g_s(u)=-\frac{4k_1}{\pi}\frac{x_y(u-x_y)}{u}\qquad\text{for } u>x_y$$

Presuming that the response is narrowband and Gaussian, we now use the Rayleigh probability density function for $A(t)$:

$$p_{A(t)}(u)=\frac{u}{\sigma_X^2}\exp\left(\frac{-u^2}{2\sigma_X^2}\right)$$

and evaluate the expectations of Eqs. 10.63 and 10.67. The integral in Eq. 10.63 has the form

$$b_1=\frac{2}{\sigma_X^2}\Bigg(\int_0^{x_y}k_1u^2\frac{u}{\sigma_X^2}\exp\left[\frac{-u^2}{2\sigma_X^2}\right]du+\int_{x_y}^{\infty}\Bigg[\frac{k_1u}{\pi}\cos^{-1}\left(\frac{u-2x_y}{u}\right)-$$
$$\frac{2k_1}{\pi}\left(\frac{u-2x_y}{u}\right)[x_y(u-x_y)]^{1/2}\Bigg]\frac{u}{\sigma_X^2}\exp\left[\frac{-u^2}{2\sigma_X^2}\right]du\Bigg)$$

and Caughey (1960b) showed that it can be converted into

$$b_1=k_1\left(1-\frac{8}{\pi}\int_1^{\infty}\left[\frac{1}{z^3}+\frac{x_y^2}{2\sigma_X^2\,z}\right](z-1)^{1/2}\exp\left[\frac{-x_y^2z^2}{2\sigma_X^2}\right]dz\right)$$

No closed form has been found for this integral, but it can be easily evaluated numerically if one first chooses a value for the dimensionless ratio σ_X/x_y. Note that σ_X/x_y can viewed as a root-mean-square (rms) ductility ratio, because it represents the rms response displacement divided by the yield displacement. The evaluation of b_2 is somewhat easier. Writing out the integral for the expected value in Eq. 10.67 gives

$$b_2 = \frac{8k_1}{\pi\omega_a\sigma_X^2}\int_{x_y}^{\infty} x_y(u - x_y)\frac{u}{\sigma_X^2}\exp[-u^2/(2\sigma_X^2)]\,du$$

and this reduces to

$$b_2 = \left(\frac{2}{\pi}\right)^{1/2}\frac{k_1 x_y}{\omega_a\sigma_X}\left(1 - \Phi(x_y/\sigma_X)\right)$$

in which $\Phi(\cdot)$ denotes the cumulative distribution function for a standardized Gaussian random variable and is related to the error function (see Example 2.7). The following sketches show normalized values of b_1 and b_2 for a fairly wide range of σ_X/x_y values.

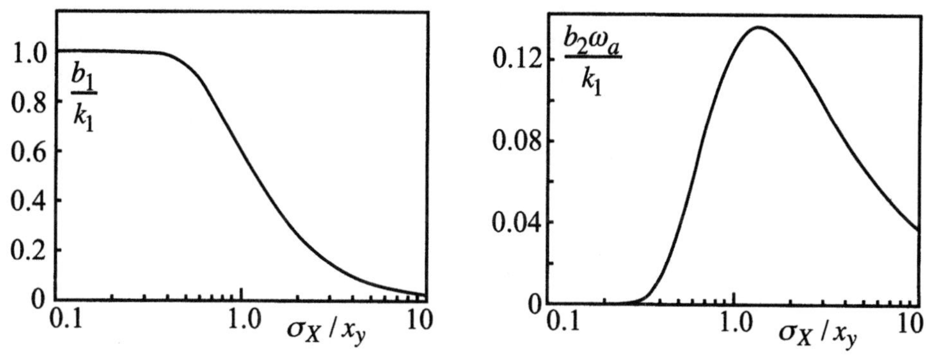

The decay of b_1 from its initial value of k_1 for $\sigma_X << x_y$ to its final value of zero for $\sigma_X >> x_y$ corresponds to ω_a^2 varying from $(k_1+k_2)/m$ to k_2/m. If the excitation is very small, then $\sigma_X << x_y$ and the elastoplastic element experiences little or no yielding, so the k_1 spring does contribute effectively to the system stiffness. If the excitation is very large so that $\sigma_X >> x_y$, then the force added by the elastoplastic element is relatively small, so the total stiffness approaches k_2. The damping effect of the elastoplastic element has a rather different characteristic. The dashpot effect is the greatest when $\sigma_X \approx 1.5x_y$ and is near zero for both very large and very small values of σ_X/x_y. Of course, the energy dissipated per cycle does continue to grow as σ_X increases beyond $1.5x_y$, even though the value of the "equivalent" dashpot coefficient b_2 decreases.

One can use an iterative procedure to find the solution of Eqs. 10.63, 10.67, 10.68, and 10.69 for given values of the system parameters m, c, k_1, k_2, and x_y, and the excitation level S_0. Alternatively, one can consider fixed values of some parameters and plot the results versus other parameters. We include three plots of this form, in which σ_X/σ_F^* is plotted versus σ_F^*/x_y, with σ_F^* defined as

$$(\sigma_F^*)^2 = \frac{2S_0\omega_0}{(k_1+k_2)^2}$$

with $\omega_0 = [(k_1 + k_2)/m]^{1/2}$ = resonant frequency for the linear system that results when x_y is infinite. For each curve the values of k_2/k_1 and $\zeta_0 \equiv c/(2[(k_1+k_2)m]^{1/2})$ are held constant. Note that for given system parameters, σ_{F^*} is a measure of the level of the excitation. Thus, any curve of the type given can be considered to represent response normalized by the excitation plotted versus the excitation normalized by the yield level. The curves were obtained by choosing a value of σ_X / x_y, evaluating b_1/k_1 and $\omega_a b_2/k_1$ as for the preceding plots, using Eq. 10.68 for the ω_a term in the latter of these expressions, and finally using Eq. 10.69 to give the response variance.

The plots include three different values of k_2/k_1, to represent three different situations. For $k_2/k_1 = 9$, it is reasonable to consider the nonlinearity to be small so that the assumptions used in the linearization procedure should be appropriate. For $k_2/k_1 = 1$ the system can lose up to half of its stiffness due to yielding and there is considerable hysteretic energy dissipation, so it is questionable whether the narrowband and Gaussian assumptions are appropriate for $\{X(t)\}$. The situation with $k_2/k_1 = 0.05$ goes far beyond the range in which it is reasonable to expect the assumptions to be valid. Numerical data for the response levels of the bilinear hysteretic oscillator with $k_2/k_1 = 1$ and $k_2/k_1 = 0.05$ were presented by Iwan and Lutes (1968), and some of these data are reproduced in the plots. In particular, the plots include data for $\zeta_0 = 0$ and $\zeta_0 = 0.01$ for $k_2/k_1 = 1$ and for $\zeta_0 = 0$ and $\zeta_0 = 0.05$ for $k_2/k_1 = 0.05$. Somewhat surprisingly, the data show that the linearization results are generally quite good for $k_2/k_1 = 1$, even though this is hardly a situation with a small amount of nonlinearity. For $k_2/k_1 = 0.05$, on the other hand, there are sometimes significant discrepancies between the results from the bilinear hysteretic system and those from linearization.

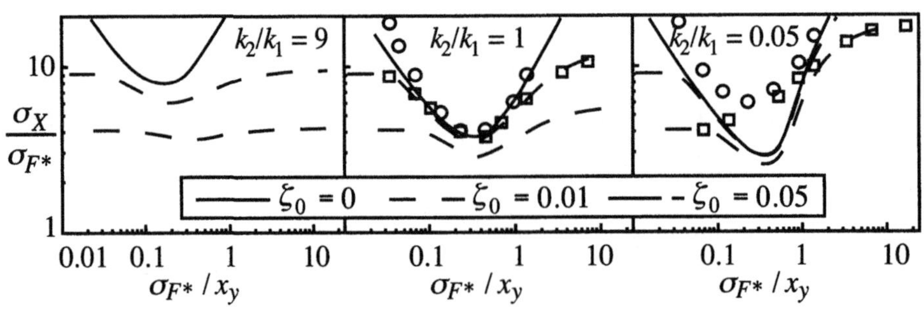

****

Note that the narrowband assumption used in the preceding analysis imposes some limitations on the technique. It may be appropriate for most SDF systems, but it is surely not appropriate for an MDF system in which more than one mode contributes significantly to the response. Also, it could be

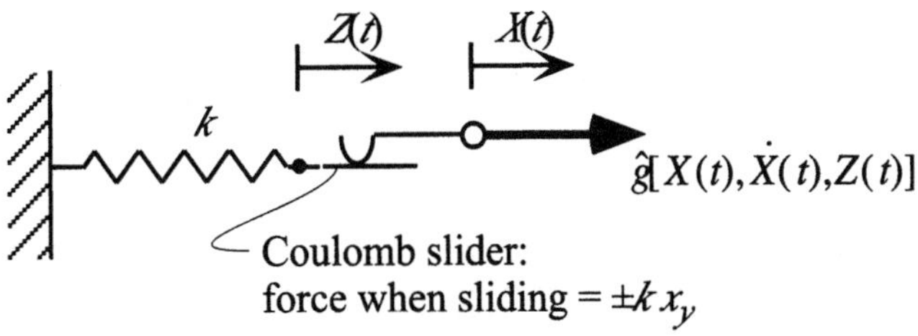

Figure 10.3 Nonhysteretic description of the elastoplastic force.

inappropriate for an SDF problem in which the response spectral density is bimodal because of a dominant frequency in the excitation that is different from the resonant frequency of the oscillator.

For some hysteretic systems, it is possible to approach the problem of linearization in a different way that eliminates the need for the narrowband approximation used in obtaining Eqs. 10.62 and 10.66. This approach involves the introduction of one or more additional state variables such that the $g[\{X(t)\}]$ restoring force can be written as a nonhysteretic function of the new set of state variables. This idea is easily illustrated for the elastoplastic element that contributes the hysteresis to the bilinear oscillator of Example 10.10. An appropriate additional state variable for this example is shown as $Z(t)$ in Fig. 10.3. Using this variable allows one to express the force across the elastoplastic element as

$$\hat{g}[X(t),\dot{X}(t),Z(t)] = k\,Z(t) \tag{10.70}$$

One notes that the state variables $X(t)$ and $\dot{X}(t)$ do not appear explicitly in this equation for the force but that they are necessary if one is to determine the rate of change of the force:

$$\dot{\hat{g}}[X(t),\dot{X}(t),Z(t)] = k\,\dot{Z}(t)$$

with

$$\dot{Z}(t) = \dot{X}(t)\Big(1 - U[Z(t)-x_y]\,U[\dot{X}(t)] - U[-Z(t)-x_y]\,U[-\dot{X}(t)]\Big) \tag{10.71}$$

That is, when $Z(t)$ reaches the level $\pm x_y$, we know that $\dot{Z}(t)$ changes from $\dot{X}(t)$ to zero, so $Z(t)$ remains unchanged until $\dot{X}(t)$ reverses its sign. When this sign reversal occurs for $\dot{X}(t)$, the value of $\hat{g}[X(t),\dot{X}(t),Z(t)]$ begins to move back from $\pm k\,x_y$ toward zero.

Because Eq. 10.70 is already linear, there is no reason to apply statistical linearization to it; however, definition of an appropriate linear model in this formulation involves replacing Eq. 10.71 with a linear relationship of the form

$$\dot{Z}(t) \approx b_0 + b_1 X(t) + b_2 \dot{X}(t) + b_3 Z(t) \tag{10.72}$$

We will now make the major assumption that $\{X(t)\}$ and $\{Z(t)\}$ can be approximated as being mean-zero and jointly Gaussian processes. This makes $X(t)$, $\dot{X}(t)$, and $Z(t)$ jointly Gaussian random variables, so we can conveniently use Eq. 10.24 in evaluating the coefficients of linearization. The results are

$$b_0 = E[\dot{Z}(t)] = 0 \tag{10.73}$$

$$b_1 = E\left(\frac{\partial \dot{Z}(t)}{\partial X(t)}\right) = 0 \tag{10.74}$$

$$b_2 = E\left(\frac{\partial \dot{Z}(t)}{\partial \dot{X}(t)}\right) = E\Big(1 - U[Z(t) - x_y]U[\dot{X}(t)]\Big) - E\Big(U[-Z(t) - x_y]U[-\dot{X}(t)]\Big)$$

and

$$b_3 = E\left(\frac{\partial \dot{Z}(t)}{\partial Z(t)}\right) = -E\Big(\dot{X}(t)\delta[Z(t) - x_y]U[\dot{X}(t)]\Big) + E\Big(\dot{X}(t)\delta[-Z(t) - x_y]U[-\dot{X}(t)]\Big)$$

which yield

$$b_2 = 1 - 2\int_{x_y}^{\infty}\int_0^{\infty} p_{\dot{X}(t)Z(t)}(v,w)\,dv\,dw \tag{10.75}$$

and

$$b_3 = -2\int_0^{\infty} v\, p_{\dot{X}(t)Z(t)}(v,x_y)\,dv \tag{10.76}$$

In addition to Eqs. 10.73–10.76 relating the values of the linearization coefficients to the response levels, it is necessary to have other equations relating the response levels to the linearization parameters. These equations come from stochastic analysis of the equation of motion of a system containing this linearized elastoplastic element. This can be illustrated by considering the bilinear hysteretic system of Example 10.10. Using Eq. 10.70 to linearize the elastoplastic element in this system gives an equation of motion of

$$m\,\ddot{X}(t) + c\,\dot{X}(t) + k_2 X(t) + k_1 Z(t) = F(t) \tag{10.77}$$

and the fact that b_0 and b_1 are zero allows the auxiliary relationship of Eq. 10.72 to be written as

$$\dot{Z}(t) = b_2 \dot{X}(t) + b_3 Z(t) \tag{10.78}$$

One can find the response levels of this system as functions of the parameters m, c, k_1, k_2, b_2, and b_3 by any of the methods appropriate for a linear system. If the $\{F(t)\}$ excitation is stationary white noise, then it is particularly convenient to use the state-space moment or cumulant methods of Chapter 9 to derive simultaneous algebraic equations giving the response variances and covariances. In particular, one can take the state vector as $\vec{Y}(t) = [X(t), \dot{X}(t), Z(t)]^T$ so that the response variances and covariances are given by the components of a symmetric matrix of dimension 3×3. One of the six simultaneous equations turns out to be trivial, simply giving the well-known fact that the covariance of $X(t)$ and $\dot{X}(t)$ is zero. The other five equations can be solved simultaneously with Eqs. 10.75 and 10.76 in order to find not only the five remaining terms describing the response levels but also the b_2 and b_3 coefficients.

****

Example 10.11: Find the linearization parameters and the mean-squared response levels predicted by Eqs. 10.70–10.76 for stationary response of the bilinear hysteretic oscillator shown in Fig. 10.2 with an $\{F(t)\}$ excitation that is a stationary, mean-zero, Gaussian white noise with autospectral density S_0.

As previously noted, the system is described by Eqs. 10.77 and 10.78 and the b_2 and b_3 linearization parameters are related to the response levels by Eqs. 10.75 and 10.76. Writing these two expressions as integrals of the jointly Gaussian probability density function gives

$$b_2 = 1 - 2\int_{x_y/\sigma_Z}^{\infty} \frac{e^{-r^2/2}}{(2\pi)^{1/2}} \Phi\left(\rho_{\dot{X},Z}\, r / [1 - \rho_{\dot{X}Z}^2]^{1/2}\right) dr$$

and

$$b_3 = -2 \frac{\sigma_{\dot{X}}}{\sigma_Z} \hat{b}_3$$

with

$$\hat{b}_3 = \frac{\rho_{\dot{X}Z}\, x_y}{(2\pi)^{1/2}\sigma_Z} \exp\left(\frac{-x_y^2}{2\sigma_Z^2}\right) \Phi\left(\frac{\rho_{\dot{X}Z}\, x_y}{\sigma_Z (1-\rho_{\dot{X}Z}^2)^{1/2}}\right) +$$

$$\frac{(1-\rho_{\dot{X}Z}^2)^{1/2}}{2\pi} \exp\left(\frac{-x_y^2}{2\sigma_Z^2 (1-\rho_{\dot{X}Z}^2)}\right)$$

Note that b_2 and $\hat{b}_3$ are dimensionless quantities that depend only on the two parameters $\rho_{\dot{X}Z}$ and x_y / σ_Z, whereas b_2 has units of frequency and depends on the value $\sigma_{\dot{X}} / \sigma_Z$.

To evaluate parameters and response levels, we must use expressions giving the response levels of the linearized system of Eqs. 10.77 and 10.78. We will do this by using the state-space formulation of the covariance equations, as given in Section 9.5. Thus, we rewrite the equations of motion in the form of $\dot{\vec{Y}}(t) + \mathbf{B}\,\vec{Y}(t) = \vec{Q}(t)$, with $\vec{Y}(t) = [X(t), \dot{X}(t), Z(t)]^T$, $\vec{Q}(t) = [0, F(t)/m, 0]^T$, and

$$\mathbf{B} = \begin{bmatrix} 0 & -1 & 0 \\ k_2/m & c/m & k_1/m \\ 0 & -b_2 & -b_3 \end{bmatrix}$$

Because we are interested only in stationary response, we can simplify Eq. 9.18 to be

$$\mathbf{B}\,\mathbf{K}_{YY} + \mathbf{K}_{YY}\mathbf{B}^T = 2\pi\mathbf{S}_0$$

with the only nonzero element of the $\mathbf{S}_0$ matrix being the scalar S_0/m^2 in the (2,2) position. After some very minor simplification, the six distinct components of this relationship can be written as

$$K_{X\dot{X}} = 0$$

$$c\,K_{X\dot{X}} + k_1 K_{XZ} + k_2 K_{XX} - m\,K_{\dot{X}\dot{X}} = 0$$

$$b_2 K_{X\dot{X}} + b_3 K_{XZ} + K_{\dot{X}Z} = 0$$

$$k_2 K_{X\dot{X}} + c\,K_{\dot{X}\dot{X}} + k_1 K_{\dot{X}Z} = \pi S_0 / m$$

$$k_2 K_{XZ} + (c - b_3 m)\,K_{\dot{X}Z} - b_2 m\,K_{\dot{X}\dot{X}} + k_1 K_{ZZ} = 0$$

and

$$-b_2 K_{\dot{X}Z} - b_3 K_{ZZ} = 0$$

The first of these equations is simply a well-known property of any stationary process, and this can be used to give a slight simplification in the second, third, and fourth equations. Similarly, Eq. 10.78 shows that the sixth equation is the corresponding condition that $K_{Z\dot{Z}} = 0$. Using the linearization condition of $b_3 = -2(\sigma_{\dot{X}}/\sigma_Z)\,\hat{b}_3$ along with this stationarity condition gives

$-b_2\rho_{\dot{X}Z}+2\hat{b}_3=0$ as a necessary condition for the joint probability density function of $\dot{X}(t)$ and $Z(t)$. One can use this relationship along with the expressions for b_2 and $\hat{b}_3$ to derive a unique relationship between $\rho_{\dot{X}Z}$ and x_y/σ_Z. In particular, for a selected value of x_y/σ_Z one can use an initial guess for $\rho_{\dot{X}Z}$, evaluate estimates of b_2 and $\hat{b}_3$, reevaluate $\rho_{\dot{X}Z}$, and so on. After finding these values for the selected x_y/σ_Z value, one can use the third response equation to write

$$K_{XZ}=-K_{\dot{X}Z}/b_3=\rho_{\dot{X}Z}\sigma_{\dot{Z}}^2/(2\hat{b}_3)=\sigma_Z^2/b_2$$

Rewriting $K_{\dot{X}\dot{X}}$ and K_{ZZ} as $\sigma_{\dot{X}}^2$ and σ_Z^2, respectively, allows relatively simple solution of the fourth and fifth response equations to find these quantities. The second response equation then gives the value of $K_{XX}\equiv\sigma_X^2$. Although there are multiple solution branches for these equations, there seems to be only one solution that gives real positive values of σ_X^2, $\sigma_{\dot{X}}^2$, and σ_Z^2 for any given value of x_y/σ_Z. The following sketches show the results of this linearization for the situations with $k_2=k_1$ and $k_2=0.05k_1$. The form of the plots is the same as in Example 10.10. For $k_2=0.05k_1$ and intermediate values of σ_F*/x_y, it is observed that the fit of this linearization to the numerical data is significantly better than for the linearization of Example 10.10. For larger values of σ_F*/x_y, though, this linearization is less accurate than the one of Example 10.10, for either value of k_2/k_1.

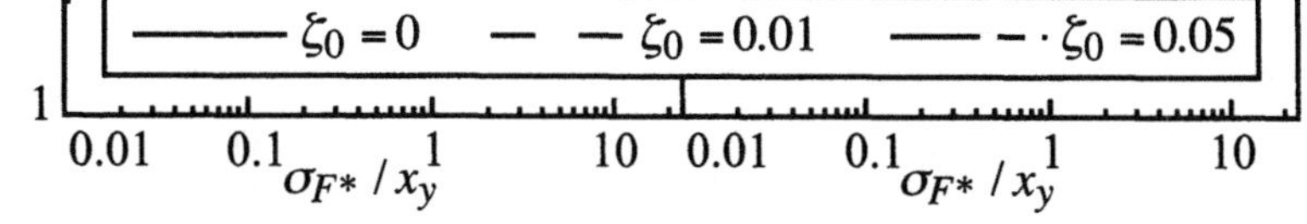

****

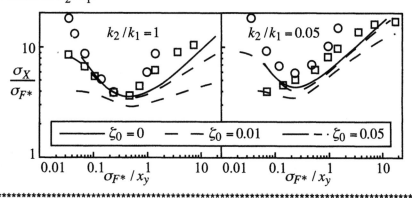

Another version of the linearization of the elastoplastic element was offered by Asano and Iwan (1984), who noted that Eq. 10.70 allows the possibility of the absolute value of the restoring force exceeding kx_y, whereas that should be impossible. To avoid this situation they replaced the right-hand side of Eq. 10.70 with an expression that truncates the restoring force at the levels $\pm kx_y$. This can be written as

$$\hat{g}[X(t),\dot{X}(t),Z(t)] = k\,Z(t)\Big(1 - U[Z(t)-x_y]U[\dot{X}(t)] - U[-Z(t)-x_y]U[-\dot{X}(t)]\Big) + \\ k\,x_y\Big(U[Z(t)-x_y]U[\dot{X}(t)] - U[-Z(t)-x_y]U[-\dot{X}(t)]\Big) \tag{10.79}$$

Note that Eq. 10.79 is identical to Eq. 10.70 whenever $|Z(t)| \le x_y$. Furthermore, the behavior of the truly elastoplastic element never allows the system to enter the region of $|Z(t)| > x_y$, where Eqs. 10.70 and 10.79 differ from each other. Thus, both Eq. 10.70 and Eq. 10.79 exactly describe truly elastoplastic behavior. When one linearizes the equations in the usual way, though, the results are not identical. In particular, one must consider Eq. 10.79 to give a nonlinear relationship between the elastoplastic force and the state variables. Linearization of this model, then, involves not only using Eq. 10.78 in place of Eq. 10.71 but also using an approximation of

$$\hat{g}[X(t),\dot{X}(t),Z(t)] \approx a_0 + a_1 X(t) + a_2 \dot{X}(t) + a_3 Z(t) \tag{10.80}$$

in place of Eq. 10.79. Using the jointly Gaussian assumption for the state variables gives the coefficients as

$$a_0 = E\Big(\hat{g}[X(t),\dot{X}(t),Z(t)]\Big) = 0 \tag{10.81}$$

$$a_1 = E\left(\frac{\partial\,\hat{g}[X(t),\dot{X}(t),Z(t)]}{\partial X(t)}\right) = 0 \tag{10.82}$$

$$\begin{aligned} a_2 &= E\left(\frac{\partial\,\hat{g}[X(t),\dot{X}(t),Z(t)]}{\partial \dot{X}(t)}\right) \\ &= k\,E\Big[Z(t)\,\delta[\dot{X}(t)]\Big(-U[Z(t)-x_y] + U[-Z(t)-x_y]\Big)\Big] + \\ &\qquad k\,x_y E\Big[\,\delta[\dot{X}(t)]\Big(U[Z(t)-x_y] + U[-Z(t)-x_y]\Big)\Big] \\ &= -2k\int_{x_y}^{\infty} (w - x_y)\, p_{\dot{X}(t)Z(t)}(0,w)\,dw \end{aligned} \tag{10.83}$$

$$\begin{aligned} a_3 &= E\left(\frac{\partial\,\hat{g}[X(t),\dot{X}(t),Z(t)]}{\partial Z(t)}\right) \\ &= k\,E\Big(1 - U[Z(t)-x_y]U[\dot{X}(t)] - U[-Z(t)-x_y]U[-\dot{X}(t)]\Big) \end{aligned} \tag{10.84}$$

One may also note that Eqs. 10.84 and 10.75 give a_3 as being identical to $k b_2$.

It is somewhat disturbing to find that Eqs. 10.70 and 10.79 are both exact descriptions of the elastoplastic nonlinearity, but they give different results when analyzed by statistical linearization. For this system there is not a unique answer even when one adopts the standard procedure of minimizing the mean-squared error and uses the standard assumption that the response quantities have a jointly Gaussian probability distribution. By using a different linearization procedure or by assuming a different probability distribution, of course, it is possible to obtain any number of other results, but there is some ambiguity even within the standard procedure. In this particular instance, the additional ambiguity arose because the new $Z(t)$ state variable was bounded, so it was undefined on certain regions of the state space. The Gaussian assumption, though, does assign probability to the entire state space, so the linearization results are affected by assumptions about behavior in the regions that really should have zero probability. One way to avoid this particular difficulty is to analyze only models that have a finite probability density everywhere. This is one of the advantages of a smooth hysteresis model that has gained popularity in recent years. In this model the new $Z(t)$ state variable is proportional to the hysteretic portion of the restoring force, as in Eq. 10.77, and its rate of change is governed by the differential equation

$$\dot{Z}(t) = -c_1 [\,|\dot{X}(t) Z^{c_3 - 1}(t)|\, Z(t)] - c_2 |Z^{c_3}(t)|\, \dot{X}(t)$$

for some parameters c_1, c_2, and c_3. This gives $Z(t)$ as unbounded, so there is not any region of zero probability. This form is a generalization by Wen (1976) of a model used by Bouc (1968) with $c_3 = 1$. Wen achieved generally good results using statistical linearization for this hysteretic system, and the model seems to be capable of approximating many physical problems.

In some problems it is not as obvious how to avoid having regions of zero probability in the state space. For example, when modeling damaged structures it may be desirable to consider the tangential stiffness at time t to be a function of the maximum distortion of the structure at any time in the past. This can be accommodated by defining a state variable $Y(t)$ that is the maximum absolute value of the $X(t)$ displacement at any prior time. This new $Y(t)$ variable is not bounded, but there should be zero probability on the regions $X(t) > Y(t)$ and $X(t) < -Y(t)$. Statistical linearization has been used for such problems (Senthilnathan and Lutes, 1991), but the difficulties and ambiguities involved are similar to those for elastoplastic behavior.

****

Example 10.12: Find the linearization parameters and the mean-squared response levels predicted by Eqs. 10.80–10.84 for stationary response of the bilinear hysteretic oscillator shown in Fig. 10.2 with an $\{F(t)\}$ excitation that is a stationary, mean-zero, Gaussian white noise with autospectral density S_0.

The procedure is the same as in Example 10.11. In addition to the b_2 and b_3 parameters evaluated there, the linearization now also uses $a_3 = k_1 b_2$ and the parameter a_2 from Eq. 10.83, which can be rewritten as $a_2 = -2k_1(\sigma_Z / \sigma_{\dot X})\hat{a}_2$ with

$$\hat{a}_2 = \frac{(1-\rho_{\dot X Z}^2)^{1/2}}{2\pi} \exp\left(\frac{-x_y^2}{2\sigma_Z^2 (1-\rho_{\dot X Z}^2)} \right) - \frac{x_y}{(2\pi)^{1/2}\sigma_Z} \Phi\left(\frac{-x_y}{\sigma_Z (1-\rho_{\dot X Z}^2)^{1/2}} \right)$$

Note that $\hat{a}_2$, like b_2 and $\hat{b}_3$, is a function only of $\rho_{\dot X Z}$ and x_y / σ_Z, so it can be evaluated independently from the analysis of the dynamics of the oscillator. Substituting Eq. 10.80, with $a_0 = 0$, $a_1 = 0$, and $a_3 = k_1 b_2$ into Eq. 10.56, gives the equation of motion for the linearized system as

$$m\ddot X(t) + (c + a_2)\dot X(t) + k_2 X(t) + k_1 b_2 Z(t) = F(t)$$

This equation must be solved in conjunction with Eq. 10.78, giving the rate of change of $Z(t)$. Note that the new equation of motion has the same form as Eq. 10.77, but it has different stiffness and damping values. In particular, the presence of b_2, which is a number between zero and unity, results in a reduced stiffness for this linearization of the problem. Similarly, a_2 is a negative quantity, so it causes a reduction in the damping of the system. Thus, we can anticipate that the response of this linear model will be greater than that for Eq. 10.70. The details of the state-space analysis to find the response quantities as functions of the linearization parameters will be omitted, because the procedure is fundamentally the same as in Example 10.11. The σ_X / σ_{F^*} results are shown in the following plot.

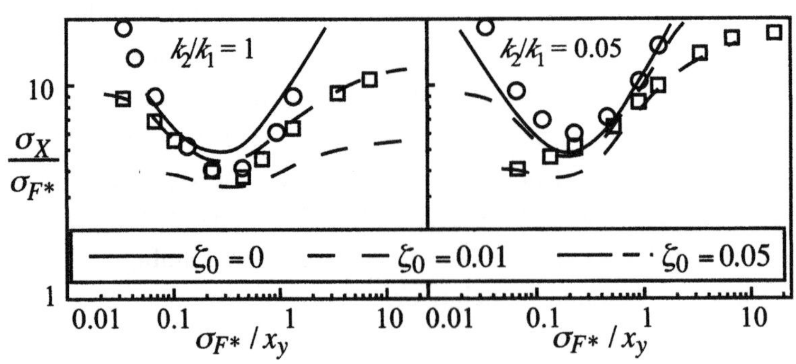

It is observed that this linearization does a better job of fitting the numerical simulation data than does the method of either Example 10.10 or 10.11. The difference between the linearization shown here and that of Example 10.11 is most obvious for relatively high values of $\sigma_F * / x_y$, because this is the situation where Eqs. 10.70 and 10.79 are most significantly different.

****

10.6 State-Space Moment and Cumulant Equations

We will now consider how the state-space methods of Chapter 9 can be applied to nonlinear problems. For this purpose we presume that the nonlinear equation of motion is written as in Eq. 10.2:

$$\dot{\vec{Y}}(t) + \vec{g}[\vec{Y}(t)] = \vec{Q}(t) \tag{10.85}$$

This general equation is adequate for analysis of nonlinear systems without hysteresis. However, with the introduction of additional state variables, as illustrated by $Z(t)$ in Eqs. 10.70 and 10.71, it can also be applied to many problems that are generally classified as hysteretic.

The procedure of Section 9.3 can now be used to derive state-space mean and covariance equations. The mean-value equation of

$$\dot{\vec{\mu}}_Y(t) + E\Big(\vec{g}[\vec{Y}(t)]\Big) = \vec{\mu}_Q(t) \tag{10.86}$$

results from simply taking the expectation of Eq. 10.85. Recalling that

$$\frac{d}{dt}\mathbf{K}_{YY}(t,t) = \mathbf{K}_{\dot{Y}Y}(t,t) + \mathbf{K}_{Y\dot{Y}}(t,t)$$

we can take cross-covariances involving terms of Eq. 10.85 and the state vector to give

$$\frac{d}{dt}\mathbf{K}_{YY}(t,t) + \mathbf{K}_{gY}(t,t) + \mathbf{K}_{Yg}(t,t) = \mathbf{K}_{QY}(t,t) + \mathbf{K}_{YQ}(t,t) \tag{10.87}$$

in which $\mathbf{K}_{Yg}(t,t)$ is as defined in Eq. 10.50, and $\mathbf{K}_{gY}(t,t)$ is its transpose.

One use of the exact mean and covariance relationships given in Eqs. 10.86 and 10.87 is to provide valuable information about the assumptions of the statistical linearization procedure introduced in Section 10.4. In particular, a comparison of Eq. 10.86 and the approximate linearization result given in Eq. 10.53 shows that integrating the two expressions will give exactly the same $\vec{\mu}_Y(t)$ values if $\vec{b}(t)$ is identically the same as $E(\vec{g}[\vec{Y}(t)])$. This latter relationship, though, is exactly the linearization condition given in Eq. 10.48. Similarly, one can compare Eqs. 10.54 and 10.87 and note that they are identical if $\mathbf{B}(t)\mathbf{K}_{YY}(t,t)=\mathbf{K}_{gY}(t,t)$, which is exactly equivalent to the linearization condition given in Eq. 10.49. Thus, if one can manage to choose $\vec{b}(t)$ and $\mathbf{B}(t)$ in such a way that the conditions of Eqs. 10.48 and 10.49 are exactly satisfied, then statistical linearization will give exact values of the mean and covariance of response of the nonlinear system. In one sense, this provides a justification for use of the statistical linearization procedure that is much stronger than any argument that we were able to make in Section 10.4. On the other hand, one must realize that it is generally impossible to satisfy Eqs. 10.48 and 10.49. To satisfy these linearization conditions it would be necessary to know the probability distribution of the response vector $\vec{Y}(t)$, but this probability distribution is precisely the unknown quantity in the problem. Nonetheless, it is useful to know that any mean and covariance error of statistical linearization lies entirely in the evaluation of the linearization coefficients, and not in the difference between the linearized equation and the original nonlinear equation. For higher-order cumulants, of course, the situation is more complicated. For example, if $\{\vec{Q}(t)\}$ is Gaussian then the linearized system will give $\{\vec{Y}(t)\}$ as also being Gaussian, so its higher-order cumulants are zero. Obviously, this is not true for the nonlinear problem.

Returning to consideration of the nonlinear state-space equations, note that the right-hand side of Eq. 10.87 is essentially the same as that of Eq. 9.9. In particular, the right-hand side of the equation is not affected by the nonlinearity that has been introduced into the system. Furthermore, the methods used in Section 9.4 to simplify this right-hand side are not in any way dependent on the system being linear. Thus, if the $\{\vec{Q}(t)\}$ excitation vector is delta-correlated with a nonstationary covariance of $\mathbf{K}_{QQ}(t,s)=2\pi\mathbf{S}_0(t)\delta(t-s)$, then one can write the nonlinear version of Eq. 9.17 as

$$\frac{d}{dt}\mathbf{K}_{YY}(t,t)+\mathbf{K}_{gY}(t,t)+\mathbf{K}_{Yg}(t,t)=2\pi\mathbf{S}_0(t) \tag{10.88}$$

By switching to Kronecker notation we can also give the corresponding equations for higher-order cumulants. The general relationship that is analogous to Eq. 9.48 for the linear problem is

$$\frac{d}{dt}\kappa_j^{\otimes}[\vec{Y}(t),\cdots,\vec{Y}(t)] + \sum_{l=1}^{j}\kappa_j^{\otimes}[\underbrace{\vec{Y}(t),\cdots,\vec{Y}(t)}_{l-1},\vec{g}[Y(t)],\underbrace{\vec{Y}(t),\cdots,\vec{Y}(t)}_{j-l}] =$$
$$\sum_{l=1}^{j}\kappa_j^{\otimes}[\underbrace{\vec{Y}(t),\cdots,\vec{Y}(t)}_{l-1},\vec{Q}(t),\underbrace{\vec{Y}(t),\cdots,\vec{Y}(t)}_{j-l}] \quad (10.89)$$

For the special case of a delta-correlated excitation, we will write, as in Section 9.7, the jth-order cumulant array for the excitation as $\kappa_j^{\otimes}[Q(t_1),\cdots,Q(t_j)] = (2\pi)^{j-1}S_j(t_j)\delta(t_1 - t_j)\cdots\delta(t_{j-1} - t_j)$, which allows Eq. 10.89 to be rewritten as

$$\frac{d}{dt}\kappa_j^{\otimes}[\vec{Y}(t),\cdots,\vec{Y}(t)] + \sum_{l=1}^{j}\kappa_j^{\otimes}[\underbrace{\vec{Y}(t),\cdots,\vec{Y}(t)}_{l-1},\vec{g}[Y(t)],\underbrace{\vec{Y}(t),\cdots,\vec{Y}(t)}_{j-l}] = (2\pi)^{j-1}\vec{S}_j(t) \quad (10.90)$$

The state-space cumulant relationships of Eqs. 10.87 and 10.89 give exact descriptions of the general situation in which the excitation of the nonlinear system is not delta-correlated, but they are less useful than the expressions in Eqs. 10.88 and 10.90 for delta-correlated excitations. This, of course, is because of the difficulty in evaluating the terms on the right-hand side of Eq. 10.89 for a general stochastic excitation. If the excitation can be modeled as a filtered delta-correlated process, then useful cumulant equations can be formulated by using the technique of analyzing the filter and the nonlinear system of interest as one composite system, as was mentioned in Chapter 9 for linear problems.

The difficulty in using Eq. 10.86, 10.88, or 10.90 to estimate response levels is in finding adequate approximations for cumulants involving $\vec{Y}(t)$ and the nonlinear function $\vec{g}[\vec{Y}(t)]$. Not surprisingly, if $\{\vec{Q}(t)\}$ is a Gaussian process then it is common to approximate $\{\vec{Y}(t)\}$ as also being Gaussian. This is called *Gaussian closure*.[2] With this assumption, the mean and cross-covariance terms involving $\vec{g}[\vec{Y}(t)]$ can be written as functions of the mean and covariance of $\vec{Y}(t)$, so it is possible to find solutions to the equations. In fact, these mean and covariance solutions obtained by considering $\{\vec{Y}(t)\}$ to be Gaussian are identical to those obtained by statistical linearization with the same assumption about the distribution of $\{\vec{Y}(t)\}$. For example, in state-space analysis the value of $E(\vec{g}[\vec{Y}(t)])$ is taken as the coefficient in Eq. 10.86; in statistical linearization it is used as the value of the vector $\vec{b}(t)$ of linearization coefficients, but using this

[2]The term *closure* comes from consideration of systems with polynomial nonlinearities, which are discussed in the following paragraphs.

value for $\vec{b}(t)$ makes Eq. 10.53 identical to Eq. 10.86. Similarly, the $\mathbf{K}_{gY}$ and $\mathbf{K}_{Yg}$ matrices can be taken either as the coefficients in the state-space formulation of Eq. 10.88 or as the values of $\mathbf{B}(t)\,\mathbf{K}_{YY}(t,t)$ and $\mathbf{K}_{YY}(t,t)\,\mathbf{B}^T(t)$, which are the corresponding coefficients in Eq. 10.55 for statistical linearization. Thus, state-space cumulant analysis and statistical linearization are completely equivalent when both are applied with the Gaussian assumption. This is despite the fact that there seem to be more assumptions involved in the statistical linearization approach.

In many nonlinear problems it is possible to write, at least approximately, the nonlinear function as a polynomial of the state variables. For example, one can do this by writing a power series expansion for the nonlinearity. Such a polynomial nonlinearity leads to an infinite set of coupled state-space equations governing the response cumulants or moments. To illustrate this situation in the simplest possible situation, we will consider the moment equations when the state vector of Eq. 10.85 is a scalar and the nonlinear function $g[Y(t)]$ is a polynomial. The equation of motion is then

$$\dot{Y}(t)+\sum_{k=0}^{K} a_k Y^k(t)=Q(t) \tag{10.91}$$

and the moment equations can be found by multiplying this equation by powers of $Y(t)$ and then taking the expected value. In particular, multiplying by $Y^{j-1}(t)$ gives

$$E[Y^{j-1}(t)\,\dot{Y}(t)]+\sum_{k=0}^{K} a_k E[Y^{k+j-1}(t)]=E[Q(t)\,Y^{j-1}(t)]$$

which can be rewritten as

$$\frac{1}{j}\frac{d}{dt}E[Y^j(t)]+\sum_{k=0}^{K} a_k E[Y^{k+j-1}(t)]=E[Q(t)\,Y^{j-1}(t)] \tag{10.92}$$

The derivative term in Eq. 10.92 demonstrates that this equation governs the evolution of the jth moment of $Y(t)$. The summation, though, includes moments up to the order $(K+j-1)$, where K is the highest-order term included in the polynomial of Eq. 10.91. Taking $j=1$ shows that the behavior of the mean-value term $E[\vec{Y}(t)]$ depends on the values of $E[Y^2(t)],\cdots,E[Y^K(t)]$.

Similarly, using $j=2$ shows that the evolution of $E[Y^2(t)]$ depends on the values of $E[Y(t)],\cdots,E[Y^{K+1}(t)]$, and so forth. Because the evolution of any moment term depends on the values of still higher moment terms, the set of coupled equations continues without limit. Methods for truncating this infinite set of equations are commonly called *closure methods*.

Gaussian closure is the most common technique for truncating the set of equations that result from consideration of a polynomial nonlinearity. For the scalar situation of Eq. 10.92, this involves assuming that any moment of $Y(t)$ can be written in terms of the mean and variance in the same way as for a Gaussian random variable. In particular, it is assumed that

$$E[Y^m(t)] = \sum_{r=0,2,4}^{[m]} \frac{m!}{r!(m-r)!}(1)(3)\cdots(r-1)\mu_Y^{m-r}(t)\sigma_Y^r(t) \qquad (10.93)$$

in which $[m]$ denotes either m or $m-1$, depending on which is an even integer (see Eq. 3.10 and Example 3.8). With this assumption, $\mu_Y(t)$ and $\sigma_Y(t)$ are the only unknowns in Eq. 10.92. Furthermore, by choosing two particular values of j in this equation, one can obtain two simultaneous equations from which, in principle, one can solve for the values of $\mu_Y(t)$ and $\sigma_Y(t)$. One must expect that the values obtained for the unknowns will depend on the choices of j, because the other moment equations will generally not be satisfied. The usual procedure in applying such a closure method is to use the lowest-order moment equations that will serve the purpose. Thus, Gaussian closure for the scalar problem involves using Eq. 10.92 with $j=1$ and with $j=2$, giving the equations that directly govern the evolution of $\mu_Y(t)$ and $\sigma_Y(t)$. If symmetry dictates that $\mu_Y(t)$ is zero for the problem of interest, then the equation for $j=1$ is trivial and the equation for $j=2$ gives the evolution of the variance.

It may be noted that Gaussian closure of the state-space moment equations with a Gaussian excitation gives exactly the same results as simply assuming that $\{Y(t)\}$ is a Gaussian process, even though the formulation of the assumption is slightly different. In particular, Gaussian closure is often presented as an assumption that certain moments of $Y(t)$ are the same as they would be if $Y(t)$ were Gaussian, rather than as an assumption that $Y(t)$ truly is Gaussian, but the results are the same. Furthermore, we have demonstrated that the nonlinear state-space moment or cumulant equations are exactly the same as those for the system obtained by statistical linearization when $\{Y(t)\}$ is assumed to be Gaussian in both approaches. Thus, we see that the most straightforward implementations of

Gaussian closure and statistical linearization make the two methods exactly equivalent when the excitation is Gaussian.

A more general method to close the infinite set of moment equations for a problem with a polynomial nonlinearity is to assume that all cumulants of $Y(t)$ beyond some specified order are so small that they can be neglected (Crandall, 1980; Wu and Lin, 1984). If one chooses to consider only the first J cumulants of $Y(t)$, for example, then it is necessary to consider only the moment expressions from Eq. 10.92 for $j=1,\cdots,J$, because there are only J unknowns to be evaluated. When J is chosen to be 2, this general cumulant neglect closure approach reduces to Gaussian closure. Cumulant neglect closure with $J>2$ generally provides better approximations of the mean and the variance of the response and also gives at least crude approximations of other cumulants up to order J. The method does not necessarily converge as J is made larger, however. The reason seems to be related to the fact that it is not possible for a stochastic process to have zero values for all cumulants beyond order J unless $J=1$ or 2.[3] Gaussian closure can be viewed as approximating $\{Y(t)\}$ by a Gaussian process, but it is not possible to view any higher-order cumulant neglect closure scheme as corresponding to use of such a substitute process.

A more promising closure method is called *quasi-moment* or *Hermite-moment closure.* It is based on approximating the probability distribution of the dynamic response with a truncated type A Gram-Charlier series. In principle, it is possible to exactly represent any probability distribution for a random variable Y in the form

$$p_Y(u) = p_{YG}(u)\left(1+\sum_{l=3}^{\infty} a_l H_l\left[\frac{u-\mu_Y}{\sigma_Y}\right]\right) \tag{10.94}$$

in which $p_{YG}(u)$ is a Gaussian approximation of the probability density for Y

$$p_{YG}(u) = \frac{1}{(2\pi)^{1/2}\sigma_Y} \exp\left(-\frac{(u-\mu_Y)^2}{2\sigma_X^2}\right) \tag{10.95}$$

The $H_l(\cdot)$ term is the lth order Hermite polynomial, which can be written as

---

[3]Kendall and Stuart (1977) attribute this result to Marcinkiewicz in 1938.

$$H_l(u) = (-1)^l e^{u^2/2} \frac{d^l}{du^l} e^{-u^2/2}$$

and the a_l coefficients are

$$a_l = \frac{1}{l!} E\left(H_l\left[\frac{Y-\mu_Y}{\sigma_Y}\right]\right)$$

in which the expectation term is sometimes called the lth *Hermite moment* of Y. Thus, truncating the series to an upper limit of L amounts to neglecting all Hermite moments above order L. Because the lth quasi-moment of Y is simply $l!a_l(\sigma_X)^l$, this truncation is identical to quasi-moment neglect closure.

Note that there are L unknown values in $p_Y(u)$: μ_Y, σ_Y, $a_3,\cdots,a_L$. These values can be found for a nonlinear dynamics problem from simultaneous solution of a set of L moment equations for the system. In particular, for a scalar equation of motion of $\dot{Y}(t)+g[Y(t)]=Q(t)$, the jth moment equation corresponding to Eq. 10.89 is

$$\frac{d}{dt}E[Y^j(t)] + jE\left(Y^{j-1}(t)\, g[Y(t)]\right) = jE[Y^{j-1}(t)\, Q(t)]$$

in which the right-hand side is identical to the $\psi(t)$ in Eq. 9.57 for the corresponding linear problem. Using a truncated form of Eq. 10.94 for the probability density of $Y(t)$ converts each of the expectations on the left-hand side to a series of expectations in terms of the Gaussian form of Eq. 10.95. An interesting and valuable property of this system of moment equations is that they are linear in the $a_3,\cdots,a_L$ unknown Hermite coefficients, even though they are nonlinear in the unknown μ_Y and σ_Y terms. This has been used to give simplified solution procedures for the L equations. Cacciola et al. (2003) used simulated time histories of the dynamic response to estimate μ_Y and σ_Y, then solved the linear equations for the other unknowns, and Falsone and Rundo Sotera (2003) used an iterative approach in which estimated values of μ_Y and σ_Y were updated after solving for the other unknowns. Falsone and Rundo Sotera specifically used this closure method for the same non-Gaussian problems that they investigated by equivalent linearization. They achieved significantly improved results by using the closure approach, which included non-Gaussian effects due to the nonlinearity as well as due to the excitation.

An undesirable aspect of the truncated type-A Gram-Charlier approximation is that it can give negative values for the approximation of the probability density function. Also, it may be necessary to include 10 to 20 terms in the summation to obtain a good approximation of a significantly non-Gaussian distribution. One method of avoiding these difficulties is to use a similar approach with a truncated type-C Gram-Charlier series in which

$$p_Y(u) \approx \exp\left(\sum_{l=0}^{L} b_l H_l\left[\frac{u-\mu_Y}{\sigma_Y}\right]\right)$$

Muscolino et al. (1997) found that good approximations could be achieved for quite small values of L in this approach; however, evaluation of the unknown coefficients is more difficult than for Eq. 10.94. Also it is necessary that L be even and that $b_L < 0$ in order for the approximate probability density to converge to zero for $|u| \rightarrow \infty$. Note that the type-C series amounts to a polynomial approximation of $\log[p_Y(u)]$, which is an approach that has been investigated in a somewhat different way by Di Paola et al. (1995). Development of accurate and efficient closure methods is an area of ongoing research.

The generalization to a vector process of the polynomial nonlinearity of Eq. 10.91 has the form

$$\dot{\vec{Y}}(t) + \sum_{k=0}^{K} \mathbf{A}_k \vec{Y}^{[k]}(t) = \vec{Q}(t)$$

in which $\vec{Y}^{[k]}(t)$ is a Kronecker power of the vector and $\mathbf{A}_k$ is a matrix of dimension $n_Y \times n_Y^k$. Equation 10.90 then becomes

$$\frac{d}{dt}\kappa_j^{\otimes}[\vec{Y}(t),\cdots,\vec{Y}(t)] + \sum_{l=1}^{j} \kappa_j^{\otimes}[\underbrace{\vec{Y}(t),\cdots,\vec{Y}(t)}_{l-1}, \sum_{k=0}^{K} \mathbf{A}_k \vec{Y}^{[k]}(t), \underbrace{\vec{Y}(t),\cdots,\vec{Y}(t)}_{j-l}] = (2\pi)^{j-1}\vec{S}_j(t)$$

which can be rewritten (see Eqs. 9.3 and 9.42) as

$$\frac{d}{dt}\kappa_j^{\otimes}[\vec{Y}(t),\cdots,\vec{Y}(t)] + \sum_{l=1}^{j}\sum_{k=0}^{K}(I^{[l-1]} \otimes \mathbf{A}_k \otimes I^{[j-l]}) \times \kappa_j^{\otimes}[\underbrace{\vec{Y}(t),\cdots,\vec{Y}(t)}_{l-1}, \vec{Y}^{[k]}(t), \underbrace{\vec{Y}(t),\cdots,\vec{Y}(t)}_{j-l}] = (2\pi)^{j-1}\vec{S}_j(t) \tag{10.96}$$

The corresponding moment equation is similar but has a more complicated right-hand side:

$$\frac{d}{dt}E[\vec{Y}^{[j]}(t)]+\sum_{l=1}^{j}\sum_{k=0}^{K}(I^{[l-1]}\otimes \mathbf{A}_k\otimes I^{[j-l]})\,E[\vec{Y}^{[k+j-1]}(t)]=$$
$$\sum_{l=1}^{j}E[\vec{Y}^{[l-1]}(t)\otimes Q(t)\otimes \vec{Y}^{[j-l]}(t)] \qquad (10.97)$$

The primary difficulty in using Eq. 10.96 concerns the fact that the Kronecker cumulant term on the left-hand side of the equation includes the Kronecker product $\vec{Y}^{[k]}(t)$, in addition to $\vec{Y}(t)$ terms. Use of cumulant neglect closure with Eq. 10.97 is somewhat more complicated, because the relationship between moments and cumulants becomes somewhat complicated for higher-order terms, in addition to the complexity of the right-hand side, even if the $\{Q(t)\}$ excitation is delta-correlated. Despite the difficulties, procedures have been developed for the implementation of cumulant neglect closure for both the moment analysis of Eq. 10.97 (Di Paola and Muscolino, 1990; Di Paola et al., 1992; Di Paola and Falsone, 1993) and the cumulant analysis of Eq. 10.96 (Papadimitriou and Lutes, 1994; Papadimitriou, 1995).

At least one other general approach has also been used to find approximate solutions to Eq. 10.90 or 10.96. This approach is based on the fact that one can often anticipate not only that certain state variables will have non-Gaussian distributions but also the general sense in which these distributions will differ from Gaussian. In particular, one can often predict that some variable will have a greater than Gaussian or a smaller than Gaussian probability of large excursions. In such a situation, it may be possible to assume a particular non-Gaussian form for the probability distribution of the state variable. One of the easiest ways to do this is to assume a Gaussian distribution for a new variable that is defined as a nonlinear function of the non-Gaussian state variable. This method was used by Iyengar and Dash (1978), who rewrote a state variable $Z(t)$ that was bounded by $\pm b$ as $Z(t)=(2b/\pi)\tan^{-1}[\hat{Z}(t)]$, and then assumed that $\hat{Z}(t)$ was Gaussian. In essence, this amounts to rewriting the equation of motion for the system in terms of a new state variable $\hat{Z}(t)$ in place of $Z(t)$, then using Gaussian closure. In fact, Iyengar and Dash referred to the technique as Gaussian closure, even though it can also be considered a method of non-Gaussian closure for the problem formulated in terms of the original state variables. In a somewhat similar vein, it has been shown (Lutes, 2000) that improved results for the bilinear hysteretic system of Example 10.10 can be obtained by modifying Caughey's procedure to assume that it is the hysteretic restoring force $g[\{X(t)\}]$, rather than $X(t)$ itself, that is a narrowband, Gaussian process. It should be noted, though, that use of the narrowband approximation limits the applicability of this particular technique, as it did in Caughey's original method. The general method of assuming a Gaussian distribution for some auxiliary process does seem to have considerable potential for use, and it avoids a major difficulty of cumulant neglect and quasi-moment neglect closure schemes. That is, the new state variable, as illustrated by $\hat{Z}(t)$ or $g[\{X(t)\}]$, is a physically meaningful quantity, whereas a random variable with only J nonzero cumulants or quasi-moments generally is not physically meaningful. No general procedures have as yet been formulated, though, for achieving non-Gaussian closure in this way.

****

Example 10.13: Let $\{X(t)\}$ denote the response of the system of Examples 10.1 and 10.5 with an equation of motion

$$\dot{X}(t) + k_1 X(t) + k_3 X^3(t) = F(t)$$

in which $\{F(t)\}$ is a mean-zero, stationary, delta-correlated process with autospectral density S_0. Write the state-space equation for the second moment of the system response, verify that the result agrees with that of statistical linearization, and apply Gaussian closure to simplify the equation.

This scalar system is described by Eq. 10.85 with $\vec{Y}(t) = X(t)$, $g[\vec{Y}(t)] = k_1 X(t) + k_3 X^3(t)$, and $\vec{Q}(t) = F(t)$. Because the system is symmetric and the excitation has a mean value of zero, the response gives both $X(t)$ and $g[X(t)]$ as also being mean-zero. The state-space covariance relationship of Eq. 10.88 can then be written as

$$\frac{d}{dt} E[X^2(t)] + 2E\big(X(t)\, g[X(t)]\big) = 2\pi S_0$$

or

$$\frac{d}{dt} E[X^2(t)] + 2k_1 E[X^2(t)] + 2k_3\, E[X^4(t)] = 2\pi S_0$$

One may note that this equation is also identical to the result of simply multiplying $X(t)$ times each term of the equation of motion and then taking the expected value, because

$$\frac{d}{dt} E[X^2(t)] = 2E[X(t)\, \dot{X}(t)]$$

and $E[F(t)\, X(t)] = \pi S_0$. One always has the option of deriving state-space moment or cumulant equations by this direct procedure. The advantage of Eq. 10.88 is that it may help clarify exactly which expressions are needed for the evaluation of given moments or cumulants in a more complicated problem.

Recall that in Example 10.5 we introduced a term b_1 that can be written as

$$b_1 = \frac{E\big(X(t)\, g[X(t)]\big)}{E[X^2(t)]} = k_1 + k_3 \frac{E[X^4(t)]}{E[X^2(t)]}$$

If we now introduce that term into our current analysis of the nonlinear system, the moment equation takes the form

$$\frac{d}{dt} E[X^2(t)] + 2b_1 E[X^2(t)] = 2\pi S_0$$

This relationship, though, is identical to the second-moment state-space equation for the linear system described by $\dot{X}(t) + a_1 X(t) = F(t)$, which is the linearized model in Example 10.5. Inasmuch as no approximations have been used in deriving this state-space relationship for the nonlinear problem, it exactly describes the second moment of the response. Thus, this example confirms the fact that statistical linearization also gave an exact relationship for the second moment of the nonlinear response. The difficulty in practice, of course, is that one cannot exactly evaluate the b_1 linearization coefficient, because it involves another moment of the response. The approximate stationary solution based on a Gaussian assumption was given in Example 10.5.

To implement Gaussian closure, we now evaluate all the terms in the state-space moment equations in terms of the unknown variance (or second moment) of the response. We do this by assuming that all expected values are the same as they would be if $X(t)$ were Gaussian. In our state-space moment equations for the current problem, $E[X^4(t)]$ is the only term that is not already written in terms of the second moment of the response. Thus, we use the Gaussian assumption to obtain $E[X^4(t)] = 3\sigma_X^4 = 3(E[X^2(t)])^2$. With this result, the state-space moment equation becomes

$$\frac{d}{dt} E[X^2(t)] + 2k_1 E[X^2(t)] + 6k_3 \Big(E[X^2(t)]\Big)^2 = 2\pi S_0$$

One could numerically solve this nonlinear differential equation to find an approximation of the nonstationary $E[X^2(t)]$. For the special case of stationary response, the equation becomes algebraic and the solution is as found in Example 10.5.

****

Example 10.14: Let $\{X(t)\}$ denote the response of the oscillator of Example 10.8 with an equation of motion

$$m\, \ddot{X}(t) + c_1 \dot{X}^3(t) + c_2\, X^2(t)\, \dot{X}(t) + k\, X(t) = F(t)$$

in which $\{F(t)\}$ is a mean-zero, delta-correlated process with autospectral density S_0. Write the general state-space equations for the second moments of the system response, verify that the results agree with those of nonstationary statistical linearization, and apply Gaussian closure to simplify the equations.

Using the usual state vector of $\vec{Y}(t)=[X(t),\dot{X}(t)]^T$, the component equations of motion are

$$\dot{Y}_1(t)-Y_2(t)=0$$

and

$$m\dot{Y}_2(t)+c_1Y_2^3(t)+c_2Y_1^2(t)Y_2(t)+kY_1(t)=F(t)$$

This agrees with Eq. 10.85 if we write the nonlinear restoring force as

$$\vec{g}[\vec{Y}(t)]=\begin{pmatrix}-Y_2(t)\\(c_1/m)Y_2^3(t)+(c_2/m)Y_1^2(t)Y_2(t)+(k/m)Y_1(t)\end{pmatrix}$$

and the excitation as $\vec{Q}(t)=[0,F(t)/m]^T$. Because the mean values are equal to zero, the cross-covariances are the same as cross-products and the $\mathbf{K}_{Yg}(t,t)$ matrix in Eq. 10.88 is

$$\mathbf{K}_{Yg}(t,t)=\begin{pmatrix}-E[Y_1(t)Y_2(t)] & \frac{c_1}{m}E[Y_1(t)Y_2^3(t)]+\frac{c_2}{m}E[Y_1^3(t)Y_2(t)]+\frac{k}{m}E[Y_1^2(t)]\\-E[Y_2^2(t)] & \frac{c_1}{m}E[Y_2^4(t)]+\frac{c_2}{m}E[Y_1^2(t)Y_2^2(t)]+\frac{k}{m}E[Y_1(t)Y_2(t)]\end{pmatrix}$$

Of course, $\mathbf{K}_{gY}(t,t)$ is the transpose of this expression. The $\mathbf{S}_0(t)$ matrix is simply

$$\mathbf{S}_0(t)=\begin{pmatrix}0 & 0\\0 & S_0/m^2\end{pmatrix}$$

Thus, the three distinct component equations from Eq. 10.88 can be written as

$$\frac{d}{dt}E[X^2(t)]-2E[X(t)\dot{X}(t)]=0$$

$$\frac{d}{dt}E[X(t)\dot{X}(t)]+\frac{c_1}{m}E[X(t)\dot{X}^3(t)]+\frac{c_2}{m}E[X^3(t)\dot{X}(t)]+\frac{k}{m}E[X^2(t)]-E[\dot{X}^2(t)]=0$$

and

$$\frac{d}{dt}E[\dot{X}^2(t)]+2\frac{c_1}{m}E[\dot{X}^4(t)]+2\frac{c_2}{m}E[X^2(t)\dot{X}^2(t)]+2\frac{k}{m}E[X(t)\dot{X}(t)]=\frac{2\pi S_0}{m^2}$$

Note that the first of these three state-space moment equations is trivial, because it states only a well-known property of the derivative. Furthermore, the second and third equations could have been derived in an alternative direct manner. In particular, we can obtain two related equations by multiplying the original equation of motion by $X(t)$ and $\dot{X}(t)$, respectively, and then taking the expected value of the products. These equations are

$$mE[X(t)\ddot{X}(t)]+c_1E[X(t)\dot{X}^3(t)]+c_2E[X^3(t)\dot{X}(t)]+kE[X^2(t)]=0$$

and

$$mE[\dot{X}(t)\ddot{X}(t)]+c_1E[\dot{X}^4(t)]+c_2E[X^2(t)\dot{X}^2(t)]+kE[X(t)\dot{X}(t)]=\frac{2\pi S_0}{m}$$

Noting that

$$\frac{d}{dt}E[X(t)\,\dot{X}(t)] = E[X(t)\,\ddot{X}(t)] + E[\dot{X}^2(t)], \qquad \frac{d}{dt}E[\dot{X}^2(t)] = 2E[\dot{X}(t)\,\ddot{X}(t)]$$

allows these expressions to be converted into the two state-space moment equations already obtained. The right-hand sides of the equations, of course, have been simplified by using the relationships given in Section 9.4 for delta-correlated excitations.

As in Example 10.8, the linearized equation can be written as $m\,\ddot{X}(t) + b_2\dot{X}(t) + (b_1 + k)\,X(t) = F(t)$. The values of b_1 and b_2 for stationary response were obtained in Example 10.8, but more general results are required for comparison with the general nonstationary state-space equations. From Eq. 10.23, the general linearization relationship of $g[\vec{X}(t)] = c_1\dot{X}^3(t) + c_2\,X^2(t)\,\dot{X}(t)$ can be written as

$$\mathbf{K}_{XX}\,\vec{b} = \text{Cov}\Big(\vec{X}(t), g[\vec{X}(t)]\Big) = \begin{pmatrix} c_1\text{Cov}[X(t),\dot{X}^3(t)] + c_2\text{Cov}[X(t),X^2(t)\dot{X}(t)] \\ c_1\text{Cov}[\dot{X}(t),\dot{X}^3(t)] + c_2\text{Cov}[\dot{X}(t),X^2(t)\dot{X}(t)] \end{pmatrix}$$

Because the processes are mean-zero, though, the component equations can be written as

$$b_1 E[X^2(t)] + b_2 E[X(t)\,\dot{X}(t)] = c_1 E[X(t)\,\dot{X}^3(t)] + c_2 E[X^3(t)\,\dot{X}(t)]$$

and

$$b_1 E[X(t)\,\dot{X}(t)] + b_2 E[\dot{X}^2(t)] = c_1 E[\dot{X}^4(t)] + c_2 E[X^2(t)\,\dot{X}^2(t)]$$

The state-space equations for the linearized system are

$$\frac{d}{dt}E[X(t)\,\dot{X}(t)] - E[\dot{X}^2(t)] + \frac{b_2}{m}E[X(t)\,\dot{X}(t)] + \frac{b_1 + k}{m}E[X^2(t)] = 0$$

and

$$\frac{d}{dt}E[\dot{X}^2(t)] + 2\frac{b_2}{m}E[\dot{X}^2(t)] + 2\frac{b_1 + k}{m}E[X(t)\,\dot{X}(t)] = \frac{2\pi S_0}{m^2}$$

and substituting from the equations for b_1 and b_2 makes these equations identical to the component equations derived for the nonlinear system.

As in Example 10.13, we note that no approximations have been used in deriving these state-space relationships of the nonlinear problem, so they exactly describe the moments of the response. Thus, this example also confirms the fact that statistical linearization gives exact relationships for the second moments of the nonlinear response, but with the practical limitation that one cannot exactly evaluate the b_1 and b_2 linearization coefficients. The approximate stationary solution based on a Gaussian assumption was given in Example 10.8.

To obtain the Gaussian closure approximation of the nonstationary nonlinear state-space moment equations, we also use relationships that would be true if $X(t)$ and $\dot{X}(t)$ were mean-zero and jointly Gaussian. In particular, this gives

$$E[X(t)\,\dot{X}^3(t)] = 3E[X(t)\,\dot{X}(t)]\,E[\dot{X}^2(t)]$$

$$E[X^3(t)\,\dot{X}(t)] = 3E[X(t)\,\dot{X}(t)]\,E[X^2(t)]$$

$$E[\dot{X}^4(t)] = 3\Big(E[\dot{X}^2(t)]\Big)^2$$

and

$$E[X^2(t)\,\dot{X}^2(t)] = E[X^2(t)]\,E[\dot{X}^2(t)] + 2\Big(E[X(t)\,\dot{X}(t)]\Big)^2$$

Substituting these relationships gives the state-space equations as

$$\frac{d}{dt}E[X^2(t)] - 2E[X(t)\,\dot{X}(t)] = 0$$

$$\frac{d}{dt}E[X(t)\,\dot{X}(t)] + \frac{3c_1}{m}E[X(t)\,\dot{X}(t)]\,E[\dot{X}^2(t)] + \frac{3c_2}{m}E[X(t)\,\dot{X}(t)]\,E[X^2(t)] + \frac{k}{m}E[X^2(t)] - E[\dot{X}^2(t)] = 0$$

and

$$\frac{d}{dt}E[\dot{X}^2(t)] + \frac{6c_1}{m}\Big(E[\dot{X}^2(t)]\Big)^2 + \frac{2c_2}{m}E[X^2(t)]\,E[\dot{X}^2(t)] + \frac{4c_2}{m}\Big(E[X(t)\,\dot{X}(t)]\Big)^2 + 2\frac{k}{m}E[X(t)\,\dot{X}(t)] = \frac{2\pi S_0}{m^2}$$

In principle, one can now solve these three simultaneous differential equations for the three unknown second-moment terms: $E[X^2(t)]$, $E[\dot{X}^2(t)]$, and $E[X(t)\,\dot{X}(t)]$. The problem is made somewhat complicated, of course, by the fact that two of the equations are nonlinear.

Restricting attention to the special case of stationary response simplifies the situation by making all the derivative terms in the equations be zero and gives $E[X(t)\,\dot{X}(t)] = 0$. The values of $E[X^2(t)]$ and $E[\dot{X}^2(t)]$ are then exactly as found in Example 10.8.

****

Example 10.15: Consider the nonlinear oscillator described by

$$\ddot{X}(t) + c\,\dot{X}(t)\exp[-a\,X^2(t)] + k\,X^3(t)\,\dot{X}^2(t) = F(t)$$

in which $\{F(t)\}$ is a mean-zero, stationary, delta-correlated process with autospectral density S_0. Write the general state-space moment equations for the system, and find the approximate stationary solution that results from Gaussian closure.

Using the alternative approach presented in Example 10.14, we can obtain appropriate state-space moment equations by multiplying the equation of motion by the components of the $\vec{Y}(t) \equiv [X(t), \dot{X}(t)]^T$ state vector, then taking the expected values. The resulting equations are

$$E[X(t)\,\ddot{X}(t)] + c\,E\Big(X(t)\,\dot{X}(t)\exp[-a\,X^2(t)]\Big) + k\,E[X^4(t)\,\dot{X}^2(t)] = E[X(t)\,F(t)] = 0$$

and

$$E[\dot{X}(t)\,\ddot{X}(t)]+c\,E\Big(\dot{X}^2(t)\exp[-a\,X^2(t)]\Big)+k\,E[X^3(t)\,\dot{X}^3(t)]=E[\dot{X}(t)\,F(t)]=\pi\,S_0$$

For stationary response, though, we can use the simplifications that $E[\dot{X}(t)\,\ddot{X}(t)]=0$, $E[X(t)\,\ddot{X}(t)]=-E[\dot{X}^2(t)]$, and $E(X(t)\dot{X}(t)\exp[-a\,X^2(t)])=0$. The last of these three relationships follows from the fact that the term has the form $E(\dot{X}(t)\,f'[X(t)])$, which is exactly the derivative with respect to t of $E(f[X(t)])$. Using these conditions gives the exact state-space equations for stationary response as

$$-E[\dot{X}^2(t)]+k\,E[X^4(t)\,\dot{X}^2(t)]=0$$

and

$$c\,E\Big(\dot{X}^2(t)\exp[-a\,X^2(t)]\Big)+k\,E[X^3(t)\,\dot{X}^3(t)]=\pi\,S_0$$

Now we introduce the Gaussian assumption to further simplify the equations. In particular, we presume that $X(t)$ and $\dot{X}(t)$ are jointly Gaussian, which requires that they also be independent inasmuch as they are uncorrelated. This gives

$$E[X^4(t)\,\dot{X}^2(t)]=E[X^4(t)]\,E[\dot{X}^2(t)]=3\sigma_X^4\sigma_{\dot{X}}^2$$

$$E[X^3(t)\,\dot{X}^3(t)]=E[X^3(t)]\,E[\dot{X}^3(t)]=0$$

and

$$E\Big(\dot{X}^2(t)\exp[-a\,X^2(t)]\Big)=E[\dot{X}^2(t)]\,E\Big(\exp[-a\,X^2(t)]\Big)$$

and the Gaussian distribution allows the final term to be evaluated as

$$E\Big(\exp[-a\,X^2(t)]\Big)=\frac{1}{(2\pi)^{1/2}\sigma_X}\int_{-\infty}^{\infty}\exp\left(-u^2\left[\frac{2\sigma_X^2\,a^2+1}{2\sigma_X^2}\right]\right)du$$

$$=\frac{1}{(2\pi)^{1/2}(2\sigma_X^2\,a^2+1)^{1/2}}\int_{-\infty}^{\infty}e^{-w^2/2}\,dw=\frac{1}{(2\sigma_X^2\,a^2+1)^{1/2}}$$

Thus, the state-space equations become

$$-\sigma_{\dot{X}}^2+3k\,\sigma_X^4\,\sigma_{\dot{X}}^2=0$$

and

$$\frac{c\,\sigma_{\dot{X}}^2}{(2\sigma_X^2\,a^2+1)^{1/2}}=\pi\,S_0$$

From the first of these equations, we determine that $\sigma_X^2=(3k)^{-1/2}$; the second equation then gives

$$\sigma_{\dot{X}}^2=\frac{\pi\,S_0}{c}(2\sigma_X^2\,a^2+1)^{1/2}=\frac{\pi\,S_0}{c}\left(\frac{2a^2}{(3k)^{1/2}}+1\right)^{1/2}$$

****

Exercises

****

Fokker-Planck Equation

****

10.1 Consider the response of the nonlinear system governed by the first-order differential equation

$$\dot{X}(t) + a \frac{X(t)}{|X(t)|} = F(t)$$

in which $\{F(t)\}$ is a mean-zero, Gaussian, white noise process with an autospectral density of S_0.

(a) Give the Fokker-Planck equation governing the probability density of the first-order Markov process $\{X(t)\}$, and evaluate the two nonzero coefficients in that equation:

$$C^{(1)}(u) = \lim_{\Delta t \to 0} \frac{1}{\Delta t} E[\Delta X | X(t) = u], \quad C^{(2)}(u) = \lim_{\Delta t \to 0} \frac{1}{\Delta t} E[(\Delta X)^2 | X(t) = u]$$

(b) Verify that the Fokker-Planck equation is satisfied by the stationary probability density function

$$p_{X(t)}(u) = A \exp\left(-\frac{a|u|}{\pi S_0} \right)$$

and evaluate A for this density function.

(c) Evaluate the variance and the kurtosis of $X(t)$ for this stationary response.

****

10.2 Consider the response of the nonlinear system governed by the first-order differential equation

$$\dot{X}(t) + a\, X^5(t) = F(t)$$

in which $\{F(t)\}$ is a mean-zero, Gaussian, white noise process with an autospectral density of S_0.

(a) Give the Fokker-Planck equation governing the probability density of the first-order Markov process $\{X(t)\}$, and evaluate the two nonzero coefficients in that equation: $C^{(1)}(u)$ and $C^{(2)}(u)$.

(b) Verify that the Fokker-Planck equation is satisfied by the stationary probability density function

$$p_{X(t)}(u) = A \exp\left(-\frac{a\, u^6}{6\pi S_0} \right)$$

and evaluate A for this density function.

(c) Evaluate the variance and the kurtosis of $X(t)$ for this stationary response.

****

10.3 Consider the response of the nonlinear system governed by the first-order differential equation

$$\dot{X}(t)+k_1 X(t)+k_5 X^5(t)=F(t)$$

in which $\{F(t)\}$ is a mean-zero, Gaussian, white noise process with an autospectral density of S_0.

(a) Give the Fokker-Planck equation governing the probability density of the first-order Markov process $\{X(t)\}$, and evaluate the two nonzero coefficients in that equation: $C^{(1)}(u)$ and $C^{(2)}(u)$.

(b) Verify that the Fokker-Planck equation is satisfied by the stationary probability density function

$$p_{X(t)}(u)=A\exp\left(-\frac{1}{\pi S_0}\left[k_1\frac{u^2}{2}+k_5\frac{u^6}{6}\right]\right)$$

in which A is a constant chosen to make the integral of the expression with respect to u be unity. [You need not evaluate A.]

****

10.4 Consider the nonlinear system governed by the second-order differential equation

$$m\ddot{X}(t)+c\,\dot{X}(t)+k\frac{X(t)}{|X(t)|}=F(t)$$

in which $\{F(t)\}$ is a mean-zero, Gaussian, white noise process with an autospectral density of S_0.

(a) Using the state vector $\vec{Y}(t)=[X(t),\dot{X}(t)]^T$, give the Fokker-Planck equation governing the probability density $p_{\vec{Y}(t)}(\vec{u})$ for the vector Markov process $\{\vec{Y}(t)\}$ and evaluate the first-order and second-order coefficients in that equation:

$$C_1^{(1)}(\vec{u})=\lim_{\Delta t\to 0}\frac{1}{\Delta t}E[\Delta Y_1|\vec{Y}(t)=\vec{u}],\quad C^{(2,0)}(\vec{u})=\lim_{\Delta t\to 0}\frac{1}{\Delta t}E[(\Delta Y_1)^2|\vec{Y}(t)=\vec{u}]$$

$$C_2^{(1)}(\vec{u})=\lim_{\Delta t\to 0}\frac{1}{\Delta t}E[\Delta Y_2|\vec{Y}(t)=\vec{u}],\quad C^{(0,2)}(\vec{u})=\lim_{\Delta t\to 0}\frac{1}{\Delta t}E[(\Delta Y_2)^2|\vec{Y}(t)=\vec{u}]$$

$$C^{(1,1)}(\vec{u})=\lim_{\Delta t\to 0}\frac{1}{\Delta t}E[\Delta Y_1\Delta Y_2|\vec{Y}(t)=\vec{u}]$$

(b) Verify that the Fokker-Planck equation has a stationary solution of

$$p_{\vec{Y}(t)}(\vec{u})=A\exp\left(-\frac{c}{\pi S_0}\left[k|u_1|+\frac{m u_2^2}{2}\right]\right)$$

and evaluate A in that expression.

(c) Find stationary variance values for $X(t)$ and $\dot{X}(t)$.

(d) Find stationary values of the kurtosis of $X(t)$ and $\dot{X}(t)$.

****

10.5 Consider the nonlinear system governed by the second-order differential equation

$$m\ddot{X}(t) + c\,\dot{X}(t) + k\,|X(t)|\,X(t) = F(t)$$

in which $\{F(t)\}$ is a mean-zero, Gaussian, white noise process with an autospectral density of S_0.

(a) Using the state vector $\vec{Y}(t) = [X(t), \dot{X}(t)]^T$, give the Fokker-Planck equation governing the probability density $p_{\vec{Y}(t)}(\vec{u})$ for the vector Markov process $\{\vec{Y}(t)\}$, and evaluate the first-order and second-order coefficients in that equation.

(b) Verify that the Fokker-Planck equation has a stationary solution of

$$p_{\vec{Y}(t)}(\vec{u}) = A \exp\left(-\frac{c}{\pi S_0}\left[\frac{k\,|u_1^3|}{3} + \frac{m\,u_2^2}{2}\right]\right)$$

and evaluate A in that expression.

(c) Find stationary variance values for $X(t)$ and $\dot{X}(t)$.

(d) Find stationary values of the kurtosis of $X(t)$ and $\dot{X}(t)$.

****

10.6 Consider the nonlinear system governed by the second-order differential equation

$$m\ddot{X}(t) + c_2\,|\dot{X}(t)|\,\dot{X}(t) + k_1 X(t) + k_5 X^5(t) = F(t)$$

in which $\{F(t)\}$ is a mean-zero, Gaussian, white noise process with an autospectral density of S_0. Using the state vector $\vec{Y}(t) = [X(t), \dot{X}(t)]^T$, give the Fokker-Planck equation governing the probability density $p_{\vec{Y}(t)}(\vec{u})$ for the vector Markov process $\{\vec{Y}(t)\}$, and evaluate the first-order and second-order coefficients in that equation.

****

10.7 The Fokker-Planck equation for a certain vector Markov process $\{\vec{Y}(t)\}$ has the form

$$\frac{\partial}{\partial t} p_{\vec{Y}(t)}(\vec{u}) + u_2 \frac{\partial}{\partial u_1} p_{\vec{Y}(t)}(\vec{u}) - \left(2b\,u_1 u_2 + 3c\,u_1^2 u_2^2\right) p_{\vec{Y}(t)}(\vec{u}) -$$

$$\left(b\,u_1 u_2^2 + c\,u_1^2 u_2^3\right) \frac{\partial}{\partial u_2} p_{\vec{Y}(t)}(\vec{u}) = \pi S_0 \frac{\partial^2}{\partial u_2^2} p_{\vec{Y}(t)}(\vec{u})$$

Furthermore, this $\{\vec{Y}(t)\}$ process is the state vector $[X(t), \dot{X}(t)]^T$ for the solution of a differential equation $\ddot{X}(t) + g[X(t), \dot{X}(t)] = F(t)$ in which $\{F(t)\}$ is a mean-zero, Gaussian, white noise process with an autospectral density of S_0. Find the appropriate $g[X(t), \dot{X}(t)]$ function in this equation of motion.

****

Statistical Linearization

****

10.8 Let $\{X(t)\}$ denote the response of the system of Exercise 10.1 with an equation of motion

$$\dot{X}(t) + a\,\frac{X(t)}{|X(t)|} = F(t)$$

in which $\{F(t)\}$ is a mean-zero, Gaussian, white noise process with an autospectral density of S_0.

(a) Using statistical linearization, find an expression for the b_1 parameter of Eqs. 10.26 and 10.27.

(b) Evaluate b_1 in terms of σ_X^2 using the assumption that $\{X(t)\}$ is a stationary Gaussian process.

(c) Verify that this stationary value of b_1 agrees with the results of Eq. 10.24.

(d) Estimate the value of σ_X^2 for stationary response.

****

10.9 Let $\{X(t)\}$ denote the response of the system of Exercise 10.2 with an equation of motion

$$\dot{X}(t) + a\,X^5(t) = F(t)$$

in which $\{F(t)\}$ is a mean-zero, Gaussian, white noise process with an autospectral density of S_0.

(a) Using statistical linearization, find an expression for the b_1 parameter of Eqs. 10.26 and 10.27.

(b) Evaluate b_1 in terms of σ_X^2 using the assumption that $\{X(t)\}$ is a stationary Gaussian process.

(c) Verify that this stationary value of b_1 agrees with the results of Eq. 10.24.

(d) Estimate the value of σ_X^2 for stationary response.

****

10.10 Let $\{X(t)\}$ denote the response of the system of Exercise 10.3 with an equation of motion

$$\dot{X}(t) + k_1 X(t) + k_5 X^5(t) = F(t)$$

in which $\{F(t)\}$ is a mean-zero, Gaussian, white noise process with an autospectral density of S_0.

(a) Using statistical linearization, find an expression for the b_1 parameter of Eqs. 10.26 and 10.27.

(b) Evaluate b_1 in terms of σ_X^2 using the assumption that $\{X(t)\}$ is a stationary Gaussian process.

(c) Verify that this stationary value of b_1 agrees with the results of Eq. 10.24.

(d) Find a cubic algebraic equation that could be solved to obtain an estimate of the value of σ_X^2 for stationary response. [You need not solve this equation.]

****

10.11 Let $\{X(t)\}$ denote the response of the system of Exercise 10.4 with an equation of motion

$$m\ddot{X}(t)+c\dot{X}(t)+k\frac{X(t)}{|X(t)|}=F(t)$$

in which $\{F(t)\}$ is a mean-zero, Gaussian, white noise process with an autospectral density of S_0.

(a) Using statistical linearization, find expressions for the b_1 and b_2 parameters of Eqs. 10.31, 10.34, and 10.35.

(b) Evaluate b_1 and b_2 in terms of σ_X^2 and $\sigma_{\dot{X}}^2$ using the assumption that $\{X(t)\}$ is a stationary Gaussian process.

(c) Verify that these stationary values of b_1 and b_2 agree with the results of Eq. 10.24.

(d) Estimate the values of σ_X^2 and $\sigma_{\dot{X}}^2$ for stationary response.

****

10.12 Let $\{X(t)\}$ denote the response of the system of Exercise 10.5 with an equation of motion

$$m\ddot{X}(t)+c\dot{X}(t)+k|X(t)|X(t)=F(t)$$

in which $\{F(t)\}$ is a mean-zero, Gaussian, white noise process with an autospectral density of S_0.

(a) Using statistical linearization, find expressions for the b_1 and b_2 parameters of Eqs. 10.31, 10.34, and 10.35.

(b) Evaluate b_1 and b_2 in terms of σ_X^2 and $\sigma_{\dot{X}}^2$ using the assumption that $\{X(t)\}$ is a stationary Gaussian process.

(c) Verify that these stationary values of b_1 and b_2 agree with the results of Eq. 10.24.

(d) Estimate the values of σ_X^2 and $\sigma_{\dot{X}}^2$ for stationary response.

****

10.13 Let $\{X(t)\}$ denote the response of the system of Exercise 10.6 with

$$m\ddot{X}(t)+c_2|\dot{X}(t)|\dot{X}(t)+k_1X(t)+k_5X^5(t)=F(t)$$

in which $\{F(t)\}$ is a mean-zero, Gaussian, white noise process with autospectral density of S_0.

(a) Using statistical linearization, find expressions for the b_1 and b_2 parameters of Eqs. 10.31, 10.34, and 10.35.

(b) Evaluate b_1 and b_2 in terms of σ_X^2 and $\sigma_{\dot{X}}^2$ using the assumption that $\{X(t)\}$ is a stationary Gaussian process.

(c) Verify that the stationary values of b_1 and b_2 agree with the results of Eq. 10.24.

(d) Estimate the value of $\sigma_{\dot{X}}^2$ for stationary response, and find a cubic algebraic equation that could be solved to obtain a corresponding estimate of the value of σ_X^2. [You need not solve this equation.]

****

State-Space Moment Equations

****

10.14 Let $\{X(t)\}$ denote the response of the system of Exercises 10.1 and 10.8 with an equation of motion

$$\dot{X}(t) + a\,\frac{X(t)}{|X(t)|} = F(t)$$

in which $\{F(t)\}$ is a mean-zero, Gaussian, white noise process with an autospectral density of S_0.

(a) Derive an exact state-space equation for the second moment of the response.

(b) Verify that the state-space moment equation in part (a) is the same as that for the linearized system of Exercise 10.10.

(c) Use Gaussian closure to estimate the value of σ_X^2 for stationary response.

****

10.15 Let $\{X(t)\}$ denote the response of the system of Exercises 10.2 and 10.9 with an equation of motion

$$\dot{X}(t) + a\,X^5(t) = F(t)$$

in which $\{F(t)\}$ is a mean-zero, Gaussian, white noise process with an autospectral density of S_0.

(a) Derive an exact state-space equation for the second moment of the response.

(b) Verify that the state-space moment equation in part (a) is the same as that for the linearized system of Exercise 10.11.

(c) Use Gaussian closure to estimate the value of σ_X^2 for stationary response.

****

10.16 Let $\{X(t)\}$ denote the response of the system of Exercises 10.3 and 10.10 with an equation of motion

$$\dot{X}(t) + k_1 X(t) + k_5 X^5(t) = F(t)$$

in which $\{F(t)\}$ is a mean-zero, Gaussian, white noise process with an autospectral density of S_0.

(a) Derive an exact state-space equation for the second moment of the response.

(b) Verify that the state-space moment equation in part (a) is the same as that for the linearized system of Exercise 10.13.

(c) Use Gaussian closure to find a cubic algebraic equation that could be solved to obtain an estimate of the value of σ_X^2 for stationary response. [You need not solve this cubic equation.]

****

10.17 Let $\{X(t)\}$ denote the response of the system of Exercises 10.4 and 10.11 with an equation of motion

$$m\,\ddot{X}(t) + c\,\dot{X}(t) + k\,\frac{X(t)}{|X(t)|} = F(t)$$

in which $\{F(t)\}$ is a mean-zero, Gaussian, white noise process with an autospectral density of S_0.

(a) Derive exact state-space equations for the second moments of the response of the system.

(b) Verify that the state-space moment equations in part (a) are the same as those for the linearized system of Exercise 10.14.

(c) Use Gaussian closure to estimate the values of σ_X^2 and $\sigma_{\dot{X}}^2$ for stationary response.

****

10.18 Let $\{X(t)\}$ denote the response of the system of Exercises 10.5 and 10.12 with an equation of motion

$$m\ddot{X}(t) + c\,\dot{X}(t) + k\,|X(t)|\,X(t) = F(t)$$

in which $\{F(t)\}$ is a mean-zero, Gaussian, white noise process with an autospectral density of S_0.

(a) Derive exact state-space equations for the second moments of the response of the system.

(b) Verify that the state-space moment equations in part (a) are the same as those for the linearized system of Exercise 10.15.

(c) Use Gaussian closure to estimate the values of σ_X^2 and $\sigma_{\dot{X}}^2$ for stationary response.

****

10.19 Let $\{X(t)\}$ denote the response of the system of Exercises 10.6 and 10.13 with an equation of motion

$$m\ddot{X}(t) + c_2\,|\dot{X}(t)|\,\dot{X}(t) + k_1 X(t) + k_5 X^5(t) = F(t)$$

in which $\{F(t)\}$ is a mean-zero, Gaussian, white noise process with an autospectral density of S_0.

(a) Derive exact state-space equations for the second moments of the response of the system.

(b) Verify that the state-space moment equations in part (a) are the same as those for the linearized system of Exercise 10.16.

(c) Use Gaussian closure to estimate the value of $\sigma_{\dot{X}}^2$ for stationary response, and to find a cubic algebraic equation that could be solved to obtain an estimate of the corresponding value of σ_X^2. [You need not solve this cubic equation.]

****

Chapter 11
Failure Analysis

11.1 Modes of Failure

One can generally separate the various modes of failure of structural or mechanical systems into two broad categories. The first category can be called *first-passage failure*, because failure is anticipated the first time that stress or displacement exceeds some critical level. Brittle fracture and buckling are two modes of failure that can be approximated in this way. The other broad category of failure is *fatigue*, in which failure occurs due to an accumulation of damage, rather than due to first passage of a critical level. The intent in this chapter is to introduce some of the more common methods for estimating the likelihood of either type of failure when the dynamic response is a stochastic process.

Clearly the occurrence of failure typically is related to the occurrence of large stresses or displacements, so a stochastic analysis of failure requires study of the extreme values of a process. In fact, two different types of extreme problems are relevant. The two types of extremes might be classified as "local" and "global," although we mean global in a somewhat limited sense in this instance. As in Chapter 7, we will use the term *peak* to refer to the purely local extrema of a time history—the points where the first derivative of the time history is zero and the second derivative is negative. The more global sort of extreme problem that we will consider involves the extreme value of some stochastic process $\{X(t)\}$ during a fixed time interval, such as $0 \le t \le T$. Henceforth, we will use the term *extreme distribution* to refer only to the probabilities for this more global problem, and not for the probability distribution of peaks.

Clearly the prediction of first-passage failure is related to the occurrence of large global extrema, whereas the accumulation of fatigue damage is generally approximated as depending only on the sequence of local extrema of the stress or strain process. Note that we have not analyzed the probability distribution of either local or global extrema in the preceding chapters; hence we must begin with this analysis here. We will develop exact expressions describing the peak distribution for some problems, but we will generally not be able to obtain such expressions for the joint probability distribution of all the peaks and valleys of a process, as would be desirable for fatigue analysis. We will also consider approximations for the extreme probability distribution, because it is generally not feasible to find an exact expression for it either.

We will not consider dynamic equations of motion in this chapter. That is, we will assume that the failure analysis can be uncoupled from the dynamic analysis, first using methods from Chapters 6–10 to find the characteristics of the response process $\{X(t)\}$, then performing a failure analysis based on those characteristics. This, of course, is an approximation, because it neglects the possibility that the system dynamics may be influenced by an occurrence of a large extremum or an accumulation of damage. It should be kept in mind, though, that failure generally only means unsatisfactory performance. Thus, we might consider a structure to have failed in the first-passage sense if the displacement or stress reached a value that we consider unsafe, even if no fracture or buckling has yet occurred, or to have failed in fatigue as soon as a crack has been observed. This provides a stronger basis for neglecting the interaction of failure and dynamics, because the dynamic characteristics of a structure generally do not change drastically until very considerable damage has occurred.

11.2 Probability Distribution of Peaks

The probability distribution of the peaks of $\{X(t)\}$ can be found by a procedure that is basically the same as that used in Section 7.2 in deriving the rates of occurrence of crossings or peaks. That is, we will derive the probability distribution from consideration of occurrence rates. First we define $\nu_P[t; X(t) \le u]$ as the expected rate of occurrence of peaks not exceeding the level u. Next we recall that for an infinitesimal time interval Δt, we can say that the expected number of occurrences in the interval is the same as the probability of one occurrence in the interval, because we can neglect the probability of two or more occurrences. Thus, we say that

$$\nu_P[t; X(t) \le u]\,\Delta t = P\big(\text{peak} \le u \text{ during } [t, t+\Delta t]\big) \tag{11.1}$$

just as

$$\nu_P(t)\,\Delta t = P\big(\text{peak during } [t, t+\Delta t]\big) \tag{11.2}$$

in which $\nu_P(t)$ is the total expected rate of peak occurrences, which is the limit as u goes to infinity of $\nu_P[t; X(t) \le u]$. Furthermore, we can say that

$$P(\text{peak} \le u \text{ during } [t,t+\Delta t]) = P(\text{peak during } [t,t+\Delta t])\, P(\text{peak} \le u \,|\, \text{peak during } [t,t+\Delta t])$$

The final, conditional probability, term in this expression is precisely what we consider to be the cumulative distribution function for a peak at time t:

$$F_{P(t)}(u) \equiv P(\text{peak} \le u \,|\, \text{peak during } [t,t+\Delta t])$$

From Eqs. 11.1 and 11.2 we then solve for this cumulative distribution function as

$$F_{P(t)}(u) = \frac{\nu_P[t; X(t) \le u]}{\nu_P(t)} \tag{11.3}$$

Thus, we see that determining the probability distribution of the peaks depends on finding the rate of occurrence of peaks below any level u, and this is relatively straightforward. First, we note that $U[-\dot{X}(t)]$ is a process that has a positive unit step at each peak of $X(t)$ and has a negative unit step at each valley of $X(t)$. Thus, the $-\ddot{X}(t)\,\delta[-\dot{X}(t)]$ derivative of this process has positive and negative unit Dirac delta functions at the peaks and valleys, respectively. By multiplying by $U[-\ddot{X}(t)]$, we can eliminate the negative Dirac delta functions in order to count only peaks. Similarly, we can multiply by $U[u - X(t)]$ in order to eliminate all peaks above the level u. In this way, we obtain the rate of occurrence of peaks not exceeding the level u as

$$\nu_P[t; X(t) \le u] = E\left(-\ddot{X}(t)\,\delta[-\dot{X}(t)]\,U[-\ddot{X}(t)]\,U[u - X(t)]\right) \tag{11.4}$$

Substituting into Eq. 11.3 gives

$$F_{P(t)}(u) = \frac{E\left(-\ddot{X}(t)\,\delta[-\dot{X}(t)]\,U[-\ddot{X}(t)]\,U[u - X(t)]\right)}{E\left(-\ddot{X}(t)\,\delta[-\dot{X}(t)]\,U[-\ddot{X}(t)]\right)}$$

which can be rewritten in terms of joint probability density functions as

$$F_{P(t)}(u) = \frac{\int_{-\infty}^{\infty}\int_{-\infty}^{\infty}\int_{-\infty}^{\infty} (-z)\,\delta(-v)\,U(-z)\,U(u-w)\, p_{X(t)\dot{X}(t)\ddot{X}(t)}(w,v,z)\,dw\,dv\,dz}{\int_{-\infty}^{\infty}\int_{-\infty}^{\infty} (-z)\,\delta(-v)\,U(-z)\, p_{\dot{X}(t)\ddot{X}(t)}(v,z)\,dv\,dz}$$

or

$$F_{P(t)}(u) = \frac{\int_{-\infty}^{0} \int_{-\infty}^{u} |z| \, p_{X(t)\dot{X}(t)\ddot{X}(t)}(w,0,z) \, dw \, dz}{\int_{-\infty}^{0} |z| \, p_{\dot{X}(t)\ddot{X}(t)}(0,z) \, dz} \tag{11.5}$$

Taking a derivative with respect to u now gives the probability density function for the peak distribution as

$$p_{P(t)}(u) = \frac{\int_{-\infty}^{0} |z| \, p_{X(t)\dot{X}(t)\ddot{X}(t)}(u,0,z) \, dz}{\int_{-\infty}^{0} |z| \, p_{\dot{X}(t)\ddot{X}(t)}(0,z) \, dz} \tag{11.6}$$

Either Eq. 11.5 or 11.6 describes the probability distribution of any peak that occurs within the vicinity of time t. The probability that the peak is within any given interval can be found directly from Eq. 11.5 or from integration of Eq. 11.6, and Eq. 11.6 is also convenient for evaluating other quantities such as the mean value

$$\mu_P(t) \equiv E[P(t)] = \int_{-\infty}^{\infty} u \, p_{P(t)}(u) \, du$$

the mean-squared value

$$E[P^2(t)] = \int_{-\infty}^{\infty} u^2 \, p_{P(t)}(u) \, du$$

the variance, $[\sigma_P(t)]^2 = E[P^2(t)] - [\mu_P(t)]^2$, and so forth. A word of caution about the notation may be in order at this point. We write the various equations describing $P(t)$ in exactly the same way as we do for a continuously parametered process, but there is no continuously parametered $\{P(t)\}$ process. We presume that it is possible for a peak to occur at any t value, but there may or may not actually be a peak in the vicinity of a particular t. What we have derived is the conditional probability distribution and conditional moments of a peak $P(t)$ in the vicinity of t, given that such a peak exists.

From Eqs. 11.5 and 11.6 we note that in order to find the probability distribution of the peak $P(t)$, one must know the joint probability distribution of $X(t)$, $\dot{X}(t)$, and $\ddot{X}(t)$. This is as expected, because the occurrence of a peak $P(t)$ at level u requires the intersection of the events $X(t) = u$, $\dot{X}(t) = 0$, and $\ddot{X}(t) < 0$. The need for the joint probability distribution of three random variables, though, can make these expressions a little more difficult than most of the expressions we have considered so far. One special case in which the expressions are relatively simple is that in which the $\{X(t)\}$ process is Gaussian and stationary. In particular, we note (just as we did in Section 7.4) that $\dot{X}(t)$ is independent of the pair $[X(t),\ddot{X}(t)]$. Thus, the only parameters in the joint distribution of the three Gaussian random variables are the three standard deviations and the correlation coefficient between $X(t)$, and $\ddot{X}(t)$. We found in Section 7.3, though, that this correlation coefficient is exactly the negative of the α_2 bandwidth parameter. Thus, we see that α_2, in addition to its other interpretations, is a parameter governing the distribution of the peaks of $\{X(t)\}$. The details of the peak distribution for a stationary Gaussian process are worked out in Example 11.1.

****

Example 11.1: Find the cumulative distribution function and the probability density function for the peaks of a stationary Gaussian process $\{X(t)\}$.

$\dot{X}(t)$ is independent of $[X(t),\ddot{X}(t)]$, so we can factor $p_{\dot{X}(t)}(0)$ out of both the numerator and the denominator of Eq. 11.6, giving

$$p_P(u) = \frac{\int_{-\infty}^{0} |z|\, p_{X\ddot{X}}(u,z)\, dz}{\int_{-\infty}^{0} |z|\, p_{\ddot{X}}(z)\, dz} = \frac{(2\pi)^{1/2}}{\sigma_{\ddot{X}}} \int_{-\infty}^{0} |z|\, p_{X\ddot{X}}(u,z)\, dz$$

Using a conditional probability density function, this can be rewritten as

$$p_P(u) = \frac{(2\pi)^{1/2}}{\sigma_{\ddot{X}}} p_X(u) \int_{-\infty}^{0} |z|\, p_{\ddot{X}}(z | X=u)\, dz$$

We know that the conditional probability distribution is also Gaussian, so we can write it as

$$p_{\ddot{X}}[z | X=u] = \frac{1}{(2\pi)^{1/2}\sigma_*} \exp\left(-\frac{1}{2}\left[\frac{z-\mu_*}{\sigma_*}\right]^2 \right)$$

in which the conditional mean and standard deviation of $\ddot{X}(t)$ are

$$\mu_* \equiv E[\ddot{X}(t) | X(t)=u] = \rho_{X(t)\ddot{X}(t)} \left(\frac{\sigma_{\ddot{X}}}{\sigma_X} \right)(u-\mu_X) = -\alpha_2 \frac{\sigma_{\ddot{X}}}{\sigma_X}(u-\mu_X)$$

and

$$\sigma_* = \sigma_{\ddot{X}} \left(1-\rho_{X(t)\ddot{X}(t)}^2 \right)^{1/2} = \sigma_{\ddot{X}} (1-\alpha_2^2)^{1/2}$$

Substitution of this Gaussian form gives

$$p_P(u) = \frac{-1}{\sigma_{\ddot{X}}\sigma_*} p_X(u) \int_{-\infty}^{0} z \exp\left(-\frac{1}{2}\left[\frac{z-\mu_*}{\sigma_*}\right]^2 \right) dz$$

which can written as

$$p_P(u) = p_X(u)\left(\frac{\sigma_*}{\sigma_{\ddot{X}}}\exp\left[-\frac{\mu_*^2}{2\sigma_*^2}\right] - (2\pi)^{1/2}\frac{\mu_*}{\sigma_{\ddot{X}}}\Phi\left[-\frac{\mu_*}{\sigma_*}\right]\right)$$

or

$$p_P(u) = p_X(u)\left((1-\alpha_2^2)^{1/2}\exp\left[-\frac{\alpha_2^2(u-\mu_X)^2}{2(1-\alpha_2^2)\sigma_X^2}\right] + \right.$$

$$\left.(2\pi)^{1/2}\frac{\alpha_2(u-\mu_X)}{\sigma_X}\Phi\left[\frac{\alpha_2(u-\mu_X)}{(1-\alpha_2^2)^{1/2}\sigma_X}\right]\right)$$

Finally, substituting the Gaussian form for $p_X(u)$ gives the probability density function as

$$p_P(u) = \frac{(1-\alpha_2^2)^{1/2}}{(2\pi)^{1/2}\sigma_X}\exp\left(-\frac{\alpha_2^2(u-\mu_X)^2}{2(1-\alpha_2^2)\sigma_X^2}\right) +$$

$$\frac{\alpha_2(u-\mu_X)}{\sigma_X^2}\exp\left(-\frac{(u-\mu_X)^2}{2\sigma_X^2}\right)\Phi\left(\frac{\alpha_2(u-\mu_X)}{(1-\alpha_2^2)^{1/2}\sigma_X}\right)$$

The corresponding cumulative distribution function can be written in the slightly simpler form of

$$F_P(u) = \Phi\left(\frac{u-\mu_X}{(1-\alpha_2^2)^{1/2}\sigma_X}\right) - \alpha_2\exp\left(-\frac{(u-\mu_X)^2}{2\sigma_X^2}\right)\Phi\left(\frac{\alpha_2(u-\mu_X)}{(1-\alpha_2^2)^{1/2}\sigma_X}\right)$$

These formulas are commonly referred to as the *S.O. Rice distribution*, in recognition of Rice's pioneering work on this problem in 1945.

The limiting forms of this distribution for $\alpha_2 = 1$ and $\alpha_2 = 0$ yield interesting results regarding the peak distribution. For the narrowband situation with α_2 approaching unity, we see that some of the arguments in $p_P(u)$ and $F_P(u)$ tend to infinity. For the $\Phi(\cdot)$ function we must take proper account of the sign of the infinite argument, because $\Phi(\infty) = 1$ and $\Phi(-\infty) = 0$. Thus, we obtain quite different results for $u > \mu_X$ than we do for $u < \mu_X$. For $\alpha_2 = 1$, we obtain

$$p_P(u) = \frac{(u-\mu_X)}{\sigma_X^2}\exp\left(-\frac{(u-\mu_X)^2}{2\sigma_X^2}\right)U(u-\mu_X)$$

and

$$F_P(u) = \left(1-\exp\left[-\frac{(u-\mu_X)^2}{2\sigma_X^2}\right]\right)U(u-\mu_X)$$

In the special case when $\mu_X = 0$, this is exactly the Rayleigh distribution that in Section 7.4 we found to describe the amplitude of the Gaussian process. When $\mu_X \neq 0$, we see that the peak distribution has the same shape as the Rayleigh amplitude distribution, but it is shifted to make μ_X be the smallest possible peak value. The agreement of the peak distribution and the amplitude distribution of the limiting narrowband process is consistent with our prior observations that a narrowband process can be considered a harmonic function with slowly varying amplitude and phase. Because the narrowband amplitude varies slowly, we can say that each peak of the narrowband process is equal to the amplitude of the process at that instant of time, so it is not surprising that the two quantities have the same probability distribution.

For the opposite extreme situation with $\alpha_2 = 0$, the probability distribution of the peaks becomes

$$p_P(u) = \frac{1}{(2\pi)^{1/2}\sigma_X} \exp\left(-\frac{(u-\mu_X)^2}{2\sigma_X^2}\right), \qquad F_P(u) = \Phi\left(\frac{u-\mu_X}{\sigma_X}\right)$$

which is simply the Gaussian distribution of $X(t)$. In this broadband situation, we find that the distribution of peaks is the same as the distribution of the process itself. This may seem to be a surprising result, but it is consistent with the result we have previously obtained for the rate of occurrence of peaks. In particular, it is consistent with the fact found in Section 7.3, that α_2 is the same as the irregularity factor for a Gaussian process. Thus, if $\alpha_2 = 0$ and the process has finite crossing rates, then the rate of occurrence of peaks is infinite, as was shown in a particular case in Example 7.6. However, if the rate of occurrence of peaks is infinite, then it is reasonable to think that there may be peaks everywhere along the process, which should be expected to give the distribution of peaks to be the same as the distribution of $X(t)$. The sketch at the right shows the probability density function for peaks for several values of α_2.

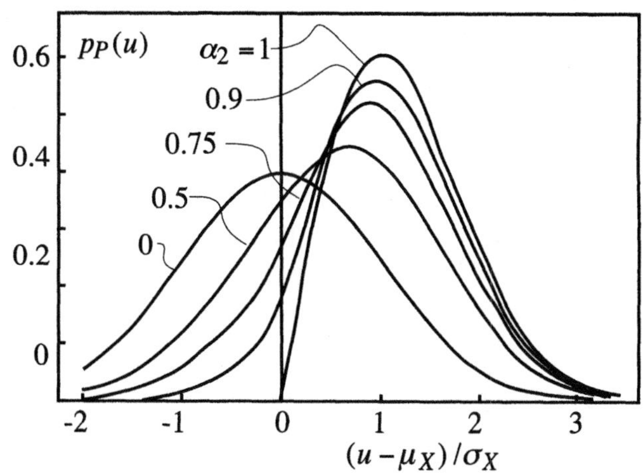

A more common form of the S.O. Rice distribution is with $\mu_X = 0$ in the preceding expressions, which can be considered to be the distribution for $P - \mu_X$ if $\mu_X \neq 0$.

A convenient feature of the normalized S.O. Rice distribution is that it also describes a random variable $R = \alpha_2 R_1 + [1-(\alpha_2)^2]^{1/2} R_2$, in which R_1 and R_2 are independent random variables with R_1 having the Rayleigh distribution of the amplitude of $X(t) - \mu_X$ and R_2 having the Gaussian distribution of $X(t) - \mu_X$.

****

Example 11.2: Find the probability that any peak of a stationary $\{X(t)\}$ process is below the level μ_X, provided that the distribution of $X(t)$ is symmetric about the level μ_X.

Simply substituting $u=\mu_X$ into Eq. 11.5 gives an expression for this probability, but we can also present the information in a somewhat different way. In particular, based on consideration of any continuous time history, we can say that any valley of $X(t)$ below the level μ_X is followed either by a peak below μ_X or by an upcrossing of the level μ_X. Thus, the rates of occurrence of these three events must satisfy the following relationship:

$$\nu_V(X<\mu_X)=\nu_P(X<\mu_X)+\nu_X^+(\mu_X)$$

Because $\nu_P(X<\mu_X)=\nu_P\, P[P(t)<\mu_X]$ and $\nu_V(X<\mu_X)=\nu_V\, P[V(t)<\mu_X]$, in which $V(t)$ denotes a valley of $X(t)$, and also $\nu_V=\nu_P$, we can divide by ν_P to obtain

$$P[V(t)<\mu_X]=P[P(t)<\mu_X]+IF$$

The specified symmetry of the distribution of $X(t)$ now gives $P[V(t)<\mu_X]=P[P(t)>\mu_X]=1-P[P(t)<\mu_X]$, so

$$P[P(t)<\mu_X]\equiv F_P(\mu_X)=\frac{1-IF}{2}$$

Somewhat surprisingly, it is possible to obtain this one point on the cumulative distribution function for $P(t)$ from knowledge only of the occurrence rates for crossings and peaks. Of course, we find that the probability that a peak is below the mean value of $X(t)$ is nearly zero for a narrowband process with $IF\approx 1$.

****

The distribution of peaks of a narrowband process is sometimes approximated by a function that can be obtained from knowledge only of the crossing rates of $X(t)$. The rationale is that in the narrowband case we can ignore the possibility of peaks below μ_X or valleys above μ_X. With this simplification, we can say that a peak occurs within the interval $[u,u+\Delta u]$ if and only if there is an upcrossing of the level u that is not followed by an upcrossing of the level $u+\Delta u$. This approximation then implies that the expected number of peaks within an interval $[u,u+\Delta u]$ is the difference between the number of upcrossings of the level u and the number of upcrossings of the level $u+\Delta u$, and the expected rate of occurrence of peaks in the interval is the difference between the upcrossing rates

$$\nu_P(u\le X\le u+\Delta u)\approx \nu_X^+(u)-\nu_X^+(u+\Delta u)$$

Saying that $p_{P(t)}(u)\,\Delta u$ is approximately $\nu_P(u \le X \le u+\Delta u)/\nu_P$, then taking the limit as Δu tends to zero, gives

$$p_{P(t)}(u) \approx \frac{-1}{\nu_P}\frac{d\nu_X^+(u)}{du}U(u-\mu_X)$$

Integrating this equation over the set of possible values, though, does not generally give unity, so the function cannot truly be a probability density function. This problem is easily remedied, however, by noting that our choice to neglect peaks below μ_X and valleys above μ_X is consistent with saying that the rate of peak occurrences is the same as the rate of upcrossings of the level μ_X. This gives

$$p_{P(t)}(u) = \frac{-1}{\nu_X^+(\mu_X)}\frac{d\nu_X^+(u)}{du}U(u-\mu_X) \qquad (11.7)$$

which is properly normalized for a probability density function.

For the special case of a stationary Gaussian process, one can use $\nu_X^+(u)$ from Example 7.1 to evaluate the $p_P(u)$ approximation in Eq. 11.7. The result is exactly the Rayleigh distribution that was shown in Example 11.1 to be the true answer in the limiting case with $\alpha_2 = 1$. Thus, the approximation does give the correct answer for the limiting narrowband process, but we can also see from the plot in Example 11.1 that the approximation can be in significant error for processes that still seem to be quite narrowband, such as when $\alpha_2 = 0.9$. The most notable error of the approximation may be the neglect of the peaks below the level μ_X, which was shown in Example 11.2 to be $(1-IF)/2$ for a symmetric distribution. Thus, for a Gaussian process with $\alpha_2 = 0.9$, one should have 5% of the peaks occurring below the level μ_X, and these are neglected in the approximation. For a non-Gaussian process, as well, we can anticipate that Eq. 11.7 will be asymptotically correct for α_2 approaching unity but may be significantly inaccurate for other situations.

11.3 Extreme Value Distribution and Poisson Approximation

A simple way to formulate the extreme value problem is to define a new stochastic process $\{Y(t)\}$ that is the extreme value of $\{X(t)\}$ during the past. Specifically, we let

$$Y(t) = \max_{0 \le s \le t} X(s) \qquad (11.8)$$

The extreme value distribution for $\{X(t)\}$ is then simply the distribution of the $Y(t)$ random variable. Note that even for a stationary $\{Y(t)\}$ process, one must expect that $\{Y(t)\}$ will be nonstationary, because larger and larger values of $X(t)$ will generally occur if we extend the period of our observation. Letting $L_X(u,t)$ denote the cumulative distribution function of $Y(t)$ gives

$$L_X(u,t) = F_{Y(t)}(u) \equiv P[Y(t) \le u] \equiv P[X(s) \le u : 0 \le s \le t] \tag{11.9}$$

in which the notation on the final term means that the $X(s) \le u$ inequality holds for all the given s values. This $L_X(u,t)$ is sometimes called the probability of survival, which is certainly appropriate if u denotes a critical value for $\{X(t)\}$ corresponding to some failure mode of the system. The probability density function for the extreme value, of course, is simply the derivative

$$p_{Y(t)}(u) = \frac{\partial}{\partial u} L_X(u,t) \tag{11.10}$$

and from this information one can also calculate the mean, variance, and so forth, of the extreme value.

An alternative problem that is almost equivalent to extreme value analysis involves the random quantity called first-passage time. Let $T_X(u)$ denote the first time (after time zero) at which $X(t)$ has an upcrossing of the level u. That is, $X[T_X(u)] = u$, $\dot{X}[T_X(u)] > 0$, and there has been no upcrossing in the interval $0 \le t < T_X(u)$. For any given u value, this $T_X(u)$ quantity is a random variable, and one can consider the family of all such variables to constitute a form of stochastic process $\{T_X(u)\}$, although the index set u is not time or frequency in this instance, as it has been in our other stochastic processes.

The relationship between the first-passage time and the extreme value distribution becomes evident when we consider the event $\{X(s) \le u : 0 \le s \le t\}$ that appears in Eq. 11.9. We see that this event can also be written as $\{X(0) \le u, T_X(u) \ge t\}$, because $X(s)$ can be less than u throughout the time interval only if it starts below u and does not have an upcrossing during the time interval. Taking probabilities then gives

$$\begin{aligned} L_X(u,t) &= P[X(0) \le u]\, P[T_X(u) \ge t \mid X(0) \le u] \\ &= L_X(u,0)\, P[T_X(u) \ge t \mid X(0) \le u] \end{aligned} \tag{11.11}$$

and the final term is related to a conditional form of the cumulative distribution of $T_X(u)$. In many practical problems, we can further simplify the relationship by completely neglecting the conditioning on this final term. For example, in some problems of interest we may have $P[X(0) \le u] = 1$, such as when the system is known to start at $X(0) = 0$, and conditioning by a sure event can always be neglected. In other situations we may not have any specific information that the distribution of $T_X(u)$ is independent of $X(0)$, but we anticipate that the effect of the conditioning will be significant only for small values of time.

Taking the derivative of Eq. 11.11 with respect to t gives

$$p_{T_X}[t \mid X(0) \le u] = \frac{-1}{L_X(u,0)} \frac{\partial}{\partial t} L_X(u,t) \tag{11.12}$$

which, along with Eq. 11.10, shows that the function governs the conditional distribution of the first-passage time and the distribution of the extreme value in very similar ways. The primary difference is that one probability density function involves a partial derivative with respect to u while the other involves a partial derivative with respect to t. The reader is cautioned that this close relationship between the extreme value problem and the first-passage problem is not always mentioned in the literature, with some authors using only one terminology and some using only the other. It should also be mentioned that we are omitting one advanced method of analysis that specifically relates to the first-passage formulation of the problem. In particular, the moments of $T_X(u)$ for some simple systems can be found by recursively solving a series of differential equations, called the generalized Pontryagin equations (Lin and Cai, 1995).

It is often convenient to write the probability of survival as some sort of exponential function of time. In particular, we use the form

$$L_X(u,t) = L_X(u,0) \exp\left(-\int_0^t \eta_X(u,s)\, ds \right) \tag{11.13}$$

Clearly it is always possible to find a function $\eta_X(u,s)$ such that one can write Eq. 11.13, even though it is not immediately obvious why this is desirable. However, we will now derive an interpretation of $\eta_X(u,s)$ that will motivate some useful approximations of the problem. First, we note that the derivative of Eq. 11.13 gives

$$\frac{\partial}{\partial t} L_X(u,t) = -L_X(u,t)\, \eta_X(u,t)$$

which can be solved for $\eta_X(u,t)$ as

$$\eta_X(u,t) = \frac{-1}{L_X(u,t)} \frac{\partial}{\partial t} L_X(u,t) = \lim_{\Delta t \to 0} \frac{1}{\Delta t} \frac{L_X(u,t) - L_X(u,t+\Delta t)}{L_X(u,t)}$$

This can then be written in terms of probabilities of first passage as

$$\eta_X(u,t) = \lim_{\Delta t \to 0} \frac{1}{\Delta t} \frac{P[t \le T_X(u) < t + \Delta t \mid X(0) \le u]}{P[T_X(u) \ge t \mid X(0) \le u]} \tag{11.14}$$

Now we note that the numerator of Eq. 11.14 relates to the event that the first upcrossing is in the specified time interval. This event, though, is the intersection of the event that there is no upcrossing prior to t and the event that there is an upcrossing in the interval:

$$\{t \le T_X(u) < t + \Delta t\} = \{T_X(u) \ge t\} \cap \{\text{upcrossing in } [t, t+\Delta t\}$$

Inasmuch as the denominator of Eq. 11.14 is the probability of one of these two events, we can see that the ratio is the conditional probability

$$\eta_X(u,t) = \lim_{\Delta t \to 0} \frac{1}{\Delta t} P\big(\text{upcrossing in } [t, t+\Delta t] \mid X(0) \le u, \text{no upcrossing prior to } t\big)$$

or

$$\eta_X(u,t) = \lim_{\Delta t \to 0} \frac{1}{\Delta t} \times$$
$$E\big(\text{number of upcrossings in } [t, t+\Delta t] \mid X(0) \le u, \text{no upcrossing prior to } t\big) \tag{11.15}$$

Thus, $\eta_X(u,t)$ can be regarded as an occurrence rate. It is the conditional rate of upcrossings of the level u, given the initial condition and the fact that there has been no prior upcrossing. If exceedance of the level u is considered to correspond to a "failure" of the system, then $\eta_X(u,t)$ is what is called the *hazard function* in reliability theory.

Unfortunately, it is not easy to calculate the $\eta_X(u,t)$ conditional rate of upcrossings. In fact, we have no rigorous relationship between $\eta_X(u,t)$ and unconditional probability density functions for $X(t)$ and $\dot{X}(t)$ random variables.[1] One can use a conditional probability density function to write an expression for the conditional crossing rate $\eta_X(u,t)$ in the same way as we previously did for the unconditional crossing rate $\nu_X(u,t)$. In particular,

$$\eta_X(u,t) = \int_0^\infty v\, p_{X(t)\dot{X}(t)}\big(u,v \mid X(0) \le u, \text{ no upcrossings in } [0,t]\big)\, dv$$

but this expression is not very useful for calculating values of $\eta_X(u,t)$, because the necessary conditional probability density function is generally unknown. Nonetheless, we will use it to gain some general information about the behavior of $\eta_X(u,t)$.

First we note that most physical processes have only a finite memory, in the sense that $X(t)$ and $X(t-\tau)$ can generally be considered independent if $\tau > T$ for some large T value. In this case, one can argue that

$$p_{X(t)\dot{X}(t)}(u,v \mid X(0) \le u, \text{ no upcrossings in } [0,t]) \approx$$
$$p_{X(t)\dot{X}(t)}(u,v \mid \text{no upcrossings in } [t-T,t])$$

for $t > T$. Some conditioning events are ignored in the second form, but they occurred prior to time $t-T$ and are out of the memory of $X(t)$. If $\{X(t)\}$ is a stationary process, though, this new form of conditional probability density is stationary, because it is independent of the choice of the origin of the time axis, except for the restriction that $t > T$. This means that $\eta_X(u,t)$ tends asymptotically to a stationary value $\eta_X(u)$ as $p_{X(t)\dot{X}(t)}(u,v \mid X(0) \le u$, no upcrossings in [0,t]) tends to $p_{X(t)\dot{X}(t)}(u,v \mid \text{no upcrossings in } [0,t])$. This asymptotic behavior of $\eta_X(u,t)$, then, implies that one can approximate $L_X(u,t)$ as

$$L_X(u,t) \approx L_0\, e^{-\eta_X(u)t} \quad \text{for large } t \tag{11.16}$$

This limiting behavior for large t will also apply if $\{X(t)\}$ is a nonstationary process that has finite memory and that becomes stationary with the passage of time.

The value L_0 in Eq. 11.16 is related to the behavior of $\eta_X(u,t)$ for small values of t. One extreme situation arises when $\{X(t)\}$ is a nonstationary "zero-

---

[1] The inclusion-exclusion series in Section 11.5 rigorously describes the effect of the conditioning by the event of no prior upcrossings, but it ignores the initial condition at time zero.

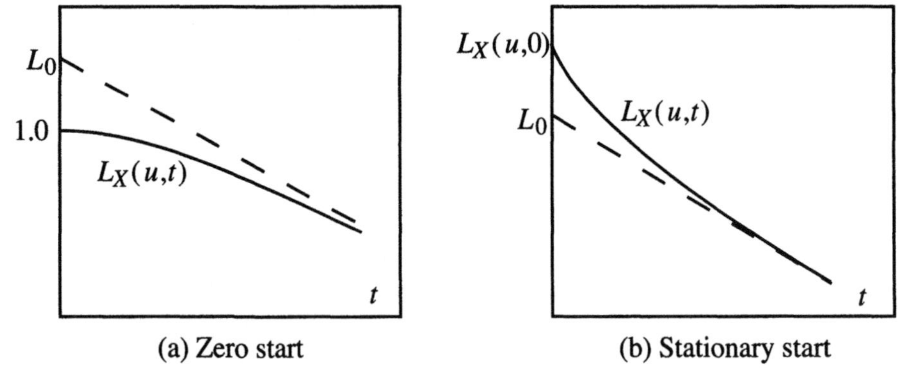

Figure 11.1 Effect of initial conditions on L_0 for Eq. 11.16.

start" process with $X(0)=0$ and $\dot{X}(0)=0$. This is the case, for example, if $\{X(t)\}$ represents the response of an oscillator that starts from a condition of rest. In this case, we can see that $\eta_X(u,0)=0$, and $\eta_X(u,t)$ increases from this initial value as $\sigma_X(t)$ and $\sigma_{\dot{X}}(t)$ grow, especially if $\mu_X(t)$ also grows. Another limiting condition is the "stationary-start" problem in which $\{X(t)\}$ is stationary for all time t. This stationary initial condition will usually give $\eta_X(u,t)$ for small time as being greater than the stationary value $\eta_X(u)$, because the unconditional $p_{X(t)\dot{X}(t)}(u,v)$ is larger than the stationary conditional probability density $p_{X(t)\dot{X}(t)}(u,v\,|\,\text{no upcrossings in }[t-T,t])$. Thus, for the zero-start problem $\eta_X(u,t)$ will generally grow from zero toward its stationary value, and for the stationary-start problem it will generally decay toward the stationary value. This behavior gives the L_0 multiplier in Eq. 11.16 as being greater than unity for the zero-start problem and as being less than $L_X(u,0)=P[X(0)\le u]$ for the stationary-start problem, as illustrated in Fig. 11.1.

It is not easy to calculate the $\eta_X(u,t)$ conditional rate of upcrossings, so we must use approximations in solving practical problems. We will present the most commonly used approximation here and discuss methods giving somewhat better estimates in the following section.

Inasmuch as $\eta_X(u,t)$ is a conditional upcrossing rate of the level u, we obviously expect it to be related to the unconditional upcrossing rate $\nu_X^+(u,t)$ that we studied in Section 7.2. In fact, the most widely used approximation of the extreme distribution problem results from simply neglecting the conditioning event in $\eta_X(u,t)$ and replacing it with $\nu_X^+(u,t)$:

$$\eta_X(u,t)\approx\nu_X^+(u,t) \tag{11.17}$$

giving

$$L_X(u,t) \approx L_X(u,0)\exp\left(-\int_0^t \nu_X^+(u,s)\,ds\right) \qquad (11.18)$$

and if $\{X(t)\}$ is a stationary process, this becomes

$$L_X(u,t) \approx L_X(u,0)\exp[-\nu_X^+(u)\,t] \quad \text{for stationary process} \qquad (11.19)$$

which is exactly of the form of Eq. 11.16. The approximation of Eqs. 11.17, 11.18, and 11.19 is commonly called the *Poisson approximation* of the extreme value, or first-passage, problem. This name comes from the fact that if the crossing rate is independent of the past history of the process, then the lengths of time between upcrossings will be independent; this makes the integer-valued process that counts the number of upcrossings a Poisson process.

From Eqs. 11.12 and 11.19, one can see that use of the Poisson approximation gives an exponential distribution for the first-passage time of a stationary $\{X(t)\}$ process:

$$p_{T_X}(t) \approx \nu_X^+(u)\exp[-\nu_X^+(u)\,t] \quad \text{for stationary process} \qquad (11.20)$$

The mean of this exponentially distributed random variable is

$$E[T_X(u)] \approx [\nu_X^+(u)]^{-1} \quad \text{for stationary process} \qquad (11.21)$$

and the standard deviation also has this same value. One of the characteristics of the Poisson process is that the time between occurrences is exponentially distributed, and the Poisson approximation of the extreme value problem gives the same exponential distribution for the first-passage time as for the time between upcrossings for a stationary $\{X(t)\}$ process. (Example 4.13 gives more information on the Poisson process.)

The Poisson approximation is most seriously in error when the $\{X(t)\}$ process is very narrowband. In that situation, an upcrossing of level u at time t is very likely to be associated with another upcrossing approximately one period later, due to the slowly varying amplitude of $\{X(t)\}$. Such a relationship between the upcrossing times is inconsistent with the Poisson approximation that the times between upcrossings are independent. On the other hand, when u is very large it is found that the independence assumption seems to be better. It is difficult to find general results that apply to all $\{X(t)\}$ processes, but for Gaussian processes it has been demonstrated that $\eta_X(u,t)$ does tend asymptotically to $\nu_X^+(u,t)$ as u tends to infinity.[2] Thus, the Poisson approximation is best when the $\{X(t)\}$ process is very broadband and/or the level u is very large. In some other situations, it may be significantly in error.

From the form of Eq. 11.13 it is clear that an overestimation of $\eta_X(u,t)$ will result in an underestimation of $L_X(u,t)$. An error of this type is usually considered to be conservative, because it overestimates the probability of failure due to large excursions. Furthermore, it is usually assumed that $\eta_X(u,t) \le \nu_X^+(u,t)$, so the Poisson approximation will underestimate $L_X(u,t)$. There are situations, though, in which this is not true, so some caution is required. In particular, we can see that if the level u is so small that $P[X(t) < u]$ is very small, but we are given the initial condition that $X(0) < u$, then it is very likely that $X(t)$ will quickly have an upcrossing of u. Mathematically this requires that $L_X(u,t)$ approach zero as $u \to -\infty$, for any finite t value. This is ensured for any choice of $L_X(u,0)$, though, only if $\eta_X(u,t)$ tends to infinity for $u \to -\infty$, at least for $t \approx 0$. This unbounded behavior of $\eta_X(u,t)$ surely violates the usual assumption that $\eta_X(u,t) \le \nu_X^+(u,t)$. On the other hand, the usual assumption does seem to be justified for the larger u values that are usually of primary importance.

In some problems, failure may occur due to large excursions of $X(t)$ in either the positive or negative direction, whereas all the development up until now has been concerned only with the probability that $X(t)$ remains below $+u$. However, the event of $X(t)$ remaining between $-u$ and $+u$ is exactly the same as the event of $|X(t)|$ remaining below the level u, and this allows us to apply the results to the new problem. Thus, we can write

$$L_{|X|}(u,t) = L_{|X|}(u,0) \exp\left(-\int_0^t \eta_{|X|}(u,s)\, ds \right) \tag{11.22}$$

for the new probability. Particularly in the study of first-passage time, the terms *double-barrier problem* and *single-barrier problem* are often used to distinguish between the consideration of upcrossings by $|X(t)|$ and $X(t)$, respectively. Of course, one can also consider double-barrier problems in which the levels of interest are not symmetric, or even problems in which the constant level u is replaced by a given function $u(t)$. The Poisson approximation of the symmetric double-barrier problem of Eq. 11.22 is simply to replace $\eta_{|X|}(u,s)$ with $\nu_{|X|}^+(u,s) = \nu_X^+(u,s) + \nu_X^-(-u,s)$. If the distribution of $X(t)$ and $\dot{X}(t)$ is

---

[2]Nigam (1983) attributes this result to Cramer (1966).

symmetric, this then gives $\nu_{|X|}^{+}(u,s)=2\nu_X^{+}(u,s)$, indicating that the decay of L with increasing t is twice as fast as for the single-barrier problem.

Note that very little information is needed to estimate the probability of first-passage failure by using the Poisson approximation. In particular, Eqs. 10.18 and 10.19 show that for the single-barrier problem one needs only the nonstationary or stationary unconditional rate of upcrossings and the initial condition of $L_X(u,0)\equiv F_{X(0)}(u)$. Using Eq. 10.22, or its stationary version, for the double-barrier problem requires the same information for $|X(t)|$.

****

Example 11.3: Use the Poisson approximation to estimate the probability of first-passage failure of a linear oscillator excited by stationary, Gaussian, white noise that is mean-zero and has autospectral density S_0. The oscillator has resonant frequency ω_0 and damping $\zeta=0.01$, and failure occurs if $X(t)$ exceeds the level $4\sigma_{stat}=4(\pi S_0)^{1/2}/(2m^2\zeta\,\omega_0^3)^{1/2}$ within the time interval $0\le\omega_0 t\le 250$.

We begin by considering the $\{X(t)\}$ response process to have zero initial conditions. This provides a slight simplification inasmuch as it makes $L_X(u,0)=1$, but it requires consideration of a nonstationary rate of upcrossings in Eq. 11.18. In particular, Example 7.2 gives

$$\nu_X^{+}(u,t)=\frac{\sigma_{\dot X}(t)}{\sigma_X(t)}\exp\left(\frac{-u^2}{2\sigma_X^2(t)}\right)\left(\frac{\rho_{X\dot X}(t,t)\,u}{(2\pi)^{1/2}\sigma_X(t)}\Phi\left[\frac{\rho_{X\dot X}(t,t)\,u}{[1-\rho_{X\dot X}^2(t,t)]^{1/2}\sigma_X(t)}\right]+\frac{[1-\rho_{X\dot X}^2(t,t)]^{1/2}}{2\pi}\exp\left[\frac{-\rho_{X\dot X}^2(t,t)\,u^2}{2[1-\rho_{X\dot X}^2(t,t)]\sigma_X^2(t)}\right]\right)$$

and the time-varying response quantities in this expression are given in Eqs. 5.50, 5.54, and 5.55. Rather than use these rather complicated expressions, though, we simplify the problem by using the approximation of Eq. 5.52 and a corresponding nonoscillatory approximation for the velocity:

$$\sigma_X^2(t)\approx\frac{\pi S_0}{2m^2\zeta\,\omega_0^3}\left(1-e^{-2\zeta\omega_0 t}\right),\quad \sigma_{\dot X}^2(t)\approx\frac{\pi S_0}{2m^2\zeta\,\omega_0}\left(1-e^{-2\zeta\omega_0 t}\right)$$

The covariance of $X(t)$ and $\dot X(t)$ is given in Eq. 5.55. If we ignore the oscillatory term in this expression, as well as for the variance terms, we have

$$K_{X\dot X}(t,t)\approx\frac{\pi S_0}{2m^2\omega_d^2}e^{-2\zeta\omega_0 t},\quad \rho_{X\dot X}(t,t)\approx\frac{\zeta\, e^{-2\zeta\omega_0 t}}{1-e^{-2\zeta\omega_0 t}}$$

This is a reasonable approximation of the correlation coefficient except that it tends to infinity for small values of t. In Example 5.8 we found that the initial value of the correlation coefficient is $3^{1/2}/2$, so we will truncate the correlation coefficient at that value, giving

$$\rho_{X\dot{X}}(t,t) \approx \rho(t) \equiv \min\left(\frac{3^{1/2}}{2}, \frac{\zeta\, e^{-2\zeta\omega_0 t}}{1-e^{-2\zeta\omega_0 t}}\right)$$

With the approximations of the time-varying terms in $\nu_X^+(u,t)$, one can substitute into Eq. 10.18 and integrate numerically to obtain $L(u,250/\omega_0)=0.9939$, indicating that the probability of first-passage failure is $P_f \approx 0.0061$.
We can also easily obtain the result for a stationary $\{X(t)\}$ process. In this situation we use Eq. 11.19, with the stationary crossing rate of

$$\nu_X^+(u) = \frac{\omega_0}{2\pi} e^{-4^2/2} = 5.34\times 10^{-5}$$

This then gives $L(u,50/\omega_0) = \Phi(4)\, e^{-250\nu_X^+(u)} = 0.9867$ as the probability of survival, and $P_f \approx 0.0133$. This is a problem for which the initial condition on $\{X(t)\}$ has a major effect on the prediction of the probability of failure.

****

11.4 Improved Estimates of the Extreme Value Distribution

First we will consider a modification of the usual Poisson estimate to account, in an approximate way, for the effect of the given information that $X(0) \le u$. This modification will improve our estimate of $\eta_X(u,t)$ for small values of u but will have little effect for large u values. In particular, the modification will address the difficulty noted in Section 11.3 for situations in which u is so low that the average time between upcrossings may be large, but this does not give a good estimate of the time until the first upcrossing. Recall that the usual Poisson assumption gives the probability distribution of the first-passage time $T_X(u)$ for a stationary process as being the same as that of a random variable T_{CR} representing the time interval between successive upcrossings of the level u. However, the time interval T_{CR} is composed of two segments: the time T_1 between the upcrossing and the following downcrossing and the time T_2 between the downcrossing and the next upcrossing. Of these two segments, only T_2 is spent below the level u. Inasmuch as $T_X(u)$ is also a time interval spent below u, it seems more appropriate to approximate it by T_2 instead of T_{CR}. Obviously, this will reduce our estimate of the time until first passage and give more conservative estimates of failure probabilities.

If we consider T_2 to be governed by a Poisson process, then the only parameter needed to describe its distribution is its arrival rate. Furthermore, we know that the arrival rate is the inverse of $E(T_2)$, as in Eq. 11.21, and because T_2 represents the portion of T_{CR} spent below the level u, we can say that

$E(T_2) = E(T_{CR}) P(X < u)$. Unfortunately, we do not have an exact result for $E(T_{CR})$ in this situation. If T_{CR} were governed by a Poisson process, then we would know that $E(T_{CR})$ was the same as the $[\nu_X^+(u)]^{-1}$ value, but the assumption that T_2 is governed by a Poisson process precludes the possibility that T_{CR} also has this Poisson property. Nonetheless, we will use this value as an approximation of $E(T_{CR})$ in order to obtain $E(T_2) = [\nu_X^+(u)]^{-1} P(X < u)$. Approximating $\eta_X(u)$ by the arrival rate for T_2 then gives $\eta_X(u) = \nu_X^+(u) / F_X(u)$ for a stationary process, and a consistent modification for a nonstationary $\{X(t)\}$ is

$$\eta_X(u,t) \approx \frac{\nu_X^+(u,t)}{F_{X(t)}(u)} \tag{11.23}$$

Clearly this approximation is almost identical to Eq. 11.17 for large values of u, because $F_{X(t)}(u)$ is almost unity in that situation. For low values of u, however, it gives $\eta_X(u,t)$ as having a very large value and ensures that $L_X(u,t)$ tends to zero as u tends to negative infinity, as desired. It may also be noted that the approximation of Eq. 11.23 was obtained by Ditlevsen (1986) by a somewhat different method of reasoning.

Next we will consider modifications to the Poisson approximation based directly on the narrowband limitation we previously noted. In particular, for a narrowband process with a slowly varying amplitude $\{A(t)\}$, a single upcrossing of the level u by $\mu_X(t) + A(t)$ is likely to be associated with several (almost uniformly spaced) upcrossings of u by $X(t)$. Because this is inconsistent with a Poisson assumption that the times between upcrossings are independent, a better approximation is desirable. The narrowband approximations that we will consider are based on consideration of the behavior of the $\{A(t)\}$ process.

Note that the developments until now have placed little restriction on the probability distribution of $\{X(t)\}$. In particular, there has been no restriction that it be mean-zero. Now, though, we are going to consider models based on the behavior of $\{A(t)\}$, and this amplitude is associated with the deviation of $X(t)$ from its mean value. Thus, it is more convenient to switch to a mean-zero random process, and henceforth we assume that $\mu_X(t) = 0$. Some information about a more general $\{Y(t)\}$ process can be obtained by writing it as $Y(t) = X(t) + \mu_Y(t)$. If $\{Y(t)\}$ is mean-value stationary, then it is simple to account for the constant μ_Y. In particular, we can say that the extreme of $Y(t)$ is simply μ_Y plus the extreme of $X(t)$. If $\mu_Y(t)$ is not a constant, then care must be used, because we have $P[Y(s) \le u : 0 \le s \le t] = P[X(s) \le u - \mu_Y(s) : 0 \le s \le t]$, which amounts to a problem with a variable barrier level. We will not explicitly address this more complicated problem.

The simplest extreme value approximation based on the amplitude process amounts to assuming that the extreme value of the mean-zero $\{X(t)\}$ process is the same as that of $\{A(t)\}$, giving

$$L_X(u,t) \approx L_A(u,t) = L_A(u,0)\exp\left(-\int_0^t \eta_A(u,s)\,ds\right) \tag{11.24}$$

Because $X(t) \le A(t)$, we know that $L_X(u,t) \ge L_A(u,t)$ for all u and t. Thus, using Eq. 11.24 will always be conservative in the sense that it overestimates the probability of large extreme values. Of course we still have the problem of determining $L_A(u,t)$, but it is generally assumed that the Poisson assumption is much better for $\{A(t)\}$ than for $\{X(t)\}$, and using it gives

$$\eta_A(u,t) \approx \nu_A^+(u,t) \tag{11.25}$$

The integration in Eq. 11.24, of course, is almost trivial if $\nu_A^+(u)$ is stationary.

It is also reasonable to apply an initial condition correction to $\nu_A^+(u,t)$, similar to what we did in obtaining Eq. 11.23. That is, if u is very small, then the initial condition of $A(0) < u$ implies that the time until first passage by $A(t)$ is likely to be short. By exactly the same reasoning as was used in obtaining Eq. 11.23, we say that

$$\eta_A(u,t) \approx \frac{\nu_A^+(u,t)}{F_{A(t)}(u)} \tag{11.26}$$

This correction seems to be much more significant in this case than it was in Eq. 11.23, though. In particular, the $F_{A(t)}(u)$ term approaches zero as u approaches zero. Thus, the correction in Eq. 11.26 is evident at values of u that may be of practical importance.

It should be noted that using the Poisson amplitude crossings model of Eqs. 11.24–11.26 is not exactly the same as introducing a new estimate of $\eta_X^+(u,t)$ in the general extreme value formula of Eq. 11.13. In particular, the multiplier of $L_X(u,0) \equiv P[X(0) \le u]$ in Eq. 11.13 has been replaced by $P[A(0) \le u]$. We will now investigate the implication of this difference. Consider an ensemble of possible time histories of $\{X(t)\}$. The term $P[X(0) \le u]$ represents the fraction of the ensemble that is below the level u at time $t = 0$, excluding only samples with $X(0) > u$, while $P[A(0) \le u]$ also excludes samples with $A(0) > u$, even if $X(0) \le u$. For a narrowband process, though, we can see that it is very likely that the samples with $A(0) > u$ and $X(0) \le u$ will have an upcrossing of u during the first cycle of the process. That is, if the narrowband process has average frequency ω_c, then it is very likely that an $X(t)$ sample with $A(0) > u$ will exceed u prior to time $2\pi/\omega_c$. Using an initial condition of $P[A(0) \le u]$ on the probability distribution of the extreme values is equivalent to counting these time histories that are particularly likely to cross u during the first cycle as having crossed at time zero. This discrepancy should be significant only if one is particularly interested in the details of the first cycle. The $\eta_X(u,t)$ true conditional rate of upcrossings is particularly large during the first cycle of a narrowband process, due to these samples that started with $A(0) > u$ having upcrossings, then it settles down to a lower value. By using an initial condition of $P[A(0) \le u]$, one can obtain an approximate extreme value distribution that is good for $t \ge 2\pi/\omega_c$, without explicit consideration of the high early value of $\eta_X(u,t)$. Because the Poisson approximation ignores this high early value of the upcrossing rate, it seems appropriate that it should be used with an initial condition of $P[A(0) \le u]$ if $\{X(t)\}$ is narrowband.

****

Example 11.4: Compare the $\eta_X(u)$ values from Eqs. 11.17, 11.23, 11.25, and 11.26 for a stationary, mean-zero, Gaussian, narrowband process with $\alpha_1 = 0.995$. (One particular process with this value of α_1 is the response to white noise of an SDF oscillator with approximately 0.8% of critical damping.)

The basic upcrossing rate $\nu_X^+(u)$ for this process was found in Example 7.1 to be

$$\nu_X^+(u) = \frac{\omega_{c2}}{2\pi} \exp\left(\frac{-u^2}{2\sigma_X^2}\right)$$

in which $\omega_{c2} = \sigma_{\dot{X}} / \sigma_X$. This, then, is the approximation of $\eta_X(u)$ for Eq. 11.17. Because this is the most commonly used approximation, we will present our comparisons for each of the other three approximations as the ratio $\eta_X(u)/\nu_X^+(u)$. (This normalization of the results has been commonly used since Crandall et al. in 1966.) The $\eta_X(u)/\nu_X^+(u)$ ratio for Eq. 11.23, of course, is simply $[F_X(u)]^{-1}$, which is $[\Phi_X(u/\sigma_X)]^{-1}$ for a mean-zero Gaussian process.
The amplitude upcrossing rates for this process were found in Examples 7.12 and 7.13, respectively, for both the Cramer and Leadbetter and the energy-based definitions. The result for the energy-based amplitude, though, depends on α_2, which is not given in the statement of this problem and cannot be determined

from the given value of α_1. (Recall, for example, that $\alpha_2 = 0$ for the response of any SDF oscillator excited by white noise.) The Cramer and Leadbetter result for Eq. 11.25 is

$$\frac{\eta_X(u)}{\nu_X^+(u)} = \frac{\nu_{A_1}^+(u)}{\nu_X^+(u)} = \frac{u}{\sigma_X}[2\pi(1-\alpha_1^2)]^{1/2} \approx 0.250\frac{u}{\sigma_X}$$

Because the amplitude has the Rayleigh distribution, the additional term needed to obtain the approximation of Eq. 11.26 is

$$F_A(u) = 1 - \exp\left(\frac{-u^2}{2\sigma_X^2}\right)$$

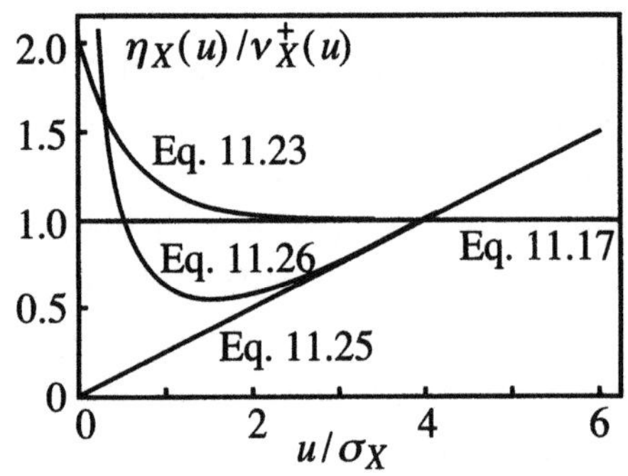

The accompanying sketch shows the values of $\eta_X(u)/\nu_X^+(u)$ versus u/σ_X for Eqs. 11.17 and 11.23, and for Eqs. 11.25 and 11.26 using the Cramer and Leadbetter amplitude.

From the plot we see that the approximations give quite different results for small values of u, which may not be of too much practical significance because failure is generally related to large values of u. We also see, though, that the X crossing and the amplitude-crossing results behave very differently for the crucial region with large values of u. In particular, the $\nu_A^+(u)/\nu_X^+(u)$ ratio grows linearly with u, so the amplitude-crossing result eventually becomes more conservative (i.e., predicts larger probabilities of upcrossings) than the original Poisson approximation.

****

For very large u values, it is generally found that many of the upcrossings of u by $A(t)$ are not accompanied by upcrossings by $X(t)$. This leads to $\nu_A^+(u,t)$ being much larger than $\nu_X^+(u,t)$, as was found in Example 11.4 for the special case of a Gaussian process. On the other hand, we know that the $\eta_X(u,t)$ conditional crossing rate tends to $\nu_X^+(u,t)$ as u becomes very large, so there must be limits to the usefulness of Eqs. 11.25 and 11.26 for large u values. An improved approximation of the extreme value distribution can be obtained by estimating the fraction of the upcrossings by $A(t)$ that are accompanied by upcrossings by $X(t)$. The $\eta_X(u,t)$ conditional crossing rate can then be taken to represent the rate of occurrence of this subset of the amplitude upcrossings. Vanmarcke introduced such a scheme in 1972, and the following paragraph uses some of his approximations in deriving a very similar result, although the derivation given here follows that of Madsen et al. (1986) much more closely than it does that of Vanmarcke. It should also be noted that extensions of the scheme were presented by Corotis et al. in 1972 and by Vanmarcke in 1975.

Using the same notation as we used before for the $\{X(t)\}$ process, let the random variable T_1 denote the time between an upcrossing of u by $A(t)$ and the subsequent downcrossing by $A(t)$. Then T_1 represents the duration of an interval with $A(t) > u$. If T_1 is large, then it seems almost certain that $X(t)$ will have an upcrossing of u within the interval, but if T_1 is small, then it seems quite likely that no upcrossing by $X(t)$ will occur. Following Vanmarcke, we approximate this relationship by

$$P[\text{no upcrossing by } X(t) \mid T_1 = \tau] \approx [1 - \nu_X^+(0,t)\,\tau]\,U[1 - \nu_X^+(0,t)\,\tau] \tag{11.27}$$

Considering $[\nu_X^+(0,t)]^{-1}$ to represent the period of an average cycle of the $\{X(t)\}$ process, this approximation amounts to saying that an upcrossing by $X(t)$ is sure if T_1 exceeds the period, and the probability of its occurrence grows linearly with T_1 for T_1 less than the period. Clearly this is a fairly crude approximation, but it should be substantially better than simply assuming that an upcrossing by $X(t)$ occurs in connection with every upcrossing by $A(t)$. To calculate the unconditional probability of an upcrossing in the T_1 interval, it is necessary to have a probability distribution for T_1. Consistent with the arrival times in the Poisson process, we use Vanmarcke's approximation that this is the exponential distribution

$$p_{T_1}(\tau) = \frac{e^{-\tau / E(T_1)}}{E(T_1)}$$

and we also use the approximation that $E(T_1) = E(T_{CR})\,P[A(t) > u] \approx P[A(t) > u] / \nu_A^+(u,t)$, with T_{CR} being the time between successive upcrossing of u by $A(t)$. This gives

$$P[\text{upcrossing by } X(t) \text{ during } T_1] \approx \frac{P[A(t) > u]\,\nu_X^+(0,t)}{\nu_A^+(u,t)} \left(1 - \exp\left[\frac{-\nu_A^+(u,t)}{P[A(t) > u]\,\nu_X^+(0,t)}\right]\right)$$

and taking $\eta_X(u,t) \approx \nu_A^+(u,t)\,P[\text{upcrossing by } X(t) \text{ during } T_1]$ gives

$$\eta_X(u,t) \approx P[A(t) > u]\, \nu_X^+(0,t) \left(1 - \exp\left[\frac{-\nu_A^+(u,t)}{P[A(t) > u]\, \nu_X^+(0,t)} \right] \right) \tag{11.28}$$

Note that as u tends to zero, the results of Eq. 11.28 approach those of Eq. 11.25 based on considering each upcrossing by $A(t)$ to correspond to a crossing by $X(t)$. The limiting behavior for large values of u may be of more interest. In this case we find that $P[A(t) > u]$ is very small, so $\nu_A^+(u,t) >> P[A(t) > u]\, \nu_X^+(0,t)$ and Eq. 11.28 gives $\eta_X(u,t) \approx P[A(t) > u]\, \nu_X^+(0,t)$. For the special case of a Gaussian process it can be shown that this is identical to $\nu_X^+(u,t)$, so the approximation agrees with the results from the assumption of Poisson crossings by $X(t)$. For a non-Gaussian process, these two results for large u values may not be identical, although they are expected to be quite similar.

As in our other approximations, we can expect to obtain better results for small u values by including the effect of the initial condition. Because this estimate of $\eta_X(u,t)$ is a modified version of the amplitude-crossing rate, we do this by dividing by $P[A(t) < u]$. This gives

$$\eta_X(u,t) \approx \frac{P[A(t) > u]\, \nu_X^+(0,t)}{P[A(t) < u]} \left(1 - \exp\left[\frac{-\nu_A^+(u,t)}{P[A(t) > u]\, \nu_X^+(0,t)} \right] \right) \tag{11.29}$$

and it is easily verified that Eq. 11.29 agrees with Eq. 11.26 in the limit for u near zero.

It should be noted that for a general non-Gaussian $\{X(T)\}$ process, Eqs. 11.28 and 11.29 are not identical to Vanmarcke's results and that his derivation uses somewhat more sophisticated assumptions about the behavior of $\{X(T)\}$. For the special case of the Gaussian process, though, it can be shown that $P[A_1(t) > u]\, \nu_X^+(0,t) = \nu_X^+(u,t)$, so Eq. 11.29 with the Cramer and Leadbetter definition of amplitude does become identical to Vanmarcke's form of

$$\eta_X(u,t) \approx \nu_X^+(u,t) \left(1 - \exp\left[\frac{-\nu_{A_1}^+(u,t)}{\nu_X^+(u,t)} \right] \right) \left(1 - \frac{\nu_X^+(u,t)}{\nu_X^+(0,t)} \right)^{-1} \tag{11.30}$$

It can be expected that the two approximations will also give similar results for other processes that do not differ greatly from the Gaussian distribution.

****

Example 11.5: Compare the $\eta_X(u)$ values from Eqs. 11.28–11.30 with those obtained by other methods in Example 11.4 for a stationary, mean-zero, Gaussian, narrowband process with $\alpha_1 = 0.995$.

Using the Gaussian relationships and the Cramer and Leadbetter definition of amplitude in Eq. 11.28 gives

$$\frac{\eta_X(u,t)}{\nu_X^+(u,t)} \approx 1-\exp\left(\frac{-\nu_A^+(u,t)}{\nu_X^+(u,t)}\right) = 1-\exp\left(-[2\pi(1-\alpha_1^2)]^{1/2}\frac{u}{\sigma_X}\right) \approx 1-e^{-0.250\,u/\sigma_X}$$

and Eq. 11.29 and 11.30 give identical results of

$$\frac{\eta_X(u,t)}{\nu_X^+(u,t)} \approx \frac{1-\exp\left(-(2\pi)^{1/2}(1-\alpha_1^2)^{1/2}\dfrac{u}{\sigma_X}\right)}{1-\exp\left(\dfrac{-u^2}{2\sigma_X^2}\right)} \approx \frac{1-e^{-0.250\,u/\sigma_X}}{1-\exp\left(\dfrac{-u^2}{2\sigma_X^2}\right)}$$

The plot confirms that for very large values of u, Eqs. 11.28 and 11.29 (or the identical Eq. 11.30) both tend to the original Poisson approximation of Eq. 11.17, whereas for small u values, Eq. 11.28 tends to Eq. 11.25 and Eq. 11.29 (or 11.30) tends to Eq. 11.26.

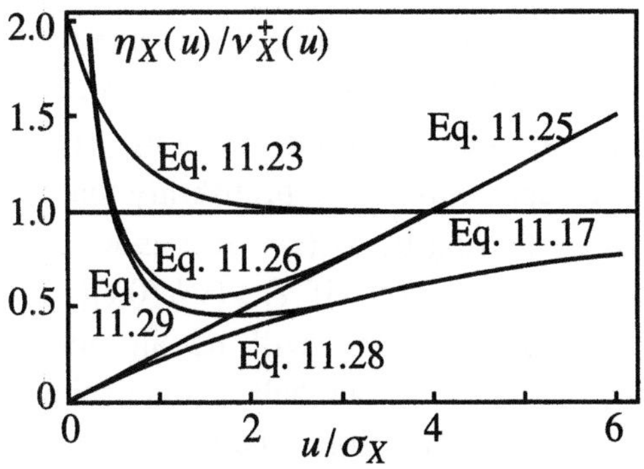

****

We indicated in Eq. 11.22 how the original Poisson crossings approximation should be applied to the double-barrier problem, in which one is concerned with the probability distribution of the extreme value of $|X(t)|$. Now we need to consider the same issue for the various modifications of the Poisson model that have been introduced in this section. Note that replacing X by $|X|$ has no effect at all on Eqs. 11.25 and 11.26, because these approximations are wholly based on the occurrence of amplitude crossings. That is, for the mean-zero process, the extreme distribution of $A(t)$ can be considered an approximation of the extreme distribution of $|X(t)|$, just as we previously assumed that it approximated the extreme distribution of $X(t)$. In fact, it seems likely that the approximation will be better for $|X(t)|$ than for $X(t)$.

The approximations of Eqs. 11.23 and 11.28–11.30 are all affected by the change from single to double barrier, because these expressions all include probability or crossing rate information regarding $X(t)$. Looking first at the initial condition effect approximated by Eq. 11.23, we see that for the double-barrier problem we should use

$$\eta_{|X|}(u,t) \approx \frac{\nu_{|X|}^{+}(u,t)}{P[|X(t)| \le u]} \tag{11.31}$$

The modification of the numerator is as in the original Poisson approximation of Eq. 11.22, but the change in the denominator is also significant. In particular, $P[|X(t)| \le u]$ will tend to zero for u approaching zero, so this approximation will give $\eta_{|X|}(u,t)$ tending to infinity in this situation. Thus, consideration of the $|X(0)| \le u$ condition gives small u behavior that resembles that of Eq. 11.26 with the initial condition of $A(0) \le u$.

One must give a little more thought to the modification of Eqs. 11.28–11.30 to describe the double-barrier problem. Recall that these equations were designed to include the probability that an upcrossing by $A(t)$ is accompanied by an upcrossing by $X(t)$. For the double-barrier problem, then, we must approximate the probability that an upcrossing by $A(t)$ is accompanied by an upcrossing by $|X(t)|$. Just as we assumed that there would be an upcrossing by $X(t)$ during T_1 if $T_1 > [\nu_X^{+}(0,t)]^{-1}$ = one period of the process, we will now assume that there will be an upcrossing by $|X(t)|$ during T_1 if $T_1 > [\nu_X^{+}(0,t)]^{-1}/2$ = one-half period of the process. Thus, we replace $\nu_X^{+}(0,t)$ in Eq. 11.27 with $2\nu_X^{+}(0,t)$, and this gives exactly the same replacement in Eqs. 11.28 and 11.29:

$$\eta_{|X|}(u,t) \approx 2\,P[A(t) > u]\,\nu_X^{+}(0,t)\left(1 - \exp\left[\frac{-\nu_A^{+}(u,t)}{2\,P[A(t) > u]\,\nu_X^{+}(0,t)}\right]\right) \tag{11.32}$$

neglecting the initial condition at t = 0, and

$$\eta_{|X|}(u,t) \approx \frac{2\,P[A(t) > u]\,\nu_X^{+}(0,t)}{P[A(t) < u]}\left(1 - \exp\left[\frac{-\nu_A^{+}(u,t)}{2\,P[A(t) > u]\,\nu_X^{+}(0,t)}\right]\right) \tag{11.33}$$

when the initial condition is included. The corresponding modification of Vanmarcke's formula in Eq. 11.30 is

$$\eta_{|X|}(u,t) \approx \nu_{|X|}^{+}(u,t)\left(1-\exp\left[\frac{-\nu_{A_1}^{+}(u,t)}{\nu_{|X|}^{+}(u,t)}\right]\right)\left(1-\frac{\nu_X^{+}(u,t)}{\nu_X^{+}(0,t)}\right)^{-1} \tag{11.34}$$

For the case of a mean-zero Gaussian $\{X(t)\}$ process, using the Cramer and Leadbetter amplitude in Eq. 11.33 gives a result that is identical to Eq. 11.34. This is consistent with the results obtained for the single-barrier situation.

For the special case of a stationary, mean-zero, Gaussian $\{X(t)\}$ process, Vanmarcke (1975) has also offered an empirical correction that improves the approximation of the conditional crossing rate. For this situation, the ratio $\nu_A^{+}(u)/\nu_X^{+}(u)$ of Eq. 11.30 is given by

$$\frac{\nu_A^{+}(u)}{\nu_X^{+}(u)} = [2\pi(1-\alpha_1^2)]^{1/2}\frac{u}{\sigma_X}$$

in which the term $(1-\alpha_1^2)$ introduces the effect of the bandwidth of the process. Similarly, the $\nu_A^{+}(u)/\nu_{|X|}^{+}(u)$ ratio of Eq. 11.34 is one-half this amount. As an empirical correction for effects not included in the derivation of these equations, Vanmarcke has suggested replacing the $(1-\alpha_1^2)$ term with $(1-\alpha_1^2)^{1.2}$. This gives the modified Vanmarcke approximations as

$$\frac{\eta_X(u)}{\nu_X^{+}(u)} \approx \left(1-\exp\left[-(1-\alpha_1^2)^{0.6}(2\pi)^{1/2}\frac{u}{\sigma_X}\right]\right)\left(1-\exp\left[\frac{-u^2}{2\sigma_X^2}\right]\right)^{-1}$$

and

$$\frac{\eta_X(u)}{\nu_{|X|}^{+}(u)} \approx \left(1-\exp\left[-(1-\alpha_1^2)^{0.6}(\pi/2)^{1/2}\frac{u}{\sigma_X}\right]\right)\left(1-\exp\left[\frac{-u^2}{2\sigma_X^2}\right]\right)^{-1} \tag{11.35}$$

for the single-barrier and double-barrier problems, respectively.

If one chooses to use the modified Vanmarcke approximation for the stationary situation, then it is desirable also to modify the nonstationary approximation to achieve consistent results. The modification amounts to a reduction in the bandwidth of the process, and the bandwidth parameter α enters into first passage calculation only as it affects the $\nu_A^{+}(u,t)$ rate of upcrossings by the envelope. For the nonstationary situation, we have seen in Example 7.14 that the $(1-\alpha_1^2)$ term in the stationary upcrossing rate by $A_1(t)$ is replaced by

$(1-\alpha_1^2-\rho_{X\dot{X}}^2)$. One can obtain a consistent modified version of the nonstationary crossing rate by applying the power 1.2 either to $(1-\alpha_1^2)$ or to $(1-\alpha_1^2-\rho_{X\dot{X}}^2)$ in the expression for the upcrossing rate. Both of these options converge to Vanmarcke's modified result in the stationary situation, in which the correlation coefficient is zero, and it is not completely clear which is better. Inasmuch as the correlation coefficient is not a measure of bandwidth, though, it seems more appropriate to modify the bandwidth effect by applying the power 1.2 only to $(1-\alpha_1^2)$.

11.5 Inclusion-Exclusion Series for Extreme Value Distribution

Finally, we will derive the infinite series called the *inclusion-exclusion relationship* between $\eta_X(u,t)$ and probability density functions for $X(t)$ and $\dot{X}(t)$ random variables.[3] This cumbersome equation rigorously includes the fact that $\eta_X(u,t)$ is conditioned by the event of no prior upcrossings of the level u by $X(t)$. It ignores the additional conditioning in Eq. 11.15 by the event of $X(0)\le u$, however. To emphasize this distinction, we use the notation $\hat{\eta}_X(u,t)$ for the conditional rate of upcrossings given only the fact that there have been no prior upcrossings. The presentation is simplified by introducing a notation for upcrossing events as follows: $B(t)$ = {event of an upcrossing of level u occurring in $[t,t+\Delta t]$} and $B^*(t)$ = {event of the first upcrossing of u occurring in $[t,t+\Delta t]$}, giving $B^*(t)$ as a subset of $B(t)$. We also introduce new terms $\nu_X^+(u,t,s_1,\cdots,s_j)$ and $\hat{\eta}_X(u,t,s_1,\cdots,s_j)$ that give the probabilities of intersections of these events as

$$P[B(t)\,B(s_1)\cdots B(s_j)]=\nu_X^+(u,t,s_1,\cdots,s_j)\;\Delta t\,\Delta s_1\cdots\Delta s_j \qquad (11.36)$$

and

$$P[B(t)\,B(s_1)\cdots B^*(s_j)]=\hat{\eta}_X(u,t,s_1,\cdots,s_j)\;\Delta t\,\Delta s_1\cdots\Delta s_j \qquad (11.37)$$

in which $s_j\le s_{j-1}\le\cdots\le s_1\le t$. In the first relationship there is no requirement that there has not been an upcrossing (or many upcrossings) prior to time s_j, but the second relationship does include this restriction. If the first upcrossing is not at time s_j, then it must be at some time s_{j+1} that is prior to s_j, so we can say

$$\nu_X^+(u,t,s_1,\cdots,s_j)=\hat{\eta}_X(u,t,s_1,\cdots,s_j)+\int_0^{s_j}\hat{\eta}_X(u,t,s_1,\cdots,s_j,s_{j+1})\,ds_{j+1}$$

[3]The inclusion-exclusion relationship is attributed to S.O. Rice in 1944, but the derivation given here is due to Madsen et al. in 1986.

or

$$\hat{\eta}_X(u,t,s_1,\cdots,s_j)=\nu_X^+(u,t,s_1,\cdots,s_j)-\int_0^{s_j}\hat{\eta}_X(u,t,s_1,\cdots,s_j,s_{j+1})\,ds_{j+1} \quad (11.38)$$

The corresponding relationship for $j=0$ is

$$\hat{\eta}_X(u,t)=\nu_X^+(u,t)-\int_0^t\hat{\eta}_X(u,t,s_1)\,ds_1 \quad (11.39)$$

describing the fact that the occurrence of an upcrossing at time t implies that the first upcrossing is either at t or at some time $s_1<t$. Now we can substitute for $\hat{\eta}_X(u,t,s_1)$ in Eq. 11.39 from Eq. 11.38 with $j=1$ to give

$$\hat{\eta}_X(u,t)=\nu_X^+(u,t)-\int_0^t\left(\nu_X^+(u,t,s_1)-\int_0^{s_1}\hat{\eta}_X(u,t,s_1,s_2)\,ds_2\right)ds_1$$

into which we can substitute for $\hat{\eta}_X(u,t,s_1,s_2)$ from Eq. 11.38 with $j=2$. Repetition of this procedure gives

$$\hat{\eta}_X(u,t)=\nu_X^+(u,t)+\sum_{j=1}^{\infty}(-1)^j\int_0^t\int_0^{s_1}\cdots\int_0^{s_{j-1}}\nu_X^+(u,t,s_1,\cdots,s_j)\,ds_j\cdots ds_2\,ds_1 \quad (11.40)$$

which is the general inclusion-exclusion relationship.

To use the inclusion-exclusion relationship, of course, one must have knowledge of the rate of multiple upcrossings term $\nu_X^+(u,t,s_1,\cdots,s_j)$ defined in Eq. 11.36. It is not difficult to write an integral expression for this term. In fact, a simple generalization of Eq. 7.2 gives

$$\nu_X^+(u,t,s_1,\cdots,s_j)=\int_0^{\infty}\cdots\int_0^{\infty}\int_0^{\infty}(v_j\cdots v_1\,v)\times$$
$$p_{X(s_j),\dot{X}(s_j),\cdots,X(s_1),\dot{X}(s_1),X(t),\dot{X}(t)}(u,v_j,\cdots,u,v_1,u,v)\,dv_j\cdots dv_1\,dv \quad (11.41)$$

Unfortunately, it is usually difficult to carry out the integration in Eqs. 11.40 and 11.41 unless j is quite small.

It should be noted that using only the first term ($j=0$) in Eq. 11.40 gives exactly the Poisson approximation. Furthermore, the alternating signs of the terms in Eq. 11.40 show that each truncation of the series gives either an upper or a lower bound on $\hat{\eta}_X(u,t)$. Thus, the Poisson approximation of $\nu_X^+(u,t)$ is an upper bound on $\hat{\eta}_X(u,t)$; including only the $j=1$ term from the summation in Eq. 11.40 gives a lower bound; and so on. It should be kept in mind, though, that we are not assured that the truncations of Eq. 11.40 will provide upper and lower bounds on the original $\eta_X(u,t)$, because $\hat{\eta}_X(u,t)$ omits conditioning by the event $X(0) \le u$. In fact, we have already argued that $\nu_X^+(u,t)$ surely does not give an upper bound on $\eta_X(u,t)$ for very low u values.

11.6 Extreme Value of Gaussian Response of SDF Oscillator

Inasmuch as we have no exact solution for the extreme value distribution, it is appropriate to use simulation data to verify which, if any, of the various approximate solutions may give accurate estimates. We will present this comparison only for the particular problem of stationary response of the SDF oscillator excited by mean-zero, Gaussian, white noise. We will consider the double-barrier problem, related to crossings of the level u by $|X(t)|$. One reason for the choice of this particular problem is the existence of multiple sets of simulation data that can be considered to provide reliable estimates of the true solution of the problem. The other reason is that this is a mathematical model often used in the estimation of first-passage failure in practical engineering problems.

Figure 11.2 shows analytical and simulation results for the situation of an oscillator with 1% of critical damping, which gives $\alpha_1 \approx 0.9937$. The plot includes the double-barrier versions of most of the analytical approximations we have considered for $\eta_{|X|}(u,t)/\nu_X^+(u,t)$. It is seen that the simulated values of this ratio are smaller than the predictions from any of the approximations, except when u is less than about $1.2\sigma_X$. Also, the simulation data clearly show that the ratio has a minimum value when u is approximately $2\sigma_X$. Of all the approximations, it appears that the analytical form due to Vanmarcke comes closest to fitting the simulation data. In particular, the modified Vanmarcke form of Eq. 11.35 gives a minimum value of $\eta_{|X|}(u,t)/\nu_X^+(u,t)$ at approximately the right u value, and the values it gives for the ratio in this vicinity are better than those of any of the other approximations that have reasonable behavior for smaller u values. Nonetheless, it must be noted that there is sometimes a significant discrepancy between Vanmarcke's formula and the simulation data. For $u=2\sigma_X$, for example, the approximation is about 70% above the value of 0.10 or 0.11 obtained from simulation, even though the modified approximation agrees almost perfectly with the data point from simulation for $u=4\sigma_X$. Any overestimation of $\eta_{|X|}(u,t)$ gives $L_X(u,t)$ values that decay more rapidly with increasing t than do the values from simulation. For $u=2\sigma_X$ and large values of

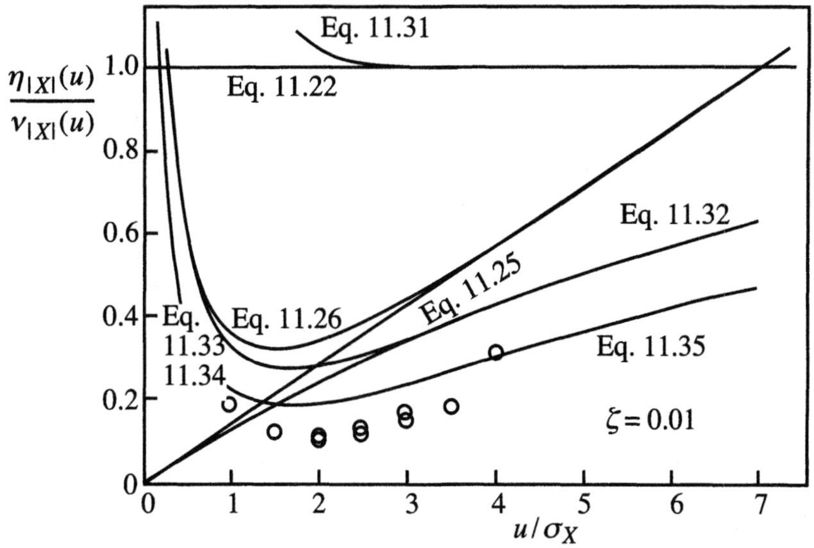

Figure 11.2 Simulation and various approximations.

time, Vanmarcke's formula will significantly overpredict the probability that $|X(t)|$ has ever reached the level u. This discrepancy, though, is smaller when u is large, and this is often the region of primary interest in predicting failure.

Simulation data also exist for larger values of damping in the SDF system, and Fig. 11.3 presents some of these values. The only analytical result shown for comparison is Vanmarcke's modified formula given in Eq. 11.35, because this seems to be better than any of the other approximate methods that we have considered. It is noted that for $\zeta \ge 0.05$ the Vanmarcke approximation fits the simulation data very well, and the error consistently increases as the damping is decreased below this level. In all cases it appears that the Vanmarcke approximation is significantly better than the value of unity predicted by assuming that a Poisson process describes the crossings of the level u by $|X(t)|$.

****

Example 11.6: Use the modified Vanmarcke approximation to estimate the probability of first-passage failure of the linear oscillator considered in Example 11.3. It has resonant frequency ω_0 and damping $\zeta = 0.01$, and it is excited by stationary, Gaussian, white noise with mean zero and autospectral density S_0. Failure occurs if $X(t)$ exceeds the level $4\sigma_{stat} = 4(\pi S_0)^{1/2} / (2m^2 \zeta \, \omega_0^3)^{1/2}$ within the time interval $0 \le \omega_0 t \le 250$.

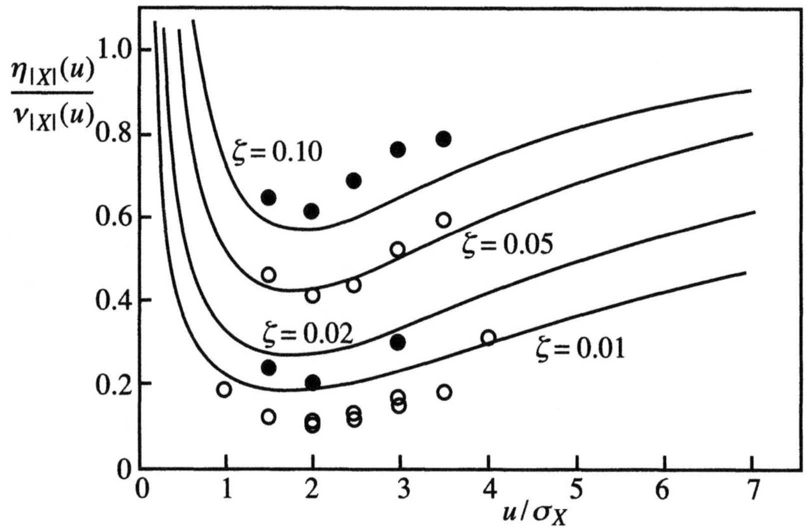

Figure 11.3 Simulation and Vanmarcke's approximation.

Considering $\{X(t)\}$ to have zero initial conditions requires that we use the nonstationary $\eta_X(u,t)$ from Eq. 11.30. In addition to the nonstationary crossing rate for $\{X(t)\}$, we must now use the corresponding crossing rate for $\{A_1(t)\}$. This was given in Example 7.14, but we now modify it for the current purpose by adding a power of 1.2 to the $(1-\alpha_1^2)$ term, as discussed in Section 11.4, giving

$$\nu_{A_1}^+(u,t) = \frac{u}{\sigma_X(t)} \exp\left(-\frac{u^2}{2\sigma_X^2(t)} \right) U(u) \frac{\sigma_{\dot{X}}(t)}{\sigma_X(t)} \times$$

$$\left(\frac{\rho_{X\dot{X}}(t)\, u}{\sigma_X(t)} \Phi\left[\frac{\rho_{X\dot{X}}(t)\, u}{\sigma_X(t)\left([1-\alpha_1^2(t)]^{1.2} - \rho_{X\dot{X}}^2(t)\right)^{1/2}} \right] + \right.$$

$$\left. \left(\frac{[1-\alpha_1^2(t)]^{1.2} - \rho_{X\dot{X}}^2(t)}{2\pi} \right)^{1/2} \exp\left[-\frac{1}{2}\left(\frac{\rho_{X\dot{X}}^2(t)\, u^2}{\sigma_X^2(t)\left([1-\alpha_1^2(t)]^{1.2} - \rho_{X\dot{X}}^2(t)\right)} \right) \right] \right)$$

Most of the time-varying terms in this expression were given in Example 11.3 and are the same here. The new information that is needed is the nonstationary bandwidth parameter $\alpha_1(t)$, which was investigated in Example 7.10. Evaluation of $\alpha_1(t) = \lambda_1(t)/[\lambda_0(t)\lambda_2(t)]^{1/2}$ requires a formula for $\lambda_1(t)$, and the expression for this term was given in Example 7.10 as

$$\lambda_1(t)=2S_0\Bigg([1+e^{-2\zeta\omega_0 t}]\int_0^\infty |H_x(\omega)|^2\,\omega\,d\omega + \omega_d^{-1}e^{-\omega_0 t}\sin(\omega_d t)\times$$
$$\left[\omega_0^2\int_0^\infty |H_x(\omega)|^2\sin(\omega t)\,d\omega+\int_0^\infty |H_x(\omega)|^2\,\omega^2\sin(\omega t)\,d\omega\right]-$$
$$2e^{-\zeta\omega_0 t}\cos(\omega_d t)\int_0^\infty |H_x(\omega)|^2\,\omega\cos(\omega t)\,d\omega\Bigg)$$

Using this formula along with the approximations in Example 11.3 allows evaluation of the conditional rate of upcrossings in Eq. 11.30:

$$\eta_X(u,t)\approx\nu_X^+(u,t)\left(1-\exp\left[\frac{-\nu_{A_1}^+(u,t)}{\nu_X^+(u,t)}\right]\right)\left(1-\frac{\nu_X^+(u,t)}{\nu_X^+(0,t)}\right)^{-1}$$

Numerical integration of Eq. 11.13 then gives $L(u,250/\omega_0)=0.99663$, and the probability of survival is $P_f\approx 0.0034$. This value is approximately 55% of that obtained in Example 11.3 by the Poisson approximation.

Comparison of $\eta_X(u,t)$ and $\nu_X^+(u,t)$ reveals a possible simplified approximation of this problem. In particular, the accompanying sketch shows that $\eta_X(u,t)$ is almost proportional to $\nu_X^+(u,t)$. Thus, one might use an approximation of $\eta_X(u,t)\approx \nu_X^+(u,t)[\eta_X(u,\infty)/\nu_X^+(u,\infty)]$. For the current problem the ratio of stationary crossing rates is 0.5166, so $\eta_X(u,t)\approx 0.5166\,\nu_X^+(u,t)$ and Eq. 11.13 yields

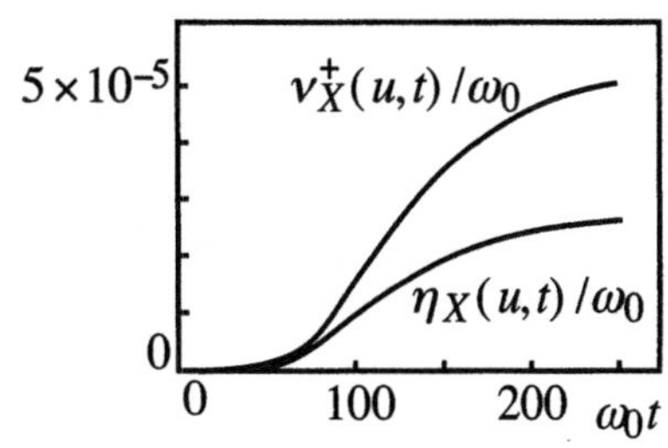

$$L_{Vanm}\approx(L_{Poisson})^{0.5166}=0.9968,\qquad P_{f,Vanm}\approx 0.0032$$

in which we have used the Poisson result from Example 11.3.The result is seen to be quite accurate, and it can be obtained with the much simpler integration of the unconditional nonstationary crossing rate, in place of the conditional rate.

****

The results in Example 11.6 indicate that it is sometimes acceptable to approximate the nonstationary Vanmarcke result for the response of the linear oscillator by using

$$\eta_X(u,t)\approx\frac{\eta_{X,Vanm}(u)}{\nu_X^+(u)}\nu_X^+(u,t)$$

in which $\nu_X^+(u)$ and $\eta_{X,Vanm}(u)$ denote the stationary values resulting for large time. This approximation then gives

$$P_{f,Vanm} \approx 1-(1-P_{f,Poisson})^{\eta_{X,Vanm}(u)/\nu_X^+(u)} \tag{11.42}$$

In the example this approximation is about 6% lower than the value obtained using Vanmarcke's modified formula for the nonstationary conditional rate of upcrossings. It can also be shown that this error depends on $c\,t/m \equiv 2\zeta\,\omega_0\,t$. In fact, the error is approximately proportional to $(c\,t/m)^{-2}$. It is not surprising that the error is small when t is very large, because that is a situation in which the $\{X(t)\}$ process is stationary throughout most of the time until failure. What is possibly surprising is that the error is only 6% for the situation in Example 11.6, in which $\nu_X^+(u)$ and $\eta_X(u)$ are clearly nonstationary. It should be noted that the approximation of Eq. 11.42 is nonconservative, in the sense that it predicts a smaller value for the probability of failure than is obtained directly from Vanmarcke's formulation. Nonetheless, it may be found useful, because it greatly simplifies the mathematical computations for a nonstationary process.

It can be expected that the approximation of Eq. 11.42 may also be used for some other problems, in addition to the response of the SDF oscillator. Crucial requirements for its application, of course, are that $\{X(t)\}$ is tending to a stationary process and that the time until failure is sufficiently large that stationarity is nearly achieved.

11.7 Accumulated Damage

Analysis of fatigue is typically based on the concept of a function $D(t)$ that represents the accumulated damage due to stresses and strains occurring prior to time t. This function is presumed to increase monotonically, and failure is expected when the accumulated damage reaches some critical level. Usually the damage function is normalized so that it reaches unity at failure—$D(T)=1$ if T is time of failure. This reasonable concept is very compatible with stochastic modeling, with $\{D(t)\}$ becoming a stochastic process when stresses are stochastic. The problem of predicting time of failure then becomes one of studying the $\dot{D}(t)$ rate of growth of $\{D(t)\}$.

The practical difficulty in implementing the accumulated damage model is that $D(t)$ is not observable by currently available means. $D(t)$ is known at only two points; it is presumed to be at least approximately zero for a specimen that has not yet been subjected to loads and it is unity when failure occurs. For a specimen that has been subjected to loads but has not yet failed, we have no method of measuring the amount of accumulated damage or its rate of growth. The nonobservability of damage is probably most relevant prior to the formation of an observable crack but also can be considered to be true during crack growth. Even if crack size is observable, there is no inherent reason to assume that $D(t)$ is proportional to that size. For example, if crack size is denoted by s, then $D(t)$ could equally well be proportional to s^a, in which a is any positive number. Some analysts consider the initial phase of crack formation simply to be the microscopic portion of the crack growth, whereas others make a greater distinction between the crack initiation and the crack propagation phases. In this introductory discussion we will use the term *fatigue* to describe both of the mentioned situations, and the concept of accumulated damage measured by a damage function $D(t)$ will be used to denote progress toward failure whether or not a crack has formed. It should also be noted that there are many definitions of the term *failure* in common usage. These vary from "appearance of a visible crack" to "complete fracture."

The rational analysis of fatigue is further complicated by the fact that there is generally a very considerable scatter in fatigue data. That is, apparently identical specimens will give greatly varying failure times when tested with identical loadings. In fact, it is not unusual to have the largest failure time be as much as 5 to 10 times greater than the smallest value from a set of identical tests. This variability is presumably due to microscopic differences between apparently identical specimens. Although this is surely a problem amenable to probabilistic analysis, it does not fall within the scope of the present study of the effect of stochastic loadings on failure prediction.

Current fatigue analysis methods are based on various approximations and assumptions about the $D(t)$ function. The goal in formulating such approximations is to achieve compatibility with the results of experiments. These experiments are sometimes performed with quite complicated time histories of loading, but they more typically involve simple periodic (possibly harmonic) loads in so-called *constant-amplitude* tests. It is presumed that each period of the motion contains only one peak and one valley. In this situation, the number of cycles until fatigue failure is usually found to depend primarily on the amplitude of the cycle, although this is usually characterized with the alternative nomenclature of *stress range*, which is essentially the double amplitude, being equal to a peak value minus a valley value. There is a secondary dependence on the mean stress value, which can be taken as the average of the peak and valley stresses. The exact shape of the cycle of periodic loading has been found not to be important. Neither is the frequency of loading important, except for either very high frequencies or situations of corrosion fatigue.

We will use the notation S_r to denote the stress range of a cycle and let N_f designate the number of cycles until failure in a constant-amplitude test.[4] A typical experimental investigation of constant-amplitude fatigue for specimens of a given configuration and material involves performing a large number of tests including a number of values of S_r, then plotting the (S_r,N_f) results. This is called an *S/N curve,* and it forms the basis for most of our information and assumptions about $D(t)$. We will emphasize the dependence of fatigue life on stress range by writing $N_f(S_r)$ for the fatigue life observed for a given value of the stress range. In principle, the S/N curve of $N_f(S_r)$ versus S_r could be any nonincreasing curve, but experimental data commonly show that a large portion of that curve is well approximated by an equation of the form

$$N_f(S_r) = K\, S_r^{-m} \tag{11.43}$$

in which K and m are positive constants whose values depend on both the material and the geometry of the specimen. This S/N form plots as a straight line on log-log paper. If S_r is given either very small or very large values, then the form of Eq. 11.43 will generally no longer be appropriate, but fatigue analysis usually consists of predicting failure due to moderately large loads, so Eq. 11.43 is often quite useful.

Because of the considerable scatter in S/N data, it is necessary to use a statistical procedure to account for specimen variability and choose the best $N_f(S_r)$ curve. For the common power law form of Eq. 11.43, the values of K and m are usually found from so-called linear regression, which minimizes the mean-squared error in fitting the data points. It should be noted that this linear regression of data does not involve any use of a probability model, so is distinctly different from the probabilistic procedure with the same name, which was discussed briefly in Section 3.3.

As noted, the fatigue life also has a secondary dependence on the mean stress. Although this effect is often neglected in fatigue prediction, there are

---

[4]It should be noted that S_r is a completely different concept from spectral density, although the symbol S is used for both. It is hoped that the subscript r on the stress range will assist in avoiding confusion about the notation.

simple empirical formulas that can be used to account for it in an approximate way. The simplest such approach is called the *Goodman correction*, and it postulates that the damage done by a loading $x(t)$ with mean stress x_m and stress range S_r is the same as would be done by an "equivalent" mean-zero loading with an increased stress range of S_e, given by $S_e = S_r / (1 - x_m / x_u)$, in which x_u denotes the ultimate stress capacity of the material. The *Gerber correction* is similar, $S_e = S_r / [1 - (x_m / x_u)^2]$, and has often been determined to be in better agreement with empirical data. Note that use of the Goodman or Gerber correction allows one to include both mean stress and stress range in fatigue prediction while maintaining the simplicity of a one-dimensional S/N relationship.

11.8 Stochastic Fatigue

As previously noted, our fundamental problem in stochastic analysis of fatigue is to define a stochastic process $\{D(t)\}$ to model the accumulation of damage caused by a stochastic stress time history $\{X(t)\}$. Some rather mathematically sophisticated formulations have been postulated for this purpose. Among these is one in which $\{D(t)\}$ is modeled as a Markov process, such that the increment of damage during any "duty cycle" at time t is a random variable that depends on both $D(t)$ and the stress $X(t)$ but is independent of prior values (Bogdanoff and Kozin, 1985). This is a general approach, but it has been found to be very difficult to identify the conditional probability distribution of the damage increment based on experimental data (Madsen et al., 1986; Sobczyk and Spencer, 1992).

Other approaches to modeling damage focus exclusively on crack growth, treating crack size as an observable $D(t)$ function. Experimental data have been used, for example, in developing models in which the rate of crack growth depends on the values of prior overloads, which cause residual stresses affecting the state of stress and strain at the crack tip (see Sobczyk and Spencer, 1992). In many practical situations, though, a specimen may have very little remaining fatigue life after the first appearance of a visible crack. Furthermore, it is not obvious whether the models that describe visible crack growth also apply to earlier parts of the fatigue process. Clearly, there is limited benefit in having an accurate model of visible crack growth if that is only a small fraction of the total fatigue life. Overall, it seems that most predictions of fatigue life in practical problems are based on simpler models in which information about the accumulated damage is obtained from the constant-amplitude S/N curve. This is the approach that we will present here.

Because experimental fatigue data are typically characterized by the number of cycles to failure rather than time, we will define the accumulated damage to be the sum of a number of discrete quantities:

$$D(t) = \sum_{j=1}^{N(t)} \Delta D_j$$

in which ΔD_j denotes the increment of damage during cycle j and $N(t)$ is the number of applied cycles of load up to time t. Furthermore, let T = failure time so that $N_f = N(T)$ is the number of cycles to failure. This gives $D(T) = 1$, so

$$1 = \sum_{j=1}^{N(T)} \Delta D_j \qquad (11.44)$$

This model is very reasonable for the constant-amplitude situation, although it does require devising a method to relate the magnitude of ΔD_j to the characteristics of the stress time history. Applying it to a stochastic problem also requires a method for defining cycles within a complicated time history of stress. Both of these problems are addressed with fairly crude approximations.

First we note that Eq. 11.44 indicates that the average value of ΔD_j over an entire constant-amplitude fatigue test at constant stress level S_r is $1/N_f(S_r)$. Let us now assume that the conditional expected value of ΔD_j for all the cycles of stress level S_r within a complicated or stochastic time history will have this same average level of damage per cycle: $E(\Delta D | S_r = u) = 1/N_f(u)$. This then gives the expected value of damage per cycle for any cycle, with random stress range S_r, within the time history as

$$E(\Delta D) = \int_0^\infty p_{S_r}(u)\, E(\Delta D | S_r = u)\, du = E[1/N_f(S_r)] \qquad (11.45)$$

and the expected number of cycles to failure is given by

$$E[N(T)] = 1/E[1/N_f(S_r)] \qquad (11.46)$$

Note that Eqs. 11.45 and 11.46 are identical to the results that one would obtain by assuming that the damage caused by any cycle of range S_r was exactly $1/N_f(S_r)$, whether that cycle was in a constant-amplitude or a stochastic time history. However, our derivation of the relationship does not require such a drastic assumption. It requires only that the damage due to cycles of range S_r distributed throughout a stochastic time history be similar to that for cycles of the same range distributed throughout a constant-amplitude time history. The results in Eqs. 11.45 and 11.46 are also equivalent to the common Palmgren-Miner hypothesis that

$$1 = \sum_{j=1}^{N(T)} [N_f(S_{r,j})]^{-1} \tag{11.47}$$

When $D(t)$ is approaching unity, we expect the number $N(t)$ of cycles to be large, because it approaches the fatigue life $N(T)$ in the limit. Furthermore, it is generally true that $\sigma_D(t)/\mu_D(t)$ decays like $[N(t)]^{-1/2}$ as $N(t)$ becomes large, in which $\sigma_D(t)$ represents only the uncertainty about damage due to the uncertainty about the stochastic time history. Precisely, the necessary condition for this decay is that the damage increments ΔD_j and ΔD_k become uncorrelated when j and k are well separated such that

$$\sum_{l=-\infty}^{\infty} \text{Cov}[\Delta D_j \, \Delta D_{j+l}] < \infty$$

which can also be stated as the requirement that the sequence of values be ergodic in mean value (see Section 4.7). Under these conditions, the variations of the random ΔD_j values tend to average out so that D has relatively less uncertainty than do the individual ΔD_j values. This has also been verified by much more rigorous techniques (see Crandall and Mark, 1963). As in constant-amplitude testing, there is considerable uncertainty about T, and the gist of the statement that $\sigma_D(t) << \mu_D(t)$ for stochastic time histories is simply that having a stochastic time history results in very little increase in the uncertainty about T, compared with a constant-amplitude situation. In fact, experimental data often suggest that there is less statistical scatter in stochastic fatigue than in deterministic fatigue.

The fact that $\sigma_D(t)/\mu_D(t)$ becomes very small as t approaches T in most fatigue problems assures us that σ_T/μ_T is also very small, which allows us to use the approximation

$$E[D(\mu_T)] \approx 1 \tag{11.48}$$

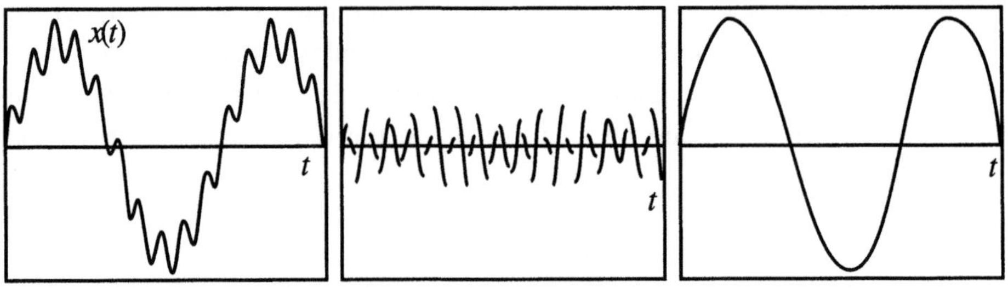

(a) Original time history (b) Half-cycles based on subsequent extrema (c) Cycles based on largest extrema

Figure 11.4 Inadequate cycle identification schemes.

That is, we say that mean time to failure, μ_T, is approximately the time at which the mean value of damage, $\mu_D(t)$, reaches unity. Furthermore, we have little need to estimate $\sigma_D(t)$ or the probability density function if $\sigma_D(t) << \mu_D(t)$. Any random variable with $\sigma << \mu$ is almost deterministic at the value μ. Thus, we will concentrate on estimating μ_T as approximated by Eq. 11.48.

The remaining problem in developing the model is to divide a complicated time history into cycles and to predict the rate of occurrence of those cycles. One of the most obvious cycle identification schemes is to consider the segment of a stress time history $x(t)$ between any two subsequent local extrema (from a peak to a valley or from a valley to a peak) to be a half cycle. In this scheme, the number of cycles is the same as the number of peaks. Another simple scheme ignores all but the largest extrema between any two subsequent upcrossings of zero, then proceeds as discussed. This generally gives a significantly smaller number of cycles—namely, the number of upcrossings of the level zero. Neither of these schemes, though, seems to give appropriate answers in some quite simple situations. In particular, consider the effect of adding a high-frequency component to a basically low-frequency time history, as shown in Fig. 11.4. Using the subsequent peaks and valleys identifies the half cycles shown in part (b) of the figure. This gives a large number of cycles, but no large half cycles are identified, because there are many peaks and valleys between any high peak and any low valley of the original time history shown in part (a). Using only the largest peaks and valleys does give large half cycles but completely ignores most of the effect of the high-frequency component, as illustrated in part (c) of the figure. Thus, it becomes apparent that more elaborate techniques are needed for identifying cycles within a complicated time history.

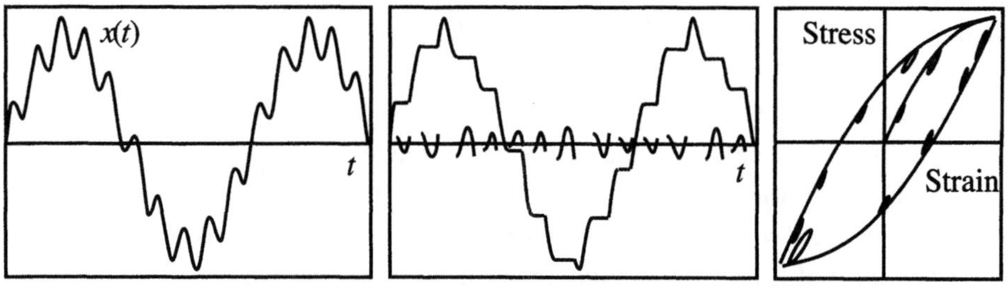

(a) Original time history (b) Rainflow cycles (c) Hysteresis loops

Figure 11.5 Rainflow cycle identification scheme.

Clearly, for identifying cycles within a complicated time history one would like to use a scheme that does count all the cycles but that does not lose the large cycles that happen to be interrupted by small cycles. The most commonly adopted cycle identification scheme of this type is called the *rainflow* method.[5] When a rainflow analysis is complete, every segment of $x(t)$ belongs to exactly one of the identified half cycles, and there is one half cycle terminating and another beginning at each extremum (peak or valley) of the time history. Furthermore, the peaks and valleys are paired in such a way as to give the largest possible half cycle, then the largest possible remaining half cycle, and so forth. If one assumes that the stress-strain behavior is slightly nonlinear and satisfies Masing's hypothesis (Masing, 1926),[6] then a rainflow cycle can be identified with a closed stress-strain hysteresis loop. Various other interpretations and several algorithms for rainflow analysis of given time histories have been published, with one of the simplest and most efficient algorithms being that of Downing and Socie (1982).

Figure 11.5 illustrates the cycles identified by the rainflow method for the time history of Fig. 11.4 and also illustrates the concept of relating rainflow cycles with closed hysteresis loops for a nonlinear stress-strain relationship. Although the plot shows a very pronounced nonlinearity, this is only for purposes of illustration. It is presumed that the material really behaves in an almost linear manner, but an accurate plot of this type would hide the hysteresis loops that we seek to illustrate. The hysteresis loops are shown with $x(t)$ being the stress time

[5]This method is generally attributed to Matsuishi and Endo in 1968. It was thoroughly studied by Dowling (1972) and is available in standard reference books such as the one by Fuchs and Stephens (1980).

[6]This hypothesis states that either half of any closed hysteresis loop has the same shape as the initial loading curve for the specimen, but it is magnified by a factor of 2.

history, but for a material that is almost linear it makes little practical difference whether $x(t)$ is associated with the stress or the strain axis in the plot.

Because the rainflow method gives one cycle for each peak of the time history, it is easy to calculate the rate of occurrence of rainflow cycles by using the formula for ν_P from Eq. 7.5. The other necessary calculation is the evaluation of $E[1/N_f(S_r)]$ in Eq. 11.46. If the S/N curve is taken to have the power law form of Eq. 11.43, though, then Eq. 11.46 reduces to

$$\frac{1}{E[N(T)]} \equiv E(\Delta D) = K^{-1} E(S_r^m) \qquad (11.49)$$

so the crucial step in estimating the expected value of the fatigue life is simply the evaluation of the mth moment of the stress range. It should be kept in mind, though, that m may not be an integer, because it is determined on the basis of obtaining a best fit of experimental data.

Ideally, we would next evaluate $E(S_r^m)$ for rainflow cycles. Lindgren and Rychlik (1987) have derived the relationship between the probability distribution of the rainflow cycles and that of the sequence of extrema of the stress process. The difficulty, though, is in finding or estimating the joint probability distribution of the extrema. Formulas for the marginal distributions were given in Section 11.2, but it is generally not feasible to obtain the joint distributions. Rychlik (1989) has presented an approximate result based on the assumption that the extrema of the stress process form an n-step Markov chain with a finite number of states, and he has presented some numerical results for the cases of $n=1$ and $n=2$. However, this involves using a numerical procedure to derive the needed conditional probability distributions of the extrema of a given time history.

Thus, the problem of finding a closed-form solution or simple approximation for the probability distribution of rainflow ranges remains unsolved. Nonetheless, there are certain limiting cases in which one can evaluate the mth moment of S_r for rainflow cycles. The most notable of these is when the $\{X(t)\}$ process is Gaussian and very narrowband, and this situation will be investigated in the following section. It should also be noted that one always has the option of studying stochastic fatigue by simulating long time histories of samples from $\{X(t)\}$ and then performing deterministic rainflow analysis of these samples. This, however, is somewhat awkward for use as a routine design tool. More easily implemented approximate methods are discussed in the following two sections.

11.9 Rayleigh Fatigue Approximation

Several analytical approximation techniques exist whereby one can estimate the stochastic fatigue life based only on the knowledge of the autospectral density function of the stress process. In particular, these methods typically make use of the values of certain spectral moments, as defined in Section 7.3. These techniques are classified as spectral methods. The first spectral method that we will discuss, called the Rayleigh approximation method, is one of the simplest and most widely used analytical techniques for stochastic fatigue analysis. This method was originally developed to predict the fatigue life under a narrowband Gaussian loading.

For a very narrowband $\{X(t)\}$ stress process with $\mu_X(t)=0$, it is reasonable to say that the value of the stress range S_r is twice the value of either the peak or amplitude of the process. Of course, the peak and amplitude distributions are essentially identical for such a very narrowband mean-zero process. Furthermore, the number of cycles per unit time of the process can be taken as either the rate of occurrence of peaks or the rate of occurrence of upcrossings of the mean value of $\{X(t)\}$. Calculation of the mth moment of S_r in this situation is particularly simple if $\{X(t)\}$ is also Gaussian, because the amplitude then has the Rayleigh distribution. Neglecting any effect of mean stress, one finds that (Miles, 1954)

$$E(\Delta D)=K^{-1}E(S_r^m)=K^{-1}2^m\int_0^\infty \frac{u^{m+1}}{\sigma_X^2}\exp\left(\frac{-u^2}{2\sigma_X^2}\right)du$$
$$=K^{-1}(2)^{3m/2}\sigma_X^m\,\Gamma\left(1+\frac{m}{2}\right)$$

or

$$E(\Delta D)=K^{-1}(2)^{3m/2}\lambda_0^{m/2}\,\Gamma\left(1+\frac{m}{2}\right) \tag{11.50}$$

in which $\Gamma(\cdot)$ is the gamma function and λ_0 denotes the zero-order spectral moment. The simplest form for the appropriate rate of cycle occurrence for this mean-zero narrowband process is given by the rate of upcrossings of the mean value, $\nu_X^+(\mu_X)$. Because the process is Gaussian (see Example 7.1), this is given by $\sigma_{\dot{X}}/(2\pi\sigma_X)$, so the estimate of failure time can be written as

$$E(T) = \frac{2\pi\sigma_X}{E(\Delta D)\sigma_{\dot{X}}} = \frac{2\pi}{E(\Delta D)}\left(\frac{\lambda_0}{\lambda_2}\right)^{1/2} \tag{11.51}$$

If $\mu_X \neq 0$ is constant throughout the loading process, one can correct for this fact by using the Goodman or Gerber correction to obtain an equivalent stress range, which merely introduces a constant scale factor for each range.

The so-called *Rayleigh approximation* of fatigue damage consists of using Eqs. 11.50 and 11.51 to predict $E(T)$. Its simplicity is, no doubt, one of the strong motivations for its widespread use. To predict $E(T)$ by using this method, one needs to find only σ_X and $\nu_X^+(\mu_X)$ for the $\{X(t)\}$ process. Furthermore, these two characteristics are easily evaluated either from an autospectral density curve, by using the λ_0 and λ_2 spectral moments, or from a representative sample time history. In addition, it should be noted that the results of the Rayleigh approximation tend to those of rainflow analysis in the limiting situation of a very narrowband Gaussian process. A common approach is to take the rainflow results as the basis of comparison for other approximate techniques, and on this basis the Rayleigh method becomes perfect in the limit as the bandwidth of a Gaussian stress process tends to zero.

Although Eqs. 11.50 and 11.51 were obtained from narrowband assumptions, they are also very commonly used to predict the fatigue life for stress processes that are not narrowband. In fact, the Rayleigh approximation is probably the most widely used analytical method for predicting fatigue life for any sort of stochastic stress process. It must be kept in mind, though, that the assumptions in the Rayleigh method are not necessarily appropriate for broadband processes, and simulation results confirm that Rayleigh predictions can differ significantly from rainflow predictions when stress processes are not narrowband. This is true for stress processes with broadband autospectral densities, but it is much more noticeable for a so-called bimodal autospectral density, having two narrowband peaks at different frequency values (Wirsching and Light, 1980; Ortiz and Chen, 1987; Lutes and Larsen, 1990).

Using Eqs. 11.50 and 11.51 for a stress process that is not narrowband may be viewed as finding the damage for an "equivalent narrowband process." That is, rather than attempting to define cycle ranges in $\{X(t)\}$, it is presumed that the damage done by $\{X(t)\}$ is the same as would be done by a narrowband Gaussian process with the same rate of upcrossings of its mean value. It is difficult to assess the validity of this choice of an "equivalent" process. Using the $\nu_X^+(\mu_X)$ rate of mean crossings as the rate of cycle occurrence seems to neglect the effect of high-frequency components. Using S_r as twice the value of the peaks of the narrowband process, however, gives unduly large magnitudes of the ranges for many time histories, such as ones similar to those in Figs. 11.4 and 11.5. Thus, the two errors in the Rayleigh approximation at least partially cancel each other.

****

Example 11.7: Compare the Rayleigh and rainflow predictions of the fatigue life for the special case of a stationary, mean-zero, Gaussian stress process and an S/N curve given by Eq. 11.43 with $m=1$.

From Eqs. 11.50 and 11.51 the Rayleigh approximation for $m=1$ can be written as

$$E(T)=\frac{2\pi\sigma_X}{K^{-1}(2)^{3/2}\sigma_X\,\Gamma\left(\frac{3}{2}\right)\sigma_{\dot X}}=\frac{(2\pi)^{1/2}K}{\sigma_{\dot X}}$$

The situation with $m=1$ is a special case in which we can also exactly evaluate the rainflow prediction of the fatigue life. In particular, Eqs. 11.43 and 11.47 give the failure condition as

$$1=K^{-1}\sum_{j=1}^{N(T)}S_{r,j}$$

However, we can rewrite this summation of S_r values by noting the contributions to the summation from each time increment of length dt. In particular, there is an excursion $|\dot X(t)|dt$ during the time increment, and this increment of excursion becomes a part of some $S_{r,j}$ stress range. Thus, it adds directly to the summation of all those stress ranges, and we can say that

$$\sum_{j=1}^{N(T)}S_{r,j}=\frac{1}{2}\int_0^T|\dot X(t)|\,dt$$

in which the factor of 1/2 comes from the fact that a full cycle with range S_r corresponds to a total excursion of $2S_r$. Substituting this relationship and taking the expected value gives $2=K^{-1}E(T)\,E(|\dot X(t)|)$. For the Gaussian process, we find that

$$E(|\dot X(t)|)=2\int_0^\infty\frac{u}{(2\pi)^{1/2}\sigma_{\dot X}}\exp\left(\frac{-u^2}{2\sigma_{\dot X}^2}\right)du=\left(\frac{2}{\pi}\right)^{1/2}\sigma_{\dot X}$$

and this gives

$$E(T)=\frac{2K}{E(|\dot X(t)|)}=\frac{(2\pi)^{1/2}K}{\sigma_{\dot X}}$$

This, however, is identical to the result of the Rayleigh approximation. Note, also, that we have used no assumption that the stress is narrowband. This is the only situation in which the Rayleigh approximation is in perfect agreement with rainflow analysis regardless of the form of the autospectral density of the stress.

****

Example 11.8: A particular mechanical bracket has been subjected to constant-amplitude fatigue tests at the single level of $S_r = 150$ MPa, and the observed fatigue life was $N_f = 10^6$ cycles. Based on experience with similar devices, it is estimated that the parameter m of the S/N curve is in the range of $3 \le m \le 5$. Based on this limited information, it is necessary to choose a level for the standard deviation σ_X of the narrowband, mean-zero, Gaussian stress that will be applied to the bracket in actual service. Using the limiting values of $m = 3$ and $m = 5$, find the level of σ_X meeting each of two possible design situations: $E[N(T)] = 10^6$ and $E[N(T)] = 10^8$.

Beginning with $m = 3$ we find the value of K in the S/N curve of Eq. 11.43 such that $10^6 = K/(150)^3$, giving $K = 3.375 \times 10^{12}$. For $E[N(T)] = 10^6$, Eq. 11.50 then gives $10^{-6} = K^{-1} 2^{4.5} (\sigma_X)^3 (3\pi^{1/2}/4)$. Solving this gives the answer of $\sigma_X = 48.2$ MPa. Similarly, $E[N(T)] = 10^8$ gives $\sigma_X = 10.4$ MPa.
For $m = 5$, we proceed in the same way and find that $K = 7.594 \times 10^{16}$. For $E[N(T)] = 10^6$, Eq. 11.50 becomes $10^{-6} = K^{-1} 2^{7.5} (\sigma_X)^5 (15\pi^{1/2}/8)$, which gives $\sigma_X = 41.7$ MPa, and $E[N(T)] = 10^8$ gives $\sigma_X = 16.6$ MPa.
Given the uncertainty about the value of m, the safe choice is to use $\sigma_X = 41.7$ MPa for a design condition of $E[N(T)] = 10^6$ and $\sigma_X = 10.4$ MPa for a design condition of $E[N(T)] = 10^8$. That is, one should use the smaller stress for the given design condition, to avoid failure due to uncertainty about the value of m.

Note that one must be cautious in choosing conservative approximations for this type of problem. In particular, one cannot conclude that either choice of m value is generally more conservative than the other. For the design condition of $E[N(T)] = 10^6$, it is more conservative to assume that $m = 5$, because this gives an allowable stress that is smaller than the value one would obtain by using $m = 3$. For the alternative design condition of $E[N(T)] = 10^8$, though, it is more conservative to assume that $m = 3$, because this gives the smaller allowable stress value. In general, one must consider the range of possible m values when dealing with such limited experimental data rather than simply using the largest possible or smallest possible m value.

****

11.10 Other Spectral Fatigue Methods for Gaussian Stress

One attempt to improve on the Rayleigh approximation technique can be called the *peak-approximation method.* The fundamental idea of this approach is to estimate the probability distribution of the stress ranges according to the probability distribution for the peaks of a Gaussian $\{X(t)\}$ process, as given in Example 11.1, and to calculate the number of cycles according to the rate of occurrence of peaks. For a process that is not very narrowband, these assumptions sound more plausible than the amplitude distribution and the mean crossing rate used in the Rayleigh approximation. Actually both of these peak-approximation assumptions are slightly altered in practice, because some peaks of $\{X(t)\}$ are negative, and these peaks cannot correspond to values of the nonnegative S_r quantity. The modification to neglect the negative peaks consists of using an appropriate conditional probability distribution. It was shown in Example 11.2 that the fraction of negative peaks for any continuous $\{X(t)\}$ process is $(1-IF)/2$, in which IF is the irregularity factor, so the relevant conditional probability density function for positive peaks for a mean-zero process becomes

$$p_{P(t)}(u \mid P(t) \ge 0) = \left(\frac{2}{1+IF}\right) p_{P(t)}(u)\, U(u) \tag{11.52}$$

One cannot generally perform a simple analytical evaluation of $E[P^m(t)]$ with m not being an integer for this distribution because of the Gaussian cumulative distribution function that appears in $p_{P(t)}(u)$, but numerical integration can be used fairly easily. Also, various analytical approximations can be obtained by curve fitting the results of numerical integration. One such simple approximation was given by Lutes et al. (1984):

$$E[P^m(t) \mid P(t) \ge 0] \approx (2)^{m/2} \sigma_X^m \left(1 + \frac{IF(1-IF)}{4}\right)^{(-1+m/2)} \frac{\Gamma\left(\dfrac{m+1+IF}{2}\right)}{\Gamma\left(\dfrac{1+IF}{2}\right)} \tag{11.53}$$

This equation agrees exactly with the results of Eq. 11.52 for the special cases of $IF = 0$ and $IF = 1$, for any positive m value. Furthermore, it is within 3% of results from numerical integration for all IF values for m values varying from one to nine.

Because the peak-approximation method for stress ranges is based only on positive peaks, it is natural to use the rate of occurrence of positive peaks as the rate of occurrence of cycles in this approach. Thus, $(1+IF)n_P/2$ becomes the rate of occurrence of stress ranges in this approximation, and one can also use the fact that $\nu_P = \nu_X^+(\mu_X)/IF$. The final result of using these expressions in conjunction with $S_r = 2P(t)$ for nonnegative peaks can be written as

$$E(T) \approx \frac{2\pi K}{2^m E[P^m(t) \mid P(t) \ge 0]} \frac{2(IF)}{(1+IF)} \frac{\sigma_X}{\sigma_{\dot X}}$$
$$= \frac{4\pi K(IF)}{2^m (1+IF) E[P^m(t) \mid P(t) \ge 0]} \left(\frac{\lambda_0}{\lambda_2} \right)^{1/2} \tag{11.54}$$

Equations 11.53 and 11.54 can then be used to obtain results for the peak-approximation method. The information required about $\{X(t)\}$ is the value of the irregularity factor, in addition to the σ_X and $\sigma_{\dot X}$ values that are needed for the Rayleigh approximation. In terms of spectral moments, one now needs λ_0, λ_2, and λ_4, because *IF* for a Gaussian process is the same as α_2, which is computed from these three spectral moments (see Eq. 7.12).

The basic concept of the peak-approximation method seems to be very sound for a narrowband process. Using the conditional distribution for $P(t)$ seems to be somewhat more reasonable than using the Rayleigh distribution, and using the rate of occurrence of positive peaks gives a correction for the inherent underestimation of the number of cycles when only crossings of the mean value are counted. Numerical results show, though, that these two corrections mostly cancel out for a narrowband process with $IF \approx 1$, so $E(T)$ is changed very little. Using the more accurate probability distribution for S_r reduces the value for $E(S_r^m)$, and this almost offsets the effect of the increased rate of cycle occurrence. For a narrowband process the peak-approximation method is reasonable, but it gives the same results as the simpler Rayleigh approximation. For some other problems, the peak-approximation method and the Rayleigh approximation may give very different results, but one cannot generally conclude that one is better than the other. Examples 11.10 and 11.11 will illustrate two such situations.

Another simple spectral technique is called the *single-moment method* (Lutes and Larsen, 1990). It is particularly easy to apply when one is given an autospectral density curve, because it depends on only one spectral moment. It has the general form of $E(T) = c(\lambda_a)^b$, in which a, b, and c are positive constants that may depend on the specimen but not on the autospectral density. One way to determine the appropriate values for a, b, and c is on the basis of making this formula agree with the rainflow and Rayleigh methods for the limiting situation of a very narrowband process. In particular, consider an autospectral density of

$$S_{XX}(\omega)=\frac{\sigma_X^2}{2}[\delta(\omega+\omega_0)+\delta(\omega-\omega_0)]$$

so that the λ_a spectral moment is given by $\lambda_a=(\sigma_X)^2(\omega_0)^a$ and the single-moment approximation gives

$$E(T)=c\,\sigma_X^{2b}\,\omega_0^{ab}$$

Equations 11.50 and 11.51 give the Rayleigh approximation for this autospectral density, as

$$E(T)=\frac{2\pi}{\omega_0\,E(\Delta D)}=\frac{2\pi K}{(2)^{3m/2}\Gamma\left(1+\frac{m}{2}\right)\omega_0\,\sigma_X^m}$$

Thus, the single-moment method agrees with the Rayleigh method in the narrowband situation only if

$$a=\frac{2}{m},\qquad b=\frac{-m}{2},\qquad c=\frac{2\pi K}{(2)^{3m/2}\Gamma\left(1+\frac{m}{2}\right)}$$

With these parameter values, the single-moment method can be written as

$$E(T)=\frac{2\pi K}{(2)^{3m/2}\Gamma\left(1+\frac{m}{2}\right)(\lambda_{2/m})^{m/2}}\tag{11.55}$$

The development of Eq. 11.55 ensures that the single-moment expression agrees with the Rayleigh approximation for a narrowband process, and we know that this also assures agreement with the rainflow method for this type of autospectral density. The rather surprising fact is that the results of Eq. 11.55 also are in relatively good agreement with simulation results from the rainflow method for a great variety of autospectral densities. Some of the data that support this conclusion are included in Example 11.11. It should be noted that $2/m$ is generally not an integer, so this method uses a noninteger spectral moment.

Although this may seem unusual, it is consistent with the fact that simulation data have indicated that rainflow damage seems to be more closely related to noninteger spectral moments than to integer spectral moments such as λ_1, λ_2, and so on. (Lutes et al., 1984).

Ortiz and Chen (1987) have presented another spectral method using noninteger spectral moments and have demonstrated that it gives very good results for some spectral densities. It may be written as

$$E(T) = \frac{2\pi K (\lambda_{2+2/m})^{m/2}}{(2)^{3m/2}\Gamma\left(1+\dfrac{m}{2}\right)(\lambda_{2/m})^{m/2}\lambda_2^{(m-1)/2}\lambda_4^{1/2}} \tag{11.56}$$

which is slightly more complicated to use than the single-moment method because it involves four spectral moments. Rainflow results from simulation have shown that this method is sometimes more accurate than the single-moment method but that in the worst case its error is substantially greater than for any situation with the single-moment method. This is illustrated in Examples 11.8 and 11.9.

Recall that the Rayleigh approximation is simple and widely known and involves assumptions that are reasonable for a narrowband process. Thus, it may be helpful to characterize any other method for analyzing a broadband process in terms of how much its fatigue life prediction differs from that of the Rayleigh approximation. For this purpose, we define a Rayleigh ratio term as

$$RR = \frac{E(T) \text{ by alternative approximation}}{E(T) \text{ by Rayleigh approximation}} \tag{11.57}$$

An idea of this type was apparently first introduced by Wirsching and Light (1980), who characterized the difference between the Rayleigh approximation and the rainflow method by using a correction factor CF of the form $CF = (RR)^{-1}$.

The Rayleigh ratio may be viewed as a method of interpreting results, but if this RR factor for the rainflow method were known as a function of some bandwidth parameter, then it would also provide a simple method of estimating the fatigue life. One could find the rainflow prediction of $E(T)$ by calculating the Rayleigh approximation and then multiplying by RR. In addition to the problem of predicting RR, though, there is no simple way to know what bandwidth parameter will work best for this purpose. Wirsching and Light obtained empirical correction factor values by simulating time histories of response, performing rainflow analysis, and plotting the results versus the *IF* irregularity factor. For their autospectral density curves and *m* values of 3, 4, 5, and 6, they found that their simulation results gave values of CF that could be approximated by

$$CF = a + (1-a)\left(1 - [1-(IF)^2]^{1/2}\right)^b \tag{11.58}$$

in which

$$a = 0.926 - 0.033m, \quad b = 1.587m - 2.323 \tag{11.59}$$

This relationship implies that different spectral density curves having nearly the same value of *IF* give *CF* values that are approximately the same, but Wirsching and Light found that this approximation was not accurate for $m = 10$. In particular, their *CF* values for $m = 10$ were found to show considerable scatter when plotted versus *IF*. Examples 11.10 and 11.11 demonstrate other situations in which the empirical relationship is not adequate.

For very narrowband processes, there is complete agreement between the results of all of the spectral prediction methods that we have discussed. This, of course, is appropriate. Our Palmgren-Miner approach involves making stochastic fatigue predictions based on the results of constant-amplitude tests, and the very narrowband process is the stochastic process that most closely resembles a constant-amplitude loading. It has been found, though, that there are sometimes significant differences between the results of the various fatigue prediction methods for stress processes that are not narrowband. The following three examples provide limited comparisons among the predictions of the various approximation schemes.

****

Example 11.9: Let the mean-zero, Gaussian stress process $\{X(t)\}$ have an autospectral density function of $S_{XX}(\omega) = S_0\, U(\omega_0 - |\omega|)$. For a specimen with $m = 4$, compare the fatigue lives predicted by the Rayleigh approximation, the peak-approximation method, the single-moment method, the formula of Ortiz and Chen, and the formula of Wirsching and Light.

For this spectral density, we find that the spectral moments are given by

$$\lambda_j = 2 S_0 (\omega_0)^{j+1} / (j+1)$$

From Eqs. 11.50 and 11.51 we obtain the Rayleigh approximation as

$$E(T_{Ray}) = 2\pi\left(\frac{\lambda_0}{\lambda_2}\right)^{1/2} \frac{K}{2^6 \lambda_0^2 (2)} = \frac{\pi K}{64 \lambda_0^{3/2} \lambda_2^{1/2}} = \frac{3^{1/2} \pi K}{256 S_0^2 \omega_0^3} \approx \frac{K}{47.05 S_0^2 \omega_0^3}$$

The irregularity factor is $IF = \lambda_2 / (\lambda_0 \lambda_4)^{1/2} = 5^{1/2}/3$, and Eqs. 11.53 and 11.54 then give the result of the peak-approximation method as

$$E(T_{peak}) \approx K/(47.15 S_0^2 \omega_0^3)$$

Similarly, Eqs. 11.58 and 11.59 give $a \approx 0.794$, $b \approx 4.025$, and $RR = (CF)^{-1} \approx 1.256$. To evaluate the single-moment approximation, we first find $\lambda_{2/m} = \lambda_{1/2} = 4 S_0 (\omega_0)^{3/2}/3$; Eq. 11.55 then gives

$$E(T_{sm}) \approx K/(36.22 S_0^2 \omega_0^3)$$

Finally, we also evaluate $\lambda_{2+2/m} = \lambda_{5/2} = 4 S_0 (\omega_0)^{7/2}/7$, and Eq. 11.56 gives the Ortiz and Chen approximation as

$$E(T_{oc}) \approx K/(38.18 S_0^2 \omega_0^3)$$

Comparing the results, we see that the peak-approximation method predicts a fatigue life that is 0.2% smaller than the Rayleigh approximation. The single-moment method, the approximation of Ortiz and Chen, and the Wirsching and Light correction factor all predict a fatigue life that is greater than the Rayleigh prediction, with the increases being 30%, 23%, and 26%, respectively. The spectral density used here is of a type considered by Wirsching and Light in developing their empirical correction factor from rainflow simulation data, so their formula should be expected to be quite accurate for this problem. Thus, it is safe to conclude that the Rayleigh and peak-approximation approximations significantly underpredict the failure time for this problem, as compared with the rainflow method.

****

Example 11.10: Let the mean-zero, Gaussian stress process $\{X(t)\}$ be the response of an SDF oscillator excited by white noise. For a specimen with $m = 4$, compare the fatigue lives predicted by the Rayleigh approximation, the peak-approximation method, the single-moment method, the formula of Ortiz and Chen, and the formula of Wirsching and Light.

From the results in Section 6.8, we can write the autospectral density for $\{X(t)\}$ as

$$S_{XX}(\omega) = S_0 m^{-2} \left([\omega_0^2 - \omega^2]^2 + [2\zeta \omega_0 \omega]^2\right)^{-1}$$

in which ω_0 and ζ are the resonant frequency and the damping of the oscillator and S_0 is the autospectral density of the white noise excitation. From the results in either Chapter 5 or 6, we know the values of λ_0 and λ_2 to be

$$\lambda_0 \equiv \sigma_X^2 = \frac{\pi S_0}{2m^2 \zeta \omega_0^3}, \qquad \lambda_0 \equiv \sigma_{\dot{X}}^2 = \frac{\pi S_0}{2m^2 \zeta \omega_0}$$

Using these two spectral moments in Eqs. 11.50 and 11.51 with $m=4$ gives the Rayleigh approximation result as

$$E(T_{Ray})=\frac{2\pi K}{2^6\sigma_X^4\Gamma(3)\,\omega_0}=\frac{K\,m^4\zeta^2\omega_0^5}{16\pi S_0^2}$$

It is not possible to apply some of the other spectral methods to this particular problem. In particular, this $S_{XX}(\omega)$ gives $\lambda_4=\infty$, as was noted in Example 7.9. Thus, we have $IF=0$, for which Eq. 11.54 gives $E(T_{peak})=0$, and Eq. 11.58 gives $RR=(CF)^{-1}=a^{-1}\approx 1.26$ for the approximation of Wirsching and Light. Similarly, Eq. 11.56 gives $E(T_{oc})=0$ for the Ortiz and Chen approximation, because $\lambda_4=\infty$ and the other pertinent spectral moments are finite.

To apply the single-moment method to a situation with $m=4$, we must calculate the $\lambda_{2/m}=\lambda_{0.5}$ spectral moment. For our autospectral density, this can be written as

$$\lambda_{0.5}=2\int_0^\infty \omega^{1/2}\,S_{XX}(\omega)\,d\omega=\frac{2S_0}{m^2\omega_0^{5/2}}\int_0^\infty \frac{\Omega^{1/2}\,d\Omega}{(1-\Omega^2)^2+(2\zeta\,\Omega)^2}$$

in which the dimensionless variable Ω is ω/ω_0. This integral can be evaluated numerically to obtain values that can be used in Eq. 11.55 to give single-moment results. One convenient way to present these numerical results is as the RR Rayleigh ratio of Eq. 11.57. These RR values have been computed for ζ values varying from zero to unity, and it is found that they agree with $RR=1+\zeta$, which implies that

$$\lambda_{0.5}=\frac{\pi S_0}{2m^2\omega_0^{5/2}\zeta(1+\zeta)^{1/2}}$$

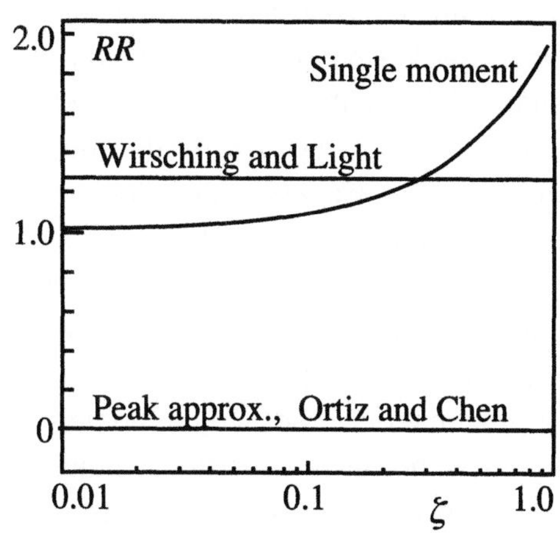

For $IF=0$, the Wirsching and Light formulas give a value of RR that depends only on m, not on the properties of the autospectral density function. This empirical approximation, thus, gives the rather unusual prediction that the rainflow fatigue life will be 26% greater than the Rayleigh approximation, regardless of the amount of damping in the oscillator. This value of RR agrees with the single-moment result only for $\zeta\approx 0.26$. The various results are sketched versus damping value.

This is an example in which the single-moment method is the only one of the alternative spectral methods that gives what one might consider to be a bandwidth correction for the Rayleigh approximation. That is, the bandwidth of the oscillator response is generally considered to depend on ζ, and only the single-moment method gives an RR value that depends on ζ.

****

Example 11.11: Let the mean-zero, Gaussian stress process $\{X(t)\}$ be the sum of two very narrowband processes, such that $\{X(t)\}$ has a bimodal autospectral density. For a specimen with $m=3$, compare the fatigue lives predicted by the Rayleigh approximation, the peak-approximation method, the single-moment method, the formula of Ortiz and Chen, and the formulas of Wirsching and Light.

We write the $\{X(t)\}$ stress process as $X(t)=X_1(t)+X_2(t)$, with $\{X_1(t)\}$ and $\{X_2(t)\}$ being independent processes with autospectral densities of $S_1(\omega)$ and $S_2(\omega)$, respectively. For simplification we also choose $S_2(\omega)$ to have the form $S_2(\omega)=(b/r)S_1(\omega/r)$ so that it has the same shape as $S_1(\omega)$. The parameters r and b are the relative frequency and the relative variance of $S_2(\omega)$ as compared with $S_1(\omega)$. That is, if $\{X_1(t)\}$ is narrowband at a frequency of ω_0 and has variance σ_0^2, then $\{X_2(t)\}$ is narrowband at frequency $r\omega_0$ and has variance $b\sigma_0^2$. This gives any spectral moment of $\{X(t)\}$ as having the form $\lambda_j=(\lambda_j)_0(1+br^j)$, in which $(\lambda_j)_0$ denotes the corresponding spectral moment of $\{X_1(t)\}$. The spectral method comparisons can be made even simpler by letting the autospectral density of $\{X_1(t)\}$ be the limiting narrowband function, giving

$$S_1(\omega)=(\sigma_0^2/2)[\delta(\omega+\omega_0)+\delta(\omega-\omega_0)], \quad \lambda_j=\sigma_0^2\,\omega_0^j(1+br^j)$$

One can now use these spectral moments in finding all the spectral predictions of $E(T)$, and the Rayleigh ratio values for the other three methods.

Note that the bimodal autospectral density reduces to a narrowband unimodal shape if b is either very small or very large, and also if $r\approx 1$. As expected, one can show that the RR values do tend to unity in these limiting situations. Furthermore, we also know that the Rayleigh approximation will be a good approximation of the rainflow analysis for these narrowband situations. For more general values of b and r, for which the autospectral density is bimodal, there can be notable differences in the various predictions. The peak-approximation method always predicts a shorter fatigue life than does the Rayleigh approximation. The single-moment method and the Wirsching and Light formulas, however, consistently predict that the fatigue life should be greater than the Rayleigh approximation. The Ortiz and Chen predictions are sometimes above and sometimes below the Rayleigh results. In fact, the primary difference between the results of Ortiz and Chen and those of the single-moment method is that the RR values are sometimes less than unity for the former method. Although the Wirsching and Light and single-moment results agree to the extent that they always give RR greater than unity, there are major differences in the numerical values. It is found that the differences among the various approximations become more pronounced when r takes on larger values.

The comparison of the various spectral methods, of course, does not necessarily demonstrate which method is most accurate. This assessment is usually based on the results of rainflow analysis of simulated time histories. One of the more extensive such studies (Larsen and Lutes, 1991) has included 180 bimodal autospectral densities, and we can compare Rayleigh ratio values from these results to those from the spectral approximations. We will do this, though, only for the largest r value included in the bimodal simulations, because this gives the greatest differences between the various predictions. Thus, for our situation with $m=3$, we plot RR values versus b for $r=15$. The plot at the right includes curves for all the spectral methods and the Wirsching and Light approximation, as well as data points obtained from rainflow analysis.

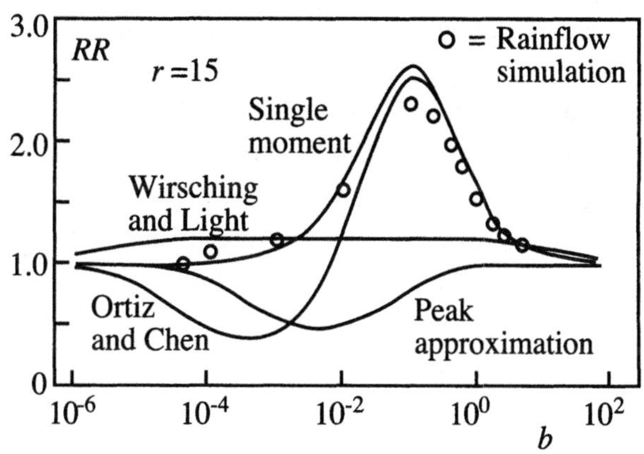

Clearly, the single-moment predictions are in the best agreement with the rainflow values. In particular, RR values less than unity, as predicted by the peak-approximation method and the Ortiz and Chen approximation, are not found by the rainflow method. Also, the rainflow RR values do substantially exceed the maximum value of 1.20 predicted by Wirsching and Light.

****

These three examples cover only a small fraction of the possible spectral density functions that one could investigate, but they do illustrate the similarities and critical differences that have been found among the various spectral methods. Example 11.9 is typical of problems in which the autospectral density is unimodal and limited to a finite frequency range. In such situations one generally finds that the predictions of the single-moment method, Ortiz and Chen's formula, and Wirsching and Light's formula are in reasonably good agreement, predicting fatigue life values that are somewhat above the Rayleigh prediction, whereas the result of the peak-approximation method is quite close to the

Rayleigh value.[7] The autospectral densities in Examples 11.10 and 11.11 emphasize the differences among the various methods. In particular, Example 11.10 removes the limitation of a bounded frequency range, and Example 11.11 introduces a bimodal autospectral density. Both of these situations must be considered common occurrences, so the differences revealed by these examples have serious implications regarding the reliability of the various spectral methods.

The peak-approximation method differs significantly from the Rayleigh approximation only for IF values smaller than about 0.5, but the difference can be significant for these nonnarrowband processes. However, the idea of using $S_r = 2P(t)$ with $P(t)$ representing every positive peak in a nonnarrowband process is not very reasonable. For a stochastic time history resembling Fig. 11.4(a), for example, it is obvious that $IF << 1$ and most of the additional peaks correspond to fairly small oscillations in the time history, not to completely reversed cycles with $S_r = 2P(t)$. For such a process with a broadband or bimodal autospectral density, it is not surprising that the peak-approximation method significantly overpredicts the rate of damage accumulation. Overall, the peak-approximation has never been found to be an improvement over the Rayleigh method, and in some broadband situations it is much less satisfactory than the Rayleigh method.

The major difficulty with the Ortiz and Chen procedure is that it sometimes predicts RR values that are significantly less than unity, whereas simulation has never demonstrated this to be true for the rainflow method. In Example 11.11 it was seen that the Ortiz and Chen prediction of $E(T)$ for $m = 3$ was sometimes only about 40% of the rainflow value for bimodal spectral densities, and Example 11.10 illustrated a situation in which the Ortiz and Chen prediction of $E(T)$ was zero. Thus, this formula can give major errors in the predicted fatigue life. The Wirsching and Light formula does always give RR values greater than unity, but these values are sometimes in significant disagreement with rainflow results, as shown in Example 11.11. Furthermore, its RR prediction for the SDF system of Example 11.10 is independent of the damping in the oscillator, whereas the bandwidth of the response process is generally acknowledged to vary with damping. The Wirsching and Light prediction that $E(T)$ is 26% greater than the Rayleigh value even if the damping tends to zero is not only inaccurate but also unconservative.

---

[7]It should be noted that an example of this type is also given by Wirsching et al. (1995).

Example 11.10 showed that the single-moment method is the only one of the considered spectral techniques that is capable of providing a consistent bandwidth correction to the Rayleigh approximation for the unimodal response of an SDF oscillator excited by white noise. Similarly, Example 11.11 showed that it is the only one that is consistent with rainflow analysis of bimodal time histories. Although there can be significant discrepancies between the single-moment method and rainflow analysis, these errors are much smaller than the maximum error for any of the other spectral methods. Overall, the single-moment method is particularly attractive, based on simplicity, general consistency with the idea of bandwidth, and agreement with rainflow analysis of simulated time histories.

Based on the available data, it appears that one should avoid using high-order spectral moments in any attempt to find a better spectral approximation method. In particular, a small high-frequency component in the autospectral density can cause a significant increase in the λ_4 spectral moment that enters into the Ortiz and Chen formulation. Furthermore, the major error of this method in Examples 11.8 and 11.9 occurs in situations in which there is a small high-frequency component that is contributing substantially to λ_4 and thereby reducing the predicted value of $E(T)$. The rainflow fatigue calculations, on the other hand, are not very sensitive to small high-frequency components, and the single-moment method, which uses only $\lambda_{2/m}$, also seems to avoid such difficulties. Inasmuch as the irregularity factor for a Gaussian process also depends on λ_4, one can anticipate difficulty when using any method that predicts fatigue life based on the *IF* value.

11.11 Non-Gaussian Fatigue Effects

Because fatigue damage is a significantly nonlinear function of stress, we anticipate that it may be sensitive to variations in the probability distribution of that stress. For example, if the stress process $\{X(t)\}$ has a greater-than-Gaussian probability of taking on large values, then this is likely to cause large stress ranges and these may cause significantly accelerated fatigue damage. We will summarize one simple approach for approximating this non-Gaussian effect for a narrowband stress process.

One method that has proved useful for modeling a non-Gaussian process $\{X(t)\}$ is that of using

$$X(t) = \mu_X(t) + \sigma_X(t) g[Y(t)] \tag{11.60}$$

in which $\{Y(t)\}$ is a mean-zero, Gaussian process and $g(\cdot)$ is a monotonic nonlinear function with $g(0)=0$. The restrictions placed on the $g(\cdot)$ function assure that the peaks, valleys, and mean-value crossings of $X(t)$ will be at the same time values as those of $Y(t)$. This can be considered as giving a very loose equivalence between the frequency content of the two processes. In particular, it ensures that they have the same values of $\nu_X^+(\mu_X)$, ν_P, and irregularity factor *IF*. In principle, it is always possible to choose the $g(\cdot)$ function so that $X(t)$ has any desired probability distribution, but this condition is often only approximated in practice. In particular, $g(\cdot)$ may be chosen to give the correct values only for the first few moments of $X(t)$.

We will now use Eq. 11.60 to estimate the fatigue damage accumulation for the non-Gaussian $\{X(t)\}$ process. In particular, we will use this idea to generalize the Rayleigh approximation method. In the Rayleigh approximation, we calculate the expected fatigue damage per cycle by taking the stress range as $S_r=2A(t)$, in which $\{A(t)\}$ is the amplitude of $\{X(t)\}$. Recall, though, that this involves the observation that the probability distribution of $A(t)$ is essentially the same as that of a peak $P(t)$ for a mean-zero, narrowband process. If we now let $P_X(t)$ and $P_Y(t)$ designate the peaks of $\{X(t)\}$ and $\{Y(t)\}$, respectively, then we can say that

$$P_X(t)=\mu_X(t)+\sigma_X(t)\,g[P_Y(t)]$$

because $g(\cdot)$ is a monotonic function. Similarly, the relationship between the corresponding valleys of the processes is

$$V_X(t)=\mu_X(t)+\sigma_X(t)\,g[V_Y(t)]$$

We now impose the condition that $\{Y(t)\}$ is narrowband, which gives $P_Y(t)=A_Y(t)$ and $V_Y(t)=-A_Y(t)$. Of course, the peak and valley do not occur simultaneously, but the narrowband assumption implies that $A(t)$ varies slowly, so we can consider it to have the same value at the times of the peak and the valley. This gives

$$S_{r,X}=P_X(t)-V_X(t)=\sigma_X(t)\big(g[A_Y(t)]-g[-A_Y(t)]\big)$$

Because $\{Y(t)\}$ is Gaussian and narrowband, we know that its amplitude $A_Y(t)$ has the Rayleigh distribution, so we can write the following integral for the expected damage per cycle:

$$E(\Delta D) = K^{-1} \sigma_X^m \, E\Big([g(A_Y) - g(-A_Y)]^m\Big)$$
$$= K^{-1} \lambda_0^{m/2} \int_0^\infty [g(u) - g(-u)]^m \, u e^{-u^2/2} \, du \qquad (11.61)$$

This can be used with Eq. 11.51 to give an estimate of the expected fatigue life. Consider now the special case in which the $g(\cdot)$ function is antisymmetric so that $g(-u) = -g(u)$. This further simplifies Eq. 11.61 to give

$$E(\Delta D) = 2^m K^{-1} \lambda_0^{m/2} \int_0^\infty g^m(u) \, u e^{-u^2/2} \, du \qquad (11.62)$$

There are many possible choices for the $g(\cdot)$ function (e.g., Lutes et al., 1984; Sarkani et al., 1994), but we limit our attention to cubic Hermite polynomial forms studied by Winterstein (1988):

$$X(t) = \sigma_X \sum_{j=1}^{3} b_j H_j[Y(t)] \quad \text{for } kurtosis > 3 \qquad (11.63)$$

and

$$Y(t) = \sum_{j=1}^{3} c_j H_j[X(t)/\sigma_X] \quad \text{for } kurtosis < 3 \qquad (11.64)$$

in which $H_j(\cdot)$ denotes the order j Hermite polynomial, as was briefly introduced in Section 10.6. It may be noted that Eq. 11.63 gives $g(\cdot)$ as a cubic function, whereas Eq. 11.64 actually gives a cubic form for the inverse of the $g(\cdot)$. The distinction is necessary in order to have $g(\cdot)$ be monotonic for the full range of kurtosis values.

Winterstein showed that the coefficients in Eqs. 11.63 and 11.64 can be approximated with the simplified relationships of

$$b_1 = (1 + 2\beta_2^2 + 6\beta_3^2)^{-1/2}, \quad b_2 = b_1 \beta_2, \quad b_3 = b_1 \beta_3$$

in which

$$\beta_3 = \frac{1}{18}\left(\left[1 + \frac{3}{2}(kurtosis - 3)\right]^{1/2} - 1\right), \quad \beta_2 = \frac{skewness}{6(1 + 6\beta_3)}$$

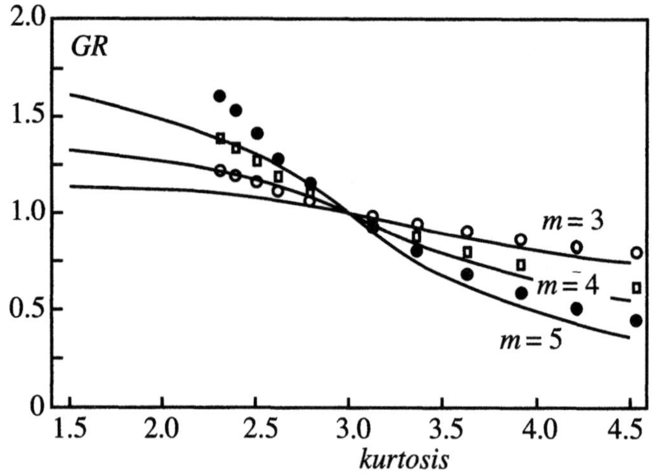

Figure 11.6 Non-Gaussian effect for narrowband process—Eq. 11.62.

and

$$c_1 = 1, \quad c_2 = -\frac{skewness}{6}, \quad c_3 = \frac{(3 - kurtosis)}{24}$$

Of course, it is necessary to invert Eq. 11.64 before we can use it to calculate the expected value of the damage increment in Eq. 11.61. One form in which the inverse can be written is

$$g(u) = \frac{c_2}{3c_3} + \left(Q_2 + [Q_2^2 - Q_1^3]^{1/2}\right)^{1/3} + \left(Q_2 - [Q_2^2 - Q_1^3]^{1/2}\right)^{1/3}$$

with

$$Q_1 = 1 - \frac{c_1}{3c_3} + \left(\frac{c_2}{3c_3}\right)^2, \quad Q_2 = \frac{u}{2c_3} + \frac{c_1 c_2}{6c_3^2} - \left(\frac{c_2}{3c_3}\right)^3$$

Numerical integration can then be used for the evaluation of $E(\Delta D)$.

To illustrate the effect of non-Gaussianity on fatigue, we introduce a Gaussian ratio term defined as

$$GR = \frac{E(T) \text{ for alternative distribution}}{E(T) \text{ for Gaussian distribution}} \tag{11.65}$$

Note that this is very similar in concept to the Rayleigh ratio RR introduced to present the effect of the autospectral density. Fig. 11.6 shows numerical values of

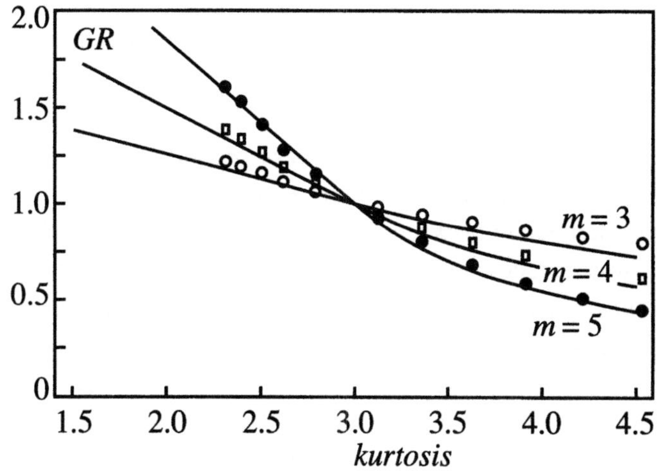

Figure 11.7 Non-Gaussian effect for narrowband process—Eq. 11.66.

GR for the symmetric situation with $skewness = 0$ and $c_2 = 0$. In addition to the results from numerical integration of Eq. 11.62, we include some results obtained from rainflow analysis of simulated time histories (Lutes et al., 1984). It should be noted that the simulation results represent the average of the results for several different stress processes, not all of which were very narrowband. Thus, the simulation data may show some effect of bandwidth as well as of kurtosis. Nonetheless, they do confirm that Eq. 11.62 gives the proper trend in predicting the non-Gaussian effect, particularly for $kurtosis > 3$.

Winterstein (1985, 1988) also offered a simplification whereby $(GR)^{-1}$ is approximated as a linear function of the kurtosis, matching the slope of the curve from Eq. 11.62 at the point $kurtosis = 3$. It has been found, though, that this approximation seems to overestimate *GR* for small kurtosis values and that the simulation data are better approximated by taking *GR* as a linear function of the kurtosis in that situation. Thus, a reasonable approximation seems to be

$$
\begin{aligned}
GR &= \left(1 + \frac{m(m-1)(kurtosis - 3)}{24}\right)^{-1} \qquad \text{for } kurtosis > 3 \\
GR &= 1 + \frac{m(m-1)(3 - kurtosis)}{24} \qquad \text{for } kurtosis < 3
\end{aligned}
\tag{11.66}
$$

Figure 11.7 shows that this approximation is in very good agreement with the simulation data. Rather surprisingly, it seems that these particular data are better fitted by Eq. 11.66 than by Eq. 11.62. There is no obvious reason to expect this to be true in general, but it does appear that the simple formulas in Eq. 11.66 may be adequate for predicting the non-Gaussian effect for moderately non-Gaussian processes.

One practical situation in which stress processes are known to be non-Gaussian relates to the stresses in offshore structures subjected to wave loadings. The nonlinear relationship between water velocity and the hydrodynamic loading can cause the stress process to have significantly elevated kurtosis values, even if the water waves are assumed to be Gaussian (Lutes and Wang, 1993). The skewness of the stress process has also been investigated for this offshore problem, but it appears that skewness is much less significant than kurtosis for predicting fatigue values (Wang and Lutes, 1993). These offshore investigations have also provided some evidence that if the stress is neither Gaussian nor narrowband, it is sometimes acceptable to use correction factors for both the autospectral density effect and the non-Gaussian effect. This approximation gives the final estimate of $E(T)$ as the Rayleigh approximation multiplied by $(RR)(GR)$.

It should be noted that at least one recent study found that applying the *GR* correction of Eq. 11.66 to the results of the single-moment method did not always give accurate results for a problem with a significantly non-Gaussian, but relatively narrowband, stress process (Yang et al., 2003). The authors made a detailed study of the distribution of peaks of their $\{X(t)\}$ process and derived an improved non-Gaussian approximation for fatigue. Their results can be summarized by the statement that if the probability density of $\{X(t)\}$ was written in the form $p_X(u) = C_1 \exp[-G(u)]$, then an improved fatigue analysis involves using $S_r = 2A$ with $p_A(u) = C_2 G'(u) \exp[-G(u)] U(u)$. It may be noted for the special case of a quadratic $G(u)$, these $p_X(u)$ and $p_A(u)$ relationships become exactly Gaussian and Rayleigh, respectively. The authors also corrected the frequency ν_0 of occurrence of fatigue cycles. Instead of using the value of $\sigma_{\dot{X}} / (2\pi\sigma_X)$ that is appropriate for a Gaussian process, they considered the special case of $m = 1$ and chose ν_0 to make $E(T) = K / [\nu_0 E(S_r)]$ match the exact result of $E(T) = 2K / E(|\dot{X}|)$ (see Example 11.7). This latter calculation was simplified by the fact that their $\{X(t)\}$ process represented the response of an oscillator with linear damping and a nonlinear spring, for which $\dot{X}(t)$ is Gaussian (see Eq. 10.15). Very good results were obtained for this problem by using the single-moment method with a *GR* value based on the derived $p_A(u)$ distribution and ν_0 value. Further study is needed to determine the range of problems for which this approach may be better than Winterstein's approximations.

****

Example 11.12: Consider the fatigue life of a structural joint for which the S/N curve has been found to be $N_f = K(S_r)^{-4}$. In actual use, the joint will be subjected to a narrowband stochastic stress $\{X(t)\}$ that is mean-zero and has standard deviation σ_X. This narrowband stress can be written as $X(t) = A(t)\cos[\omega_c t + \theta(t)]$, with $A(t)$ and $\theta(t)$ being independent and $\theta(t)$ uniformly distributed on the set of possible values. Rather than being Rayleigh distributed, it is believed that $A(t)$ has the one-sided Gaussian distribution

$$p_{A(t)}(u) = \frac{1}{\pi^{1/2}\sigma_X}\exp\left(\frac{-u^2}{4\sigma_X^2}\right)U(u)$$

Compare the fatigue life predicted by the Rayleigh approximation, the use of $S_r = 2A(t)$ and $p_{A(t)}(u)$ in Eq. 11.49, and the use of Eq. 11.66.

Before starting on the fatigue computation, it is appropriate to check for consistency of the model. In particular, we find that

$$E[A^2(t)] = \int_0^\infty u^2\, p_{A(t)}(u)\, du = 2\sigma_X^2$$

so the narrowband process gives $E[X^2(t)] = E[A^2(t)]/2 = \sigma_X^2$ (see Example 4.3). Thus, the model is consistent with σ_X being the standard deviation of the stress. Now we note that the Rayleigh approximation from Eqs. 11.50 and 11.51 is $E(T_{Ray}) = \pi K/[64(\lambda_0)^{3/2}(\lambda_2)^{1/2}]$, as in Example 11.9. For this narrowband process we can say that $\lambda_2 = \omega_c^2\lambda_0$, and we also know that $\lambda_0 = \sigma_X^2$. This allows the Rayleigh approximation to be rewritten as

$$E(T_{Ray}) = \pi K/(64\omega_c\sigma_X^4)$$

Using $S_r = 2A(t)$ in Eq. 11.49 gives

$$E(\Delta D) = K^{-1}E(S_r^4) = 2^4 K^{-1}\int_0^\infty u^4\, p_{A(t)}(u)\, du = 192\, K^{-1}\sigma_X^4$$

and Eq. 11.51 then gives the expected fatigue life as $E(T) = \pi K/[96\omega_c(\sigma_X)^4]$. Thus, this gives a prediction of $E(T)$ that is 33% smaller than the Rayleigh approximation.

In order to use Eq. 11.66, we must find the kurtosis of the stress process. To do this, we first integrate over the possible values for $\theta(t)$ to find that

$$E[X^4(t)] = E[A^4(t)]\int_0^{2\pi}\frac{\cos^4(\psi)}{2\pi}\, d\psi = \frac{3}{8}E[A^4(t)]$$

From the integral used in the previous step, we find that $E[A^4(t)] = 12\sigma_X^4$, so $E[X^4(t)] = 4.5\sigma_X^4$. Thus this process has $kurtosis = 4.5$, and Eq. 11.66 then gives $GR = 24/[24 + 12(1.5)] \approx 0.571$, which predicts a 43% reduction in fatigue life as compared with the Rayleigh approximation (which would be appropriate for a Gaussian process). This value is relatively consistent with the 33% reduction calculated in the previous step. To neglect the non-Gaussian effect would be a more serious error, because it would result in a fatigue life prediction that was significantly too large.

****

Exercises

****

Peak Distribution

****

11.1 Let $\{X(t)\}$ be a covariant stationary mean-zero Gaussian process with standard deviation σ_X, energy-based characteristic frequency ω_{c2}, and irregularity factor $IF = 0.5$.

(a) Evaluate $P[P(t) > b\sigma_X]$ for $b = -1$, 1, 3, and 5, in which $P(t)$ is any peak of $X(t)$.

(b) Compare the values in part (a) with those that would apply for the limiting cases of $IF = 0$ and $IF = 1$.

(c) Find the expected rate of occurrence of peaks with value greater than $3\sigma_X$.

(d) Find the expected rate of occurrence of peaks with value less than $-\sigma_X$.

****

11.2 Let $\{X(t)\}$ be a covariant stationary mean-zero Gaussian process with standard deviation σ_X, energy-based characteristic frequency ω_{c2}, and irregularity factor $IF = 0.6$.

(a) Evaluate $P[P(t) > b\sigma_X]$ for $b = -1.5$, 1.5, 3, and 4, in which $P(t)$ is any peak of $X(t)$.

(b) Compare the values in part (a) with those that would apply for the limiting cases of $IF = 0$ and $IF = 1$.

(c) Find the expected rate of occurrence of peaks with value greater than $3\sigma_X$.

(d) Find the expected rate of occurrence of peaks with value less than $-1.5\sigma_X$.

****

Extreme Value Distribution—Poisson Approximation

****

11.3 Each of the four formulas gives the joint probability density of $X(t)$ and $\dot{X}(t)$ for a particular covariant stationary stochastic process $\{X(t)\}$. For each process, estimate $L_X(u,t)$ by using the Poisson approximation. Also give a qualitative sketch showing the shape of $L_X(u,t)$ versus u, both for $t = 0$ and for some $t > 0$. [Note: The upcrossing rates for these problems were found in Exercise 7.2.]

(a) $p_{X(t)\dot{X}(t)}(u,v) = \dfrac{1}{\pi a^2} U(a^2 - u^2 - v^2)$

(b) $p_{X(t)\dot{X}(t)}(u,v)=\dfrac{1}{2(2\pi)^{1/2}\,b\,\sigma_2}\exp\left(-\dfrac{v^2}{2\sigma_2^2}\right)U(b-|u|)$

(c) $p_{X(t)\dot{X}(t)}(u,v)=\dfrac{1}{2\pi\sigma_1\sigma_2}\exp\left(-\dfrac{u^2}{2\sigma_1^2}-\dfrac{v^2}{2\sigma_2^2}\right)$

(d) $p_{X(t)\dot{X}(t)}(u,v)=\dfrac{3\alpha^{1/3}}{2\Gamma(1/3)}\left(\dfrac{\gamma}{\pi}\right)^{1/2}e^{-\alpha|u^3|-\gamma v^2}$

****

11.4 Consider the stationary $\{X(t)\}$ process of Exercise 7.3 with

$$p_{X(t)\dot{X}(t)}(u,v)=A\exp\left(-\lambda|u|-\frac{v^2}{2\sigma_2^2}\right)$$

Use the Poisson approximation to estimate the values of $L_X(3\sigma_X,t)$ and $L_Y(3\sigma_X,t)$ for $\{X(t)\}$ and the stationary Gaussian process $\{Y(t)\}$ with the same mean value and autocovariance function as $\{X(t)\}$.

****

11.5 Consider the stationary $\{X(t)\}$ process of Exercise 7.4 with

$$p_{X(t)\dot{X}(t)}(u,v)=A\exp\left(-\alpha u^4-\frac{v^2}{2\sigma_2^2}\right)$$

Use the Poisson approximation to estimate the values of $L_X(3\sigma_X,t)$ and $L_Y(3\sigma_X,t)$ for $\{X(t)\}$ and the stationary Gaussian process $\{Y(t)\}$ with the same mean value and autocovariance function as $\{X(t)\}$.

****

11.6 Let $\{X(t)\}$ be a covariant stationary, Gaussian process with the autospectral density function of Exercises 7.6 and 7.9, with $b=8$:

$$S_{XX}(\omega)=A|\omega|^8 e^{-c|\omega|}$$

(a) Use the Poisson approximation to obtain an estimate of the probability that the energy-based amplitude $A_2(t)$ will ever cross the level $u=4\sigma_X$ during a time interval of length $50c/8$.

(b) Perform the same calculation for the Cramer and Leadbetter amplitude $A_1(t)$.

****

11.7 Let $\{X(t)\}$ be a covariant stationary, Gaussian process with the autospectral density function of Exercise 7.19:

$$S_{XX}(\omega)=S_0\,U(\omega_0-|\omega|)+S_0\left|\frac{\omega_0}{\omega}\right|^4 U(-|\omega|-\omega_0)\qquad \text{with } \omega_0>0$$

(a) Use the Poisson approximation to estimate the probability that the Cramer and Leadbetter amplitude $A_1(t)$ will ever cross the level $u = 4\sigma_X$ during a time interval of length $100/\omega_0$.

(b) Is it feasible to perform the same calculation for the energy-based amplitude $A_2(t)$? Briefly explain your answer.

****

11.8 Let $\{X(t)\}$ represent the dynamic response of an oscillator governed by

$$\ddot{X}(t) + 20\zeta\,\dot{X}(t) + 100\,X(t) = F(t)$$

with $\{F(t)\}$ being mean-zero, Gaussian white noise with autospectral density $S_{FF}(\omega) = 400/\pi$ for all ω.

Use the Poisson approximation to estimate $L_X(u,t)$ for the following three situations:

(a) $\zeta = 0.20$, $u = 4$

(b) $\zeta = 0.20$, $u = 2$

(c) $\zeta = 0.05$, $u = 4$

(d) For which of the three situations {(a), (b), or (c)} does the Poisson approximation give the best estimate?

(e) For which of the three situations {(a), (b), or (c)} does the Poisson approximation give the worst estimate?

(f) Briefly explain your answers to parts (d) and (e).

****

11.9 Let $\{X(t)\}$ be a mean-zero Gaussian process with autospectral density given by

$$S_{XX}(\omega) = 4\,U(|\omega| - 10 + b)\,U(10 + b - |\omega|) \text{ mm}^2/(\text{rad/s})$$

Use the Poisson approximation to estimate $L_X(u,t)$ for the following three situations:

(a) $b = 4$ rad/s, $u = 32$ mm

(b) $b = 4$ rad/s, $u = 16$ mm

(c) $b = 1$ rad/s, $u = 16$ mm

(d) For which of the three situations {(a), (b), or (c)} does the Poisson approximation give the best estimate?

(e) For which of the three situations {(a), (b), or (c)} does the Poisson approximation give the worst estimate?

(f) Briefly explain your answers to parts (d) and (e).

****

11.10 Consider the problem of predicting the probability of structural yielding during a wind storm that has been modeled as a stationary segment of Gaussian stochastic force. It has been determined that the critical response is the $X(t)$ distortion in the first story and that yielding will occur if $|X(t)| \ge 60$ mm. The unyielded structural response $\{X(t)\}$ can be modeled as the response of a linear

SDF system with a 1 second period ($\omega_0 = 2\pi$ rad/s) and 2% of critical damping, and the stationary response has been found to have $\mu_X = 0$ and $\sigma_X = 11$ mm. The duration of the wind storm is 2 hours. Estimate the probability that yielding will occur during the 2-hour wind storm by using the Poisson approximation for the double-barrier problem.

****

11.11 You wish to estimate the probability that two buildings will collide during an earthquake. Let $X_1(t)$ be the response at the top of building 1 (the shorter building), $X_2(t)$ be the response of building 2 at the level of the top of building 1, and $Z(t) = X_1(t) - X_2(t)$. Collision will occur if $Z(t) > 20$ mm (i.e., there is 20 mm clearance in the static position). For your design earthquake, you find that the standard deviations and average frequencies of the individual stationary responses are:

$$\sigma_{X_1} = 2.5\,\text{mm},\ \nu^+_{X_1}(0,t) = 1.0 \text{ zero-upcrossings/s}$$

$$\sigma_{X_2} = 5.0\,\text{mm, and } \nu^+_{X_2}(0,t) = 0.25 \text{ zero-upcrossings/s}$$

Consider $\{X_1(t)\}$ and $\{X_2(t)\}$ to be mean-zero, independent, Gaussian processes.

(a) Find the standard deviation and the average frequency of zero-upcrossings by $\{Z(t)\}$.

(b) Find the rate of upcrossings by $\{Z(t)\}$ of the level $Z = 20$ mm.

(c) Assume that the probability of collision is the probability of at least one upcrossing of $Z = 20$ mm during 20 seconds of stationary response. Estimate this probability by using the Poisson approximation.

****

Extreme Value Distribution—Vanmarcke's Approximation

****

11.12 Let $\{X(t)\}$ be a covariant stationary Gaussian stochastic process with zero mean value and the autospectral density of Exercise 7.12:

$$S_{XX}(\omega) = |\omega|^5 e^{-\omega^2}$$

(a) Estimate $L_{|X|}(u,t)$ for $u = 10$ and $t = 10$ by using the Poisson approximation.

(b) Estimate $L_{|X|}(u,t)$ for $u = 10$ and $t = 10$ by using Vanmarcke's modified approximation (Eq. 11.35).

****

11.13 Let $\{X(t)\}$ be a covariant stationary Gaussian stochastic process with zero mean value and the autospectral density of Exercise 7.13:

$$S_{XX}(\omega) = |\omega| \exp(-\omega^2)$$

(a) Estimate $L_{|X|}(u,t)$ for $u = 3$ and $t = 75$ by using the Poisson approximation.

(b) Estimate $L_{|X|}(u,t)$ for $u = 3$ and $t = 75$ by using Vanmarcke's modified approximation (Eq. 11.35).

****

Rayleigh Fatigue Approximation

****

11.14 Use the Rayleigh approximation to estimate the expected number of cycles until failure for a mean-zero stress process $\{X(t)\}$ with $\sigma_X = 25\,\text{MPa}$, using each of the following S/N curves.

(a) $N_f = (13{,}000 / S_r)^{3.3}$

(b) $N_f = (7{,}000 / S_r)^{4.0}$

[Note: Properties of the gamma function are given in Example 3.7.]

****

11.15 For a material with an ultimate stress of $x_u = 516\,\text{MPa}$, use the Rayleigh approximation to estimate the expected number of cycles until failure for a stress process $\{X(t)\}$ with $\mu_X = 100\,\text{MPa}$ and $\sigma_X = 25\,\text{MPa}$, using the Gerber correction and the following S/N curves:

(a) $N_f = (13{,}000 / S_r)^{3.3}$

(b) $N_f = (7{,}000 / S_r)^{4.0}$

****

11.16 A structural joint has been subjected to constant-amplitude fatigue tests at the single level of $S_r = 200\,\text{MPa}$, and the observed fatigue life was $N_f = 4 \times 10^5$ cycles. Based on experience with similar joints, it is estimated that the m parameter of the S/N curve is in the range of $3.5 \le m \le 4.5$. Based on this limited information, choose acceptable levels for the standard deviation σ_X of the narrowband, mean-zero, Gaussian stress that will be applied to the joint in actual service. You need consider only the two limiting values of $m = 3.5$ and $m = 4.5$, not all the possible m values. Solve the problem for the design conditions of:

(a) $E[N(T)] = 10^6$ cycles

(b) $E[N(T)] = 10^8$ cycles

****

11.17 A machine part has been subjected to constant-amplitude fatigue tests at the single level of $S_r = 100\,\text{MPa}$, and the observed fatigue life was $N_f = 5 \times 10^6$ cycles. Based on experience with similar parts it is estimated that the m parameter of the S/N curve is in the range of $3.0 \le m \le 4.5$. Based on this limited information, choose acceptable levels for the standard deviation σ_X of the narrowband, mean-zero, Gaussian stress that will be applied to the joint in actual service. You need consider only the two limiting values of $m = 3.0$ and $m = 4.5$, not all the possible m values. Solve the problem for the design conditions of:

(a) $E[N(T)] = 10^7$ cycles

(b) $E[N(T)] = 10^9$ cycles

****

11.18 The S/N curve for a particular connection has been found to be $N_f = (10{,}000/S_r)^{3.5}$. In service it is subjected to a stress process $\{X(t)\}$ that is mean zero and has $\sigma_X = 12\,\text{MPa}$. Furthermore, $\{X(t)\}$ can be modeled as the response of an SDF oscillator subjected to a Gaussian white noise excitation. The natural frequency and damping ratio of the oscillator are $\omega_0 = 3$ rad/s and $\zeta = 0.02$. Use the Rayleigh approximation to estimate the expected hours of service prior to fatigue failure.

****

Single-Moment Fatigue Method

****

11.19 Let the mean-zero, Gaussian stress process $\{X(t)\}$ have an autospectral density function of $S_{XX}(\omega) = S_0 \exp(-\beta|\omega|)$. For a specimen with $m = 3$, compare the fatigue lives predicted by the Rayleigh approximation and the single-moment method.

****

11.20 Let the mean-zero, Gaussian stress process $\{X(t)\}$ have an autospectral density function of

$$S_{XX}(\omega) = S_0 \left(1 - \frac{1}{5}\left[\frac{|\omega|}{\omega_0} - 1\right]\right) U(|\omega| - \omega_0)\, U(6\omega_0 - |\omega|)$$

For a specimen with $m = 4$, compare the fatigue lives predicted by the Rayleigh approximation and the single-moment method.

****

11.21 Let the mean-zero, Gaussian stress process $\{X(t)\}$ have an autospectral density function of

$$S_{XX}(\omega) = S_0\, U(\omega_0 - |\omega|) + S_0 \left(\frac{\omega_0}{|\omega|}\right)^4 U(|\omega| - \omega_0)$$

For a specimen with $m = 3$, compare the fatigue lives predicted by the Rayleigh approximation and the single-moment method.

****

11.22 Let the mean-zero, Gaussian stress process $\{X(t)\}$ represent the response of a 2DF system with lightly damped modes. In particular, the fundamental mode has damping $\zeta_1 = 0.01$ and frequency $\omega_1 = 6$ rad/s, and the second mode has $\zeta_2 = 0.02$ and frequency $\omega_2 = 80$ rad/s. The standard deviations of the modal stress values are $\sigma_{X_1} = 100\,\text{MPa}$ and $\sigma_{X_2} = 7.5\,\text{MPa}$. Make an approximate comparison of the fatigue lives predicted by the Rayleigh approximation and the single-moment method for a specimen with $m = 4$. Do this by considering the two modal responses to be independent and concentrated at their natural frequencies ω_1 and ω_2 so that the autospectral density can be approximated by the form used in Example 11.11.

****

Non-Gaussian Fatigue Effects

****

11.23 Consider the fatigue life of a structural joint for which the S/N curve has been found to be $N_f = K(S_r)^{-3}$. In actual use the joint will be subjected to a non-Gaussian narrowband stochastic stress $\{X(t)\}$ that is mean zero and has a standard deviation σ_X. This narrowband stress can be written as $X(t) = A(t)\cos[\omega_c t + \theta(t)]$, with $A(t)$ and $\theta(t)$ being independent, $\theta(t)$ being uniformly distributed on the set of possible values, and $A(t)$ having an exponential distribution: $p_{A(t)}(u) = (\sigma_X)^{-1} \exp(-u/\sigma_X) U(u)$. Evaluate the fatigue life predicted by:

(a) using the Rayleigh approximation.
(b) using $S_r = 2A(t)$ and $p_{A(t)}(u)$ in Eq. 11.49.
(c) using Eq. 11.66.

****

11.24 Consider the fatigue life of a structural joint for which the S/N curve has been found to be $N_f = K(S_r)^{-4}$. In actual use the joint will be subjected to a non-Gaussian narrowband stochastic stress $\{X(t)\}$ that is mean zero and has a standard deviation σ_X. This narrowband stress can be written as $X(t) = A(t)\cos[\omega_c t + \theta(t)]$, with $A(t)$ and $\theta(t)$ being independent, $\theta(t)$ being uniformly distributed on the set of possible values, and $A(t)$ having a Weibull distribution of the form

$$p_{A(t)}(u) = \left(\frac{\pi u^3}{4\sigma_X^4}\right) \exp\left(\frac{-\pi u^4}{16\sigma_X^4}\right) U(u)$$

Evaluate the fatigue life predicted by:

(a) using the Rayleigh approximation.
(b) using $S_r = 2A(t)$ and $p_{A(t)}(u)$ in Eq. 11.49.
(c) using Eq. 11.66.

****

Chapter 12
Effect of Parameter Uncertainty

12.1 Basic Concept

All the preceding chapters have been based on the presumption that the equation of motion of the dynamic system is described deterministically. The only source of randomness has been the stochastic excitation. In most cases the systems considered have had constant coefficients, but even when we considered the parameters to vary with time, that variation was deterministic. It is rather obvious that this is a major simplifying assumption in many practical situations, because there may be considerable uncertainty about the parameter values in many problems, especially structural engineering problems. Taking the multi-degree-of-freedom (MDF) system as the prototypical problem, there may be uncertainty about the appropriate values of the **m**, **c**, and **k** matrices during a particular dynamic event, even though it may be appropriate to consider them not to vary throughout the dynamic analysis. Similarly, in the state-space formulation of the equations of motion in Section 8.6 and Chapter 9, there is often uncertainty about the appropriate values for the **A** and **B** matrices in the equation of motion.

We have previously noted that there is almost always uncertainty about the energy dissipation during dynamic motion of a mechanical or structural system, so the components of the **c** matrix may vary significantly from their "design" values. The mass of a system is generally known much more accurately, and the same is generally true for the stiffness of a mechanical system. There may be significant uncertainty about both mass and stiffness, though, in structural engineering systems, particularly buildings. The mass magnitude and distribution within a building depend on the details of the usage of the building. While a building is in use, the mass distribution can surely be ascertained reasonably well by careful study of the contents, but this can be time-consuming and costly. If one wishes to do a *post mortem* of a building after a disaster, such as an earthquake, there may be much more uncertainty about the mass matrix. Probably the greatest uncertainty about the stiffness matrix for a structure has to do with the effect of "nonstructural" components. Testing has shown that partitions within a building do add significant stiffness, at least during small deflections, even though they are not considered load-bearing components. Thus, both **k** and **m** must also be considered uncertain in many practical problems.

One of the first issues is to decide what one wishes to learn about the effect of parameter uncertainty. To be more specific, let Q denote the response quantity of interest and a vector $\vec{r}$ contain all the uncertain parameters. A complete description of the problem, then, may be considered to be the finding of Q as a function of $\vec{r}$. If $\vec{r}$ has only one or two components, for example, then one could plot $Q(\vec{r})$. If $\vec{r}$ has many components, though, this may be too much information. Furthermore, the $Q(\vec{r})$ function may be so complicated that even numerical evaluation is prohibitive. For practical purposes, one often needs some simplified formulation that quantifies the effect of parameter uncertainty without direct consideration of the full $Q(\vec{r})$ function. Such simplified formulations are the focus of this chapter.

First consider the problem of determining how the response would be altered if a single parameter were changed by a small amount. Approximating this as a linear problem gives the key factors as being the so-called *sensitivity coefficients*, defined as first derivatives:

$$\beta_l \equiv \left(\frac{\partial Q}{\partial r_l}\right)_{\vec{r}=\vec{r}_0} \tag{12.1}$$

in which $\vec{r}_0$ is the original design point in parameter space. Roughly speaking, one can say that a Δr_l change in parameter r_l will cause a change of $\beta_l\,\Delta r_l$ in the value of Q. Of course, this proportionality may not hold for finite changes in the parameter if Q depends on r_l in a significantly nonlinear way. This sensitivity analysis has been carried out for a number of deterministic problems, in particular in connection with stability studies. In this chapter we are particularly interested in focusing on problems of stochastic response, and we also wish to consider more than infinitesimal amounts of uncertainty about the parameters. Nonetheless, the sensitivity coefficients in Eq. 12.1 will play a key role.

The most common approach for modeling uncertainty about the parameters of a problem is to use probability analysis, in which each uncertain parameter is treated as being a random variable. There are other approaches, such as fuzzy sets, but we will not consider these here. At first glance it may seem that we are now introducing a new problem for which none of our previous work will be applicable. Fortunately, though, the situation is not this bad. In particular, any analysis using deterministic values of the parameters may be considered to give a conditional solution, given those specific parameter values. For example, let $\mathcal{M}$, $\mathcal{C}$, and $\mathcal{K}$ be matrices of random variables denoting the uncertain mass, damping, and stiffness, respectively, of an MDF system. Then it can be said that each analysis with fixed values of the $\mathbf{m}$, $\mathbf{c}$, and $\mathbf{k}$ matrices gives a solution conditioned on the event that $\mathcal{M}=\mathbf{m}$, $\mathcal{C}=\mathbf{c}$, and $\mathcal{K}=\mathbf{k}$. Thus, the analyses from the preceding chapters are not irrelevant, but they are inadequate for a complete description of the problem. The new problem is to find the effect of variations of the parameters away from the deterministic values we have used in our analyses.

Rather than beginning with a modeling of $\mathcal{M}$, $\mathcal{C}$, and $\mathcal{K}$, it is often convenient to consider there to be a random vector $\vec{\mathcal{R}}$ of uncertain parameters with $\mathcal{M}$, $\mathcal{C}$, and $\mathcal{K}$ each being functions of $\vec{\mathcal{R}}$. This can be particularly useful when the components of $\vec{\mathcal{R}}$ can be considered to be independent random variables, whereas $\mathcal{M}$, $\mathcal{C}$, and $\mathcal{K}$ do not have independent components and may not be independent of each other. For example, if a building owner installs additional nonstructural walls within the building, this modification will almost surely increase both the mass and the stiffness of the structure and will probably also increase the damping. In principle, introducing the new parameter vector $\vec{\mathcal{R}}$ does add much additional complication to the problem. If one knows the probability distribution for the vector $\vec{\mathcal{R}}$ and knows how $\mathcal{M}$, $\mathcal{C}$, and $\mathcal{K}$ each depends on $\vec{\mathcal{R}}$, then one can find the probability distribution of $\mathcal{M}$, $\mathcal{C}$, and $\mathcal{K}$. In practice, this step is simple if the dependence on $\vec{\mathcal{R}}$ is linear, and less so if it is nonlinear. Approximations are often used.

Before proceeding further it is worthwhile to identify the notation that will be used. The symbols $\vec{r}$, $\mathbf{m}$, $\mathbf{c}$, $\mathbf{k}$, and so on, will denote general values of parameters and may be considered as variables in our problems. The specific values $\vec{r}_0$, $\mathbf{m}_0$, $\mathbf{c}_0$, $\mathbf{k}_0$, and so forth, will denote the "design point" used when uncertainty is ignored. Finally, $\vec{\mathcal{R}}$, $\mathcal{M}$, $\mathcal{C}$, $\mathcal{K}$, and so on, will denote random variables defined on the selected ranges for $\vec{r}$, $\mathbf{m}$, $\mathbf{c}$, $\mathbf{k}$, and so forth. Throughout the following analysis we will adopt a common convention in which the uncertain parameters are modeled such that the mean value of the vector $\vec{\mathcal{R}}$ is the design point $\vec{r}_0$. This choice is certainly not essential, but it does sometimes simplify both the results and the comparisons with the solutions without parameter uncertainty.

Modeling the parameters as random variables gives the response quantity $\mathcal{Q}=\mathcal{Q}(\vec{\mathcal{R}})$ as also being a random variable. Thus, one might consider the probability distribution of Q to be a complete probabilistic description of the effect of random parameter uncertainty. This general problem of finding the probability distribution of a function of random variables was studied in Section 2.6, but it can become very complicated for the problems of interest here. A less complete, but more practical, approach is to study the mean and variance of Q. Roughly speaking, knowledge of the first moment of Q provides information about the extent to which consideration of parameter uncertainty may shift the most likely value of the response to a higher or lower value than predicted from the design point. Similarly, the variance, or standard deviation, of Q indicates the extent to which the actual value of Q may differ significantly from its own mean value.

The general problem of finding the moments of a function of random variables was considered in Section 3.3. In principle, one can always find the moments of $Q(\vec{\mathcal{R}})$ by either analytical or numerical integration over the chosen probability distribution of $\vec{\mathcal{R}}$. We will include such "exact" results in some examples presented in this chapter. The practical difficulty is that the functional relationship is sometimes quite complicated, making it necessary to use approximate methods of analysis that will simplify computations. Another approach, which we will not pursue here, is the so-called *Monte Carlo method*, in which independent samples of $\vec{\mathcal{R}}$ are simulated and the value of Q is found for each of these (Astill and Shinozuka, 1972; Shinozuka, 1972; Chang and Yang, 1991; Zhang and Ellingwood, 1995). Statistical averages over the resulting samples of Q give estimates of the moments or probability distribution of Q. The difficulty with the Monte Carlo method, as with direct evaluation of the quantities from integration and differentiation, is the time and cost involved in the procedures when the $Q(\vec{\mathcal{R}})$ relationship is complicated.

Much of the presentation here, as in the research work on this problem, will focus on so-called *perturbation methods* based on derivatives of Q with respect to the uncertain parameters, including the sensitivity coefficients (Contreras, 1980; Ghanem and Spanos, 1991; Kleiber and Hien, 1992). The primary purpose in the perturbation approaches is to reduce the computational task as compared with direct evaluation of the moments and/or probability distribution of Q. A secondary benefit of perturbation is the derivation of some approximate formulas for finding moments of Q without selection of a probability distribution for the uncertain parameters.

It should be noted that the effect of uncertain parameters applies to each of the types of dynamic analysis that we have considered. For example, the impulse response function used in traditional time-domain analysis will become random when the input parameters are random, as will the frequency response function used in traditional frequency-domain analysis. The response moments or cumulants will then be random, whether determined from time-domain, frequency-domain, or direct analysis. Furthermore, the probabilities of first-passage or fatigue failure will also have uncertainty due to the uncertain parameters. Note also that the uncertainty about response values is not limited to random vibration problems, or even to dynamic problems. Within this chapter, though, we will focus on the random vibration topics that form the basis of this book.

12.2 Modeling Parameter Uncertainty

The analyst rarely has much empirical evidence available to indicate an appropriate probability distribution for physical parameters such as $\mathcal{M}$, C, and $\mathcal{K}$ or the parameters in $\vec{\mathcal{R}}$ that affect $\mathcal{M}$, C, and $\mathcal{K}$. One response to this situation is to use the "maximum-entropy" principle to choose probability distributions that are consistent with this ignorance. In particular, Jaynes (1957) gave results for choosing a one-dimensional probability distribution that is consistent with a given set of constraints while maximizing Shannon's measure of uncertainty about the random variable. The basic idea is that maximizing the uncertainty minimizes the biases introduced by the modeling process. The simplest situation for this approach is that in which the constraints are a smallest possible and a largest possible value for the random variable, for which the maximum-entropy distribution is uniform on this set of possible values. When mean and variance, as well as smallest and largest values are known, the solution is a truncated Gaussian distribution on the set of possible values. The uniform and truncated Gaussian distributions have commonly been used in studying the effect of parameter uncertainty (Udwadia, 1987).

One difficulty with using the uniform and truncated Gaussian distributions may be the choice of the smallest and largest possible values of the parameters. These extreme values are generally not well known. Furthermore, these two distributions (particularly the uniform) assign significant probability of being near these limiting values, so the results may be sensitive to the choice of smallest and largest values. In the extreme, one might consider that the smallest and largest values are unknown but that the mean and variance are known. In this case it seems reasonable to use the usual untruncated Gaussian distribution, but this may cause significant difficulty. For example, it is known that mass damping and stiffness must all be nonnegative, but using a Gaussian distribution for any of them allows the possibility that the quantity will take on negative values. Another difficulty arises if the response quantity of interest varies inversely with the parameter being modeled. For example, if the response is inversely proportional to $\mathcal{R}$ and $\mathcal{R}$ has the Gaussian distribution, then the moments of the response will not exist because the probability density giving the likelihood of $\mathcal{R}^{-1}$ being in the vicinity of v will decay only as v^{-2} as $|v| \rightarrow \infty$. Thus, use of the uniform or Gaussian distribution to model the uncertainty about many parameters does require the knowledge or estimation of smallest and largest possible values.

One way in which the Gaussian distribution can be used is to assume that it applies to the logarithms of the parameters. In particular, if one considers the logarithms of the parameters to be the uncertain quantities of interest, then one can consider them to be Gaussian. For example, if $\log(\mathcal{R})$ is considered Gaussian, then $\mathcal{R}$ has what is called the *log-normal distribution*. The probability distribution can be written in terms of the mean and variance of $\log(\mathcal{R})$ or in terms of the mean and variance of $\mathcal{R}$. Using the log-normal distribution for $\mathcal{R}$ gives it as nonnegative, and $\mathcal{R}^{-1}$ also has a log-normal distribution. Another advantage of the log-normal model has to do with "symmetry" with respect to the uncertainty. For example, modeling $\mathcal{R}$ as having a symmetric distribution seems to imply that $1.5\mu_{\mathcal{R}}$ has the same likelihood as $0.5\mu_{\mathcal{R}}$. It sometimes seems more logical, though, to assume that $2\mu_{\mathcal{R}}$ and $\mu_{\mathcal{R}}/2$ have the same likelihood, representing a symmetric uncertainty about the percentage error. This latter type of symmetry is provided by considering $\log(\mathcal{R})$ to have a symmetric distribution, as in the log-normal modeling. For very small ranges of uncertainty, there is no practical difference between considering $\mathcal{R}$ or $\log(\mathcal{R})$ to be symmetric, but the difference is significant when the uncertainty is large. Which is correct, of course, depends on the particular problem being considered and the given information about the parameters.

12.3 Direct Perturbation Method

One of the simplest approaches to calculating the effect of parameter uncertainty is to approximate the response as a polynomial in the uncertain parameter, or in some parameter that can be used to define the magnitude of the uncertainty. This polynomial is generally a Taylor series about the mean value of the parameters. For example, if the vector $\vec{\mathcal{R}}$ has components that are all the uncertain parameters and the response of interest is $Q = Q(\vec{\mathcal{R}})$, then the first-order approximation is

$$Q_{lin} = Q(\vec{\mu}_{\mathcal{R}}) + \sum_{l=1}^{R} \left(\frac{\partial Q(\vec{r})}{\partial r_l} \right)_{\vec{r}=\vec{\mu}_{\mathcal{R}}} (\mathcal{R}_l - \mu_{\mathcal{R}_l})$$

in which R is the number of uncertain parameters. Note that the first-order approximation gives a linear relationship between Q_{lin} and each of the uncertain parameters. Similarly, a quadratic approximation is obtained by using the second-order polynomial

$$Q_{quad} = Q_{lin} + \frac{1}{2} \sum_{l=1}^{R} \sum_{k=1}^{R} \left(\frac{\partial^2 Q(\vec{r})}{\partial r_l \partial r_k} \right)_{\vec{r}=\vec{\mu}_{\mathcal{R}}} (\mathcal{R}_l - \mu_{\mathcal{R}_l})(\mathcal{R}_k - \mu_{\mathcal{R}_k})$$

Higher-order approximations are easily written but are generally difficult to use. The linear relationship is often used in practice because of its simplicity.

We now introduce into the approximation formulas the restriction that $\vec{\mu}_{\mathcal{R}}$ is the design point $\vec{r}_0$. This makes the first-order derivatives in the linear approximation be exactly the β_l sensitivity coefficients defined in Eq. 12.1. For convenience we also use the same symbol with a double subscript to denote the second derivatives at the design point:

$$\beta_{lk} \equiv \left(\frac{\partial^2 Q}{\partial r_l \partial r_k} \right)_{\vec{r}=\vec{r}_0} \tag{12.2}$$

With this notation we can rewrite the linear and quadratic approximations as

$$Q_{lin} = Q_0 + \sum_{l=1}^{R} \beta_l (\mathcal{R}_l - r_{l,0}) \tag{12.3}$$

and

$$Q_{quad} = Q_{lin} + \frac{1}{2} \sum_{l=1}^{R} \sum_{k=1}^{R} \beta_{lk} (\mathcal{R}_l - r_{l,0})(\mathcal{R}_k - r_{k,0}) \tag{12.4}$$

in which we have also used the notation $Q_0 \equiv Q(\vec{r}_0)$.

For the linear relationship of Eq. 12.3, the mean value and variance of Q are

$$E[Q_{lin}] = Q_0, \quad \mathrm{Var}[Q_{lin}] = \sum_{l=1}^{R}\sum_{k=1}^{R} \beta_l \beta_k \,\mathrm{Cov}[\mathcal{R}_l, \mathcal{R}_k] \tag{12.5}$$

whereas the quadratic approximation gives

$$E[Q_{quad}] = E[Q_{lin}] + \frac{1}{2}\sum_{l=1}^{R}\sum_{k=1}^{R} \beta_{lk} \,\mathrm{Cov}[\mathcal{R}_l, \mathcal{R}_k] \tag{12.6}$$

and

$$\begin{aligned}\mathrm{Var}[Q_{quad}] = {} & \mathrm{Var}[Q_{lin}] + \sum_{l=1}^{R}\sum_{k=1}^{R}\sum_{j=1}^{R} \beta_j \beta_{lk} E[(\mathcal{R}_l - r_{l,0})(\mathcal{R}_k - r_{k,0})(\mathcal{R}_j - r_{j,0})] + \\ & \frac{1}{4}\sum_{l=1}^{R}\sum_{k=1}^{R}\sum_{j=1}^{R}\sum_{i=1}^{R} \beta_{ji}\beta_{lk}\Big(E[(\mathcal{R}_l - r_{l,0})(\mathcal{R}_k - r_{k,0})(\mathcal{R}_j - r_{j,0})(\mathcal{R}_i - r_{i,0}] - \\ & \mathrm{Cov}[\mathcal{R}_l, \mathcal{R}_k]\mathrm{Cov}[\mathcal{R}_j, \mathcal{R}_i]\Big)\end{aligned} \tag{12.7}$$

The results are considerably simplified if the components of $\vec{\mathcal{R}}$ are independent and therefore uncorrelated. In particular, Eqs. 12.5–12.7 become

$$E[Q_{lin}] = Q_0, \quad \mathrm{Var}[Q_{lin}] = \sum_{l=1}^{R} \beta_l^2 \sigma_{\mathcal{R}_l}^2 \tag{12.8}$$

$$E[Q_{quad}] = E[Q_{lin}] + \frac{1}{2}\sum_{l=1}^{R} \beta_{ll} \sigma_{\mathcal{R}_l}^2 \tag{12.9}$$

$$\begin{aligned}\mathrm{Var}[Q_{quad}] = {} & \mathrm{Var}[Q_{lin}] + \sum_{l=1}^{R} \beta_l \beta_{ll} E[(\mathcal{R}_l - r_{l,0})^3] + \\ & \frac{1}{4}\sum_{l=1}^{R} \beta_{ll}^2 \Big(E[(\mathcal{R}_l - r_{l,0})^4] - \sigma_{\mathcal{R}_l}^4\Big) + \sum_{l=1}^{R}\sum_{k=1}^{l-1} \beta_{lk}^2 \sigma_{\mathcal{R}_l}^2 \sigma_{\mathcal{R}_k}^2\end{aligned} \tag{12.10}$$

Note that including the quadratic term in the variance calculation requires knowledge of the third and fourth cumulants of the input parameters. If one presumes a given probability distribution for the uncertain parameters, then these quantities can be calculated, but one usually has relatively little confidence in the estimates of these skewness and kurtosis values for an uncertain parameter. One way of avoiding this difficulty is to use a somewhat inconsistent approach (e.g., see Benajamin and Cornell, 1970) in which the first-order approximation of Eq. 12.5 or 12.8 is used for the variance and the second-order approximation of Eq. 12.6 or 12.9 is used for the mean. In this way, both estimates are based only on the first and second moments of the uncertain parameters. We will call this a *mixed-order perturbation method.*

One major advantage of the first-order and mixed-order approximations is that one can estimate the mean and variance of the response quantity without choosing a specific probability distribution for the uncertain parameters. Calculation of the probability distribution of the response quantity, however, generally does require choice of a specific probability distribution for the uncertain parameters. Furthermore, this calculation may become complicated even for relatively simple distributions of the uncertain parameters, such as uniform distributions. One alternative approach might be the selection of a form for the probability distribution of the response quantity, then a fitting of this form to moments calculated from perturbation. If the response quantity must be nonnegative, then the previously mentioned log-normal distribution or a gamma distribution may be reasonable. On the other hand, this is not always appropriate. For example, the probability of failure must always be between zero and unity, so a beta distribution might be more appropriate for modeling this response quantity.

It should also be noted that one can sometimes simplify the differentiation required within the perturbation method by using the chain rule. For example, assume that an eigenvalue of the system is given as $\lambda = g(r)$ and the response quantity of interest is given as $Q = f(\lambda, r)$. Then one can write

$$\frac{dQ}{dr} = \frac{\partial f}{\partial r} + \frac{\partial f}{\partial \lambda} g'(r)$$

This sort of procedure can be carried on for as many steps as necessary. Taking the derivative of this expression gives a relationship for the second derivative with respect to r, if that is desired.

It is often helpful to normalize the Q and $\vec{\mathcal{R}}$ terms in the analysis. One convenient way of doing this is to divide each uncertain quantity by its value obtained when uncertainty is ignored. For example, if one takes $\mathcal{R}_l = \mathcal{K} / k_0$ with $E(\mathcal{K}) = k_0$, then $E(\mathcal{R}_l) = 1$. Similarly, if Q is equal to a physical response quantity divided by the conditional value when $\mathcal{R} = E(\mathcal{R}) = \vec{r}_0$, then $Q_0 = 1$ and deviations of $E(Q)$ from unity indicate the extent to which the response quantity is affected by nonlinear relationships between Q and $\vec{\mathcal{R}}$. Also, these normalizations give dimensionless sensitivity coefficients that give the percentage change in the response due to a small change in the uncertain parameter. These ideas will be illustrated in the applications in subsequent sections.

The major shortcoming of the perturbation approach is that the linear approximation may not adequately capture the nonlinear relationship between the response and the input parameter (Igusa and Der Kiureghian, 1988; Katafygiotis and Beck, 1995; Gupta and Joshi, 2001). This is particularly true when the uncertainty is large. In principle, one can improve the method by using the quadratic or higher-order polynomial approximations, but this is not necessarily an improvement. Including higher-order terms does give a better fit near the point of expansion of the Taylor series, but this does not ensure a better fit for large variations in the parameter. In addition, there is the difficulty of having the mean and variance of the response quantity depend on the details of the assumed probability distribution of the parameter, as was mentioned regarding the variance computed from the quadratic model.

12.4 Logarithmic Perturbation Method

We will now consider an alternative approximation in which we consider $\log(Q)$ as a function of the $\log(\mathcal{R}_l)$ variables and use a perturbation method on the logarithms. In general, we want our approximation to match Q at the design point used in an analysis with no parameter uncertainty. That is, we want a perfect fit at the point $\vec{\mathcal{R}} = E(\vec{\mathcal{R}}) = \vec{r}_0$. This suggests that our logarithmic perturbation should be about the point $\log(\vec{r}) = \log(\vec{r}_0)$, rather than the point $\log(\vec{r}) = E[\log(\vec{\mathcal{R}})]$.

Let the linear approximation in logarithmic space be written as

$$[\log(Q)]_{lin} = \log(Q_0) + \sum_{l=1}^{R} \chi_l [\log(\mathcal{R}_l) - \log(r_{l,0})] \qquad (12.11)$$

in which χ_l is the derivative of $\log(Q)$ with respect to $\log(r_l)$, evaluated at the design point. Similarly, a quadratic approximation can be written as

$$[\log(Q)]_{quad} = [\log(Q)]_{lin} + \frac{1}{2}\sum_{l=1}^{R}\sum_{k=1}^{R}\chi_{lk}[\log(\mathcal{R}_l) - \log(r_{l,0})][\log(\mathcal{R}_k) - \log(r_{k,0})] \tag{12.12}$$

in which χ_{lk} is the mixed second derivative in the logarithmic variables. The logarithmic derivatives can be evaluated by the chain rule to give

$$\frac{\partial \log(Q)}{\partial \log(\partial r_l)} = \frac{\partial \log(Q)}{\partial Q}\frac{\partial Q}{\partial \log(r_l)} = \frac{\partial \log(Q)}{\partial Q}\frac{\partial Q}{\partial r_l}\frac{\partial r_l}{\partial \log(r_l)} = \frac{r_l}{Q}\frac{\partial Q}{\partial r_l}$$

and

$$\frac{\partial^2 \log(Q)}{\partial \log(r_l)\partial \log(r_k)} = \frac{\partial}{\partial \log(r_k)}\left(\frac{r_l}{Q}\frac{\partial Q}{\partial r_l}\right) = r_k\frac{\partial}{\partial r_k}\left(\frac{r_l}{Q}\frac{\partial Q}{\partial r_l}\right)$$

$$= \frac{r_k}{Q}\frac{\partial r_l}{\partial r_k}\frac{\partial Q}{\partial r_l} - \frac{r_l r_k}{Q^2}\left(\frac{\partial Q}{\partial r_k}\frac{\partial Q}{\partial r_l}\right) + \frac{r_l r_k}{Q}\frac{\partial^2 Q}{\partial r_l \partial r_k}$$

This allows the χ coefficients to be written in terms of the β coefficients defined in Eqs. 12.1 and 12.2 as

$$\chi_l = \frac{r_{l,0}}{Q_0}\beta_l, \qquad \chi_{lk} = \frac{r_{l,0}}{Q_0}\delta_{lk}\,\beta_l - \frac{r_{l,0}r_{k,0}}{Q_0^2}\beta_l\,\beta_k + \frac{r_{l,0}r_{k,0}}{Q_0}\beta_{lk} \tag{12.13}$$

in which δ_{lk} is unity if $l = k$ and zero otherwise (the Kronecker delta).

The linear approximation in the logarithmic space gives a simple power-law form for the dependence of Q on $\vec{\mathcal{R}}$:

$$Q_{log\text{-}lin} = Q_0\prod_{l=1}^{R}\left(\frac{\mathcal{R}_l}{r_{l,0}}\right)^{\chi_l} \tag{12.14}$$

and the quadratic approximation gives a more complicated form that can be written as

$$Q_{log\text{-}quad} = Q_{log\text{-}lin}\prod_{l=1}^{R}\prod_{k=1}^{R}\left(\frac{\mathcal{R}_l}{r_{l,0}}\right)^{\chi_{lk}\log(\mathcal{R}_k/r_{k,0})/2} \tag{12.15}$$

Finding the mean and variance of Q from either Eq. 12.14 or 12.15 requires use of a particular probability distribution for the uncertain $\mathcal{R}_l$ parameters, inasmuch as one must find the expected values of powers of the parameters. This is a disadvantage of the logarithmic formulation as compared with perturbation without logarithms. The logarithmic approximations, however, include forms that are not well approximated by the linear and quadratic approximations in the original variables. Of course, one can use numerical integration to obtain response moments for Q using the logarithmic approximations, and analytical evaluations are also easy for Eq. 12.14 if the components of $\vec{\mathcal{R}}$ have uniform distributions.

Another situation in which it is convenient to use the logarithmic linear approximation is that in which one assumes that the $\mathcal{R}_l$ terms are independent and log-normal, which requires that the $\log(\mathcal{R}_l)$ terms are independent and Gaussian. The linear approximation of Eq. 12.11 then gives $\log(Q)$ as also being Gaussian, so Q has the log-normal distribution given by

$$p_Q(u)=\frac{1}{u\left(2\pi\,\text{Var}[\log(Q)]\right)^{1/2}}\exp\left(-\frac{\left(\log(u)-E[\log(Q)]\right)^2}{2\,\text{Var}[\log(Q)]}\right)U(u) \qquad (12.16)$$

The relationships between the moments of log-normal quantities and their Gaussian logarithms are relatively simple (see, e.g., Benjamin and Cornell, 1970), and using them along with Eq. 12.11 gives

$$\mu_Q=Q_0\prod_{l=1}^{R} r_{l,0}^{\chi_l}\left(1+\frac{\sigma_{\mathcal{R}_l}^2}{r_{l,0}^2}\right)^{\chi_l(\chi_l-1)/2}, \qquad \sigma_Q^2=\mu_Q^2\left(-1+\prod_{l=1}^{R}\left[1+\frac{\sigma_{\mathcal{R}_l}^2}{r_{l,0}^2}\right]^{\chi_l^2}\right) \qquad (12.17)$$

Note, also, that these expressions become much simpler when one uses a normalization that gives $Q_0=1$ and each $r_{l,0}=1$.

Thus, it is quite easy to obtain approximations for the mean value, variance, and probability distribution of the response quantity if one uses the logarithmic-linear approximation along with the assumptions that the uncertain parameters are independent and log-normal. Analytical expressions for the mean and variance of Q can also be obtained for the logarithmic-quadratic expansion with the independent and log-normal assumption for the parameters. These expressions, though, are quite long and complicated and will be omitted here. Of course, numerical integration can be used to find moments of Q using the logarithmic-quadratic approximation of $Q(\vec{r})$ and any chosen distribution for the uncertain parameters.

The gamma distribution has also been used for modeling uncertain parameters (Katafygiotis and Beck, 1995). It has the form

$$p_{\mathcal{R}_l}(u) = \frac{b_l^{a_l}}{\Gamma(a_l)} u^{a_l - 1} e^{-b_l u} U(u) \tag{12.18}$$

in which the a_l and b_l constants can be written in terms of the mean and variance of the parameters as $a_l = (\mu_{\mathcal{R}_l} / \sigma_{\mathcal{R}_l})^2$ and $b_l = \mu_{\mathcal{R}_l} / (\sigma_{\mathcal{R}_l})^2$. This distribution works particularly well with the logarithmic-linear perturbation of Eq. 12.14 inasmuch as it gives

$$E(Q^n) = Q_0^n \prod_{l=1}^{R} \frac{\Gamma(a_l + n\chi_l)}{\Gamma(a_l)(r_{l,0}\, b_l^n)^{\chi_l}} \tag{12.19}$$

from which it is easy to compute the mean and the variance of Q.

12.5 Stationary Response Variance for Delta-Correlated Excitation

The simplest stochastic problem we can consider is the variance of the $\{X(t)\}$ response of an SDF system with a delta-correlated force excitation. As shown in Eqs. 5.63 and 5.64, the conditional variance of the displacement given m, c, and k is

$$\sigma_X^2 = \frac{\pi S_0}{2m^2 \zeta \omega_0^3} = \frac{\pi S_0}{kc} \tag{12.20}$$

in which we have used $G_0 = 2\pi S_0$ based on the results in Section 6.5. Note that this equation is for the force excitation, with S_0 being the autospectral density of the exciting force. Similarly, the results for a base acceleration with constant autospectral density S_{aa} is given by

$$\sigma_X^2 = \frac{\pi S_{aa}}{2\zeta \omega_0^3} = \frac{m^2 \pi S_{aa}}{kc} \tag{12.21}$$

One can now generalize this to the situation with uncertain parameters by replacing m, c, and k with random variables $\mathcal{M}$, $\mathcal{C}$, and $\mathcal{K}$. One has the choice of performing the analysis of parameter uncertainty either by using the first forms, written in terms of $\mathcal{M}$, ζ, and ω_0, or the second forms, written in terms of $\mathcal{M}$, $\mathcal{C}$, and $\mathcal{K}$. If one does use ζ and ω_0, then they must be considered as random variables given by

$$\zeta = \frac{\mathcal{C}}{2(\mathcal{K}\mathcal{M})^{1/2}}, \qquad \omega_0 = \left(\frac{\mathcal{K}}{\mathcal{M}}\right)^{1/2} \tag{12.22}$$

We will now analyze the case of Eq. 12.21, rewritten as

$$\sigma_X^2 = \frac{\pi S_{aa}}{2\zeta\omega_0^3} = \frac{\mathcal{M}^2 \pi S_{aa}}{\mathcal{K}\mathcal{C}} \tag{12.23}$$

We choose this base-excited case because it can be considered the more general of the two. In particular, general results for Eq. 12.23 can be applied to the force-excited problem by letting $\mathcal{M} = m$ have no uncertainty and replacing the constant $m^2 S_{aa}$ with S_0. We will consider $\mathcal{M}$, $\mathcal{C}$, and $\mathcal{K}$ to be independent and to have mean values of m, c, and k, respectively.

We now introduce dimensionless random variables for the parameter uncertainties as

$$\mathcal{R}_1 = \mathcal{M}/m_0, \qquad \mathcal{R}_2 = \mathcal{C}/c_0, \qquad \mathcal{R}_3 = \mathcal{K}/k_0 \tag{12.24}$$

Similarly, the nondimensional response quantity will be the random result of Eq. 12.23 divided by the deterministic value of Eq. 12.21. The simplest form is

$$Q = \frac{\sigma_X^2}{(\sigma_X^2)_0} = \frac{\mathcal{M}^2 \pi S_{aa}}{\mathcal{K}\mathcal{C}} \frac{k_0 c_0}{m_0^2 \pi S_{aa}} = \mathcal{R}_1^2 \mathcal{R}_2^{-1} \mathcal{R}_3^{-1} \tag{12.25}$$

but one also has the option of using a form written in terms of ζ and ω_0, as defined in Eqs. 12.22:

$$Q = \frac{\sigma_X^2}{(\sigma_X^2)_0} = \frac{\pi S_{aa}}{2\zeta\omega_0^3} \frac{k_0 c_0}{m_0^2 \pi S_{aa}} = \frac{k_0 c_0}{2 m_0^2} \frac{1}{\zeta\omega_0^3} \tag{12.26}$$

The form in Eq. 12.26 is introduced only to illustrate the use of the chain-rule differentiation for the perturbation method. For example, using this equation we can write

$$\frac{\partial Q}{\partial m}=\frac{k_0 c_0}{2m_0^2}\left(\frac{-1}{\zeta^2\omega_0^3}\frac{\partial \zeta}{\partial m}-\frac{3}{\zeta\omega_0^4}\frac{\partial \omega_0}{\partial m}\right)=\frac{-Q}{\zeta}\frac{\partial \zeta}{\partial m}-\frac{3Q}{\omega_0}\frac{\partial \omega_0}{\partial m}$$

Differentiating Eq. 12.22 gives

$$\frac{\partial \zeta}{\partial m}=\frac{-C}{4k^{1/2}m^{3/2}}=\frac{-\zeta}{2m}, \qquad \frac{\partial \omega_0}{\partial m}=\frac{-C}{4k^{3/2}m^{1/2}}=\frac{-\zeta}{2k}$$

so

$$\frac{\partial Q}{\partial r_1}=\frac{\partial Q}{\partial m}\frac{\partial m}{\partial r_1}=\left(\frac{Q}{\zeta}\frac{\zeta}{2m}+\frac{3Q}{\omega_0}\frac{\omega_0}{2m}\right)m=\frac{2Q}{r_1}$$

Of course, this result could have been obtained more easily by using Eq. 12.25, which gives all the first derivatives as

$$\beta_1\equiv\left(\frac{\partial Q}{\partial r_1}\right)_{\vec{r}=\vec{r}_0}=2r_{1,0}r_{2,0}^{-1}r_{3,0}^{-1}=\frac{2Q_0}{r_{1,0}}=2$$

$$\beta_2\equiv\left(\frac{\partial Q}{\partial r_2}\right)_{\vec{r}=\vec{r}_0}=\frac{-Q_0}{r_{2,0}}=-1, \quad \beta_3\equiv\left(\frac{\partial Q}{\partial r_3}\right)_{\vec{r}=\vec{r}_0}=\frac{-Q_0}{r_{3,0}}=-1$$

Similarly, the second derivatives are

$$\beta_{11}=\frac{2Q_0}{r_{1,0}^2}=2, \quad \beta_{22}=\frac{2Q_0}{r_{2,0}^2}=2, \quad \beta_{33}=\frac{2Q_0}{r_{3,0}^2}=2$$

$$\beta_{12}=\frac{-2Q_0}{r_{1,0}r_{2,0}}=-2, \quad \beta_{13}=\frac{-2Q_0}{r_{1,0}r_{3,0}}=-2, \quad \beta_{23}=\frac{Q_0}{r_{2,0}r_{3,0}}=1$$

Substituting $Q_0=1$ and the first derivatives into the linear perturbation approximation of Eq. 12.8 gives

$$E(Q) \approx 1, \quad \mathrm{Var}(Q) \approx 4\sigma^2_{\mathcal{R}_1} + \sigma^2_{\mathcal{R}_2} + \sigma^2_{\mathcal{R}_3} = 4\frac{\sigma^2_{\mathcal{M}}}{m_0^2} + \frac{\sigma^2_{\mathcal{C}}}{c_0^2} + \frac{\sigma^2_{\mathcal{K}}}{k_0^2} \qquad (12.27)$$

In the same way, the second-order perturbation approximation for the mean value is

$$E(Q) \approx 1 + \sigma^2_{\mathcal{R}_1} + \sigma^2_{\mathcal{R}_2} + \sigma^2_{\mathcal{R}_3} = 1 + \frac{\sigma^2_{\mathcal{M}}}{m_0^2} + \frac{\sigma^2_{\mathcal{C}}}{c_0^2} + \frac{\sigma^2_{\mathcal{K}}}{k_0^2} \qquad (12.28)$$

As previously noted, one cannot compute the second-order approximation of the variance without first choosing probability distributions for $\mathcal{M}$, $\mathcal{C}$, and $\mathcal{K}$.

Next let us consider the common choice of using a uniform distribution for each of the uncertain parameters. This can be written in terms of $\mathcal{M}$, $\mathcal{C}$, and $\mathcal{K}$, but it is equally easy to work directly with the $\mathcal{R}_l$ components. Because each of these components has a unit mean value, an appropriate uniform distribution is on the set $1-\gamma_l \le \mathcal{R}_l \le 1+\gamma_l$. This gives

$$\sigma^2_{\mathcal{R}_l} = \gamma_l^2/3, \quad E[(\mathcal{R}_l - \mu_{\mathcal{R}_l})^3] = 0, \quad E[(\mathcal{R}_l - \mu_{\mathcal{R}_l})^4] = \gamma_l^4/5 \qquad (12.29)$$

and the second-order approximation of the variance with these uniform distributions can be written as

$$\mathrm{Var}(Q) \approx \frac{4}{3}\gamma_1^2 + \frac{1}{3}\gamma_2^2 + \frac{1}{3}\gamma_3^2 + \frac{4}{45}(\gamma_1^4 + \gamma_2^4 + \gamma_3^4) + \frac{4}{9}\gamma_1^2\gamma_2^2 + \frac{4}{9}\gamma_1^2\gamma_3^2 + \frac{1}{9}\gamma_2^2\gamma_3^2 \qquad (12.30)$$

Alternatively, one may wish to choose a given value for the variance of $\mathcal{R}_l$, then find an appropriate value for γ_l. In particular, this gives

$$\gamma_l^2 = 3\sigma^2_{\mathcal{R}_l}, \quad E[(\mathcal{R}_l - \mu_{\mathcal{R}_l})^4] = 9\sigma^4_{\mathcal{R}_l}/5 \qquad (12.31)$$

For the first-order approximation, in particular, it is also relatively easy to find an approximation of the probability distribution for Q for this problem once one has chosen values for γ_1, γ_2, and γ_3. In addition, the form of Q and the uniform distribution for the uncertain parameters make it feasible to find the exact probability distribution of Q. Both of these ideas will be illustrated in Examples 12.1 and 12.2.

Consider now the logarithm of Eq. 12.25:

$$\log(Q) = 2\log(\mathcal{R}_1) - \log(\mathcal{R}_2) - \log(\mathcal{R}_3)$$

Note that this equation is exactly linear in the $\log(\mathcal{R}_l)$ terms, so the first-order form of Eq. 12.11 is exact. That is, no perturbation is necessary. Of course, this is not true for most problems. If we also assume that the $\log(\mathcal{R}_l)$ random variables are independent and Gaussian, then the mean, variance, and probability density function of Q are found from Eqs. 12.16 and 12.17.

****

Example 12.1: Consider the stationary response to delta-correlated excitation of a simplified problem in which the mass and stiffness are considered deterministic, but the damping has a uniform distribution on the set $(1-\gamma)c_0 \le C \le (1+\gamma)c_0$.

This problem can be considered a special case of the general formulation, with

$$\sigma^2_{\mathcal{R}_1} = \sigma^2_{\mathcal{R}_3} = 0, \quad \sigma^2_{\mathcal{R}_2} = \gamma^2/3$$

Equation 12.27 then gives the first-order perturbation approximations as

$$\mu_Q \approx 1, \quad \sigma^2_Q \approx \sigma^2_{\mathcal{R}_2} = \gamma^2/3$$

and Eqs. 12.28 and 12.30 give the second-order approximations as

$$\mu_Q \approx 1 + \frac{\gamma^2}{3}, \quad \sigma^2_Q \approx \frac{\gamma^2}{3} + \frac{4\gamma^4}{45} = \frac{\gamma^2}{3}\left(1 + \frac{4\gamma^2}{15}\right)$$

For this uniform distribution of C it is also easy to evaluate the exact moments as

$$\mu_Q = \int_{(1-\gamma)}^{(1+\gamma)} \frac{du}{2\gamma u} = \frac{\log[(1+\gamma)/(1-\gamma)]}{2\gamma}$$

$$E(Q^2) = \int_{(1-\gamma)}^{(1+\gamma)} \frac{du}{2\gamma u^2} = \frac{(1-\gamma)^{-1} - (1+\gamma)^{-1}}{2\gamma}$$

from which the exact variance can be calculated as $E(Q^2) - (\mu_Q)^2$.
For the log-normal approximation we use $\chi_2 = \beta_2 = -1$ in Eq. 12.17 to obtain

$$\mu_Q = 1 + \gamma^2/3, \quad \sigma^2_Q = \left(1 + \frac{\gamma^2}{3}\right)^2 \left(\frac{\gamma^2}{3}\right)$$

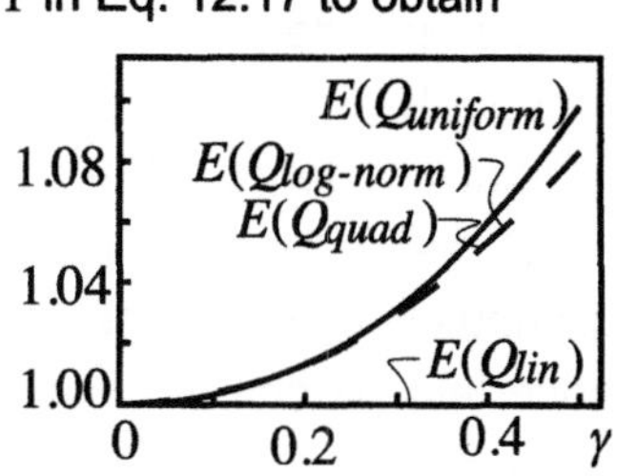

The accompanying sketches show the various calculations of mean value and variance for a range of γ values.

From the sketches we see that all but one of the curves are quite close to each other for $\gamma \le 0.3$. The one exception is the first-order perturbation estimate of the mean value. For larger values of γ, we see some deviation. For this problem, the log-normal result for the mean value is identical to that of the second-order perturbation, and these results are quite close to the exact value for a uniform distribution of C, even for relatively large values of γ. The variance value from the log-normal method is closer to the exact value for the uniform distribution than is either of the perturbation results. It should be kept in mind that the log-normal results are actually exact under the condition that the uncertain parameter has a log-normal distribution. Comparing the log-normal results with the exact results for a uniform distribution of the uncertain parameter is a comparison of the effect of the choice of model, not of the accuracy of the results. It is seen that the choice of model does not have a great effect on the mean value and variance results for this particular problem.

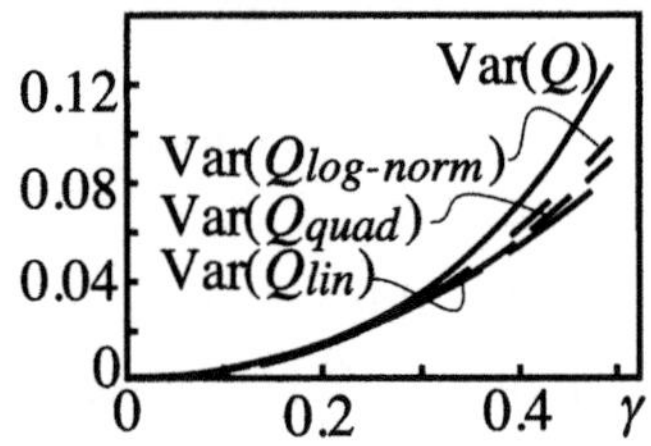

It is also easy to investigate the probability distribution of Q for this problem with a uniform distribution for the uncertain parameter. The exact result is

$$p_Q(u) = \frac{1}{u^2} p_{\mathcal{R}_2}(u^{-1}) = \frac{1}{u^2} U\left(u - [1+\gamma]^{-1}\right) U\left([1-\gamma]^{-1} - u\right)$$

The linear first-order perturbation relationship predicts the same type of probability distribution for Q as for $\mathcal{R}_2$. Thus, the uniform distribution for $\mathcal{R}_2$ gives

$$p_Q(u) \approx \left(\frac{1-\gamma^2}{2\gamma}\right) U\left(u - [1+\gamma]^{-1}\right) U\left([1-\gamma]^{-1} - u\right)$$

The log-normal probability density function is given by Eq. 12.16.

The sketch to the right compares these three results for the case of $\gamma = 0.3$. The results for the uniform distribution of C show that the first-order approximation may not be an adequate approximation of the probability distribution when the uncertainty is relatively large. The comparison of the exact results for the uniform and log-normal situations shows that the probability distribution can be quite sensitive to the assumption about the distribution of the uncertain parameter, even when the mean and variance are in reasonable agreement.

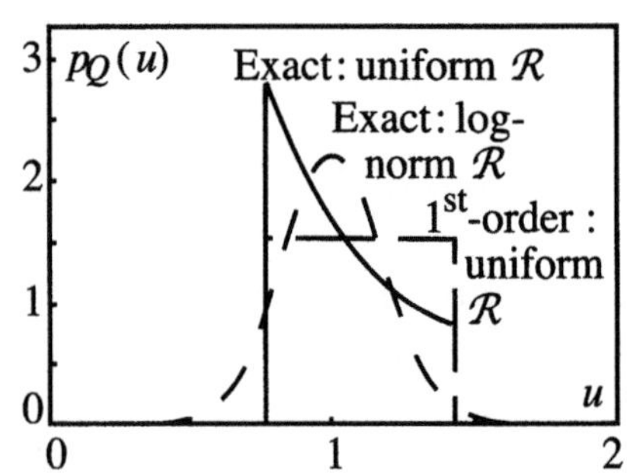

****

Example 12.2: Consider the stationary response to delta-correlated force excitation of a system in which there is uncertainty about mass, stiffness, and damping parameters. The uncertainty is greatest for the damping, and least for the mass, with $\sigma_{\mathcal{K}}/k_0 = 0.5\sigma_C/c_0$ and $\sigma_{\mathcal{M}}/m_0 = 0.25\sigma_C/c_0$. The mass, damping, and stiffness are independent with mean values of m_0, c_0, and k_0.

Because the response variance to delta-correlated force excitation does not depend on the magnitude of $\mathcal{M}$, we can use a dimensionless response quantity of the form

$$Q = \frac{\sigma_X^2}{(\sigma_X^2)_0} = \frac{\pi S_0}{\mathcal{K} C}\frac{k_0 c_0}{\pi S_0} = \mathcal{R}_2^{-1}\mathcal{R}_3^{-1}$$

We will begin by choosing the pertinent parameters to be uniformly distributed with $1-\gamma_l \le \mathcal{R}_l \le 1+\gamma_l$. The given restriction on the standard deviations then requires that $\gamma_3 = 0.5\gamma_2$. Eq. 12.27 gives the first-order approximations as

$$\mu_Q \approx 1, \quad \sigma_Q^2 \approx \sigma_{\mathcal{R}_2}^2 + \sigma_{\mathcal{R}_3}^2 = 1.25\sigma_{\mathcal{R}_2}^2 = (1.25/3)\gamma_2^2$$

and Eqs. 12.28 and 12.30 give the second-order approximation as

$$\mu_Q \approx 1 + \sigma_{\mathcal{R}_2}^2 + \sigma_{\mathcal{R}_3}^2 = 1 + \frac{1.25}{3}\gamma_2^2$$

$$\sigma_Q^2 \approx \frac{\gamma_2^2+\gamma_3^2}{3} + \frac{4(\gamma_2^4+\gamma_3^4)}{45} + \frac{\gamma_2^2\gamma_3^2}{9} = \frac{1.25}{3}\gamma_2^2 + \frac{5.5}{45}\gamma_2^4$$

For the uniform distribution of C and $\mathcal{K}$ it is also easy to evaluate the exact moments as

$$\mu_Q = E(\mathcal{R}_2^{-1})\,E(\mathcal{R}_3^{-1}) = \frac{\log[(1+\gamma_2)/(1-\gamma_2)]}{2\gamma_2}\,\frac{\log[(1+0.5\gamma_2)/(1-0.5\gamma_2)]}{\gamma_2}$$

$$E(Q^2) = E(\mathcal{R}_2^{-2})\,E(\mathcal{R}_3^{-2}) = \frac{(1-\gamma_2)^{-1}-(1+\gamma_2)^{-1}}{2\gamma_2}\,\frac{(1-0.5\gamma_2)^{-1}-(1+0.5\gamma_2)^{-1}}{\gamma_2}$$

from which the exact variance can be calculated as $E(Q^2)-(\mu_Q)^2$.

For the log-normal approximation we use Eq. 12.17, along with the sensitivity values of $\beta_1 = 0$ and $\beta_2 = \beta_3 = -1$ to obtain

$$\mu_Q = \left(1+\frac{\gamma_2^2}{3}\right)\left(1+\frac{\gamma_2^2}{12}\right)$$

$$\sigma_Q^2 = \left(1+\frac{\gamma^2}{3}\right)^2\left(1+\frac{\gamma^2}{12}\right)^2\left(\left[1+\frac{\gamma^2}{3}\right]\left[1+\frac{\gamma^2}{12}\right]-1\right)$$

The accompanying sketches show the values of mean value and variance for a range of γ_2 values.

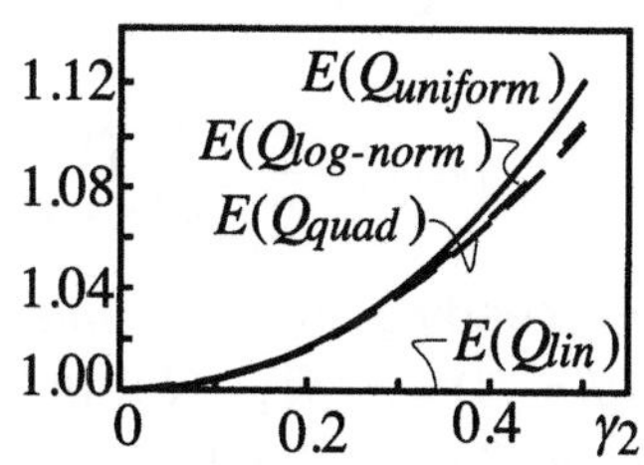

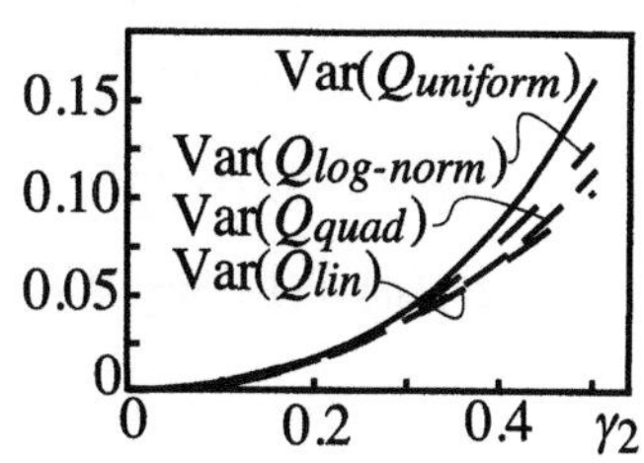

These results are very similar to those in Example 12.1. In the current problem the $E(Q)$ values from the log-normal approximation and the second-order perturbation are not identical, but they are almost indistinguishable on the plot. The curve for the log-normal distribution is slightly higher. Again, both are quite close to the exact curve for the uniform distribution. The other variance values are all below, but relatively close to the exact values for the uniform distribution. The variance values for the first-order and second-order perturbations are quite close to each other, indicating that a mixed-order perturbation method is almost as good as the second-order method.

The exact probability distribution of Q, for uniform distributions of $\mathcal{R}_2$ and $\mathcal{R}_3$, can be found most easily by first considering the cumulative distribution

$$P(Q \le u) = P(\mathcal{R}_2 \mathcal{R}_3 \ge u^{-1})$$

then differentiating to find the probability density function. For the given ranges of $\mathcal{R}_2$ and $\mathcal{R}_3$ this gives three distinct ranges that must be investigated separately. For $(1+\gamma_2)^{-1}(1+\gamma_2/2)^{-1} \le u \le (1+\gamma_2)^{-1}(1-\gamma_2/2)^{-1}$ we have

$$P(Q \le u) = \int_{(1+\gamma_2)^{-1}u^{-1}}^{1+\gamma_2/2} \int_{u^{-1}w^{-1}}^{(1+\gamma_2)} \frac{dv}{2\gamma_2^2} dw$$

$$p_Q(u) = \frac{1}{2\gamma_2^2 u^2} \int_{(1+\gamma_2)^{-1}u^{-1}}^{1+\gamma_2/2} \frac{dw}{w} = \frac{\log[(1+\gamma_2)(1+\gamma_2/2)\,u]}{2\gamma_2^2 u^2}$$

For $(1+\gamma_2)^{-1}(1-\gamma_2/2)^{-1} \le u \le (1-\gamma_2)^{-1}(1+\gamma_2/2)^{-1}$

$$P(Q \le u) = \int_{1-\gamma_2/2}^{1+\gamma_2/2} \int_{u^{-1}w^{-1}}^{1+\gamma_2} \frac{dv}{2\gamma_2^2} dw$$

$$p_Q(u) = \frac{1}{2\gamma_2^2 u^2} \int_{1-\gamma_2/2}^{1+\gamma_2/2} \frac{dw}{w} = \frac{1}{2\gamma_2^2 u^2} \log\left(\frac{1+\gamma_2/2}{1-\gamma_2/2}\right)$$

and for $(1-\gamma_2)^{-1}(1+\gamma_2/2)^{-1} \le u \le (1-\gamma_2)^{-1}(1-\gamma_2/2)^{-1}$, the result is

$$P(Q \le u) = 1 - \int_{1-\gamma_2/2}^{(1-\gamma_2)^{-1}u^{-1}} \int_{1-\gamma_2}^{u^{-1}w^{-1}} \frac{dv}{2\gamma_2^2} dw$$

$$p_Q(u) = \frac{1}{2\gamma_2^2 u^2} \int_{1-\gamma_2/2}^{(1-\gamma_2)^{-1}u^{-1}} \frac{dw}{w} = \frac{1}{2\gamma_2^2 u^2} \log[(1-\gamma_2)^{-1}(1-\gamma_2/2)^{-1}u^{-1}]$$

Proceeding in the same way with the linear model of $Q \approx 1-(\mathcal{R}_1-1)-(\mathcal{R}_2-1) = 3-\mathcal{R}_1-\mathcal{R}_2$ gives $P(Q \le u) = P(\mathcal{R}_1+\mathcal{R}_2 \ge 3-u)$. After integrating over $\mathcal{R}_2$ and $\mathcal{R}_3$, then differentiating with respect to u, one obtains

$$p_Q(u) \approx (u-1+3\gamma_2/2)/(2\gamma_2^2) \quad \text{for } 1-3\gamma_2/2 \le u \le 1-\gamma_2/2$$

$$p_Q(u) \approx \gamma_2/(2\gamma_2^2) \quad \text{for } 1-\gamma_2/2 \le u \le 1+\gamma_2/2$$

$$p_Q(u) \approx (1+3\gamma_2/2-u)/(2\gamma_2^2) \quad \text{for } 1+\gamma_2/2 \le u \le 1+3\gamma_2/2$$

Finally, the probability density function of the log-normal model is given by Eq. 12.16. These various results are compared in the accompanying sketch for the special case of $\gamma_2 = 0.4$.

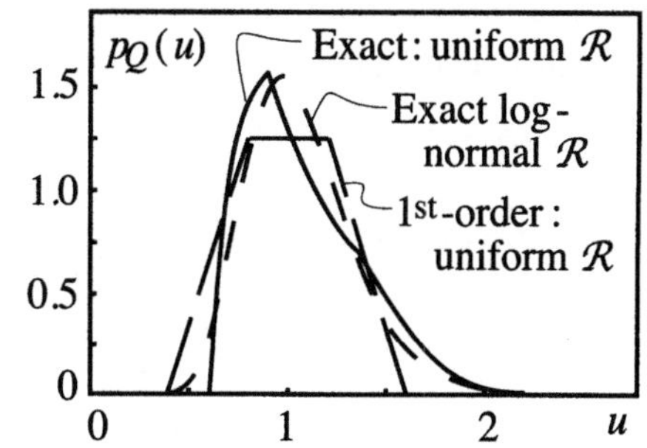

We see that the probability density function of the response quantity is not well approximated by the linearized form of the dependence on the uncertain parameters. In particular, the approximation does not accurately predict the largest possible value of Q.

Recall that there is no reason that the log-normal curve should agree with the other curves in the figure, inasmuch as it is based on a different set of assumptions about the distribution of C and $\mathcal{K}$. It is seen, however, that the log-normal result is better than the linear approximation in fitting the exact distribution for the uniform problem. This may be viewed as evidence that a logarithmic approximation may be better than a linear one in some problems even when the uncertain parameters are not necessarily log-normal. It should also be kept in mind, though, that this particular problem is especially appropriate for application of the log-normal approach, because using log-normal parameters gives the Q response quantity as having exactly a log-normal distribution. This is not true in most problems.

****

Example 12.3: Consider the stationary response to delta-correlated base excitation of a system in which there is uncertainty about mass, stiffness, and damping parameters. The uncertainty is greatest for the damping, and least for the mass, with $\sigma_{\mathcal{K}} / k_0 = 0.5 \sigma_C / c_0$ and $\sigma_{\mathcal{M}} / m_0 = 0.25 \sigma_C / c_0$. The mass, damping, and stiffness are independent with mean values of m_0, c_0, and k_0.

First we will consider the perturbation method. In the preceding section we found that the quadratic approximation of the variance was little better than the linear. On the other hand, it was significantly better than the linear model for predicting the mean value. Thus, we will use the mixed-perturbation method, with the quadratic approximation of Eq. 12.9 for the mean value and the linear approximation of Eq. 12.8 for the variance. This gives

$$\mu_Q \approx 1 + \sigma_{\mathcal{R}_1}^2 + \sigma_{\mathcal{R}_2}^2 + \sigma_{\mathcal{R}_3}^2 = 1 + \frac{21}{16} \sigma_{\mathcal{R}_2}^2$$

$$\sigma_Q^2 \approx 4 \sigma_{\mathcal{R}_1}^2 + \sigma_{\mathcal{R}_2}^2 + \sigma_{\mathcal{R}_3}^2 = \frac{3}{2} \sigma_{\mathcal{R}_2}^2$$

For log-normal distributions for the uncertain parameters, Eq. 12.17 with the sensitivity coefficients of $\beta_1 = 2,\ \beta_2 = \beta_3 = -1$ gives

$$\mu_Q = (1+\sigma_{\mathcal{R}_2}^2/16)(1+\sigma_{\mathcal{R}_2}^2)(1+\sigma_{\mathcal{R}_2}^2/4)$$

$$\sigma_Q^2 = (1+\sigma_{\mathcal{R}_2}^2/16)^2(1+\sigma_{\mathcal{R}_2}^2)^2(1+\sigma_{\mathcal{R}_2}^2/4)^2 \times \left((1+\sigma_{\mathcal{R}_2}^2/16)^4(1+\sigma_{\mathcal{R}_2}^2)(1+\sigma_{\mathcal{R}_2}^2/4)-1\right)$$

For a uniform distribution we use Eq. 12.31 to find γ_l for each $\mathcal{R}_l$ and obtain

$$E(Q) = E(\mathcal{R}_1^2)\,E(\mathcal{R}_2^{-1})\,E(\mathcal{R}_3^{-1}) = \frac{(1+3^{1/2}\sigma_{\mathcal{R}_2}/4)^3-(1-3^{1/2}\sigma_{\mathcal{R}_2}/4)^3}{1.5(3^{1/2})\sigma_{\mathcal{R}_2}}\,\frac{\log[(1+3^{1/2}\sigma_{\mathcal{R}_2})/(1-3^{1/2}\sigma_{\mathcal{R}_2})]}{2(3^{1/2})\sigma_{\mathcal{R}_2}} \times \frac{\log[(1+3^{1/2}\sigma_{\mathcal{R}_2}/2)/(1-3^{1/2}\sigma_{\mathcal{R}_2}/2)]}{3^{1/2}\sigma_{\mathcal{R}_2}}$$

$$E(Q^2) = E(\mathcal{R}_1^4)\,E(\mathcal{R}_2^{-2})\,E(\mathcal{R}_3^{-2}) = \frac{(1+3^{1/2}\sigma_{\mathcal{R}_2}/4)^5-(1-3^{1/2}\sigma_{\mathcal{R}_2}/4)^5}{2.5(3^{1/2})\sigma_{\mathcal{R}_2}} \times \frac{(1-3^{1/2}\sigma_{\mathcal{R}_2})^{-1}-(1+3^{1/2}\sigma_{\mathcal{R}_2})^{-1}}{2(3^{1/2})\sigma_{\mathcal{R}_2}}\,\frac{(1-3^{1/2}\sigma_{\mathcal{R}_2}/2)^{-1}-(1+3^{1/2}\sigma_{\mathcal{R}_2}/2)^{-1}}{3^{1/2}\sigma_{\mathcal{R}_2}}$$

The following sketches show the various calculations of mean value and variance for a range of $\sigma_{\mathcal{R}_2}$ values. Again we see that the perturbation and log-normal results are in good agreement, and both are smaller than the exact results for a uniform distribution.

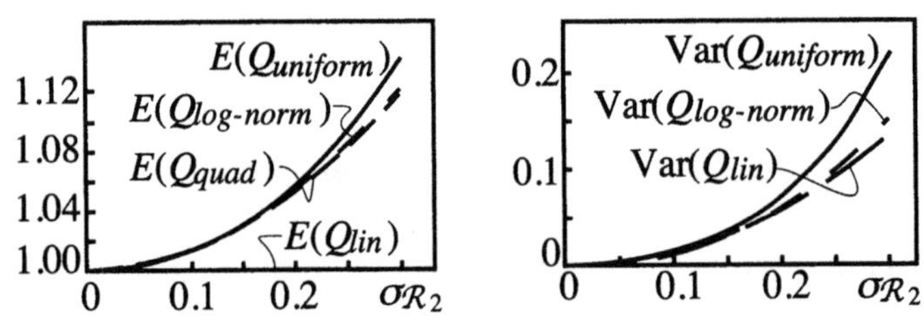

****

12.6 Nonstationary Response Variance for Delta-Correlated Excitation

Consider the linear SDF system with a stationary, mean-zero, delta-correlated excitation force suddenly applied at time $t=0$:

$$m\,\ddot{X}(t) + c\,\dot{X}(t) + k\,X(t) = F(t)\,U(t)$$

The general result for the variance was given in Eq. 5.50 as

$$\sigma_X^2(t) \equiv K_{XX}(t,t) = \frac{\pi S_0}{2m^2 \zeta \omega_0^3}\left(1 - e^{-2\zeta\omega_0 t}\left[\frac{\omega_0^2}{\omega_d^2} + \frac{\zeta\omega_0}{\omega_d}\sin(2\omega_d t) - \frac{\zeta^2 \omega_0^2}{\omega_d^2}\cos(2\omega_d t)\right]\right)U(t) \qquad (12.32)$$

in which S_0 is the autospectral density of $\{F(t)\}$. Recall, though, that except for $t \approx 0$, this expression is well approximated by the much simpler Eq. 5.52, which can be rewritten as

$$\sigma_X^2(t) \approx \frac{\pi S_0}{kc}\left(1 - e^{-ct/m}\right)U(t) \qquad (12.33)$$

We use the simpler form and define our dimensionless response quantity as the value from using Eq. 12.33 with random parameters $\mathcal{M}$, $\mathcal{C}$, and $\mathcal{K}$ divided by the value with deterministic m_0, c_0, and k_0:

$$Q \equiv \frac{c}{\mathcal{C}}\frac{k_0}{\mathcal{K}}\frac{1 - e^{-\mathcal{C}t/\mathcal{M}}}{1 - e^{-c_0 t/m_0}}$$

Using $\vec{\mathcal{R}} = [\mathcal{M}/m_0, \mathcal{C}/c_0, \mathcal{K}/k_0]^T$ then gives

$$Q \equiv \mathcal{R}_2^{-1}\mathcal{R}_3^{-1}\frac{1 - e^{-(c_0 t/m_0)(\mathcal{R}_2/\mathcal{R}_1)}}{1 - e^{-c_0 t/m_0}}$$

Because Q depends only on t and the deterministic parameters in the form of $c_0 t/m_0 \equiv 2\zeta\omega_0 t$, we simplify the equations by introducing a normalized time of the form

$$T = c_0 t/m_0 \qquad (12.34)$$

which gives

$$Q = \mathcal{R}_2^{-1}\mathcal{R}_3^{-1}\frac{1 - e^{-T(\mathcal{R}_2/\mathcal{R}_1)}}{1 - e^{-T}} \qquad (12.35)$$

We also use the common simplification that $\mu_{\mathcal{M}} = m_0$, $\mu_{\mathcal{C}} = c_0$, and $\mu_{\mathcal{K}} = k_0$, so $E(\vec{\mathcal{R}}) = (1,1,1)^T$.

For the perturbation method we need the derivatives with respect to each of the uncertain parameters, which we take to be independent. The derivatives with respect to $\mathcal{R}_3$ are simple (and the same as in the preceding section):

$$\beta_3 \equiv \left(\frac{\partial Q}{\partial r_3} \right)_{\vec{r}=\vec{r}_0} = -1, \quad \beta_{33} = \left(\frac{\partial^2 Q}{\partial r_3^2} \right)_{\vec{r}=\vec{r}_0} = 2$$

but the derivatives with respect to r_1 and r_2 are somewhat more complicated. In particular,

$$\frac{\partial Q}{\partial r_1} = -r_2^{-1} r_3^{-1} \frac{T r_2}{r_1^2} \frac{e^{-T(r_2/r_1)}}{1-e^{-T}} = -\frac{T r_2}{r_1^2} \frac{e^{-T(r_2/r_1)}}{1-e^{-T(r_2/r_1)}} Q$$

$$\frac{\partial^2 Q}{\partial r_1^2} = \left(\frac{2 T r_2}{r_1^3} - \frac{T^2 r_2^2}{r_1^4} \right) \frac{e^{-T(r_2/r_1)}}{1-e^{-T(r_2/r_1)}} Q$$

$$\frac{\partial Q}{\partial r_2} = \left(\frac{T e^{-T(r_2/r_1)}}{r_1 [1-e^{-T(r_2/r_1)}]} - \frac{1}{r_2} \right) Q$$

and

$$\frac{\partial^2 Q}{\partial r_2^2} = \left(\frac{2}{r_2^2} - \frac{2 T e^{-T(r_2/r_1)}}{r_1 r_2 [1-e^{-T(r_2/r_1)}]} - \frac{T^2 e^{-T(r_2/r_1)}}{r_1^2 [1-e^{-T(r_2/r_1)}]} \right) Q$$

so

$$\beta_1 = \frac{-T e^{-T}}{1-e^{-T}}, \quad \beta_2 = \frac{T e^{-T}}{1-e^{-T}} - 1$$

$$\beta_{11} = \frac{(2T - T^2) e^{-T}}{1-e}, \quad \beta_{22} = 2 - \frac{(2T + T^2) e^{-T}}{1-e^{-T}}$$

Thus, the first-order approximation for the variance is

$$\sigma_Q^2 \approx T^2 \left(\frac{e^{-T}}{1-e^{-T}} \right)^2 \sigma_{\mathcal{R}_1}^2 + \left(\frac{T e^{-T}}{1-e^{-T}} - 1 \right)^2 \sigma_{\mathcal{R}_2}^2 + \sigma_{\mathcal{R}_3}^2 \tag{12.36}$$

and the second-order approximation for the mean value is

$$\mu_Q \approx 1 + \frac{T(1-T/2)e^{-T}\sigma_{\mathcal{R}_1}^2}{1-e^{-T}} + \left(1 - \frac{T(1+T/2)e^{-T}}{1-e^{-T}}\right)\sigma_{\mathcal{R}_2}^2 + \sigma_{\mathcal{R}_3}^2 \tag{12.37}$$

The situation with the logarithmic model is not as desirable as in Section 12.5. In particular, we note from Eq. 12.35 that $\log(Q)$ is now a nonlinear function of $\log(\mathcal{R}_1)$ and $\log(\mathcal{R}_2)$, although it is linear with respect to $\log(\mathcal{R}_3)$. Thus, we must use a perturbation approximation in the logarithmic variables, in contrast to the situation in Section 12.5. If we assume log-normal distributions for the parameters, then our estimates of the mean and variance of Q will still be given by Eq. 12.17, but these will be only approximately correct, even if the uncertain parameters truly have log-normal distributions, due to the approximation of the linear model in the logarithmic quantities.

****

Example 12.4: Consider the nonstationary response to delta-correlated force excitation of a system in which there is uncertainty about mass, stiffness, and damping parameters. The uncertainty is greatest for the damping, and least for the mass, with $\sigma_{\mathcal{K}}/k_0 = 0.5\sigma_{\mathcal{C}}/c_0$ and $\sigma_{\mathcal{M}}/m_0 = 0.25\sigma_{\mathcal{C}}/c_0$. The mass, damping, and stiffness are independent with mean values of m_0, c_0, and k_0.

The following sketches show numerical values from Eqs. 12.36 and 12.37 for direct mixed-order perturbation and for the log-normal model with first-order perturbation of $\log(Q)$ on the $\log(\mathcal{R}_l)$ components. The results are plotted versus $T = c_0 t/m_0$ for $\sigma_{\mathcal{R}_2} = 0.3$.

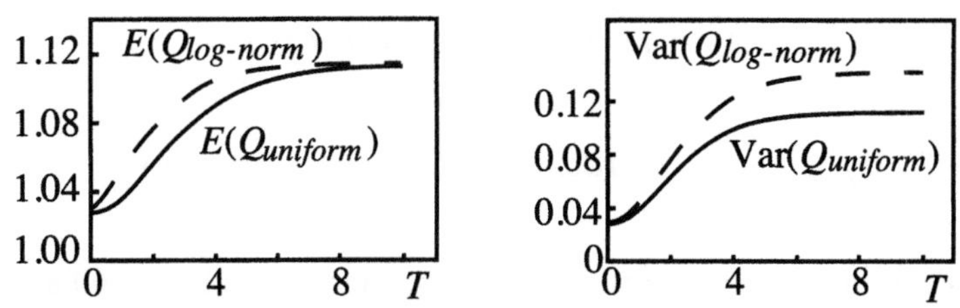

The log-normal form gives a higher value than direct perturbation for both the mean and the variance of Q for this problem. It should be kept in mind that both are approximations. That is, neither is exact for any particular probability distribution of the uncertain parameters.
To find the mean or variance of σ_X^2 one simply reverses the procedure used in defining Q to obtain

$$\sigma_X^2(t) \approx \frac{\pi S_0}{\mathcal{K} C}\left(1 - e^{-C t/\mathcal{M}}\right) = \frac{\pi S_0}{k_0 c_0}\left(1 - e^{-c_0 t/m_0}\right) Q = \frac{\pi S_0}{k_0 c_0}\left(1 - e^{-T}\right) Q$$

so that

$$E[\sigma_X^2(t)] \approx \frac{\pi S_0}{k_0 c_0}(1 - e^{-T})\,\mu_Q, \quad \text{Var}[\sigma_X^2(t)] \approx \left(\frac{\pi S_0}{k_0 c_0}(1 - e^{-T})\right)^2 \sigma_Q^2$$

****

Consider now the possibility of using the exact variance expression in Eq. 12.32 instead of the smooth approximation in Eq. 12.33. Of course, the first thing one notes is that the exact expression is much more complicated. This will be a problem if derivatives are evaluated "by hand," but not if a symbolic processor is used to allow a computer to perform the task. Beyond the matter of work, though, the presence of the sine and cosine terms in the expression will lead to additional uncertainty. In particular, uncertainty about the values of m and k will lead to uncertainty about ω_0 and ω_d. Even a small change in ω_d can cause a large change in the $\sin(2\omega_d t)$ and $\cos(2\omega_d t)$ terms as t grows. That is, a small change in mass or stiffness can cause these terms to be anywhere in the range of -1 to 1, at any particular large value of t. Unless t is quite small, this additional uncertainty is not large for Eq. 12.32 because of the exponential decay of the harmonic terms, but it is one of the more difficult issues in some forms of dynamic analysis with uncertain parameters.

12.7 Stationary Response Variance for General Stochastic Excitation

If the excitation of a linear system is not delta-correlated, one can find the response variance either from Eq. 5.30 or Eqs. 6.22 and 6.31 giving expressions of

$$\sigma_X^2 = \int_{-\infty}^{\infty} G_{FF}(\tau)\, h_{x,var}(\tau)\, d\tau \tag{12.38}$$

and

$$\sigma_X^2 = \int_{-\infty}^{\infty} S_{XX}(\omega)\, d\omega = \int_{-\infty}^{\infty} S_{FF}(\omega) |H(\omega)|^2\, d\omega \tag{12.39}$$

For an SDF system the necessary functions are found from Eqs. 5.60 and 6.43 as

$$h_{x,var}(\tau) = \frac{e^{-\zeta \omega_0 |\tau|}}{4 m^2 \zeta \omega_0^3}\left(\cos(\omega_d \tau) + \frac{\zeta \omega_0}{\omega_d}\sin(\omega_d |\tau|)\right)$$

and

$$|H(\omega)|^2 = \frac{1}{(k - m\omega^2)^2 + (c\,\omega)^2} = \frac{1}{m^2[(\omega_0^2 - \omega^2)^2 + (2\zeta \omega_0 \omega)^2]}$$

In principle, there is no particular difficulty in applying the perturbation method to Eq. 12.38 or 12.39. For example, one can find the *j*th derivative with respect to an uncertain input parameter r_l as either

$$\frac{\partial^j}{\partial r_l^j}\sigma_X^2 = \int_{-\infty}^{\infty} G_{FF}(\tau)\frac{\partial^j}{\partial r_l^j} h_{x,var}(\tau)\, d\tau \tag{12.40}$$

or

$$\frac{\partial^j}{\partial r_l^j}\sigma_X^2 = \int_{-\infty}^{\infty} S_{FF}(\omega)\frac{\partial^j}{\partial r_l^j} |H(\omega)|^2\, d\omega \tag{12.41}$$

Thus, applying the perturbation method requires that an integral must be performed for the evaluation of each derivative term. Essentially, this means that there is as much effort involved (either analytically or numerically) in evaluating each derivative as there is in evaluating the conditional value of the response variance given deterministic input parameters. Once the derivatives are evaluated, the procedure is identical to that for the problems studied in Sections 12.5 and 12.6.

The time-domain expression in Eqs. 12.38 and 12.40 are analytically equivalent to the frequency-domain results in Eqs. 12.39 and 12.41, but the latter form is generally preferred for numerical calculations. This is because both $G_{FF}(\tau)$ and $h_{x,var}(\tau)$ are generally oscillating functions, whereas $S_{FF}(\omega)$ and $|H(\omega)|^2$ are usually somewhat better behaved. This leads to more accurate numerical integration in the frequency domain.

A difficulty with the perturbation approach to this problem is that there are situations in which the response variance is very sensitive to changes in the physical parameters. The formulation in Eq. 12.40 is most useful for visualizing this phenomenon. For typical values of damping, we know that $|H(\omega)|^2$ has a sharp peak near the resonant frequency and that uncertainty about the values of *m* and *k*, in particular, will give uncertainty about the location of this peak. If the excitation has a broadband autospectral density, then this does not cause any serious difficulty. That is, an uncertainty about the value of ω_0 will not introduce much uncertainty about $S_{FF}(\omega_0)$ if the spectral density has little variation. The delta-correlated situation considered in Sections 12.5 and 12.6 is the prototypical case of this type. If the autospectral density of $\{F(t)\}$ has one or more sharp peaks, however, the response variance can be quite sensitive to uncertainty about *m* and *k* due to uncertainty about whether or not $|H(\omega)|^2$ is "tuned" to a peak in $S_{FF}(\omega)$. This problem is never as severe for a stochastic excitation, though, as it is for a harmonic excitation, which may be considered the ultimate narrowband excitation. The following examples will explore this problem numerically.

****

Example 12.5: Consider the response variance for a lightly damped SDF system with a relatively narrowband excitation that is perfectly tuned with the oscillator. In particular, let the oscillator have nominal natural frequency ω_0 and damping $\zeta = 0.01$ and the excitation have an autospectral density of

$$S_{FF}(\omega) = S_0 \exp\left(\frac{-(|\omega| - \omega_0)^2}{2b^2\omega_0^2}\right)$$

with $b = 0.1$. The autospectral density and autocorrelation functions for the excitation are shown in the accompanying sketch.

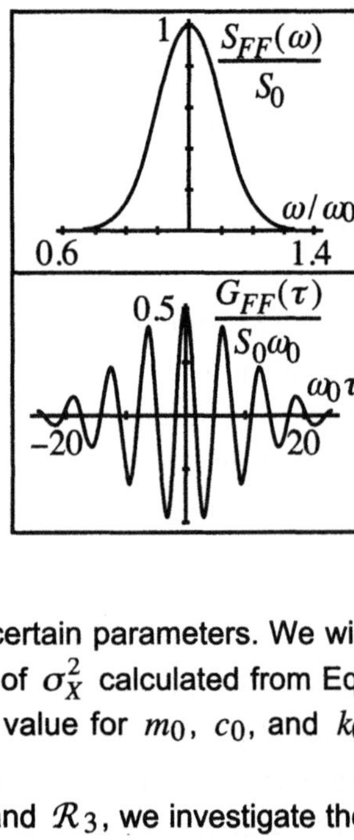

Note that any discrepancy between the true value of $\mathcal{M}$ or $\mathcal{K}$ and the nominal value of m_0 or k_0, respectively, will result in a detuning. Thus, this is a situation that is likely to be quite sensitive to variations in the parameters. Discrepancies between the true damping C and the nominal value c_0 will also have an effect on the response, but little effect on detuning.

We use the frequency domain formulation with the uncertain frequency response function term as

$$|H(\omega)|^2 = \frac{1}{(k - m\omega^2)^2 + (c\,\omega)^2}$$

and choose $\vec{\mathcal{R}} = [\mathcal{M}/m_0, C/c_0, \mathcal{K}/k_0]^T$, with m_0, c_0, and k_0 being the expected values of the three uncertain parameters. We will take the response random variable Q to be the value of σ_X^2 calculated from Eq. 12.38 with parameters $\mathcal{M}$, C, and $\mathcal{K}$ divided by the value for m_0, c_0, and k_0 so that $Q_0 = 1$.

Before considering any random variation of $\mathcal{R}_1$, $\mathcal{R}_2$, and $\mathcal{R}_3$, we investigate the sensitivity of Q to variations in each of these parameters. In each case we will compare the Q values with those obtained from linear and quadratic approximations of the form of Eqs. 12.3 and 12.4, but with only one uncertain parameter at a time. The derivatives needed for the linear and quadratic approximations have been numerically evaluated from Eq. 12.41, and the sensitivity coefficients have been found to be $\beta_1 = 0.0356$, $\beta_{11} = -23.12$, $\beta_2 = -1.076$, $\beta_{22} = 2.160$, $\beta_3 = -0.9600$, and $\beta_{33} = -21.22$. The β_{lk} coefficients for $k \neq l$ have not been evaluated, because they do not enter into calculation of moments by a mixed-order perturbation method. The following sketches show numerical results for both the linear and quadratic approximations, along with direct numerical evaluation of Q for situations with only one uncertain parameter at a time.

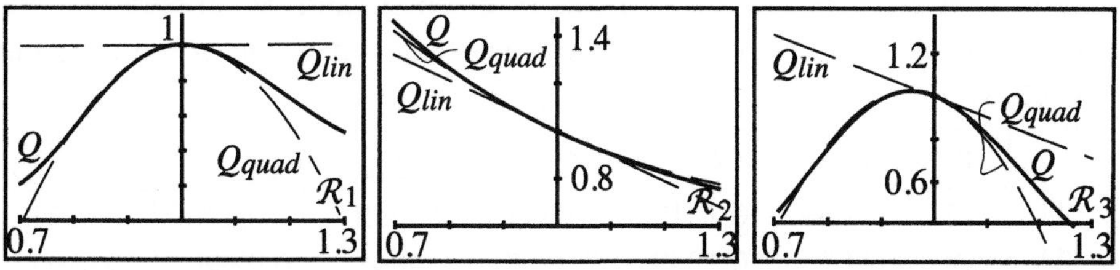

From the sketches it is clear that the linear approximation is close to the computed value for Q only for very small variations in m or k, although it is significantly better for variations in c. This is in agreement with the prior observation that either positive or negative variations in m or k can cause considerable reduction of the response variance, due to detuning, and also with the large values of the second-order sensitivity coefficients β_{11} and β_{33}. The dependence of the response variance on c, though, is monotonic. The quadratic approximations are seen to be reasonably good for a much broader range of parameters, especially for m and k. Nonetheless, they seriously overpredict the response reduction due to large changes in mass or stiffness, especially when the change is positive.

Let us now consider perturbation results for simple cases with only one random parameter. First we will take the damping and stiffness to be known and the mass to be uniformly distributed: $\mathcal{R}_2 = \mathcal{R}_3 = 1$ and $\mathcal{R}_1$ uniformly distributed on $1-\gamma \le \mathcal{R}_1 \le 1+\gamma$. Next the mass and damping are held constant, and stiffness is taken to be uniformly distributed: $\mathcal{R}_1 = \mathcal{R}_2 = 1$, $1-\gamma \le \mathcal{R}_3 \le 1+\gamma$. Mean and standard deviation results have been obtained for three different γ values. Also, the results from a logarithmic linear perturbation and a log-normal parameter distribution (labeled $Q_{log\text{-}lin}$) have been obtained by using Eq. 12.17 and the sensitivity coefficients from the Q_{quad} formulas. Results from direct integration, the mixed-order direct perturbation, and the linear log-normal perturbation are shown in the following table.

The results for the quadratic approximation of the mean value of Q are quite good for $\gamma \le 0.2$, and are reasonable approximations for $\gamma = 0.3$, when compared with the results of numerical integration with a uniform parameter distribution. The log-normal approximation of the mean is generally quite close to the value of unity that would be obtained by a direct linear perturbation. These values are generally acceptable only for very small amounts of uncertainty. The results for the variance of Q are in reasonable agreement between the three methods for an uncertain $\mathcal{K}$ but show major discrepancies for uncertain $\mathcal{M}$. This is not surprising based on the plot of Q versus $\mathcal{R}_1$. The β_1 sensitivity coefficient is very small because Q is near a maximum at the design point of $\mathcal{R}_1 = 1$. A linear approximation cannot capture the dependence of Q on $\mathcal{R}_1$.

The logarithmic-linear formulation of this problem provides little benefit because of the form of the nonlinear dependence of Q on $\mathcal{K}$ and $\mathcal{M}$. In particular, the dependence does not even vaguely resemble that of a power of the uncertain parameters.

	γ	μ_Q	σ_Q	$\mu_{Q_{quad}}$	$\sigma_{Q_{lin}}$	$\mu_{Q_{log\text{-}lin}}$	$\sigma_{Q_{log\text{-}lin}}$
$\mathcal{M}$	0.3	0.7338	0.2243	0.6531	0.0062	0.9995	0.0061
$\mathcal{M}$	0.2	0.8623	0.1239	0.8458	0.0041	0.9998	0.0041
$\mathcal{M}$	0.1	0.9625	0.0340	0.9615	0.0021	0.9999	0.0021
$\mathcal{K}$	0.3	0.7505	0.2141	0.6817	0.1663	1.0282	0.1709
$\mathcal{K}$	0.2	0.8733	0.1318	0.8585	0.1109	1.0125	0.1122
$\mathcal{K}$	0.1	0.9656	0.0587	0.9646	0.0554	1.0031	0.0556

The poor fit of Q versus $\mathcal{R}_1$ for any of the distributions considered so far prompts consideration of the logarithmic-quadratic approximation of Eqs. 12.12 and 12.15. The following sketch shows that this approximation is almost indistinguishable from Q for the same range of $\mathcal{R}_1$ as was used before. The mean and standard deviation of $Q_{log\text{-}quad}$ have also been evaluated using the uniform distribution for $\mathcal{R}_1$. The answers of $\mu_{Q_{log\text{-}quad}} = 0.7343$ and $\sigma_{Q_{log\text{-}quad}} = 0.2187$ are seen to be very close to those evaluated directly from Q. Note that this is only a use of the logarithmic-quadratic form for curve fitting and is not a log-normal approximation. It does illustrate some of the versatility of this form, though.

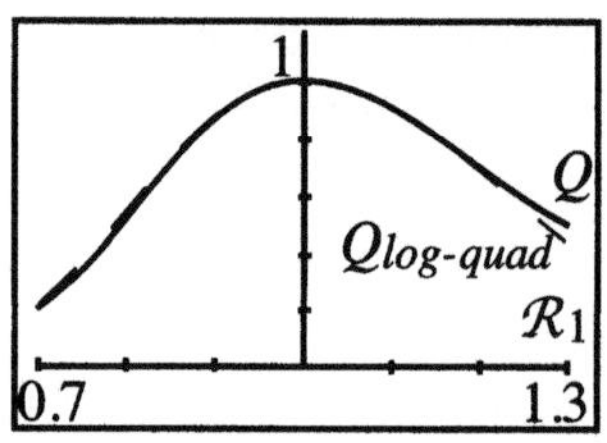

****

Example 12.6: Consider the same lightly damped system as in Example 12.5, but with an excitation that is less narrowband. In particular, the oscillator has parameters ω_0 and $\zeta = 0.01$ and the excitation has an autospectral density of

$$S_{FF}(\omega) = S_0 \exp\left(\frac{-(|\omega| - \omega_0)^2}{2b^2 \omega_0^2} \right)$$

with $b = 0.5$. The autospectral density and autocorrelation functions for the excitation are shown in the preceding sketch.

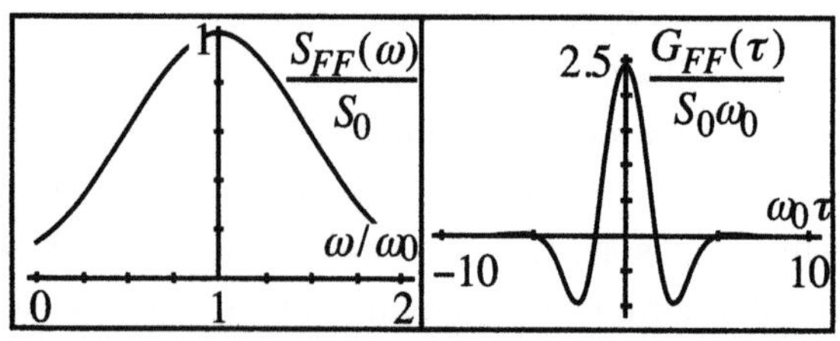

Proceeding as in the prior example, we find the sensitivity coefficients as $\beta_1 = 0.00848$, $\beta_{11} = -1.008$, $\beta_2 = -1.016$, $\beta_{22} = 2.033$, $\beta_3 = -0.9921$, and $\beta_{33} = 0.9774$.

The following sketches show numerical results for the linear and quadratic approximations, along with direct numerical evaluation of Q, and the table that follows shows the mean and standard deviations obtained for Q for the uniform distributions, the mixed-order direct perturbation, and the logarithmic linear perturbation.

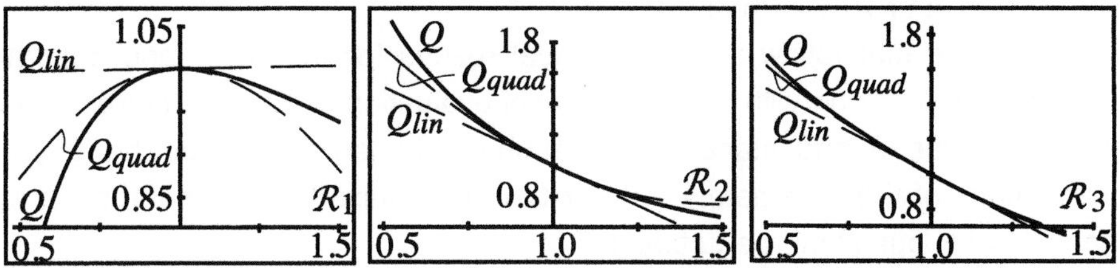

	γ	μ_Q	σ_Q	$\mu_{Q_{quad}}$	$\sigma_{Q_{lin}}$	$\mu_{Q_{log\text{-}lin}}$	$\sigma_{Q_{log\text{-}lin}}$
$\mathcal{M}$	0.5	0.9471	0.0685	0.9580	0.0024	0.9997	0.0024
$\mathcal{M}$	0.2	0.9930	0.0071	0.9933	0.0010	0.9999	0.0010
$\mathcal{K}$	0.5	1.0436	0.3012	1.0407	0.2864	1.0823	0.3099

The $\mathcal{M}$ and C results are much like those in Example 12.5, except that $\mathcal{M}$ now has an even smaller linear sensitivity coefficient, which leads to larger discrepancies in predicting the standard deviation of Q. The $\mathcal{K}$ result is much more nearly linear than it was in Example 12.5. Again the mean values are predicted quite well by the quadratic perturbation, and the linear logarithmic model provides little improvement over a direct linear perturbation. As shown in the sketch at the right, using the logarithmic-quadratic form of Eqs. 12.12 and 12.15 provides an improved approximation, as it did in Example 12.5. The mean and standard deviation values computed in this way for a uniform distribution on [0.5, 1.5] are 0.9538 and 0.0508, which agree quite well with the direct computation for $\gamma = 0.5$.

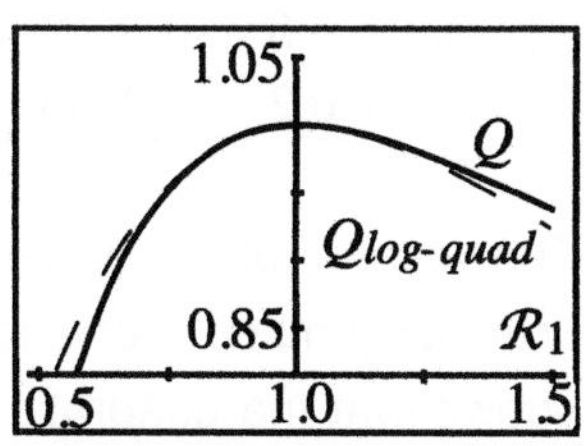

****

Example 12.7: Consider the same lightly damped system and narrowband excitation as in Example 12.5, but with some detuning between the excitation and oscillator. In particular, the oscillator has nominal natural frequency ω_0 and damping $\zeta = 0.01$ and the excitation has an autospectral density of

$$S_{FF}(\omega) = S_0 \exp\left(\frac{-(|\omega| - 1.1\omega_0)^2}{2b^2\omega_0^2}\right)$$

with $b = 0.1$. The autospectral density for the excitation is shown in the sketch to the right.

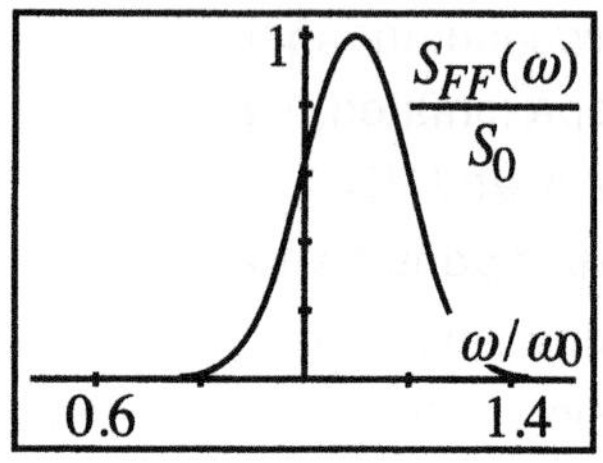

Proceeding as in the prior examples, we find the sensitivity coefficients as $\beta_1 = -4.570$, $\beta_{11} = 4.682$, $\beta_2 = -1.046$, $\beta_{22} = 2.093$, $\beta_3 = 3.616$, and $\beta_{33} = -10.55$.

The sketches that follow show numerical results for the linear and quadratic approximations, along with direct numerical evaluation of Q. The following table shows the mean and standard deviations obtained for Q for the uniform distributions of the parameters and for the direct mixed-order perturbation and the first-order logarithmic perturbation.

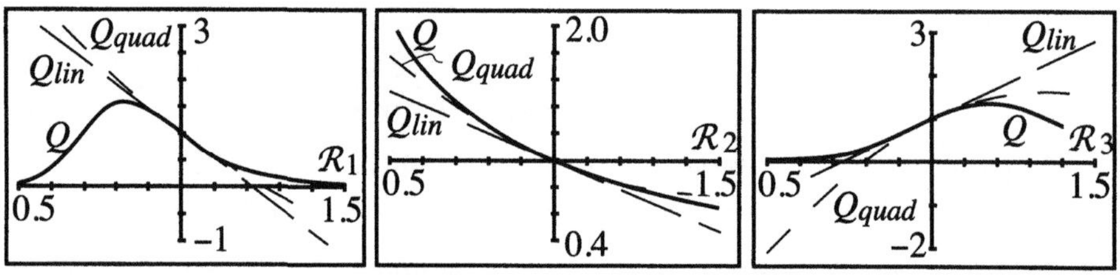

	γ	μ_Q	σ_Q	$\mu_{Q_{quad}}$	$\sigma_{Q_{lin}}$	$\mu_{Q_{log\text{-}lin}}$	$\sigma_{Q_{log\text{-}lin}}$
$\mathcal{M}$	0.5	0.6594	0.5288	1.1951	1.3193	0.9997	0.0024
$\mathcal{M}$	0.2	0.9973	0.4266	1.0312	0.5277	0.9999	0.0010
$\mathcal{M}$	0.1	1.0055	0.2512	1.0078	0.2639	1.0000	0.0005
$\mathcal{K}$	0.5	0.6947	0.4722	0.5603	1.0438	1.0823	0.3099
$\mathcal{K}$	0.2	0.9344	0.3484	0.9296	0.4175	1.0132	0.1161

The results for uncertain $\mathcal{M}$ (or $\mathcal{R}_1$) demonstrate an undesirable behavior not previously noted. For a large uncertainty such as $\gamma_1 = 0.5$, the preceding sketch shows that the quadratic approximation for Q may be considered less accurate than the linear approximation. In particular, the second derivative of Q with respect to $\mathcal{R}_1$ is positive at $\mathcal{R}_1 = 1$, giving the quadratic approximation a positive curvature. This, however, completely misses the negative curvature associated with the peak of Q at $\mathcal{R}_1 \approx 0.8$. This problem is also reflected in the value of $E(Q_{quad})$ in the table for $\gamma_1 = 0.5$. For $\gamma_1 = 0.1$, however, the peak at $\mathcal{R}_1 \approx 0.8$ has little effect and the quadratic approximation gives good results. Neither of the perturbation methods gives very good results when the uncertainty about $\mathcal{M}$ or $\mathcal{K}$ is large.

Because the dependence of Q on $\mathcal{R}_1$ and $\mathcal{R}_3$ has not been reasonably approximated, consider again the logarithmic-quadratic form used in Examples 12.5 and 12.6. The accompanying sketches show that it does give a much better fit than was obtained from any of the other forms, even though it does have some error. The mean and variance of Q were also computed both for $\mathcal{R}_1$ uniformly distributed on [0.5, 1.5] and for $\mathcal{R}_3$ with the same distribution. For $\mathcal{R}_1$ variation, the mean and standard deviation values are 0.7377 and 0.5643, respectively, and for $\mathcal{R}_3$ variation they are 0.7470 and 0.5201. Again, these do provide reasonable approximations of those obtained from direct evaluation, even for these very nonlinear functions and wide ranges of the parameters.

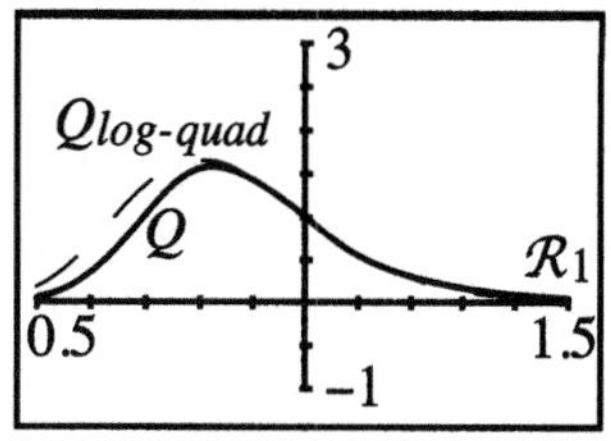

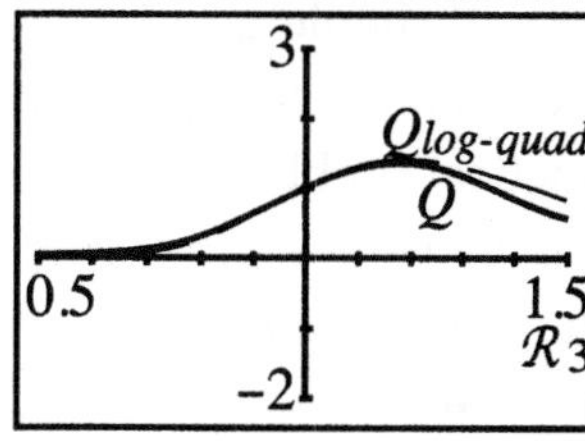

****

The preceding examples illustrate that the usual perturbation methods may not give good results for this problem unless the uncertainty is small. This difficulty can be anticipated in any problem in which the response quantity has a highly nonlinear dependence on the uncertain parameters. In some instances the quadratic approximation is appropriate for such nonlinear situations, but this is not always the case, as illustrated by the $\mathcal{R}_1$ and $\mathcal{R}_3$ dependence of Q in Example 12.7. In these examples, numerical integration was used to evaluate the mean and variance of Q based on the actual nonlinear dependence on the uncertain parameters and an assumed probability distribution, but this can be a somewhat costly procedure when there are multiple uncertain parameters. The logarithmic-quadratic perturbation has been shown to give a reasonable approximation of some nonlinear relationships that cannot be approximated with simpler methods. Its disadvantage is that it requires numerical integration for evaluation of the moments of Q, which may be somewhat inconvenient for some choices of the distribution of the uncertain parameters.

Recall that the only difference between the problem considered in this section and that for force-excitation in Section 12.5 is the change of the autospectral density of $\{F(t)\}$ to be variable rather than constant. That is, if we let $S_{FF}(\omega_0) = S_0$ in this section, then we would reduce the problem to that of Section 12.5, in which there was no dependence on $\mathcal{M}$ (or $\mathcal{R}_1$). Thus, using a more broadband excitation in the current problem should give Q as having less dependence on $\mathcal{R}_1$. This is confirmed by the results in Examples 12.5 and 12.6, which are identical except for the bandwidth of $\{F(t)\}$. For example, using $\gamma_1 = 0.2$ for the narrowband excitation in Example 12.5 gave μ_Q as being 14% below that for deterministic $\mathcal{R}_1$ and σ_Q as being 0.12, whereas the corresponding situation with the more broadband excitation of Example 12.6 gave μ_Q as only 0.7% below that for deterministic $\mathcal{R}_1$ and σ_Q as 0.007. Nonetheless, the plot of Q versus $\mathcal{R}_1$ in Example 12.6 shows that there is still considerable dependence on $\mathcal{R}_1$ when it varies significantly from the deterministic value of 1.0.

It should also be noted that the examples have considered extremely large uncertainties for k and m—up to ±50%. These are not anticipated values for practical problems but are included to emphasize the difficulties that can occur due to very nonlinear relationships between the response quantity and the uncertain parameters. Perturbation analyses generally work better when uncertainties are smaller, but one must always be aware that they can fail to give acceptable answers in particular problems.

A somewhat more complicated problem that will not be explored here is a combination of Sections 12.6 and 12.7—the nonstationary response variance for an excitation that is not delta-correlated. The basic formula for this situation can be taken as either the time-domain integral of Eq. 5.35 or the integral of the evolutionary spectral density in Eq. 6.47. The latter equation is the same as Eq. 12.39, but with a modulating term reflecting the fact that the oscillator response begins at zero. One can anticipate that the major difficulty will relate to the matter of tuning/detuning between the oscillator and excitation, as in the current section. The effect of time-variation, as shown in Section 12.6, is less severe and is better modeled by perturbation.

12.8 First-Passage Failure Probability

In studying the effect of uncertain parameters on the probability of first-passage failure, we will focus on the response of an SDF oscillator. We consider the excitation to be a mean-zero, stationary, Gaussian white noise force. The oscillator is taken to be initially at rest, so it must start below the failure level u, and Eq. 11.13 gives the failure probability as

$$P_f(t) = 1 - \exp\left(-\int_0^t \eta_X(u,s)\, ds\right) \qquad (12.42)$$

Consider first the Poisson approximation of $\eta_X(u,t) = \nu_X^+(u,t)$, which is found from Example 7.2 as

$$\nu_X^+(u,t) = \frac{\sigma_{\dot{X}}(t)}{\sigma_X(t)} \exp\left(\frac{-u^2}{2\sigma_X^2(t)}\right)\left(\frac{\rho_{X\dot{X}}(t,t)\,u}{(2\pi)^{1/2}\sigma_X(t)} \Phi\left[\frac{\rho_{X\dot{X}}(t,t)\,u}{[1-\rho_{X\dot{X}}^2(t,t)]^{1/2}\sigma_X(t)}\right] + \frac{[1-\rho_{X\dot{X}}^2(t,t)]^{1/2}}{2\pi} \exp\left[\frac{-\rho_{X\dot{X}}^2(t,t)\,u^2}{2[1-\rho_{X\dot{X}}^2(t,t)]\sigma_X^2(t)}\right]\right)$$

We use the nonoscillatory approximations for the growth of the variances of $\{X(t)\}$ and $\{\dot{X}(t)\}$:

$$\sigma_X^2(t) \approx \frac{\pi S_0}{kc}\left(1-e^{-ct/m}\right)U(t), \qquad \sigma_{\dot{X}}^2(t) \approx \frac{\pi S_0}{mc}\left(1-e^{-ct/m}\right)U(t)$$

as we did for the variance of $\{X(t)\}$ in Section 12.5. Also, the nonoscillatory approximation of their correlation coefficient, as used in Example 11.3, is used:

$$\rho_{X\dot{X}}(t,t) \approx \min\left(\frac{3^{1/2}}{2}, \frac{c}{2(km)^{1/2}} \frac{e^{-ct/m}}{1-e^{-ct/m}}\right)$$

Note that all of the preceding expressions are written in terms of m, c, and k in order to reveal more clearly the dependence of the probability of failure on these parameters. It is relatively easy to evaluate directly and plot P_f versus m, c, or k using the given expressions, as was done in Examples 12.5–12.7. One can also evaluate the derivatives needed for a perturbation approximation. For example, if we let r denote m, c, or k, then we find that the derivative of Eq. 12.42 is

$$\frac{\partial}{\partial r} P_f(t) = P_f(t) \int_0^t \frac{\partial}{\partial r} \eta_X(u,s)\, ds \tag{12.43}$$

and

$$\frac{\partial^2}{\partial r^2} P_f(t) = P_f(t)\left(\left[\frac{\partial}{\partial r} P_f(t)\right]^2 + \int_0^t \frac{\partial^2}{\partial r^2} \eta_X(u,s)\, ds\right) \tag{12.44}$$

The derivatives of $\eta_X(u,t) = \nu_X^+(u,t)$, for the Poisson approximation, then involve derivatives of $\sigma_X(t)$, $\sigma_{\dot X}(t)$, and $\rho_{X\dot X}(t,t)$. The first two of these are obviously straightforward, as is the third after one notes that

$$\frac{\partial}{\partial r}\min\left(\frac{3^{1/2}}{2}, \frac{c}{2(k\,m)^{1/2}}\frac{e^{-ct/m}}{1-e^{-ct/m}}\right) =$$

$$U\left(\frac{3^{1/2}}{2}, \frac{c}{2(k\,m)^{1/2}}\frac{e^{-ct/m}}{1-e^{-ct/m}}\right)\frac{\partial}{\partial r}\left(\frac{c}{2(k\,m)^{1/2}}\frac{e^{-ct/m}}{1-e^{-ct/m}}\right)$$

In obtaining numerical results for a given set of parameters, it is useful to note that one can first perform a numerical integration of Eq. 12.42 to obtain the value of P_f at the design point, then perform the numerical integration of Eq. 12.43 to obtain the first derivative, and finally do the same for Eq. 12.44 to obtain the second derivative. Thus, the effort required for a second-order or mixed-order perturbation approximation is essentially the same as for evaluating P_f for three sets of parameters. That is, it involves only three numerical integrations over t. The linear log-normal approximations can also be found from the first-order derivatives.

****

Example 12.8: Investigate the effects of uncertain parameters on the probability of first-passage failure for a linear oscillator by using the Poisson approximation. The oscillator is the same as the one considered in Example 11.3, with resonant frequency ω_0 and damping $\zeta = 0.01$. The oscillator has zero initial conditions, and it is excited by stationary, Gaussian, white noise with mean zero and autospectral density S_0. Failure occurs if $X(t)$ exceeds the level $4\,\sigma_{stat} = 4(\pi\,S_0)^{1/2}/(c_0\,k_0)$ within the time interval $0 < \omega_0\,t \le 250$.

Letting $\vec{\mathcal{R}} = [\mathcal{M}/m_0, \mathcal{C}/c_0, \mathcal{K}/k_0]^T$, as in the preceding examples, numerical results have been obtained for $Q = P_f(\mathcal{M},\mathcal{C},\mathcal{K})/P_f(m_0,c_0,k_0)$, as well as for the first and second derivatives with respect to each parameter. The sensitivity coefficients are found as $\beta_1 = -1.565$, $\beta_{11} = 2.374$, $\beta_2 = -7.277$, $\beta_{22} = 51.92$, $\beta_3 = 0.4972$, and $\beta_{33} = -0.2486$.

Note that P_f is very sensitive to changes in $\mathcal{C}$. In particular, a 1% increase in $\mathcal{C}$ results more than a 7% decrease in P_f. The logarithmic approximation from Eq. 12.14 is also easily obtained for each single-parameter variation as $Q = (\mathcal{R}_l)^{\beta_l}$. The following sketches show the numerical evaluations of Q for rather large parameter variations. The logarithmic-linear curve is omitted for $\mathcal{K}$ variation because it is essentially indistinguishable from the curve from direct evaluation.

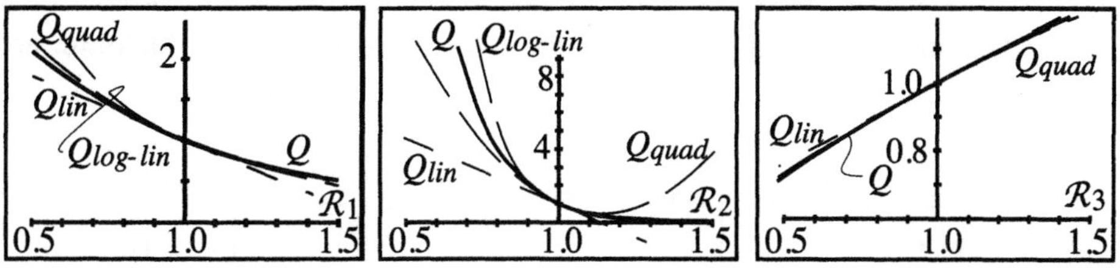

It is seen that the dependence of P_f on C is far from linear. This, though, was quite predictable, inasmuch as the linear approximation would predict Q being negative for $\mathcal{R}_2 > 1.14$, which is obviously impossible. The sketch does reveal, though, that P_f is greatly reduced by increasing C. For $C = 1.5c_0$, the probability of failure is only 2.3% of the value for $C = c_0$. The dependence of P_f on $\mathcal{M}$ and $\mathcal{K}$ is much simpler and is quite well approximated by the linear or quadratic functions. The logarithmic-linear approximation is not generally better than the others, but it does give the proper behavior for large C values. The logarithmic-quadratic approximation was not found to be significantly better than the logarithmic-linear one for this problem.

The following table gives sample values of the mean and variance computed from direct numerical integration with the uniform parameter distribution, the mixed-order perturbation, and the logarithmic approximation with the log-normal parameter distribution.

	γ	μ_Q	σ_Q	$\mu_{Q_{quad}}$	$\sigma_{Q_{lin}}$	$\mu_{Q_{log-lin}}$	$\sigma_{Q_{log-lin}}$
$\mathcal{M}$	0.5	1.1062	0.4963	1.0989	0.4518	1.1743	0.5466
C	0.2	1.3748	1.0723	1.3461	0.8403	1.4902	1.5025
$\mathcal{K}$	0.5	0.9890	0.1468	0.9896	0.1435	0.9900	0.1400

For the $\mathcal{M}$ and $\mathcal{K}$ dependence, the approximations are in reasonable agreement even for $\gamma = 0.5$. The plots have already shown that the C dependence is not well approximated over this wide range, so the results shown are for $\gamma = 0.2$. Even for this more reasonable level of uncertainty, the mean and variance values computed from the logarithmic-linear approximation with log-normal variables are significantly higher than for the uniform distribution. Also, the first-order variance approximation in the mixed-order perturbation is significantly low.

****

Next we consider the approximation for the probability of first passage as given by Eq. 11.42:

$$P_{f,Vanm} \approx 1 - (1 - P_{f,Pois})^{\eta_{X,Vanm}(u)/\nu_X^+(u)} \tag{12.45}$$

in which $P_{f,Pois}$ is computed exactly as before, using $\eta_X(u,t) = \nu_X^+(u,t)$. Using Vanmarcke's modified bandwidth effect, the exponent is given in Eq. 11.35 as

$$\frac{\eta_{X,Vanm}(u)}{\nu_X^+(u)} \approx \left(1-\exp\left[-(1-\alpha_1^2)^{0.6}(2\pi)^{1/2}\frac{u}{\sigma_X}\right]\right)\left(1-\exp\left[\frac{-u^2}{2\sigma_X^2}\right]\right)^{-1}$$

The other quantity needed is the stationary α_1, which was given in Example 7.9:

$$\alpha_1 = \frac{1}{(1-\zeta^2)^{1/2}}\left(1-\frac{2}{\pi}\tan^{-1}\left[\frac{\zeta}{(1-\zeta^2)^{1/2}}\right]\right)$$

with $\zeta = c/[2(k\,m)^{1/2}]$.

To find the β_I and β_{II} sensitivity parameters for this formulation, one can use

$$\frac{\partial P_{f,Vanm}}{\partial r} \approx -(1-P_{f,Pois})^{\eta_{X,Vanm}(u)/\nu_X^+(u)} \times$$
$$\left(\log[(1-P_{f,Pois})]\frac{\partial}{\partial r}\frac{\eta_{X,Vanm}(u)}{\nu_X^+(u)} - \frac{1}{1-P_{f,Pois}}\frac{\eta_{X,Vanm}(u)}{\nu_X^+(u)}\frac{\partial P_{f,Pois}}{\partial r}\right)$$

and

$$\frac{\partial^2 P_{f,Vanm}}{\partial r^2} \approx -(1-P_{f,Pois})^{\eta_{X,Vanm}(u)/\nu_X^+(u)} \times$$
$$\left(\left[\log[(1-P_{f,Pois})]\frac{\partial}{\partial r}\frac{\eta_{X,Vanm}(u)}{\nu_X^+(u)} - \frac{1}{1-P_{f,Pois}}\frac{\eta_{X,Vanm}(u)}{\nu_X^+(u)}\frac{\partial P_{f,Pois}}{\partial r}\right]^2 + \right.$$
$$\log[(1-P_{f,Pois})]\frac{\partial^2}{\partial r^2}\frac{\eta_{X,Vanm}(u)}{\nu_X^+(u)} - \frac{2}{1-P_{f,Pois}}\frac{\partial P_{f,Pois}}{\partial r}\frac{\partial}{\partial r}\frac{\eta_{X,Vanm}(u)}{\nu_X^+(u)} -$$
$$\left.\frac{1}{(1-P_{f,Pois})^2}\frac{\eta_{X,Vanm}(u)}{\nu_X^+(u)}\left[\frac{\partial P_{f,Pois}}{\partial r}\right]^2 - \frac{1}{1-P_{f,Pois}}\frac{\eta_{X,Vanm}(u)}{\nu_X^+(u)}\frac{\partial^2 P_{f,Pois}}{\partial r^2}\right)$$

The new derivatives of the $\eta_{X,Vanm}(u)/\nu_X^+(u)$ term in these expressions do not involve integrals over time, and the derivatives of $P_{f,Pois}$ are the same as one would use for perturbation analysis using the Poisson approximation.

In principle, one can proceed in the same way for the complete nonstationary Vanmarcke approximation of the first-passage failure, as explored in Example 11.6. The derivative expressions needed for perturbation, though, are very lengthy. The key to such evaluations seems to be the use of an appropriate mix of analytical and numerical operations in order to achieve reasonable efficiency in the calculations. Direct numerical evaluation of Vanmarcke's approximation of P_f over a range of parameters is also feasible, and numerical differentiation can give estimates of the perturbation relationships. Similarly, numerical integration can give direct estimates of the mean and variance of P_f for a given distribution of the uncertain parameters.

****

Example 12.9: Find the linear and quadratic perturbation formulas for Vanmarcke's modified approximation of the probability of first-passage failure for the oscillator of Example 12.8. The oscillator has 1% of critical damping, stationary, white noise excitation, and zero initial conditions.

The numerical results from the preceding formulas with the approximation of Eq. 12.45 give $\beta_1 = -1.769$, $\beta_{11} = 3.233$, $\beta_2 = -6.548$, $\beta_{22} = 40.87$, $\beta_3 = 0.2964$, and $\beta_{33} = -0.2320$.

One notes that the values of the sensitivity parameters are generally similar to those from the Poisson perturbation formulas in Example 12.8. This in contrast to the probability of failure, which is reduced by 48%, from 6.1×10^{-3} for the Poisson approximation to 3.2×10^{-3} for the modified Vanmarcke formula. Clearly the sensitivity to parameter uncertainty is less affected by the choice of analysis method than is the actual estimate of P_f.

****

Note that the results in this section show that the probability of first-passage failure is very sensitive to changes in the damping value in the system, as indicated by the large values of β_2 and β_{22}. In a sense this might be anticipated, inasmuch as changes in damping directly influence the level of response variance, which is crucial in predicting the probability of first passage. The probability of first passage, however, is much more sensitive than the response variance to uncertainty in the damping. Also, the response variance, for white noise excitation, is as sensitive to changes in stiffness as to changes in damping, and this is clearly not true for the probability of first-passage failure.

If one considers a nonwhite excitation of the SDF oscillator, then the sensitivity of first-passage failure to uncertain parameters becomes much more complicated. Sections 12.5 and 12.7 illustrate this for the simpler problem of studying stationary response variance, and the same general effects must be anticipated for the study of first-passage failure. If uncertainty about mass or stiffness can affect the tuning/detuning of the oscillator and a peak in the excitation autospectral density, then this uncertainty will have a major effect on the probability of first-passage failure. Although the method of analysis is the same for this more general excitation, it is clear that the computations become much more involved.

12.9 Fatigue Life

The results in Eqs. 11.54 and 11.55 give the most commonly used estimate of stochastic fatigue life as

$$E(T_{Ray}) = \frac{2\pi K}{2^{3\breve{m}/2} \sigma_X^{\breve{m}-1} \sigma_{\dot{X}} \Gamma(1+\breve{m}/2)} \tag{12.46}$$

in which K and $\breve{m}$ are parameters from constant-amplitude fatigue tests. (Note that the brev has been added over the symbol m used in Chapter 11 to avoid confusion with mass, which appears in various applications in this chapter.) The approximation In Eq. 12.46 is generally considered appropriate when the $\{X(t)\}$ stress process is narrowband and Gaussian. To include bandwidth effects and non-Gaussianity, this result was modified by RR and GR factors, respectively. For the present study we will focus only on the single-moment method for the RR term and Winterstein's approximations for GR, giving

$$E(T) = E(T_{Ray})(RR)(GR)$$

with

$$RR = \frac{\sigma_X^{\breve{m}-1} \sigma_{\dot{X}}}{(\lambda_{2/\breve{m}})^{\breve{m}/2}}$$

and

$$GR = \left(1 + \frac{\breve{m}(\breve{m}-1)(\kappa-3)}{24}\right)^{-1} U(\kappa-3) + \left(1 + \frac{\breve{m}(\breve{m}-1)(3-\kappa)}{24}\right)[1 - U(\kappa-3)] \tag{12.47}$$

in which κ denotes the kurtosis of $\{X(t)\}$. The spectral moment in RR is given by

$$\lambda_{2/\breve{m}} = 2\int_0^\infty \omega^{2/\breve{m}}\, S_{XX}(\omega)\, d\omega \tag{12.48}$$

Analysis of the prediction including the correction factors can be simplified by rewriting the failure time as

$$E(T) = \frac{2\pi K(GR)}{2^{3\hat{m}/2}(\lambda_{2/\hat{m}})^{\hat{m}/2}\,\Gamma(1+\hat{m}/2)} \tag{12.49}$$

Two of the major sources of uncertainty about the expected fatigue life in Eqs. 12.46–12.49 may be considered to be the two parameters K and $\breve{m}$ of the S/N curve. At best these parameters are based on experiments with specimens similar to the prototype of interest, and in other situations they may be extrapolated from tests of rather different components. The effect of uncertainty about mass, damping, and stiffness in the prototype enters into Eq. 12.46 with GR correction as uncertainty about the kurtosis and standard deviation of stress and about the standard deviation of the derivative of stress. Uncertain structural parameters affect Eq. 12.49 by introducing uncertainty about the stress kurtosis and the value of the $\lambda_{2/\breve{m}}$ spectral moment. It is not easy, though, to give any general relationships between these quantities in the fatigue equation and the structural parameters. In principle, one can generally consider the $\{X(t)\}$ process to be proportional to the dynamic response of a linear or nonlinear dynamic equation, but the details of the response will depend strongly on the form of the excitation process. In many situations, that excitation will have a predominant frequency that is not near the resonant frequency of the structure. This situation commonly leads to the autospectral density of the response being multimodal, having peaks both at the dominant frequency of the excitation and at one or more resonant frequencies of the structure. Furthermore, non-Gaussianity of the stress process may be due to non-Gaussianity of the excitation, nonlinearity of the equation of motion, or both.

Taking the derivative of Eq. 12.49 with respect to the fatigue parameter K is essentially trivial, because $E(T)$ is proportional to K. The derivative with respect to $\breve{m}$, though, is more interesting:

$$\frac{\partial E(T)}{\partial \breve{m}} = E(T)\left(\frac{1}{GR}\frac{\partial (GR)}{\partial \breve{m}} - \frac{3}{2}\log(2) - \frac{1}{2}\log(\lambda_{2/\breve{m}}) - \frac{\breve{m}}{2\lambda_{2/\breve{m}}}\frac{\partial(\lambda_{2/\breve{m}})}{\partial \breve{m}} - \frac{1}{\Gamma(1+\breve{m}/2)}\frac{\partial \Gamma(1+\breve{m}/2)}{\partial \breve{m}}\right)$$

in which

$$\frac{\partial (GR)}{\partial \breve{m}} = -\left(1+\frac{\breve{m}(\breve{m}-1)(\kappa-3)}{24}\right)^{-2} \frac{(2\breve{m}-1)(\kappa-3)}{24} U(\kappa-3) + \frac{(2\breve{m}-1)(3-\kappa)}{24}[1-U(\kappa-3)]$$

Using

$$\Gamma\left(1+\frac{\breve{m}}{2}\right) = \int_0^\infty u^{\breve{m}/2} e^{-u}\, du$$

gives its derivative as

$$\frac{\partial}{\partial \breve{m}} \Gamma\left(1+\frac{\breve{m}}{2}\right) = \frac{1}{2}\int_0^\infty u^{\breve{m}/2} e^{-u} \log(u)\, du$$

Finally, Eq. 12.48 gives

$$\frac{\partial \lambda_{2/\breve{m}}}{\partial \breve{m}} = \frac{-4}{\breve{m}^2}\int_0^\infty \omega^{2/\breve{m}} \log(\omega)\, S_{XX}(\omega)\, d\omega$$

Another sensitivity that may be of interest is that due to an uncertainty about κ for which

$$\frac{\partial E(T)}{\partial \kappa} = \frac{E(T)}{GR}\frac{\partial GR}{\partial \kappa}$$

with

$$\frac{\partial GR}{\partial \kappa} = -\left(1+\frac{\breve{m}(\breve{m}-1)(\kappa-3)}{24}\right)^{-2} \frac{\breve{m}(\breve{m}-1)}{24} U(\kappa-3) - \frac{\breve{m}(\breve{m}-1)}{24}[1-U(\kappa-3)]$$

The second derivatives needed for a quadratic approximation can be obtained by the same procedures. Not all the details will be written out here, but the two terms involving derivatives of integrals are

$$\frac{\partial^2}{\partial \breve{m}^2} \Gamma\left(1+\frac{\breve{m}}{2}\right) = \frac{1}{4}\int_0^\infty u^{\breve{m}/2} e^{-u} [\log(u)]^2\, du$$

and

$$\frac{\partial^2 \lambda_{2/\breve{m}}}{\partial \breve{m}^2} = \frac{8}{\breve{m}^3} \int_0^\infty \omega^{2/\breve{m}} \log(\omega)\, S_{XX}(\omega)\, d\omega + \frac{8}{\breve{m}^4} \int_0^\infty \omega^{2/\breve{m}} [\log(\omega)]^2\, S_{XX}(\omega)\, d\omega$$

and the general formula is

$$\frac{\partial^2 E(T)}{\partial \breve{m}^2} = E(T) \Bigg(-\frac{1}{(GR)^2} \left[\frac{\partial (GR)}{\partial \breve{m}} \right]^2 + \frac{1}{GR} \frac{\partial^2 (GR)}{\partial \breve{m}^2} - \frac{1}{\lambda_{2/\breve{m}}} \frac{\partial (\lambda_{2/\breve{m}})}{\partial \breve{m}} +$$
$$\frac{\breve{m}}{2(\lambda_{2/\breve{m}})^2} \left[\frac{\partial (\lambda_{2/\breve{m}})}{\partial \breve{m}} \right]^2 - \frac{\breve{m}}{2\lambda_{2/\breve{m}}} \frac{\partial^2 (\lambda_{2/\breve{m}})}{\partial \breve{m}^2} +$$
$$\frac{1}{[\Gamma(1+\breve{m}/2)]^2} \left[\frac{\partial \Gamma(1+\breve{m}/2)}{\partial \breve{m}} \right]^2 - \frac{1}{\Gamma(1+\breve{m}/2)} \frac{\partial^2 \Gamma(1+\breve{m}/2)}{\partial \breve{m}^2} +$$
$$\Bigg[\frac{1}{GR} \frac{\partial (GR)}{\partial \breve{m}} - \frac{3}{2} \log(2) - \frac{1}{2} \log(\lambda_{2/\breve{m}}) - \frac{\breve{m}}{2\lambda_{2/\breve{m}}} \frac{\partial (\lambda_{2/\breve{m}})}{\partial \breve{m}} -$$
$$\frac{1}{\Gamma(1+\breve{m}/2)} \frac{\partial \Gamma(1+\breve{m}/2)}{\partial \breve{m}} \Bigg]^2 \Bigg)$$

Note that the evaluation of the sensitivity of $\lambda_{2/\breve{m}}$ with respect to $\breve{m}$ treats the spectral density $S_{XX}(\omega)$ as given. If one wishes also to evaluate the sensitivity of $E(T)$ to uncertainty about the spectral density, then it is necessary to express the spectral density as a function of a finite number of parameters. For example, one might use an approximation of $S_{XX}(\omega) = \Sigma\, a_j\, f_j(\omega)$, with given $f_j(\omega)$ functions and uncertain a_j coefficients, then investigate the sensitivity to the uncertain coefficients. In this approach it is essential that the $f_j(\omega)$ functions be chosen such that

$$\int_0^\infty \omega^{2/\breve{m}}\, f_j(\omega)\, d\omega < \infty$$

exists for each j value. If the a_j coefficients can be related to the mass, damping, and stiffness of the structure, then this gives a method for investigating the effects on $E(T)$ of those parameters. For example, if the structure is modeled as a SDF system and if the dominant frequency of the excitation is well separated from the resonant frequency of the structure, then one might use a crude approximation of $f_1(\omega)$ being $|H(\omega)|^2$ at the design point for the structure and a_1 being proportional to $k^{-1}c^{-1}$, inasmuch as response variance is proportional to this quantity. The other portion of the spectral density would then come from an $f_2(\omega)$ that has the form of the autospectral density of the excitation. If the dominant frequency of the excitation is lower than the resonant frequency of the structure, then one might consider a_2 to be proportional to k^{-1}. Similarly, a dominant frequency of the excitation that is higher than the resonant frequency would give a_2 approximately proportional to m^{-1}.

The examples here will focus on the sensitivity of $E(T)$ to the parameters that appear explicitly in Eqs. 12.46–12.49 rather than on mass, damping, and stiffness, because there are so many ways that the spectral density could depend on those structural parameters.

****

Example 12.10: Find the sensitivity coefficients for the effect of the $\breve{m}$ and κ parameters on the expected fatigue life using the fatigue prediction of Eq. 12.49. The nominal values of the parameters are $\breve{m}_0 = 3$ and $\kappa_0 = 4$, and the autospectral density of the $\{X(t)\}$ stress is one for which it has been shown in Example 11.11 that the single-moment method gives $E(T)$ values significantly greater than those from the Rayleigh approximation. In particular, it has a limiting bimodal form of

$$S_{XX}(\omega) = B[\delta(\omega+\omega_0)+\delta(\omega-\omega_0)]+(B/10)[\delta(\omega+15\omega_0)+\delta(\omega-15\omega_0)]$$

We let $\mathcal{A}$ and $\breve{\mathcal{M}}$ be random variables denoting uncertainty about the values of κ and $\breve{m}$, respectively, and define dimensionless parameters as $\mathcal{R}_1 = \mathcal{A}/\kappa_0$ and $\mathcal{R}_2 = \breve{\mathcal{M}}\ /\breve{m}_0$. Also we define the normalized response quantity Q as the ratio of $E(T)$ for parameters $\mathcal{A}$ and $\mathcal{M}$ to the value for parameters κ_0 and $\breve{m}_0$.
Note that the simplified autospectral density for $\{X(t)\}$ reduces the need for some numerical integration, because it gives $\lambda_{2/\breve{m}} = 2B(\omega_0)^{2/\breve{m}}[1+15^{2/\breve{m}}/10]$. Carrying through the formulas now gives $\beta_1 = -0.8$, $\beta_{11} = 1.28$, $\beta_2 = -5.402$, and $\beta_{22} = 26.58$.

Thus, in this problem the fatigue life is quite sensitive to the value of $\breve{m}$ but only moderately sensitive to κ. The accompanying sketch shows the dependence of Q on $\breve{m}$ for a relatively wide range of the parameter. The shape of the plot is quite similar to that in Example 12.8 for the dependence of the first-passage probability on the amount of damping in the system. As in that problem, none of the simple perturbation approximations match the actual value very well for such a wide range of the parameter.

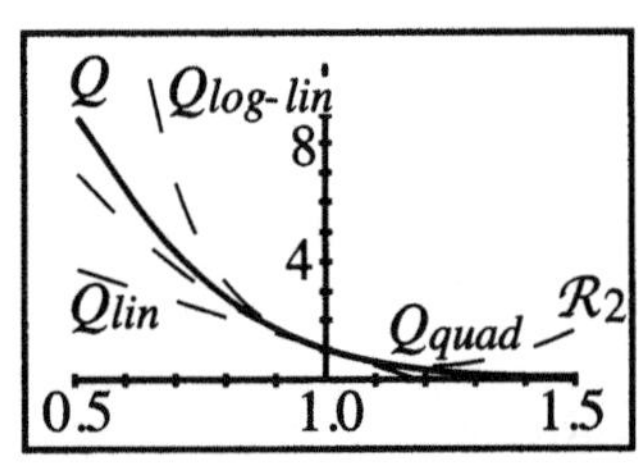

****

12.10 Multi-Degree-of-Freedom Systems

The specific situations considered in Sections 12.5–12.8 have involved the dynamic response of SDF systems, even though the methods introduced in the earlier sections and used here are general. For more complex systems the computational effort is considerably greater, and there are additional factors to consider. Some of these items will be discussed in this section, although it is not feasible to give as much detail as was provided for the SDF problems.

One of the most common ways of reducing the effort for dynamic analysis of MDF systems is to use truncated modal summations, as mentioned in Section 8.4. Typically, the truncation is to include only the modes with the lowest resonant frequencies. Furthermore, exactly the same concept can be used in the state-space formulation of the equations, since the imaginary parts of the eigenvalues represent frequencies. The effect of uncertain parameters in such modal analysis studies is most easily analyzed by first considering the effect of the parameters on the eigensolutions that give modal frequencies and mode shapes, then investigating the effect of the uncertainties in eigensolutions on modal responses.

Relatively straightforward procedures have been developed for evaluating the first and second derivatives of eigenvalues and eigenvectors in terms of the derivatives of the **m** and **k** matrices for an MDF system or in terms of the derivatives of the corresponding **A** and **B** matrices in the state-space formulation. These results have been particularly used in perturbation studies of finite element models, which typically yield MDF equations of motion (Ghanem and Spanos, 1991; Kleiber and Hien, 1992). The simplest form of the uncertain eigenvalue problem (Fox and Kapoor, 1968) will be outlined here. We will use the **m** and **k** matrices, but exactly the same procedure can be used for **A** and **B** matrices provided that they are both symmetric. Recall the notation of Chapter 8 in which the eigenvectors of $\mathbf{m}^{-1}\mathbf{k}$ are the columns of a matrix $\boldsymbol{\theta}$ and the eigenvalues are the elements of a diagonal matrix $\boldsymbol{\lambda}$, so $\mathbf{m}^{-1}\mathbf{k}\boldsymbol{\theta}=\boldsymbol{\theta}\boldsymbol{\lambda}$ or $\mathbf{k}\boldsymbol{\theta}=\mathbf{m}\boldsymbol{\theta}\boldsymbol{\lambda}$. Writing out only the *j*th column of this result gives $\mathbf{k}\vec{\theta}_j=\mathbf{m}\vec{\theta}_j\,\lambda_{jj}$ or

$$(\mathbf{k}-\lambda_{jj}\mathbf{m})\vec{\theta}_j=\vec{0}$$

Let r denote an uncertain parameter, for which we wish to find the sensitivity coefficient. Taking the derivative gives

$$\left(\frac{\partial \mathbf{k}}{\partial r}-\frac{\partial \lambda_{jj}}{\partial r}\mathbf{m}-\lambda_{jj}\frac{\partial \mathbf{m}}{\partial r}\right)\vec{\theta}_j+(\mathbf{k}-\lambda_{jj}\mathbf{m})\frac{\partial \vec{\theta}_j}{\partial r}=0 \qquad (12.50)$$

Premultiplying by $(\theta_j)^T$ makes the last term zero, though, because $(\theta_j)^T(\mathbf{k}-\lambda_{jj}\mathbf{m}) = [(\mathbf{k}-\lambda_{jj}\mathbf{m})\vec{\theta}_j]^T = \vec{0}^T$. Using the notation $\hat{\mathbf{m}}$ for the diagonal matrix $\boldsymbol{\theta}^T\mathbf{m}\boldsymbol{\theta}$ then gives the derivative of the jth eigenvalue as

$$\frac{\partial\lambda_{jj}}{\partial r} = \frac{1}{\hat{m}_{jj}}\vec{\theta}_j^T\left(\frac{\partial\mathbf{k}}{\partial r} - \lambda_{jj}\frac{\partial\mathbf{m}}{\partial r}\right)\vec{\theta}_j \tag{12.51}$$

Finding the derivative of the eigenvector is a little more difficult. From Eq. 12.50 we see that we want the solution of the equation

$$(\mathbf{k}-\lambda_{jj}\mathbf{m})\frac{\partial\vec{\theta}_j}{\partial r} = -\left(\frac{\partial\mathbf{k}}{\partial r} - \frac{\partial\lambda_{jj}}{\partial r}\mathbf{m} - \lambda_{jj}\frac{\partial\mathbf{m}}{\partial r}\right)\vec{\theta}_j \tag{12.52}$$

in which all the terms on the right-hand side are now presumed to be known. The problem, of course, is that the $(\mathbf{k}-\lambda_{jj}\mathbf{m})$ matrix on the left-hand side is singular, so its inverse does not exist. An equation of the type $\mathbf{A}\vec{x}=\vec{y}$ with $|\mathbf{A}|=0$ presents two difficulties. The first is that there is no solution for $\vec{x}$ for most $\vec{y}$ vectors. The fact that $|\mathbf{A}|=0$ implies that one can write $A_{nl} = \Sigma\alpha_j A_{jl}$ in which the summation is over j, but with $j \ne n$. That is, the last row (or any specific row) can be written as a linear combination of the other rows. It is easily shown, then, that a solution exists only if $y_n = \Sigma\alpha_j y_j$. This is not really an issue with our problem, because we know that Eq. 12.52 must have a solution, so it must meet this condition. The remaining difficulty is that the solution of the equation is not unique. We know that $(\mathbf{k}-\lambda_{jj}\mathbf{m})\vec{\theta}_j = \vec{0}$, so if $\vec{v}$ represents one solution of Eq. 12.52, then $\vec{v}+b\vec{\theta}_j$ is also a solution for any value of the scalar b. The problem now is first to find a solution $\vec{v}$, then find a condition to give us a unique value of b in the form

$$\frac{\partial\vec{\theta}_j}{\partial r} = \vec{v} + b\vec{\theta}_j \tag{12.53}$$

Nelson (1976) has presented a very useful way to handle this problem. First one selects a value of one component of $\vec{v}$ and deletes one row from Eq. 12.52 to give a reduced equation with a unique solution for the other components of $\vec{v}$. Using a subscript $[R]$ to denote reduced matrices with any one row removed then gives a particular solution $\vec{v}$ of Eq. 12.52 as the selected element combined with the elements of the reduced vector solution of

$$(\vec{v})_{[R]} = -(\mathbf{k} - \lambda_{jj}\mathbf{m})^{-1}_{[R]} \left(\frac{\partial \mathbf{k}}{\partial r} - \frac{\partial \lambda_{jj}}{\partial r}\mathbf{m} - \lambda_{jj}\frac{\partial \mathbf{m}}{\partial r} \right)_{[R]} \vec{\theta}_j$$

The problem of finding a unique value for the derivative of the eigenvector is similar to the basic eigenvalue problem, for which the eigenvectors are not unique unless one stipulates some normalization condition. Various normalizations are possible and in use. One approach is to require that the diagonal matrix $\hat{\mathbf{m}} \equiv \boldsymbol{\theta}^T \mathbf{m} \boldsymbol{\theta}$ be the identity matrix. Probably a more common choice is to say that $\boldsymbol{\theta}^T \boldsymbol{\theta} = \mathbf{I}$. Whatever the chosen normalization, we now want it also to apply to $\vec{\theta}_j + d\vec{\theta}_j$, so the derivative of the normalization condition with respect to r must be zero. Using $\boldsymbol{\theta}^T \boldsymbol{\theta} = \mathbf{I}$ gives

$$\vec{\theta}_j^T \frac{\partial \vec{\theta}_j}{\partial r} + \frac{\partial \vec{\theta}_j^T}{\partial r} \vec{\theta}_j = 0$$

Substituting $\vec{v} + b\vec{\theta}_j$ for the derivative then gives

$$\vec{\theta}_j^T (\vec{v} + b\vec{\theta}_j) + (\vec{v}^T + b\vec{\theta}_j^T)\vec{\theta}_j = 0$$

from which one can find that $b = -\vec{v}^T \vec{\theta}_j$. For other normalizations of the eigenvectors the result for b is slightly less simple, but straightforward.

This procedure for the derivative of the eigenvector has an advantage in common with the Fox and Kapoor evaluation of the derivative of the eigenvalue in Eq. 12.51. Namely, both derivatives can be found without knowledge of any other eigenvalues or eigenvectors. This is in contrast to some other techniques in which one uses expansions of the eigenvector derivative in terms of other eigenvectors (Adelman and Haftka, 1986). Thus, one can use these formulas in a truncated modal summation approach without evaluating other eigenvalues or eigenvectors.

Consider now the relatively general problem in which the stationary variance of response at a particular point in an MDF system is written as a truncated modal summation:

$$\sigma_X^2 = \sum_{j=1}^{J} \sigma_{X_j}^2 \tag{12.54}$$

in which J is the number of modes being included. For example, let $X(t)$ be taken as $X_l(t)$ for a system with uncoupled modes so that using $\vec{X}=\boldsymbol{\theta}\vec{Y}$ converts

$$\mathbf{m}\ddot{\vec{X}}+\mathbf{c}\dot{\vec{X}}+\mathbf{k}\vec{X}=\vec{F}$$

into

$$\ddot{\vec{Y}}+\boldsymbol{\gamma}\dot{\vec{Y}}+\boldsymbol{\lambda}\vec{Y}=\vec{G}$$

with $\boldsymbol{\gamma}$ and $\boldsymbol{\lambda}$ being diagonal. If the damping is small and the spectral density of the excitation is smooth in the neighborhood of the resonant frequency $\omega_j=(\lambda_{jj})^{1/2}$, then one can approximate the variance of the Y_j response as

$$\sigma_{Y_j}^2\approx\frac{\pi S_{GG}(\omega_j)}{\hat{m}_{jj}^2\gamma_{jj}\lambda_{jj}}$$

Provided that the modal frequencies are well separated, one can neglect modal correlations and higher-frequency modes to obtain

$$\sigma_{X_l}^2\approx\sum_{j=1}^{J}\theta_{lj}^2\sigma_{Y_j}^2$$

so the modal contribution in Eq. 12.54 is

$$\sigma_{X_j}^2\approx\theta_{lj}^2\sigma_{Y_j}^2\approx\theta_{lj}^2\frac{\pi S_{GG}(\omega_j)}{\hat{m}_{jj}^2\gamma_{jj}\lambda_{jj}} \tag{12.55}$$

Note that the jth modal contribution involves only the jth eigenvector and jth eigenvalue. Thus, one can write

$$\frac{d\sigma_X^2}{dr}=\sum_{j=1}^{J}\frac{d\sigma_{X_j}^2}{dr}=\sum_{j=1}^{J}\left(\frac{d\sigma_{X_j}^2}{d\lambda_j}\frac{d\lambda_j}{dr}+\sum_{l=1}^{n}\frac{d\sigma_{X_j}^2}{d\theta_{jl}}\frac{d\theta_{jl}}{dr}\right)$$

in which θ_{jl} is the lth component of the jth eigenvector. The evaluation of the derivatives of the eigenvalue and eigenvector has already been presented, so the remaining issue is the derivative of the modal responses with respect to the eigenvalue and eigenvector. Each of these modal responses, though, is like the response of an SDF system and is therefore similar to the problems considered in Sections 12.5–12.7. The parameters considered in the earlier sections were mass, damping, and stiffness, but there is no particular difficulty in adapting that coverage to the current situation. It must be kept in mind that $\hat{m}_{jj}$, γ_{jj}, and $S_{GG}(\omega)$ all depend on the eigenvectors because of their definitions: $\hat{\mathbf{m}} = \boldsymbol{\theta}^T \mathbf{m} \boldsymbol{\theta}$, $\hat{\mathbf{m}}\boldsymbol{\gamma} = \boldsymbol{\theta}^T \mathbf{c} \boldsymbol{\theta}$, and $\vec{G} = \hat{\mathbf{m}}^{-1} \boldsymbol{\theta}^T \vec{F}$. Nonetheless, the problem is basically the same as for the SDF system, and similar results can be expected for such broadband excitations as were assumed in deriving Eq. 12.55.

Recall that the major difficulty for the perturbation method for the SDF system was in Section 12.7, in which the excitation was not broadband. One must anticipate that this same difficulty will appear for MDF systems if there are peaks in the spectral density of the excitation. In fact, the problem may be somewhat more complicated than in the SDF situation, because now there will be a number of resonant frequencies that can be affected by tuning/detuning due to variations in the parameters. This difficulty is particularly severe when there is a nonmassive "secondary" system attached to a massive "primary" system, because small changes in parameters can have major effects on the tuning/detuning of the primary-secondary system (Igusa and Der Kiureghian, 1988; Jensen and Iwan, 1992). In these situations with complicated and very nonlinear dependence of the response on the input parameters, it may also be difficult or costly to obtain results from direct calculation of the response at various parameter values, particularly when there are many uncertain parameters, as there typically are in an MDF system. Approximate methods can sometimes be used to obtain acceptable results for these complicated situations.

Jensen and Iwan (1992) formulated one simplified approach for studying the parameter dependence of MDF systems and other higher-order linear systems described by the state-space equations of Chapter 9. In particular, they studied response variance, which in state-space analysis must be found as a component of a matrix $\mathbf{K}$ of response covariances. As shown in Eq. 9.17, the response covariance is found from solution of a first-order matrix differential equation involving $\mathbf{A}$ and $\mathbf{B}$ matrices that depend on system parameters. Simplifying their presentation to the situation of a single uncertain parameter, denoted by $\mathcal{R}$, it was assumed that the response covariance could be written as a series of the form $\mathbf{K} = \Sigma \mathbf{K}_j(t) H_j(\mathcal{R})$. The $H_j(\mathcal{R})$ were taken as functions that are orthonormal with respect to the probability density of $\mathcal{R}$, so $E([H_j(\mathcal{R})]^2) = 1$ and $E[H_j(\mathcal{R}) H_k(\mathcal{R})] = 0$. In this way they found uncoupled differential equations for the $\mathbf{K}_j(t)$ coefficient matrices in the expansion. They obtained good results with only a few terms in the expansion when considering two uncertain parameters having uncertainty levels characterized by $\sigma_{\mathcal{R}}$ being 30% and 40% of $\mu_{\mathcal{R}}$.

Katafygiotis and Beck (1995) have also presented an approach for studying the dependence of the response variance on uncertainty about the damping and stiffness matrices in an MDF system. They concluded that the perturbation approach was adequate for predicting the effect of parameter uncertainty on the modal damping and frequency but not for predicting the dependence of response variance on these modal parameters. They used the Duhamel convolution form for the response of each mode and the fast Fourier transform to evaluate efficiently the convolution integrals. They also used gamma distributions to model the uncertainty of the modal parameters, matching the mean and variance determined from perturbation analysis. The gamma distribution allowed analytical evaluation of certain integrals, avoiding the cost of numerical integration, and providing a very efficient method of analysis.

Research continues on the development of efficient and general methods for analyzing the effect of many uncertain parameters in situations where the nonlinearity of the response dependence makes perturbation results inaccurate.

Exercises

****

Stationary Response Variance for Delta-Correlated Excitation

****

12.1 Let $\{X(t)\}$ be the stationary response of an SDF system with white noise excitation: $m\ddot{X}(t)+C\dot{X}(t)+\mathcal{K}X(t)=F(t)$, with $S_{FF}(\omega)=S_0$. The uncertain damping and stiffness coefficients are independent random variables with mean values $\mu_C=c_0$ and $\mu_{\mathcal{K}}=k_0$ and standard deviations $\sigma_C=0.2c_0$ and $\sigma_{\mathcal{K}}=0.1k_0$. Let the normalized response be $Q=c_0k_0\sigma_X^2/(\pi S_0)$ and the normalized parameters be $\mathcal{R}_2=C/c_0$ and $\mathcal{R}_3=\mathcal{K}/k_0$.

(a) Find the estimates of the mean and variance of Q based on mixed-order perturbation.

(b) Find the mean and variance of Q based on the assumption that C and $\mathcal{K}$ have uniform distributions.

(c) Find estimates of the mean and variance of Q based on the logarithmic-linear perturbation and the assumption that C and $\mathcal{K}$ have log-normal distributions.

****

12.2 Let $\{X(t)\}$ be the stationary response of an SDF system with white noise excitation: $m\ddot{X}(t)+\mathcal{C}\dot{X}(t)+\mathcal{K}X(t)=F(t)$, with $S_{FF}(\omega)=S_0$. The uncertain damping and stiffness coefficients are independent random variables with $\mathcal{C}$ uniformly distributed on $0.7c_0 \le \mathcal{C} \le 1.3c_0$ and $\mathcal{K}$ uniformly distributed on $0.75k_0 \le \mathcal{K} \le 0.25k_0$. Let the normalized response be $Q=c_0k_0\sigma_X^2/(\pi S_0)$ and the normalized parameters be $\mathcal{R}_2=\mathcal{C}/c_0$ and $\mathcal{R}_3=\mathcal{K}/k_0$.

(a) Find the mean and variance of Q.

(b) Find the estimates of the mean and variance of Q based on mixed-order perturbation.

(c) Estimate the mean and variance of Q by using the logarithmic-linear perturbation and replacing the uniform distributions of $\mathcal{C}$ and $\mathcal{K}$ with log-normal distributions.

****

12.3 Let $\{X(t)\}$ be the stationary response of an SDF system with white noise base acceleration: $\mathcal{M}\ddot{X}(t)+\mathcal{C}\dot{X}(t)+\mathcal{K}X(t)=-m\,a(t)$, with $S_{aa}(\omega)$ being a constant. The uncertain mass, damping, and stiffness coefficients are independent random variables with $\mu_{\mathcal{M}}=m_0$, $\mu_{\mathcal{C}}=c_0$, $\mu_{\mathcal{K}}=k_0$, $\sigma_{\mathcal{M}}=0.1m_0$, $\sigma_{\mathcal{C}}=0.2c_0$, and $\sigma_{\mathcal{K}}=0.1k_0$. Let the normalized parameters be $\mathcal{R}_1=\mathcal{M}/m_0$, $\mathcal{R}_2=\mathcal{C}/c_0$, and $\mathcal{R}_3=\mathcal{K}/k_0$ and the normalized response be $Q=c_0k_0\sigma_X^2/(\pi S_0)$.

(a) Find the estimates of the mean and variance of Q based on mixed-order perturbation.

(b) Find the mean and variance of Q based on the assumption that $\mathcal{M}$, $\mathcal{C}$, and $\mathcal{K}$ have uniform distributions.

(c) Find estimates of the mean and variance of Q based on the logarithmic-linear perturbation and the assumption that $\mathcal{M}$, $\mathcal{C}$, and $\mathcal{K}$ have log-normal distributions.

****

Nonstationary Response Variance for Delta-Correlated Excitation

****

12.4 Let $\{X(t)\}$ be the nonstationary response of an SDF system with zero initial conditions and white noise excitation: $\mathcal{M}\ddot{X}(t)+\mathcal{C}\dot{X}(t)+\mathcal{K}X(t)=F(t)$, with $S_{FF}(\omega)=S_0$. The uncertain mass, damping, and stiffness coefficients are independent random variables with $\mu_{\mathcal{M}}=m_0$, $\mu_{\mathcal{C}}=c_0$, $\mu_{\mathcal{K}}=k_0$, $\sigma_{\mathcal{M}}=0.1m_0$, $\sigma_{\mathcal{C}}=0.2c_0$, and $\sigma_{\mathcal{K}}=0.1k_0$. The damping and frequency at the nominal values of the parameters are $\omega_0=(k_0/m_0)^{1/2}=10$rad/s and $\zeta=c_0/[2(k_0m_0)^{1/2}]=0.05$. Let the normalized parameters be $\mathcal{R}_1=\mathcal{M}/m_0$, $\mathcal{R}_2=\mathcal{C}/c_0$, and $\mathcal{R}_3=\mathcal{K}/k_0$, and let Q be the ratio of $[\sigma_X(4\,s)]^2$ for parameters $\mathcal{M}$, $\mathcal{C}$, and $\mathcal{K}$ to its value for parameters m_0, c_0, and k_0.

(a) Find the estimates of the mean and variance of Q based on mixed-order perturbation.

(b) Find the estimates of the mean and variance of Q based on the logarithmic-linear perturbation and the assumption that $\mathcal{M}$, $\mathcal{C}$, and $\mathcal{K}$ have uniform distributions.

(c) Find estimates of the mean and variance of Q based on the logarithmic-linear perturbation and the assumption that $\mathcal{M}$, $\mathcal{C}$, and $\mathcal{K}$ have log-normal distributions.

****

12.5 Let $\{X(t)\}$ be the nonstationary response of an SDF system with zero initial conditions and white noise excitation: $\mathcal{M}\ddot{X}(t)+\mathcal{C}\dot{X}(t)+\mathcal{K}X(t)=F(t)$, with $S_{FF}(\omega)=S_0$. The uncertain mass, damping, and stiffness coefficients are independent random variables with $\mu_{\mathcal{M}}=m_0$, $\mu_{\mathcal{C}}=c_0$, $\mu_{\mathcal{K}}=k_0$, $\sigma_{\mathcal{M}}=0.05\,m_0$, $\sigma_{\mathcal{C}}=0.25\,c_0$, and $\sigma_{\mathcal{K}}=0.1\,k_0$. The damping and frequency at the nominal values of the parameters are $\omega_0=(k_0/m_0)^{1/2}=6\text{rad/s}$ and $\zeta=c_0/[2(k_0 m_0)^{1/2}]=0.01$. Let the normalized parameters be $\mathcal{R}_1=\mathcal{M}/m_0$, $\mathcal{R}_2=\mathcal{C}/c_0$, and $\mathcal{R}_3=\mathcal{K}/k_0$, and let Q be the ratio of $[\sigma_X(50\,s)]^2$ for parameters $\mathcal{M}$, $\mathcal{C}$, and $\mathcal{K}$ to its value for parameters m_0, c_0, and k_0.

(a) Find the estimates of the mean and variance of Q based on mixed-order perturbation.

(b) Find estimates of the mean and variance of Q based on the logarithmic-linear perturbation and the assumption that $\mathcal{M}$, $\mathcal{C}$, and $\mathcal{K}$ have log-normal distributions.

(c) Find estimates of the mean and variance of Q based on the logarithmic-linear perturbation and the assumption that $\mathcal{M}$, $\mathcal{C}$, and $\mathcal{K}$ have uniform distributions.

****

Stationary Response Variance for Nonwhite Stochastic Excitation

****

12.6 Let $\{X(t)\}$ be the stationary response of the oscillator considered in Example 12.5. The oscillator has resonant frequency ω_0 and damping $\zeta=0.01$. The excitation has autospectral density $S_{FF}(\omega)=S_0\exp[-(\omega-\omega_0)^2/(0.02\omega^2)]$. The uncertain mass, damping, and stiffness coefficients are independent random variables with $\mu_{\mathcal{M}}=m_0$, $\mu_{\mathcal{C}}=c_0$, $\mu_{\mathcal{K}}=k_0$, $\sigma_{\mathcal{M}}=0.1\,m_0$, $\sigma_{\mathcal{C}}=0.3c_0$, and $\sigma_{\mathcal{K}}=0.1\,k_0$. Let the normalized parameters be $\mathcal{R}_1=\mathcal{M}/m_0$, $\mathcal{R}_2=\mathcal{C}/c_0$, and $\mathcal{R}_3=\mathcal{K}/k_0$, and let Q be the ratio of σ_X^2 for parameters $\mathcal{M}$, $\mathcal{C}$, and $\mathcal{K}$ to its value for parameters m_0, c_0, and k_0.

(a) Find the estimates of the mean and variance of Q based on mixed-order perturbation.

(b) Find the estimates of the mean and variance of Q based on the logarithmic-linear perturbation and the assumption that $\mathcal{M}$, $\mathcal{C}$, and $\mathcal{K}$ have uniform distributions.

(c) Find estimates of the mean and variance of Q based on the logarithmic-linear perturbation and the assumption that $\mathcal{M}$, $\mathcal{C}$, and $\mathcal{K}$ have log-normal distributions.

****

12.7 Let $\{X(t)\}$ be the stationary response of the oscillator considered in Example 12.6. The oscillator has resonant frequency ω_0 and damping $\zeta = 0.01$. The excitation has autospectral density $S_{FF}(\omega) = S_0 \exp[-(\omega - \omega_0)^2 / (0.5\omega^2)]$. The uncertain mass, damping, and stiffness coefficients are independent random variables with $\mu_{\mathcal{M}} = m_0$, $\mu_{\mathcal{C}} = c_0$, $\mu_{\mathcal{K}} = k_0$, $\sigma_{\mathcal{M}} = 0.1 m_0$, $\sigma_{\mathcal{C}} = 0.3 c_0$, and $\sigma_{\mathcal{K}} = 0.2 k_0$. Let the normalized parameters be $\mathcal{R}_1 = \mathcal{M} / m_0$, $\mathcal{R}_2 = \mathcal{C} / c_0$, and $\mathcal{R}_3 = \mathcal{K} / k_0$, and let Q be the ratio of σ_X^2 for parameters $\mathcal{M}$, $\mathcal{C}$, and $\mathcal{K}$ to its value for parameters m_0, c_0, and k_0.

(a) Find the estimates of the mean and variance of Q based on mixed-order perturbation.

(b) Find estimates of the mean and variance of Q based on the logarithmic-linear perturbation and the assumption that $\mathcal{M}$, $\mathcal{C}$, and $\mathcal{K}$ have log-normal distributions.

(c) Find estimates of the mean and variance of Q based on the logarithmic-linear perturbation and the assumption that $\mathcal{M}$, $\mathcal{C}$, and $\mathcal{K}$ have uniform distributions.

****

12.8 Let $\{X(t)\}$ be the stationary response of the oscillator considered in Example 12.7. The oscillator has frequency ω_0 and damping $\zeta = 0.01$. The excitation has autospectral density $S_{FF}(\omega) = S_0 \exp[-(\omega - 1.1\omega_0)^2 / (0.02\omega^2)]$. The uncertain mass, damping, and stiffness coefficients are independent random variables with $\mu_{\mathcal{M}} = m_0$, $\mu_{\mathcal{C}} = c_0$, $\mu_{\mathcal{K}} = k_0$, $\sigma_{\mathcal{M}} = 0.1 m_0$, $\sigma_{\mathcal{C}} = 0.2 c_0$, and $\sigma_{\mathcal{K}} = 0.1 k_0$. Let the normalized parameters be $\mathcal{R}_1 = \mathcal{M} / m_0$, $\mathcal{R}_2 = \mathcal{C} / c_0$, and $\mathcal{R}_3 = \mathcal{K} / k_0$, and let Q be the ratio of σ_X^2 for parameters $\mathcal{M}$, $\mathcal{C}$, and $\mathcal{K}$ to its value for parameters m_0, c_0, and k_0.

(a) Find the estimates of the mean and variance of Q based on mixed-order perturbation.

(b) Find estimates of the mean and variance of Q based on the logarithmic-linear perturbation and the assumption that $\mathcal{M}$, $\mathcal{C}$, and $\mathcal{K}$ have log-normal distributions.

(c) Find the estimates of the mean and variance of Q based on the logarithmic-linear perturbation and the assumption that $\mathcal{M}$, $\mathcal{C}$, and $\mathcal{K}$ have uniform distributions.

****

First-Passage Failure Probability

****

12.9 Let $\{X(t)\}$ be the stationary response of the oscillator considered in Examples 12.8 and 12.9. The oscillator has frequency ω_0, damping $\zeta = 0.01$, and zero initial conditions. The excitation is white noise with autospectral density S_0. The uncertain mass, damping, and stiffness coefficients are independent random variables with $\mu_{\mathcal{M}} = m_0$, $\mu_C = c_0$, $\mu_{\mathcal{K}} = k_0$, $\sigma_{\mathcal{M}} = 0.1\, m_0$, $\sigma_C = 0.1 c_0$, and $\sigma_{\mathcal{K}} = 0.1\, k_0$. Let the normalized parameters be $\mathcal{R}_1 = \mathcal{M} / m_0$, $\mathcal{R}_2 = C / c_0$, and $\mathcal{R}_3 = \mathcal{K} / k_0$, and let Q be the ratio of the probability of first-passage failure for parameters $\mathcal{M}$, C, and $\mathcal{K}$ to its value for parameters m_0, c_0, and k_0, in which failure occurs if $X(t)$ exceeds the level $4(\pi S_0)^{1/2} / (c_0 k_0)$ within the time interval $0 < \omega_0 t \le 250$.

(a) Find the estimates of the mean and variance of Q based on mixed-order perturbation and the Poisson approximation of first-passage probability.
(b) Find estimates of the mean and variance of Q based on the logarithmic-linear perturbation and the assumption that $\mathcal{M}$, C, and $\mathcal{K}$ have log-normal distributions with the Poisson approximation of first-passage probability.
(c) Find the estimates of the mean and variance of Q based on mixed-order perturbation and the approximation of Eq. 11.42 for the probability of first passage.
(d) Find estimates of the mean and variance of Q based on the logarithmic-linear perturbation and the assumption that $\mathcal{M}$, C, and $\mathcal{K}$ have log-normal distributions with the approximation of Eq. 11.42 for the probability of first passage.

****

12.10 Let $\{X(t)\}$ be the stationary response of the oscillator considered in Examples 12.8 and 12.9. The oscillator has frequency ω_0, damping $\zeta = 0.01$, and zero initial conditions. The excitation is white noise with autospectral density S_0. The uncertain mass, damping, and stiffness coefficients are independent random variables with $\mu_{\mathcal{M}} = m_0$, $\mu_C = c_0$, $\mu_{\mathcal{K}} = k_0$, $\sigma_{\mathcal{M}} = 0.05\, m_0$, $\sigma_C = 0.1 c_0$, and $\sigma_{\mathcal{K}} = 0.05\, k_0$. Let the normalized parameters be $\mathcal{R}_1 = \mathcal{M} / m_0$, $\mathcal{R}_2 = C / c_0$, and $\mathcal{R}_3 = \mathcal{K} / k_0$, and let Q be the ratio of the probability of first-passage failure for parameters $\mathcal{M}$, C, and $\mathcal{K}$ to its value for parameters m_0, c_0, and k_0, in which failure occurs if $X(t)$ exceeds the level $4(\pi S_0)^{1/2} / (c_0 k_0)$ within the time interval $0 < \omega_0 t \le 250$.

(a) Find the estimates of the mean and variance of Q based on mixed-order perturbation and the Poisson approximation of first-passage probability.

(b) Find estimates of the mean and variance of Q based on the logarithmic-linear perturbation and the assumption that $\mathcal{M}$, C, and $\mathcal{K}$ have log-normal distributions with the Poisson approximation of first-passage probability.

(c) Find the estimates of the mean and variance of Q based on mixed-order perturbation and the approximation of Eq. 11.42 for the probability of first passage.

(d) Find estimates of the mean and variance of Q based on the logarithmic-linear perturbation and the assumption that $\mathcal{M}$, C, and $\mathcal{K}$ have log-normal distributions with the approximation of Eq. 11.42 for the probability of first passage.

****

Fatigue Life

****

12.11 Let $\{X(t)\}$ be the stationary stress process considered in Example 12.10. It consists of two very narrowband frequency components with their dominant frequencies differing by a factor of 15. The low-frequency component contributes 10 times as much to the variance as does the high-frequency component. The uncertain parameter of the stress process is its kurtosis, which is a random variable $\mathcal{A}$ with $\mu_{\mathcal{A}} = \kappa_0$ and $\sigma_{\mathcal{A}} = 0.2\kappa_0$. The uncertain parameters of the S/N curve are independent random variables $\breve{\mathcal{M}}$ and $\mathcal{K}$ with $\mu_{\breve{\mathcal{M}}} = \breve{m}_0$, $\mu_{\mathcal{K}} = K_0$, $\sigma_{\breve{\mathcal{M}}} = 0.1\breve{m}_0$, and $\sigma_{\mathcal{K}} = 0.5K_0$. Let the normalized parameters be $\mathcal{R}_1 = \mathcal{A}/\kappa_0$, $\mathcal{R}_2 = \breve{\mathcal{M}}/\breve{m}_0$, and $\mathcal{R}_3 = \mathcal{K}/K_0$, and let Q be the ratio of the expected fatigue life for parameters $\mathcal{A}$, $\mathcal{M}$, and $\mathcal{K}$ to its value for parameters κ_0, $\breve{m}_0$, and K_0.

(a) Find the estimates of the mean and variance of Q based on mixed-order perturbation.

(b) Find estimates of the mean and variance of Q based on the logarithmic-linear perturbation and the assumption that $\mathcal{A}$, $\breve{\mathcal{M}}$, and $\mathcal{K}$ have log-normal distributions.

(c) Find the estimates of the mean and variance of Q based on the logarithmic-linear perturbation and the assumption that $\mathcal{A}$, $\breve{\mathcal{M}}$, and $\mathcal{K}$ have uniform distributions.

****

12.12 Let $\{X(t)\}$ be the stationary stress process considered in Example 12.10. It consists of two very narrowband frequency components with their dominant frequencies differing by a factor of 15. The low-frequency component contributes 10 times as much to the variance as does the high-frequency component. The uncertain parameter of the stress process is its kurtosis, which is a random variable $\mathcal{A}$ with $\mu_{\mathcal{A}} = \kappa_0$ and $\sigma_{\mathcal{A}} = 0.1\kappa_0$. The uncertain parameters of the S/N curve are independent random variables $\mathcal{M}$ and $\mathcal{K}$ with $\mu_{\breve{\mathcal{M}}} = \breve{m}_0$,

$\mu_{\mathcal{K}} = K_0$, $\sigma_{\breve{\mathcal{M}}} = 0.1\breve{m}_0$, and $\sigma_{\mathcal{K}} = 0.3K_0$. Let the normalized parameters be $\mathcal{R}_1 = \mathcal{A}/\kappa_0$, $\mathcal{R}_2 = \breve{\mathcal{M}}/\breve{m}_0$, and $\mathcal{R}_3 = \mathcal{K}/K_0$, and let Q be the ratio of the expected fatigue life for parameters $\mathcal{A}$, $\breve{\mathcal{M}}$, and $\mathcal{K}$ to its value for parameters κ_0, $\breve{m}_0$, and K_0.

(a) Find the estimates of the mean and variance of Q based on mixed-order perturbation.

(b) Find estimates of the mean and variance of Q based on the logarithmic-linear perturbation and the assumption that $\mathcal{A}$, $\breve{\mathcal{M}}$, and $\mathcal{K}$ have log-normal distributions.

(c) Find the estimates of the mean and variance of Q based on the logarithmic-linear perturbation and the assumption that $\mathcal{A}$, $\breve{\mathcal{M}}$, and $\mathcal{K}$ have uniform distributions.

****

Appendix A
Dirac Delta Function

The fundamental properties of the so-called *Dirac delta function* are

$$\delta(x) = 0 \qquad \text{for } x \neq 0 \tag{A.1}$$

and

$$\int_{-\infty}^{\infty} \delta(x - x_0)\, f(x)\, dx = f(x_0) \tag{A.2}$$

for any function $f(\cdot)$ that is finite and continuous at the point $x = x_0$. Strictly speaking, $\delta(x)$ is not a function, because it is not finite at one point on the real line. Of course, saying that $\delta(0) = \infty$ is not an adequate definition of the behavior of the function at the origin, inasmuch as infinity is not a number. For example, ∞/∞ can have any value from zero to infinity. The definition of the critical property of $\delta(\cdot)$ at the origin is given by Eq. A.2. We will use Dirac delta functions only in situations where the quantity of real interest is to be obtained from an integral involving the $\delta(\cdot)$ function rather than from the value of $\delta(\cdot)$ at any single argument.

Another way of viewing the Dirac delta function is as the formal derivative of the unit step function. This interpretation follows directly from the fundamental properties of $\delta(\cdot)$. In particular, let a function $U(x)$ be defined as the integral of $\delta(x)$:[1]

$$U(x) = \int_{-\infty}^{x} \delta(u)\, du \tag{A.3}$$

with an initial condition of $U(-\infty) = 0$. Based on Eqs. A.1 and A.2, we then obtain

$$U(x) = 0 \quad \text{for } x < 0, \qquad U(x) = 1 \quad \text{for } x > 0$$

---

[1]The reader is reminded of the equivalence of the indefinite integral of $f(x)$ and the definite integral of $f(\cdot)$ with x appearing only as the upper limit of the integral.

This specification of $U(x)$, though, is identical to that given in Section 2.3 for the unit step function, except for uncertainty about the value of $U(x)$ at the point $x=0$.[2] If the derivative of Eq. A.3 existed, then it would be given by

$$\frac{dU}{dx}=\delta(x) \tag{A.4}$$

The difficulty with this procedure, of course, is that the unit step function is not truly differentiable at the point $x=0$, because it is discontinuous at that point.

To define $\delta(x)$ precisely it is necessary to consider a sequence of functions, because $\delta(x)$ is not truly a function. One way to do this is to consider a sequence of functions that asymptotically approach the condition of Eq. A.1, with each member satisfying Eq. A.2. For example,

$$\delta_j(x)=0 \text{ for } x\le -2^{-j},\quad \delta_j(x)=0 \text{ for } x\ge 2^{-j},\quad \delta_j(x)=2^{j-1} \text{ for } |x|<2^{-j} \tag{A.5}$$

or

$$\delta_j(x)=0 \text{ for } x\le -2^{-j},\ \delta_j(x)=0 \text{ for } x\ge 2^{-j},\ \delta_j(x)=2^j(1-2^j\,|x|) \text{ for } |x|<2^{-j} \tag{A.6}$$

Clearly, each member of these sequences of rectangles and triangles does exactly satisfy Eq. A.2, and as j increases toward infinity, either sequence comes closer and closer to meeting the condition of Eq. A.1. The term *generalized function* is sometimes used for a relationship like $\delta(x)$, which is singular but can be approached asymptotically by a sequence of functions.

It may be noted that each sequence that asymptotically approaches $\delta(x)$ can be integrated, term by term, to give a sequence that converges to $U(x)$, except possibly at the point $x=0$. For the two preceding sequences, the $U_j(x)$ integrals are exactly 1/2 at $x=0$. This can be remedied by shifting the "pulses" to the left of the origin, such as replacing Eq. A.5 by

$$\delta_j(x)=0 \text{ for } x\le -2^{-j+1},\ \delta_j(x)=0 \text{ for } x\ge 0,\ \delta_j(x)=2^{j-1} \text{ for } -2^{-j+1}<x<0$$

---

[2]In Section 2.3 we defined $U(x)$ to be continuous from the right, because this simplifies its usage in the description of cumulative distribution functions. This choice, though, is somewhat arbitrary.

(A.7)

but this also has certain disadvantages. In particular, the Dirac delta function is inherently symmetric [$\delta(-x) = \delta(x)$], whereas the sequence in Eq. A.7 is always asymmetric for any finite j. The integral of Eq. A.5 or A.7 gives a sequence of piecewise linear functions that tend to the unit step function, whereas integration of Eq. A.6 gives a corresponding sequence that is piecewise quadratic. If needed, one can go so far as to use a sequence of analytic (i.e., infinitely differentiable) functions that asymptotically approaches $\delta(x)$. One example is

$$\delta_j(x) = \frac{2^j}{(2\pi)^{1/2}} \exp\left(-2^{2j-1} x^2\right) \tag{A.8}$$

Term-by-term differentiation of such a sequence can be used also to provide a precise generalized function definition of the derivatives of a Dirac delta function. It is good to know that such a definition is possible, because there are situations in which it is convenient to use this derivative concept.

Appendix B
Fourier Analysis

The basic idea of Fourier analysis is to represent a function $x(t)$ as a sum, or linear combination, of harmonic components in order to simplify its analysis. The simplest such situation occurs when $x(t)$ is defined only on a finite region. We take this region to be of length T, and we take it to be symmetric about the origin, $-T/2 \le t \le T/2$, because this symmetry will lead to some simplifications later. In this case one can write

$$x(t) = \sum_{j=0}^{\infty} a_j \cos\left(\frac{2\pi j t}{T}\right) + \sum_{j=1}^{\infty} b_j \sin\left(\frac{2\pi j t}{T}\right) \qquad \text{(B.1)}$$

Note that the frequencies of the harmonic terms have been taken such that each term contains an integer number of cycles of oscillations during the interval $-T/2 \le t \le T/2$. Also, note that the a_0 term is simply a constant, so the equation could equally well be written as

$$x(t) = a_0 + \sum_{j=1}^{\infty} a_j \cos\left(\frac{2\pi j t}{T}\right) + \sum_{j=1}^{\infty} b_j \sin\left(\frac{2\pi j t}{T}\right)$$

The problem, now, is to evaluate all the a_j and b_j coefficients in Eq. B.1. This task is made quite easy, though, by the orthogonality of the harmonic terms included. In particular,

$$\begin{aligned}\int_{-T/2}^{T/2} \cos\left(\frac{2\pi j t}{T}\right) \cos\left(\frac{2\pi k t}{T}\right) dt &= 0 && \text{for } j \ne k \\ &= T/2 && \text{for } j = k \ne 0 \\ &= T && \text{for } j = k = 0\end{aligned}$$

$$\begin{aligned}\int_{-T/2}^{T/2} \sin\left(\frac{2\pi j t}{T}\right) \sin\left(\frac{2\pi k t}{T}\right) dt &= 0 && \text{for } j \ne k \\ &= T/2 && \text{for } j = k \ne 0\end{aligned}$$

and

$$\int_{-T/2}^{T/2} \cos\left(\frac{2\pi\, j\, t}{T}\right) \sin\left(\frac{2\pi\, k\, t}{T}\right) dt = 0$$

First we multiply Eq. B.1 by $\cos(2\pi\, k\, t/T)$, for any integer k, then integrate both sides of the equation from $t = -T/2$ to $t = T/2$:

$$\int_{-T/2}^{T/2} x(t) \cos\left(\frac{2\pi\, k\, t}{T}\right) dt =$$
$$\int_{-T/2}^{T/2} \cos\left(\frac{2\pi\, k\, t}{T}\right) \left[\sum_{j=0}^{\infty} a_j \cos\left(\frac{2\pi\, j\, t}{T}\right) + \sum_{j=1}^{\infty} b_j \sin\left(\frac{2\pi\, j\, t}{T}\right)\right] dt$$

The order of summation and integration can be reversed on the right-hand side of this equation, and the orthogonality relationships cause all but one of these integral terms to be zero. In particular, the only nonzero term is the one $\cos(2\pi\, j\, t/T)$ term with $j = k$. Thus, one finds that

$$\int_{-T/2}^{T/2} x(t) \cos\left(\frac{2\pi\, k\, t}{T}\right) dt = a_k \left(\frac{T}{2}\right) \qquad \text{for } k \neq 0$$

$$\int_{-T/2}^{T/2} x(t)\, dt = a_0\, T$$

Rewriting these expressions with an index variable of *j*, instead of *k*, gives

$$a_0 = \frac{1}{T} \int_{-T/2}^{T/2} x(t)\, dt \tag{B.2}$$

$$a_j = \frac{2}{T} \int_{-T/2}^{T/2} x(t) \cos\left(\frac{2\pi\, j\, t}{T}\right) dt \qquad \text{for } j \neq 0 \tag{B.3}$$

This gives the values for all the a_j coefficients in Eq. B.1. Performing the same operations using a multiplier of $\sin(2\pi\, k\, t/T)$ gives

$$b_j = \frac{2}{T} \int_{-T/2}^{T/2} x(t) \sin\left(\frac{2\pi\, j\, t}{T}\right) dt \qquad \text{for } j \neq 0 \tag{B.4}$$

for the b_j coefficients in Eq. B.1.

The usefulness of the Fourier series represented in Eqs. B.1–B.4 depends on the fact that the series does converge, so a truncated series of the form

$$x_N(t) = \sum_{j=0}^{N} a_j \cos\left(\frac{2\pi\, j t}{T}\right) + \sum_{j=1}^{N} b_j \sin\left(\frac{2\pi\, j t}{T}\right) \tag{B.5}$$

can be used as an approximation of $x(t)$. In particular, the truncated series converges to $x(t)$ as N goes to infinity at every point t where $x(t)$ is continuous. If $x(t)$ is discontinuous at the value t, then the truncated series converges to the average of the limits from the right and the left. The fact that Eq. B.1 can be used to represent any continuous function on $-T/2 \le t \le T/2$ can be viewed as a statement that the harmonic functions form a complete basis for the space of these continuous functions. Thus, Eqs. B.1–B.4 give a complete representation of $x(t)$ as a sum of harmonic components. Alternatively, one can say that Eq. B.1 is the harmonic decomposition of $x(t)$.

Sometimes it is more convenient to use an alternative form of Eq. B.1, based on the complex exponential representation of the harmonic functions. In particular, if one uses the identities

$$\cos\left(\frac{2\pi\, j t}{T}\right) = \frac{e^{2\pi i\, j t/T} + e^{-2\pi i\, j t/T}}{2}, \qquad \sin\left(\frac{2\pi\, j t}{T}\right) = \frac{e^{2\pi i\, j t/T} - e^{-2\pi i\, j t/T}}{2i}$$

then Eqs. B.1–B.4 become

$$x(t) = \sum_{j=-\infty}^{\infty} c_j e^{2\pi i\, j t/T} \tag{B.6}$$

with

$$c_j = \frac{1}{T} \int_{-T/2}^{T/2} x(t)\, e^{-2\pi i\, j t/T}\, dt \tag{B.7}$$

in which $c_0 = a_0$ and

$$c_j = \frac{a_j}{2} + \frac{b_j}{2i} = \frac{a_j - i b_j}{2}, \qquad c_{(-j)} = \frac{a_j}{2} - \frac{b_j}{2i} = \frac{a_j + i b_j}{2} \qquad \text{for } j > 0$$

The validity of this exponential form is also easily confirmed by direct evaluation of the coefficients in Eq. B.6. That is, rather than using Eqs. B.2–B.4 to find the coefficients in Eq. B.6, one can use the orthogonality of the complex exponential functions, which can be written as

$$\int_{-T/2}^{T/2} e^{2\pi i\, j t/T}\, e^{2\pi i k t/T}\, dt = 0 \quad \text{for } j+k \neq 0$$
$$= T \quad \text{for } j+k = 0$$

Thus, multiplying Eq. B.6 by $\exp(2\pi i\, k t/T)$ and integrating directly gives the coefficient values given in Eq. B.7.

It should be noted that the Fourier series representation converges to $x(t)$ only within the interval $-T/2 \le t \le T/2$. In fact, it was assumed in Eqs. B.1, B.5, and B.6 that $x(t)$ was defined only on that finite region. In some situations $x(t)$ is actually defined on a broader domain, but the series converges only within the $[-T/2, T/2]$ interval used in evaluating the coefficients according to Eqs. B.2–B.4 or Eq. B.7. In fact, the functions of Eqs. B.1, B.5, and B.6 are periodic. For example, $x_j(t \pm T) = x_j(t)$, so the series repeats itself with period *T*. This finite period representation is not adequate for many problems in which we wish to consider $x(t)$ to be aperiodic and to extend from $-\infty$ to ∞. A direct way of extending our Fourier analysis to include this situation is to let the period *T* tend to infinity. In investigating this limiting situation, we will use the exponential form of Eqs. B.6 and B.7, because it is somewhat simpler than the form using sine and cosine representations of the harmonic components.

First, we introduce a notation for the frequency of the *j* term in Eqs. B.6 and B.7:

$$\omega_j = j\,\Delta\omega, \qquad \Delta\omega = 2\pi/T$$

so that the equations can be written as

$$x(t) = \sum_{j=-\infty}^{\infty} c_j e^{i\omega_j t}, \qquad c_j = \frac{\Delta\omega}{2\pi}\int_{-T/2}^{T/2} x(t)\, e^{-i\omega_j t}\, dt$$

Letting *T* tend to infinity now gives $\Delta\omega \to d\omega$ and the summation tending to an integral, so one can introduce a renormalized function $\tilde{x}(\omega_j) = c_j/\Delta\omega$ and write

$$x(t) = \int_{-\infty}^{\infty} \tilde{x}(\omega)\, e^{i\omega t}\, d\omega \qquad \text{(B.8)}$$

with

$$\tilde{x}(\omega) = \frac{1}{2\pi} \int_{-\infty}^{\infty} x(t)\, e^{-i\omega t}\, dt \tag{B.9}$$

We will call the function $\tilde{x}(\omega)$ given in Eq. B.9 the Fourier transform of $x(t)$, and Eq. B.8 is then the inverse Fourier transform formula. Other terms, such as integral Fourier transform, exponential Fourier transform, and so forth, are also used for Eq. B.9. Note that the frequency decomposition of the aperiodic, infinite period $x(t)$ function generally contains all frequencies. That is, the Fourier transform $\tilde{x}(\omega)$ is defined for all ω values, and Eq. B.8 is a superposition of $\tilde{x}(\omega)\, e^{i\omega t}$ terms for all ω values.

One also has the option of deriving Eq. B.9 directly, without any consideration of the Fourier series introduced earlier. In particular, if one assumes that an $\tilde{x}(\omega)$ transform exists such that it is possible to write Eq. B.8, then one can use the orthogonality of the complex exponential functions to derive Eq. B.9. In particular,

$$\int_{-\infty}^{\infty} e^{i\psi t}\, dt = 2\pi\, \delta(\psi) \tag{B.10}$$

so multiplying Eq. B.8 by $e^{i\eta t}$ and integrating gives

$$\int_{-\infty}^{\infty} x(t)\, e^{i\eta t}\, dt = \int_{-\infty}^{\infty} \tilde{x}(\omega) \int_{-\infty}^{\infty} e^{i\eta t}\, e^{i\omega t}\, dt d\omega$$

$$= 2\pi \int_{-\infty}^{\infty} \tilde{x}(\omega)\, \delta(\eta + \omega)\, d\omega = 2\pi\, \tilde{x}(-\eta)$$

which is easily rewritten as Eq. B.9. The orthogonality relationship of Eq. B.10 appears quite frequently in applications of Fourier analysis. It can be viewed as the Fourier transform of the function $f(t) = 1$. Written in the usual notation with ω, rather than η, representing frequency, this gives $\tilde{f}(\omega) = \delta(\omega)$, showing that the only harmonic component of a constant is the term with zero frequency.

References

Abramowitz, M. and Stegun, I.A. (1965). *Handbook of Mathematical Functions,* Dover Publications, Inc., New York.

Adelman, H.M. and Haftka, R.T. (1986). "Sensitivity analysis of discrete structural systems," *AIAA Journal,* 24, 823-832.

Asano, K. and Iwan, W.D. (1984). "An alternative approach to the random response of bilinear hysteretic systems," *Earthquake Engineering and Structural Dynamics,* **12,** 229-236.

Astill, J., and Shinozuka, M. (1972). "Random eigenvalue problem in structural mechanics," *AIAA Journal,* **10**(4), 456-462.

Atalik, T.S. and Utku, S. (1976). "Stochastic linearization of multi-degree-of-freedom non-linear systems," *Earthquake Engineering and Structural Dynamics,* **4,** 411-420.

Bendat, J.S. and Piersol, A.G. (1966). *Measurement and Analysis of Random Data,* Wiley, New York.

Benjamin, J.R. and Cornell, C.A. (1970). *Probability, Statistics, and Decision for Civil Engineers,* McGraw-Hill, New York.

Bogdanoff, J.L. and Kozin, F. (1985). *Probabilistic Models of Cumulative Damage,* Wiley, New York.

Bouc, R. (1968). "Forced vibration of a mechanical system with hysteresis," Abstract, in *Proceedings of the Fourth Conference on Nonlinear Oscillations,* Academia Publishing, Prague, Czechoslovakia, p. 315.

Cacciola, P., Muscolino, G., and Ricciardi, G. (2003). "On some closure methods in nonlinear stochastic dynamics," *Computational Stochastic Mechanics,* Eds. Spanos and Deodatis, Millpress, Rotterdam, pp. 79-88.

Caddemi, S. and Di Paola, M. (1997). "Nonlinear system response for impulsive parametric input," *Journal of Applied Mechanics,* **64,** 642-648.

Cai, G.Q. and Lin, Y.K. (1988). "A new approximate solution technique for randomly excited nonlinear oscillators," *International Journal of Non-Linear Mechanics,* **23,** 409-420.

Caughey, T.K. (1960a). "Classical normal modes in damped linear dynamic systems," *Journal of Applied Mechanics,* ASME, **27,** 269-271.

Caughey, T.K. (1960b). "Random excitation of a system with bilinear hysteresis," *Journal of Applied Mechanics,* **27,** 649-652.

Caughey, T.K. (1965). "On the response of a class of nonlinear oscillators to stochastic excitation," *Les vibrations forcées dans les systèmes non-linéaires,* (Éditions du Centre National de la Recherche Scientifique), Paris, France, pp. 393-405.

Caughey, T.K. (1986). "On response of non-linear oscillators to stochastic excitation," *Probabilistic Engineering Mechanics,* **1,** 2-4.

Caughey, T.K. and O'Kelly, M.E.J. (1965). "Classical normal modes in damped linear dynamic systems," *Journal of Applied Mechanics,* ASME, **32,** 583-588.

Caughey, T.K. and Stumpf, H.S. (1961). "Transient response of a dynamic system under random excitation," *Journal of Applied Mechanics,* ASME, **28,** 563-566.

Chang, C.C. and Yang, H.T.Y. (1991). "Random vibration of flexible uncertain beam element," *Journal of Engineering Mechanics,* ASCE, **117**(10), 2329-2350.

Chen, D.C-K. and Lutes, L.D. (1994). "First-passage time of secondary system mounted on a yielding structure," *Journal of Engineering Mechanics,* ASCE, **120,** 814-834.

Choi, D-W., Miksad, R.W., Powers, E.J., and Fischer, F.J. (1985). "Application of digital cross-bispectral analysis techniques to model the nonlinear response of a moored vessel system in random seas," *Journal of Sound and Vibration,* **99,** 309-326.

Contreras, H. (1980). "The stochastic finite element model," *Computers and Structures,* **12,** 341-348.

Corotis, R.B., Vanmarcke, E.H., and Cornell, C.A. (1972). "First passage of nonstationary random processes," *Journal of the Engineering Mechanics Division,* ASCE, **98,** 401-414.

Cramer, H. (1966). "On the intersection between the trajectories of a normal stationary stochastic process and a high level," *Arkiv för Matematik,* **6,** 337-349.

Cramer, H. and Leadbetter, M.R. (1967). *Stationary and Related Stochastic Processes,* Wiley, New York.

Crandall, S.H. (1980). "Non-Gaussian closure for random vibration excitation," *International Journal of Non-Linear Mechanics,* **15,** 303-313.

Crandall, S.H., Chandaramani, K.L., and Cook, R.G. (1966). "Some first passage problems in random vibration," *Journal of Applied Mechanics,* ASME, **33,** 532-538.

Crandall, S.H. and Mark, W.D. (1963). *Random Vibrations in Mechanical Systems,* Academic Press, New York.

Der Kiureghian, A. (1980). "Structural response to stationary excitation," *Journal of the Engineering Mechanics Division,* ASCE, **106,** 1195-1213.

Di Paola, M. (1985). "Transient spectral moments of linear systems," *Solid Mechanics Archives,* **10,** 225-243.

Di Paola, M. and Falsone, G. (1993). "Stochastic dynamics of nonlinear systems driven by non-normal delta-correlated processes," *Journal of Applied Mechanics,* ASME, **60,** 141-148.

Di Paola, M., Falsone, G., and Pirotta, A. (1992). "Stochastic response analysis of nonlinear systems under Gaussian inputs," *Probabilistic Engineering Mechanics,* **7,** 15-21.

Di Paola, M. and Muscolino, G. (1990). "Differential moment equations of FE modeled structures with geometrical non-linearities," *International Journal of Non-Linear Mechanics,* **25,** 363-373.

Di Paola, M., Ricciardi, G., and Vasta, M. (1995). "A method for the probabilistic analysis of nonlinear systems," *Probabilistic Engineering Mechanics,* **10,** 1-10.

Ditlevsen, O. (1986). "Duration of Gaussian process visit to critical set," *Probabilistic Engineering Mechanics,* **1,** 82-93.

Dowling, N.E. (1972). "Fatigue failure predictions for complicated stress-strain histories," *Journal of Materials,* **7,** 71-87.

Downing, S.D. and Socie, D.F. (1982). "Simple rainflow counting algorithms," *International Journal of Fatigue,* **4,** 31-40.

Elishakoff, I. (1995). "Random vibration of structures: A personal perspective," *Applied Mechanics Reviews,* **48,** Part 1, 809-825.

Falsone, G. and Rundo Sotera, M. (2003). "About the evaluation of $E[f(X)X]$ when X is a non-Gaussian random variable," *Computational Stochastic Mechanics,* Eds. Spanos and Deodatis, Millpress, Rotterdam, pp. 189-198.

Fox, R.L. and Kapoor, M.P. (1968). "Rate of change of eigenvalues and eigenvectors," *AIAA Journal,* **6,** 2426-2429.

Fuchs, H.O. and Stephens, R.I. (1980). *Metal Fatigue in Engineering,* Wiley, New York.

Ghanem, R. and Spanos, P.D. (1991). *Stochastic Finite Element: A Spectral Approach,* Springer, New York.

Grigoriu, M. (1995). *Applied Non-Gaussian Processes,* PTR Prentice Hall, Englewood Cliffs, NJ.

Gupta, I.D. and Joshi, R.G. (2001). "Response spectrum superposition for structures with uncertain properties," *Journal of Engineering Mechanics,* **127,** 233-241.

Ibrahim, R.A. (1985). *Parametric Random Vibration,* Wiley, New York.

Ibrahim, R.A. (1995). "Recent results in random vibrations of nonlinear mechanical systems," *50th Anniversary of the Design Engineering Division,* Transactions of the ASME, pp. 222-233.

Igusa, T., Der Kiureghian, A., and Sackman, J.L. (1984). "Modal decomposition method for stationary response of non-classically damped systems," *Earthquake Engineering and Structural Dynamics,* **12,** 121-136.

Igusa, T. and Der Kiureghian, A. (1988). "Response of uncertain systems to stochastic excitation," *Journal of Engineering Mechanics,* **114,** 812-832.

Iwan, W.D. and Lutes, L.D. (1968). "Response of the bilinear hysteretic system to stationary random excitation," *Journal of the Acoustical Society of America,* **43,** 545-552.

Iyengar, R.N. and Dash, P.K. (1978). "Study of the random vibration of nonlinear systems by the Gaussian closure technique," *Journal of Applied Mechanics,* ASME, **45,** 393-399.

Jaynes, E.T. (1957). "Information theory and statistical mechanics," *Physical Review,* **106**(4), 620-630 and **108**(2), 171-190.

Jensen, H. and Iwan, W.D. (1992). " Response of systems with uncertain parameters to stochastic excitation," *Journal of Engineering Mechanics,* **118,** 1012-1025.

Kareem, A., Zhao, J., and Tognarelli, M.A. (1995). "Surge response statistics of tension leg platforms under wind and wave loads: A statistical quadratization approach," *Probabilistic Engineering Mechanics,* **10,** 225-240.

Katafygiotis, L.S. and Beck, J.L. (1995). "A very efficient moment calculation method for uncertain linear dynamic systems," *Probabilistic Engineering Mechanics,* **10,** 117-128.

Kazakov, I. E. (1965). "Generalization of the method of statistical linearization to multidimensional systems," *Automation and Remote Control,* **26,** 1201-1206.

Kendall, M. and Stuart, A. (1977). *The Advanced Theory of Statistics,* Macmillan Publishing, New York.

Kleiber, M., and Hien, T.D. (1992). *The Stochastic Finite Element Method,* Wiley, Chichester.

Larsen, C.E. and Lutes, L.D. (1991). "Predicting the fatigue life of offshore structures by the single-moment spectral method," *Probabilistic Engineering Mechanics,* **6,** 96-108.

Lebedev, V.L. (1958). *Random Processes in Electrical and Mechanical Systems,* State Publishing House of Physical and Mathematical Literature, Moscow (Russian). Translation: Flancreich, J., and Segal, M., Office of Technical Services, U.S. Dept. of Commerce, 1961.

Li, X-M., Quek, S-T., and Koh, C-G. (1995). "Stochastic response of offshore platforms by statistical cubicization," *Journal of Engineering Mechanics,* ASCE, **121,** 1056-1068.

Lin, Y.K. (1967). *Probabilistic Theory of Structural Dynamics,* McGraw-Hill, New York. Reprinted in 1976 by Krieger Publishing, Melbourne, FL.

Lin, Y.K. and Cai, G.Q. (1995). *Probabilistic Structural Dynamics: Advanced Theory and Applications,* McGraw-Hill, New York.

Lindgren, G., and Rychlik, I. (1987). "Rain flow cycle distribution for fatigue life prediction under Gaussian load processes," *Fatigue Fracture Engineering Materials and Structures,* **10,** 251-260.

Lutes, L.D. (2000). "The Gaussian assumption in equivalent linearization," *Proceedings of the 8th ASCE Specialty Conference on Probabilistic Mechanics and Structural Reliability* (CD ROM), Notre Dame, IN, Paper PMC2000-334, 6 pp.

Lutes, L.D. (2002). "Integral representations of increments of stochastic processes," *Meccanica,* **37,** 193-206.

Lutes, L.D. and Chen, D.C.K. (1992). "Stochastic response moments for linear systems," *Probabilistic Engineering Mechanics,* **7,** 165-173.

Lutes, L.D., Corazao, M., Hu, S-L.J., and Zimmerman, J.J. (1984). "Stochastic fatigue damage accumulation," *Journal of Structural Engineering,* ASCE, **110,** 2585-2601.

Lutes, L.D. and Larsen, C.E. (1990). "Improved spectral method for variable amplitude fatigue," *Journal of Structural Engineering,* ASCE, **116,** 1149-1164.

Lutes, L.D. and Wang, J. (1993). "Kurtosis effects on stochastic structural fatigue," in *Proceedings of the International Conference on Structural Safety and Reliability* (ICOSSAR '93), Innsbruck, Austria, pp. 1091-1097.

Madsen, H.O., Krenk, S., and Lind, N.C. (1986). *Methods of Structural Safety,* Prentice Hall, Englewood Cliffs, NJ.

Marcinkiewicz, J. (1938). "Sur une propriété de la loi de Gauss," *Mathematische Zeitschrift,* **44,** 612-618.

Marple, Jr., S.L. (1987). *Digital Spectral Analysis: With Applications,* Prentice Hall, Englewood Cliffs, NJ.

Masing, G. (1926). "Eigenspannungen und verfestigung beim messing," *Proceedings of the 2nd International Congress of Applied Mechanics,* Zurich, Switzerland, pp. 332-335 (German).

Matsuishi, M. and Endo, T. (1968). "Fatigue of metals subject to varying stress," presesented to the Japan Society of Mechanical Engineers, Fukuoka, Japan.

Michaelov, G., Sarkani, S., and Lutes, L.D. (1999a). "Spectral characteristics of nonstationary random processes—A critical review," *Structural Safety,* **21,** 223-244.

Michaelov, G., Sarkani, S., and Lutes, L.D. (1999b). "Spectral characteristics of nonstationary random processes—Response of a simple oscillator," *Structural Safety,* **21,** 245-267.

Miles, J.W. (1954). "On structural fatigue under random loading," *Journal of the Aeronautical Society,* **21,** 753-762.

Moyal, J.E. (1949). "Stochastic processes and statistical physics," *Journal of the Royal Statistical Society* (London), **B11,** 150-210.

Muscolino, G., Ricciardi, G., and Vasta, M. (1997). "Stationary and non-stationary probability density function for non-linear oscillators," *International Journal of Non-Linear Mechanics,* **32,** 1051-1064.

Naess, A. and Ness, G. M. (1992). "The statistical distribution of second-order, sum-frequency response statistics of tethered platforms in random waves," *Applied Ocean Research,* **14**(2), 23-32.

Nelson, R.B. (1976). "Simplified calculation of eigenvector derivatives," *AIAA Journal,* **14,** 1201-1205.

Nigam, N.C. (1983). *Introduction to Random Vibrations,* M.I.T. Press, Cambridge, MA.

Ortiz, K. and Chen, N.K. (1987). "Fatigue damage prediction for stationary wideband random stresses," in *Proceedings of the Fifth International Conference on Applications of Statistics and Probability in Soil and Structural Engineering* (ICASP 5), Vancouver, B.C., pp. 309-316.

Papadimitriou, C. (1995). "Stochastic response cumulants of MDOF linear systems," *Journal of Engineering Mechanics,* ASCE, **121,** 1181-1192.

Papadimitriou, C. and Lutes, L.D. (1994). "Approximate analysis of higher cumulants for MDF random vibration," *Probabilistic Engineering Mechanics,* **9,** 71-82.

Priestly, M.B. (1988). *Non-Linear and Non-Stationary Time Series Analysis,* Academic Press, London.

Rice, S.O. (1944, 1945). "Mathematical analysis of random noise," *Bell System Technical Journal,* **23** and **24.** Reprinted in *Selected Papers on Noise and Stochastic Processes,* Ed. N. Wax (1954), Dover, New York.

Roberts, J.B. and Spanos P.D. (1986). "Stochastic averaging: An approximate method for solving random vibration problems," *International Journal of Non-linear Mechanics,* **21,** 111-134.

Roberts, J.B. and Spanos, P.D. (1990). *Random Vibration and Statistical Linearization,* Wiley, New York.

Rugh, W.J. (1981). *Nonlinear System Theory: The Volterra/Wiener Approach,* Johns Hopkins University Press, Baltimore.

Rychlik, I. (1989). "Simple approximations of the rain-flow-cycle distribution for discretized random loads," *Probabilistic Engineering Mechanics,* **4,** 40-48.

Sarkani, S., Kihl, D.P., and Beach, J.E. (1994). "Fatigue of welded joints under narrowband non-Gaussian loadings," *Probabilistic Engineering Mechanics,* **9,** 179-190.

Schetzen, M. (1980). *The Volterra and Wiener Theories of Nonlinear Systems,* Wiley Interscience, New York.

Senthilnathan, A. and Lutes, L.D. (1991). "Nonstationary maximum response statistics for linear structures," *Journal of Engineering Mechanics,* ASCE, **117,** 294-311.

Shinozuka, M. (1972). "Monte Carlo simulation of structural dynamics," *Computers and Structures,* **2,** 865-874.

Sobczyk, K. and Spencer, Jr., B.F. (1992). *Random Fatigue: From Data to Theory,* Academic Press, Boston.

Soong, T.T. and Grigoriu, M. (1993). *Random Vibration of Mechanical and Structural Systems,* PTR Prentice Hall, Englewood Cliffs, NJ.

Spanos, P-T.D. (1983). "Spectral moments calculation of linear system output," *Journal of Applied Mechanics,* ASME, **50,** 901-903.

Spanos, P.D. and Donley, M.G. (1991). "Equivalent statistical quadratization for nonlinear systems," *Journal of Engineering Mechanics,* ASCE, **117,** 1289-1309.

Stratonovich, R.L. (1963). *Topics in the Theory of Random Noise,* Gordon and Breach, New York. Translated from the Russian by Silverman, R.A.

To, C.W.S. and Li, D.M. (1991). "Equivalent nonlinearization of nonlinear systems to random excitations," *Probabilistic Engineering Mechanics,* **6,** 184-192.

Udwadia, F.E. (1987). "Response of uncertain dynamic systems," *Journal of Applied Mathematics and Computation,* **22,** 115-187.

Vanmarcke, E.H. (1972). "Properties of spectral moments with applications to random vibration," *Journal of the Engineering Mechanics Division,* ASCE, **98,** 425-446.

Vanmarcke, E.H. (1975). "On the distribution of the first-passage time for normal stationary random processes," *Journal of Applied Mechanics,* ASME, **42,** 215-220.

Veletsos, A.S. and Ventura, C.E. (1986). "Modal analysis of non-classically damped linear systems," *Earthquake Engineering and Structural Dynamics,* **14,** 217-243.

Wang, J. and Lutes, L.D. (1993). "Current induced skewness of dynamic response for offshore structures," in *Proceedings of the International Conference on Structural Safety and Reliability* (ICOSSAR '93), Innsbruck, Austria, pp. 545-549.

Wen, Y.K. (1976). "Method for random vibration of hysteretic systems," *Journal of the Engineering Mechanics Division,* ASCE, **102,** 249-263.

Wilson, E.L., Der Kiureghian, A., and Bayo, E.P. (1981). "A replacement for the SRSS method in seismic analysis," *Earthquake Engineering and Structural Dynamics,* **9,** 187-192.

Winterstein, S.R. (1985). "Non-normal responses and fatigue damage," *Journal of Engineering Mechanics,* ASCE, **111,** 1291-1295.

Winterstein, S.R. (1988). "Nonlinear vibration models for extremes and fatigue," *Journal of Engineering Mechanics,* ASCE, **114,** 1772-1790.

Winterstein, S.R. and Cornell, C.A. (1985). "Energy fluctuation scale and diffusion models," *Journal of Engineering Mechanics,* ASCE, **111,** 125-142.

Winterstein, S.R., Ude, T.C., and Marthinsen, T. (1994). "Volterra models of ocean structures: Extreme and fatigue reliability," *Journal of Engineering Mechanics,* ASCE, **120,** 1369-1385.

Wirsching, P.H. and Light, M.C. (1980). "Fatigue under wide band random stress," *Journal of the Structural Division,* ASCE, **106,** 1593-1607.

Wirsching, P.H., Paez, T.L., and Ortiz, K. (1995). *Random Vibrations: Theory and Practice,* Wiley Interscience, New York.

Wu, W.F. and Lin, Y.K. (1984). "Cumulant-neglect closure for non-linear oscillators under random parametric and external excitations," *International Journal of Non-Linear Mechanics,* **19,** 349-362.

Yang, B., Mignolet, M.P., and Spottswood, S.M. (2003). "Modeling of damage accumulation for Duffing-type systems under severe random excitations," *Computational Stochastic Mechanics,* Eds. Spanos and Deodatis, Millpress, Rotterdam, pp. 663-673.

Yang, J.N. (1972). "Nonstationary envelope process and first excursion probability," *Journal of Structural Mechanics,* **1,** 231-249.

Zhang, J. and Ellingwood, B. (1995). "Effect of uncertain material properties on structural stability," *Journal of Structural Engineering,* ASCE, **121,** 705-716.

Zhu, W.Q. and Yu, J.S. (1989). "The equivalent non-linear system method," *Journal of Sound and Vibration,* **129,** 385-395.

Author Index

Subject Index